Employees' Intellectual Property Rights

AIPPI Law Series

VOLUME 1

Series Editors

Sanna Wolk, Kacper Szkalej

Introduction & Contents/Subjects

Books in this series are developed within the framework of the International Association for the Protection of Intellectual Property (AIPPI), a non-affiliated, non-profit organization dedicated to improving and promoting the protection of intellectual property at both national and international levels.

Objective

The aim is to publish innovative work appealing to practitioners and academics alike.

Readership

Academics and practitioners dealing with IP law.

The titles in this series are listed at the back of this volume.

Employees' Intellectual Property Rights

Second Edition

Edited by

Sanna Wolk
Kacper Szkalej

Wolters Kluwer

Published by:
Kluwer Law International B.V.
PO Box 316
2400 AH Alphen aan den Rijn
The Netherlands
Website: lrus.wolterskluwer.com

Sold and distributed in North, Central and South America by:
Wolters Kluwer Legal & Regulatory U.S.
7201 McKinney Circle
Frederick, MD 21704
United States of America
Email: customer.service@wolterskluwer.com

Sold and distributed in all other countries by:
Air Business Subscriptions
Rockwood House
Haywards Heath
West Sussex
RH16 3DH
United Kingdom
Email: international-customerservice@wolterskluwer.com

ISBN 978-90-411-9230-1

e-Book: ISBN 978-90-411-9265-3
web-PDF: ISBN 978-90-411-9297-4

List of Contributors

Michel Abello, lawyer at the Paris Bar specialising in IP law, engineer in mechanics, European patent attorney and former French and Community Trademark and Design Attorney. He is the managing partner of an IP Law firm in France and an IP litigator. He is Professor in charge of law courses at Ecole Centrale Paris, University of Science. He has written several publications in the field of IP.

Gusztáv Bacher, Dr, attorney-at-law in Hungary, partner at Szecskay Attorneys at Law, specialising in IP law, litigation and competition law. Former President of LIDC (the International League of Competition Law) (2012-2014), Secretary General of the Hungarian Group of AIPPI. He has published various publications in the fields of intellectual property, competition law, and advertising law, co-author of commentaries, and was head of several working groups which prepared national reports for AIPPI and LIDC.

Yaniv Benhamou (PhD and lecturer at the University of Geneva) teaches, conducts research and publishes in the areas of Intellectual Property, E-Commerce and Entertainment Law. In addition to his academic activities, he is attorney-at-law in a Swiss law firm, where he advises and represents clients before courts with regard to Commercial Law, in particular to Intellectual Property, Technology, Media & Telecoms and Data Protection. Link to the institutional website: http://www.unige.ch/droit/collaborateur/cema/benhamou-yaniv.html.

Martín Bensadon, partner and head of the patent department at Marval, O'Farrell & Mairal. Martin specialises in industrial property and more particularly in patent law, having advised many national and foreign firms in this area. Martin has written various articles on subjects related to his area of specialty in national and international publications as well as a book on Argentine patent law.

Jeppe Brinck-Jensen, attorney and partner at the Danish law firm Plesner within the field of intellectual property. He has extensive litigation experience and has contributed to a number of books and articles within the field. Further, he gives courses for patent agents on patent law, litigation and intellectual property rights agreements arranged by the Danish Association for Intellectual Property Law.

Niklas Bruun, Professor of Commercial Law, especially Intellectual Property and Labour Law, at Hanken School of Economics. He has published a number of academic publications on intellectual property and labour law. He is also the Director of IPR University Centre, an institute of six universities which coordinates and promotes education of and research into issues of intellectual property rights.

Ella Cheong, lawyer, founder of Ella Cheong & Alan Chiu, Solicitors & Notaries (formerly known as Ella Cheong Law Office) and Ella Cheong LLC in Singapore. Apart from being on the Roll of Honour of the HK Law Society, she has also received many international accolades (including awards of Honour from AIPPI, FICPI, APAA, INTA), and featured in numerous IP professional publications for all aspects of IP.

Christian Czychowski, attorney at law, Certified IT Lawyer and Copyright and Media Law Lawyer, is a partner at the IP law firm Boehmert & Boehmert. He specialises *inter alia* in IT law, patent and licensing, contract law. He took part in developing the German Model Agreements for Research and Development Cooperation's between Industry and University which was conceived in the Council for Innovation of the Chancellor's Office/Federal Ministry of Economics. Christian was appointed Honorary Professor at the Faculty of Law at the University of Potsdam.

Mr. Korcan Dericioğlu, *D.E.A.*, LL.M, CLP, is a Board Member of Ankara Patent Bureau and Managing Partner of Dericioglu & Eren Law Office. He is an attorney-at law, registered mediator and patent & trademark attorney. His main focus is on domain names, trademarks and copyrights. He is also actively involved in policy and IP management matters such as valuation, commercialisation, taxation and licensing of IP. He actively represents both firms in national and international platforms such as AIPPI, FICPI, LES and ICC. He is continuously involved in IP&IT based projects, publishes articles. He is fluent in English and French.

Luis-Alfonso Durán, managing partner at the Spanish IP firm Durán-Corretjer, S.L.P. as well as President of the Association of Spanish Industrial Property Attorneys before International Intellectual Property Organisations (AGESORPI), President of the Reporting Group for the reform of the European Patent Institute (epi), and delegate of the International Federation of Intellectual Property Attorneys (FICPI). He holds a degree in Industrial Engineering, a Diploma in Spanish and European Patents from the University of Barcelona and a Diploma on European Patents Litigation by the University of Strasbourg He is a Spanish industrial property attorney and a European patent, trade mark and design attorney.

Elisabeth Kasznar Fekete, Attorney at Law, SjD in Commercial Law, University of São Paulo, Brazil. Co-founder partner of the firm Kasznar Leonardos Intellectual Property. Member of the Board, INTA, 2016-2017. Past President of AIPPI's Brazilian Group and of ABPI – the Brazilian Intellectual Property Association. Published a Treatise on Trade

Secrets and more than 30 papers. Co-Coordinator of the Specialisation in Intellectual Property Course of the São Paulo Bar and professor in several courses in this area.

Gustavo Fischer, President of the Uruguayan Group of AIPPI, Past President of the Uruguayan Patent and Trademark Attorneys Association, Judge of the Tribunal of ASIPI, Professor of Public Law Litigation at the Master Program in Economic Administrative Law, University of Montevideo. Partner of Fischer Abogados, Intellectual Property and Technology Law, Montevideo, Uruguay.

Jorge Gomez, partner and head of the Trademark, Litigation and Corporate & Regulatory Affairs practice at DBB. He has 17 years of experience in foreign trade, IP, litigation, corporate and related regulatory matters. Jorge is an active member of several important professional associations such as AMPPI, ASIPI, AIPF, INTA and AIPPI.

Joseph Jeffrey Hawley, Distinguished Fellow in Intellectual Property at the University of New Hampshire School of Law. After a 37-year career at Eastman Kodak, he began teaching part time in 2007. He was active in several professional organisations and was President of IPO in 2004-2005.

José Roberto Herrera Díaz, LLM Munich Intellectual Property Law Centre, is Partner at Herrera Díaz Abogados, Intellectual Property Boutique Law firm located in Colombia. He is arbitrator for IP-related matters before the Chamber of Commerce of Bogota. In addition, he is an Intellectual Property Professor at Javeriana University and College of Administration in Superior Studies (Pontificia Universidad Javeriana, Colegio de Estudios Superiores de Administración).

Willem Hoorneman, heads the IP/TMT team of CMS Netherlands. With some 25 years of experience in intellectual property law, and author of the patent chapters in the leading handbook on software protection in the Netherlands, Willem is considered a pre-eminent intellectual property lawyer, and one of the very best in the field of patent law.

Belinda Isaac, Dr, is the founder and Principal Solicitor at Isaac & Co Solicitors, a specialist Intellectual Property practice based in Oxford, England. She has written a number of articles and contributed to several practitioner publications on Intellectual Property topics. Belinda advises clients on a range of intellectual property matters and is currently working on the text for an international and comparative law book focussing on the issue of compensation for employee inventors.

Edita Ivanauskiene, LLM, attorney at law, patent attorney and European trade mark and design attorney practising intellectual property law in Lithuania. She is also a member of the Commission of Copyright and Related Rights of the Republic of Lithuania, part time expert in the field of intellectual property law at the Faculty of Law, Vytautas Magnus University. She is also involved in the activities of the INTA Copyright Committee and the Association of Patent Attorneys (the APP).

Ronald Kakungulu-Mayambala is a Senior Lecturer with the Human Rights and Peace Centre (HURIPEC) and formerly a Graduate Teaching Assistant at The University of British Columbia, Vancouver, Canada (2007-2008). R. Kakungulu-Mayambala lectures human rights in the domestic perspective, consumer law and protection, introducing law and legal methods, intellectual property law, equity and trusts to undergraduate students and international environmental law, and computers & the law to graduate students at Makerere University School of Law. He has extensive knowledge on international and comparative indigenous peoples law, rights of ethnic minorities, oil and gas law, economic, social and cultural rights and has also done work on the rights of indigenous communities in East Africa. He is an Advocate of the High Court of Uganda. Dr Kakungulu-Mayambala's doctoral thesis at The University of Arizona was on *Indigenous People, Human Rights, and the African Problem: The Case of the Twa, Ogiek and Maasai*. He has consulted for many international and national agencies including IDRC, CIDA, DFID, IGAD, SIDA, DANIDA, NORAD, EASSI, KCK and ACODE.

Tsugizo Kubo, Professor and Vice Dean, specialises in the field of intellectual property law at the Senshu University Law School located in Tokyo, Japan, since 2008. He worked with Nissan Motor Corporation as a corporate intellectual property practitioner, especially handling patent, industrial design, trademark and licensing matters for 28 (years from 1971 to 1999). He participated in the WIPO Internet Domain Name Process as a member of the Panel of Experts (its final WIPO Report was released on 30 April 1999). He also served as a WIPO Consultant of the Madrid Protocol system for its promotion in Japan 2000 to 2003. He began teaching IP laws at the Kagoshima University and its Law School located in Kagoshima, Japan between 2003 and 2008.

Gyooho Lee, Professor of law at Chung-Ang University School of Law in Seoul, Republic of Korea. Currently he is serving as the president of the Association of Contents Property for Next Generation, of the International Cultural Property Law Association, and as the vice president of the Korea Association for Informedia Law and of the Korea Computer Game Law Association. Also, he is an arbitrator of KCAB (The Korean Commercial Arbitration Board), a mediator of Seoul Central District Court, and the vice president of KOSA (Korean Society of Authors). In addition, he is an editor of the *Korean Yearbook of International Law* and a member of the International Law Association's Committee on Intellectual Property and Private International Law.

Susanna Leong, Professor of Law (Department of Strategy & Policy), is a faculty at the NUS Business School, National University of Singapore with research interests in intellectual property and technology-related laws. She has published in several international and local academic journals, and is also the author of *Intellectual Property Law of Singapore*, Academy Publishing (Singapore). She is a Senior Fellow at the Intellectual Property Academy of Singapore; a member of the World Intellectual Property Organization (WIPO) Arbitration and Mediation Centre's Domain Name Panel; a member of Asian Domain Name Dispute Resolution Centre (Hong Kong) and also a member of the Singapore Copyright Tribunal.

Li Luo, Associate Professor, at the Law School of Renmin University of China. Her research interests mainly focus on intellectual property law, but also include civil law and legal history. Before joining Renmin Law School she had been teaching at China University of Political Science and Law for more than ten years. She is a council member of China Intellectual Property Law Association.

Marja-Leena Mansala, LLM, was the Secretary General of IPR University Centre. She has published a number of publications about intellectual property. She has been the Chief Editor of IPRinfo magazine for 15 years.

Špelca Mežnar, Associate Professor, is a researcher in the field of intellectual property at the International School for Social and Business Studies, Celje, Slovenia and an attorney specialising in business law (including copyright and trademarks). She has authored a number of academic publications in the field of intellectual property, media and internet law.

Ann Monotti is a Professor in the Faculty of Law at Monash University. She was the inaugural Director of the Centre for Commercial Law and Regulatory Studies in the Faculty of Law from 2014 - July 2017. Ann is an author of many publications in the area of intellectual property law including *Universities and Intellectual Property: Ownership and Exploitation* (OUP, 2013), *Australian Intellectual Property Law* (CUP, 2008, 2012, 2017). Her edited book in the Edward Elgar Publishing Critical Concepts in Intellectual Property Law series, *Employment Law and Intellectual Property Law,* is to be published in early 2018.

Jan Bernd Nordemann, attorney-at-law, Certified Copyright and Media Law Lawyer and Certified IP lawyer, is a partner at the IP Law Firm Boehmert & Boehmert. He specialises in several areas of intellectual property and is today a recognised expert in these fields. Jan has written numerous publications. He was appointed Honorary Professor at the Faculty of Law at Humboldt University Berlin, where he teaches IP law.

Gerardo Naranjo Ormaza is specialised in Intellectual Property as well as in IP administrative and judicial litigation. His knowledge specialty fields are trademarks, patents, and copyright; yet, he also gives advice in other related areas. He has over 5 years of experience in intellectual property litigation. He is a legal issues teacher in several areas, including IP, at a top technological level.

Andrea Ottolia, Associate Professor of Business Law at the University of Genoa, Italy. He holds an LLM and a doctorate from the University of California at Berkeley, Boalt Hall School of Law. His fields of research include copyright, patent, biotech law and data protection. He is a member of the Genoa Centre for Law and Finance and the Aida Journal.

Cristóbal Porzio, lawyer (J.D Catholic University of Chile, 1994), partner at PORZIO RIOS GARCIA, current. Immediate Past-President of Asociación Chilena de la Propiedad Intelectual (ACHIPI) – the Chilean Group of AIPPI - Professor of Law of the

Catholic University of Chile, Santiago, Chile. He has written a number of academic publications in the fields of intellectual property, distribution law and franchising.

Jonathan Rangel, lawyer graduated with honours from Universidad Nacional Autónoma de México (UNAM). His area of practice focuses mainly on IP litigation, trademark and patent prosecution, IP contracts, copyright litigation, trademark and patent infringement, well-known and famous trademarks, and domain names disputes.

Cyrill Rieder, Dr. iur, attorney-at-law, is associate partner at FMP Fuhrer Marbach. His practice is focused on all relevant aspects of intellectual property, including litigation in all Swiss courts. Furthermore he works in the fields of unfair competition law as well as IT, data and advertising law. His thesis examines the question of ownership and the exploitation of university inventions.

Morten Rosenmeier, Professor, is a researcher within the field of copyright law at the Faculty of Law, University of Copenhagen. He has written a number of academic publications in his research field.

Ilmārs Šatovs, attorney-at-law at law firm 'Šatovs', Riga, Latvia, Latvian and European Trade Mark and Design Attorney. He obtained Mag. iur. at the University of Latvia and LLM at the University of Stockholm in European Intellectual Property Law. His main field of interest is intellectual property litigation.

Helene Egede Scotwin, assistant attorney at the Danish law firm Accura Advokatpartnerselskab.

Norman Siebrasse, Professor of Law at the University of New Brunswick. He has published extensively, particularly in the areas of pharmaceutical patent law, intellectual property remedies, and the intersection of intellectual property law and commercial law.

Lígia Gutierrez Setúbal, associate lawyer of the Intellectual Property team at Vieira de Almeida & Associates, LLM in Intellectual Property and Competition Law, the Munich Intellectual Property Law Centre – MIPLC (Academic centre jointly run by the Max-Planck-Institut für Immaterialgüter und Wettbewerbsrecht [Max Planck Institute for Intellectual Property and Competition Law], the University of Augsburg, the Technische Universität München and The George Washington University Law School in Washington, D.C. [USA]); she has written some academic publications in this field.

Ekrem Soylu is registered Turkish Patent and Trademark Attorney since 2002. After 5 years patent law firm experience, he started the patent department of Bosch Home Appliances in Turkey and managed it for 11 years. He is the founding partner of Soylu Patent, Inc. currently counselling R&D centres on patent matters and now Patent Division Manager under IP & Law Directorate of Sanovel, one of Europe's top patent filers in pharmaceutical industry.

Kacper Szkalej, LLB (Hons), LLM (IT Law), LLM (IP Law), PhD Researcher at the Faculty of Law, Uppsala University currently conducts research focusing on the intersection of copyright, competition and consumer protection law. Active at the Institute of Intellectual Property, Marketing and Competition Law (IMC) at Uppsala University. Course Director a\ for several intellectual property law courses at the Faculty of Law, Uppsala University and lecturer at Uppsala University and KTH – Royal Institute of Technology.

Jérôme Tassi, lawyer at the Paris Bar since 2007 in the field on Intellectual Property. He is also a lecturer in IP law at Ecole Centrale Paris, University of Science and he has written several publications in this field.

Marketa Trimble, Samuel S. Lionel Professor of Intellectual Property Law at the William S. Boyd School of Law, the University of Nevada, Las Vegas. She conducts research and teaches in the areas of intellectual property law, private international law/conflict of laws, and internet law. She has authored numerous works in the areas of her research.

Benedetta Ubertazzi, Aggregate Professor of International Law, School of Law, University of Macerata, Italy; Fellow, Von Humboldt Foundation Fellowship for Experienced Researchers: host Institute Max Planck Institute for Competition and Innovation, Munich, Germany.

Hendrik Vanhees, Professor at the Faculty of Law at the University of Antwerp and at the Faculty of Law at Ghent University. He teaches in both faculties courses on intellectual property. His research is focused on intellectual property rights. He has written several books and articles on different aspects of intellectual property rights. He is also editor of the *International Encyclopaedia of Intellectual Property Law*.

Estelle Vard, specialises in intellectual property law. She completed her Master's degree in Industrial Property Law at the Panthéon-Assas University (2014) and received her LLM from the Humboldt University of Berlin (2012). She wrote a Master's thesis on employees' inventions under French and German law.

Maria Rosa Fabara Vera, founding partner and General Manager of Fabara & Compañía Abogados. Throughout her almost twenty-five years of successful professional career, she has provided advice to a wide range of national and international clients on unfair competition practices, competition, mergers & acquisitions, IP cases and mining, among other specialisation fields. She is the head of the highly regarded IP Boutique practice, which is part of the firm, acknowledged for having a very active and professional multidisciplinary team specialised in life science matters and patents preparation and litigation, including engineering fields such as bio-chemical, chemical, pharmaceutical amidst other sectors. She has outstanding negotiation, litigation, and settlement skills and knowledge.

Pablo Fabara Vera is a head lawyer specialised in IP and in administrative and procedural law litigation. With over 12 years of experience, Pablo is renowned for his skills in Trademarks and in Patents Development as well as for effectively protecting his clients' intellectual property assets, both through advice and litigation. He has sound knowledge in authors' right protection, and in unfair competition practices linked to IP. He is a settlement lawyer experienced in alternative dispute resolution methods.

Rogier de Vrey, after obtaining his PhD (unfair competition law) with Utrecht University he started as an attorney, currently in the IP/TMT practice group of CMS Netherlands. Rogier is also university lecturer at the Department of Private Law, Centre of Intellectual Property Rights (CIER) of Utrecht University. He still lectures and publishes regularly, in particular on the enforcement of IP rights.

Darryl C. Wilson is the Attorneys Title Fund Professor of Law at Stetson University College of Law in Gulfport Florida. He directs students in the intellectual property internship program and is the faculty advisor to the student intellectual property law organisation. Darryl has practised and taught in the field of property law for 30 years both domestically and internationally. Darryl is the author of numerous publications and his primary area of focus within the intellectual property field is trademark and unfair competition law in both real space and cyberspace.

Sanna Wolk, Professor, is a researcher within the field of intellectual property at the Faculty of Law, Uppsala University. She has written a number of academic publications in her research field. President of the national Swedish AIPPI group. She is also co-chair of the AIPPI Copyright Committee. Sanna is further appointed EU copyright expert by the European Commission and arbitrator at the Swedish Industrial Inventions Board. In addition, Sanna is the initiative taker to establishing the Institute of Intellectual Property, Marketing and Competition Law (IMC) at the Faculty of Law.

Mateusz M. Wozniak, lawyer practicing for more than a decade in intellectual property and new technology law. PhD candidate at Jagiellonian University in Cracow. Former academic. Author or co-author of number of academic publications in the field of international public law and IP and IT law.

Elizabeta Zirnstein, Assistant Professor at the Faculty of Management, University of Primorska, Slovenia. She is one of the leading experts in the field of legal aspects of innovativeness in Slovenia. Elizabeta's research interest includes different areas of law, mainly intellectual property law, corporate law and labour law. She has authored a number of academic publications in her research field. Currently she is also Vice Dean at the Faculty.

Summary of Contents

List of Contributors v

Foreword lix

Preface lxi

List of Abbreviations lxiii

Part I
A Legislative Overview 1

Chapter 1
Introduction to Employees Intellectual Property Rights
Sanna Wolk 3

Chapter 2
Choice of Law
Gyooho Lee 5

Chapter 3
Jurisdiction
Benedetta Ubertazzi 29

Chapter 4
Recognition and Enforcement of Foreign Judgments
Marketa Trimble 49

Part II
National Legal Rules 53

A In Europe 53

CHAPTER 5
Ownership and Harmonisation Efforts at EU Level
Sanna Wolk 55

CHAPTER 6
Belgium
Hendrik Vanhees 63

CHAPTER 7
Denmark
Jeppe Brinck-Jensen, Morten Rosenmeier & Helene Egede Scotwin 81

CHAPTER 8
Finland
Niklas Bruun & Marja-Leena Mansala 105

CHAPTER 9
France
Michel Abello, Jérôme Tassi & Estelle Vard 121

CHAPTER 10
Germany
Jan Bernd Nordemann & Christian Czychowski 155

CHAPTER 11
Hungary
Gusztáv Bacher 205

CHAPTER 12
Italy
Andrea Ottolia 221

CHAPTER 13
Latvia
Ilmārs Šatovs 249

CHAPTER 14
Lithuania
Edita Ivanauskiene 259

CHAPTER 15
The Netherlands
Rogier de Vrey & Willem Hoorneman 283

Chapter 16
Poland
Kacper Szkalej 297

Chapter 17
Portugal
Lígia Gutierrez Setúbal 321

Chapter 18
Slovenia
Elizabeta Zirnstein & Špelca Mežnar 337

Chapter 19
Spain
Luis-Alfonso Durán & Laura Nadal 355

Chapter 20
Sweden
Sanna Wolk 369

Chapter 21
Switzerland
Yaniv Benhamou & Cyrill Rieder 391

Chapter 22
Turkey
Ekrem Soylu & Korcan Dericioğlu 419

Chapter 23
The United Kingdom
Belinda Isaac 435

B In Asia 469

Chapter 24
China
Ella Cheong & Li Luo 471

Chapter 25
Hong Kong
Ella Cheong 485

Chapter 26
Japan
Tsugizo Kubo 501

CHAPTER 27
Republic of Korea
Gyooho Lee 525

CHAPTER 28
Singapore
Susanna Leong 557

C In Oceania 589

CHAPTER 29
Australia
Rights in Intellectual Property Created in Employment
Ann Monotti 591

D In North America 631

CHAPTER 30
Canada
Norman Siebrasse 633

CHAPTER 31
United States of America
Joseph Jeffrey Hawley, Marketa Trimble & Darryl C. Wilson 649

E In Latin America 663

CHAPTER 32
Argentina
Martín Bensadon 665

CHAPTER 33
Brazil
Elisabeth Kasznar Fekete 679

CHAPTER 34
Chile
Cristóbal Porzio 699

CHAPTER 35
Colombia
José Roberto Herrera Díaz 721

CHAPTER 36
Ecuador
Maria Rosa Fabara Vera, Pablo Fabara Vera & Gerardo Naranjo Ormaza 737

CHAPTER 37
Mexico
Jorge Gomez, Liliana Arellano & Jonathan Rangel 747

CHAPTER 38
Uruguay
Gustavo Fischer 757

F In Africa 777

CHAPTER 39
Uganda
Ronald Kakungulu-Mayambala 779

Employees' Intellectual Property Rights: An Epilogue
Kacper Szkalej & Sanna Wolk 803

Table of Contents

List of Contributors v

Foreword lix

Preface lxi

List of Abbreviations lxiii

PART I
A Legislative Overview 1

CHAPTER 1
Introduction to Employees Intellectual Property Rights
Sanna Wolk 3

1 Background 3
2 Different Approaches in Different Law Systems 3
2.1 The Common Law System 4
2.2 The Civil Law System 4
3 Concluding Remarks 4

CHAPTER 2
Choice of Law
Gyooho Lee 5

1 Current Issues 5
2 Characterisation of Employment Relationships 6
3 Choice of Law 7
3.1 Public Policy or Mandatory Rules 17
3.2 Academic Proposal 18
3.2.1 ALI Principles 18
3.2.2 CLIP Principles 20

3.2.3 Joint Japanese-Korean Proposal 23
3.3 Some Remarks 24
Bibliography 25

CHAPTER 3
Jurisdiction
Benedetta Ubertazzi 29

1 International Framework 29
1.1 Harmonising International Jurisdiction Rules 29
1.2 Mosaic Approach Originated in the Territoriality Principle 30
1.3 Consolidation of Claims under the Brussels System 32
1.3.1 Exclusive Jurisdiction 32
1.3.2 General Jurisdiction 35
1.3.3 Infringement Jurisdiction 36
1.4 Consolidation of Claims under the Sets of Principles 39
2 Employees' Intellectual Property Rights 41
2.1 Consolidation of Claims under the Protocol on Jurisdiction 41
2.2 Consolidation of Claims under the Brussels System 42
Bibliography 43

CHAPTER 4
Recognition and Enforcement of Foreign Judgments
Marketa Trimble 49

1 Introduction 49
2 National Rules 49
3 Different Approaches 50
4 Conditions of Recognition and Enforcement 50
5 Enforcement of a Foreign Judgment 51

PART II
National Legal Rules 53

A In Europe 53

CHAPTER 5
Ownership and Harmonisation Efforts at EU Level
Sanna Wolk 55

1 Background 55
2 Employees' Copyright 56
3 Employees' Semiconductor Chips 58
4 Employees' Inventions 58
5 Employees' Designs 59

6 Employees' Plant Varieties 60
Bibliography 61

CHAPTER 6
Belgium
Hendrik Vanhees 63

1 Background 63
2 Employees' Copyright 64
2.1 The Legal Framework 64
2.2 Ownership of Rights 65
2.3 The Parties' Duty to Inform 68
2.4 Employee Remuneration Right 68
3 Employees' Semiconductor Chips 69
3.1 The Legal Framework 69
3.2 Ownership of Rights 69
3.3 The Parties' Duty to Inform 69
3.4 Employee Remuneration Right 70
4 Employees' Inventions 70
4.1 The Legal Framework 70
4.2 Ownership of Rights 70
4.3 The Parties' Duty to Inform 71
4.4 Employee Remuneration Right 72
5 Employees' Utility Models 72
6 Employees' Design Rights 72
6.1 The Legal Framework 72
6.2 Ownership of Rights 73
6.3 The Parties' Duty to Inform 75
6.4 Employee Remuneration Right 75
7 Employees' Plant Varieties 75
7.1 The Legal Framework 75
7.2 Ownership of Rights 76
7.3 The Parties' Duty to Inform 76
7.4 Employee Remuneration Right 76
8 Instances for Disputes 76
9 Concluding Remarks 78
Bibliography 78

CHAPTER 7
Denmark
Jeppe Brinck-Jensen, Morten Rosenmeier & Helene Egede Scotwin 81

1 Background 81
2 Employees' Copyright 82
2.1 The Legal Framework 82

2.2 Ownership of Rights 83
2.2.1 What Is the 'Customary Activity' of the Employer? 83
2.2.2 Cases Where the Employee Is under a Legal Obligation to Create Works 84
2.2.3 Droit Moral 85
2.2.4 The Work Must Have Been Created as Part of an Employment Obligation 85
2.2.5 Contracts on the Copyright of Employees 86
2.2.6 Cases in Which the Employer Changes the Customary Activity Without Agreement 88
2.2.7 Computer Programs 89
2.2.8 Neighbouring Rights 89
2.3 The Parties' Duty to Inform 91
2.4 Employees' Remuneration Right 91
3 Employees' Semiconductor Chips 91
3.1 The Legal Framework 91
3.2 Ownership of Rights 91
3.3 The Parties' Duty to Inform 92
3.4 Employees' Remuneration Right 92
4 Employees' Inventions 92
4.1 The Legal Framework 92
4.2 Ownership of Rights 92
4.2.1 Employees (Not Public Researchers) 92
4.2.1.1 An Invention and an Employee 92
4.2.1.2 Duty to Assign 93
4.2.1.3 Inventions Made after Termination of Employment 93
4.2.2 Public Researchers 94
4.2.2.1 An Invention and a Public Researcher 94
4.2.2.2 Duty to Assign 94
4.2.2.3 Cooperation with Third Parties 94
4.2.2.4 Ethical Considerations 94
4.2.2.5 Assignment from Others than Public Researchers 95
4.3 The Parties' Duty to Inform 95
4.3.1 Employees (Not Public Researchers) 95
4.3.2 Public Researchers 95
4.4 Employees' Remuneration Right 96
4.4.1 Employees (Not Public Researchers) 96
4.4.1.1 Right to Remuneration 96
4.4.1.2 Amount of Any Remuneration 96
4.4.1.3 Case Law on Remuneration 97
4.4.2 Public Researchers 99
4.4.2.1 Right to Compensation 99

4.4.2.2 Amount of Any Compensation 100
5 Employees' Utility Models 101
6 Employees' Design Rights 101
6.1 The Legal Framework 101
6.2 Ownership of Rights 101
6.3 The Parties' Duty to Inform 102
6.4 Employees' Remuneration Right 102
7 Employees' Plant Varieties 102
7.1 The Legal Framework 102
7.2 Ownership of Rights 102
7.3 The Parties' Duty to Inform 102
7.4 Employees' Remuneration Right 102
8 Instances for Disputes 103
9 Concluding Remarks 103
Bibliography 103

CHAPTER 8
Finland
Niklas Bruun & Marja-Leena Mansala 105

1 Background 105
2 Employees' Copyright 105
2.1 The Legal Framework 105
2.2 Ownership of Rights 107
2.2.1 Private Sector 107
2.2.2 Exemption for Researchers 108
2.3 The Parties' Duty to Inform 108
2.4 Employee Remuneration Right 108
3 Employees' Semiconductor Chips 108
3.1 The Legal Framework 108
3.2 Ownership of Rights 109
3.2.1 Private Sector 109
3.2.2 Public Sector 109
3.3 The Parties' Duty to Inform 109
3.4 Employees' Remuneration Right 109
4 Employees' Inventions 109
4.1 The Legal Framework 110
4.2 Ownership of Rights 111
4.3 The Parties' Duty to Inform 112
4.4 Employee Remuneration Right 112
5 Employees' Utility Models 114
5.1 The Legal Framework 114
5.2 Ownership of Rights 114
5.3 The Parties' Duty to Inform 114
5.4 Employee Remuneration Right 115

6 Employees' Design Rights 115
6.1 The Legal Framework 115
6.2 Ownership of Rights 115
6.3 The Parties' Duty to Inform 115
6.4 Employee Remuneration Right 115
7 Employees' Plant Breeder's Rights 116
7.1 The Legal Framework 116
7.2 Ownership of Rights 116
7.2.1 Private Sector 116
7.2.2 Public Sector 116
7.3 The Parties' Duty to Inform 116
7.4 Employees' Remuneration Right 117
8 Instances for Disputes 117
9 Concluding Remarks 117
Bibliography 118

CHAPTER 9
France
Michel Abello, Jérôme Tassi & Estelle Vard 121

1 Background 121
2 Employees' Copyright 122
2.1 The Legal Framework 122
2.2 Ownership of Rights 123
2.2.1 Transfer of Rights 124
2.2.2 Collective Works 125
2.2.3 Public Servants 126
2.2.4 Journalists 127
2.2.5 Employees' Computer Programs 128
2.2.6 Neighbouring Rights of Database Producers 128
2.2.7 Neighbouring Performers' Rights 129
2.3 The Parties' Duty to Inform 130
2.4 Employee Remuneration Right 130
3 Employees' Semiconductor Chips 131
3.1 The Legal Framework 131
3.2 Ownership of Rights 132
3.3 The Parties' Duty to Inform 132
3.4 Employee Remuneration Right 132
4 Employees' Inventions 132
4.1 The Legal Framework 132
4.2 Ownership of Rights 134
4.2.1 Inventions Made in the Course of the Employee's Normal Duties 135
4.2.2 Inventions Made Beyond the Employee's Normal Duties 135

4.2.3 Inventions Made Beyond Employment Duties 136
4.3 The Parties' Duty to Inform 136
4.3.1 The Procedure 136
4.3.2 Failure to Comply With the Duty to Inform 137
4.3.3 New Duty to Inform of the Filing and Granting of a Patent 138
4.4 Employee Remuneration Right 138
4.4.1 Supplementary Remuneration in the Private Sector 139
4.4.2 Supplementary Remuneration in the Public Sector 141
4.4.3 Fair Amount 142
5 Employees' Utility Models 143
6 Employees' Design Rights 143
6.1 The Legal Framework 143
6.2 Ownership of Rights 144
6.2.1 Ownership of a French Registered Design 144
6.2.2 Ownership of a Community Design and French Copyright 144
6.3 The Parties' Duty to Inform 145
6.4 Employee Remuneration Right 145
7 Employees' Plant Varieties 146
8 Instances for Disputes 146
8.1 The Commission of Conciliation 146
8.2 IP Courts 148
8.3 Labour Court 150
9 Concluding Remarks 150
Bibliography 150

CHAPTER 10
Germany
Jan Bernd Nordemann & Christian Czychowski 155

1 Background 155
2 Employees' Copyright 156
2.1 The Legal Framework 156
2.2 Ownership of Rights 156
2.2.1 Transfer of Economic Rights in Employees' Works 157
2.2.2 Inner Connection: Private Sector 158
2.2.3 Inner Connection: Public Sector 159
2.2.4 Scope of Transfer 159
2.2.5 Transfer of Rights Beyond the Object of the Business 160
2.2.6 Moral Rights 160
2.2.7 Free Works: Employee's Duty to Offer 162
2.2.8 Termination 163
2.2.9 Specific Provision on Employees' Software 164

2.3 The Parties' Duty to Inform 165
2.4 Employee Remuneration Right 165
2.4.1 Contractual Remuneration Right 166
2.4.2 Statutory Remuneration Rights 169
2.4.3 Remuneration for Rental and Lending 169
2.4.4 Cable Retransmission Right 170
2.4.5 Resale Right 171
2.4.6 Compensation for Limitations on Authors' Rights 171
3 Employees' Semiconductor Chips 173
3.1 The Legal Framework 173
3.2 Ownership of Rights 174
3.2.1 Allocation of Rights to Employee-Created Topographies 174
3.2.2 Termination 175
3.3 The Parties' Duty to Inform 175
3.4 Employee Remuneration Right 176
4 Employees' Inventions 176
4.1 The Legal Framework 177
4.2 Ownership of Rights 177
4.2.1 Allocation of Rights in Employees' Inventions 177
4.2.2 Employers' Duty of Registration 178
4.2.3 Ownership of Rights in University Employees' Inventions 179
4.2.4 Termination 180
4.3 The Parties' Duty to Inform 181
4.4 Employee Remuneration Right 181
4.4.1 The Inventive Value 182
4.4.2 The Share Factor 183
4.4.3 Remuneration in Practice 184
5 Employees' Utility Models 184
5.1 The Legal Framework 184
5.2 Ownership of Rights 185
5.2.1 Allocation of Rights in Employee's Inventions 185
5.2.2 Ownership of Rights in Inventions by University Employees 185
5.2.3 Termination 186
5.3 The Parties' Duty to Inform 186
5.4 Employee Remuneration Right 187
6 Employees' Design Rights 187
6.1 The Legal Framework 187
6.2 Ownership of Rights 187
6.2.1 General Allocation of Rights to Designs 187
6.2.2 Allocation of Rights to Designs in the Public Sector 190
6.2.3 Termination 190
6.3 The Parties' Duty to Inform 191

6.4 Employee Remuneration Right 191
6.4.1 Service Designs 191
6.4.2 Free Designs 193
7 Employees' Plant Breeder's Rights 193
7.1 The Legal Framework 193
7.2 Ownership of Rights 193
7.2.1 Assignment 194
7.2.2 Licence 195
7.2.3 Termination 196
7.3 The Parties' Duty to Inform 196
7.4 Employee Remuneration Right 197
8 Instances for Disputes 197
Bibliography 200

Chapter 11
Hungary
Gusztáv Bacher 205

1 Background 205
2 Employees' Copyright 205
2.1 The Legal Framework 205
2.2 Ownership of Rights 207
2.3 The Parties' Duty to Inform 207
2.4 Employee Remuneration Right 208
3 Employees' Semiconductor Chips 208
3.1 The Legal Framework 208
3.2 Ownership of Rights 208
3.3 The Parties' Duty to Inform 209
3.4 Employee Remuneration Right 209
4 Employees' Inventions 209
4.1 The Legal Framework 209
4.2 Ownership of Rights 209
4.3 The Parties' Duty to Inform 210
4.4 Employee Remuneration Right 210
4.4.1 Service Inventions 211
4.4.1.1 Remuneration on the Basis of the Licence Analogy 212
4.4.1.2 Lump Sum Payments 213
4.4.2 Employee Inventions 214
5 Employees' Utility Models 214
5.1 The Legal Framework 214
5.2 Ownership of Rights 215
5.3 The Parties' Duty to Inform 215
5.4 Employee Remuneration Right 215
6 Employees' Design Rights 215

6.1 The Legal Framework 215
6.2 Ownership of Rights 215
6.3 The Parties' Duty to Inform 216
6.4 Employee Remuneration Right 216
7 Employees' Plant Varieties 217
7.1 The Legal Framework 217
7.2 Ownership of Rights 217
7.3 The Parties' Duty to Inform 217
7.4 Employee Remuneration Right 217
8 Instances for Disputes 217
9 Concluding Remarks 218
Bibliography 218

CHAPTER 12
Italy
Andrea Ottolia 221

1 Background 221
2 Employees' Copyright 224
2.1 The Legal Framework 224
2.2 Ownership of Rights 225
2.2.1 Employees Moral Rights 229
2.3 The Parties' Duty to Inform 230
2.4 Employee Remuneration Right 230
3 Employees' Semiconductor Chips 230
3.1 The Legal Framework 230
3.2 Ownership of Rights 230
3.3 The Parties' Duty to Inform 231
3.4 Employee Remuneration Right 231
4 Employees' Inventions 231
4.1 The Legal Framework 231
4.2 Ownership of Rights 231
4.2.1 Distinguishing Between Service and Business Inventions 233
4.2.2 The Causal Nexus Between the Employment Relation and the Invention 234
4.2.3 Defining 'Occasional Inventions' 234
4.2.4 Article 64 Beyond the Employment Relationship 236
4.2.5 University Researchers' Inventions 238
4.2.5.1 Exceptional Nature of the Norm and Scope of Application 238
4.2.5.2 Contract-Based Solutions 239
4.3 The Parties' Duty to Inform 239
4.4 Employee Remuneration Right 240
4.4.1 The Equitable Reward under Article 64.2 240

4.4.2 Equitable Remuneration and Transfer of the Invention 241
5 Employees' Utility Models 241
6 Employees' Design Rights 242
7 Employees' Plant Varieties 242
8 Instances for Disputes 243
9 Concluding Remarks 243
Bibliography 244

CHAPTER 13
Latvia
Ilmārs Šatovs 249

1 Background 249
2 Employees' Copyright 249
2.1 The Legal Framework 249
2.2 Ownership of Rights 250
2.2.1 General Rule 250
2.2.2 Employees' Computer Programs 250
2.2.3 Neighbouring Rights 250
2.3 The Parties' Duty to Inform 251
2.4 Employee Remuneration Right 251
3 Employees' Semiconductor Chips 251
3.1 The Legal Framework 251
3.2 Ownership of Rights 251
3.3 The Parties' Duty to Inform 252
3.4 Employee Remuneration Right 252
4 Employees' Inventions 252
4.1 The Legal Framework 252
4.2 Ownership of Rights 252
4.2.1 General Rule 252
4.2.2 Inventions of Scientific Personnel 253
4.3 The Parties' Duty to Inform 254
4.4 Employee Remuneration Right 254
5 Employees' Utility Models 254
6 Employees' Design Rights 254
6.1 The Legal Framework 254
6.2 Ownership of Rights 255
6.3 The Parties' Duty to Inform 255
6.4 Employee Remuneration Right 255
7 Employees' Plant Varieties 255
7.1 The Legal Framework 255
7.2 Ownership of Rights 256
7.3 The Parties' Duty to Inform 256
7.4 Employee Remuneration Right 256
8 Instances for Disputes 257

9 Concluding Remarks 257
Bibliography 258

CHAPTER 14
Lithuania
Edita Ivanauskiene 259

1 Background 259
2 Employees' Copyright 260
2.1 The Legal Framework 260
2.2 Ownership of Rights 260
2.2.1 Employees' Copyright Economic and Moral Rights 262
2.2.2 Employee's Computer Programs 264
2.2.3 Producers' Rights to Audiovisual Works 264
2.2.4 Related Rights of Performers and Producers 265
2.2.5 Sui Generis Database Producers Rights 266
2.2.6 University Researchers' Copyright 266
2.3 The Employee's Duty to Give Notice 267
2.4 Employee Remuneration Right 267
3 Employees' Semiconductor Chips 268
3.1 The Legal Framework 268
3.2 Ownership of Rights 269
3.3 The Parties' Duty to Inform 270
3.4 Employee Remuneration Right 270
4 Employees' Inventions 270
4.1 The Legal Framework 270
4.2 Ownership of Rights 271
4.2.1 Employees' Inventions 271
4.2.2 University Researchers' Inventions 272
4.3 The Parties' Duty to Inform 272
4.4 Employee Remuneration Right 272
4.4.1 Remuneration 273
5 Employees' Utility Models 274
6 Employees' Design Rights 274
6.1 The Legal Framework 274
6.2 Ownership of Rights 274
6.3 The Parties' Duty to Inform 276
6.4 Employee Remuneration Right 276
7 Employees' Plant Varieties 277
7.1 The Legal Framework 277
7.2 Ownership of Rights 277
7.3 The Parties' Duty to Inform 278
7.4 Employee Remuneration Right 278
8 Instances for Disputes 278

9 Concluding Remarks 280
Bibliography 281

Chapter 15
The Netherlands
Rogier de Vrey & Willem Hoorneman 283

1 Background 283
2 Employees' Copyright 283
2.1 The Legal Framework 283
2.2 Ownership of Rights 283
2.2.1 Copyright 283
2.2.1.1 Employment and Software 284
2.2.1.2 Directory Law 284
2.2.1.3 Moral Rights 285
2.2.2 Neighbouring Rights 285
2.2.2.1 The Remuneration of the Employee 286
2.2.2.2 Moral Rights 286
2.2.3 Database Rights 287
2.3 The Parties' Duty to Inform 287
2.4 Employee Remuneration Right 287
3 Employees' Semiconductor Chips 288
3.1 The Legal Framework 288
3.2 Ownership of Rights 289
3.3 The Parties' Duty to Inform 289
3.4 Employees' Remuneration Right 289
4 Employees' Inventions 289
4.1 The Legal Framework 289
4.2 Ownership of Rights 290
4.3 The Parties' Duty to Inform 291
4.4 Employee Remuneration Right 291
5 Employees' Utility Models 292
6 Employees' Design Rights 292
6.1 The Legal Framework 292
6.2 Ownership of Rights 292
6.2.1 Ownership of Rights: Designs and Copyright Law 293
6.3 The Parties' Duty to Inform 293
6.4 Employees' Remuneration Right 293
7 Employees' Plant Varieties 293
7.1 The Legal Framework 293
7.2 Ownership of Rights 294
7.3 The Parties' Duty to Inform 294
7.4 Employee Remuneration Right 294
8 Instances for Disputes 295

9 Concluding Remarks 295
Bibliography 296

CHAPTER 16
Poland
Kacper Szkalej 297

1 Background 297
2 Employees' Copyright 299
2.1 The Legal Framework 299
2.2 Ownership of Rights 300
2.2.1 Employees' Works and Scientific Works 300
2.2.2 Research Results Financed by the National Centre for Research and Development 302
2.2.3 Computer Programs 302
2.3 The Parties' Duty to Inform 303
2.4 Employee Remuneration Right 304
3 Employees' Semiconductor Chips 305
3.1 The Legal Framework 305
3.2 Ownership of Rights 306
3.3 The Parties' Duty to Inform 307
3.4 Employee Remuneration Right 307
4 Employees' Inventions 308
4.1 The Legal Framework 308
4.2 Ownership of Rights 309
4.2.1 The Right to File a Patent Application and Transfer of Rights 309
4.2.2 Employer's Right to Exploit the Invention 310
4.3 The Parties' Duty to Inform 310
4.4 Employee Remuneration Right 311
5 Employees' Utility Models 312
5.1 The Legal Framework 312
5.2 Ownership of Rights 312
5.3 The Parties' Duty to Inform 313
5.4 Employee Remuneration Right 313
6 Employees' Design Rights 313
6.1 The Legal Framework 313
6.2 Ownership of Rights 313
6.2.1 Copyright in Designs 314
6.2.2 Industrial Property Rights in Designs 314
6.3 The Parties' Duty to Inform 314
6.4 Employee Remuneration Right 314
7 Employees' Plant Varieties 314
7.1 The Legal Framework 314
7.2 Ownership of Rights 315

7.3 The Parties' Duty to Inform 315
7.4 Employee Remuneration Right 316
8 Instances for Disputes 316
9 Concluding Remarks 317
Bibliography 318

Chapter 17
Portugal
Lígia Gutierrez Setúbal 321

1 Background 321
2 Employees' Copyright 322
2.1 The Legal Framework 322
2.2 Ownership of Rights 322
2.2.1 General Rules 322
2.2.2 Exceptions 323
2.2.2.1 Journalistic Works 323
2.2.2.2 Photographic Works 323
2.2.3 Computer Programs 324
2.3 The Parties' Duty to Inform 324
2.4 Employee Remuneration Right 324
3 Employees' Semiconductor Chips 325
3.1 The Legal Framework 325
3.2 Ownership of Rights 325
3.3 The Parties' Duty to Inform 325
3.4 Employee Remuneration Right 325
4 Employees' Inventions 325
4.1 The Legal Framework 325
4.2 Ownership of Rights 326
4.3 The Parties' Duty to Inform 328
4.4 Employee Remuneration Right 328
4.4.1 Employment Contract With a Clause Establishing the Employee's Monetary Remuneration for Their Inventive Activity 329
4.4.2 Employment Contract Without a Clause Establishing the Employee's Monetary Remuneration for Their Inventive Activity 329
4.4.3 When the Employee's Inventive Activity Is Not Covered by the Employment Contract 330
5 Employees' Utility Models 331
5.1 The Legal Framework 331
5.2 Ownership of Rights 331
5.3 The Parties' Duty to Inform 331
5.4 Employee Remuneration Right 331
6 Employees' Design Rights 331

6.1 The Legal Framework 331
6.2 Ownership of Rights 331
6.3 The Parties' Duty to Inform 332
6.4 Employee Remuneration Right 332
7 Employees' Plant Varieties 332
7.1 The Legal Framework 332
7.2 Ownership of Rights 333
7.3 The Parties' Duty to Inform 333
7.4 Employee Remuneration Right 334
8 Instances for Disputes 334
8.1 Arbitration 334
8.1.1 Mandatory Arbitration 334
8.1.2 Voluntary Arbitration 334
8.2 Labour Courts 335
8.3 Intellectual Property Court 335
9 Concluding Remarks 335
Bibliography 336

CHAPTER 18
Slovenia
Elizabeta Zirnstein & Špelca Mežnar 337

1 Background 337
2 Employees' Copyright 338
2.1 The Legal Framework 338
2.2 Ownership of Rights 338
2.2.1 Non-obligatory Nature of the Statutory Provision 339
2.2.2 Exceptions 339
2.3 The Parties' Duty to Inform 339
2.4 Employee Remuneration Right 340
2.4.1 Compulsory Licence 340
3 Employees' Semiconductor Chips 340
3.1 The Legal Framework 340
3.2 Ownership of Rights 341
3.3 The Parties' Duty to Inform 341
3.4 Employee Remuneration Right 341
4 Employees' Inventions 341
4.1 The Legal Framework 341
4.2 Ownership of Rights 342
4.2.1 Reporting of Inventions 343
4.2.2 Claiming Rights 344
4.2.3 Released Service Inventions 345
4.2.4 Free Inventions 345
4.2.5 Insolvency Proceedings 345
4.3 The Parties' Duty to Inform 346

4.4 Employee Remuneration Right 346
4.4.1 Arbitration 348
5 Employees' Utility Models 348
6 Employees' Design Rights 348
7 Employees' Plant Varieties 349
7.1 The Legal Framework 349
7.2 Ownership of Rights 349
7.3 The Parties' Duty to Inform 349
7.4 Employee Remuneration Right 350
8 Instances for Disputes 350
8.1 Copyright-Protected Works 350
8.2 Inventions and Utility Models 350
8.3 Plant Varieties and Semiconductor Chips 351
9 Concluding Remarks 352
Bibliography 352

CHAPTER 19
Spain
Luis-Alfonso Durán & Laura Nadal 355

1 Background 355
2 Employees' Copyright 355
2.1 The Legal Framework 355
2.2 Ownership of Rights 356
2.3 The Parties' Duty to Inform 356
2.4 Employee Remuneration Right 357
3 Employees' Semiconductor Chips 357
3.1 The Legal Framework 357
3.2 Ownership of Rights 357
3.3 The Parties' Duty to Inform 357
3.4 Employee Remuneration Right 357
4 Employees' Inventions 357
4.1 The Legal Framework 357
4.2 Ownership of Rights 358
4.2.1 Private and Public Companies 358
4.2.2 Universities 359
4.2.3 Public Research Bodies 359
4.3 The Parties' Duty to Inform 360
4.3.1 Private and Public Companies 360
4.3.2 University Researchers 360
4.3.3 Public Research Bodies 361
4.4 Employee Remuneration Right 361
4.4.1 Private and Public Companies 361
4.4.2 Universities 361
4.4.3 Public Research Bodies 362

5 Employees' Utility Models 362
5.1 The Legal Framework 362
5.2 Ownership of Rights 363
5.3 The Parties' Duty to Inform 363
5.4 Employee Remuneration Right 363
6 Employees' Design Rights 363
6.1 The Legal Framework 363
6.2 Ownership of Rights 363
6.3 The Parties' Duty to Inform 363
6.4 Employee Remuneration Right 364
7 Employees' Plant Varieties 364
7.1 The Legal Framework 364
7.2 Ownership of Rights 364
7.3 The Parties' Duty to Inform 364
7.4 Employee Remuneration Right 364
8 Instances for Disputes 365
9 Concluding Remarks 365
Bibliography 366

Chapter 20
Sweden
Sanna Wolk 369

1 Background 369
2 Employees' Copyright 370
2.1 The Legal Framework 370
2.2 Ownership of Rights 370
2.2.1 Employees' Copyright and Personal Neighbouring Rights 371
2.2.2 Employees' Computer Programs 373
2.2.3 Neighbouring Producers Rights 375
2.2.4 University Researchers' Copyright 376
2.3 The Employees' Duty to Give Notice 376
2.4 Employee Remuneration Right 377
3 Employees' Semiconductor Chips 377
3.1 The Legal Framework 377
3.2 Ownership of Rights 377
3.3 Employees' Duty to Give Notice 378
3.4 Employee Remuneration Right 378
4 Employees' Inventions 378
4.1 The Legal Framework 378
4.2 Ownership of Rights 379
4.2.1 Employees in the Private and Public Sectors 379
4.2.2 University Researchers' Inventions 380
4.3 The Employees' Duty to Give Notice 380

4.4 Employee Remuneration Right 381
4.4.1 Mandatory Reasonable Remuneration 381
5 Employees' Utility Models 383
6 Employees' Design Rights 383
6.1 The Legal Framework 383
6.2 Ownership of Rights 383
6.3 The Employee's Duty to Give Notice 384
6.4 Employee Remuneration Right 384
7 Employees' Plant Varieties 384
7.1 The Legal Framework 384
7.2 Ownership of Rights 385
7.3 The Employee's Duty to Give Notice 386
7.4 Employee Remuneration Right 386
8 Instances for Disputes 386
8.1 Disputes Concerning Employees' Copyright, Semiconductor Chips, Designs and Plant Varieties 386
8.2 Disputes Concerning Employees' Inventions 386
9 Concluding Remarks 387
9.1 Overlapping Intellectual Property Rights 387
9.2 Interconnection Between Copyright Law and Patent Law 388
9.3 Interconnection Between Designs Law and Copyright Law 389
9.4 Concluding Remarks Concerning Overlapping Rights 389
Bibliography 390

CHAPTER 21
Switzerland
Yaniv Benhamou & Cyrill Rieder 391

1 Background 391
2 Employees' Copyright 393
2.1 The Legal Framework 393
2.2 Ownership of Rights 393
2.2.1 Employees' Copyright and Performers' Rights 393
2.2.2 Employees' Computer Programs 398
2.2.3 Neighbouring Producer's Rights 400
2.2.4 Neighbouring Broadcasting Rights 401
2.3 The Parties' Duty to Inform 401
2.4 Employee's Remuneration Right 401
3 Employees' Semiconductor Chips 402
3.1 The Legal Framework 402
3.2 Ownership of Rights 403
3.3 The Parties' Duty to Inform 403
3.4 Employee's Remuneration Right 403
4 Employees' Inventions 403
4.1 The Legal Framework 403

4.2 Ownership of Rights 404
4.2.1 Employees in the Private Sector 404
4.2.2 Employees in the Public Sector 405
4.3 The Parties' Duty to Inform 406
4.4 Employee's Remuneration Right 406
5 Employees' Utility Models 407
6 Employees' Design Rights 407
6.1 The Legal Framework 407
6.2 Ownership of Rights 407
6.2.1 Employees in the Private Sector 407
6.2.2 Employees in the Public Sector 408
6.3 The Parties' Duty to Inform 408
6.4 Employee's Remuneration Right 409
7 Employees' Plant Varieties 409
7.1 The Legal Framework 409
7.2 Ownership of Rights 409
7.3 The Parties' Duty to Inform 410
7.4 Employee's Remuneration Right 411
8 Instances for Disputes 411
8.1 Dispute Concerning Employees' IPR 411
8.2 Disputes Concerning Employees' Inventions 412
8.3 Disputes Concerning Public Institutions 412
8.4 Competing Jurisdiction Between IPR and Labour Courts 413
9 Concluding Remarks 413
9.1 Overlapping Intellectual Property Rights 413
9.2 Discrepancy in IPRs 414
9.3 Adaptation of the Contract Once Transfer Has Been Made 414
Bibliography 415

CHAPTER 22
Turkey
Ekrem Soylu & Korcan Dericioğlu 419

1 Background 419
2 Employees' Copyright 420
2.1 The Legal Framework 420
2.2 Ownership of Rights 420
2.2.1 Nature of the Right 421
2.2.2 Definition of Employee 421
2.2.3 Criteria Required for the Exercise of Rights by the Employer 421
2.3 The Parties' Duty to Inform 422
2.4 Employees' Remuneration Right 422
3 Employees' Semiconductor Chips 422
3.1 The Legal Framework 422

3.2 Ownership of Rights 423
3.2.1 Ownership of Rights in Employment Contracts 423
3.2.2 Ownership of Rights of University Researchers 423
3.2.3 Ownership of Rights in Other Service Contracts 424
3.3 The Parties' Duty to Inform 424
3.3.1 University Researchers 424
3.3.2 Employees 424
3.4 Employees' Remuneration Right 424
3.4.1 University Researchers 424
3.4.2 Employees 424
4 Employees' Inventions 425
4.1 The Legal Framework 425
4.2 Ownership of Rights 425
4.3 Parties' Duty to Inform 426
4.3.1 Invention Notification 426
4.3.2 Information after Notifying – Pending Potential Patent Application 426
4.3.3 Information after Patent Application 427
4.4 Employees' Remuneration 427
5 Employees' Utility Models 428
6 Employees' Design Rights 429
6.1 The Legal Framework 429
6.2 Ownership of Rights 429
6.2.1 Ownership of Rights in Employment Contracts 429
6.2.2 Ownership of Rights of University Researchers 429
6.2.3 Ownership of Rights in Other Service Contracts 429
6.3 The Parties' Duty to Inform 430
6.3.1 Duty to Inform for University Researchers 430
6.3.2 Duty to Inform for Employees 430
6.4 Employees' Remuneration Right 430
6.4.1 For University Researchers 430
6.4.2 For Employees 430
7 Employees' Plant Varieties 430
7.1 The Legal Framework 430
7.2 Ownership of Rights 431
7.2.1 Ownership of Rights for Employees of Private Companies 431
7.2.2 Ownership of Rights for Other Service Contracts 431
7.2.3 Ownership of Rights by Employees of Public Institutions 431
7.3 The Parties' Duty to Inform 432
7.4 Employees' Remuneration Right 432
7.4.1 For Employees of Private Companies 432
7.4.2 For Employees of Public Institutions 432

8 Instances for Disputes 433
9 Concluding Remarks 433
9.1 General Remarks 433
9.2 Cumulative Protection 433
Bibliography 434

CHAPTER 23
The United Kingdom
Belinda Isaac 435

1 Background 435
2 Employees' Copyright 437
2.1 The Legal Framework 437
2.2 Ownership of Rights 438
2.2.1 Works Made by Directors 440
2.2.2 Special Situations 440
2.2.3 Performer's Rights 442
2.2.4 Moral Rights 443
2.2.5 Rental & Lending Rights 443
2.2.6 Artists' Resale Right 444
2.2.7 Database Rights 445
2.3 The Parties' Duty to Inform 445
2.4 Employee Remuneration Right 446
3 Employees' Semiconductor Chips 448
3.1 The Legal Framework 448
3.2 Ownership of Rights 448
3.3 The Parties' Duty to Inform 449
3.4 Employee Remuneration Right 449
4 Employees' Inventions 450
4.1 The Legal Framework 450
4.2 Ownership of Rights 450
4.3 The Parties' Duty to Inform 452
4.4 Employee Remuneration Right 454
5 Employees' Utility Models 458
6 Employees' Design Rights 458
6.1 The Legal Framework 458
6.2 Ownership of Rights 458
6.3 The Parties' Duty to Inform 459
6.4 Employee Remuneration Right 460
7 Employees' Plant Varieties 460
7.1 The Legal Framework 460
7.2 Ownership of Rights 460
7.3 The Parties' Duty to Inform 461
7.4 Employee Remuneration Right 461
8 Instances for Disputes 462

8.1 Applications for Employee Inventor Compensation 463
9 Concluding Remarks 463
9.1 Overlapping Intellectual Property Rights 463
9.2 Interaction between Moral Rights and Copyright 464
9.3 Interaction between Copyright and Design Rights 464
9.4 Interaction between Copyright and Patents 464
9.5 Concluding Remarks 465
Bibliography 465

B In Asia 469

CHAPTER 24
China
Ella Cheong & Li Luo 471

1 Background 471
2 Employees' Copyright 472
2.1 The Legal Framework 472
2.2 Ownership of Rights 472
2.2.1 Copyright in the Employment Work Owned by the Employee 472
2.2.2 Copyright in the Employment Work Owned by the Employer 473
2.2.3 Copyright in the Employment Work Co-owned by the Employer and Employee 473
2.3 The Parties' Duty to Inform 473
2.4 Employee Remuneration Right 473
3 Employees' Semiconductor Chips 474
3.1 The Legal Framework 474
3.2 Ownership of Rights 474
3.2.1 Layout Design Made in Execution of Employee's Duty 475
3.2.2 Layout Design Made Mainly by Using Employer's Materials and Technology 475
3.3 The Parties' Duty to Inform 476
3.4 Employee Remuneration Right 476
4 Employees' Inventions 476
4.1 The Legal Framework 476
4.2 Ownership of Rights 477
4.2.1 Inventions Made in Execution of Employment Tasks 477
4.2.2 Inventions Made Mainly by Using Employer's Materials and Technical Means 478
4.3 The Parties' Duty to Inform 478
4.4 Employee Remuneration Right 478
5 Employees' Utility Models 479
6 Employees' Design Rights 479

7 Employees' Plant Varieties 480
7.1 The Legal Framework 480
7.2 Ownership of Rights 480
7.3 The Parties' Duty to Inform 481
7.4 Employee Remuneration Right 481
8 Instances for Disputes 481
8.1 Litigation 481
8.2 Arbitration 482
8.3 Mediation 482
9 Concluding Remarks 482
Bibliography 483

CHAPTER 25
Hong Kong
Ella Cheong 485

1 Background 485
2 Employees' Copyright 485
2.1 The Legal Framework 485
2.2 Ownership of Rights 486
2.2.1 Private Sector 486
2.2.2 Public Sector 486
2.2.3 Case Law 487
2.3 The Parties' Duty to Inform 488
2.4 Employee Remuneration Right 488
3 Employees' Semiconductor Chips 488
3.1 The Legal Framework 488
3.2 Ownership of Rights 489
3.3 The Parties' Duty to Inform 489
3.4 Employee Remuneration Right 489
4 Employees' Inventions 489
4.1 The Legal Framework 489
4.2 Ownership of Rights 490
4.2.1 Ownership of Employee's Inventions 490
4.2.2 Two Additional Criteria in Determining Ownership 491
4.2.3 Employee's Rights in Jointly-Made Inventions 491
4.2.4 Employee's Rights Arising from Patents Outside Hong Kong 491
4.2.5 Employee's Rights Shall Not Be Diminished 491
4.3 The Parties' Duty to Inform 492
4.4 Employee Remuneration Right 492
4.4.1 The Invention Has Always Belonged to the Employer 492
4.4.2 The Invention Originally Belonging to the Employee but Is Subsequently Assigned or Exclusively Licensed to the Employer 493

4.4.3 Limitation Period for Seeking Compensation by Employee 493
4.4.4 Amount of Compensation 493
5 Employees' Utility Models 495
5.1 The Legal Framework 495
5.2 Ownership of Rights and Termination 495
5.3 The Parties' Duty to Inform 495
5.4 Employee Remuneration Right 495
6 Employees' Design Rights 495
6.1 The Legal Framework 495
6.2 Ownership of Rights and Termination 496
6.3 The Parties' Duty to Inform 496
6.4 Employee Remuneration Right 496
7 Employees' Plant Breeder's Rights 496
7.1 The Legal Framework 496
7.2 Ownership of Rights and Termination 497
7.3 The Parties' Duty to Inform 497
7.4 Employee Remuneration Right 497
8 Instances for Disputes 497
8.1 Copyright Tribunal 498
8.2 Labour Tribunal and Minor Employment Claims Adjudication Board 498
9 Concluding Remarks 498
Bibliography 498

CHAPTER 26
Japan
Tsugizo Kubo 501

1 Background 501
2 Employees' Copyright 502
2.1 The Legal Framework 503
2.2 Ownership of Rights 503
2.2.1 Conditions for Enabling the Authorship of Employees' Works to Be Attributed to the Employer 503
2.2.2 Works other than Computer Programs 504
2.2.3 Computer Programs 504
2.2.4 Influence of the Status of a Legal Entity's Works 504
2.3 The Parties' Duty to Inform 505
2.4 Employee Remuneration Right 505
3 Employees' Semiconductor Chips 505
3.1 The Legal Framework 505
3.2 Ownership of Rights 505
3.3 The Parties' Duty to Inform 506
3.4 Employee Remuneration Right 506

4 Employees' Inventions 506
4.1 The Legal Framework 506
4.1.1 History of the Employee Invention System 507
4.1.2 Basic Structure of Article 35 of the Patent Act 507
4.1.3 Provisions of Article 35 of the Patent Act on the Employee Invention System as Applicable to Inventions Created Prior to 1 April 2016 508
4.2 Ownership of Rights 509
4.2.1 Patent Rights Originally Belong to Employees 509
4.2.2 Conditions for Transferring Employee Inventions to Employers 509
4.3 The Parties' Duty to Inform 511
4.4 Employee Remuneration Right 511
4.5 The Latest Amendment of Article 35 of the Patent Act in 2015 as Applicable to Inventions Made on or after 1 April 2016 514
4.5.1 Outline of the Amendment of Article 35 (Paragraphs 1 to7) 515
5 Employees' Utility Models 517
6 Employees' Industrial Design Rights 518
7 Employees' New Plant Varieties 518
7.1 The Legal Framework 518
7.2 Ownership of Rights 518
7.2.1 Article 8 of the PVP Act 518
7.2.2 Breeder's Rights Originally Belong to Employees 519
7.3 The Parties' Duty to Inform 520
7.4 Employee Remuneration Right 520
8 Instances for Disputes 520
9 Concluding Remarks 522
Bibliography 523

CHAPTER 27
Republic of Korea
Gyooho Lee 525

1 Background 525
2 Employers' Copyright 527
2.1 The Legal Framework 527
2.2 Ownership of Rights 527
2.2.1 Ownership of Rights at Universities or Research Institutions 531
2.2.2 Case Law Related to Employee's Copyright 531
2.3 The Parties' Duty to Inform 532
2.4 Employee Remuneration Right 532
3 Employers' Semiconductor Chips 532

3.1 The Legal Framework 532
3.2 Ownership of Rights 533
3.2.1 Ownership of Rights at Universities or Research Institutions 536
3.3 The Parties' Duty to Inform 537
3.4 Employee Remuneration Right 537
4 Employers' Inventions 537
4.1 The Legal Framework 537
4.2 Ownership of Rights 538
4.2.1 Case Law 540
4.2.2 Ownership of Rights at Universities or Research Institutions 542
4.3 The Parties' Duty to Inform 543
4.4 Employee Remuneration Right 543
4.4.1 Case Law 545
5 Employees' Utility Models 547
6 Employers' Design Rights 547
6.1 The Legal Framework 547
6.2 Ownership of Rights 547
6.2.1 Ownership of Rights at Universities or Research Institutions 548
6.3 The Parties' Duty to Inform 548
6.4 Employee Remuneration Right 548
7 Employers' Plant Breeder's Rights 548
7.1 The Legal Framework 548
7.2 Ownership of Rights 549
7.2.1 Ownership of Rights at Universities or Research Institutions 549
7.2.2 Case Law 550
7.3 The Parties' Duty to Inform 550
7.4 Employee Remuneration Right 551
8 Instances for Disputes 551
9 Concluding Remarks 553
Bibliography 553

CHAPTER 28
Singapore
Susanna Leong 557

1 Background 557
2 Employees' Copyright 558
2.1 The Legal Framework 558
2.2 Ownership of Rights 559
2.2.1 Works Produced in the Course of Employment 559
2.2.2 Employed Journalists 560

2.2.3 Commissioned Photographs, Portraits and Engravings 561
2.2.4 Works Produced in Pursuant to Contracts Varying First Ownership 562
2.3 The Parties' Duty to Inform 562
2.4 Employee Remuneration Right 563
3 Employees' Semiconductor Chips 563
3.1 The Legal Framework 563
3.2 Ownership of Rights 563
3.3 The Parties' Duty to Inform 564
3.4 Employee Remuneration Right 564
4 Employees' Inventions 564
4.1 The Legal Framework 564
4.2 Ownership of Rights 565
4.2.1 Inventor or Joint Inventors: Who is an 'Inventor'? 565
4.2.2 Employee-Inventors 567
4.2.2.1 Who Is an 'Employee'? 567
4.2.2.2 Relevance of the Terms in the Employment Contract 573
4.2.2.3 Putting the Patent Grant in Issue 575
4.3 The Parties' Duty to Inform 579
4.3.1 Obligations of Employees 579
4.4 Employee Remuneration Right 580
4.4.1 Employee-Inventors' Compensation Rights 580
5 Employees' Utility Models 581
6 Employees' Design Rights 581
6.1 The Legal Framework 581
6.2 Ownership of Rights 582
6.3 The Parties' Duty to Inform 584
6.4 Employee Remuneration Right 584
7 Employees' Plant Varieties 584
7.1 The Legal Framework 584
7.2 Ownership of Rights 584
7.3 The Parties' Duty to Inform 585
7.4 Employee Remuneration Right 585
8 Instances for Disputes 585
9 Concluding Remarks 585
Bibliography 586

C In Oceania 589

CHAPTER 29
Australia
Rights in Intellectual Property Created in Employment
Ann Monotti 591

1 Background 591
2 Employees' Copyright 593
2.1 The Legal Framework 593
2.2 Ownership of Rights 594
2.2.1 Employee or Independent Contractor 595
2.2.2 In Pursuance of the Terms of His or Her Employment 596
2.2.3 Producers and Broadcasters 597
2.2.4 Moral Rights 599
2.2.5 University (Academic) Employees' Copyright 600
2.3 The Parties' Duty to Inform 604
2.4 Employee Remuneration Right 604
3 Employees' Semiconductor Chips 605
3.1 The Legal Framework 605
3.2 Ownership of Rights 605
3.2.1 Employees Generally 605
3.2.2 University Researchers' EL Rights 606
3.3 The Parties' Duty to Inform 606
3.4 Employee Remuneration Right 606
4 Employees' Inventions 607
4.1 The Legal Framework 607
4.2 Ownership of Rights 607
4.2.1 Express Provisions in the Employment Contract 608
4.2.2 Implied Duty to Assign Inventions 609
4.2.3 Fiduciary Obligations 611
4.2.4 University Employee Researchers' Rights 612
4.3 The Parties' Duty to Inform 615
4.4 Employee Remuneration Right 615
5 Employees' Utility Models 615
6 Employees' Design Rights 616
6.1 The Legal Framework 616
6.2 Ownership of Rights 616
6.2.1 University Researchers' Rights in Registered Designs 617
6.3 The Parties' Duty to Inform 618
6.4 Employee Remuneration Right 618
7 Employees' Plant Varieties 619
7.1 The Legal Framework 619
7.2 Ownership of Rights 619
7.2.1 University Researchers' Plant Breeder's Rights 619
7.3 The Parties' Duty to Inform 620

7.4 Employee Remuneration Right 620
8 Instances for Disputes 621
8.1 Jurisdiction Conferral 621
8.2 Appeal Jurisdiction 622
9 Concluding Remarks 623
Bibliography 625

D In North America 631

CHAPTER 30
Canada
Norman Siebrasse 633

1 Background 633
2 Employees' Copyright 633
2.1 The Legal Framework 633
2.2 Ownership of Rights 634
2.2.1 Basic Rule 634
2.2.1.1 Contract 'of Service' 634
2.2.1.2 In the Course of Employment 635
2.2.1.3 Agreement to the Contrary 636
2.2.1.4 Academic Employees 637
2.2.2 Newspaper Articles 637
2.2.3 Government Works 638
2.2.4 Reversion 638
2.2.5 Moral Rights 638
2.3 The Parties' Duty to Inform 639
2.4 Employee Remuneration Right 639
3 Employees' Semiconductor Chips 639
3.1 The Legal Framework 639
3.2 Ownership of Rights 639
3.3 The Parties' Duty to Inform 640
3.4 Employee Remuneration Right 640
4 Employees' Inventions 640
4.1 The Legal Framework 640
4.2 Ownership of Rights 640
4.2.1 Government Inventions 641
4.3 The Parties' Duty to Inform 642
4.4 Employee Remuneration Right 642
5 Employees' Utility Models 642
6 Employees' Design Rights 643
6.1 The Legal Framework 643
6.2 Ownership of Rights 643
6.3 The Parties' Duty to Inform 643
6.4 Employee Remuneration Right 644

7 Employees' Plant Varieties 644
7.1 The Legal Framework 644
7.2 Ownership of Rights 644
7.3 The Parties' Duty to Inform 645
7.4 Employee Remuneration Right 645
8 Instances for Disputes 645
9 Concluding Remarks 645
Bibliography 646

CHAPTER 31
United States of America
Joseph Jeffrey Hawley, Marketa Trimble & Darryl C. Wilson 649

1 Background 649
2 Employees' Copyright 650
2.1 The Legal Framework 650
2.2 Ownership of Rights 650
2.3 The Parties' Duty to Inform 651
2.4 Employee Remuneration Right 651
3 Employees' Semiconductor Chips 652
3.1 The Legal Framework 652
3.2 Ownership of Rights 652
3.3 The Parties' Duty to Inform 652
3.4 Employee Remuneration Right 652
4 Employees' Inventions 652
4.1 The Legal Framework 652
4.2 Ownership of Rights 652
4.3 The Parties' Duty to Inform 653
4.4 Employee Remuneration Right 653
5 Employees' Utility Models 653
6 Employees' Design Rights 654
7 Employees' Plant Varieties 654
8 Instances for Disputes 654
Bibliography 655

E In Latin America 663

CHAPTER 32
Argentina
Martín Bensadon 665

1 Background 665
2 Employees' Copyright 665
2.1 The Legal Framework 665
2.2 Ownership of Rights 666
2.3 The Parties' Duty to Inform 667

2.4 Employee Remuneration Right 667
3 Employees' Semiconductor Chips 667
3.1 The Legal Framework 667
3.2 Ownership of Rights 668
3.3 The Parties' Duty to Inform 669
3.4 Employee Remuneration Right 669
4 Employees' Inventions 669
4.1 The Legal Framework 669
4.2 Ownership of Rights 671
4.3 The Parties' Duty to Inform 672
4.4 Employee Remuneration Right 672
5 Employees' Utility Models 673
6 Employees' Design Rights 673
6.1 The Legal Framework 673
6.2 Ownership of Rights 673
6.3 The Parties' Duty to Inform 674
6.4 Employee Remuneration Right 674
7 Employees' Plant Varieties 674
7.1 The Legal Framework 674
7.2 Ownership of Rights 674
7.3 The Parties' Duty to Inform 674
7.4 Employee Remuneration Right 674
8 Instances for Disputes 675
8.1 Competent Court 675
8.2 Procedure 676
9 Concluding Remarks 676
Bibliography 676

CHAPTER 33
Brazil
Elisabeth Kasznar Fekete 679

1 Background 679
2 Employees' Copyright 682
2.1 The Legal Framework 682
2.2 Ownership of Rights 682
2.2.1 Copyright 682
2.2.2 Computer Programs 684
2.2.3 Jurisprudence Concerning Computer Programs 684
2.3 The Parties' Duty to Inform 685
2.4 Employee Remuneration Right 685
3 Employees' Semiconductor Chips 685
3.1 The Legal Framework 685
3.2 Ownership of Rights 685
3.3 The Parties' Duty to Inform 686

3.4 Employee Remuneration Right 686
4 Employees' Inventions 686
4.1 The Legal Framework 686
4.2 Ownership of Rights 687
4.2.1 Service Inventions 687
4.2.2 Free Inventions 688
4.2.3 Composite Inventions 688
4.2.4 Jurisprudence 689
4.3 The Parties' Duty to Inform 690
4.4 Employee Remuneration Right 690
5 Employees' Utility Models 691
6 Employees' Design Rights 691
7 Employees' Plant Varieties 691
7.1 The Legal Framework 691
7.2 Ownership of Rights 691
7.3 The Parties' Duty to Inform 692
7.4 Employee Remuneration Right 692
8 Instances for Disputes 692
8.1 Labour Relationships 692
8.2 Relationships Other Than Labour 693
9 Concluding Remarks 693
Bibliography 695

CHAPTER 34
Chile
Cristóbal Porzio 699

1 Background 699
2 Employees' Copyright 701
2.1 The Legal Framework 701
2.2 Ownership of Rights 702
2.2.1 Preliminary Considerations 702
2.2.2 Moral Rights of the Employee-Author 703
2.2.3 Economics Rights of the Employee-Author 705
2.2.4 Computer Programs/Software 706
2.2.5 Works Created by Employee-Authors of Media Companies, News and Information Agencies or Television or Broadcaster Stations 707
2.2.6 Databases and Other Compilations 708
2.2.7 Cinematographic Works 709
2.2.8 Works Produced by Public Employees in the Performance of Their Duties 710
2.3 The Parties' Duty to Inform 710
2.4 Employee Remuneration Right 710
3 Employees' Semiconductor Chips 711

4 Employees' Inventions 711
4.1 The Legal Framework 711
4.2 Ownership of Rights 711
4.2.1 Inventions Resulting from the Inventive or Creative Activity of an Employee, Where the Provision of Services is Contracted by the Employer 712
4.2.2 Inventions Owned by the Employee or Free Inventions 713
4.2.3 Inventions Not Related to the Provision of Services Contracted by the Employer, but Produced in the Framework of it or Taking Advantage of the Media that the Employer Has Made Available 714
4.2.4 Inventions of Employees of Universities or Investigation Institutions 715
4.3 The Parties' Duty to Inform 715
4.4 Employee Remuneration Right 716
4.4.1 Employee's Remuneration for Inventions Resulting from the Inventive or Creative Activity of an Employee, Where Provision of Services Is Contracted by the Employer 716
4.4.2 Employee's Remuneration for Inventions Not Related to the Provision of Services Contracted by the Employer, but Produced in the Framework of It or Taking Advantage of the Media That the Employer Makes Available to the Employee 717
5 Employees' Utility Models 717
6 Employees' Design Rights 717
7 Employees' Plant Breeders' Rights 717
8 Instances for Disputes 717
8.1 Resolution of Conflicts Related to the Attribution of Intellectual Property Rights 717
8.2 Resolution of Conflicts Related to the Attribution of Industrial Property Rights 718
9 Concluding Remarks 718
Bibliography 719

Chapter 35
Colombia
José Roberto Herrera Díaz 721

1 Background 721
2 Employees' Copyright 723
2.1 The Legal Framework 723
2.2 Ownership of Rights 723
2.3 The Parties' Duty to Inform 726
2.4 Employee Remuneration Right 726

3 Employees' Semiconductor Chips 726
3.1 The Legal Framework 726
3.2 Ownership of Rights 726
3.3 The Parties' Duty to Inform 727
3.4 Employee Remuneration Right 727
4 Employees' Inventions 727
4.1 The Legal Framework 727
4.2 Ownership of Rights 728
4.3 The Parties' Duty to Inform 729
4.4 Employee Remuneration Right 729
5 Employees' Utility Models 729
6 Employees' Design Models 729
6.1 The Legal Framework 729
6.2 Ownership of Rights 730
6.3 The Parties' Duty to Inform 730
6.4 Employee Remuneration Right 730
7 Employees' Plant Varieties 730
7.1 The Legal Framework 730
7.2 Ownership of Rights 731
7.3 The Parties' Duty to Inform 731
7.4 Employee Remuneration Right 731
8 Instances for Disputes 731
9 Concluding Remarks 732
Bibliography 734

CHAPTER 36
Ecuador
Maria Rosa Fabara Vera, Pablo Fabara Vera & Gerardo Naranjo Ormaza 737

1 Background 737
2 Employees' Copyright 739
2.1 The Legal Framework 739
2.2 Ownership of Rights 739
2.3 The Parties' Duty to Inform 739
2.4 Employee Remuneration Right 740
3 Employees' Semiconductor Chips 740
3.1 The Legal Framework 740
3.2 Ownership of Rights 740
3.3 The Parties' Duty to Inform 740
3.4 Employee Remuneration Right 740
4 Employees' Inventions 741
4.1 The Legal Framework 741
4.2 Ownership of Rights 741
4.3 The Parties' Duty to Inform 741
4.4 Employee Remuneration Right 741

5 Employees' Utility Models 742
5.1 The Legal Framework 742
5.2 Ownership of Rights 742
5.3 The Parties' Duty to Inform 742
5.4 Employee Remuneration Right 742
6 Employees' Design Rights 742
6.1 The Legal Framework 742
6.2 Ownership of Rights 743
6.3 The Parties' Duty to Inform 743
6.4 Employee Remuneration Right 743
7 Employees' Plant Varieties 743
7.1 The Legal Framework 743
7.2 Ownership of Rights 743
7.3 The Parties' Duty to Inform 743
7.4 Employee Remuneration Right 744
8 Instances for Disputes 744
9 Concluding Remarks 744

CHAPTER 37
Mexico
Jorge Gomez, Liliana Arellano & Jonathan Rangel 747

1 Background 747
2 Employees' Copyright 748
2.1 The Legal Framework 748
2.2 Ownership of Rights 748
2.2.1 Computer Programs/Software 748
2.2.2 Government Employees 749
2.3 The Parties' Duty to Inform 750
2.4 Employee Remuneration Right 750
3 Employees' Semiconductor Chips 750
3.1 The Legal Framework 750
3.2 Ownership of Rights 751
3.3 The Parties' Duty to Inform 751
3.4 Employee Remuneration Right 751
4 Employees' Inventions 751
4.1 The Legal Framework 751
4.2 Ownership of Rights 751
4.3 The Parties' Duty to Inform 752
4.4 Employee Remuneration Right 752
5 Employees' Utility Models 752
5.1 The Legal Framework 752
5.2 Ownership of Rights 752
5.3 The Parties' Duty to Inform 753
5.4 Employee Remuneration Right 753

6 Employees' Design Rights 753
6.1 The Legal Framework 753
6.2 Ownership of Rights 753
6.3 The Parties' Duty to Inform 754
6.4 Employee Remuneration Right 754
7 Employees' Plant Varieties 754
7.1 The Legal Framework 754
7.2 Ownership of Rights 754
7.3 The Parties' Duty to Inform 755
7.4 Employee Remuneration Right 755
8 Instances for Disputes 755
9 Concluding Remarks 755
Bibliography 756

CHAPTER 38
Uruguay
Gustavo Fischer 757

1 Background 757
2 Employees' Copyright 757
2.1 The Legal Framework 757
2.2 Ownership of Rights 758
2.2.1 The General Rule 759
2.2.2 Computer Programs 760
2.2.3 Employees' Copyright in Journalism 761
2.3 The Parties' Duty to Inform 761
2.4 Employee Remuneration Right 762
3 Employees' Semiconductor Chips 762
3.1 The Legal Framework 762
3.2 Ownership of Rights 762
3.3 The Parties' Duty to Inform 762
3.4 Employee Remuneration Right 762
4 Employees' Inventions 763
4.1 The Legal Framework 763
4.2 Ownership of Rights 764
4.2.1 General Remarks 764
4.2.2 Employee Inventions Situations 765
4.2.2.1 The Employee-Researcher Case 766
4.2.2.2 The Company-Provided Knowledge or Means Case 766
4.2.2.3 The Residual Criterion 768
4.3 The Parties' Duty to Inform 768
4.4 Employee Remuneration Right 770
5 Employees' Utility Models 771
6 Employees' Design Rights 771

7 Employees' Plant Varieties 772
7.1 The Legal Framework 772
7.2 Ownership of Rights 772
7.3 The Parties' Duty to Inform 773
7.4 Employee Remuneration Right 773
8 Instances for Disputes 773
9 Concluding Remarks 774
Bibliography 775

F In Africa 777

CHAPTER 39
Uganda
Ronald Kakungulu-Mayambala 779

1 Background 779
2 Employees' Copyright 782
2.1 The Legal Framework 782
2.2 Ownership of Rights 783
2.2.1 Employee Works 783
2.2.2 Commissioned Works 784
2.3 The Parties' Duty to Inform 786
2.4 Employee Remuneration Right 786
3 Employees' Semi-conductor Chips 787
4 Employees' Inventions 787
4.1 The Legal Framework 787
4.1.1 Protection of Computer Programs 788
4.2 Ownership of Rights 788
4.2.1 Employees' Technovations 789
4.2.1.1 Infringement of Inventions and Remedies 789
4.2.1.2 Limitations on the Employer's Rights in Respect of the Invention 789
4.3 The Parties' Duty to Inform 790
4.4 Employee Remuneration Right 790
5 Employees' Utility Models 791
5.1 The Legal Framework 791
5.2 Ownership of Rights 792
5.2.1 Infringement of a Utility Model 792
5.2.2 Remedies for Infringement of a Utility Model 792
5.3 The Parties' Duty to Inform 793
5.4 Employee Remuneration Right 793
6 Employees' Design Rights 793
6.1 The Legal Framework 793
6.2 Ownership of Rights 794
6.2.1 Infringement of Industrial Design 794

6.2.2 Remedies for Infringement of Employees' Design Rights 795
6.3 The Parties' Duty to Inform 795
6.4 Employee Remuneration Right 795
7 Employees' Plant Varieties 796
7.1 The Legal Framework 796
7.2 Ownership of Rights 796
7.2.1 Rights of Plant Breeders 797
7.2.2 Infringement of Plant Breeders Rights and Remedies 797
7.3 The Parties' Duty to Inform 798
7.4 Employee Remuneration Right 798
8 Instances for Disputes 798
9 Concluding Remarks 800
Bibliography 800

Employees' Intellectual Property Rights: An Epilogue
Kacper Szkalej & Sanna Wolk 803

1 Overview 803
2 Are Uniform Rules Desirable? 804

Foreword

AIPPI, the International Association for the Protection of Intellectual Property, was founded in 1897 and is dedicated to the development, improvement, and legal protection of intellectual property. AIPPI is a non-affiliated, non-profit organisation headquartered in Switzerland, having approximately 9,000 members representing over 100 countries. The members of AIPPI include lawyers, attorneys, and agents working across all fields of intellectual property in corporate and private practice throughout the world, as well as academics, judges, government officials and other persons interested in intellectual property. AIPPI is organised into 66 National and Regional Groups.

The objective of AIPPI is to improve and promote the protection of intellectual property at both national and international levels. It does this by studying and comparing existing and proposed laws and policies relating to intellectual property, and working with both government and non-government organisations for the development, expansion and improvement of international and regional treaties and agreements, and national laws.

The current book is an example of how AIPPI in cooperation with a publisher like Kluwer can reach an even bigger audience and be of assistance in the daily work of the business, creators of works of art, inventors, lawyers, academics, of each and every one engaged in the process of creating and protecting objects of intellectual property rights.

The project was initiated by the Communications Committee of AIPPI based on an initial idea from Sanna Wolk, member of the Programme Committee of AIPPI. It was afterwards strongly supported by Kluwer. This is a joint effort to make public the substantial studies carried out by the national and regional groups of the Association in the preparation of the Resolutions adopted by its Executive Committee. The present book was inspired by Resolution on Q183 'Employers' rights to intellectual property', adopted in 2004 during the 39th Congress of AIPPI in Geneva. The time that run since then, and after ten years, the subject necessitated certain changes in the approach. The contributors to this edition are primarily AIPPI members, but there are also non-AIPPI members, who also supported the idea with their work. AIPPI is proud to stamp the final result as it is proposed to the readers of this book.

Felipe Claro
President of AIPPI
June 2015

Preface

The transfer of employees' intellectual property rights is crucial to the success of any business in today's knowledge-based and globalised economy. Given that increasingly more businesses establish entities in different countries and increasingly more employees cooperate across borders, it becomes essential for businesses to be able to establish the conditions under which ownership subsists in intellectual property created in employment relationships in various countries.

With the addition of Uganda, this revised and updated second edition of the book describes and analyses employers' acquisition of employees' intellectual property rights in 34 different jurisdictions collectively representing all habitable continents. Embarking upon a project that encapsulates contributors who are scattered around the world creates numerous challenges, particularly with regards to the structure a book such as this should take, and attributable, *inter alia*, to the amount of contributors, legal developments, structure of laws, way of writing, and academic creativity. The focal point of this book has been clear from the beginning, which is why we were forced to impose a layer of sections with a view of providing a common frame of reference and ensuring a clear and consistent body of chapters, sometimes at the cost of depriving contributors of the possibility to follow their own preferences. One size does not fit all, but we are therefore forever grateful to all contributors who agreed to become part of this project and without whom this book would not have been possible.

This publication is the result of a common effort and we wish to thank many people who contributed with their support in various ways during the course of the project. First we would like to express our deepest thanks to all contributing authors. Further, we wish to direct especial and sincere gratitude to Charters Macdonald-Brown who believed in and supported this project from the beginning to the end.

Lastly, the views expressed in this publication are those of the named authors or editors (as applicable).

Sanna Wolk
Kacper Szkalej
July 2017

List of Abbreviations

ACTA	Anti-Counterfeiting Trade Agreement
CJEU	Court of Justice of the European Union
EEC	European Economic Community
EC	European Council
EFTA	European Free Trade Association
EPC	European Patent Convention
EPO	European Patent Office
EU	European Union
ILO	The International Labour Organization
IP	Intellectual Property
IPRs	IP rights
WIPO	the World Intellectual Property Organization
WTO	The World Trade Organization

Other abbreviations and acronyms are defined in the chapters in which they appear.

PART I A Legislative Overview

Chapter 1

Introduction to Employees Intellectual Property Rights

Sanna Wolk

1 BACKGROUND

There are different types of intellectual property rights, such as trademarks, patents for inventions, designs of industrial products, copyright and neighbouring rights. Nowadays many companies are well aware of the value of intellectual property. Ownership and control of intellectual property rights is crucial to the success of any business and in order to maximise the value of intellectual property assets it is necessary to maintain and effectively manage all of their associated ownership rights. Intellectual property is created by various groups of persons, which can include contractors, employees or sub-contracted manpower. Most frequently, however, development, research and creative activities are still performed by employees. Traditionally, companies' most important assets were limited to fixed assets, such as real property, machines and equipment, and similar. Even though manpower has always been regarded as essential for a business, it has also been considered replaceable. Employees could be replaced without jeopardising the company and its future. However, in our service-oriented, technologically innovative economy, human capital, such as key persons with know-how and creative ideas regarding a company's products and services, is a valued asset.

2 DIFFERENT APPROACHES IN DIFFERENT LAW SYSTEMS

From a legal perspective, the view on the right of employers to employees' intellectual rights differs substantially between the two main legal systems in the world today, the common law system and the civil law system.

2.1 The Common Law System

In the common law system, as in the United States and in the United Kingdom, the investor (employer) benefits from its employees' intellectual creations. In those countries in many cases the employer is the initial owner of the employees' intellectual property rights produced in the course of the employment. The employer is treated as the first owner, but not deemed to be the author. Therefore the duration of copyright, e.g., is measured with reference to the life of the employed creator. Further, the national laws in these countries make it clear that contractual provisions, whether expressed or implied, can affect the employer's initial ownership. Differences in ownership can result from whether the author is an employee, or is self-employed or a contractor.

2.2 The Civil Law System

In the civil law system (otherwise called the Roman-Germanic or written law system), to which most of the countries of Continental Europe - Germany, France and the Nordic countries, e.g. - belong, a legal person such as an employer, may not generally be deemed the first holder of an intellectual property right. Those rights are normally linked to individual persons. Therefore, an employer may normally only obtain intellectual property rights by assignment by law or in contract.

3 CONCLUDING REMARKS

Employed persons often develop intellectual property rights and recent decades have brought an upsurge of interest in the legal ownership of intellectual property rights created by employees. This interest is reflected in international discussions from the beginning of the 1900s onwards. However, there are no international solutions regarding employees' intellectual property rights. At present the nature of both employees' and employers' rights is defined by national legislation.

Furthermore, there is no standard formula for the employers' right at national level. In some countries the relevant rules on the employer's right to employees' intellectual assets are to be found in the national intellectual property acts. In other countries and for certain intellectual property rights, a transfer from the employee to the employer follows from general principles of law. Consequently, the methods of identifying the owner of, and establishing rights over an employee's intellectual property assets, are relatively uncertain. Therefore, in our global economy with cross-border research and development, clarification of these questions in the employment contract or within a specific contract is vital to both parties.

CHAPTER 2

Choice of Law

Gyooho Lee

1 CURRENT ISSUES

A *choice of law provision* in an employment contract refers to the law which will be applied for the interpretation of the employment contract in a cross-border case. For instance, an IPR-related employment contract can include a choice of law provision, stating that 'the validity, interpretation, performance, and enforcement of this agreement shall be governed by and interpreted in accordance with the laws of country A'.[1] Sometimes, the choice-of-law provision can be combined with a venue provision in terms of the employment contract, although mandatory rules of the jurisdictions in which copyright subsists may nevertheless apply, despite a choice of forum and a choice of law. In particular, in some instances, which involve cases related to multi-state countries, the employment contract could embody a venue provision, and a choice of law provision as well.

Because choice of law rules vary in each country in connection with IPR-related employment contracts, employers are likely to avoid the application of the law of countries which favour the rights of employee inventors and authors over the rights of employers. Hence, it is very important for the international legal community to harmonise each country's laws to strike a balance between employee inventors and authors and their employers. It is of significant importance because about 80-90% of all inventions are made by employees.[2]

1. Steven L. Oberholtzer and Joshua E. Ney, *Licensing: Choice of Law and Venue*, at http://www.brinksgilson.com/files/landslide_mayjune_2011.pdf (accessed on 3 August 2014).
2. Christian Heinze, *Comment to Art. 3:503*, in *European Max Planck Group on Conflict of Laws in Intellectual Property, Conflict of Laws in Intellectual Property: The CLIP Principles and Commentary*, 279 (1st ed., Oxford University Press 2013) ('CLIP Principles and Commentary').

Are the courts of a forum country A expected to apply the law chosen by the parties to a dispute in question when the chosen law is the law of country B in IPR-related cross-border employment contracts? In some countries, the courts honour the parties' contract even though the chosen law is not the law of the forum. In other countries, courts prefer the application of the law of the forum to the application of the law designated in the parties' choice of law provision.

We can witness the increase in intellectual creations made by employees in the course of their duties.[3] From the perspective of private international law practitioners and scholars, the following issues need to be addressed in terms of intellectual property rights in connection with an employment relationship: (i) the establishment of an employment relationship; (ii) the initial ownership of intellectual property rights in an employment relationship; (iii) transfer and licensing of intellectual property rights in an employment relationship; (iv) infringement and remedies; and (v) security interests of intellectual property rights in an employment relationship.

2 CHARACTERISATION OF EMPLOYMENT RELATIONSHIPS

Before a court determines the applicable law for a dispute pertaining to an intellectual property right created by one or more employees, it should take a close look at the case to decide if the dispute has something to do with an employment relationship as a preliminary question.[4]

The definition and scope of *employee, employer*, and *employment contract* is a matter of characterisation. For example, the Court of Justice of the European Union has characterised an employment relationship by the fact that: (i) for a certain period of time a person performs services for, and under the direction of, another person in return for which they receive remuneration;[5] or (ii) the services create a lasting bond which brings the worker, to some extent, within the organisational framework of the business of the undertaking or employer.[6] The test applied in the courts of Member

3. Mireille M.M. van Eechoud, *Choice of Law in Copyright and Related Rights: Alternatives to the Lex Protectionis* § 6.3.3. (2003).
4. Heinze, *Comment to Art. 3:503*, in CLIP Principles and Commentary, p. 281.
5. CJEU Case 66/85 *Lawrie/Blum* [1986] ECR 2121, paragraphs 16-17; CJEU Case C-138/02 *Collins* [2004] ECR I-2703, paragraph 26; CJEU Case C-456/02 *Trojani* [2004] ECR I-7573, paragraph 15; and CJEU Case C-94/07 *Raccaneill v Max-Planck-Gesellschaft* [2008] ECR I-5939, paragraph 33. In particular, in *Raccaneill v Max-Planck-Gesellscahft* ('MPG'), the European Court of Justice held that 'Any person who pursues activities which are real and genuine, to the exclusion of activities on such a small scale as to be regarded as purely marginal and ancillary, must be regarded as a "worker".' Hence, in *Reccanelli*, the CJEU found that a researcher preparing a doctoral thesis on the basis of a grant contract concluded with MPG, must be regarded as a worker within the meaning of Art. 39 EC if he is called upon to perform as much work as a researcher preparing a doctoral thesis on the basis of a BAT/2 employment contract with MPG.
6. CJEU Case 266/85 *Schenavai v Kreischer* [1987] ECR 239, paragraph 16. In *Schenavai v Kreischer*, the European Court of Justice found that '[T]he obligation to be taken into consideration in a dispute concerning proceedings for the recovery of fees commenced by an architect commissioned to draw up plans for the building of houses is the contractual obligation which forms the actual basis of legal proceedings.'

States and other countries may or may not reflect this approach. A forum court determines whether a dispute in question arises out of an employment relationship from the perspective of private international law.

3 CHOICE OF LAW[7]

Issues concerning choice of law arising from disputes on intellectual property rights in connection with employment relationships need to be delved into at international, regional, and national levels.

Article 5(2)(2) of the Berne Convention[8] prescribes that '[T]he extent of protection, as well as the means of redress afforded to the author to protect his rights, shall be governed exclusively by the laws of the country where protection is claimed.' Article 6 of the Berne Convention provides for a choice of law rule in terms of moral rights. In accordance with Article 6 of the Berne Convention, *lex protectionis* is the law governing the means of redress for safeguarding the rights to claim authorship of the work and to object to any distortion, mutilation or other modification of, or other derogatory action in relation to, the said work.[9] In addition, Article 14*bis*(2)(a) of the Berne Convention states that 'Ownership of copyright in a cinematographic work shall be a matter for legislation in the country where protection is claimed.' Hence, Article 14*bis*(2)(a) of the Berne Convention stipulates that the law governing ownership of copyright in a cinematographic work is *lex protectionis*. Whether Article 5(2)(2) of the Berne Convention mandates the application of *lex protectionis* is highly debatable.[10]

Meanwhile, Article 60(1)(2) of the European Patent Convention (hereinafter EPC)[11] prescribes that:

> [T]he right to a European patent shall be determined in accordance with the law of the State in which the employee is mainly employed; if the State in which the employee is mainly employed cannot be determined, the law to be applied shall be that of the State in which the employer has the place of business to which the employee is attached.

Article 60(1)(2) of the EPC focuses on the employment relationship rather than on the employee in terms of a connecting factor.[12] The country in which the employee is mainly employed is the main connecting factor in accordance with this provision.

7. This part is elaborated under the auspices of projects MINECO DER 2013-44749, 'Mediacion, arbitraje y jurisdicion en el actual paradigma de justicia: integracion, "glocalizacion"', Derecho publico y ODR como alguno de sus retos.
8. It refers to Berne Convention for the Protection of Literary and Artistic Works, which was adopted on 9 September 1886 and became effective since 4 December 1887.
9. Art. 6(1) and (3) of the Berne Convention.
10. Josef Drexl, *Comment to Art. 3:201: Initial Ownership*, in CLIP Principles and Commentary, p. 244.
11. It refers to the Convention on the Grant of European Patents of 5 October 1973.
12. The American Law Institute, Intellectual Property: Principles Governing Jurisdiction, Choice of Law, and Judgments in Transnational Disputes, Section 311, Reporters' note (2008) ('ALI', 'ALI Principles').

Also, this provision goes on to state that the residual connecting factor is the country in which the employer has the place of business in that the employee may often reside their place of employment.[13] Article 60(1)(2) of the EPC appears to be inconsistent with Article 8 of Rome 1 Regulation.[14] Article 8(1) of the Rome I Regulation provides that:

> An individual employment contract shall be governed by the law chosen by the parties in accordance with Article 3. Such a choice of law may not, however, have the result of depriving the employee of the protection afforded to him by provisions that cannot be derogated from by agreement under the law that, in the absence of choice, would have been applicable pursuant to paragraphs 2, 3 and 4 of this Article.

Article 8(1) of the Rome I Regulation recognises party autonomy in terms of choice of law in an individual employment contract, for the interpretation of the contract and for determining its effect, while prohibiting such a choice of law from having the result of depriving the employee of the protection afforded to them by provisions that cannot be derogated from by agreement under the law applicable pursuant to the following fall-back provisions.

Article 8(2) to (4) of the Rome I Regulation provides fall-back provisions in the absence of choice of law made by the parties. Article 8(2)(1) of the Rome I Regulation prescribes that, in the absence of a choice of law made by the parties, the individual employment contract shall be governed by the law of the country in which or, failing that, from which the employee habitually carries out their work in performance of the contract. In this regard, the country where the work is habitually carried out shall not be deemed to have changed if they are temporarily employed in another country.[15] Where the law of the country in which or, failing that, from which the employee habitually carries out their work in performance of the contract cannot be determined, the contract shall be governed by the law of the country where the place of business through which the employee was engaged is situated.[16] Where it appears from the circumstances as a whole that the contract is more closely connected with a country other than the law of the country in which or, failing that, from which the employee habitually carries out their work in performance of the contract, or the law of the country where the place of business through which the employee was engaged is situated, the law of that other country shall apply.[17] The essential difference between the two provisions could be the possibility that Article 8(1) of the Rome I Regulation allows party autonomy in terms of choice of law for the interpretation of an

13. *Ibid.*
14. Regulation (EC) No. 593/2008 of the European Parliament and of the Council of 17 June 2008 on the law applicable to contractual obligations (Rome I).
15. Art. 8(2) 2 of the Rome I Regulation.
16. Art. 8(3) of the Rome I Regulation.
17. Art. 8(4) of the Rome I Regulation.

employment contract and to determine its effect, whereas Article 60(1)(2) EPC identifies the forum whose mandatory rules regarding the initial ownership of patents should apply.

Article 11(4) of the Plant Variety Right Regulation[18] provides that 'If the breeder is an employee, the entitlement to the Community plant variety right shall be determined in accordance with the national law applicable to the employment relationship in the context of which the variety was bred, or discovered and developed.' The national law in question may or may not be the law chosen for the interpretation of the employment contract.

In terms of a non-contractual obligation arising from an infringement of an intellectual property right of one or several employees, Article 8 of the Rome II Regulation[19] should be discussed. Under Rome II, the courts of all EU Member States other than Denmark must apply the same set of rules to determine the law governing non-contractual obligations arising between the parties in civil and commercial matters.[20]

Article 8(1) of the Rome II Regulation prescribes that 'The law applicable to a non-contractual obligation arising from an infringement of an intellectual property right shall be the law of the country for which protection is claimed'. Therefore, a non-contractual obligation arising from an infringement of an employees' intellectual property right will be governed by *lex protectionis* pursuant to Article 8(1) of the Rome II Regulation. According to Article 8(2) of the Rome II Regulation, the law of the country in which the act of infringement was committed will govern a non-contractual obligation arising from an infringement of a unitary Community intellectual property right, for any question that is not governed by the relevant Community instrument. The law applicable to a non-contractual obligation arising from an infringement of an intellectual property right may not be derogated from by an agreement based on party autonomy.[21]

Some national legislation and case law need to be explored because international or regional schemes have failed to provide for uniform and comprehensive choice of law rules in connection with intellectual property rights arising from an employment relationship.

According to Article 110(1) of the Swiss Private International Law Act, intellectual property rights are governed by *lex protectionis*. This provision also covers initial ownership of intellectual property rights.[22]

18. Regulation (EC) No. 2100/94 of the European Parliament and of the Council of 27 July 1994 on Community plant variety rights.
19. Regulation (EC) No. 864/2007 of the European Parliament and of the Council of 11 July 2007 on the law applicable to non-contractual obligations.
20. Assignment of intellectual property rights: drafting note, at http://us.practicallaw.com/2-200-8871 (accessed on 28 July 2014).
21. Arts. 8(3) and 14 of the Rome II Regulation.
22. Drexl, *Notes on Art. 3:201: Initial Ownership*, in CLIP Principles and Commentary, p. 241.

The same rules hold true in Austria,[23] Italy,[24] Belgium,[25] South Korea,[26] and China.[27] On the other hand, the laws of some other jurisdictions stick to a rule-of-origin approach. Good illustrations can be found in Greece,[28] Portugal,[29] and the US.[30] Some jurisdictions, such as the US, resort to case law in terms of a choice of law clause in the employment agreement. In the US, if a choice of law clause is included in the employment contract and there is a reasonable relationship between that state and the transaction, the choice of law provision will probably be given effect[31] unless it is not contrary to public policy of the relevant state.[32] Section 187 of the Restatement (Second) of Conflicts of Laws provides:

> (1) The law of the state chosen by the parties to govern their contractual rights and duties will be applied if the particular issue is one which the parties could have resolved by an explicit provision in their agreement directed to that issue.
> (2) The law of the state chosen by the parties to govern their contractual rights and duties will be applied, even if the particular issue is one which the parties could not have resolved by an explicit provision in their agreement directed to that issue, unless either
> (a) the chosen state has no substantial relationship to the parties or the transaction and there is no other reasonable basis for the parties' choice, or
> (b) application of the law of the chosen state would be contrary to a fundamental policy of a state which has a materially greater interest than the chosen state in the determination of the particular issue and which, under the rule of § 188, would be the state of the applicable law in the absence of an effective choice of law by the parties.

23. Art. 34(1) of Austrian Private International Law Act.
24. Art. 54(1) of Italian Private International Law Reform Act.
25. Art. 93(1) of Belgian Private International Law Code.
26. Art. 24 of Korean Private International Law Act.
27. Art. 48 of the Chinese Private International Law Act (prescribing that the ownership and contents of intellectual property rights shall be governed by *lex protectionis*), and Art. 50 of that Act (stating that the determination of tort liabilities for the infringement of intellectual property rights shall be governed by *lex protectionis*). See Mo Zhang, *Codified Choice of Law in China: Rules, Processes And Theoretic Underpinnings*, 37 N.C.J.Int'l L. and Com. Reg. 83, 143-44 (2011).
28. Art. 67 of Greek Copyright Act of 1993.
29. Art. 48(1) of Portuguese Civil Code.
30. *Itar-Tass Russian News Agency v Russian Kurier*, 153 F.3d 82 (2nd Cir. 1998).
31. Corp. Counsel's Guide to Intel. Prop. § 13:4 (2014).
32. *Howard v Nitro-Lift Technologies*, L.L.C., 2011 OK 98, 273 P. 3d 20, 331 I.E.R. Cas. (BNA) 902 (Okla. 2011), petition for cert. filed, 80 U.S.L.W. 3647 (U.S. May 14, 2012).

(3) In the absence of a contrary indication of intention, the reference is to the local law of the state of the chosen law.[33]

Therefore, if the parties opt for the law of a jurisdiction having a substantial relationship to the transaction and the governing law in such jurisdiction is not manifestly contrary to the law of another state having a substantial relationship to that transaction, their choice of law is likely to be honoured by any forum court.[34]

In this context, the Korean experience will give some guidance to establish clear-cut rules for intellectual property rights arising from an employment relationship because the Korean Private International Act provides for *lex protectionis* in connection with the protection of intellectual property rights and special choice of law for employment contracts, and because the Korean case law has made the relationship between the Berne Convention and the Korean Private International Act clear, which provides for *lex protectionis* in terms of the protection of intellectual property rights.

For instance, Article 24 of the Korean Private International Act,[35] the title of which is 'Protection of Intellectual Property Rights', stipulates that 'the protection of intellectual property rights shall be governed by the law of the place where such rights are violated'.

In terms of the governing law on the effect of trademark rights, the *'Myspace' Domain Name* case[36] ruled that:

---

33. S. 188 of the Restatement (Second) of Conflicts of Law prescribes that:

'The rights and duties of the parties with respect to an issue in contract are determined by the local law of the state which, with respect to that issue, has the most significant relationship to the transaction and the parties under the principles stated in § 6. In the absence of an effective choice of law by the parties, the contacts to be taken into account in applying the principles of § 6 to determine the law applicable to an issue include (a) the place of contracting, (b) the place of negotiation of the contract, (c) the place of performance, (d) the location of the subject matter of the contract, and (e) the domicil, residence, nationality, place of incorporation and place of business of the parties. These contracts are to be evaluated according to their relative importance with respect to the particular issue.

If the place of negotiating the contract and the place of performance are in the same state, the local law of this state will usually be applied, except as otherwise provided in §§ 189 to 199 and 203.'

S. 6 of the Restatement (Second) of Conflicts of Law reads:

'A court, subject to constitutional restrictions, will follow a statutory directive of its own state on choice of law.

When there is no such directive, the factors relevant to the choice of the applicable rule include (a) the needs of the interstate and international systems, (b) the relevant policies of the forum, (c) the relevant policies of other interested states and the relative interests of those states in the determination of the particular issue, (d) the protection of justified expectations, (e) the basic policies underlying the particular field of law, (f) certainty, predictability and uniformity of result, and (g) ease in the determination and application of the law to be applied.'

34. *Law of Electronic Commerce*, § 3.04 Conflict of Law Rules, 2013 WL 4285430 (C.C.H.) (2014).
35. Act No. 6465, comprehensively amended on 7 April 2001.
36. Judgment Rendered by Seoul District Court on 30 August 2007, Case No. 2006 Gaha.

> [t]he effect of a trademark right is governed by the registered State, which falls under the category of lex protectionis pursuant to Article 24 of the Private International Act, having the closest connection with the trademark right. Each country grants an exclusive right on a trademark throughout its application and registration process. Many countries have employed territoriality principles in terms of trademark rights. Accordingly, creation, transfer, and effect of a trademark registered in a country will be governed by the law of the country, and its effect will be limited only to the country's territory. In this regard, the State where protection is claimed is the registered State, so that the nation having closest connection with the trademark right is the registered State.

According to the *'Myspace' Domain Name* court's holding, applying the Anticybersquatting Consumer Protection Act ('ACPA') of the US to the infringement of a trademark registered in the USA, and thus imposing on a Korean defendant a duty not to use the domain name and a duty to transfer it to the plaintiff, is not acceptable from the standpoint of the fundamental ideology of the Korean Trademark Law scheme and is contrary to a social order prescribed in Article 10 of the Private International Act.[37] Hence, the court applied the Unfair Competition Act in Korea to the case in question instead of the ACPA of the USA.

The overall legal issues, including initial ownership, existence, and transfer, of intellectual property rights will be determined pursuant to the law of the state where protection is claimed. In the *Von Dutch* case[38] it was held that the law of the state where protection is claimed should govern all legal issues related to intellectual property rights inclusive of the determination as to initial ownership or existence, extinction, and assignability of the copyright even though Article 24 of the Korean Private International Act deals only with the infringement of intellectual property rights. Nonetheless, it is debatable whether the ruling in the *Von Dutch* case holds true in the context of disputes on intellectual property rights arising out of an employment relationship.

It is an important issue whether initial ownership of intellectual property rights arising from an employment relationship is governed by *lex protectionis*, the law of origin, *lex fori*, or the law applicable to the employment relationship. In terms of a choice of law rule on initial ownership of intellectual property rights arising from an employment relationship, some scholarly opinions and the Supreme Court in Korea

37. Id.
38. Judgment Rendered by Seoul High Court on 8 July 2008, Case No. 2007Na80093.

appear to favour the law applicable to the employment relationship[39] [40] while the *Von Dutch* case favours *lex protectionis*. In addition to those views there is a view that the law of the State in which an employee-inventor resides shall govern initial ownership

39. Dong-Won Kim, *The law applicable to overseas patent rights of work-for-hire invention*, Law and Technology, Vol. 10, No. 1, p. 28 (2014) (by citing the Judgment rendered by Seoul High Court on 8 December 2011 (Case No. 2011 Na 20210), which held that the grant of a free non-exclusive licence to an employer by operation of law would be governed by the law applicable to the employment relationship); Eon Suk Kim, *Invention and Work Made for Hire in Private International Law*, Korea Private International Law Journal, Vol. XVII, p. 346 (2011). See Art. 28 (Employment Contract) of the Korean Private International Act which prescribes that: 'In case of an employment contract, even if the parties choose the applicable law, the protection, given to the employee under the mandatory provisions of the country of the applicable laws designated under paragraph (2), shall not be deprived. In case the parties do not choose the applicable law, irrespective of the provisions of Art. 26, the employment contract concerned shall be governed by the law of the country where the employee habitually provides his service. In case the employee does not habitually provide his service within one country, the law of the country, where the business office of the employer who hires the employee is located, shall govern. In case of an employment contract, an employee may file a lawsuit against an employer in a country, where he habitually provides his service or even where he habitually provided his service for the last time. In case the employee does not or did not habitually provide his service within one country, he may file a lawsuit against an employer even in a country where the business office of the employer, who hired the employee, is or was located. In case of an employment contract, an employer may file a lawsuit against an employee only in the country where the habitual residence of the employee is located or where the employee habitually provides his service. The parties of an employment contract may, in writing, make an agreement on the international jurisdiction: Provided that such agreement shall be effective only in any of the following subparagraphs: 1. In case a dispute already occurred; or 2. In case filing a lawsuit with other courts in addition to the competent court under this Article is permitted to the employee.'

In addition, Article 26 (Objective Connection at Time of Decision of Applicable Law) of the Korean Private International Act stipulates that: 'In case the parties to a contract do not choose the applicable law, the contract shall be governed by the law of the country which is most closely connected with the contract. In case a party, according to the contract concerned, is to effect any performance under any of the following subparagraphs, the law of the country, where the habitual residence of the party is located at the time of the conclusion of the contract (the law of the country where the principal business office is located in case the party is a corporation or organisation), shall be presumed to be most closely connected: in case the contract is concluded by the party's occupational or business activities, the law of the country, where the business office of the party is located, shall be presumed to be most closely connected: 1. In case of assignment contracts, the performance of the assignor; 2. In case of lease contracts, the performance of the party who provides the things or the rights to use; and 3. In case of mandate contracts, contracts for work, or other similar contracts of service supplies, the performance of the service. In case of the contract subject to the right to immovables, the law of the country, where the immovables are located, shall be presumed to be most closely connected.'

40. Judgment rendered by the Korean Supreme Court on 15 January, 2015 (Case No. 2012 Da 4763), which held that the grant of a free non-exclusive licence to an employer by operation of law would be governed by the law applicable to the employment relationship. Its lower court decision was the judgment rendered by Seoul High Court on 8 December 2011 (Case No. 2011 Na 20210) mentioned in footnote 39. In this case, the issue was dealt with as a preliminary question because the plaintiff, a Korean company, sought an injunction to enjoin the defendant from interfering with the transnational business of the plaintiff before Seoul Central District Court in Korea. The Korean Supreme Court applied for the law applicable to the employment relationship even though its main issue concerns torts rather than infringement of intellectual properties. In this case, the defendant, a former employee of the plaintiff, patented his invention made in the course of his employment in his name in Canada and sent his letter to a Canadian company in order for the Canadian company not to transact with the plaintiff anymore because the goods manufactured by the plaintiff allegedly infringed his patent registered in Canada.

of the right to claim a patent, grant of licences, transferability of a patent right or a right to patent.[41] So, the relationship between Article 24 and Article 28 of the Korean Private International Act needs to be clarified. Therefore, the current Korean Private International Act needs to be amended to solve this problem. The solution to the problem is to set *lex protectionis* as the default rule pertaining to legal disputes on initial ownership of intellectual property rights while applying to initial ownership of an intellectual property right arising from an employment contract or other pre-existing relationship, the law applicable to that contract or relationship.[42]

There are two interesting cases which dealt with the relationship between Article 24 of the Private International Act and choice of law provision on the Berne Convention. The first one is the *Sonata of Temptation* case, [43] in which the facts were as follows:

Square Enix, a Japanese popular game maker, filed a civil suit against the co-defendants, Fantom Entertainment Group, and a Korean music video director before Seoul Central District Court in March 2007, alleging that Ivy's 2007 music video, entitled *Sonata of Temptation*, copied its 2005 sequel to the popular *Final Fantasy VII* franchise, *Final Fantasy VII Advent Children*. The plaintiff alleged that the defendants produced the said music video 'Sonata of Temptation' by exploiting plotlines, background, characters, and their hand movements, look, and outfitting without any authorisation of the plaintiff. The plaintiff claimed compensation for damages caused by the fact that the defendants violated its economic and moral rights, and elucidation ads necessary to recover his reputation caused by its moral right's infringement.

In addition to dealing with issues as to whether the court has international jurisdiction over the case at hand, the *Sonata of Temptation* case delved into the following issues:

(i) what law is applied to the case in question in terms of actions for damages caused by infringement of economic rights or moral rights;
(ii) whether co-defendants are liable for infringement of the plaintiff's economic right (i.e., reproduction right, right to create derivative works, broadcasting right, transmission right) and of its moral right (i.e., its right to maintain its integrity); and
(iii) whether the plaintiff is allowed to claim elucidation ads necessary to recover his reputation caused by its moral right's infringement.

The court held that it:

> ... deems the Japanese plaintiff entitled to the rights enjoyed by Korean nationals in Korea pursuant to Article 5 (1) of the Berne Convention, to which both countries

Here, *lex protectionis* is the Canadian law which does not allow the grant of a free non-exclusive licence to an employer by operation of law while the law applicable to the employment relationship is Korean law.

41. Kyung-Han Sohn, *The Law applicable to disputes on intellectual properties*, 78 Justice 183-184 (2004).
42. Kwang Hyun Suk, *Private International Law*, Pakyoung-sa, p. 289 (2013).
43. Judgment Rendered by Seoul Central District Court on 13 March 2008, Case No. 2007 Gahap53681.

> are signatories. Applying Article 5 (2) of the Berne Convention which provided that the extent of protection, as well as the means of redress afforded to the author shall be governed exclusively by the laws of the country where protection is claimed, we rule that the claim for injunction is governed by Korean law.[44]

The court went on to apply Article 6*bis*(3) of the Berne Convention which provides that 'the means of safeguarding moral rights shall be governed by the laws of the country where protection is claimed'. Hence, the court concluded that Korean law also governs the claim for recovery of the author's reputation. As to the claim for damages, the court deemed copyright infringement as tort and applied Korean law, consistent with Article 32 of the Korean Private International Law that tort cases are governed by the law of the place where the tortious act took place. The court found the defendant liable for infringement of the plaintiff's economic right and right to maintain the integrity and ordered it to pay KRW 300 million to the Japanese game maker for plagiarising the popular film *Final Fantasy VII*. However, the court dismissed the plaintiff's claim for elucidation ads because it failed to prove co-defendants' defamation against the plaintiff.

The *Sonata of Temptation* case is related to a cinematographic work which is a work for hire. In this regard the court omitted, or must have misunderstood, some issues in terms of international jurisdiction and choice of laws. I reach the following conclusions, criticising some holdings observed by the court

First of all, the court should have touched upon issues as to whether joinder of claims was available in the instant case.

Secondly, the court ought to have discussed the applicable law on the ownership of copyright insofar as the facts in the case were based on a cinematographic work as a work for hire. Unfortunately, the court seems to impliedly have followed the *lex fori* approach rather than the *lex loci protectionis* test without any further discussion in its holdings.

44. The relevant provisions of Korean Copyright Act ('KCA') were as follows: Art. 93 [Claim for Damages] [corresponding to Art. 125 of KCA of 2010]: '1.The owner's of authors' property rights or other rights protected by this Act (excluding authors' moral rights) (hereinafter referred to as 'owner of authors' property rights, etc.') may claim compensation for damages from a person who has infringed his rights intentionally or by negligence. 2. If the owner of authors' property rights, etc. claims compensation under section 1, the amount of profit which the infringing person has gained by his act of infringement shall be presumed to be the amount of damages sustained by the owner of authors' property rights, etc. 3. The owner of authors' property rights, etc. may claim, in addition to the amount of compensation for damages as referred to in section 2, an amount of compensation for damages equivalent to that which he could have earned by the ordinary exercise of his moral rights. 4. The infringing act of the person who infringed the copyright, right of publication, neighbouring right, or database producer's right shall be presumed to be carried out by negligence.

Art. 94 [Determination of the amount of compensation] [corresponding to Art. 126 of KCA of 2010]: 'When the court found that the victim's injury was incurred but that it is difficult for the court to determine the amount of compensation pursuant to Art. 93 [corresponding to Art. 125 of KCA of 2010], the court may determine reasonable amount of compensation by taking into account the gist of oral hearing and the outcome of evidential investigation.'

Art. 95 [corresponding to Art. 127 of KCA of 2010] [Right of Demanding Recovery of Honour, etc.]: 'The author or performer may demand the person who has infringed his moral rights intentionally or by negligence to take measures necessary for the recovery of his honour or reputation in return for or together with compensation for damages.'

Thirdly, the court applied Article 5(2) of the Berne Convention to injunctive relief based on infringement of moral rights whereas it applied Article 6 of the Convention to a claim for elucidation ads. I certainly agree with its holding on this claim.[45]

Fourthly, I am out of accord with the court's holding that characterised the action as an action for damages caused by torts and applied Article 32(1) of the Korean Private International Act to the case at hand.[46] I think that Article 5(2) of the Convention should be applied to damages claims based on violation of economic rights and Article 6*bis*(3) of the Convention to an action for damages incurred in violation of moral rights.

Unlike the *Sonata of Temptation* case, which applied Article 32 of the Korean International Private Act in an action to seek damages incurred by copyright infringement in terms of international copyright dispute, the *49 Things to Do in Your Life* case correctly applied Article 5(2) of the Berne Convention.[47]

The facts of *49 Things to Do in Your Life* case were as follows:

> A Chinese book called *49 Things to Do in Your Life* is a collection of moral anecdotes compiled and published by Tan Zhuoying and Wang Jing in 2003. The compilers' copyright was taken over that year by Shenyang Yuanliu. Shenyang Yuanliu then signed a publishing contract with the Beijing Industrial University's publishing firm. The Beijing Industrial University publisher later reached a deal with Wisdom House, the Korean office of the multinational publisher. Wisdom House selected 45 excerpts from the original book and added four others for the Korean book.
>
> Shenyang Yuanliu filed a compensation suit in 2007 against Wisdom House and Korea's eight largest bookstores including Kyobo Book Center, alleging that they infringed its copyright and ignored its call to suspend sales in 2007.
>
> The *49 Things to Do in Your Life* court should determine: (i) whether a Korean court has jurisdiction over this case; (ii) what law should be applied to a damages claim caused by copyright infringement; and (iii) whether the Beijing Industrial University publisher obtained a worldwide exclusive licence from Shenyang Yuanliu and Wisdom House et al., and had reason to believe that the Chinese publisher did. In this regard, jurisdiction issues will not be dealt with because they were explained above. As far as choice of law issues are concerned, the court ruled that:
>
> > Article 24 of the Korean Private International Act is supplemental to Article 5(2) of the Berne Convention ... Therefore, the former is pre-empted by the latter. As

45. Art. 5, 6*bis*, and 14*bis* of the Berne Convention.
46. Art. 2, 24, and 32 of Korean Private International Act [Act No. 6465, comprehensively amended on 7 April 2001].
47. Judgment Rendered by Seoul Central District Court on 20 June 2008, Case No. 2007 Gaha; Kwang Hyun Suk, *Comments on International Intellectual Property Law: Expressing My Regret at Some of Recent Lower Courts' Cases*, 3656 Law Times (9 June 2008); Gyooho Lee, *Some Issues on International Dispute Related to Entertainment Industries: Focused on Choice of Law*, 14 Korea Private International Law Journal 40, 65 (2008) ('Gyooho Lee, *Entertainment*'); But see Judgment Rendered by Seoul High Court on 23 September 2008, Case No. 2007Na127657 ('*Desperado* case') having applied Art. 24 of Private International Act rather than relevant provisions of the Berne Convention in terms of an action for damages incurred by infringement of economic rights and moral rights of a copyright holder. The *Desperado* court did not explain why it applied Art. 24 of Private International Act. See Sang Ick Ha, *International Dispute about the Copyright Issue and the International Private Law of Korea: A Brief Introduction about the Judgments [Rendered] by the Courts of Korea in 2008*, 14 Korea Private International Law Journal 3, 34 (2008).

a corollary, Article 5(2) of the Berne Convention will be applied in copyright infringement cases including an action to seek damages. According to its provision, Korean substantive law, the law of the country where protection is claimed, is applied in this case.

The court went on to hold that '[t]he fact that Wisdom House translated and published 45 excerpts with the approval of the Beijing Industrial University publisher and not with Shenyang Yuanliu is a violation of the latter's secondary copyright'. That is, the court observed that *49 Things to Do in Your Life*, the most sold book of 2005 in Korea, was illegal as its Korean publisher, Wisdom House, published the book via a contract with its Chinese publisher, not its copyright holder, because the Beijing Industrial University publisher had an exclusive licence only within the territory of China. The court ordered Wisdom House to recall and destroy all copies of the book and pay compensation to its Chinese copyright holder. Also, the court ruled that Wisdom House should pay KRW 300 million in compensation to the Chinese copyright holder, Shenyang Yuanliu Co., Ltd. The amount is equal to a 3% royalty on the total sales of 1.13 million copies in Korea.

3.1 Public Policy or Mandatory Rules

By virtue of Article 9 of the Rome I Regulation, the courts of the EU Member States may, in exceptional circumstances, apply exceptions on the basis of public policy and overriding mandatory provisions. In this regard, the term 'overriding mandatory provisions' should be differentiated from the phase 'provisions which cannot be derogated from by agreement' and should be interpreted more narrowly.[48] Hence, according to Article 9(1) and (2) of the Rome I Regulation the law governing an employment contract to intellectual property rights cannot restrict the application of the overriding mandatory provisions of the law of the forum, which refers to:

> [P]rovisions the respect for which is regarded as crucial by a country for safeguarding its public interests, such as its political, social or economic organisation, to such an extent that they are applicable to any situation falling within their scope, irrespective of the law otherwise applicable to the employment contract.[49]

Article 9(3) of the Rome I Regulation provides that:

> 'Effect may be given to the overriding mandatory provisions of the law of the country where the obligations arising out of the contract have to be or have been performed, in so far as those overriding mandatory provisions render the performance of the contract unlawful. In considering whether to give effect to those provisions. regard shall be had to their nature and purpose and to the consequences of their application or non-application.'

Article 9(3) of the Rome I Regulation provides for the criteria under which the overriding mandatory provisions may take precedence over application of the law

48. Recital 37 of the Rome I Regulation.
49. Art. 9(1)-(2) of the Rome I Regulation.

governing the employment contract. For example, intellectual property laws of a certain jurisdiction contain overriding mandatory provisions regarding ownership of intellectual property rights.[50]

3.2 Academic Proposal

The following academic projects have been completed to coordinate transnational intellectual property disputes. It should be noted that the following principles do not reiterate the existing rules and case laws of a certain country, which may sometimes be inconsistent with the existing international, regional, or national regimes. The following principles could aim in creating a model law that could provide influence over national,[51] regional, and international legislations.[52] Also, the following principles may assist courts and arbitrators in resolving transnational disputes relating to intellectual property rights and help parties shape their contractual and non-contractual transactions.[53]

3.2.1 ALI Principles[54]

General rules on the territoriality principle and party autonomy are prescribed under Sections 301 and 302 of the ALI Principles.

The law applicable to determine the existence, validity, duration, attributes, and infringement of intellectual property rights and the remedies for their infringement is the law of each state of registration for registered intellectual property rights and *lex protectionis* for unregistered intellectual property rights.[55] The law governing a non-obligation arising from an act of unfair competition is the law of each state in which direct and substantial damage results or is likely to result, without regard to the state or states in which the act giving rise to the damage occurred.[56]

The parties may agree at any time, including after a dispute arises, to designate a law that will govern all or part of their dispute, except for: (i) the validity and

50. Art. 32b of the German Copyright Act on adequate remuneration for employees. See Paul Torremans, *Notes on Art. 3:901: Overriding mandatory provisions*, in CLIP Principles and Commentary, p. 376; William F. Patry, *National Copyright Laws Restricting Assignments*, 7 Patry on Copyright 25:57 (2014).
51. *Principles of Private International Law on Intellectual Property Rights*, in *Studies on the Principles of Private International Law on Intellectual Property Rights – A Japanese-Korean Joint Proposal from East Asian Points of View* 125 (Shoichiro Kidana ed., The Institute of Comparative Law, Waseda University, 2012) ('Kidana (ed.), Joint Japanese-Korean Proposal').
52. Part 1 of the CLIP Principles.
53. *Id.*
54. Principles Governing Jurisdiction, Choice of Law, and Judgments in Transnational Disputes (2008) ('ALI Principles').
55. S. 301(1) of the ALI Principles.
56. S. 301(2) of the ALI Principles.

maintenance of registered rights; (ii) the existence, attributes, transferability, and duration of rights, whether or not registered; and (iii) formal requirements for recording assignments and licensees.[57]

According to Section 311(2) of the ALI Principles, when the subject matter of the registered right arises out of a contractual or other pre-existing relationship between or among the parties, initial ownership is governed by the law that governs the contract or relationship. This provision was conceived based on Article 60 of the EPC.[58] In this context, the pre-existing relationship includes employment relationships.[59] When there is a contractual or other pre-existing relationship among the parties, the law applicable to that contract or relationship will govern even initial title to an unregistered trademark or trade dress.[60] Initial ownership of other rights that do not arise out of registration, such as copyright, is governed by the law of the State that governs an employment relationship when the subject matter was created pursuant to that relationship.[61] The provision on initial ownership of copyright whose subject matter was created pursuant to the employment relationship conflicts with Article 14*bis*(2)(a) of the Berne Convention, which provides that the initial ownership of copyright of a cinematographic work is governed by *lex protectionis*.[62]

As far as transfer of title and grants of licences are concerned, the default rule is that the contract law of the State designated by agreement of the parties governs a transfer of interest in, or grant of licence of, intellectual property rights.[63] In the absence of a choice of law agreement, the contract law of the State with the closest connection to the contract governs a transfer of interest in, or grant of licence of, intellectual property rights.[64] The employment contract is presumed to be most closely connected to the State in which the assignor or the licensor resided at the time of the execution of the contract.[65] In standard form agreements of transfer or licences, the contract law of the State designated in the contract will be applicable if the choice of law provision is valid.[66] Choice of law clauses in standard form agreements are valid only if the choice of law clause was reasonable and readily accessible to the non-drafting party at the time the agreement was concluded, and is available for subsequent reference by the court and the parties.[67] The reasonableness is determined in light of: (i) the closeness of the connection between the parties, the substance of the agreement, and the State the law of which is chosen; and (ii) the parties' residences, interests, and resources, taking particular account of the resource and sophistication of the non-drafting party.[68] If the choice of law clause in standard form agreements of transfer or

57. S. 302(1)-(2) of the ALI Principles.
58. ALI, ALI Principles, Reporters' notes on § 311, p. 135.
59. *Id.*
60. *Id.* at 136-137.
61. S. 313(1)(c) of ALI Principles.
62. ALI, ALI Principles, Reporters' notes on § 313, p. 144.
63. S. 315(1) of the ALI Principles.
64. S. 315(2)(1) of the ALI Principles.
65. S. 315(2)(2) of the ALI Principles.
66. S. 315(3)(a) of the ALI Principles.
67. S. 302(5)(a) of the ALI Principles.
68. S. 302(5)(b) of the ALI Principles.

licences is not valid, the applicable law is the contract law of the State with the closest connection to the contract.[69] For registered intellectual property rights, the law of the State of registration governs transfers by operation of intellectual property law[70] whereas, for unregistered intellectual property rights, *lex protectionis* governs transfers by operation of intellectual property law.[71] The laws pertaining to the existence, attributes, and the extent of transferability of intellectual property rights are applicable in terms of secured transactions to the same extent that they apply to other transactions involving intellectual property rights.[72]

In cases where the alleged infringing activity is ubiquitous and the laws of multiple countries are pleaded, the court may opt to apply to the issues of existence, validity, duration, attributes, and infringement of intellectual property rights and remedies for their infringement, the law or laws of the country or countries with close connections to the dispute.[73] If the concern arises that a choice of law clause in an employment contract is only favourable to the party who has strong bargaining power (an employer), public policy considerations and mandatory rules can come into play.

In accordance with Section 322 of the ALI Principles, the application of particular rules of foreign law is excluded, if such application leads to a result that is repugnant to public policy. Also, the court may give effect to the mandatory rules of any State with which the dispute has a close connection.[74] When the mandatory rules of two or more countries collide one another, and all of the countries have close connections to the dispute, the court must weigh up the relative stakes of each country.[75]

3.2.2 CLIP Principles

Article 3:201 of the CLIP Principles provides for choice of law rules on initial ownership in connection with intellectual property rights. The default rule is that *lex protectionis* governs initial ownership including authorship of a copyrighted work and entitlement to intellectual property rights arising out of registration.[76] The only exception to the application of *lex protectionis* can be found in Article 3:201(3) of the CLIP Principles. According to this exception, in the context of an employment contract, the law applicable to the right to claim a registered right is the law applicable to the contract.[77] In this regard, the governing law is the law chosen by the parties pursuant to Article 3:503(1) and Article 3:501 of the CLIP Principles.[78]

Article 3:503 of the CLIP Principles provides for a choice of law rule in relation to the transfer or licence of an intellectual property right arising from employment

69. S. 315(3)(b) of the ALI Principles.
70. S. 316(1) of the ALI Principles.
71. S. 316(2) of the ALI Principles.
72. S. 317(1) of the ALI Principles.
73. S. 321 of the ALI Principles.
74. S. 323 of the ALI Principles.
75. ALI, ALI Principles, Comment to s. 323, p. 160.
76. Art. 3:201(1) of the CLIP Principles.
77. Art. 3:201(3) of the CLIP Principles.
78. Drexl, *Comment to Art. 3:201: Initial ownership*, in CLIP Principles and Commentary, at p. 240.

relationships. Article 3:503 of the CLIP Principles is modelled on Article 8(1), (2), and (4) of the Rome I Regulation.[79] Article 3:503(1) of the CLIP Principles prescribes that:

> The mutual obligations of employer and employee in relation to the transfer or license of an intellectual property right arising from the employee's efforts, in particular the right of the employer to claim the intellectual property right and the right of the employee to remuneration, shall be governed by the law chosen by the parties in accordance with Article 3:501 of the CLIP Principles, which allows party autonomy. Such a choice of law may not, however, have the result of depriving the employee of the protection afforded to her/him by the provisions that cannot be derogated from by agreement under the law that, in the absence of choice, would have been applicable pursuant to paragraphs 2 and 3 of this Article.

This provision affords, in principle, party autonomy upon the contracting parties in terms of choice of law issues. However, the CLIP Principles set certain exceptions to this principle, providing that:

> To the extent that the law has not been chosen by the parties, the mutual obligations of employer and employee in relation to the transfer or license of an intellectual property right arising from the employee's efforts, in particular the right of the employer to claim the intellectual property right and the right of the employee to remuneration, shall be governed by the law of the State in which or, failing that, from which the employee habitually carries out his work in performance of the contract. The State where the work is habitually carried out shall not be deemed to have changed if the employee is temporarily employed in another State.[80]

Also, the CLIP Principles stipulate that 'Where it appears from the circumstances as a whole that the contract is more closely connected with a State other than that indicated in paragraph 2, the law of that other State shall apply.'[81] Hence, the fallback provisions, in the absence of a law chosen by the parties, make it clear that the applicable law will be as follows:

(i) the law of the state in which and, failing that, from which the employee habitually carries out his/her work in performance of the contract; or
(ii) the law of the state with which the employment contract is more closely connected other than that indicated in (i) above in cases where the circumstances as a whole are taken into account.[82]

The transferability of intellectual property rights and the question of whether the transfer or licence can be invoked against third parties are governed by *lex protectionis*.[83] The law applicable to initial co-ownership pertaining to a person or persons getting involved in an employment contract (i.e., employees or/and employers) shall

79. Heinze, *Notes on Art. 3:503: Employment relationships,* in CLIP Principles and Commentary, at p. 284.
80. Art. 3:503(2) of the CLIP Principles.
81. Art. 3:503(3) of the CLIP Principles.
82. Art. 3:503(2) to (3) of the CLIP Principles.
83. Art. 3:301 of the CLIP Principles.

be determined by the law applicable to the contract.[84] Transferability of the shares of each co-owner is governed by *lex protectionis*.[85]

The law applicable to transfers by operation of law and the duty to grant a licence is *lex protectionis*.[86] Rules on the transfers by operation of law and the duty to grant a licence can often be found in employment relationships.[87] Also, as a default rule, *lex protectionis* is the law applicable to the infringement of an intellectual property right and remedies for its infringement.[88] In disputes pertaining to infringement carried out through ubiquitous media such as the internet, the court may apply the law of the state having the closest connection with the infringement if the infringement arguably takes place in every state in which the signals can be received.[89] The parties to a dispute pertaining to the infringement of an intellectual property right may choose a choice of law rule on remedies claimed for the infringement by an agreement entered into before or after the dispute has arisen.[90] If the infringement is closely connected with a pre-existing relationship between the parties, such as an employment contract, the law applicable to the pre-existing relationship shall also govern the remedies for the infringement, unless the parties have expressly excluded the application of the law governing the pre-existing relationship with regard to the remedies for infringement, or unless it is clear from all the circumstances of the case that the claim is more closely connected with another state.[91]

The mutual rights and obligations of the parties arising from a contract to create or transfer a security right in intellectual property arising out of an employment relationship is governed by the law chosen by the parties.[92] In the absence of a choice of law clause in an employment contract, the mutual rights and obligations of the parties are governed by the law of the country where the grantor of the security has their habitual residence at the time of the conclusion of the contract.[93] As far as this is concerned, the mutual rights and obligations of the parties arising from a contract to create or transfer a security right in intellectual property arising out of an employment relationship, where it is clear from all the circumstances of the case that the employment contract is manifestly more closely connected with another country, the law of that other country applies.[94]

The application of a governing law specified by the CLIP Principles may have certain limitations based on overriding mandatory rules and public policy consideration.

84. Art. 3:401(1) and 3:201 of the CLIP Principles.
85. Art. 3:401(2) of the CLIP Principles.
86. Art. 3:507(1) of the CLIP Principles.
87. Axel Metzger, *Notes on Art. 3:507: Transfers by operation of law and compulsory licenses*, in CLIP Principles and Commentary, at p. 297-298.
88. Art. 3:601 of the CLIP Principles.
89. Art. 3:603(1) of the CLIP Principles.
90. Art. 3:606(1) of the CLIP Principles.
91. Art. 3:606(2) of the CLIP Principles.
92. Art. 3:801(1) of the CLIP Principles.
93. Art. 3:801(2)(1) of the CLIP Principles.
94. Art. 3:801(2)(2) of the CLIP Principles.

Effect may be given to the overriding mandatory provisions of another state with which the situation has a close connection.[95] Any provision in the CLIP Principles is prohibited from restricting the application of the overriding mandatory provisions of *lex fori*.[96] The definition of 'overriding mandatory provisions' is modelled on Article 9 of the Rome I Regulation.[97] Also, an application of a provision of the law of any state specified by the CLIP Principles may be refused only if such application is manifestly incompatible with the public policy of the forum.[98]

3.2.3 Joint Japanese-Korean Proposal

Unless otherwise provided by the Joint Japanese-Korean Proposal,[99] including a provision on the initial ownership of an intellectual property right, the default rule under the Joint Japanese-Korean Proposal is that 'All matters concerning an intellectual property right as such, including its existence, validity, content, and revocation, shall be governed by *lex protectionis*.' [100] Article 301(2) of the Joint Japanese-Korean Proposal presents an example of *lex protectionis* by stating that:

> lex protectionis is the law of the state for which protection is sought. In the case of a registered intellectual property right, this state is presumed to be the state in which that right is or will be registered, or which is deemed to be a state of registration under the convention to which that state belongs or the local law of that state.

Party autonomy is allowed, so that the parties may at any time choose a law that will govern all or part of their dispute.[101] In the absence of a choice of law, the law applicable to an alleged infringement[102] and remedies, and to unjust enrichment arising out of such infringement,[103] is *lex protectionis*. The applicable law to non-contractual obligation arising out of an act or acts initiating unfair competition is the law of the country in which a direct and substantial damage caused by the act(s) takes place or is expected to take place.[104] In the case of disputes concerned with an infringement that occurs or has occurred in unspecific and multiple states, the court shall apply the law of the state that has the closest connection with that infringement in its entirety.[105] So what law should be applied to IP disputes arising out of an employment relationship when an employment contract is made between the parties?

95. Art. 3:901(1) 1 of the CLIP Principles.
96. Art. 3:901(3) of the CLIP Principles.
97. Art. 3:901(2) of the CLIP Principles; See Torremans, *Notes on Art. 3:901: Overriding mandatory provisions*, in CLIP Principles and Commentary, at p. 375.
98. Art. 3:902 of the CLIP Principles.
99. It refers to the 'Principles of Private International Law on Intellectual Property Rights,' also known as 'Joint Proposal Drafted by Members of the Private International Law Association of Japan and Korea. See Kidana (ed.), Joint Japanese-Korean Proposal, at p. 171.
100. Art. 301(1) of the Joint Japanese-Korean Proposal.
101. Art. 302(2)1 of the Joint Japanese-Korean Proposal.
102. Art. 304(1) of the Joint Japanese-Korean Proposal.
103. Art. 304(2) of the Joint Japanese-Korean Proposal.
104. Art. 304(3) of the Joint Japanese-Korean Proposal.
105. Art. 306(1) of the Joint Japanese-Korean Proposal.

When an employment contract is made but lacks a provision on applicable law, the contract shall be governed by the law of the state with which the contract is most closely connected at the time of its conclusion.[106] The court shall apply the law of the state in which the transferee or licensee, in particular an employer-transferee or an employer-licensee, habitually resides, if it finds that the state is more closely connected with that employment contract by taking into account the express or implied obligation to exploit the right, exclusiveness of the licensed right, and the connection between the place of substantial exploitation of intellectual property and habitual residence of the parties.[107] The law governing an employment contract shall be applied to any infringement of an intellectual property right made by a tortious act of any of the contracting parties.[108] Article 308(4) of the Joint Japanese-Korean Proposal prescribes, under the title 'Initial ownership', that 'If the intellectual property right was vested pursuant to an employment, or any other pre-existing relationship, the applicable law is the law governing that contract or relationship.' The initial ownership of an intellectual property right arising from an employment relationship will be governed by the law governing the employment relationship pursuant to Article 308(4) of the Joint Japanese-Korean Proposal. This provision is similar to Section 313(1)(c) of the ALI Principles. In addition, the transferability of intellectual property rights is governed by *lex protectionis*.[109]

Mandatory rules[110] and public policy considerations[111] may not be derogated from by the application of choice of law rules to the disputes on intellectual property rights arising from employment relationship.

3.3 Some Remarks

A cross-border dispute on an intellectual property right arising from an employment relationship gives rise to many legal issues which call for the application of choice of law rules, especially initial ownership of an intellectual property right arising from an employment relationship needs to be governed by the law applicable to that relationship. The outcome of each academic project contain provisions helpful for the courts and legislators of each jurisdiction. I hope the outcomes of each academic project can contribute to harmonisation of international, regional and national regimes.

106. Art. 307(1) of the Joint Japanese-Korean Proposal.
107. Art. 307(2) of the Joint Japanese-Korean Proposal.
108. Art. 307(3) of the Joint Japanese-Korean Proposal.
109. Art. 309(1) of the Joint Japanese-Korean Proposal.
110. Art. 312 of the Joint Japanese-Korean Proposal prescribes that:

'The court shall apply the absolutely mandatory rules of the forum state.The court may give effect to the mandatory rules of foreign state regardless of applicable law on account of the nature and purpose of those rules, or the consequences of their application.'

111. Art. 313 of the Joint Japanese-Korean Proposal states that:

'The application of rules of foreign law under these Principles may be excluded if such application leads to a manifestly repugnant consequence for public policy in the forum state.'

BIBLIOGRAPHY

Table of Sources

Axel Metzger, *Notes on Article 3:507: Transfers by operation of law and compulsory licenses*, in *European Max Planck Group on Conflict of Laws in Intellectual Property, Conflict of Laws in Intellectual Property: The CLIP Principles and Commentary* 297-298 (1st ed., Oxford University Press 2013).

Christian Heinze, *Comment to Article 3:503*, in *European Max Planck Group on Conflict of Laws in Intellectual Property, Conflict of Laws in Intellectual Property: The CLIP Principles and Commentary* 279 (1st ed., Oxford University Press 2013).

Corp. Counsel's Guide to Intel. Prop. § 13:4 (2014).

Dong-Won Kim, *The law applicable to overseas patent rights of work-for-hire invention*, 10(1) Law & Technology 28 (2014).

Eon Suk Kim, *Invention and Work Made for Hire in Private International Law*, XVII Korea Private International Law Journal 346 (2011).

Gyooho Lee, *Some Issues on International Dispute Related to Entertainment Industries: Focused on Choice of Law*, 14 Korea Private International Law Journal 40, 65 (2008).

Josef Drexl, *Comment to Article 3:201: Initial Ownership*, in *European Max Planck Group on Conflict of Laws in Intellectual Property, Conflict of Laws in Intellectual Property: The CLIP Principles and Commentary* (1st ed., Oxford University Press 2013).

Kwang Hyun Suk, *Comments on International Intellectual Property Law: Expressing My Regret at Some of Recent Lower Courts' Cases*, 3656 Law Times (9 June 2008).

Kwang Hyun Suk, *Private International Law*, Pakyoung-sa, p. 289 (2013).

Kyung-Han Sohn, *The Law applicable to disputes on intellectual properties*, 78 Justice 183-184 (2004).

Mireille M.M. van Eechoud, *Choice of Law in Copyright and Related Rights: Alternatives to the Lex Protectionis* § 6.3.3. (2003).

Mo Zhang, *Codified Choice of Law in China: Rules, Processes And Theoretic Underpinnings*, 37 N.C.J.Int'l L. & Com. Reg. 83, 143-144 (2011).

Paul Torremans, *Notes on Article 3:901: Overriding mandatory provisions*, in *European Max Planck Group on Conflict of Laws in Intellectual Property, Conflict of Laws in Intellectual Property: The CLIP Principles and Commentary* 376 (1st ed., Oxford University Press 2013).

Sang Ick Ha, *International Dispute about the Copyright Issue and the International Private Law of Korea: A Brief Introduction about the Judgments [Rendered] by the Courts of Korea in 2008*, 14 Korea Private International Law Journal 3, 34 (2008).

The American Law Institute, *Intellectual Property: Principles Governing Jurisdic-*

tion, Choice of Law, and Judgments in Transnational Disputes, Section 311, Reporters' note (2008).

William F. Patry, *National Copyright Laws Restricting Assignments*, 7 Patry on Copyright 25:57 (2014).

Assignment of intellectual property rights: drafting note, at http://us. practicallaw. com/2-200-8871 (accessed 28 July 2014).

Steven L. Oberholtzer and Joshua E. Ney, *Licensing: Choice of Law and Venue*, at http://www.brinksgilson.com/files/landslide_mayjune_2011.pdf (accessed 3 August 2014).

Table of Legislation

Berne Convention for the Protection of Literary and Artistic Works (adopted on 1886, revised on 28 September 1979).

Convention on the Grant of European Patents (5 October 1973).

Regulation (EC) No. 2100/94 of the European Parliament and of the Council on Community plant variety rights (27 July 1994).

Regulation (EC) No. 593/2008 of the European Parliament and of the Council on the law applicable to contractual obligations (Rome I) (17 June 2008).

Regulation (EC) No. 864/2007 of the European Parliament and of the Council on the law applicable to non-contractual obligations (11 July 2007).

Directive 98/44/EC of the European Parliament and of the Council on the legal protection of biotechnological inventions (6 July 1998).

ALI Principles, The American Law Institute, *Intellectual Property: Principles Governing Jurisdiction, Choice of Law and Judgments in Transnational Disputes* (Chestnut, ALI Publishers, 2008).

CLIP Principles, 'Principles for Conflict of Laws in Intellectual Property, prepared by the European Max Planck Group on Conflict of Laws in Intellectual Property (CLIP)' (31 August 2011).

Joint Japanese-Korean Proposal, *Principles of Private International Law on Intellectual Property Rights*, in *Studies on the Principles of Private International Law on Intellectual Property Rights – A Japanese-Korean Joint Proposal from East Asian Points of View* (Shoichiro Kidana ed., The Institute of Comparative Law, Waseda University, 2012).

Restatement (Second) of Conflicts of Law.

Table of Cases

Schenavai v Kreischer, CJEU Case 266/85 ECR 239, paragraph 16 (1987).

Lawrie/Blum, CJEU Case 66/85 ECR 2121, paragraphs 16 and 17 (1986).

Collins, CJEU Case C-138/02 ECR I-2703, paragraph 26 (2004).

Trojani, CJEU Case C-456/02 ECR I-7573, paragraph 15 (2004).

Raccaneill v Max-Planck-Gesellscahft, CJEU Case C-94/07 ECR I-5939, paragraph 33 (2008).

Itar-Tass Russian News Agency v Russian Kurier, 153 F.3d 82 (2nd Cir. 1998).
Howard v Nitro-Lift Technologies, L.L.C., 2011 OK 98, 273 P. 3d 20, 331 I.E.R. Cas. (BNA) 902 (Okla. 2011), petition for cert. filed, 80 U.S.L.W. 3647 (U.S. 14 May 2012).
Judgment Rendered by the Korean Supreme Court, Case No. 2012 Da 4763 (15 January, 2015).
Judgment Rendered by Seoul Central District Court, Case No. 2007 Gahap53681 (13 March 2008).
Judgment Rendered by Seoul High Court, Case No. 2007Na80093 (8 July 2008).
Judgment Rendered by Seoul District Court, Case No. 2006 Gaha (30 August 2007).
Judgment rendered by Seoul High Court, Case No. 2011 Na 20210 (8 December 2011).
Judgment Rendered by Seoul Central District Court, Case No. 2007 Gaha (20 June 2008).
Judgment Rendered by Seoul High Court, Case No. 2007Na127657 ('*Desperado* case') (23 September 2008).

Chapter 3
Jurisdiction

Benedetta Ubertazzi

1 INTERNATIONAL FRAMEWORK

1.1 Harmonising International Jurisdiction Rules

Whereas substantive IP law is far advanced in terms of international harmonisation, issues of international jurisdiction still remain national. In fact, even though states constituted international governmental organisations that centralise all or part of the administrative procedures necessary for the granting of certain IPRs, those procedures typically give rise to a portfolio of national or EU IPRs enforceable only as territorial rights. Yet, these procedures do not contain significant rules addressing either the international jurisdiction of the courts of the Member States to adjudicate IP related claims, or the recognition and enforcement of foreign judgments in the area of IPRs (see however the Protocol on Jurisdiction and Recognition of the European Patent Convention). Additionally, even though existing universal international instruments on IP, especially the ones concluded in the past two decades (including TRIPS and ACTA), strongly emphasise the need to effectively enforce IPRs, and though enforcement of IPRs across national borders is crucial for their protection, those international instruments focus on purely domestic issues, ignoring transnational disputes. So, while the first steps in the direction of harmonising international jurisdiction on IPRs were undertaken by the Hague Conference in 1999 when it launched its preliminary draft proposal for an international Convention on Jurisdiction and Enforcement in Civil and Commercial Matters ('Draft Convention'), followed by a new draft in 2001, which included rules on cross-border IPRs issues, this text and its IPRs rules were very contentious and determined the failure of the entire Draft Convention.

Reasons for this current frame are at least twofold. First, since the failure of the Draft Convention was mainly attributable to the lack of consensus on IPRs rules; states are discouraged from negotiating any IP cross-border litigation instrument. Secondly, as with other issues that do not have adequate attention at the international level, there is a lack of pressure by interest groups to place the matters on the agenda. The current system, in fact, obliges enforcement of IPRs on a national basis, country by country according to the so-called mosaic approach. This causes economic inequalities, where the big multinational companies are generally able to finance litigation in every relevant jurisdiction, whereas the small and medium-sized enterprises typically do not have the same financial strength to defend each national proceeding. Indeed, empirical studies have shown in recent years that cross-border cases are growing in number and increasing in proportion to the total percentage of IPRs transnational litigation. Additionally, this increase is seen in cases affecting IPRs that are particularly relevant to the national economy of the countries involved. Therefore, given the frequent exploitation of IPRs beyond national borders and the need for their cross-border enforcement, an international agreement should remain the ultimate goal to be achieved.

Thus, recent negotiations at different international fora of mainly an academic nature proposed four sets of principles that can serve as model for future state's negotiations of an international agreement on IP and PIL. These sets of principles are the ALI Principles Governing Jurisdiction, Choice of Law, and Judgments in Transnational Disputes, adopted on 14 May 2007; the CLIP Principles for Conflict of Laws in Intellectual Property, prepared by the European Max Planck Group on Conflict of Laws in Intellectual Property, published on 31 August 2011; the Transparency Proposal on Jurisdiction, Choice of Law, Recognition and Enforcement of Foreign Judgments in Intellectual Property, finalised in 2009; and the Principles of Private International Law on Intellectual Property Rights drafted by members of the Private International Law Association of Korea and Japan in 2011 ('the four sets of principles').

Furthermore, in November 2010 the ILA instituted a Committee on Intellectual Property and Private International Law. Building upon the four sets of principles, this Committee is currently working toward the adoption of an ILA Resolution on IP and PIL, which could serve as a model for future international agreements promoting a more efficient adjudication of transnational IP disputes.[1]

1.2 Mosaic Approach Originated in the Territoriality Principle

Despite the absence of harmonised rules with regard to international jurisdiction related to cross-border IP cases, states reach similar solutions. The rationale for these similarities is to be found in the territoriality principle.

The territoriality principle has a substantive nature and indicates a spatial delimitation of each IPR, which exists only on the territory of the state that grants it (for registered IPRs) or that recognises it (for unregistered IPRs) (in the interest of simplicity, 'states granting or conferring IPRs' will also be used in reference to 'states recognising unregistered IPRs'). Thus, to the same intellectual creation, invention, or

1. See [2012] *Journal of Intellectual Property, Information Technology and E-Commerce Law*, available at http://www.jipitec.eu/issues/jipitec-3-3-2012.

sign correspond as many IPRs as there are countries that have granted them. This occurs in cases of EU IPRs, that are spatially delimited to the territory of the entire EU, and of European patents, which, despite being conferred after a unitary examination process, to the same holder, on the same invention, by a single body – the EPO – provide a bundle of parallel rights, which are referred to as national portions of the European patents. IPRs are then separate and independent rights, and the territoriality principle is grounded on the principle of independence, which is at the basis of all international treaties on the substantive harmonisation of intellectual property, among which the CUP, CUB and TRIPS.

This territorial substantive nature of IPRs builds upon various rationales. First of all, it derives from the peculiar character of registered IPRs, which come to existence through administrative acts of registration: then IPRs, as their respective registration acts, would be born as spatially limited to the territory of their conferring countries. Secondly, territoriality derives from the nature of registered and unregistered IPRs, which are monopolies of the states that grant the IPRs owners not only the right to exploit the intellectual creation, but also the right to exclude any other person from this exploitation: then, this right of exclusion would be granted by states just as limited to their respective territories. Thirdly, territoriality derives from the character of registered and unregistered IPRs which aim at reaching interests of a public nature, such as the development of culture and science, and therefore express public policies of their granting states: then these interests, like all those with a public character, would be intrinsically linked just to the territory of their respective countries.

More complex is the understanding of whether the territoriality principle affects the determination of courts with international jurisdiction and the scope of their authority. A traditional opinion maintains that it does. In fact, since protection of an IPR is limited to the territory of its granting state, IPRs do not exist, cannot be infringed and do not require any protection, in foreign countries. Hence IPRs should be adjudicated solely by the courts of their granting states, which shall exercise their authority only on circumstances that occurred in their respective national territories and with effects thereby limited. In addition, since changes in the administrative acts of registration might be necessary to protect IPRs, and since according to the so-called act of state doctrine administrative acts can be modified just by the national competent authorities, only the courts of the IPR-granting countries can order variation in their respective national acts of registration.

This traditional opinion, however, in cases of parallel IPRs obliges the interested parties, IPR owners or alleged infringers, to defend as many proceedings as are IPRs included in the litigation, according to a mosaic approach. Yet, economic globalisation and rapid development of information and communication technologies potentially accessible in the entire world increase at the same time the necessity of registering IPRs in more countries and raise the consequent risks of these rights being infringed in many states. Jurisdiction criteria grounded on territorial reasons result in more and more problems and the limits of the mosaic approach become apparent, leading to a multiplication of parallel proceedings, which raises the risk of conflicting judgments and the costs of litigation, and creates inequalities between big multinational companies and small and medium-sized enterprises.

1.3 Consolidation of Claims under the Brussels System

Starting from the 1960s, attempts to overcome the traditional understanding of the territoriality principle, particularly with regard to its influence on international jurisdiction, were initiated in Europe. These attempts aimed at internationalising cross-border IPRs litigation, concentrating adjudication of disputes before a single judicial authority even in cases of multi-state infringement and reducing the risks of conflicting judgments and the inequalities between the different players. These attempts resulted in the entry into force of the Brussels Convention of 1968, then the Brussels I Regulation, and now the Brussels I bis Regulation, as well as of the Lugano Convention (Brussels system).

This system, in fact, limits the traditional authority of the courts of the IPR-granting states and grounds this authority on the administrative nature of the IPRs acts of registration, rather than on the territoriality principle. Consequently, the Brussels system limits the scope of the (exclusive) jurisdiction of the courts of the IPR-granting state to disputes that can result in changes in the administrative acts of registration (i.e. validity of registered IPRs claims); allows courts other than the ones of the IPR-granting countries to adjudicate other multi-state parallel IPRs disputes (pure infringement claims of registered IPRs or non-registered IPRs claims of any nature); does not impose any territorial limitation to the authorities of these courts; and favours consolidation of cross-border IP litigation.

The following sections will discuss this consolidation in the context of exclusive, general and infringement jurisdiction.

1.3.1 Exclusive Jurisdiction

Article 16(4) of the Brussels Convention (corresponding to Article 24(4) of the Brussels I bis Regulation) establishes the exclusive jurisdiction of the court at the place of registration in proceedings relating to registration and validity of registered IPRs. This norm has been interpreted by the CJEU in two cases. The first case, *Duijnstee*, concerned an employee's invention. For the present purpose, it suffices to state that the CJEU noted that the concept 'proceedings concerned with the registration or validity of patents' under Article 16(4) of the Brussels Convention does not cover a dispute having as its subject matter the transfer of IPRs.

In the second case, *GAT*, a German company (LuK) alleged out of court that GAT, another German company, was in breach of, *inter alia*, certain French patents of which LuK was the registered proprietor. In response GAT filed a declaratory action for non-infringement before the Landgericht Düsseldorf, asking the court to establish that GAT was not in breach of LuK's patents, as such patents were invalid. The Oberlandesgericht Düsseldorf stayed the proceeding and asked the CJEU if the exclusive jurisdiction of Article 16(4) of the Brussels Convention only applies if proceedings with *erga omnes* effect are brought to declare the patent invalid, or rather extends also to proceedings concerned with the validity of patents where the defendant in a patent infringement action, or the claimant in a declaratory action to establish that a patent is not infringed, pleads that the patent is invalid and that there is also no patent infringement for that reason. The CJEU answered that the rule of exclusive jurisdiction under Article 16(4) of the Brussels Convention concerns all proceedings relating to the registration or validity of a patent, irrespective of whether that issue is raised by way of an action or a plea in objection. This CJEU statement was then codified by Article 24(4) of the Brussels I bis Regulation, according to which:

> in proceedings concerned with the registration or validity of patents, trade marks, designs, or other similar rights required to be deposited or registered, irrespective of whether the issue is raised by way of an action or as a defence, [shall have exclusive jurisdiction, regardless of the domicile of the parties] the courts of the Member State in which the deposit or registration has been applied for, has taken place or is under the terms of an instrument of the Union or an international convention deemed to have taken place.

In both the *Duijnstee* and the *GAT* cases, the CJEU reached its conclusions on the basis of the following main arguments. First, according to the CJEU, exclusive jurisdiction is justified by the fact that the issue of patents necessitates the involvement of the national administrative authorities. Here, the CJEU referenced the Jenard Report, according to which the granting of a patent is an exercise of national sovereignty. From this statement, it may be concluded that the Jenard Report is grounded in the act of state doctrine. Accordingly, the subsequent CJEU decisions citing it are also based on that doctrine. Secondly, the act of state doctrine is adopted to ground the scope of the exclusive jurisdiction rule. In fact, the exclusive jurisdiction rule covers only registered IPRs and does not apply to unregistered IPRs. The reason for this distinction is that registered IPRs are granted through a public act of concession, which implies the intervention of the national administration and therefore the exercise of national sovereignty, whereas unregistered IPRs come into being without these formalities. Thirdly, exclusive jurisdiction rules cover the proceedings related to the validity of IPRs, but do not encompass transfer and infringement claims. In fact proceedings concerning an IPR's validity challenge the validity of its granting acts, whereas proceedings related to transfer and infringement claims concern only the activity of private subjects. Since the granting act is an act of the national administration of the conferring state, this act is an expression of the state's sovereignty and validity proceedings would lead to an examination of the sovereign activity of foreign states.

Yet, the CJEU did not ground the exclusive jurisdiction rules on the territoriality principle. This might be explained in light of the *Tod's* case, where the CJEU stated that the CUB does not relate to PIL and therefore does not determine any applicable laws, but instead establishes a national treatment and a territoriality principle of a substantive nature. This statement may be applied to the CUP and TRIPS agreements and also to international procedural law. Interpreted in this way, the CJEU decision in *Tod's* determines that the territoriality principle of the CUP, CUB, and TRIPS agreements neither refers to applicable law nor posits any (implicit) exclusive jurisdiction rule.

Even though the exclusive jurisdiction rule of the Brussels system is not based on the territoriality principle, it obliges litigating transnational IPRs disputes country by country and expresses a strict territorial conception of IPRs. To overcome that strict territoriality, favouring consolidation of cross-border IP litigation, several EU and EFTA Member States have shown resistance to the *GAT* decision and have mitigated its scope in four different ways. A first way relates to cases where the foreign-registered IPR validity issue is raised in an infringement proceeding for which the forum state court has international jurisdiction. In those cases, the question arises as to whether the court competent to address the foreign IPR infringement claim should dismiss the case and decline jurisdiction when the validity of the foreign IPR is raised. Alternatively, the seized court could stay the infringement proceedings, give the party raising the validity issue a short period of time to institute validity proceedings in a court with exclusive jurisdiction to hear the claim and then continue the infringement proceedings once a decision on validity has been reached. All of this subject to the condition that if the relevant party fails to bring invalidity proceedings, it will be deemed having recognised the validity of the IPR, allowing thereby the infringement proceedings to continue. The first option – to dismiss the case in its entirety – is adopted for instance in the UK and in Italy. In contrast, the second option – to simply stay the proceeding with regard to the infringement claim – is followed in the Netherlands and in Switzerland.

The first option is an incorrect application of the Brussels system since this system does not extend exclusive jurisdiction rules to infringement issues, but covers only validity claims. Thus, it may be argued that the second way should be favoured. This conclusion is based on a literal interpretation of the *GAT* judgment, according to which there is nothing in *GAT* to stop a plaintiff from exercising their right under the Brussels Regulation to bring an action for infringement at the defendant's domicile or in the place of the harmful effect according to Article 2 of the Brussels Regulation and respectively to Article 5(3) of the same Regulation (corresponding to Article 4 and respectively Article 7(2) of the Brussels I bis Regulation). Further support for this conclusion can be found in CJEU's *Roche* decision that essentially provided that the issue of validity and the claim for infringement will be split between the courts of different Member States.

A second way to mitigate the scope of the exclusive jurisdiction rules of the Brussels system concerns cases where the validity claim is incidentally raised in the frame of a preliminary proceeding to award provisional measures under Article 31 of Regulation 44/2001, corresponding to Article 35 of Regulation Brussels I bis. In such cases, certain states invoked the non-application of the exclusive jurisdiction rules of the Brussels system. This approach was confirmed by the CJEU in the *Solvay* case, by stating that Article 22(4) of Regulation 44/2001 must be interpreted as not precluding, in cases where the validity claim is incidentally raised in the frame of a preliminary proceeding, the application of Article 31 of that Regulation.

A third way to mitigate the scope of the exclusive jurisdiction rules of the Brussels system concerns cases where non-EU registered IPRs are at stake. Those cases deal with the issue of whether the exclusive jurisdiction rules of the Brussels system have a so-called reflexive effect and therefore are extendable also to validity issues incidentally raised with respect to IPRs registered in non-Member States. Specifically, the question arises as to whether Article 22(4) of the Brussels I Regulation, now Article 24(4) of the Brussels I bis Regulation, creates extra EU jurisdiction thereby potentially extending its scope to cases where the IPR is registered in a non-EU Member State and preventing courts in EU Member States from exercising jurisdiction over validity claims incidentally raised involving foreign non-EU IPRs. This question has not been solved by the CJEU. Meanwhile EU Member States have dealt with this question autonomously. So, *inter alia*, in France the answer to this question is negative and therefore French courts might make a decision upon validity issues incidentally raised for foreign non-EU IPRs, but this decision would have effects limited to the parties of the dispute. In the UK the same opinion was followed in *obiter dicta* by the Court of Appeal in *Lucasfilm*, according to which Article 22(4) of the Regulation provides for exclusive jurisdiction, regardless of domicile, in respect of a variety of identified matters, including validity of patent rights, but none of that would apply to non-EU matters of the type excluded by Article 22(4): if proceedings for infringement of a US patent were brought against a person domiciled in a Member State, Article 22(4) does not exclude jurisdiction over validity. Thus, the defendant could challenge the validity of the patent.

A fourth way to mitigate the scope of the exclusive jurisdiction rule regarding registered IPRs validity issues incidentally raised concerns cases where the courts ascertain that the declining of jurisdiction by virtue of exclusive jurisdiction rules could lead to a denial of justice. In those cases, courts would rule that the exclusive jurisdiction rules of the Brussels system are contrary to the human right of access to a court. This conclusion was reached, *inter alia*, by the District Court of The Hague in 2007 in *Single Buoy Moorings v Bluewater*.

To sum up, the Brussels system establishes exclusive jurisdiction rules with regard to registered IPRs validity issues principally and incidentally raised. Yet, the scope of those rules may be mitigated in at least the four different ways discussed above. Each of these mitigations reflects a tendency that overcomes a strict territorial approach with respect to IPRs and favours consolidation of cross-border IP litigation.

1.3.2 General Jurisdiction

Further attempts to overcome the traditional understanding of the territoriality principle, favouring consolidation of cross-border IP litigation, relate to norms that vest jurisdiction in the courts of the defendant's domicile or residence. Among these norms stands Article 2 of the Brussels I Regulation, now Article 4 of the Brussels I bis Regulation, according to which jurisdiction regarding registered or unregistered IPRs infringement claims, or registered or unregistered IPRs transfer claims, or unregistered IPRs validity claims principally or incidentally raised, is afforded to the courts of the state where the defendant is domiciled.

This norm does not establish any additional territorial requirement. So, it does not impose the consideration of whether the IPR in question exists or was infringed in the forum state. Thus, this rule reflects a tendency that overcomes a strict territorial approach with respect to IPRs and that favours a consolidation of claims in cross-border IP disputes.

1.3.3 *Infringement Jurisdiction*

Further attempts to overcome the traditional understanding of the territoriality principle, favouring consolidation of cross-border IP litigation, relate to norms that vest jurisdiction in IPRs infringement cases. Among these norms stands Article 5(3) of the Brussels I Regulation, now Article 7(2) of the Brussels I bis Regulation, according to which 'a person domiciled in a Member State may, in another Member State, be sued: ... in matters relating to tort, delict or quasi-delict, in the courts for the place where the harmful event occurred or may occur'.

The CJEU has interpreted Article 5(3) in several cases related to torts other than IPRs infringements. In *Mines de potasse* the CJEU applied the ubiquity theory by maintaining that where the place of the happening of the event and the place where that event results in damage are not identical, the expression 'place where the harmful event occurred' in Article 5(3) of the Brussels Convention must be interpreted as covering both the place where the damage occurred (*locus damni*) and that of the event giving rise to it (*locus actus*). Thus, the plaintiff may sue the defendant in both places.

With regard to the place where the damage occurred, in *Marinari, Dumez France* and *Kronhofer* the CJEU clarified that the direct damage is the only relevant one to ground international jurisdiction, and that indirect consequences or pecuniary losses following a direct damage shall play no role in that respect. With regard to the place of the event giving rise to the damage, in *Melzer* the CJEU emphasised that in a case where there are several supposed perpetrators of a tort, Article 5(3) of the Brussels I Regulation does not allow jurisdiction to be established on the basis of the place of the event giving rise to the damage in favour of a court within whose jurisdiction the supposed perpetrator who is being sued did not act.

In *Melzer* and *Folien Fischer*, the CJEU maintained that Article 5(3) of the Brussels Regulation is based on the existence of a particularly close connection between the dispute and the courts of the place where the damage occurred or of the place of the event giving rise to it. In fact, these courts are objectively best placed to determine whether the elements establishing the liability of the person sued are present, since within their jurisdiction the relevant connecting factor is to be found. So, reasons relating to the sound administration of justice and the efficacious conduct of proceedings justify the attribution of jurisdiction to those courts.

In *Folien Fischer* the CJEU emphasised that Article 5(3) applies not only to actions for a positive declaration seeking to establish the occurrence of an infringement, but also to actions for a negative declaration seeking to establish the absence of liability in tort.

The CJEU then restricted the jurisdiction of the seized court in cases of multi-state infringements. In fact, in the *Shevill* decision the CJEU allowed the consolidation of infringement proceedings interpreting the expression of Article 5(3) of the Brussels I Regulation, 'place where the harmful event occurred', in the sense that the victim of a libel by a newspaper article distributed in several Member States may bring an action for damages against the publisher either before the courts of the state of the place where the publisher of the defamatory publication is established or before the courts of each Contracting State in which the publication was distributed and where the victim claims to have suffered injury to his reputation. In the first case, the seized court has jurisdiction to award damages for all the harm caused by the defamation. In the second case, the seized court has jurisdiction to rule solely in respect of the harm caused in the state of the court seized.

The CJEU went even further in the *eDate* case concerning an alleged infringement of personality rights by means of content placed online on an internet website. According to the CJEU, Article 5(3) of the Brussels I Regulation must be interpreted as meaning that the person who considers that their rights have been infringed has the option of bringing an action for liability either before the courts of the Member State in which the publisher of that content is established or before the courts of the Member State in which the centre of the plaintiff's interests is based (*forum actoris*). Instead of an action for liability in respect of all the damage caused, the plaintiff may also bring their action before the courts of each Member State in the territory of which content placed online is or has been effectively accessible, irrespective of the fact that the publisher targeted the forum state. Yet, when the forum state is that of accessibility, its courts have jurisdiction only with regard to the damage caused in their respective territories.

In this framework the CJEU has, on several occasions, clarified how to apply its case-law related to general torts to IPRs infringements. In *Wintersteiger* the CJEU highlighted that the understanding of Article 5(3) of the Brussels system may vary according to the nature of the right allegedly infringed. Thus, the CJEU emphasised that the *eDate* case law does not extend to IPRs in its part related to the *forum actoris,* since this forum would protect owners of IPRs that are strong players too much, unlike owners of personality rights.

Yet, the ubiquity theory was extended to IPRs by the CJEU with the *Wintersteiger* judgment. In fact, the CJEU clarified that Article 5(3) of the Brussels I Regulation shall be interpreted as meaning that an action relating to infringement of a trademark registered in a Member State because of the use by an advertiser of a keyword identical to that trademark on a search engine website operating under a country-specific top-level domain of another Member State may be brought either before the courts of the Member State in which the trademark is registered (in their quality of courts at the place of the damage), or before the courts of the Member State of the place of establishment of the advertiser (in their quality of courts at the place of the event giving rise to the damage). The courts at the place of registration are always the best placed to ascertain whether the IPR at issue has been infringed, since the protection granted by registration is limited to the territory of their respective forum state. Yet, these courts always have jurisdiction to hear the action restricted to the territory of the forum state. In contrast the courts at the place of establishment of the advertiser have territorially unrestricted jurisdiction to adjudicate on the entire damages wherever located.

The ubiquity theory was subsequently extended to IPRs by the CJEU with the *Pinckney* judgment. In that case the author of a work brought before a French court an action to establish liability for an alleged infringement of his copyrights protected in France. The action was brought against a company established in Austria that had reproduced in that state the relevant work on a material support which was subsequently sold by companies established in the UK through an internet site in many countries including France. The CJEU was requested to establish if the French court seized had jurisdiction over that action according to Article 5(3) of the Brussels system. The CJEU concluded that the French court had jurisdiction by being the court at the place of the damage. In fact, within the jurisdiction of the court seized it was possible to obtain a reproduction of the work to which the rights relied on by the defendant pertained from an internet site accessible in France. Yet, since the jurisdiction of the French court was based on the criterion of accessibility, it was limited to determine the damage caused in France.

The ubiquity theory was further extended to IPRs by the CJEU with the *Coty* judgment. First, in *Coty* the CJEU was asked to determine if the ubiquity theory applies to an infringement of a Community trademark under Regulation No. 40/94 on the Community trademark. The CJEU excluded that the duality of linking factors, namely the place of the event giving rise to the damage and that where the damage occurred, applies to the notion of 'Member State in which the act of infringement has been committed' in Article 93(5) of Regulation 40/94. Yet, the CJEU grounded this exclusion on the wording of Article 93(5) of Regulation 40/94, which differs from that of Article 5(3) of the Brussels system. In fact, the linking factor provided for by the former refers to the Member State where the act giving rise to the alleged infringement occurred, while that provided for by the latter concerns the Member State where that infringement produces its effects. So, according to the CJEU jurisdiction under Article 93(5) of the Regulation 40/94 may be established solely in favour of the courts in the Member State in which the defendant committed the alleged unlawful act. It follows that unlike under Article 93(5) of the Regulation 40/94, jurisdiction under Article 5(3) shall be established according to the ubiquity theory.

Secondly, in *Coty* the CJEU was asked to interpret Article 5(3) of the Brussels system in a case where there were several supposed perpetrators of an infringement of an IPR protected in the forum state, so-called contributory infringement. In that case, the CJEU applied the conclusions reached in *Melzer* and emphasised that Article 5(3) of Regulation No. 44/2001 does not allow jurisdiction to be established on the basis of the place of the event giving rise to the damage in favour of a court within the jurisdiction of which the supposed perpetrator who is being sued did not act. On the contrary, the CJEU understood Article 5(3) as allowing the jurisdiction of that court to be established on the basis of the place where the alleged damage occurs, provided that the damage may occur within the jurisdiction of the court seized. In such case the court has jurisdiction only to rule on the damage caused in the territory of the Member State to which it belongs. So, even though the CJEU in *Coty* excluded consolidation of claims against the contributory infringer that did not act within the jurisdiction of the court seized, it confirmed that Article 5(3) of the Brussels Regulation shall be interpreted as establishing the jurisdiction of the court at the place where the damage occurred or at that of the event giving rise to it, according to the ubiquity theory. With regard to copyright and neighbouring rights, similar conclusions were reached by the CJEU in the *Hi Hotel* case.

The ubiquity theory was further extended to IPRs by the CJEU in *Gautzsch*, where it was requested to interpret Article 89(1)(d) of Regulation No. 6/2002 on Community designs, under which 'the law of the Member State in which the acts of infringement or threatened infringement are committed'. So, the *Gautzsch* case concerned applicable law rather than international jurisdiction. Yet, its outcome is relevant for international jurisdiction also, since the notion of commission of an IPR infringement corresponds to that of the place where the harmful event occurred under Article 5(3) of the Brussels system. The CJEU distinguished (the law of) the place/State of registration of an IPR from (that of) the place/state in which the acts of infringements have been committed. Thus, the CJEU confirmed that Article 5(3) shall be interpreted as distinguishing these places/states, according to the ubiquity theory.

The ubiquity theory was further extended to IPRs by the CJEU with the recent *Hejduk* judgment, where it was requested to apply the ubiquity theory in a situation in which the alleged tort consisted in the infringement of copyright by the placing of photographs online on a German website without the consent of the author, who brought claim in Austria. In that case the CJEU referred to *Coty* and stated that the expression 'place where the harmful event occurred or may occur' in Article 5(3) is intended to cover both the place where the damage occurred and the place of the event giving rise to it, so that the defendant may be sued, at the option of the applicant, in the courts of either of those places. The CJEU referred to *Wintersteiger* and stated that the place of the event giving rise to the damage is the place where the process for the technical display of the photographs on the relevant website by its owner is activated: this event is to be localised typically at the place where that owner has its seat, since that is where the defendant took and carried out the decision to place photographs online on a particular website, i.e. Germany. In addition, the CJEU referred to *Pinckney* and stated that the place where the damage occurred is to be localised in the Member State where the IPR whose infringement is alleged is protected and the website on which the photographs at issue were published was accessible, it being irrelevant that this website operated under a country-specific German top-level domain – 'de' – and therefore was not directed at the Member State in which the seized court is situated, i.e. Austria. The CJEU concluded that the German court had jurisdiction to rule on the damages caused in every jurisdiction involved, whereas the Austrian court had jurisdiction only to rule on the damage caused in Austria.

To sum up, Article 5(3) of the Brussels system as interpreted according to the ubiquity theory by the CJEU reflects a tendency that overcomes a strict territorial approach with respect to IPRs and that favours a consolidation of claims in cross-border IP disputes.

1.4 Consolidation of Claims under the Sets of Principles

Further attempts to overcome the traditional understanding of the territoriality principle, favouring consolidation of cross-border IP litigation, result from the four sets of principles referred to in section 1.1 and the current work of the ILA Committee in this field. In fact, all four sets of principles allow for consolidation of claims based on their general jurisdiction norms. In addition, all four sets of principles overcome exclusive jurisdiction in cases related to validity issues incidentally raised, by establishing that those issues can be brought before a court of a state other than that of registration, provided that the judgment on validity would have *inter partes* effect. Furthermore, all four sets of principles allow for the adjudication of an IPR multi-state infringement in its entirety by a single court under certain conditions.

Concerning exclusive jurisdiction, the four sets of principles differ with regard to the question of whether the validity issue raised as a principal claim could be adjudicated by a court of a state other than that of registration. In fact, the CLIP Principles and the Joint Japanese-Korean Proposal explicitly establish that claims principally related to validity should be brought only before the courts of the state of registration. In contrast, the ALI Principles and the Transparency Proposal enable courts of a state other than that of registration to decide upon validity issues concerning foreign IPRs principally raised, provided that the effects of the judgment on validity is limited to the parties of the proceedings. The ALI Principles pose the stricter conditions that the dispute should have a multi-state nature and that the claim on validity principally raised should be brought before the court at the defendant's residence. The Transparency Proposal requires only that the Japanese court seized with the validity claim principally raised related to a foreign-registered IPR has jurisdiction.

Concerning infringement jurisdiction, the four sets of principles differ with regard to the conditions that have to be met in order to allow the seized court to adjudicate a multi-state infringement in its entirety. More particularly, there are two main differences among these sets of principles. The first relates to the question of whether the centralised jurisdiction should be limited to infringements perpetrated through ubiquitous media, such as the internet, or should also cover multi-state infringements perpetrated by other means. The CLIP Principles refer to 'ubiquitous infringements'. In contrast, the other sets of principles concern infringement perpetrated by means other than the internet also, namely the ALI Principles refer to injuries 'wherever' they could occur, while the Transparency Proposal and the Joint Japanese-Korean Proposal refer to 'multiple states' infringement.

The second difference concerns the question of which court shall have jurisdiction to adjudicate multi-state infringement disputes in their entirety and under which conditions. The Transparency Proposal provides that if a Japanese court is the court of the place of the action giving rise to the damage or of the damage, it could have jurisdiction to hear multi-state infringement disputes in their entirety subject to two additional requirements: the infringement must occur in several states; and its results should be maximised in Japan. The Joint Japanese-Korean Proposal grounds jurisdiction to hear multi-state infringement disputes in their entirety on the court at the place of injury (but not at the place of the action) if the three following conditions are met: the injuries should occur in multiple states; the major part of the initial infringing activities should be located in the forum state; and the alleged infringement shall not be directed against a particular state. The ALI Principles allow a court of the state where the substantial infringing activities were committed to assert jurisdiction over multi-state infringement disputes in their entirety, unless it can be reasonably foreseen that those activities are directed to one particular state. The CLIP Principles provide that the court of the state where the alleged infringement occurs or may occur shall have jurisdiction over the ubiquitous infringement if it is reasonably foreseen that the infringer directed the activities to the forum state. Also, with regard to the extent of jurisdiction, the CLIP Principles allow the court to assert jurisdiction to hear ubiquitous infringement disputes in their entirety where four cumulative requirements are met: the infringement shall be carried out through ubiquitous media; the court shall have jurisdiction pursuant to the infringement jurisdiction rule; the activities shall have no substantial effect in the state where the infringer is habitually resident; and the substantial activities shall be located within the forum state or the harm caused by the infringement in the forum state shall be substantial in relation to the infringement in its entirety.

2 EMPLOYEES' INTELLECTUAL PROPERTY RIGHTS

2.1 Consolidation of Claims under the Protocol on Jurisdiction

Further attempts to overcome the traditional understanding of the territoriality principle, favouring consolidation of cross-border IP litigation, concern the field of employees' intellectual property rights. In fact, consolidation of claims follows both the Protocol on Jurisdiction to the EPC for European patents, and the Brussels system for all other IPRs. Concerning the Protocol on Jurisdiction, consolidation of claims derives from its Articles 2 to 6 on jurisdiction and from its Article 8 on *lis pendens*. Articles 2 to 6 of the Protocol establish a series of subsidiary jurisdiction criteria to determine where the plaintiff should bring actions related to the grant of a European patent. With particular regard to employee's inventions, when parties have concluded a choice of court agreement in writing or verbally with written confirmation, the designated court shall have jurisdiction to hear the claims, provided that the agreement at stake is legal according to the applicable labour law governing the contract of employment (Article 5). If the designated court is that of a non-contracting state, if there is no agreement, or if the agreement is not legal according to the applicable labour law, then the competent court to hear the claim is that of the state in which the employee is mainly employed; if that state cannot be determined, jurisdiction is grounded on the court of the state in which the employer has the place of business to which the employee is attached (Article 4 Protocol on Jurisdiction and 60(1) of the EPC).

Article 8 of the Protocol on Jurisdiction establishes that when proceedings based on the same claim and between the same parties are brought before courts of different contracting states, the court to which an earlier application is made takes priority. Therefore, the court to which a later application is made must of its own motion decline jurisdiction. If the jurisdiction of the court to which an earlier application is made is challenged, the courts to which a later application is made shall stay the proceeding until the first seized court decides on its jurisdiction.

To sum up, in cases related to the grant of a European patent in employment relationships, the Protocol on Jurisdiction obliges parties to litigate their transnational cases before a single court, rather than before the courts of each state for which portions of the European patent at stake were conferred. Thus, this Protocol reflects a tendency that overcomes a strict territorial approach with respect to IPRs and that favours a consolidation of claims in cross-border IP disputes.

2.2 Consolidation of Claims under the Brussels System

Further attempts to overcome the traditional understanding of the territoriality principle favouring consolidation of cross-border IP litigation, result under the Brussels system for disputes on entitlement and ownership of employees' IPRs. Particularly, in *Duijnstee,* Mr. Duijnstee, a liquidator in the dissolution of a company, applied to the Maastricht Arrondissementsrechtbank for an interlocutory injunction requiring Mr. Goderbauer, the former manager of the company, to transfer to Mr. Duijnstee all patents and pending patent applications in 22 countries for an invention that Mr. Goderbauer made while employed by the company. The proceedings came before the Hoge Raad, which then referred the preliminary question to the CJEU of 'whether the concept of proceedings "concerned with the registration or validity of patents" within the meaning of Article 16(4) of the [Brussels] Convention [corresponding to Article 24(4) of the Brussels I bis Regulation] may cover a dispute such as that concerned in the main action' (paras. 19-20); i.e. a dispute having as its subject matter a transfer of IPRs. The CJEU answered that Article 16(4) does not cover such disputes. In fact, it is true that a decision as to the ownership and entitlement of a registered IPR might lead to a change in the name of the owner in the corresponding IPR national registry. Yet, this change is not enough to bring the issue in the concept of proceedings concerned with the registration of a registered IPR, covered by the exclusive jurisdiction of Article 22(4) of the Brussels I bis Regulation, since in these cases the public authorities competent to change the registry would only record the transfer of ownership, without playing any administrative role.

In other words, in *Duijnstee* the CJEU restricted the scope of the exclusive jurisdiction rule provided for by Article 24(4) of the Brussels I bis Regulation. As already mentioned, exclusive jurisdiction rules oblige litigating transnational IPRs disputes country by country and express a strict territorial conception of IPRs. So, the *Duijnstee's* restriction of the exclusive jurisdiction rule favours consolidation of claims before a single competent court, and avoids the disadvantages of fragmentation and the mosaic approach.

Consequently, in the context of transnational employment relationships proceedings related to the transfer of IPRs, as well as to their ownership or entitlement, may be adjudicated by the courts designed by the abovementioned rules on general and infringement jurisdiction (Article 4 and Article 7(2) of the Brussels I bis Regulation), and also by Articles 20-23 of this Regulation on jurisdiction over individual contracts of employment. These last rules establish that an employee may sue an employer before the courts of the Member State in which the employer is domiciled. Alternatively the employee may sue the employer, even if not domiciled in a Member State, before the courts of the Member State for the place where or from where the employee habitually carries out their work or in the courts for the last place where they did so. Alternatively, if the employee does not, or did not, habitually carry out their work in any one country, the employee may sue the employer in the courts of the Member State for the place where the business which engaged the employee is or was situated. In contrast, an employer may bring proceedings only in the courts of the Member State in which the employee is domiciled. In any case, a court other than those just mentioned may have jurisdiction only when designated by an agreement that is concluded after the dispute has arisen or that allows only the employee to bring proceedings before courts other than the previously determined ones.

Additionally, the court having jurisdiction to determine cross-border cases related to employees' IPRs may be designated by Article 7(1) also, on jurisdiction in general contractual disputes, according to which:

> a person domiciled in a Member State may be sued in another Member State: (1)(a) in matters relating to a contract, in the courts for the place of performance of the obligation in question; (b) for the purpose of this provision and unless otherwise agreed, the place of performance of the obligation in question shall be: – in the case of the sale of goods, the place in a Member State where, under the contract, the goods were delivered or should have been delivered, – in the case of the provision of services, the place in a Member State where, under the contract, the services were provided or should have been provided; (c) if point (b) does not apply then point (a) applies.

With regard to this last provision, the CJEU emphasised in *Falco* that Article 5(1) of the Brussels system, corresponding to Article 7(1) of the Brussels I bis Regulation, shall be interpreted to the effect that a contract, under which the owner of an IPR grants its contractual partner the right to use that right in return for remuneration, is not a contract for the provision of services within the meaning of Article 5(1)(b), but rather shall be considered a general contract under Article 5(1)(a). Therefore jurisdiction shall be grounded in the courts of the Member State for the place of performance of the obligation that is litigated before the court, as interpreted by the CJEU in cases regarding contracts in general.

To sum up, in cross-border cases related to employees' IPRs, the Brussels system allows parties to litigate before a single court, rather than before the courts of each state for which the IPRs were conferred. Thus, even in employment relationships the Brussels system reflects a tendency that overcomes a strict territorial approach with respect to IPRs, and that favours a consolidation of claims in cross-border IP disputes.

BIBLIOGRAPHY

Table of Sources

Litigating Intellectual Property Rights Disputes Cross-Border: EU Regulations, ALI Principles, CLIP Project (ed. Stefania Bariatti, CEDAM 2010).

Intellectual Property in the Conflict of Laws (eds Jürgen Basedow, Josef Drexl, Annette Kur & Axel Metzger, Mohr Siebeck 2005).

European Max Planck Group on Conflict of Laws in Intellectual Property, *Conflict of Laws in Intellectual Property, The CLIP Principles and Commentary* (OU).

Pedro De Miguel Asensio, *Cross-border Adjudication of Intellectual Property Rights and Competition between Jurisdictions*, in *Annali italiani del diritto d'autore, della cultura e dello spettacolo*, 117 ff (2007).

François Dessemontet, *The ALI Principles: Intellectual Property in Transborder Litigation*, in *Intellectual Property in the Global Arena* (eds Jürgen Basedow, Toshiyuki Kono & Axel Metzger, Mohr Siebeck 2010).

Graeme Dinwoodie, *Trademarks and Territory: Detaching Trademark Law from the Nation-State,* Hous. L. Rev., 885 ff (2004).

Josef Drexl, *Lex americana ante portas – Zur extraterritorialen Anwendung nationalen Urheberrechts*, in *Urheberrecht im Informationszeitalter, Festschrift für Wilhelm Nordemann*, 429 ff (ed Ulrigh Loewenheim, Beck 2004).

Rochelle Dreyfuss, *Resolution Through Conflict of Laws: The ALI Principles on Transnational Intellectual Property Disputes: Why Invite Conflicts?*, Brooklyn J.Int'l L., 819 ff (2005).

Jane Ginsburg, *The Private International Law of Copyrights in an Era of Technological Change,* Rec. des Cours, 257 ff (1998).

Annette Kur and Benedetta Ubertazzi, *The ALI Principles and the CLIP Project– A Comparison*, in *Litigating Intellectual Property Rights Disputes Cross-Border: EU Regulations, ALI Principles, CLIP Project* (ed. Stefania Bariatti, CEDAM 2010).

Annette Kur, *Durchsetzung gemeinschaftsweiter Schutzrechte: Internationale Zuständigkeit und andwendbares Recht*, Gewerblicher Rechtsschutz und Urheberrecht, Internationaler Teil, 749 ff (2014).

Dario Moura Vicente, *La propriété intellectuelle en droit international privé,* Rec. des Cours, 424 ff (2008).

Paul Torremans & James Fawcett, *Intellectual Property and Private International Law* (2nd ed, OU).

Marketa Trimble Landova, *Global Patents. Limits of Transnational Enforcement* (OU).

Benedetta Ubertazzi, *Licence Agreements Relating to IP Rights and the EC Regulation on Jurisdiction*, IIC, 912 ff (2009).

Benedetta Ubertazzi, *Exclusive Jurisdiction in Intellectual Property* (Mohr Siebeck 2012).

Dai Yokomizo, *Intellectual Property Infringement on the Internet and Conflict of Laws*, AIPPI Journal, 105 ff (May 2011).

Table of Legislation and Agreements

International Legislation

ACTA, Anti-Counterfeiting Trade Agreement (concluded on November 15, 2010) (see text in Proposal for a Council Decision on the conclusion of the Agreement of 24 June 2011, COM(2011) 380 final).

CUB, Berne Convention for the Protection of Literary and Artistic Works (adopted on 1886, revised on 28 September 1979).

CUP, Paris Convention for the Protection of Industrial Property (adopted in 1883, revised on 14 July 1967).

The Hague Conference on Private International Law, Preliminary Draft Convention on Jurisdiction and the Effects of Judgments in Civil and Commercial Matters, (June 18, 1999, with an explanatory report by Nygh P. and Pocar F.), (HC) Prel. Doc. No. 11.

Permanent Bureau, Preliminary Document No.13, Report of the experts meeting in the intellectual property aspects of the future Convention on jurisdiction and foreign judgments in civil and commercial matters (1 February 2001).

TRIPS, Agreement on Trade Related Aspects of Intellectual Property Rights (annex 1C of the Marrakesh Agreement Establishing the World Trade Organization, signed in Marrakesh, Morocco on 15 April 1994), 1869 UNTS 299.

European Legislation

EPO, Convention on the Grant of European Patents (European Patent Convention) of 5 October 1973 as revised by the Act revising Article 63 EPC of 17 December 1991 and the Act revising the EPC of 29 November 2000.

Protocol on Jurisdiction and the Recognition of Decisions in respect of the Right to the Grant of a European Patent (Protocol on Jurisdiction and Recognition) of 5 October 1973.

The Brussels Convention of 27 September 1968 on jurisdiction and the enforcement of judgments in civil and commercial matters, communitarised by the Council Regulation (EC) 44/2001 of 22 December 2000 on jurisdiction and the recognition and enforcement of judgments in civil and commercial matters (Brussels I Regulation), revised by Regulation (EU) No. 1215/2012 of the European Parliament and of the Council of 12 December 2012 on jurisdiction and the recognition and enforcement of judgments in civil and commercial matters (recast) (Brussels I Regulation Recast), in: [2012] OJ L 351/1.

The Lugano Convention on jurisdiction and enforcement of judgments in civil and commercial matters of 16 September 1988, revised by the Lugano Convention on jurisdiction and the recognition and enforcement of judgments in civil and commercial matters concluded in Lugano on 30 October 2007, in [2007] OJ L 339/3.

Council Regulation No. 40/94 of 20 December 1993 on the Community trademark (OJ 1994 L 11, p. 1), now Council Regulation (EC) No. 207/2009 of February 26, 2009, on the Community trademark (codified version) (OJ 2009 L 78, p. 1).

Council Regulation (EC) No. 6/2002 of 12 December 2001 on Community designs (OJ 2002 L 3, p. 1).

Other Sources

ALI Principles

The American Law Institute, *Intellectual Property: Principles Governing Jurisdiction, Choice of Law and Judgments in Transnational Disputes* (Chestnut, ALI Publishers, 2008).

CLIP Principles

'Principles for Conflict of Laws in Intellectual Property, prepared by the European Max Planck Group on Conflict of Laws in Intellectual Property (CLIP)', published on 31 August 2011.

Transparency Proposal

Transparency Proposal on Jurisdiction, Choice of Law, Recognition and Enforcement of Foreign Judgments in Intellectual Property, finalised in 2009, and available in Jürgen Basedow, Toshiyuki Kono and Axel Metzger (eds), *Intellectual Property in the Global Arena* (Mohr Siebeck, 2010).

Joint Japanese-Korean Proposal

Commentary on Principles of Private International Law on Intellectual Property Rights (Joint Proposal Drafted by Members of the Private International Law Association of Korea and Japan)' (2011) 2(6) *Kigyō to Hōsōzō.*

Table of Cases

CJEU Cases

Handelskwekerij G. J. Bier BV v Mines de potasse d'Alsace SA, C-21/76, ECR 1735 (1976).
Duijnstee v Goderbauer, 288/82 ECR 3663 (1983).
Dumez France SA, C-220/88 I-49 (1990).
Shevill v Presse Alliance SA, C-68/93 ECR I-415 (1995).
Antonio Marinari, 364/93 ECR I-2719 (1995).
Rudolf Kronhofer, C-168/02 ECR I-6009 (2004).
Gesellschaft für Antriebstechnik mbH & Co. KA (GAT) v Lamellen und Kupplungsbau Beteilgungs KG (LuK), C-4/03 ECR I-6509 (2006).
Roche Nederland BV v Primus, C-539/03 ECR I-6535 (2006).
Falco Privatstiftung ECR I-3327 (2009).
eDate Advertising GmbH v X and Olivier Martinez, Robert Martinez v MGN Limited, joined Cases C-509/09 and C-161/10 OJ C370/9 (2011).
Wintersteiger AG v Products 4U Sondermaschinenbau GmbH, C-523/10 (2012) not yet published in ECR, available on the official website at <www.curia.eu> accessed 15 August 2014.
Solvay SA v Honeywell Fluorine Products Europe BV and Others, C-616/10 (2012) not yet published in ECR, available on the official website <www.curia.eu> accessed 15 August 2014.

Folien Fischer and Fofitec, C-133/11 (2012) not yet published in ECR, available on the official website < www.curia.eu > accessed 15 August 2014.

Melzer, C-228/11 (2013) not yet published in ECR, available on the official website < www.curia.eu > accessed 15 August 2014.

Peter Pinckney v KDG Mediatech AG, C-170/12 (2013) not yet published in ECR, available on the official website < www.curia.eu > accessed 15 August 2014.

Coty Germany GmbH, C-360/12 (2014) not yet published in ECR, available on the official website < www.curia.eu > accessed 15 August 2014.

Hi Hotel, C-387/12 (2014) not yet published in ECR, available on the official website < www.curia.eu > accessed 15 August 2014.

Gautzsch, C-479/12 (2014) not yet published in ECR, available on the official website < www.curia.eu > accessed 15 August 2014.

Hejduk, C-441/13 (2015) not yet published in ECR, available on the official website < www.curia.eu > accessed 15 March 2015.

The Netherlands

Single Buoy Moorings Inc./Bluewater c.s., District Court of The Hague, 11 April 2007.

UK

Lucasfilm Entertainment Co. v Ainsworth, Court of Appeal EWCA (Civ) 1328 (2009). *Lucasfilm Ltd. & Ors v Ainsworth & Anor*, Supreme Court UKSC 39 (2011).

CHAPTER 4

Recognition and Enforcement of Foreign Judgments

Marketa Trimble

1 INTRODUCTION

A judgment rendered by a court in one country must be recognised in another (foreign) country if the judgment includes a declaration of a legal status or an award of a remedy whose effects are sought outside the country in which a court granted the declaration or remedy. Enforcement of a remedy – an injunction and/or a monetary remedy – might be necessary in a foreign country if the person against whom the remedy was granted has no physical or other presence in the country where the original court granted the remedy, or if the person acts outside the country of the court that granted the remedy.

2 NATIONAL RULES

The recognition and/or the enforcement of foreign judgments are governed by countries' national laws. In some instances bilateral treaties govern recognition and/or enforcement between countries, and in the European Union, Regulation (EU) 1215/2012 on jurisdiction and the recognition and enforcement of judgments in civil and commercial matters (Brussels I Regulation recast, amended by Regulations (EU) 542/2014 and 2015/281) includes provisions on the recognition and enforcement of certain judgments issued by courts of the Member States of the European Union. In the United States, individual states' laws apply to the recognition and enforcement of foreign-country judgments; the laws of 21 U.S. states and the District of Columbia are modelled after the 2005 Uniform Foreign-Country Money Judgments Recognition Act (as of 8 March 2017), and some of the other states follow the 1962 Uniform Act.

3 DIFFERENT APPROACHES

Countries differ in the procedures they require for a foreign judgment to be recognised and enforced; the procedures range from a simple administrative act of registration of a foreign judgment (without a separate recognition procedure) to the filing of a new lawsuit with the recognising court in which the foreign judgment is attached to the complaint. In some jurisdictions, procedures for recognition vary depending on the remedy at issue; for example, different procedures may have to be followed for foreign judgments depending on whether they grant monetary or non-monetary remedies. Differences also exist among countries as to which court decisions are considered recognisable and/or enforceable, both in terms of types of decisions and stage of proceedings in which the decisions were rendered. Typically countries will allow recognition of foreign court decisions that are on the merits, final, and non-appealable. Some common law countries will not recognise foreign-granted injunctions, and no countries recognise judgments that enforce foreign public laws, such as criminal judgments, and decisions in administrative matters. Nevertheless, some countries enter into agreements to assist each other in enforcement of public laws; for example, some countries cooperate in investigations and arrests.

4 CONDITIONS OF RECOGNITION AND ENFORCEMENT

Countries' laws include a number of conditions that a foreign judgment must meet for a court to recognise and/or enforce the foreign judgment. These conditions correspond to grounds for non-recognition or non-enforcement, which are mostly legislated as mandatory grounds, meaning that courts must deny recognition or enforcement of a foreign judgment if any of the grounds are present. In some countries, some grounds are discretionary, meaning that courts may (but need not) deny recognition or enforcement of a foreign judgment based on a discretionary ground.

One ground for non-recognition or non-enforcement is the lack of personal jurisdiction of the foreign court that rendered the judgment. Countries have various approaches for assessing the adequacy of the jurisdiction of the original foreign court. Some countries have special jurisdictional rules legislated specifically for the purpose of determining the adequacy of the personal jurisdiction of a foreign court in the context of the recognition of a foreign judgment. Other countries adhere to the mirror-image principle and will assess the adequacy of a foreign court's jurisdiction based on the countries' own jurisdictional rules as they would have applied if the countries' own courts were in the same position as the original foreign court.

Other grounds for non-recognition or non-enforcement that are of a procedural character may also render a foreign judgment non-recognisable. Lack of subject matter jurisdiction may be a ground for the non-recognition or non-enforcement of a foreign judgment, as may be a lack of procedural due process in the original proceedings before the foreign court – whether it is fraud in the original proceedings, lack of proper notice to the defendant, or lack of opportunity for the defendants to present their case.

General problems in the foreign judicial system may also be grounds for non-recognition or non-enforcement, such as a lack of impartial tribunals and procedures.

Grounds for non-recognition or non-enforcement of a substantive character are limited; courts typically do not evaluate a foreign judgment on its merits, and a mere difference between national laws will not suffice for a court to deny recognition or enforcement of a foreign judgment that was issued based on a law that differs from the law of the court's jurisdiction. However, if the foreign judgment or the cause of action on which the judgment is based is repugnant to a country's important public policy the country's courts will not recognise and enforce the judgment. Nor will a foreign judgment be recognised and enforced if it conflicts with another judgment rendered between the same parties.

5 ENFORCEMENT OF A FOREIGN JUDGMENT

Once a foreign judgment is registered or recognised, the party seeking enforcement of the judgment may follow the required domestic procedure to have the foreign judgment enforced.

PART II National Legal Rules

A In Europe

CHAPTER 5

Ownership and Harmonisation Efforts at EU Level

Sanna Wolk

1 BACKGROUND

Within the European Union (EU), the European Commission has, from the 1970s onwards, adopted, and is continuing to introduce, a number of measures which seek to harmonise the ownership aspects of employee's intellectual property rights throughout the Union.

One difficulty in harmonising the legal framework is attributable to the differences between the various national models within the Member States. In particular three distinct models can be extracted. Transfer of rights may be based on autonomous legislation, provisions in statutes regulating intellectual property, or express or implied agreements.[1] It is predominantly the right to employees' inventions (and, where applicable, utility models)[2] that is regulated in specific legislation. Many European countries have, however, such provisions in their patent legislation. Regarding other exclusive rights, the majority of European countries regulate the transfer of rights in their respective intellectual property legislation framework.

The question regarding the employers' right to the products of intellectual effort are of great economic significance, and in spite of the differences, the EU has, since the 1980s, striven to harmonise the matter. This has to some extent been achieved as relevant provisions can be found in Directive 87/54/EEC on the Legal Protection of

1. For a comparative report on the three models and their relationship, see e.g. Jonczyk in IIC, 847 et seq. (1989).
2. See further COM (1997) 691 final, COM (1999) 309 final, and Opinion of the Economic and Social Committee (98/C 235/06). The European Commission communicated on 27 September 2005 that a European protection for utility models will not be effected. A number of countries in Europe protect utility models, though not e.g. the United Kingdom, Luxembourg or Sweden.

Topographies for Semiconductor Products, Directive 2009/24/EC on the Legal Protection of Computer Programs, Regulation (EC) No. 2100/94 on Community Plant Variety Rights and Regulation (EC) No. 6/2002 on Community Designs. However these directives and regulations have inconsistent approaches. The Computer Programs Directive and the Community Design Regulation clearly define the right that an employer enjoys, whereas the Semiconductor Directive and the Plant Variety Rights Regulation leave it to the Member States to decide whether they wish to introduce national provisions on the employer's acquisition of rights.

2 EMPLOYEES' COPYRIGHT

In the late 1980s and in the early 1990s, copyright was at the centre of attention in the EU. Copyright protects creations of the mind insofar as they are original and expressed in a particular form. Copyright protection covers a very broad range of creations, such as software, databases, web pages and multimedia works.

In the late 1980s the European Commission published its proposals for copyright protection within the Community.[3] The Commission's efforts led, *inter alia*, to the adoption of Directive 2009/24/EC on the Legal Protection of Computer Programs. Article 2 of the Directive deals with authorship of programs, collective works and employees' works. Article 2(1) sets out that the author of a computer program is the natural person or group of natural persons who have created the program or, where the legislation of the Member States permits, the legal person designated as the rightholder. Where collective works are recognised by national legislation the person considered by the laws of the Member State to have created the work will be deemed its author. From Article 2(2) it follows that where a computer program is created by a group of natural persons working together, the exclusive rights will be owned jointly. However, this means that Member States having jurisdiction and neither recognising that corporations can be authors, nor recognising the concept of collective works, do not have to change their laws. Article 2(3) of the Directive contains a mandatory requirement on employees' programs. Under the provision the employer shall exclusively be entitled to exercise all *economic rights* in an employee's computer program, where a program is created in the execution of the employee's duties or where the employee is following instructions given by the employer.[4] It is an automatic legal transfer of the copyright in computer programs. However, if the parties agree, the employed author of the computer program can recover the rights through a specific clause in the employment contract or a separate agreement on the exploitation of the computer program made by the employee (in which case they exclude the statutory provision of automatic transfer of rights).

3. See the Green Paper on Copyright and the Challenge of Technology - Copyright Issues Requiring Immediate Action, COM (88) 172 final and the Follow-up to the Green Paper - Working Programme of the Commission in the field of Copyright and Neighbouring Rights, COM (90) 584 final.
4. Cf. Art. 3(2) a Directive 87/54/EEC on the Legal Protection of Topographies of Semiconductor Products (a non-mandatory provision regarding employees' chips).

Article 2(3) of the Computer Program Directive was inspired by the employer provision in the Directive 87/54/EEC on the Legal Protection of Topographies of Semiconductor Products,[5] and the original proposal for the provision also mirrored the wording.[6] The European Economic and Social Committee proposed, however, that the chosen formulation should be amended so as to only apply to such computer programs created by employees whose duty was to develop programs. The European Parliament argued, on the other hand, that the employer should be entitled to more than that, and it was this proposal that eventually formed the basis of the formulation of Article 2(3) Computer Program Directive. One reason to widen the scope of the provision could have been that computer programs are created by various types of employees and it would have been inappropriate if the presumption only applied to some employees. The employer has therefore, in comparison to the original proposal, received a broad right to their employees' computer programs, which has come to be expressed in the laws of the Member States.[7]

Moral rights, however, such as the right of paternity and the right of integrity, are left outside the scope of the Computer Programs Directive and are therefore currently regulated by national provisions.[8] In the Member States belonging to the civil law system, moral rights are considered to arise directly in the author and to be inalienable even by voluntary transfer (cession) to an employer. On the other hand, in the Member States belonging to the common law system, moral rights may be waived by an instrument in writing,[9] and the right of paternity, and in some cases the right of integrity, does not extend over anything done by the employer where copyright was first vested in the employer.[10] Furthermore the integrity right does not extend over computer programs.[11]

A similar provision to Article 2(3) of the Computer Programs Directive was included in the first draft of the Directive 96/9/EC on the Legal Protection of Databases. However, it was deleted from the final version of the Directive and recital 29 only states that nothing prevents Member States from stipulating in national laws that where a database is created by an employee in the execution of the duties or following the instructions given by the employer, the employer shall exclusively be entitled to exercise the rights in the database so created.

During early 2000s the European Commission announced as regards ownership of employees copyright that:

5. See COM (88) 172 final p. 196.
6. See COM (88) final – SYN 183 p. 34.
7. See further COM (2000) 199 final.
8. See further COM (88) 172 final p. 197, COM (90)509 final – SYN 183 p. 18, COM (95) 382 final p. 65 et seq. and COM (96) 586 final Ch. 3. See also Czarnota/Hart, *Legal Protection of Computer Programs in Europe* 48 et seq. (1991); Dreier in EIPR, 319 et seq. (1991); and Tritton, *Intellectual Property in Europe* 492 (3rd ed., 2008). See also Art. 6*bis* of the Berne Convention for the Protection of Literary and Artistic Works of 1886.
9. See e.g. ss. 87(2)-(4) of the UK Copyright, Designs and Patents Act 1988.
10. See e.g. ss. 79(3) and 82(1) of the UK Copyright, Designs and Patents Act 1988.
11. See e.g. ss. 81(2) of the UK Copyright, Designs and Patents Act 1988.

> At this point, it would seem advisable to analyse the issue further and, in particular, identify specific situations where harmonisation would yield added value and address Internal Market needs.[12]

The problem is that it may at present not be possible to overcome national differences in legislation as they are, after all, based on different legal traditions. The interest to realise the internal market may, however, become relevant for harmonisation efforts even in this area in order to facilitate cross-border trade. This makes legislative action on an EU level a real possibility.

To recapitulate, today only employees' computer programs are harmonised on a European level and the question of employees' copyrighted works in general is left to the different national solutions. In some countries, such as in the Netherlands and the United Kingdom, national copyright acts regulate employees' copyright. In other countries, such as Germany, France and the Nordic countries, a transfer from the employee to the employer follows from general principles of law. Nevertheless, all Member States have included in their national copyright laws provisions implementing the mandatory requirement on employees' computer programs contained in Article 2(3) of the Computer Program Directive.[13]

3 EMPLOYEES' SEMICONDUCTOR CHIPS

The Directive on the Legal Protection of Topographies of Semiconductor Products (87/54/EEC) has a provision on the employer's right to employee's topographies of semiconductor products. From Article 3(1) of the Directive it follows that protection shall apply in favour of persons who are the creators of topographies of semiconductor products, and in Article 3(2)(a) it is stated that all rights are automatically vested in the employer, where the topography is developed in the execution of the employee's duties and nothing else is agreed. However, the provision is not mandatory and Member States may decide if their national laws should contain such a provision. The reason for the non-mandatory character of the provision is attributable to the fact that the purpose was not to thoroughly regulate the object of protection or the subject; rather the Directive was the result of a desire to obtain protection for European semiconductor products in the US.[14]

4 EMPLOYEES' INVENTIONS

During the 1970s an effort was made in the patent field to adopt a Community Patent Convention.[15] However, the opinion within the EU was then that matters regarding

12. Commission Staff Working Paper on the Review of the EC Legal Framework in the Field of Copyright and Related Rights, SEC (2004) 995, Brussels, 19.7.2004 p. 14.
13. See further the Report from the Commission to the Council, the European Parliament and the Economic and Social Committee on the implementation and effects of Directive 91/250/EEC on the Legal Protection of Computer Programs, COM (2000) 199 final.
14. For a detailed analysis of the European work, see e.g. Dreier in IIC 1988 p. 427 et seq.
15. The Convention was signed in 1975 by the nine Member States, but has never been ratified.

employee inventions should be dealt with under the national laws of the Member States, and not harmonised on European level.[16] The need for harmonisation in the field of employee inventions was not considered urgent, hence the existing differences in the laws of the Member States.

In the latter part of the 1990s and in the 2000s the EU announced that harmonisation regarding inventions created by employees is not necessary, and that the matter should continue to be regulated by the various national laws.[17] Today, employer rights to employee inventions are often regulated by national laws, whereby an employer can acquire the right to an invention made by an employee in a number of different ways.

At present, on the national level, the rights to employee inventions are regulated to a greater extent than rights to other intellectual property assets created by employees. For instance, all Member States of the EU have national provisions regarding inventions developed by employees in the course of their employment. However, there is a wide diversity of legal regimes in these countries as to employee inventions, ranging from general provisions in patent laws, e.g., in Austria, Estonia, France, Latvia, the Netherlands, Portugal, Spain and the United Kingdom, to specific subject-oriented legislation regulating employer rights to employee inventions, as in Denmark, Germany, Finland, Norway and Sweden. Further, many European countries have legal rules on compensation for employee inventors.[18]

5 EMPLOYEES' DESIGNS

The protection of industrial designs is growing in importance. A design is the ornamental or aesthetic aspect of an article. Designs are applied to a great variety of products from different industry and trade sectors, from complex instruments with a special or exclusive use, such as machines, vehicles, tools, computers, to simple or everyday articles, such as clothes, electrical appliances, web pages, toys and furniture.

In the 1990s, after having dealt with patents and with selected areas of copyright, the European Commission turned its attention to the harmonisation of industrial designs within the Community.[19] The first effort was the adoption of Directive 98/71/EC on the Legal Protection of Designs. However, since ownership of rights in

16. Cf. Art. 40(1) in European Communities, The Council, Draft Convention for the European Patent for the Common Market and Attached Annexes, 1973. Cf. also Art. 60(1) of the European Patent Convention ('EPC'). EPC has left patent ownership to the discretion of the states signatory to it and the right to a European patent is determined in accordance with the law of the State in which the employee is mainly employed. If the State in which the employee is mainly employed cannot be determined, the law to be applied will be that of the State in which the employee's employer has their place of business.
17. See further the Green Paper on the Community Patent and the Patent System in Europe, COM(97) 314 final and the Proposal for a Council Regulation on the Community Patent COM(2000) 412 final.
18. See e.g. Wolk, *Remuneration of Employee Inventors – is There a Common European Ground? A Comparison of National Laws on Compensation of Inventors in Germany, France, Spain, Sweden and the United Kingdom*, 42 IIC 272-298 (2011).
19. See Green Paper on the Legal Protection of Industrial Design (III/F/5131/91).

designs is an area where the laws of Member States differ, there is no provision in the Directive dealing with employees' designs.[20] Nevertheless, since 2002 there exists a Community-wide design right. This was established through Regulation 6/2002/EC on Community Designs. The Regulation is binding in its entirety and directly applicable in all Member States since 6 March 2002. From Article 14(1) of the Regulation it follows that the right to a Community design shall be vested in the designer or the designer's successor in title, and Article 14(3) provides that all design rights are automatically vested in the employer where the design is developed in the execution of the employee's duties, or when the employee is following instructions given by the employer.[21] The provision mirrors the rule in the Computer Program Directive, and both presumptions have principally the same meaning.[22]

However, the rule in Article 14(3) of the Regulation is subject to a pre-existing agreement, and gives priority to any national legal rules.[23] Nevertheless, CJEU held in C-32/08 *FEIA v Cul de Sac* that there was no room for national law to determine the issue of ownership of an unregistered Community design and that thus the issue of ownership had to be determined by reference to Article 14. In practice, because of the clear provision on employees' Community unregistered designs and in the absence of a national provision or a conflicting one that does not stipulate the same as in Article 14 of the Design Regulation, transfer of a Community design should be applied even when national design rights cover the design.

Since the design right is intended to be an economic right, rather than a moral right, transfer of a design right can therefore occur in its entirety (compare with copyright).[24] In the Green Paper copyright-inspired moral rights for designers were abandoned because it was not desirable or practically appropriate to introduce such rights, given the purpose of protection.[25] Nevertheless, since a design can also be protected by copyright, in those situations national provisions on copyright may be applicable simultaneously as the provisions of the Design Regulation.[26]

6 EMPLOYEES' PLANT VARIETIES

There is a provision in Regulation (EC) No. 2100/94 on Community plant variety rights relating to employees' plant breeders rights. Article 11(1) sets out that the person who

20. In the Green Paper on the Legal Protection of Industrial Design an elaborate provision for employee designs was proposed, based on Art. 60(1) of the EPC. However, the provision was abandoned in the initial 1993 Draft Proposal.
21. See also CJEU C-32/08 *FEIA v Cul de Sac* (Art. 14(3) which refers to designs being developed by an employee in the execution of his duties belonging to the employer was not relevant to the ownership of commissioned community designs). Cf. Art. 17 in Proposal of the Max Planck Institute for a European Design Law, 1990 published in IIC 523 et seq. (1991).
22. See COM (93) 342 final - COD 463 p. 17.
23. In an earlier proposal precedence was not given national rules, see e.g. COM (2000) 660 final/2, COM (1999) 310 final and COM (93) 342 final - COD 463.
24. See COM (93) 342 final - COD 463 p. 16 et seq. See also Green Paper on the Legal Protection of Industrial Design (III/F/5131/91), 1991 p. 96.
25. Cf. *6bis* of the Bern Convention.
26. See Green Paper on the Legal Protection of Industrial Design (III/F/5131/91), 1991 p. 95.

bred, or discovered and developed the variety, or their successor in title, shall be entitled to a Community plant variety right. From Article 11(2) it follows that where a variety is bred, or discovered and developed by a group of persons working together, the rights shall be vested jointly in them or their respective successors in title. Further, from Article 11(4) it follows that if the breeder is an employee, the entitlement to the Community plant variety right shall be determined in accordance with the national law applicable to the employment relationship in the context of which the variety was bred, or discovered and developed.[27]

BIBLIOGRAPHY

Table of Sources

COM 342 final - COD 463, Proposal for a European Parliament and Council Regulation on the Community Design (1993).

COM 310 final, Amended proposal for a Council regulation (EC) on Community Design (presented by the Commission pursuant to Article 250 (2) of the EC-Treaty) (1999).

COM 660 final/2, Amended proposal for a Council regulation on Community design (2000).

Commission Staff Working Paper on the Review of the EC Legal Framework in the Field of Copyright and Related Rights, SEC (2004) 995 (Brussels, 19 July 2004).

Czarnota & Hart, *Legal Protection of Computer Programs in Europe* (1991).

Gert Würtenberger, Paul van der Kooij, Bart Kiewiet & Martin Ekvad, *European Community Plant Variety Protection* (Oxford University Press 2006).

Green Paper on Copyright and the Challenge of Technology - Copyright Issues Requiring Immediate Action, COM (88).

Green Paper on the Legal Protection of Industrial Design (III/F/5131/91), (1991).

Guy Tritton, *Intellectual Property in Europe* (3rd ed., 2008).

Opinion of the Economic and Social Committee (98/C 235/06).

Proposal of the Max Planck Institute for a European Design Law, 1990 published in IIC p. 523 et. seq (1991).

Sanna Wolk, *Remuneration of Employee Inventors – is There a Common European Ground? A Comparison of National Laws on Compensation of Inventors in Germany, France, Spain, Sweden and the United Kingdom*, 42 IIC 272-298 (2011).

27. See also Würtenberger/Kooij van der/Kiewiet/Ekvad, *European Community Plant Variety Protection* 26 (2006). Cf. Art. 1.iv Convention for the Protection of New Varieties of Plants of 1961.

Table of Legislation

Directive 87/54/EEC on the Legal Protection of Topographies of Semiconductor Products.
Directive 2009/24/EC on the Legal Protection of Computer Programs.
Regulation (EC) No. 2100/94 on Community Plant Variety Rights.
Regulation (EC) No. 6/2002 on Community Designs.
The Berne Convention for the Protection of Literary and Artistic Works of 1886.
The European Patent Convention ('EPC').
The UK Copyright, Designs and Patents Act (1988).

Table of Cases

FEIA v Cul de Sac, CJEU C-32/08.

CHAPTER 6

Belgium

Hendrik Vanhees

1 BACKGROUND

With regard to the Belgian rules relating to the entitlement to intellectual property rights in case of creations made by employees, some important remarks have to been made.

First of all it has to be stressed that rules relating to the entitlement to employees' intellectual property rights must be sought in the legal regulations relating to the various intellectual property rights. In Belgium, the problem of the employer-employee relationship is not regulated in specific rules relating to employment (labour) law. As there exists no overarching regulation of the employer-employee relationship for the various intellectual property rights, the solution to the problem of creations made by employees must be found in the various legal regulations relating to the different intellectual property rights. Every regulation should be examined separately in order to determine how this problem is arranged for that specific intellectual property right.

It is important to note that, in terms of content, in the various intellectual property regulations, no identical solution is provided for the problem discussed here. Each distinct regulation contains its own specific rules on the employer-employee relationship, and sometimes no explicit rules are provided for. As this contribution will show, the Belgian regulations relating to patents do not contain rules relating to inventions made by an employee. This fragmented approach can be explained by the fact that the employer-employee relationship within the field of intellectual property is a very sensitive problem. The rules in the field of copyright, which are discussed below, only came about after long and very difficult discussions in the Belgian Parliament. During the discussion of the current rules relating to patents in Parliament no agreement could be reached with regard to the entitlement to creations made by employees.

Typical for Belgium is the contractual freedom given to the employer and the employee. As this contribution will show, even in the case where legal rules are provided for the employer-employee relationship, they can be waived. When no rules are provided, as is the case concerning patents, the employer and employee are free to arrange by contract who is entitled to the creations made by the employee.

Very recently most of the different intellectual property laws like the Copyright Act, the Act on the Protection of Software, the Act on the Protection of Databases, the Act on the Protection of Semiconductor Products and the Patent Act were brought together in one code, called Book XI: Intellectual Property. In Belgium all economic legislation is brought together in one large code, namely the 'Code of Economic Law', consisting of different books. Book XI of this code concerns in particular the intellectual property rights. The rules concerning design and trademark protection are however not included in Book XI, as design and trademark protection in Belgium, the Netherlands and Luxembourg is governed by the Benelux Treaty of 25 February 2005 on Intellectual Property (trademarks and designs). It is important to stress that the objective of Book XI was only to bring together the legislation as it existed at the moment of its introduction. As regards the entitlement to creations made by an employee, no changes were made to the existing rules.

2 EMPLOYEES' COPYRIGHT

2.1 The Legal Framework

Article XI.170 § 1 of Book XI stipulates very clearly that the original author is the physical person who created the protected work. So the physical person who creates a work protected by copyright is the author thereof. They are entitled to all the rights that exist with respect to the work, economic rights as well as moral rights. This basic rule in the Belgian copyright system applies to creations made by employees (and public servants). The employee who created the copyright-protected work is the author of it and receives all the rights, economic rights and moral rights. If an employer wishes to receive the economic rights then they need to make a contractual agreement with their employee in which a transfer of these rights is provided. Article XI.167 of Book XI contains conditions that must be met by this contractual arrangement.

With regard to software, Article 2(3) of Directive 2009/24/EC of the European Parliament and of the Council of 23 April 2009 on the legal protection of computer programs[1] stipulates that where a computer program is created by an employee in the execution of their duties or following the instructions given by their employer, the employer exclusively shall be entitled to exercise all economic rights in the program so created, unless otherwise provided by contract. This provision was implemented by Article XI.296 of Book XI. Article XI.296 stipulates that, unless otherwise provided by contract or in the charter of public servants, only the employer shall be deemed to be

1. O.J., 2009, L 111, 16 ('Computer Program Directive').

entitled to the economic rights in programs created by one or multiple employees or public servants in the execution of their duties or commissioned by their employer.

Directive 96/9/EC on the legal protection of databases contains no rules regarding the entitlement to the copyright protection of databases made by employees.[2] Recital 29 of this Directive only states that the arrangements applicable to databases created by employees are left to the discretion of the Member States, and that therefore nothing in the Directive prevents Member States from stipulating in their legislation that where a database is created by an employee in the execution of their duties or following the instructions given by their employer, the employer exclusively shall be entitled to exercise all economic rights in the database so created, unless otherwise provided by contract. The Member States are therefore free to provide rules relating to the employer-employee relationship. Belgium has chosen to provide legislation for this problem. Article XI.187 § 1 of Book XI stipulates that, unless otherwise provided by contract or in the charter of public servants, only the employer shall be deemed to be entitled to the economic rights in databases created in the non-cultural sector by one or multiple employees or public servants in the execution of their duties or following the instructions given by their employer. Paragraph 2 of Article XI.187 states that related to the presumption of the transfer of rights, contained in § 1, collective agreements can determine the extent and modalities thereof.

2.2 Ownership of Rights

In Belgium, as indicated above, it is the employee (or public servant) who created the copyright-protected work who is the author of it and receives all the rights, economic rights as well as moral rights.[3] If an employer wants to be entitled to the economic rights, they have to make a contractual agreement with their employee. It should be stressed that in Belgium an employer will not be entitled to the economic rights vested in a copyright-protected work created by their employee as a result of the mere existence of an employment agreement (or a charter in case of public servants),[4] even if the employee created the work in the execution of their duties.

Article XI.167 of Book XI contains conditions that must be met if the employer wishes to be entitled to the economic rights vested in the copyright-protected work created by their employee. This Article only applies when the work is created by the employee in the execution of their employment agreement (or charter in case of public servants). The work must be created by the employee executing their duties, arising

2. O.J., 1996, L 77, 20 ('Database Directive').
3. Court of Appeal of Brussels, 15 February 1996, A&M, 1996, 405, note M Buydens See also: Court of Appeal of Brussels, 12 November 2002, IRDI, 2003, 230, Court of First Instance of Brussels, 28 March 2007, ICIP, 2007, 357, note F Gevers and President of the Court of First Instance of Gent, 3 September 2001, IRDI, 2002, 104, note E Laevens.
4. See e.g.: Court of First Instance of Brussels, 5 May 2006, Ing Cons., 2006, 271.

out of their employment contract or contractual agreement during the employment relationship. So the work must be created by the employee executing contractual duties.[5]

The Copyright Act stipulates that the transfer of the economic rights towards the employer should be stated expressly.[6] This means that the transfer should be stated in writing.[7] This can be in the employment agreement (or in the charter in case of a public servant), or in any written agreement, even if this written agreement was concluded later on.[8] The transfer of economic rights towards the employer must be clearly defined. In case of doubt the agreement will be interpreted restrictively.[9] The agreement must very clearly indicate which economic rights are transferred, even though it is not required to list in detail each mode of exploitation transferred to the employer.[10]

An employee (public servant) can grant rights in relation to future modes of exploitation. Pursuant to Article XI.167 § 3(4) Book XI this must be done expressly. It is also necessary to stipulate that the employee (public servant) is entitled to a share in the profits received according to the future mode(s) of exploitation. It should be noted that Article XI.167 § 3(4) Book XI is silent on the proportion of this share in the profits to which the employee (public servant) is entitled. This can be determined freely between employer and employee.

The transfer of the economic rights may be agreed in the individual employment contract between employer and employee. Article XI.167 § 3(5) Book XI stipulates that collective agreements can determine the extent and modalities thereof. This provision allows that the mere fact that rights are transferred to the employer is provided for in the individual employment contract, but that the extent and modalities of this transfer are determined in a collective (employment) agreement, binding for both the employer and employee. It must however be stressed that the transfer to the employer can never just be arranged in the mentioned collective (employment) agreement. The mere fact that rights are transferred to the employer always has to be provided for in the individual employment contract between the employer and employee.[11]

Regarding computer programs protected by copyright, the very clear starting point of the copyright rules was retained. The person who created the computer program is the author of the program, even if they are an employee or a public servant.[12] Article XI.296 of Book XI, implementing Article 2(3) of the Computer Program Directive, however contains an automatic transfer of the economic rights as it

5. I Van Puyvelde, *Intellectuele rechten van werknemers* 59, No 91 (Antwerpen, Intersentia, 2012).
6. Art. XI.167, § 3(1) Book XI Court of First Instance of Brussels, 28 March 2007, ICIP, 2007, 357, note F Gevers.
7. Cf Court of Appeal of Gent, 13 March 2006, IRDI, 2006, 220.
8. In this sense also: F De Visscher and B Michaux, *Précis du droit d'auteur et des droits voisins* 337-338, No 420 (Brussels, Bruylant, 2000); F Brison and B Michaux, *De nieuwe auteurswet*, R.W 529 (1995-1996); A Strowel and B Strowel, *Titularité et règles contractuelles dans le domaine du droit d'auteur et des programmes d'ordinateur*, 34 DAOR 68 (1995); A Strowel and B Strowel, *La nouvelle législation belge sur le droit d'auteur*, J.T 129 (1995).
9. *Ibid.*
10. Cf Art XI 167, §3(3) Book XI.
11. See F De Visscher and B Michaux, *supra* note 8, 341, No 425 and I Van Puyvelde, *supra* note 5, 67-68.
12. Cass., 3 June 2010, A&M, 2011, 47, note H Vanhees.

stipulates that, unless otherwise provided by contract or in the charter of public servants, only the employer shall be deemed to be entitled to the economic rights in programs created by one or multiple employees or public servants in the execution of their duties or commissioned by their employer. The presumption of acquisition of rights in favour of the employer foreseen in Article XI.296 of Book XI[13] will not only apply when the computer program was created by one or multiple employees (or public servants) 'in the execution of their duties' (this is the case when it was the duty of the employee(s) or public servant(s), as this results out of their employment agreement (or charter of the public servant), to develop the program). It will also apply when the computer program is created on their employer's demand. The concept 'on their employer's demand' should be understood as meaning that the employee (public servant) developed the program because their employer asked them to do this, although this development is independent of the employee's duties resulting from their employment agreement.[14] It is the employee (or public servant) who has the burden of proof of showing that the legal presumption of Article XI.296 Book XI cannot apply.[15]

Looking at the literal text of Article XI.296 Book XI it is clear that this provision contains a legal presumption only for the economic rights and not for the moral rights, and that it applies both in the case of employees and public servants.

The automatic transfer of the economic rights in a protected program to the employer, as foreseen in Article XI.296 of Book XI, will only find application if no deviant agreement is provided. Such a deviant agreement can be provided in the employment contract, in any other contract between the employer and employee or in the charter of public servants.[16] It may be both to the benefit and the detriment of the employee.

The rules related to databases protected by copyright run parallel with the rules related to computer programs protected by copyright. As a result of this, also in relation to copyright-protected databases the very clear starting point of the copyright rules is retained. The physical person who created the database is the author of this database, even if they are an employee or a public servant. Article XI.187 of Book XI however contains an automatic transfer of the economic rights in the protected database as it stipulates that, unless otherwise provided by contract or in the charter of public servants, only the employer shall be deemed to be entitled to the economic rights in databases created in the non-cultural sector by one or multiple employees or public servants in the execution of their duties or following the instructions given by their employer. The legal presumption only applies to databases in the non-cultural sector. For many aspects of this legal presumption contained in Article XI.187 § 1 Book XI, please refer to the analysis of Article XI.296 Book XI above. Only those aspects of the legal presumption contained in Article XI.187 § 1 Book XI will be discussed that differ from the presumption that can be found in Article XI.296 Book XI.

---

13. Cf Cass., 3 June 2010, A&M, 2011, 47, note H Vanhees.
14. F Brison and B Michaux, *supra* note 8, 553; H Vanhees, *Auteursrechtelijk beschermde werken en software gemaakt in uitvoering van een arbeidsovereenkomst of statuut*, Or 174 (1994); I Van Puyvelde, *supra* note 5, 76.
15. Court of Appeal of Antwerp, 12 May 1997, R.W., 1997-1998, 1341.
16. I Van Puyvelde, *supra* note 5, 76, No 129.

As indicated above, the legal presumption in Article XI.187 § 1 Book XI only applies to databases in the non-cultural sector. The concept 'non-cultural sector' is very unclear and should be clarified by case law.[17] The automatic transfer of economic rights towards the employer also applies when a database is created in the non-cultural sector by one or multiple employees or public servants 'following the instructions given by their employer'. Consequently, Article XI.187 Book XI also finds application when the employee or public servant developed databases in the non-cultural sector because their employer asked them to do this and gave them instructions, even when the development is independent of their duties resulting from their employment agreement.[18]

Paragraph 2 of Article XI.187 Book XI states that related to the presumption contained in § 1 of this article, collective agreements can determine the extent and modalities thereof. This means that in a collective agreement, binding for both the employer and employee, the extent and the modalities of the automatic transfer of economic rights can be determined.

2.3 The Parties' Duty to Inform

Neither the copyright rules nor the rules relating to copyright-protected software and databases contain a provision imposing an obligation on the employee to inform their employer that they created a work, database or computer program protected by copyright. If, on the basis of an agreement (contractual arrangement) between the employer and employee or by virtue of the legal presumptions laid down in Article XI.296 and Article XI.187 of Book XI, the employer is entitled to the economic rights, this obligation to inform the employer arises from the general obligation to carry out in good faith the (employment) contract. In practice, this obligation to inform the employer probably will be set out in a contractual agreement, the labour regulation, service notes, etc.

2.4 Employee Remuneration Right

Neither the copyright rules nor the rules relating to copyright-protected software and databases provide for a remuneration right for employees. So when an employee creates a work protected by copyright, and by virtue of a contractual arrangement the employer is entitled to the economic rights vested in this work, the employee will have no right to an additional remuneration for the transfer of these rights in addition to their wage, when such an additional remuneration is not agreed with their employer. Also, when the legal presumption laid down in Article XI.296 or Article XI.187 of Book XI finds application, the employee who created a computer program or database protected by copyright in the execution of their duties or commissioned by their employer, will

17. See for critical comments in connection with this: B Michaux, *Droit des bases de données* 116-117 (Kluwer, Brussels, 2005).
18. I Van Puyvelde, *supra* note 5, 81, No 140.

not receive, and will not be able to claim, any remuneration. The employee will only have a right to a (monetary) compensation when this is agreed between the parties.

3 EMPLOYEES' SEMICONDUCTOR CHIPS

3.1 The Legal Framework

Council Directive 87/54/EEC on the legal protection of topographies of semiconductor products gives Member States the freedom to implement in their national legislation rules concerning the relation between an employer and employees (see Article 3 (2) (a) of the Semiconductor Directive that stipulates that Member States *may* provide that where a topography is created in the course of the creator's employment, the right to protection shall apply in favour of the creator's employer unless the terms of employment provide to the contrary).[19] The Belgian legislator opted to include this provision in the rules relating to semiconductor chips. Article XI.322 § 1 of Book XI stipulates that if a topography of a semiconductor product is created by an employee in the exercise of their function, unless otherwise agreed, the employer shall be deemed to be the creator.

3.2 Ownership of Rights

It follows from Article XI.322 § 1 of Book XI that if no agreement exists between employer and employee, the employer is deemed to be the creator of a semiconductor product created by their employee, executing, by creating the semiconductor product, the tasks that by virtue of their employment contract rested on them.[20] Consequently, in this situation, the employer owns all the rights on the protected semiconductor product. Article XI.322 § 1 of Book XI allows however that the employer and employee can agree otherwise in the employment contract or in any other contract. Such a deviant agreement may be both to the benefit and the detriment of the employee.[21]

3.3 The Parties' Duty to Inform

Book XI contains no provision imposing an obligation on the employee to inform their employer that they created a protected semiconductor product. Such an obligation to inform the employer arises however from the general obligation to carry out the employment contract in good faith. In practice, the obligation to inform the employer will probably be set out in a contractual agreement, the labour regulation, service notes, etc.

19. O.J., 1987, L 24, 36 ('Semiconductor Directive').
20. I Van Puyvelde, *supra* note 5, 78.
21. F Gotzen and M.C Janssens, *Wegwijs in het intellectueel eigendomsrecht* 332 (Brugge,Vanden Broele, 2012); I Van Puyvelde, *supra* note 5, 79.

3.4 Employee Remuneration Right

Article XI.322 § 1 of Book XI provides no remuneration right for the employee. So, when there is no agreement between the employer and their employee, and Article XI.322 § 1 of Book XI is applicable, the employee who created a semiconductor product in the exercise of their function will not receive, and will not be able to claim, any remuneration. The employee will only have a right to (monetary) compensation when this is agreed between the parties.[22]

4 EMPLOYEES' INVENTIONS

4.1 The Legal Framework

Book XI contains no rules with regard to the entitlement to inventions created by employees. There is only Article XI.9 § 1 of Book XI stipulating that the inventor, or their successor in title, is entitled to the patent. Article XI.9 § 3 of Book XI stipulates that the applicant for a patent is deemed to be entitled to the patent.

4.2 Ownership of Rights

As there exists no rules with regard to the entitlement to inventions created by employees, the situation in Belgium is characterised by absolute contractual freedom.[23] Having regard to this contractual freedom, the employer and employee are free to conclude an agreement stipulating who is entitled to apply for a patent if the invention was done by the employee.[24] Such an agreement can be made in the contract of employment but also in any other agreement.[25] A written agreement is not required.[26] However, an individual agreement must exist.[27]

If no contractual agreement exists, jurisprudence and legal doctrine make a distinction between three different types of inventions: service inventions, dependent or mixed inventions and free or personal inventions.

Service inventions are inventions made by employees executing research activities or inventive tasks arising from the employment contract. The invention is the direct result of research activities, and these activities can be normal labour tasks or temporary commands.[28] If the contract of employment concerns research activities or

22. *Ibid.*
23. F Gotzen and M.C Janssens, *supra* note 21, 324.
24. *Ibid.*
25. *Ibid.*
26. I Vandereken and H Van Hoogenbemt, *Werknemers en intellectuele rechten*, in *Intellectuele eigendomsrechten* 229, No 28 (Vlaams Pleitgenootschap bij de Balie te Brussel (ed.), Biblo, Kalmthout, 1995); I Van Puyvelde, *supra* note 5, 102, No 177.
27. See: I Van Puyvelde, *supra* note 5, 102-103, No 177-178.
28. F Gotzen and M.C Janssens, *supra* note 21, 325.

inventive tasks, it is presumed that the employee accepted the entitlement of their employer to the invention.[29] In case of service inventions the invention belongs to the employer.[30]

Inventions will qualify as dependent or mixed inventions if a clear link exists with the tasks the inventor performed as an employee or with the activities of the employer. Clear input of the employer should be proven.[31] This input can be an intellectual, financial or material, e.g., where the employee used the knowhow, machines or staff of their employer.[32] In Belgium there is still a discussion about who is entitled to these inventions.[33] In legal doctrine the majority considers that the employee is entitled to these inventions.[34] But legal scholars who defend this view accept that the employee's freedom to act is restricted.[35] For instance, the employee is banned from engaging in activities of unfair competition[36] and publishing trade secrets.[37] Some commentators defend the view that the employer is entitled to these inventions. This is for instance the case when there is an active participation or cooperation of the employer.[38]

Free or personal inventions are inventions made by an employee during their spare time, with their own means and on their own initiative.[39] Free or personal inventions even exist when they are related to the activities of the employer, but when there is no input on the employer's part.[40] It is undisputed that the employee is entitled to these inventions. The employee alone can decide on a patent application and the exploitation of the invention.[41]

4.3 The Parties' Duty to Inform

In case of service inventions the employee is obliged to inform the employer immediately about the fact that they made an invention.[42] Although such an obligation cannot be found in Book XI, it results from the general obligation to carry out the employment

29. M.C Janssens, *Uitvindingen in dienstverband met bijzondere aandacht voor uitvindingen aan universiteiten* 233-234, No 255 (Brussels, Bruylant, 1996); I Van Puyvelde, *supra* note 5 90, No 154.
30. F Gotzen and M.C Janssens, *supra* note 21 325.
31. F Gotzen and M.C Janssens, *supra* note 21, 325; I Van Puyvelde, *supra* note 5, 93, No 159.
32. F Gotzen and M.C Janssens, *supra* note 21 326.
33. F Gotzen and M.C Janssens, *supra* note 21, 326; I Van Puyvelde, *supra* note 5, 93 and following, No 160 and following.
34. See F Gotzen and M.C Janssens, *supra* note 21 326.
35. F Gotzen and M.C Janssens, *supra* note 21, 326; I Van Puyvelde, *supra* note 5, 95, No 165.
36. Cf Art 17, 3°(a) of the Act of 3 July 1978 on employment agreements, B.S., 22 August 1978 (erratum: B.S., 30 August 1978), as amended ('Act on employment agreements'); F Gotzen and M.C Janssens, *supra* note 21, 326; I Van Puyvelde, *supra* note 5, 95, No 165.
37. Cf Art 17, 3°(b) of the Act on employment agreements; F Gotzen and M.C Janssens, *supra* note 21, 326.
38. I Vandereken and H Van Hoogenbemt, *supra* note 26, 222-223, No 18.
39. F Gotzen and M.C Janssens, *supra* note 21, 326; I Vandereken and H Van Hoogenbemt, *supra* note 26, 221, No 15; I Van Puyvelde, *supra* note 5, 97, No 171.
40. M.C Janssens, *supra* note 29, 263, No 293.
41. F Gotzen and M.C Janssens, *supra* note 21, 326; I Van Puyvelde, *supra* note 5, 98, No 171.
42. F Gotzen and M.C Janssens, *supra* note 21, 325; M.C Janssens, *supra* note 29, 236, No 258; I Van Puyvelde, *supra* note 5, 91, No 156.

contract in good faith. The obligation to inform the employer includes the obligation to assist the employer in applying for a patent and in the exploitation of the invention.[43]

With regard to dependent or mixed inventions, if it is assumed that the employee is entitled to these inventions, there is of course no obligation for the employee to inform their employer that they made an invention.[44]

4.4 Employee Remuneration Right

In case of service inventions the employee is not entitled to any remuneration in addition to their wage.[45]

5 EMPLOYEES' UTILITY MODELS

The protection of utility models does not exist in Belgium.

6 EMPLOYEES' DESIGN RIGHTS

6.1 The Legal Framework

The protection of designs is regulated by the Benelux Treaty of 25 February 2005 on Intellectual Property (trademarks and designs) ('BTIP').[46] As a result it is not possible to receive design protection only in Belgium. If a design is registered by the Benelux Office for Intellectual Property (trademarks and designs), protection is obtained for the three countries of the Benelux, namely Belgium, the Netherlands and Luxembourg. The Benelux design is a unit. It is granted for the three countries of the Benelux, and when it expires or is declared null, the design will be null and will come to an end in the whole Benelux region. From the BTIP it follows that the designer is entitled to apply for design protection.[47] Article 3.8 § 1 BTIP however stipulates that if a design is developed by an employee in the execution of their function, unless otherwise agreed, the employer will be deemed to be the designer. If a design is also protected by copyright, Article 3.29 BTIP complements Article 3.8 § 1 by stating that if, pursuant to Article 3.8 § 1 BTIP, the employer is deemed to be the designer, the employer is also entitled to copyright protection.

Regarding Community designs, Article 14 §3 of Council Regulation (EC) No. 6/2002 on Community designs stipulates that where a design is developed by an employee in the execution of their duties or following the instructions given by their employer, the right to the Community design shall vest in the employer, unless otherwise agreed or specified under national law.[48]

43. M.C Janssens, *supra* note 29 238, No 259.
44. M.C Janssens, *supra* note 29, 255, No 284.
45. F Gotzen and M.C Janssens, *supra* note 21, 325.
46. B.S., 26 April 2006, 21867, as amended ('BTIP' or 'Benelux Treaty').
47. Cf Art 3.7, §1 BTIP.
48. O.J., 2002, L 3, 1 as amended ('Regulation on Community designs').

6.2 Ownership of Rights

Pursuant to Article 3.8 § 1 BTIP, unless otherwise agreed, the employer is deemed to be the designer and has the right to apply for design protection in the Benelux. This presumption only finds application when the design is made by the employee in the execution of their function. A design is only developed by an employee in the execution of their function if there exists a relation between the developed design on the one hand and the task of the employee and the nature of their activities, as they result from their employment agreement, on the other hand.[49] The Court of Justice of the Benelux has decided that Article 3.8 § 1 BTIP also applies if an application for a Benelux design has not (yet) been made.[50] So Article 3.8 § 1 BTIP also finds application on designs which have not yet been registered. As has already been indicated, Article 3.29 BTIP stipulates that if, pursuant to Article 3.8 § 1 BTIP, the employer is deemed to be the designer, and the design is also protected by copyright, the employer is also entitled to copyright protection. There is however doubt in the legal doctrine about the question of whether Article 3.29 BTIP also finds application if a design has not yet been registered. Jurisprudence and legal doctrine are divided on this topic.[51] In a decision of 25 October 2013 the Supreme Court of the Netherlands[52] implicitly assumed that Article 3.29 BTIP finds application on not-yet registered designs, as it declared that this Article finds application on designs, namely on the appearance of the whole or a part of a product, even if these designs do not comply with the requirements for design protection. In the view of the Supreme Court of the Netherlands, Article 3.29 BTIP finds application on the appearance of the whole or a part of a product, even if this appearance is not new and/or has no individual character.

Article 3.8 § 1 BTIP as well as Article 3.29 BTIP allow the employer and the employee to agree otherwise in the employment agreement or in any other contract. Such a deviant agreement may be both to the benefit and the detriment of the employee. The employee and employer can for instance agree that the employee will be entitled to the design when the design is developed by them in the execution of their

49. D.W.F Verkade, *Bescherming van het uiterlijk van produkten*, in *Recht en praktijk* No 40, 71, No 49 (J.M Polak (ed.), Deventer, Kluwer, 1985).
50. Court of Justice of the Benelux, 22 June 2007, ICIP, 2007, 381, note F De Visscher; R.W., 2007-08, 1448, note H Vanhees; T.B.H., 2008, 408, note J Keustermans; B.I.E., 2007, 601; I.E.R., 2007, 309, note Ch G (Electrolux t/Sofam).
51. Proponents of the application of Art 3.29 BTIP on not (yet) registered designs are for instance: F De Visscher and B Michaux, *supra* note 19, 190-191, No 229, and footnote 64; D Visser, *Opdrachtgeversauteursrecht*, 3 AMI 98 (2009) See also: Court of Appeal of Antwerp, 23 April 2012, ICIP, 2012/2, 359 and IEF 11735; Court of Appeal of Gent, 13 March 2006, IRDI, 2006, 220; Court of 's-Gravenhage, 7 December 2011, IEF 10640 and B9 10526 Opponents of this application are for instance: E Derclaye and A Strowel, *Droit d'auteur et numérique: logiciels, bases de données, multimédia* 124-125, No 135, footnote 385 (Brussels, Bruylant, 2001); A Strowel, *Droit d'auteur et copyright* 351, No 266 (Brussels, Bruylant, 1993); H Vanhees, *De bescherming van auteursrechtelijk beschermde, niet gedeponeerde modellen door het Benelux-modellenrecht*, 5 A&M 496 et seq. (2009) A Quaedvlieg has serious reservations about the application of Art. 3.29 BTIP on not (yet) registered designs: A Quaedvlieg, *Het opdrachtgever-smodelrecht als Fremdkörper*, B.I.E., 395 (2009) Case law that supports this view: Court of First Instance of Brussels, 28 March 2007, ICIP, 2007, 357, note F Gevers.
52. IEF, 13179 and IEPT 20131025.

duties. They can also agree that when the design is created by the employee in the execution of their function the employer is entitled to the design right and the employee to copyright protection.[53]

By virtue of Article 14(3) of the Regulation on Community designs, which applies both to registered and unregistered Community designs, unless otherwise agreed or specified under national law, the right to a Community design is vested in the employer if the design is developed by their employee in the execution of their duties or following the instructions given by their employer.[54] The phrase 'in the execution of their duties' should be understood as the duty of the employee as it results from their employment agreement to develop designs. The concept 'following the instructions given by their employer' should be understood as the development of designs by the employee because their employer asked them to do this, although this development is independent of their duties resulting from their employment agreement.

Article 14(3) of the Regulation on Community designs clearly allows the employer and the employee to agree otherwise in the employment agreement or in any other contract.[55] The employee and employer can for instance agree that the employee will be entitled to the design when the design is developed in the execution of their duties or following the instructions given by their employer. This has already been confirmed by the Court of Justice in a decision of 2 July 2009 in the case *FEIA v Cul de Sac*. The Court held that regarding the part of Article 14(3) which states 'unless otherwise agreed', it must be pointed out that this enables the parties to an employment contract to designate 'the employee' as the proprietor of the design developed by them in the execution of their duties.[56] Article 14(3) of the Regulation on Community designs also give the Member States the option of stipulating in their national legislation that the 'employee' is the proprietor of a Community design, when the design has been developed in the context of an employment relationship.[57] Such a stipulation does not exist in Belgium, nor in the BTIP.

As the Regulation on Community designs contains no rules concerning the entitlement to copyright protection, when a Community design is also protected by copyright, the provisions of Book XI and of the BTIP must be applied to decide this entitlement. This was recently confirmed in the already discussed decision of the Supreme Court of the Netherlands of 25 October 2013. As a result of this, if a registered or unregistered Community design is also protected by copyright, and if it is developed

53. See on this issue: H Vanhees, *Het Beneluxmodel* 100, No 229 and 153-154, No 341 (Brussels, Larcier, 2006).
54. The Court of Justice already decided that Art. 14, §3 of the regulation on Community designs only applies to the situation described in this provision, namely to the situation where designs are developed by an employee in the execution of their duties or following the instructions given by their employer This provision finds for instance no application when the designs are made on command (Court of Justice, 2 July 2009, C-32/08, FEIA c/ Cul de Sac, recital 51, ICIP, 2009/3, 481, note H Vanhees).
55. D Musker, *Community Design Law Principles and Practice* 112-113 (London, Sweet and Maxwell, 2002).
56. Court of Justice, 2 July 2009, C-32/08, *FEIA v Cul de Sac*, recital 50, ICIP, 2009/3, 481, note H Vanhees.
57. Cf Court of Justice, 2 July 2009, C-32/08, *FEIA v Cul de Sac*, recital 50, ICIP, 2009/3, 481, note H Vanhees.

by an employee in the execution of their function and the design protection rights are vested in the employer due to the legal presumption in Article 14(3) of the Regulation on Community designs, the employer is also entitled to the copyright protection rights. This solution is the result of the application of Article 3.29 BTIP. As was indicated above in this contribution, Article 3.29 BTIP allows the employer and the employee to agree otherwise in the employment agreement or in any other contract. They can agree for instance that when the design is created by the employee in the execution of their function, the employer is entitled to the design right and the employee to copyright protection rights.

6.3 The Parties' Duty to Inform

Neither the BTIP nor the Regulation on Community designs contains a provision imposing the obligation on the employee to inform their employer that they developed a design. Such an obligation to inform the employer arises, however, from the general obligation to carry out the employment contract in good faith. In practice the obligation to inform the employer will most likely be set out in a contractual agreement, the labour regulation, service notes, etc.

6.4 Employee Remuneration Right

Neither the BTIP nor the Regulation on Community designs provides a remuneration right for the employee. So, when there is no agreement between the employer and their employee, and Article 3.8 § 1 BTIP or Article 14(3) of the Regulation on Community designs finds application, the employee who developed a design in the exercise of their function or following the instructions given by their employer will not receive, and will not be able to claim, any remuneration. The employee will only have a right to a (monetary) compensation when this is agreed between the parties. This is also the case when Article 3.29 BTIP finds application.

7 EMPLOYEES' PLANT VARIETIES

7.1 The Legal Framework

Article XI.111 §3 of Book XI stipulates that when a new variety was bred, or discovered and developed by an employee by carrying out (by implementing) their employment agreement, the entitlement to the plant variety right belongs to the employer, unless otherwise agreed.

Of course, one can always choose to obtain a Community plant variety right. Article 11(4) of Council Regulation (EC) No. 2100/94 on Community plant variety rights[58] stipulates that if the breeder is an employee, the entitlement to the Community

58. O.J., 1994, L 227, 1 (erratum: O.J., 2001, L 111, 31) as amended This regulation will be referred to as 'the regulation on Community plant variety rights'.

plant variety right shall be determined in accordance with the national law applicable to the employment relationship in the context of which the variety was bred, or discovered and developed.

7.2 Ownership of Rights

Article XI.111 § 3 of Book XI clearly provides that, unless otherwise agreed, the employer has an entitlement to plant variety rights when the new variety was bred or discovered and developed by an employee by carrying out (by implementing) their employment agreement. Article XI.111 § 3 of Book XI permits the employer and the employee to agree otherwise in the employment agreement or in any other contract. Such a deviant agreement may be both to the benefit and the detriment of the employee.

Regarding Community plant variety rights, if Article 11(4) of the Regulation on Community plant variety rights finds application, and if Belgian law is applicable to the employment relationship in the context of which the variety was bred, or discovered and developed, Article XI.111 § 3 of Book XI will determine the entitlement to the Community plant variety right.

7.3 The Parties' Duty to Inform

Book XI contains no provision imposing an obligation on the employee to inform their employer that they created a new plant variety right. Such an obligation to inform the employer arises however from the general obligation to carry out the employment contract in good faith. In practice, the obligation to inform the employer will most likely be set out in a contractual agreement, the labour regulation, service notes, etc.

7.4 Employee Remuneration Right

Book XI does not provide for the employee a remuneration right. So, when there is no agreement between employer and employee, and Article XI.111 § 3 of Book XI finds application, the employee who created a new plant variety in the exercise of their function will not receive, and will not be able to claim, any remuneration. The employee will only have a right to (monetary) compensation when this is agreed between the parties.

8 INSTANCES FOR DISPUTES

Pursuant to Article 574 of the Belgian Judicial Code,[59] the Commercial Court is competent for the following claims, even when the parties to the proceedings are not undertakings:

59. The Belgian Judicial Code is referred to as 'JC'.

Claims referred to in Article XI.342 of Book XI.[60] Article XI.342 of Book XI stipulates that the Commercial Courts are competent for all claims relating to semiconductor chips even when the parties to the proceedings are not merchants and independently of the amount of the claim. Pursuant to Article 633quinquies § 1(2) JC only a Commercial Court having their seat at a Court of Appeal is competent for claims relating to semiconductor products. As a result of this only the Commercial Courts of Antwerp, Gent, Brussels, Mons and Liège are competent. Article XI.342 § 3(2) of Book XI permits claims referred to here to be brought before an arbitral tribunal.

Claims referred to in Article XI.337 of Book XI.[61] Article XI.337 of Book XI stipulates that the Commercial Court of Brussels is competent for all claims relating to patents, even when the parties to the proceedings are not merchants and independently of the amount of the claim. Pursuant to Article 633quinquies §1(1) JC only the Commercial Court of Brussels is competent. Article XI.337 § 2(2) of Book XI stipulates that disputes relating to the entitlement of a patent application or a patent can be brought before an arbitral tribunal.

Claims relating to Benelux designs.[62] Pursuant to Article 633quinquies, § 1(2) JC only the Commercial Courts of Antwerp, Gent, Brussels, Mons or Liège are competent. By virtue of Article 93(1) of the Regulation on Community designs these five Commercial Courts are also competent for actions relating to the ownership or entitlement to Community designs.

Claims referred to in Article XI.339 of Book XI.[63] Article XI.339, §1 of Book XI stipulates that the Commercial Courts are competent for all claims relating to plant variety rights even when the parties to the proceedings are not merchants and independently of the amount of the claim. Pursuant to Article 633quinquies § 1(2) JC only the Commercial Courts of Antwerp, Gent, Brussels, Mons or Liège are competent. By virtue of Article 103 of the Regulation on Community plant variety rights these five Commercial Courts are also competent for actions relating to the ownership or entitlement to Community plant variety rights.

The situation with regard to copyright is somewhat more complex. The rules discussed here relating to copyright also apply in the case of databases and software protected by copyright. If both parties are undertakings the Commercial Courts of Antwerp, Gent, Brussels, Mons or Liège have competence.[64] This will also be the case if the defendant is an undertaking and the plaintiff is not, but the plaintiff chooses to bring the action before the Commercial Court.[65] If both the plaintiff and the defendant are not undertakings or if the defendant is not an undertaking, or if the defendant is an undertaking and the plaintiff is not and the plaintiff does not choose the Commercial Court, and the amount of the claim exceeds EUR 2,500, then the Courts of First Instance

60. Art. 574, 17° JC.
61. Art. 574, 15° JC.
62. Art. 574, 18° JC.
63. Art. 574, 16° JC.
64. Art. 575, §1(1) and Art. 633quinquies §1(3) JC.
65. Art. 575, §1(2) and Art. 633quinquies §1(3)JC

of Antwerp, Gent, Brussels, Mons or Liège are competent.[66] If the amount of the claim does not exceed EUR 2,500, the latter claims fall within the competence of the Justice of the Peace.[67]

9 CONCLUDING REMARKS

As has become clear, the solution to the problem of creations made by employees must be found in the various rules relating to the different intellectual property rights. No answers can be found in specific rules relating to employment (labour) law. The various rules relating to the different intellectual property rights provide no consistent solution for the problem discussed here. Sometimes a legal presumption is provided in favour of the employer, sometimes the employer is obliged to secure a contract if they wish to obtain rights relating to creations made by their employees. The regulation relating to patents does not even contain rules relating to the employer-employee relationship. The consequence of all this is that the legal situation in Belgium is characterised by the contractual freedom given to the employer and employee. Even when legal rules are provided for the employer-employee relationship, these rules can be waived.

BIBLIOGRAPHY

Table of Sources

M. Buydens, *Le régime des créations et inventions salariés*, (Orientations 2013/7, 2-17).

F. Gotzen & M.C. Janssens, *Wegwijs in het intellectueel eigendomsrecht*, 357p (Brugge, Vanden Broele 2012).

M.C. Janssens, *Uitvindingen in dienstverband met bijzondere aandacht voor uitvindingen aan universiteiten*, 671 p (Brussel, Bruylant 1996).

B. Michaux, *La propriété des droits intellectuels sur les créations du salarié*, in Le patrimoine intellectuel de l'entreprise, 7-52 (Brussel, Larcier 2010).

I. Vandereken & H. Van Hoogenbemt, *Werknemers en intellectuele rechten*, in *Intellectuele eigendomsrechten* 209-246 (ed Vlaams Pleitgenootschap bij de Balie te Brussel, Biblo, Kalmthout 1995).

H. Vanhees & I. Van Puyvelde, *Intellectuele eigendomsrechten: de relatie werknemer werkgever*, in *Sociaal en fiscaal recht: 'Elck wat wils'*, 1-36. (Mechelen, Kluwer 2013).

I. Van Puyvelde, *Intellectuele rechten van werknemers*, 172p (Antwerpen, Intersentia 2012).

66. Art. 568(1) and 633quinquies §1(3) JC.
67. Art. 590(1) JC.

Table of Legislation

Benelux Treaty of 25 February 2005 on Intellectual Property (trademarks and designs).
Book XI 'Intellectual Property' of the Code Economic Law.
Council Regulation (EC) No. 2100/94 of 27 July 1994 on Community plant variety rights.
Council Regulation (EC) No. 6/2002 of 12 December 2001 on Community designs.
Council Directive 87/54/EEC of 16 December 1986 on the legal protection of topographies of semiconductor products.
Directive 96/9/EC of the European Parliament and of the Council of 11 March 1996 on the legal protection of databases.
Directive 2009/24/EC of the European Parliament and of the Council of 23 April 2009 on the legal protection of computer programs.

Table of Cases

FEIA c/Cul de Sac, Court of Justice (2 July 2009), C-32/08, ICIP, 2009/3, 481, note H. VANHEES.
Electrolux t/Sofam, Court of Justice of the Benelux (22 June 2007), ICIP, 2007, 381, note F. DE VISSCHER; R.W., 2007-2008, 1448, note H. VANHEES; T.B.H., 2008, 408, note J. KEUSTERMANS; B.I.E., 2007, 601; I.E.R., 2007, 309, note Ch. G.
Cass., (3 June 2010), A&M, 2011, 47, note H. VANHEES.
Supreme Court of the Netherlands (25 October 2013), IEF 13179 and IEPT 20131025.
Court of Appeal of Antwerp (23 April 2012), ICIP, 2012/2, 359 and IEF 11735.
Court of Appeal of Gent (13 March 2006), IRDI, 2006, 220.
Court of Appeal of Brussel (12 November 2002), IRDI, 2003, 230.
Court of Appeal of Antwerp (12 May 1997), R.W., 1997-1998, 1341.
Court of Appeal of Brussel (15 February 1996), A&M, 1996, 405, note M. BUYDENS.
Court of First Instance of Brussel (28 March 2007), ICIP, 2007, 357, note F. GEVERS.
Court of First Instance of Brussel (5 May 2006), Ing. Cons., 2006, 271.
President of the Court of First Instance of Gent (3 September 2001), IRDI, 2002, 104, note E. LAEVENS.

CHAPTER 7

Denmark

Jeppe Brinck-Jensen, Morten Rosenmeier & Helene Egede Scotwin[*]

1 BACKGROUND

Whether intellectual property rights created by employees as part of their employment belong to the employee or to their employer is an important question. In Danish law the question is dealt with by the following rules.

Employees' rights to *copyright* in *computer programs* are covered by Section 59 of the Danish Copyright Act. However, the Copyright Act does not regulate employees' rights to *copyright other than in computer programs*. Copyright other than in computer programs is regulated by legal principles derived from case law and the legal literature; see section 2 below.

Rights to semiconductor chips created by employees are dealt with in Section 3 of the Danish Act on the Protection of Semiconductor Chips; see section 3 of this chapter below.

Under Danish law, the right to *inventions* created by employees is dealt with in two acts: the Act on Inventions at Public Research Institutions concerns inventions made by employees at public research institutions such as universities; and the Employees' Inventions Act concerns inventions made by persons employed elsewhere, e.g., in private firms. See section 4 below which deals with patent rights and section 5 which deals with utility models.

---

* Sections 1-3 and 5-9 have been written by Professor Morten Rosenmeier, University of Copenhagen, Centre for Information and Innovation Law Section 4 has been written by Attorney Jeppe Brinck-Jensen and Assistant Attorney Helene Egede Scotwin, both from the Danish law firm Accura Advokatpartnerselskab The parts of paragraph 4 on the rights of public researchers has been written by Helene Egede Scotwin and has not previously been published The remaining parts are an abridged translation from Danish of Ch 2, 3, and 5 in Brinck-Jensen, Rosenmeier, Schovsbo and Sommer, *Ansattes immaterielle rettigheder* (Copenhagen 2013).

The extent to which *design rights* relating to products made by employees are owned by them or by their employer has not been expressly regulated for national design rights covered by the Danish Design Act. However, for EU design rights this is regulated in the EU Design Regulation; see section 6 below.

The extent to which an employer owns the rights to *plant varieties* created by their employees is governed by unwritten legal principles; see section 7 below.

2 EMPLOYEES' COPYRIGHT

2.1 The Legal Framework

Many copyright works are created by authors working for an employer, e.g., newspaper articles written by journalists working for newspapers and magazines, computer programs written by software engineers working for software firms, and graphic material created by graphic designers working in art departments. Both for the authors and their employers it is important to clarify the extent to which the employer owns the copyright in such works.

As a starting point, the issue is subject to the general principle of freedom of contract. Accordingly, the parties may settle by agreement to whom the rights shall belong. The question is, however, what shall apply in the absence of such a contract.

The Danish Copyright Act regulates this question expressly only in relation to *computer programs*. Section 59 of the Danish Copyright Act states that the copyright in computer programs created by employees is automatically assigned to the employer; see section 2.2.7 below.

However, the Danish Copyright Act contains no rules on the extent to which an employer owns the copyright in works other than computer programs created by employees. This is dealt with by a legal principle derived from a number of court decisions and the legal literature. The principle states as follows:

Unless there is an agreement to the contrary, the employer is only entitled to those parts of the employee's copyright that relate to a necessary part of the employer's customary activity at the time when the employee created the work. It is a condition that the creation of the work is part of the employee's job description.[1]

This principle can be derived from a number of court decisions but it is not expressly stated in the Danish Copyright Act. From time to time attempts have been made to include an express provision on employees' copyright in the Copyright Act.[2] So far these attempts have been unsuccessful.

1. See Brinck-Jensen, Rosenmeier, Schovsbo and Sommer, *Ansattes immaterielle rettigheder* Ch 2 (2014); and Schønning, *Ophavsretsloven med kommentarer* 502 f and 536 f (Copenhagen 2011).
2. See Brinck-Jensen, Rosenmeier, Schovsbo and Sommer, *Ansattes immaterielle rettigheder*, 32 f.

2.2 Ownership of Rights

2.2.1 What Is the 'Customary Activity' of the Employer?

As explained above, unless an agreement has been made to the contrary, the employer is only entitled to those parts of the employee's copyright that relates to a necessary part of the employer's *customary activity* at the time when the employee created the work.

The Danish Supreme Court's first decision on this was in 1978.[3] In this case, a local newspaper had used parts of an article from the Danish national newspaper *Politiken* without permission. The journalist who had written the article while working at *Politiken* took legal action against the local newspaper for copyright infringement. The local newspaper and *Politiken*, which intervened in the case to support the local newspaper, argued that the copyright was owned by *Politiken* rather than the journalist, and thus only *Politiken*, and not the journalist, was entitled to demand damages from the local newspaper. The Supreme Court stated: 'It cannot be assumed that, because of his employment at *Politiken*, the journalist's copyright has been assigned to *Politiken* to a greater extent than was necessary for the customary activity of *Politiken*. Consequently, he cannot be prevented from claiming infringement or demanding damages for the unwarranted copying of the article.'

This decision has subsequently been followed in a number of other cases such as:

Ugeskrift for Retsvæsen 1982.926 Ø

The customary activity of the Danish news agency *Ritzau* was to sell news articles to media which subscribed to *Ritzau* and continuously received news articles according to their subscriptions. *Ritzau* sold an article to a newspaper which had no subscription. This was found to infringe the copyright of the *Ritzau* journalist since the selling of articles to non-subscribers was not part of the customary activity of *Ritzau*. In other words, that part of the journalist's copyright which concerned the sale of articles to subscribers had been assigned to *Ritzau*, but that part of the journalist's copyright which concerned other forms of sale had not been assigned.

Ugeskrift for Retsvæsen 1993.180 Ø

Two journalists employed by a television company had made a programme about the author Isabel Allende, including an interview with her. The television company later allowed a film company to make a television programme including a part of the interview. This was found to infringe the copyright of the television journalists since it was not 'established that it was … necessary for the customary activity of the television company to dispose of [the first programme] by assigning it to another television producer'.

3. *Ugeskrift for Retsvæsen* 1978.901/2 H.

Ugeskrift for Retsvæsen 1994.349 H

Some television journalists employed by a television company had produced a number of programmes, and the company reran the programmes without their permission. This was not found to infringe their copyright since their employment was so permanent that they were found to have assigned their rights to the television company to the extent that this was a necessary part of the television company's customary activity.[4]

When an employee is employed in part of a group of companies under a parent company, the focus will be on the customary activity of the particular entity to which the employee is attached. For example, if a corporate group owns three newspapers and one publishing house, there will be a question as to the extent to which the copyright of a journalist working for one of the newspapers is assigned to the employer. When deciding this it is relevant to look at the customary activity of the newspaper for which the employee works; the customary activities of the other newspapers and the publishing house are irrelevant.

When determining what is the customary activity of an employer under Danish law, it is not clear whether account should be taken of the activities of similar employers in the same kind of business or whether the court should focus narrowly on the customary activity of the employer in question.[5]

2.2.2 *Cases Where the Employee Is under a Legal Obligation to Create Works*

As noted above, the decisive criterion for establishing whether the copyright vests with the employer or the employee is the extent to which the exploitation of a work created by an employee is part of the customary activity of the employer. On the other hand, it is not decisive whether the employee is under a *legal obligation* to create works as part of the employment. If an employee has such an obligation but the exploitation of their work does not reflect the customary activity of the employer, the copyright will not be assigned to the employer. An employer cannot set aside this principle simply by *ordering* an employee to create a work.

Among other things this means that the copyright in books and articles produced by researchers working at universities normally belongs to the researchers themselves. It is not part of the customary activity of Danish universities to exploit such works. Nor can universities unilaterally order researchers to publish their articles in institutional repositories as part of Open Access.[6]

As regards *teaching materials* created by teachers, such as examination papers, grading instructions, PowerPoint hand-outs etc., the teacher's copyright is to some extent assigned to the educational institution for which they work. For example, it is part of the customary activity of educational institutions to hand out examination papers to their students or to send grading instructions to examiners. Teachers

4. Further Danish court decisions on this topic are discussed in Brinck-Jensen, Rosenmeier, Schovsbo and Sommer, *Ansattes immaterielle rettigheder* 35 f.
5. See Brinck-Jensen, Rosenmeier, Schovsbo and Sommer, *Ansattes immaterielle rettigheder* 37.
6. See Brinck-Jensen, Rosenmeier, Schovsbo and Sommer, *Ansattes immaterielle rettigheder* 39 f.

therefore have to accept that part of their copyright concerning such use of their teaching materials is assigned to the educational institutions, but only the part of their copyright that concerns such use is assigned. However, institutions cannot commercially publish the teaching materials of their teachers without the explicit consent of the teachers.

2.2.3 *Droit Moral*

Under the Danish Copyright Act authors have both economic rights and moral rights. Economic rights give authors an exclusive right to make copies or to communicate their work to the public by distributing or displaying copies or by public performance. Moral rights are divided into the right to be identified as the author of a work (paternity rights) (Section 3(1)) and a right of integrity (Section 3(2)). Under the paternity right the author must be named when someone copies the work or communicates it to the public. Under the right of integrity an author may prevent others from modifying their work or using it in ways or in contexts which harm their literary or artistic reputation.

There is a debate in Danish copyright law about whether the principles governing employees' copyright apply both to economic rights and to moral rights. Some think that the principles apply only to economic rights, and that an employer can only acquire the moral rights of their employees if this is expressly agreed. In our opinion, however, the principles governing employees' copyright apply to both the economic rights and moral rights. For example, the principles can entitle a newspaper to systematically crop photographs taken by employed photographers even if this is not expressly regulated by a contract.[7]

2.2.4 *The Work Must Have Been Created as Part of an Employment Obligation*

An employer acquires the copyright of employees pursuant to the above principles only if the work has been created as part of an employment obligation. This means, that the relationship between the parties must be an employment relationship in a legal sense. A number of conditions can indicate the existence of an employment relationship. Among other things, it depends on whether the employer is entitled to direct how the work must be carried out, and which of the parties bears the financial risk.[8]

In Danish copyright law it is doubtful if the employment relationship must be characterised as 'fixed and lasting'. In our opinion the conditions apply even if the employee is only temporarily employed.[9]

However, the principles do not apply to works created by *freelancers and independent consultants* etc. The extent to which their copyright is assigned to the client depends on the agreement between the parties.

7. See Brinck-Jensen, Rosenmeier, Schovsbo and Sommer, *Ansattes immaterielle rettigheder* 43 f.
8. For further reading see Brinck-Jensen, Rosenmeier, Schovsbo and Sommer, *Ansattes immaterielle rettigheder*, p. 46 f.
9. See Brinck-Jensen, Rosenmeier, Schovsbo and Sommer, *Ansattes immaterielle rettigheder* 48.

It is a widespread perception in Danish copyright law that the principles governing the copyright of employees do not apply to works which employees create in their *spare time*. However, this cannot be stated categorically. It is probably more correct to say that what matters is whether the creation of the work reflects the tasks of the employee and whether the work is created in the execution of these tasks.[10] For example, if a teacher prepares an examination paper at the weekend, they must accept that the paper will subsequently be handed out to the students by the teaching institution. The part of the teacher's copyright that concerns the handing out of the work to the students is assigned to the employer even though the paper was prepared in the teacher's spare time.

The condition that the work must be created as part of an employment obligation means that the creation of works must be *part of the employee's job description*. For example, if a messenger working in an advertising agency invents a good slogan, the copyright is not automatically assigned to the agency even though it is part of the agency's customary activity to use slogans created by its employees.[11]

2.2.5 Contracts on the Copyright of Employees

Unless there is an agreement to the contrary, an employer is only entitled to those parts of an employee's copyright that relate to necessary parts of the employer's customary activity at the time when the employee created the work. Consequently, an employer who wishes to *change* their customary activity and use their employees' works in new ways must enter into a special agreement with the employees. Such agreements can take several forms. For example, it is possible to insert a provision on copyright in an existing employment contract, or a separate contract on copyright could be made. It is also possible to make a collective agreement, where at least one of the parties is an organisation representing employers or employees. For example, in Denmark the Danish Union of Journalists has made many collective agreements on the copyright of newspaper or television journalists.[12]

Organisations may only enter into agreements on the copyright of their members if the members have authorised them to do so. In Danish law this authority may be given indirectly by the regulations of the organisation in question. It may also be simply implied by the circumstances. For example, in some cases it may be sufficient that it is the practice of the organisation to enter into agreements on their members' copyright and that the members know this or ought to know it. Normally, collective agreements are only binding on the parties to the agreement. However in some cases a collective agreement may also be binding on employees who are not members of the relevant organisation.[13]

10. See Wolk, *Arbetstageres immaterialrätter* 130 (Stockholm 2008); and Brinck-Jensen, Rosenmeier, Schovsbo and Sommer, *Ansattes immaterielle rettigheder* 49.
11. See Brinck-Jensen, Rosenmeier, Schovsbo and Sommer, *Ansattes immaterielle rettigheder* 49.
12. See Brinck-Jensen, Rosenmeier, Schovsbo and Sommer, *Ansattes immaterielle rettigheder* 50.
13. See Hasselbalch and Rosenmeier, in *Ugeskrift for Retsvæsen* 299 (2013 B).

Contractual obligations may also arise other than by express provisions. For instance, if an employer begins to exploit the works of their employees in new ways and the employees do not object, in some cases their mere acceptance may amount to an implied licence making it legal for the employer to go on exploiting the works in the new ways.[14]

Agreements on the copyright of employees are covered by fundamental contractual principles and can be declared invalid pursuant to the Danish Contracts Act. Among other things, under Section 36 of the Act the courts may set aside contracts in whole or in part if it would be unreasonable or fraudulent to enforce them.[15]

Also, copyright contracts must be interpreted according to the basic principles of contract interpretation. According to 'the draftsman principle', contracts that have been drafted by one of the parties should be interpreted to the advantage of the other party. According to another contractual principle, if a contract has been drafted by only one of the parties, any provisions that are to the disadvantage of the counterparty must be emphasised in order to be valid.[16]

Finally, contracts on copyright are covered by the 'principle of speciality' in Section 53(3) of the Danish Copyright Act. According to this principle, contracts for the assignment of copyright must generally be interpreted to the advantage of the author. If there is doubt about how to interpret an agreement which assigns the copyright of an employee (author) to their employer, the starting point is to interpret any doubts in the way that benefits the author (employee).[17]

Agreements which assign the copyright of an employee to their employer should therefore be reasonable and balanced, otherwise they can be set aside under Section 36 of the Danish Contracts Act or they can be interpreted to the advantage of the author (employee) according to fundamental contractual principles or the principle of speciality.[18]

14. See Brinck-Jensen, Rosenmeier, Schovsbo and Sommer, *Ansattes immaterielle rettigheder* 51 and 53.
15. See e.g. Bryde Andersen, *Grundlæggende aftaleret* 423 f (Copenhagen 2008).
16. For further reading on the principles for the interpretation of contracts, see Lynge Andersen and Madsen, *Aftaler og mellemmænd* Ch 9 (Copenhagen 2012); and Bryde Andersen, *Grundlæggende aftaleret* Ch 5.
17. For further reading, see Schovsbo, Rosenmeier and Salung Petersen, *Immaterialret*, 133 f and 613 f (Copenhagen 2013); Schønning, *Ophavsretsloven med kommentarer* 502 f and 536 f.; and Blomqvist, *Overdragelse af ophavsrettigheder* 147 f (Copenhagen 1987) See also Betænkning 1197/90 om revision af ophavsretslovgivningen - slutbetænkning 41 f and 217 f.
18. See Brinck-Jensen, Rosenmeier, Schovsbo and Sommer, *Ansattes immaterielle rettigheder* 52 If the employees refuse to enter into an agreement with their employer the employer can to some extent impose an agreement in accordance with employment law principles where an employer can alter the terms of employment with an individual subject to giving notice In such cases employees' organisations have certain legal options; see Hasselbalch and Rosenmeier, in *Ugeskrift for Retsvæsen* 299 f (2013 B).

2.2.6 Cases in Which the Employer Changes the Customary Activity Without Agreement

If an employer wishes to change (i.e. expand) their activities in a way that involves new ways of exploiting the works of their employees, the employer must agree on this with the employees or their representative organisation, otherwise there will be infringement of their copyright.

As already mentioned, such an agreement may be implied by the circumstances, so there may be some areas of business where an employer is entitled to alter their activities without negotiating with the employees. For example, it seems likely that a design company which normally exports applied art works created by its employees to countries in Northern and Central Europe would be allowed to start exporting to countries in Southern Europe without the express agreement of the employees.[19] The continuous expansion of a market is a customary activity of an undertaking and thus the employer has an implied licence to do so. However, the starting point is that an employer must negotiate with their employees if they want to exploit the employees' works in new ways.

Not all employers are aware of this. In fact many employers alter their customary activities without seeking the agreement of the employees.

Initially, the new activity will infringe the copyright of the employees. However, for the employees to maintain their copyright they must claim their rights without waiting too long because at some point the new and illegal activities will become the customary activities of the employer. Time heals all wounds, also in copyright law.

In Danish copyright law it is unclear how much time must elapse before a new activity becomes a legitimate customary activity.[20]

When in time the new activity has become a legitimate customary activity, the part of the copyright of the employees related to the new activity will be assigned to the employer. However, the employees may still claim damages for the unlawful exploitation of their works which took place before the new activity became customary. Under the Danish Statute of Limitations such a claim will be statute-barred after three years. The limitation period starts when the claimant discovers or ought to have discovered the infringement but it always expires ten years from the time of the infringement irrespective of whether the claimant has discovered them or not; see Section 3 of the Statute of Limitations. In any case a claim may lapse if it is not brought within a reasonable time according to basic legal principles on the passivity of rights holders.[21] These principles apply independently of the Statute of Limitations

19. See Hasselbalch and Rosenmeier, in *Ugeskrift for Retsvæsen* 296 (2013 B); and Brinck-Jensen; Rosenmeier, Schovsbo and Sommer, *Ansattes immaterielle rettigheder* 53
20. See Brinck-Jensen, Rosenmeier, Schovsbo and Sommer, *Ansattes immaterielle rettigheder* 53 f.
21. See Brinck-Jensen, Rosenmeier, Schovsbo and Sommer, *Ansattes immaterielle rettigheder* 54 f.; and Andersen, Carlquist and Rubinstein, *Retshåndhævelse af immaterialrettigheder* 493 f (Copenhagen 2006).

2.2.7 *Computer Programs*

The principles described above only apply to works other than computer programs. Pursuant to Section 59 of the Danish Copyright Act, copyright in a computer program created by an employee in the execution of their duties or following the instructions of their employer is automatically assigned to the employer. This provision implements the EU Directive on the legal protection of computer programs ('Software Directive') (91/250/EEC, now replaced by a codified version 2009/24/EC).

There has been a debate in Danish law as to whether Section 59 applies to programs created by university researchers.[22]

Section 59 only concerns computer programs and not other forms of digital works.[23]

According to Article 2(3) of the Software Directive, the employer only owns only the economic rights of the employed programmers, not their moral rights. However, according to the *travaux préparatoires* of Section 59 in the Danish Copyright Act the employer obtains the whole of the employee's copyright, including their moral rights. This is probably not in accordance with the Berne Convention, Article 6*bis*(1).[24]

Section 59 may be waived by agreement.

2.2.8 *Neighbouring Rights*

The legal rules and principles described above govern the extent to which copyright in original works is assigned to the employer.

Apart from protecting original works, the Danish Copyright Act also includes a number of provisions giving a shorter 'neighbouring right' to a number of phenomena which cannot – or cannot always – be considered original works.

These neighbouring rights concern the performances of performing artists (Section 65); sound recordings (Section 66); films (Section 67); radio and television broadcasts (Section 69); photographs (Section 70); compilations (Section 71); press releases (Section 72); titles, pseudonyms and signatures (Section 73); and first publication of previously unpublished works for which protection has lapsed (Section 64).

The works protected by these provisions are often made by employees, so the question arises as to whether and to what extent the protection given to them is assigned to their employer.

22. See further in Schovsbo, Rosenmeier and Salung Petersen, *Immaterialret* 144 f.
23. For the purposes of copyright computer programs are the codes the programmer means to be fed into the processor of a computer in order to have it perform certain tasks Both 'source codes' and 'object codes' are considered programs Programs are also considered to include the preparatory design work leading to the development of a computer program, provided the nature of the preparatory work is such that a computer program can be developed from it at a later stage; see recital 7 of the Software Directive The term 'program', however, does not include the user interface of programs, including pictures, screen images, texts and so forth See Case C-393/09 *BSA* Nor does the term 'program' include digital photographs, digital databases, etc.
24. For further reading see *Bruun in Vennebog til Mogens Koktvedgaard* 160 f (Stockholm 1993); Rosén in *NIR* 370 f (1993); Nordell in *NIR* 378 (1993); and *Wolk, Arbetstagares immaterialrätter* 82 f.

Protection pursuant to Section 65 (the performance of performing artists) and Section 70 (photographs) is probably covered by a principle similar to that which concerns original works. Unless there is an agreement to the contrary the employer is only entitled to those parts of the employee's copyright that were necessary for the employer's customary activity at the time the employee created the protected material.

In our opinion, the same applies for the right pursuant to Section 73 which prohibits communication to the public of a work under a title, pseudonym or signature which can lead to confusion with a previously published work or its author. Thus, the right to give permission pursuant to Section 73 belongs to the employee.

The protection accorded by Section 66 (sound recordings) and Section 67 (films) belongs to the 'maker', i.e. the natural or legal person responsible for making them. If an employee makes a sound recording or a film as part of their job, their employer will probably often be considered the maker. In such situations the employer is the first owner of the copyright. However, if a recording or a film is made by researchers working at a university or other institute of higher education it is possible that the rights initially belong to the researcher. Whether the rights are then assigned to the employer will be governed by the same legal principles as apply to works.[25]

The right accorded by Section 69 (radio and television broadcasts) belongs to the 'radio or TV enterprise', meaning the employer.[26]

The right accorded by Section 71 (compilations) belongs to the maker of the compilations. The maker can be a legal or natural person. If a compilation is considered to be created by an employee, their right must be assignable to the employer according to a principle similar to that which applies to works. This means that, according to Section 71, the employer is only entitled to those parts of the employee's copyright that relate to a necessary part of the employer's customary activities at the time when the employee created the compilation. However, if the employer is considered the maker of a compilation, they own all rights accorded by Section 71.

Section 72 prohibits the unauthorised publication of press releases from foreign news agencies or foreign correspondents, without permission from the authorised recipient of a press release, until 12 hours after their first publication in Denmark. Normally, the recipient will be a news organisation, not its employees, and the right accorded by Section 72 will therefore belong to the news organisation.

Similarly, the rights relating to publications of old, previously unpublished works referred to in Section 64 will normally belong to the publishing house which publishes the works rather than its employees.

25. See sections 2.2.1-2.2.6 above.
26. See Schønning, *Ophavsretsloven med kommentarer* 592.

2.3 The Parties' Duty to Inform

The Danish Copyright Act contains no provision obliging employees who create works or other protected material to inform their employer. However, there may be such an obligation under an implied or explicit contract between the employer and the employee.

2.4 Employees' Remuneration Right

The Danish Copyright Act does not give the employees whose copyrights are assigned to their employer a right to remuneration.

3 EMPLOYEES' SEMICONDUCTOR CHIPS

3.1 The Legal Framework

In Denmark the rules on the protection of semiconductor chips are to be found in the Act of the Protection of Semiconductor Chips (lovbekendtgørelse 2012-01-24 No. 105). According to the Act, depending on registration at the Danish Patent and Trademark Office, it is possible to obtain protection of 'the topography of a semiconductor chip'; see Section 2(1). It is a requirement that what is protected should 'result from the creator's intellectual effort' and should not be 'ordinary within the semiconductor chip industry'; see Section 2(2).[27]

3.2 Ownership of Rights

According to Section 2 of the Act of the Protection of Semiconductor Chips, the creator of a semiconductor chip topography or the person to whom the right has been assigned can obtain an exclusive right to use the topography. The first owner must, by definition, be a natural person, i.e. an employee.

The extent to which the rights are assigned to the employer is regulated by Section 4(1)(2), which states that the exclusive right can only be obtained by 'a natural person who is citizen of or who is resident in a Member State or a legal person carrying on business in a Member State when the topography is created by a person who is employed by them or created according to their directions, unless it is or must be considered as having been agreed that the exclusive right is owned by the creator.' Thus, the rights are owned by the employer unless there is an agreement to the contrary. The rights are assigned in their entirety. This is the broadest 'work for hire' rule in Danish IP legislation.

There is no Danish case law on this provision.

27. For further reading see Schovsbo, Rosenmeier and Salung Petersen, *Immaterialret* 465 f.; and Bryde Andersen, *IT-retten* Ch 10.4 (Copenhagen 2005).

3.3 The Parties' Duty to Inform

The Act contains no provision requiring employees who create semiconductor chips to inform their employer. However, there may be such an obligation under an implied or explicit contract between the employer and the employee.

3.4 Employees' Remuneration Right

The Act does not give employees who have created a semiconductor chip a right to remuneration.

4 EMPLOYEES' INVENTIONS

4.1 The Legal Framework

The Employees' Inventions Act[28] ('EIA') came into force 9 May 1955. The Act regulates employers' and employees' rights and obligations in detail.

Employees in both the private and the public sector are covered by the Act. However, the EIA does not apply to inventions made by teachers and other scientific personnel at universities and other institutions of higher education ('public researchers'). Since 1 January 2000 the inventions of employees at such institutions have been regulated by the Act on Inventions at Public Research Institutions[29] ('AIP').

4.2 Ownership of Rights

4.2.1 Employees (Not Public Researchers)

4.2.1.1 An Invention and an Employee

The EIA applies to inventions that are patentable in Denmark and to utility models that are registrable in Denmark (hereinafter referred to collectively as 'inventions'). Whether a company chooses to patent an invention is irrelevant.

The EIA applies to employees. An employee is anyone who would be considered an employee under Danish labour laws. As described above under section 2.2.4, numerous conditions can indicate the existence of an employment relationship, including whether the employer is entitled to direct how the work must be carried out. Consequently, there must be a superior-subordinate relationship between the management and the employee. A managing director is generally not regarded as an employee if they have overall management responsibility. Other directors, such as finance directors or commercial directors, are usually considered to be employees.

28. Lovbekendtgørelse 2012-01-24 No 104 om arbejdstageres opfindelser.
29. Lovbekendtgørelse 2009-03-17 No 210 om opfindelser ved offentlige forskningsinstitutioner.

4.2.1.2 Duty to Assign

According to Section 1 of the Danish Patents Act and Section 1 of the Danish Utility Models Act, the legal starting point is that the employee owns all rights to the invention. However, an employer may require the rights for an employee's invention to be assigned to them if the relevant criteria are fulfilled. Assignment means that the employer has the right to dispose of the invention as they see fit, including choosing not to apply for a patent.

The employer's right to have the right to the invention assigned to them is subject to certain conditions. The employee must be considered to have made the invention in the course of their employment, and the employer has the burden of proving that the invention has been made in the course of employment. However, the *travaux préparatoires* to the Act and the (limited) case law seem to indicate that this requirement is not interpreted very strictly.

Further, the exploitation of the invention must either fall within the scope of the employer's customary activities, or the invention must relate to a specific task assigned to the employee by the employer. What is decisive is not the ownership structure of the undertaking but how the work areas function in practice. The scope of the employer's 'customary activities' is to be interpreted quite widely.

4.2.1.3 Inventions Made after Termination of Employment

If a former employee makes an invention it belongs to the employee and they are under no obligation to give notice of the invention to their former employer. However, Section 9 of the EIA contains a presumptive rule to protect employers against employees leaving their employ and immediately making an invention. The burden of proof is reversed if a former employee applies for a patent or for registration of a utility model within six months of leaving their employment. In such cases, the inventions are considered having been made during the employment unless the former employee can show that it is probable that the invention was made after the termination of their employment. The date of the termination of employment is the date when the employment contract formally ends.

Inventions made after six months from the termination of employment do not fall under the presumptive rule. In these cases the employer must prove that the invention is covered by the EIA. The parties can agree to extend the presumptive rule beyond the six months (subject to general contractual principles).

An agreement between an employer and an employee to limit the employee's right to dispose of inventions made more than one year after the termination of their employment will be invalid.

4.2.2 *Public Researchers*

4.2.2.1 *An Invention and a Public Researcher*

The AIP applies to inventions that are patentable under the Danish Patents Act and utility models that are registrable under the Danish Utility Models Act ('inventions').

A public researcher is defined as an employee of the institutions mentioned in Section 6 of the AIP. These are the universities covered by the Universities Act, government research institutions, public hospitals, regional health research institutions, and the Geological Survey of Denmark and Greenland ('GEUS'). Both permanent and temporary employees are covered, as are PhD students employed at these institutions and student employees. Students enrolled at an institution who are not employed at the institution are not covered.

4.2.2.2 *Duty to Assign*

The starting point is that the rights to inventions made by a researcher belong to the researcher. However, a public research institution may require an invention to be assigned to it. For an institution to require an invention to be assigned to it, the invention must have been made by a researcher as part of their work at the institution.

The work of a researcher is defined as 'all activities performed as part of their employment'. This applies regardless of whether a research, development or education programme is carried out as part of a given assignment or whether the invention has been made in a situation where the subject and method are left to the researcher to decide. Inventions made outside the scope of employment are not due to the institution. The distinction can be difficult to make, especially as the scientific staff of institutions do not have specific work hours and in some cases not even an obligation to attend the institution. The burden of proof lies with the party who requires the rights to the invention.

4.2.2.3 *Cooperation with Third Parties*

If an institution has projects carried out in cooperation with or financed by parties that are not covered by the AIP, the institution may, by prior written agreement on its own behalf and the researcher's behalf, renounce the right to any inventions that may result from the project.

4.2.2.4 *Ethical Considerations*

Where, according to an employee, particular ethical considerations apply, the institution may accept that an invention should not be commercially exploited or patented. Ethical questions can be relevant to biotechnological inventions, in particular inventions regarding human material (genes, stem cells, etc.).

4.2.2.5 Assignment from Others than Public Researchers

The AIP also applies to the commercial exploitation of rights referred to in Sections 12 and 14-14b. In addition to rights to an employee's inventions, an institution may acquire rights to inventions and computer program copyrights from non-employees for the purposes of commercial exploitation. An institution can enter into such an agreement with a student who is not employed by the institution, or with another party as part of their cooperation (when wholly or partially financed by public funds). It is a condition that the other party does not intend to exploit these rights commercially. The institution can also enter into such an agreement with another of the institutions referred to in Section 6.

4.3 The Parties' Duty to Inform

4.3.1 Employees (Not Public Researchers)

An employee who has made an invention covered by the EIA must notify their employer of their invention without undue delay. The employee must provide information about the invention so as to enable the employer to assess the importance of the invention.

If the employer wishes to acquire the right to the invention they must inform the employee of this within four months of having received the notification from the employee. The employee may not dispose of the invention before the expiry of the employer's time limit, nor disclose it to others in such a way that it may be published or exploited by others without the employer's written consent.

As soon as they have given notification, the employee may apply for a patent for the invention if they give advance notice to the employer.

4.3.2 Public Researchers

A public researcher who has made an invention as part of their work must notify the institution thereof in writing without undue delay. The researcher must provide the institution with all necessary information, as directed by the institution.

The institution may lay down specific rules for how the employee's notification must be provided, and it may decide that the notification obligation does not apply in areas where prior experience has shown that inventions are not made. The institution must evaluate the possibility of exploiting the invention commercially and of protecting the rights to the invention, and it must discuss with the researcher how the rights to the invention may be commercially exploited. This must be done within two months from the date of the researcher's notification or subsequently by agreement between the institution and the researcher.

A public researcher may neither publicise nor dispose of the invention before the institution has given written confirmation of receipt of notification. The institution must send the confirmation as soon as possible. However, the institution may order the researcher not to publish or dispose of an invention for up to two months from receipt of the notification, in order to carry out the assessment referred to. This period can be extended beyond the two months if the parties so agree.

Within the time limit (two months or longer if agreed), the institution must decide whether it wants to have the rights to the invention assigned to it, or whether the researcher may retain their rights. Should the institution fail to inform the researcher of its decision within the time limit, all rights to the invention remain with the researcher.

4.4 Employees' Remuneration Right

4.4.1 Employees (Not Public Researchers)

4.4.1.1 Right to Remuneration

An employee who has made an invention has a right to reasonable remuneration if the employer requires the right to the invention to be assigned to it.

An employee has a right to separate remuneration if the value of the invention exceeds what the employee, given their terms of employment as a whole, may reasonably be expected to produce. This rule takes into consideration that many inventions will be of little financial value and that, by virtue of their job descriptions, many employees are expected to be inventive. Such employees are already considered to be remunerated through their salaries.

What an employee may reasonably be expected to produce depends on whether the job description specifically includes development and invention or whether the job function is outside the area of development, e.g., sales. The more the job description entails invention, development or solving technical tasks, the greater the value of an invention must be for the employee to be entitled to remuneration, especially when the invention relates to a specific task assigned to the employee by the employer. It is for the employer to prove that value of the invention does not exceed what is expected when taking the job description into account.

4.4.1.2 Amount of Any Remuneration

The EIA lists a number of factors to be taken into consideration when determining the employee's remuneration. These factors include the value of the invention and its importance for the company, the employee's terms of employment and the influence of the employee's service on the invention.

The value of the invention must be determined objectively, taking into consideration its industrial and commercial significance. The valuation of the invention must not be based solely on its value to the specific company. However, its importance to the company is also one of the factors to be taken into consideration. If the invention is of great value to the company where the employee is employed, but does not have equivalent value to other companies, this should not affect the amount of the remuneration negatively.

The company's decision to keep the invention secret and not to patent it does not affect the value of the invention.

The employee's terms of employment are also relevant. There must be regard for whether the employee is wholly or partly employed to perform inventive work and whether they have already been remunerated for the assignment of the invention through their normal salary. The amount of the employee's salary will probably also be taken into consideration. If the employee receives a higher salary than similar employees this will probably have a negative impact on the amount of remuneration.

The influence of the employee's service on the invention is also relevant for assessing the value of the invention. If the employee has contributed significantly to the invention, e.g., where the employee has made the invention without assistance from anyone else or where the company has not carried out any preliminary work, the employee will be entitled to greater remuneration than if, for instance, the employee was only one of several originators of an idea that the company has had to develop further in order to make it patentable.

The EIA's list of factors to be taken into consideration is not exhaustive. For example, the remuneration can be affected by the countries in respect of which the employer chooses to have the invention assigned. If an employer only chooses to acquire the rights to the invention for Denmark, then the employee will be free to commercialise the invention in the rest of the world or to assign the rights for the rest of the world. In assessing remuneration for pharmaceutical patents account should also be taken of whether an SPC and a marketing authorisation for the patent are likely to be obtained, since these will increase the value of the patent. In order to provide the employee with an incentive to invent, any agreement (including provisions in the employment contract) which an employee enters into before making the invention and which deprives the employee of the right to reasonable remuneration will be deemed invalid.

4.4.1.3 Case Law on Remuneration

Disputes on an employee's right to remuneration can be brought before the ordinary courts. There are no special tribunals for such disputes.

In a decision from the Danish Maritime and Commercial High Court in 2005 a development engineer had invented accessories for a specific hearing aid. The Court found that the engineer had the specified task of developing accessories for the hearing aid, and that the market for the invention was very limited. Consequently, the value of the invention did not exceed the value that the employee could reasonably be expected to produce, even though the invention could be and would probably be used for future generations of hearing aids.

In a decision from the Western High Court a carpenter, who had originally been employed as a carpenter on an hourly basis, had been involved in inventing a 'Do It

Yourself' concept. The carpenter claimed that he was only employed to control a certain machine and only participated in customer visits as a normal carpenter. However, the High Court found that the carpenter's participation in business travel abroad showed that the employment had been expanded so as to include the development of new types of windows. Consequently, the High Court found that the invention was covered by the EIA. The next question was whether the carpenter could reasonably be expected to produce an invention of the value that the concept represented. The High Court found that it was impossible to assess the value of the invention but assumed that it did not exceed the value of what the employee could reasonably be assumed to provide in his employment.

In a case before the Eastern High Court in 2012 a development engineer claimed remuneration for the assignment of his rights to three inventions: a temperature sensor, a humidity sensor and a water flowmeter. The engineer had made the inventions while employed at the company and had assigned them in accordance with his employment contract. The High Court found that there was an element of invention activity in the engineer's job description and that the value of the inventions did not exceed what could reasonably be expected from the employee as part of his regular work.

There are only a limited number of cases where the courts have decided on a figure for remuneration.

In a case before the Eastern High Court in 1977 a mechanical engineer had made two inventions: a mixing machine and a method for twin casting. With regard to the mixing machine, the company had obtained a patent in Denmark and had applied for patents in Japan, Canada and Yugoslavia. The company had used the invention in Denmark and Yugoslavia. With regard to the method for twin casting, the company had obtained patent in Denmark but had not used the invention in Denmark or elsewhere. The parties agreed that the company was entitled to require the rights to the invention to be assigned to it, but they disagreed on whether the employee was entitled to reasonable remuneration. Among other things the company claimed that the development of the inventions was directly connected to the tasks that the company had assigned to the employee, and that the company had attached importance to the employee's experience in the area of technology concerned when hiring the employee. Also, the company claimed that under the terms of employment, including the salary, it was a condition that the employee had the ability to make improvements for the company without receiving further remuneration. Finally, the company claimed that the inventions did not have industrial or commercial significance for the company.

These arguments were all rejected by the Court, which found that the company had not proved that the value of the inventions for which the company had taken out a patent did not exceed what the employee could reasonably be expected to produce. In determining the remuneration, the Court stated that the invention had already been used in two of the company's factories in Denmark and in a system sold abroad. Further, the Court stated that the company might have derived an advantage from referring to the patent rights in discussions with prospective buyers. The Court calculated a discretionary remuneration of DKK 10,000, equivalent to about two months' pay for the employee.

In an advisory declaration of the Eastern High Court in 1995 the head of a new ship department at the shipping company A.P. Møller had invented a system to increase the container capacity of new ships. The Court based its assessment of the value of the invention and its significance to A.P. Møller on a loose estimate. The Court found that A.P. Møller had used the invention in the production of 27 new ships, that the invention had been patented in a number of countries and that the increased container capacity was important for the ships' cargo-carrying capacity and stability. However, the Court also stressed that it should be taken into consideration that the invention was made possible by the insight and expertise that the employee had acquired through his long employment with A.P. Møller and his position as head of the new ship department. Also, the fact that the invention had been developed and tested at A.P. Møller, which had paid for all the costs involved, had an impact on the remuneration. The Court found that the remuneration should be fixed at DKK 2.5 million.

In a 2004 case before the Supreme Court a skilled electrician who was employed as a sales consultant in a company producing valves (among other things) had invented a new valve. The sales consultant presented the valve to the company which initiated development work leading to patent applications being made in a number of countries. In the applications the sales consultant was listed as the inventor together with two employees from the company's development department. The sales consultant's work primarily consisted in selling valves to housing associations and not in development work. The Supreme Court found that, given the employee's job description and the other circumstances, the employee's contribution to the invention was outside the scope of what he could reasonably be expected to produce in his position as a sales consultant. Consequently, he was entitled to reasonable remuneration. In assessing the value of the invention, the Court noted that the company had found the idea sufficiently viable to pay development costs of DKK 1 million and to take out patent for the invention. However, the Court also noted that, according to expert opinion in the case, the employee's contribution to the invention consisted of a presentation of ideas where the company's already patented Flow Concept was combined with an ordinary radiator valve. The presentation was far from being a fully developed product. Further, the possibility of commercial exploitation was uncertain and impossible to estimate at the time when the invention had not been developed into an actual product. Accordingly, the Court found that the remuneration should be DKK 50,000 which was a little less than two months' pay for the sales consultant.

4.4.2 *Public Researchers*

4.4.2.1 Right to Compensation

If the rights to an invention which has been assigned to an institution are exploited commercially, the employee is entitled to reasonable compensation from the institution.

However, this principle goes both ways. If, with the agreement of the institution, the rights to an invention are commercially exploited by the employee, then the institution is entitled to reasonable compensation.

4.4.2.2 Amount of Any Compensation

According to the *travaux préparatoires* for the AIP, the obligation to pay compensation only arises if the exploitation produces a net profit (i.e. if the revenue exceeds the cost).

The assessment of the amount of the compensation is discretionary. The compensation must reflect the actual market value of the invention. The starting point is that the compensation must be calculated on basis of the revenues from the commercial exploitation of the invention. Thus, the employee's personal circumstances, such as their salary or the terms under which the invention was developed, are immaterial for determining the compensation.

Employees' claims for reasonable compensation under the AIP differ significantly from employees' claims for reasonable remuneration under the EIA (see section 4.4.1.2). A claim under the EIA is conditional, as an employer's right to inventions developed by their employees can arise because the development of the invention can be part of the work performed by the employee in exchange for their salary. Only when an invention exceeds what the employee can reasonably be expected to produce are they entitled to remuneration. In contrast, since it is not part of an institution's purpose to commercially exploit rights to inventions developed by its employees, the salary of an employee under the AIP is not assumed to cover the assignment of such rights to the institution. Consequently the employee should be compensated for such assignment, provided the commercial exploitation generates a net profit.

The rules for calculating the amount of compensation are laid down by the institution. The AIP does not contain any further provisions for calculating compensation but, according to the *travaux préparatoires* for the AIP, the person who takes the commercial risk by exploiting the rights to an invention should benefit most by getting the biggest share of the future net profit. Further, the rules for calculating compensation must provide a clear financial incentive for research-based entrepreneurs and others who wish to exploit their own inventions.

The institutions seem to have a generally agreed model where an institution's external costs (e.g. costs for filing patent applications) are paid first, after which the net profits are divided with one-third going to the inventor and two-thirds going to the institution when the institution has acquired the rights. Conversely, when the employee is the one commercially exploiting the invention, the net profits are divided with one-third going to the institution and two-thirds going to the inventor.

Disputes about decisions made under the AIP fall within the jurisdiction of the ordinary courts.

5 EMPLOYEES' UTILITY MODELS

The Employees' Inventions Act applies to utility models which are registrable in Denmark. Please refer to the above section on Employees' Inventions.

6 EMPLOYEES' DESIGN RIGHTS

6.1 The Legal Framework

The extent to which an employer owns the rights to designs that are created by their employees and are protected by the Danish Design Act is not dealt in the Act, but is governed by a legal principle which can be derived from legal theory; see section 6.2 below.

The extent to which an employer owns the rights to designs that are created by their employees and are covered by the EU Design Regulation is covered by the Regulation; see section 6.2 below.

6.2 Ownership of Rights

In Danish law the principles governing the design rights of employees may be summarised as follows:

Design rights pursuant to the Danish Design Act are automatically owned by the employer unless the design is also protected by copyright. In that case the design rights are treated in the same way as copyright. Thus, whether an employer obtains the design rights of their employees is governed by a principle similar to that which applies in copyright law.[30]

Example: An employee working in a design firm specialising in lamps and electrical goods designs a socket outlet and a lamp. The firm registers both as designs in the Danish Patent and Trademark Office. The lamp is original in a copyright sense and thus protected by both design rights and copyright. The socket outlet, on the other hand, is only protected by design right.

The design right to the socket outlet will belong to the employer. The design right to the lamp is treated in the same way as the copyright, i.e. the employer is only entitled to those parts of the employee's design rights that relate to a necessary part of the employer's customary activity at the time the lamp was created. On the other hand, under the EU Design Regulation, EU design rights belong to the employer; see Article 14(3) which states that 'where a design is developed by an employee in the execution of his duties or following the instructions given by his employer, the right to the Community design shall vest in the employer, unless otherwise agreed or specified under national law.' This provision only deals with actual employment relationships, not cases where designs have been produced on commission; see Case C-32/08 *FEIA*.

30. See Schovsbo, Rosenmeier and Salung Petersen, *Immaterialret* 458 f.

6.3 The Parties' Duty to Inform

Under Danish design law, employees who create designs do not have a legal obligation to inform their employer. However, in some cases there is probably such an obligation under an implied contract between the parties.

6.4 Employees' Remuneration Right

Danish design law does not give an employee who has created a design a right to remuneration.

7 EMPLOYEES' PLANT VARIETIES

7.1 The Legal Framework

Plant varieties are not patentable; see Article 53(b) of the European Patent Convention which has been implemented in Denmark by Section 1(4) of the Danish Patent Act. However, plant varieties can be protected under the Act on the Protection of Plant Varieties (adopted in its most recent form in Lovbekendtgørelse No. 190 of 12 March 2009). The Act protects varieties and strains of plants that are distinct, uniform, stable and new (Section 1 of the Act). Protection is given both to new varieties created by improvement and new varieties found in nature.

The Act does not clarify who owns the rights to plant varieties created by employees. There is no Danish case law on the subject.

7.2 Ownership of Rights

It is argued in Danish legal literature that the employer will normally own the right because of an implied agreement with the employee. However, it is thought that the rights to plant varieties created by researchers working at universities and similar research institutions will normally belong the researcher unless there is an agreement to the contrary.

7.3 The Parties' Duty to Inform

The Danish Plant Variety Protection Act does not require employees who create plant varieties to inform their employer. However, in some cases there is probably such an obligation under an implied or explicit contract between the parties.

7.4 Employees' Remuneration Right

Danish Plant Variety Protection Act does not give employees a right to remuneration.

8 INSTANCES FOR DISPUTES

In Danish law there is no specific tribunal for disputes about the intellectual property rights of employees. Such cases are dealt with by the normal courts.

9 CONCLUDING REMARKS

For further reading on the intellectual property rights of employees in Danish law, see Brinck-Jensen, Rosenmeier, Schovsbo & Sommer, *Ansattes immaterielle rettigheder.*

BIBLIOGRAPHY

Table of Sources

Andersen, Carlquist and Rubinstein, *Retshåndhævelse af immaterialrettigheder* (Copenhagen 2006).
Blomqvist, *Overdragelse af ophavsrettigheder* (Copenhagen 1987).
Brinck-Jensen, Rosenmeier, Schovsbo and Sommer, *Ansattes immaterielle rettigheder* (Copenhagen 2014).
Bruun in Vennebog til Mogens Koktvedgaard 160 ff. (Stockholm 1993).
Bryde Andersen, *Grundlæggende aftaleret* (Copenhagen 2008).
Bryde Andersen, *IT-retten* (Copenhagen 2005).
Hasselbalch and Rosenmeier, in *Ugeskrift for Retsvæsen* 299 (2013 B).
Lynge Andersen and Madsen, *Aftaler og mellemmænd* (Copenhagen 2012).
Nordell in *NIR* 378 ff. (1993).
Rosén in *NIR* 370 ff. (1993).
Schovsbo, Rosenmeier and Salung Petersen, *Immaterialret* (Copenhagen 2013).
Schønning, *Ophavsretsloven med kommentarer (*Copenhagen 2011).
Wolk, *Arbetstageres immaterialrätter* (Stockholm 2008).
Official Reports etc.
Betænkning 1197/90 om revision af ophavsretslovgivningen – slutbetænkning.

Table of Legislation

Act on Inventions at Public Research Institutions (lovbekendtgørelse 2009-03-17 No. 210).
Act on the Protection of Plant Varieties (lovbekendtgørelse 2009-03-12 No. 190).
Act of the Protection of Semiconductor Chips (lovbekendtgørelse 2012-01-24 No. 105).
Copyright Act (Lovbekendtgørelse 2014-10-23 No. 1144).
Design Act LBK (lovbekendtgørelse 2012-01-24 No. 102).

Directive 2009/24/EC on the legal protection of computer programs (The Software Directive).
Employees' Inventions Act (lovbekendtgørelse 2012-01-24 No. 104).
Patents Act (lovbekendtgørelse 2012-01-24 No. 108).
Regulation 6/2002/EC on Legal Community Design.
Statute of Limitations (lovbekendtgørelse 2013-08-28 No. 1063).
The Berne Convention for the Protection of Literary and Artistic Works, 1886 with subsequent revisions.
The Contracts Act (lovbekendtgørelse 1996-08-26 No. 781).
The European Patent Convention, 1973.
Utility Models Act (lovbekendtgørelse 2012-01-24 No. 106).

Table of Cases

Advisory declaration of the Eastern High Court 1995, *10. afd. Nr. B-3233-88.*
Decision of the Eastern High Court, *Ugeskrift for Retsvæsen* 1977.600 Ø.
Decision of the Eastern High Court, *Ugeskrift for Retsvæsen* 1982.926 Ø.
Decision of the Eastern High Court, *Ugeskrift for Retsvæsen* 1993.180 Ø.
Decision of the Eastern High Court, *Ugeskrift for Retsvæsen* 2012.420 Ø.
Decision of the European Court of Justice, C-32/08 *FEIA*.
Decision of the European Court of Justice, C-393/09 *BSA*.
Decision of the Maritime and Commercial High Court F-15-02, 2005.
Decision of the Supreme Court, *Ugeskrift for Retsvæsen* 1978.901/2 H.
Decision of the Supreme Court, *Ugeskrift for Retsvæsen* 1994.349 H.
Decision of the Supreme Court, *Ugeskrift for Retsvæsen* 2004.1018 H.
Decision of the Western High Court, *Ugeskrift for Retsvæsen* 2006.3250 V.

CHAPTER 8
Finland

Niklas Bruun & Marja-Leena Mansala

1 BACKGROUND

The legal framework in Finland concerning intellectual property rights is very similar to other Nordic Countries. Legislation concerning intellectual property rights has its basis, to a significant extent, in collaborative preparatory work. Legislation, especially that concerning employee inventions, is harmonised in the Nordic countries. EU legislation has brought new elements, e.g., concerning design rights.

2 EMPLOYEES' COPYRIGHT

2.1 The Legal Framework

The current Copyright Act entered into force on 1 September 1961 and has been amended several times since then.[1] However, the basic principles are still the same as in the original law. Section 1 of the Copyright Act provides that a person who has created a work shall have copyright thereto. The author has moral rights and economic rights to their work. The author has the right to be acknowledged as the author and they have the right to object to derogatory treatment. The economic rights provide that the author has the exclusive right to control the work by making copies thereof (reproduction right) and by making it available to the public (communication to the public, public performance, distribution to the public, public display). This means that the author has in practice all the relevant means to utilise their copyright.

It is a generally accepted principle that the initial ownership of copyright is always held by one or several individuals. A corporation may therefore obtain

1. Copyright Act 404/1961, last amendments 2016.

copyright only by assignment. The Copyright Act is based on the principle of freedom of contract, and the author is entitled to freely transfer their copyright. However the contracts are interpreted in favour of the author. Section 28 of the Act states that in the absence of an agreement to the contrary, the person to whom a copyright has been transferred may not transfer the copyright further to third parties or alter the work.

The author is entitled to transfer all the economic rights but moral rights can only be transferred for limited use of the piece of work.

There is no demand for a specified form of the transfer of copyright. Works created during employment fall under the general principle that rights are borne to the author and there is no general rule in the Copyright Act about employers' rights to the works. Thus a contract of employment can be one way of transferring copyright. The above-mentioned interpretation principle means in practice that the transfer of copyright needs a written contract.

Finland implemented the Directive on legal protection of computer programs in 1991. A new phenomenon was then introduced to the Copyright Act by stating that if a computer program and a work directly associated with it has been created in the scope of duties in an employment relationship, the copyright in the computer program and the work shall pass to the employer. The same shall apply to a computer program and a work directly associated therewith, created within the scope of a civil service post. However, the provisions shall not apply to a computer program, or to a work directly associated therewith, created by an author independently engaged in teaching or research in an institution of higher education, with the exception of institutions of military education. The same principles have subsequently been introduced to correspondingly apply to a database which is created in the scope of duties in private and public employment relationships.

The person who has made a catalogue, a table, a program or any other product in which a large number of information items are compiled, or a database the obtaining, verification or presentation of which has required substantial investment, has the *sui generis* right to the catalogue or database. The right includes the exclusive right to control the whole or, in qualitative or quantitative terms, a substantial part thereof, by making copies of it and by making it available to the public. This right is borne to the employer directly if the catalogue or database is a result of duties in employment.

The Employment Contracts Act applies to (employment) contracts entered into by an employee, or jointly by several employees as a team, agreeing personally to perform work for an employer under the employer's direction and supervision in return for a salary or other remuneration. An employment contract may be oral, written or electronic.[2]

A collective agreement within the meaning of the Collective Agreements Act from 1946 (the Act has been amended several times) is any agreement concluded by one or more employers or registered associations of employers and one or more registered associations of employees, concerning the conditions to be complied with in contracts of employment or in employment generally.

2. Employment Contracts Act 55/2001.

The contracts are interpreted in the light of the Contracts Act (13.6.1929/228) and general contractual principles. An amendment to the Copyright Act was made in 2015 concerning unreasonable terms in a copyright transfer agreement. If a term in the transfer agreement made by the original copyright holder is unreasonable compared with the terms used commonly in the field or otherwise unreasonable or if the result of the agreement would be unreasonable, the term can be mediated or ignored.

2.2 Ownership of Rights

As stated in the previous section there are no general rules regulating employees' copyright in the Copyright Act. The legislative differences between employees in private and public sectors in copyright matters are minor.

2.2.1 Private Sector

The right to a work made during employment can be transferred in the contract of employment. Also the duties assigned to the employee might be such that some transformation of copyright is essential to the employer in order to fulfil the meaning of the employment. In these cases it is commonly acknowledged that copyright is transferred to the employer in order to allow the employer to use the result as intended. This can be referred to as a right to use the piece of work: the right usually covers one-time use or publication, and republishing of the piece of work must, therefore, be agreed on separately. Other rights remain with the creator, including the right to use for other purposes and the right to alterations. However, without express agreement of the scope of the transfer, the interpretation can cause some obscurity.

Some collective agreements include definitions of how and where the employer can use the economic rights of a certain piece of work in their business; how pieces of work can be adjusted by the employer; and how, if at all, these economic rights can be transferred on to a third party. Collective agreements are of special importance in industries that include and handle plenty of copyright-protected material, such as publishing, journalism and photography, and the artistic work of, e.g., performing artists and graphic designers.

The agreement transferring the copyright can be written, spoken or even a tacit agreement. The relationship is clearer, however, if the parties conclude a written agreement. Common contractual principles are applied to the transfer of rights in this area. The only exception concerns the interpretation of the possibility to transfer the copyright further or to alter the work.

As stated above, rights to a computer program, a catalogue and a database, which have been created in the course of fulfilling the obligation of the employment, shall automatically pass to the employer.

According to the Copyright Act, forwarding rights or altering the original piece of work is prohibited, if the parties have not agreed otherwise. This also has to be taken into account in employee contracts.

The rules are the same for public officials and bureaucrats as for employees in the private sector.

2.2.2 *Exemption for Researchers*

The above-mentioned exception for computer software created while fulfilling the duties of employment does not concern researchers at universities or research institutions. In those cases no rights are transferred to the universities without a contract. The same also applies to databases. However, the *sui generis* protection for databases does not contain an exemption for researchers. The line between database protection under copyright and the *sui generis* right might be difficult to draw.

2.3 The Parties' Duty to Inform

The Finish Copyright Act contains no provision obliging employees who create works or other protected material to inform their employer.

2.4 Employee Remuneration Right

With regard to computer programs, an employer has no obligation based on the law to compensate an employee for creating a computer program during their employment since the creation of programs constitutes part of the employee's duties.

There are no statutes about compensation for authors in the Copyright Act. However, reasonable remuneration can be agreed on at the same time as transfer of rights if the employer receives a wider right than is necessary for using the work for its original purpose.

3 EMPLOYEES' SEMICONDUCTOR CHIPS

3.1 The Legal Framework

Semiconductor chips, integrated circuits and circuit designs are regulated with a specific law in Finland (Act on the Exclusive Right to the Layout-Design of an Integrated Circuit, 11.1.1991/32).

The Act protects the layout-design of inventive and original semiconductor chips. Layout-design is defined in the law as a three-dimensional placing of the parts of an integrated circuit in all of its possible patterns.

3.2 Ownership of Rights

3.2.1 Private Sector

According to the Act, the creator of the circuit can acquire exclusive rights by registration. The exclusive rights give the holder the right to determine the use of the layout-design.

If a circuit is created by an employee in the course of their duties, the employer has the right to acquire by registration an exclusive right to the circuit unless otherwise agreed. If the holder of the exclusive rights relinquishes those rights, the registration authority must remove the semiconductor chip from the registry and declare the exclusive rights expired. The duration of the exclusive rights is ten years counted from the date on which the application for registration was made.

3.2.2 Public Sector

The same applies for public officials and bureaucrats. This paragraph does not, however, apply to those teaching or doing research at universities or polytechnics governed by the Universities Act and Polytechnics Act, apart from military education institutions governed by the National Defence University Act. At those institutions, therefore, the employer does not presumably have a right to register exclusive rights over employee's semiconductor chip. The right mentioned can of course be agreed on.

3.3 The Parties' Duty to Inform

The Act contains no provision requiring employees who create semiconductor chips to inform their employer. However, there may be such an obligation under an implied contract between the employer and the employee.

3.4 Employees' Remuneration Right

The Act does not give employees who have created the semiconductor chip a right to remuneration.

4 EMPLOYEES' INVENTIONS

In Finland, patent rights on inventions made during employment or research are regulated by two acts; the Act on the Rights to Employee Inventions (No. 656/1967, Employee Inventions Act) and the Act on the Rights in Inventions made at Higher Education Institutions (369/2006, University Inventions Act).

4.1 The Legal Framework

Inventions made in employment relationships are regulated by the Employee Inventions Act. The Act is supplemented by the Decree on the Right to Employees' Inventions (No. 527/1988), which provides provisions especially on remuneration. The Employee Inventions Act is applied to inventions which have been made by employees either in private or public employment, but if a person is employed at a Finnish higher education unit, the University Invention Act will be applied. The purpose of the Employee Inventions Act is to reconcile the inventor's rights to his/ her invention and the employers' right to the fruits of their employees' labour. In general the principles in both Acts are similar.

For the law to be applicable, the invention has to be patentable in Finland. The law does not require that a patent is actually sought or issued. Instead the law is applicable when the employee-inventor's right to seek a patent is transferred to the employer or restricted by the employer (Section 1). If the employment has been terminated and the employee files a patent application in the subsequent six months, the invention will be considered to have been made under the duration of the employment if the employee does not provide reasonable evidence supporting the contrary (Section 8). However, a contract which restricts employees' rights to use their inventions made more than one year after the employment has ended is invalid.

The law is to a large extent non-mandatory, with the exception of significant employee rights, such as the right to a reasonable remuneration and the initial rights of inventors to their inventions.

Finland has its own law governing the right to inventions made at universities and polytechnics or by persons holding a research position funded by the Academy of Finland, the University Inventions Act. The University Inventions Act is applicable to inventions, which are patentable in Finland and made under employment. Similarly, the law becomes applicable when an institution takes over a right to the invention thus restricting the inventor's ability to seek for a patent. The purpose of the law is to promote the recognition, protection and exploitations of inventions made within the institutions.

The institutions regulated by this law are derived from other legislative instruments; the Universities Act, Polytechnics Act and the National Defence University Act.

Article 3 of the University Inventions Act defines two types of research depending on the amount of the involvement of the external party. Open research is research, which is either conducted without the involvement of funding from external parties, conducted with external funding but without provisions regulating the publishing of the results or if the parties have specifically agreed that the research is conducted as open research. Contract research is defined as research conducted as a chargeable service or research which does not fall within other categories and at least one external party is involved and there are obligations concerning the outcome and its publishing. Inventions that fall outside the aforementioned situations are considered to be inventions created in other circumstances.

4.2 Ownership of Rights

In the private sector the rights to an invention belong initially to the inventor or inventors, which by Finnish law have to be natural persons. The Employee Inventions Act does not recognise a corporation as an inventor. The law provides that the employer may have the right to acquire all or part of the rights to an invention from the employee if certain requirements are fulfilled. The key requirement is that there is a connection between the invention and the employer. For the employer to gain extensive rights to the invention, the requirements are that the invention has been made under a specific duty assigned by the employer or that the invention has been made by using know-how gained from the employer. In addition, the invention should belong to the corporation's sphere of operations.

If there is a weak connection between the employer and the invention but where it still falls within the field of operations, the employer might acquire the right to exploit the invention. When the connection between employee and employment is vague, the employer may only have the right to make the first offer concerning the possible utilisation of the invention.

If the employer has the right to acquire rights to the invention, the employee has to sign a written form of transfer of rights. The employer may be entitled to present it to the patent authorities when filing the application for patent protection.

The University Inventions Act stipulates that the inventor has the same rights to their invention as any inventor, if not otherwise dictated by the law. The inventor has the right to be recognised as the inventor (Article 4). The institution may acquire rights to any invention but the 'rules of procedure' differ in open research and contract research. The basic principle is that the institution has to be informed of all inventions made during the research.

The institution may acquire rights to an invention made under open research if certain requirements are fulfilled. First is that the inventor has not in the six months following the notice of invention published the invention or announced the will to exploit the invention. Before acquiring these rights by a written form, the institution has to inquire if the inventor wishes to utilise the invention themselves. If the institution does not notify of the acquisition of these rights, the institution is presumed to have waived its right.

If the invention has been made while conducting contract research, the institution may take the rights to the invention with a written announcement within six months following the notice of invention. If the institution has not filed such an announcement within the time limit, the institution will be presumed to have waived these rights. If the institution has acquired the rights, the inventor has to, upon request, immediately sign a written form of transfer.

If the invention has been made under other circumstances (i.e. not under open or contract research) the institution has the right to negotiate the utilisation of the invention. If this invention is necessary for the operations of the institution, the institution has the right to acquire a right to use the invention for adequate compensation.

4.3 The Parties' Duty to Inform

In the private sector, after creating an invention (within the meaning of the Act), the employees must inform their employer without a delay (notice of invention, Section 5). The notice shall be made in written form and the employer should be provided with sufficient information of the invention in order to understand the invention. The inventor should simultaneously notify the employer whether or not they intend to file a patent application.

After the first notification the inventors are obliged to keep their invention as a secret and refrain from filing the patent application for one month. The inventor shall provide the employer with any further information relating to the invention, if the employer needs this information in order to determine possible prospects for utilising the invention.

After receiving the notice, the employer has a period of four months to consider whether or not to acquire rights to the invention (Section 6). The employer has to handle all information relating to the invention confidentially both during and after this period, regardless of whether or not the employer will decide to utilise the invention. The employer shall during this period inform the employee whether or not they will acquire rights to the invention and what rights they will acquire.

Many corporations today have their own employment invention guidelines and use a specific form for information about an employment invention. Guidelines may extent time limits, such as the four-month period set for the employer to gain rights to an invention.

From the University Inventions Act it follows that when an invention (within the meaning of this law) has been made, the inventor has to submit a notice of the invention to the institution without delay. The notice has to be made in written form and has to include an opinion on whether the invention has been made in open or contractual research or in other circumstances. The notice has to include sufficient information concerning the invention and it has to list all inventors. The inventor has to provide the institution with other necessary information upon request.

The institution has to notify the inventor in writing and without a delay of the rights which the institution is entitled to by the Act. This notification has to be made not later than two months after receiving adequate information of the invention. Simultaneously the institution has to inform the inventors whether or not it affirms the stipulated type of research in which the invention was made.

Article 12 of the Act regulates burden of proof. The institution has the burden of proof when indicating that it has the right to acquire the rights to the invention.

4.4 Employee Remuneration Right

The employees' right for reasonable remuneration is mandatory, and a contractual term waiving this right is null and void even if included in the original employment contract. The parties cannot bindingly agree upon a reasonable compensation for future inventions, even though they would have agreed in advance that the rights to a future invention will be transferred by the inventor. Reasonable remuneration should be determined so that it consists partly of a lump sum and partly of a royalty. With minor or small-scale inventions the payment can be made as a non-recurrent compensation only. In most cases reasonable remuneration is determined as a lump sum.

When the employer has informed the inventor of acquiring rights to the invention, a period of ten years for filing a claim concerning reasonable remuneration will begin to run. However, the Act stipulates that an inventor can always file a remuneration claim after one year from a patent grant even if the aforementioned period of ten years has expired.

When calculating remuneration, all relevant, objective factors constituting the economic value of the invention should be taken into account. The employer must inform the employee of all the relevant information in order to determine the right amount of remuneration (Section 7). Relevant information includes production amounts, price and possible licence fees. Also possible savings which the invention brings to the required amount of energy, employment or raw materials will be considered to form part of the invention's value. If these do not constitute a suitable way by which to calculate remuneration, the value can be determined from licence fees, which the employer would have had to pay when buying the rights to a similar invention from outside the company. If neither of these ways for calculating the remuneration is possible, the reasonable remuneration can also be estimated.

Also, the extent of rights acquired by the employer as well as terms of the employment contract and other factors connected to the employment will be considered as forming part of the invention's economic value. If the rights or part of the rights are transferred back to the inventor, this should be taken into account as a reduction in the remuneration. The employer can deduct the necessary costs that follow the use of the invention from the compensation.

Before coming up with a sum, the parties may request an opinion concerning the reasonable amount of compensation from the Employee Inventions Board (*Työsuhdekeksintölautakunta*). Opinions issued by the board are not binding, but the parties should inform the Board two months after receiving the opinion whether or not they will comply with it.

Further, as regards university researchers inventions, when the institution acquires rights to an invention, the inventors are entitled to adequate remuneration. A contractual term concerning waiving these rights is not binding if made before the invention. The circumstances where the invention was made and the profit the institution gains from the invention shall both be taken into account when calculating the compensation (Article 9). The institution has to grant the inventor all necessary information in order to determine the right amount of compensation. If circumstances have changed significantly and the contract, decision or judgment concerning the remuneration might have become unjust, a court can re-evaluate the amount of suitable compensation.

The parties may also request an opinion concerning the reasonable amount of invention made by university researchers from the special division of the Employee Inventions Board.

The Ministry of Economic Affairs and Employment has set a working group to revise the University Employee Act and the report should be published in May 2017. Great alterations are not expected.

5 EMPLOYEES' UTILITY MODELS

5.1 The Legal Framework

The Finnish Utility Model Act of 1991 protects minor technical inventions which do not fulfil the requirements of patents. The Finnish law does not govern utility models which have been invented under employment with any specific provisions. The Act on the Rights to Employee Inventions does not include utility models.

Finnish utility models are also called 'minor patents', as the requirement for the level of inventive step is less for utility models than for patents and the maximum duration of a utility right is only ten years (4 + 4 + 2). Utility models must be new and differ clearly from what was known earlier. Processes, plants or animal species cannot get protection as utility models.

The registration of a utility model differs from a patent as that the registration authority only investigates whether or not the formal preconditions have been met and that the invention is not against public morals or public order. The authority does not inspect the inventive step or originality of the invention if the applicant does not specifically ask for such assessment. The registration of a utility model is simpler, faster and more cost-effective than the corresponding procedure for patents.

The application procedure gives the applicant and third parties the right of appeal. As the examination is not thorough, it is also possible to file utility models which do not satisfy the legal requirements. A utility model can thus be declared null and void by the Patent Office. Any person may at any time initiate such a procedure by submitting a claim and paying a fee.

5.2 Ownership of Rights

The rights to a utility model belong to the person who invented the utility model or someone to whom the rights have been transferred. The law has no specific provisions on the employers' rights to inventions which could be possibly registered as utility models. A grant of a utility right gives the inventor the exclusive right to commercial exploitation of the invention.

5.3 The Parties' Duty to Inform

There are no requirements for the parties to inform about possible utility models. However, the requirement for the employer to inform about possibly patentable inventions as stipulated in the Employee Inventions Act should be taken into account. The employer is also entitled to use a utility model as a way to protect an invention.

5.4 Employee Remuneration Right

The utility model right may be transferred. The transferee may not transfer the right onwards if such right is not specifically granted by the original inventor. Since there are no provisions concerning employers' rights to utility models, contracts by which the rights would be transferred to the employer are viewed in the same way as other transactions made under the Contracts Act (228/1929).

6 EMPLOYEES' DESIGN RIGHTS

6.1 The Legal Framework

The Registered Designs Act (221/1971) entered into force in 1971. It is based on the so-called Nordic patent approach. The Act protects 'the prototype of an article's appearance or of an ornament'. The Act was harmonised in 2002 in order to fulfil the requirements of the Directive on the legal protection of designs (98/71/EC). According to the Council Regulation on Community Designs (6/2002/EC) 'where a design is developed by an employee in the execution of his duties or following the instructions given by his employer, the right to the Community design shall vest in the employer, unless otherwise agreed or specified under national law' (Article 14). The Finnish Registered Designs Act has no specific provisions governing the assignment of a design right within an employment relationship, thus the provision of the Community Regulation become applicable if the employer applies for registered Community design protection.

6.2 Ownership of Rights

The right to a design belongs to the person named in the application who has created the design. In addition what has been said about Community Designs above, the right to use a design, if created by an employee while performing duties under an employment agreement, is considered to belong to the employer. A design right is subject to the general rules on transfer and licensing.

6.3 The Parties' Duty to Inform

Under Finnish design law, employees who create designs do not have a legal obligation to inform their employer.

6.4 Employee Remuneration Right

Contracts concerning the transfer of design rights will fall under the general Contracts Act, which provides for the possibility to alter an unjust contractual term in court.

7 EMPLOYEES' PLANT BREEDER'S RIGHTS

7.1 The Legal Framework

Employees' plant breeder's rights are regulated in the Plant Breeder's Right Act (22.12.2009/1279). The Act is applied to protect plant breeder's rights, which can be granted to varieties of all botanical genera and species. The European Council Regulation (EC) No. 2100/84 on Community plant variety rights is also of relevance. The same protection as laid down by the Plant Breeder's Act applies to the right holder, regardless of whether it is provided by the national Act or the EC Regulation.

7.2 Ownership of Rights

7.2.1 *Private Sector*

A holder of plant breeder's rights has the exclusive right to exploit the propagating material of the variety in professional production and marketing. Some acts require the authorisation of the rightholder, such as propagating, conditioning for the purpose of propagation, marketing, export and import and storage for use for purposes referred to before. Plant breeder's rights do not, however, restrict the use of a variety for the following purposes:

(1) privately for other than commercial purposes;
(2) for experimental and research purposes;
(3) for breeding new varieties.

The law does not regulate private sector plant breeder's rights concerning plants developed by an employee during working time. Regulation in this area is the same for both plant breeder's rights acquired during and outside of working hours or duties. Besides the specific law, common contractual principles are applied to transferring rights in this area.

7.2.2 *Public Sector*

The law does not regulate public sector plant breeder's rights concerning plants developed by an employee during working time.

7.3 The Parties' Duty to Inform

The Finnish Act does not require employees who create plant varieties to inform their employer. However, in some cases there is probably such an obligation under an implied contract between the parties.

7.4 Employees' Remuneration Right

The Finnish Act does not give employees a right to remuneration.

8 INSTANCES FOR DISPUTES

Disputes concerning civil matters on copyright, as well as administrative and civil matters in patents, designs, utility models and semiconductor chip rights will be heard by the Market Court, which is a special court for intellectual property right disputes in Finland. Appeals can be made in administrative matters to the Supreme Administrative Court of Justice and in civil matters to the Supreme Court of Justice if the Court in question grants leave.

In addition to courts, the Copyright Council regulated by the Copyright Act assists and provides expert opinion on copyright matters. The Council is administered by the Ministry of Education and Culture of Finland. It is important to note, however, that the Copyright Council does not give opinions on contracts in any form. Its opinions are non-binding and can be requested by anyone (private persons, businesses, organisations, authorities and courts of law, whether or not they have a personal interests in the matter).

Regarding employees' inventions, the Employee Inventions Board is a statutory body appointed by the Council of State. The Board gives advisory and non-binding opinions and may act as an arbitral tribunal, if the parties have so agreed. The Board has a special unit concerning university inventions. Disputes concerning the rights of the parties will be heard by the Market Court.

If the parties have agreed upon remuneration for the invention, but an essential change of the circumstances means that the value of the invention is more or less than what has been agreed on, the discontented party can bring the contract to court for adjustment. However, the requirements are strict – parties cannot plead to matters which they could have known and been able to take into account at the time of the contract. These adjustments can only be made in court and the employee cannot be ordered to return remunerations which have already been paid out.

9 CONCLUDING REMARKS

There are very few binding rules concerning how copyright created during employment is transferred to the employer. To avoid any obscurities concerning the matter, it is highly recommended for the employer and employee to make a contract of transfer of copyright before entering into a contract of employment or include necessary clauses in the employment contract. However, representatives of some industries have in recent years demanded legislation concerning rights to works created during employment. There have been attempts to create such legislation but the issue has been controversial and the attempts have failed.

In companies, minor inventions covered by the Utility Model Act are usually included in the company's guidelines concerning the procedure on employee inventions. Sometimes legislation may be overlapping and the interpretation of the inventive step required for patentable inventions might cause some difficulties.

In spite of the fact that the Registered Designs Act does not regulate the transformation of the design right when the design is created in the course of fulfilling the obligations of employment, it is commonly agreed that the employer is entitled to use the design for the purpose for which it has been designed. The scope of the rights transferred to the employer can however be unclear without a contract.

The legislation on semiconductor chips came into force on 1 July 1991. There have been altogether just four registrations and no disputes. Rights to semiconductor chips play no role in employment relationships.

Regarding inventions by university researchers, unless otherwise agreed, the inventor cannot publish the outcome of the research in a way which might endanger the possibility to protect or exploit the invention. Before the end of the six-month period when the institution may acquire the rights, the inventor cannot submit a patent application or otherwise govern an invention created under contractual research, unless the institution has agreed not to acquire the rights or permitted the inventor to file a patent.

The institution has to keep information concerning the invention as a secret until the invention has been adequately protected and there are no other reasons for maintaining secrecy.

It is recommended that companies and universities make their own guidelines about the procedure for employee inventions in the company or at the university. Generally the rules of procedure in both above-mentioned acts function well and possible disagreements are usually attributable to different opinions about the remuneration.

BIBLIOGRAPHY

Table of Sources

Niklas Bruun & Mikko Välimäki, *Korkeakoulukeksinnöt* (IPR University Center, Helsinki 2007).

Pirkko-Liisa Haarmann, *Immateriaalioikeus* (Talentum Helsinki 2014).

Kristiina Harenko, *Valtteri Niiranen, Pekka Tarkela, Tekijänoikeus – kommentaari ja käsikirja* (WSOYpro Helsinki 2016)

Timo Kivi-Koskinen, *Työsuhdekeksinnöt* (Kauppakaari, Helsinki 2002).

Rainer Oesch, Heli Pihlajamaa & Sami Sunila, *Patenttioikeus* (Talentum Helsinki 2014).

Mikko Välimäki, *Oikeudet tietokoneohjelmistoihin* (Talentum Helsinki 2009).

Table of Legislation

Copyright Act (8.7.1961, 404/1961).

Patent Act (15.12.1967, 550/1967).

Act of Utility Models (10.5.1991, 800/1991).
Act on Registered Designs (12.3.1971, 221/1971).
Act on Plant Breeder's Right (22.12.2009/1279).
Act on the Exclusive Right to the Layout-Design of an Integrated Circuit (11.1.1991, 32/1991).
Act on the Rights to Employee Inventions (29.12.1967, 656/1967).
Act on the Rights in Inventions made at Higher Education Institutions (19.5.2006, 369/2006) (221/1971).
Council Regulation on Community Designs (6/2002/EC).
Directive on the legal protection of designs (98/71/EC).
Employment Contracts Act (26.1.2001, 55/2001).
Collective Agreements Act (7.6.1946, 436/1946).
Contracts Act (13.6.1929, 228/1929).
The European Council Regulation (EC) No. 2100/84 on Community plant variety rights.

Table of Cases

Supreme Court KKO 1987:427.
Supreme Court KKO:1991:514.
Market Court MaO:2014:360.

CHAPTER 9
France

*Michel Abello, Jérôme Tassi & Estelle Vard**

1 BACKGROUND

The protection of employees' creations is at the junction of employment and intellectual property law. French rules about employees' creations are set in the Intellectual Property Code, not in the Labour Code. However, employment law still has an influence on this matter (e.g. possibility to depart from the legal regime only if the rules are more favourable to the employee).

Most creations are performed in the course of an employment contract and the ownership of IP rights on such creations raises much debate in France. This issue is very sensitive in many companies, which argue that the ownership of the IP rights should vest in the employer, as the counterpart of the salary paid to employees. However, the assignment of IP rights on employees' creations to the benefit of the employer is not a principle of French law.

The rules regarding employees' creations differ according to the intellectual property rights which are involved. This chapter will show that the regime of employees' creations is not homogeneous and particularly complicated to implement in day-to-day business life. The number of legal proceedings regarding this issue has increased since the year 2000.

To sum up:

- Copyright is intended to protect the authors, irrespective of their status as employees; only a natural person can be an author. An employment contract does not imply assignment of copyright to the employer. Furthermore, only a natural person can be an author and the moral rights always belong to the

* With the help of Guillaume Dubos, Lawyer at the Paris Bar.

employee since they are not assignable. Except in the case of collective work, the initial owner is always the employee.
- Industrial property rights provide an enhanced protection for the employer (and sometimes initial ownership) when an invention or a design is created by the employee in the course of their duties.

While recent reforms have partly harmonised the procedural rules governing IP rights (Laws of 29 October 2007 and 11 March 2014), no harmonisation is considered with regard to employees' creations.

2 EMPLOYEES' COPYRIGHT

2.1 The Legal Framework

The legislation on employees' copyright[1] puts in place a protective regime, which has occasionally been criticised as being too protective for the author and far from the realities of the process of creation as it exists at the present time.

The basic principles regarding copyright can be found in Article L.111-1 of the French Intellectual Property Code ('IPC'). This Article mentions in its first paragraph that 'the author of an intellectual work shall enjoy in that work, by the mere fact of its creation, an exclusive incorporeal property right which shall be enforceable against all'.

While the first paragraph of Article L.111-1 explains that the author is always the first right holder, paragraph 3 of the same Article specifically mentions the case of employees' works:

> The existence or the conclusion of a contract for the lease and hire of work or of services by the author of an intellectual work does not derogate from the enjoyment of the right recognised by the first paragraph, subject to the exceptions mentioned in the present code. With the same reservations, it is also not derogated from this same right when the author of an intellectual work is an agent of the state, a territorial authority, a public administrative body with legal personality, an independent administrative authority or the Bank of France.

Unlike industrial property rights, French copyright does not require registration, and the right belongs to the author from the moment they create it, as mentioned in Article L.111-1, para. 1 IPC. The only requirement for a creation to be protected under French law is that it shall be original. By virtue of French case law originality is understood as 'the print of the personality of the creator'.[2] This applies to works of fine art as well as works of applied art.[3] Indeed, Article L. 112-1 IPC establishes the principle

1. For convenience, we refer to 'copyright', although the proper English translation of '*droit d'auteur*' is 'right of the author'.
2. Cass Civ.1ère, 14 February 2008, No 07-12176.
3. Cass Ass Plén., 7 March 1986 No 85-91465.

of unicity of art: 'The provisions of the present Code protect authors' right on intellectual works, regardless of the gender, the form of expression, the merit or the destination'.

The originality criterion also means that if a work has been executed exclusively by a person following instructions, the creation does not bear the print of the personality of that person, and they should not enjoy any copyright on the creation.

France adopted a dualistic system concerning the rights of the author, distinguishing between the moral and the patrimonial rights on the creation. Those two rights are separable and different rules apply to them.

The moral right is regulated in Article L.121-1 to 9. This right is meant to protect the particular link between the author and the creation, and cannot be assigned nor licensed:

> The author enjoys the respect of his name, of his quality and of his work. This right is attached to his person.
>
> It is perpetual, inalienable and imprescriptible.
>
> It is transferable on death to the heirs of the author.
>
> The exercise can be conferred to a third party by testamentary provisions.[4]

The moral rights include the right to claim paternity of the work or to maintain anonymity, the right to disclose the work, the right of withdrawal and the right to respect of the work.

Like any right, moral rights are limited in case of abuse. For example, Article L.121-3 IPC states that the Court can hear a case about abuse (whether positive or negative acts) of the right of disclosure (publication) exerted by legal representative of the deceased author, and prescribe appropriate measures.

The economic rights are listed in Article L.122-1 IPC et seq. (they essentially include the right of performance and the right of reproduction).[5] They are not attached to the person of the author and can be assigned or licensed.

2.2 Ownership of Rights

Even in the case of an employee's creation, the right is first vested in the author. It should be mentioned that the author can only be a natural person and not a legal body,[6] except in the case of a collective work.[7]

However, the fact that a legal body cannot be author of the work does not mean that it cannot own rights to the work as rights can be assigned by the author.

4. Art. L 121-1 IPC.
5. The right of distribution is implicitly derivable from the right to put on the market the material exemplars of the work, without prejudice to the principle of exhaustion of rights in the European Economic Area (Art. L122-3-1 IPC).
6. Cass Civ. 1, 15 January 2015, No 13-23.566 concerning databases.
7. The qualification of collective work is the only way for the employer to obtain the rights *ab initio* to the work.

2.2.1 Transfer of Rights

The transfer is subject to specific formalities and remuneration, as we will see below. The employee's refusal to transfer their rights to their employer is not regarded as a wrongful act of the employee according to case law, except in case of a malicious intent or a breach of their duty of loyalty.[8] However, if an employee has transferred their economic rights to their employer, case law has held that publication of the work by the employee is a fault justifying a dismissal for real and serious cause.[9]

As mentioned above, the mere existence of an employment contract does not imply any transfer of the ownership of intellectual property rights of the author, unless there is a specific agreement for assigning these rights. Executing such a specific agreement is thus the safest way for an employer to secure its ownership on the economic rights over the works of its employees. Indeed the qualification of collective works (see hereunder 2.2.2), can be re-qualified in case of litigation, when the judge establishes that the legal conditions have not been fulfilled.

The specific formality for the assignment must be respected as the assignment may otherwise be nullified. The drafting of this type of clause should be made with care because a contract is always interpreted in favour of the author in case of ambiguity. This is subject to relative nullity, which means that it can only be invoked by the parties: an alleged infringer cannot use the non-conformity with formality requirements as a defence.

The conditions for the transfer of rights are mentioned in Article L.131-3 IPC:

> The transmission of rights of the author is subjected to the condition that each of the transferred rights be mentioned distinctively in the act of transfer and that the field of exploitation of the transferred rights be limited as for its extent and destination, as for its location and its duration.

The contract must mention all the rights to be transferred. The rights that have not been explicitly mentioned in the contract remain with the author. The field of exploitation of each of the transferred rights should also be mentioned as well as its extent and destination, including the place and the duration of the exploitation.

Article L.131-3 IPC only refers to contracts where one of the parties is the author. This is not applicable to subsequent contracts concerning the same work.[10]

Since 2006, case law states that the formality requirements of Article 131-3 IPC only apply to contracts listed in this Article, i.e. contracts for the performance, publishing, and audiovisual production. An implicit assignment of economic rights in the case of other contracts would thus be possible.[11] Such contracts must be made in writing (Article L.131-2 IPC). This is not a condition for the validity of the contract, but for evidentiary purposes.[12]

8. Cass Soc. 18 January 2011, No 09-41636 in the case of an employee refusing to give to their employer access to their personal website, containing pages created in behalf of the employer.
9. CA Paris 10 November 2014, No 14/00024.
10. Cass Civ. 1, 13 October 1993, No 91-11241.
11. Cass Civ. 1, 21 November 2006, No 05-19294.
12. Cass Civ. 1, 12 April 1976, No 74-12149.

Articles 1359 to 1362 of the Civil Code, concerning other ways to prove a contract, are applicable to all other contracts.[13] Proof by the production of a written contract is intended to protect the author, and cannot be used against them by the other party.[14]

In order to ensure correct transfer of rights, it advisable to prepare a written contract in any case.

It should also be mentioned that the global transfer of future intellectual works is invalid (Article L.131-1 IPC). This provision was meant to avoid hindering authors' freedom to create. It is thus impossible to state in a contract that all (present and future) works of the employee will be transferred to the employer.

Even though the transfer occurs under strict rules, case law has developed a presumption of ownership on the basis of Article L.113-1 IPC: if a work of art has been disclosed under a certain name, this person is presumed to be the author of the work or to have acquired the corresponding rights. This presumption has been used by legal entities in cases where the work has been created by an employee, and where a third party challenged ownership rights. This presumption is only valid against third parties when the authorship of the work is not the principal claim. It does not apply when the employee is challenging the rights of the employer.

There is also a presumption of transfer of the exclusive rights when a contract has been concluded between a producer and the author of an audiovisual work other than a musical composition (Article L. 132-24 IPC).

2.2.2 *Collective Works*

An exception to the rule that authors must be natural persons concerns the authorship of a collective work (*oeuvre collective*). In this case, copyright is vested in the legal or natural person who disclosed the creation under their name[15]. This is the only possibility for a legal body to be the initial owner of copyright to a work created by its employees (including moral rights). It should be noted that the court has the power to requalify the work as a collective work, and is not bound by the qualification chosen by the parties (although the anticipation of this qualification in an employment contract may create a rebuttable presumption of collective work[16]). A work can be qualified as a 'collective work' if three conditions are simultaneously fulfilled: (i) a person discloses the work under their name, (ii) the work was created under the direction of this person, and (iii) the personal contributions of each author are merged in the final work, without it being possible to attribute to each co-author a specific share of their contribution to the creation of the work.

The employer must prove that the above three conditions are fulfilled by any appropriate means (minutes/protocols of meetings concerning the creation process,

13. It is possible to prove by testimonies for those contracts.
14. See footnote 9.
15. Art. L.113-4 IPC.
16. CA Lyon 10 June 2016, n°15/01025.

affidavit, instructions of the employer, etc.). The collective work is the property of the natural or legal person under the name of which it has been disclosed, until proved otherwise (Article L.113-5 IPC).

In the case of a collective work, there is no need for a proportional remuneration of an employee who contributed to the work, as they are not considered the author.

Part of the doctrine considers that the regime of collective work should not be extended, while another part advocates for a broader application of this regime, allowing more legal security in the case of employees' works. For the past few years, case law has admitted more easily the qualification of collective work in case of works made by several employees.

2.2.3 Public Servants

Another exception to the basic regime governing ownership of rights concerns employees in the public sector. Indeed, the IPC distinguishes between employees in the private and the public sector. This distinction finds its justification in the mission of the public service to act in society's general interest, the result of which is derogation from the rules applicable in the private sector.

The right to a work made by a public servant was not specifically regulated before 2006. At that time, the Highest French Administrative Court (*Conseil d'Etat*) established in OFRATEM[17] that the right belongs to the State:

> The necessities of the service require that the Administration be vested with the author's right [copyright] in the IPC as defined by the law related to literary and artistic property, for those works the creation of which arises from public service. ... By accepting their functions, the public servants and public officials have put their creative activity, or the rights which may stem from it, at the disposal of the service, to the full extent necessary for the exercise of the said functions.

With the law of 1 August 2006, the legislator created a specific regime for public servants, more balanced than the OFRATEM opinion. The basic rules for employees' creations are applicable to public servants, which means the rights are invested in the author; however, there are exceptions to this regime.

The exploitation right of a work created by a public servant in the course of their duties, or following the instructions of the (public) employer, is, from the outset, assigned to the state (Article L.131-3-1 IPC). In such case the public servant remains the author, but the rights are automatically transferred to the state. However, this only applies when it is 'strictly necessary for the accomplishment of a duty of public service'. In case of a commercial exploitation of the work 'the state only enjoys a preferential right over the author'. This provision is not applicable in the case of scientific research activities of a technological and scientific public institution or of a public institution with scientific, cultural and professional activity, when these activities are subject to a contract with a legal entity of private law.

Concerning the moral rights, Article L. 121-7-1 IPC states that:

17. CE 21 November 1972, Ofratem, n°309.721.

> [T]he right of disclosure recognized to the public servant mentioned in the third paragraph of Article L. 111-1, who has created an intellectual work in the exercise of its functions or according to the instructions received, is exerted in the respect of the rules to which he is subject as an agent and of the rules governing the functioning, and the activity of the hierarchical authority employing him.
>
> The agent cannot:
>
> (1) Object to modifications of the work of art decided in the interest of the public service by the authority vested with a hierarchical power, when this modification does not offend their honor or their reputation;
>
> (2) *Exercise their right to reconsider and withdraw, except with agreement of the authority vested with a hierarchical power.*

This regime applies to public servants with the exception of authors of works for which the disclosure is not subject to previous control of the hierarchical authority, as mentioned in Article L.111-1 IPC. This is the case for independent agents, such as university professors.

2.2.4 *Journalists*

The situation of journalists is also specifically regulated in the IPC and in the Labour Code. However, this only applies to journalists of the written press (Article L.132-35(1) IPC excludes audiovisual communication services).

Article L.7111-3(1) of the Labour Code defines a professional journalist as '*any person whose main, regular and paid activity is the exercise of his profession in one or several media companies, or daily and periodic publications, or press agencies, and for whom such work is the main source of income*'. The collaboration between a press agency and a professional journalist includes all media types of the press publication, except if stipulated otherwise in the employment contract or any other agreement of temporary collaboration.

Unless agreed otherwise, a contract between a professional journalist and employer implies the transfer of exploitation rights on the journalist's works made within the framework of the publication, exclusively and irrespective of whether or not the work is published (Article L.132-36 IPC). This is thus an exception to the general rule that the transfer of rights should be explicitly stipulated in a written contract. This means that even though the provisions concern the written press, the rights transferred are not only those on written works but on all the copyrighted works created by the journalist in this framework. However, Article L.132-41 IPC mentions specific rules concerning the professional journalist author of a fixed image, who draw the majority of their revenue from such works, and occasionally works for a press publication. In this case, the transfer of rights as mentioned in Article L. 132-6 IPC is only applicable when this intellectual work was commissioned by the media company.

In addition, the exploitation of the work of the journalist in different media within the framework of the press publication in return for solely the salary is possible during a period determined by a company-level or a collective agreement (Article L.132-37 IPC).

If the employer wishes to exploit the work other than in the press publication they will have to receive authorisation from the journalist.

2.2.5 Employees' Computer Programs

Although the copyright system applies uniformly to both works of fine art and works of applied art, a derogative provision has been provided for computer programs. Computer programs are considered an intellectual work in France, but the applicable rules are inspired by the patent system. Indeed, as early as in the 1970s the issue of the classification of computer programs as an intellectual work or as an invention was raised. This debate is justified by the specific character of a computer program, which is both a technical creation and a literary work.

Article L.113-9 IPC is thus inspired by the regime of employees' inventions (see section 4 below). It states as a basic principle that the economic rights to computer programs and their documentation created by one or many employees, during the course of their employment or according to the instructions of their employer, belong to the employer. As indicated by the Court of Cassation in a recent decision[18], this does not mean that the employer can be the author of the software; as for other intellectual works, only natural persons can be authors (except in case of collective work). Thus, the employee is the holder of the intellectual property rights to the computer program *ab initio*, but the economic rights - only - are automatically transferred to the employer.

In respect of the remuneration of the employee-creator of a computer program, case law considers that it is not necessary,[19] even though some authors do not share this view.[20]

2.2.6 Neighbouring Rights of Database Producers

The right of the database producer is a *sui generis* right. It protects the producer against the extraction or the re-utilisation of the whole, or a substantial part, of the content of the database (Article L.342-1 IPC). The database is '*a collection of independent works, data or other materials arranged in a systematic methodical way and individually accessible by electronic or other means*' (Article 1 of Directive 96/9/EC). This protection comes in addition to the protection of the structure of the database by the copyright. The right to the data should be distinguished from the right to the database.

The right to the database is protected, regardless of the legal protection of the data, which can be protected by copyright or personal data. The right of the database producer is granted when there is a qualitative or quantitative substantial investment in either the obtaining, verification or presentation of the contents. This does not

18. Cass Civ. 1, 15 January 2015, No 13-23566.
19. CA Rennes 1 July 2014, No 12/06459.
20. M. ABELLO et J. TASSI, Comm. Comm. Elec., April 2011, Etude No 8.

include the investment for the creation of the data, but only for the creation of the database[21]. It is expressly stated in Article L.341-2 2° IPC that legal entities can benefit from this protection.

In case of a database created by an employee, the investment is usually made by the employer, for which reason the rights should vest in them and not the employee.

2.2.7 *Neighbouring Performers' Rights*

Article L.212-1 IPC provides a definition of the performer-interpreter (*'artiste-interprète'*)[22]. This is a natural person who performs an intellectual work. A similar definition of the performing artist can be found in the Labour Code, which provides a non-exhaustive list of such artists.

A performer holds moral rights to their interpretation as mentioned in Article L.212-2 IPC: the respect for their name, status as performer-interpreter and their interpretation. Like author's moral rights, the performer's moral rights cannot be transferred, and are subject to no limitation period. Rights concerning the protection of the interpretation and the memory of the (deceased) author are transferable to heirs. However, performers' moral rights remain limited compared to authors' moral rights.

Under Article L.212-3 IPC the registration of the performance, its reproduction and its communication to the public are subject to written authorisation of the performer. There is a presumption of transfer of rights to the interpretation in Article L.212-4 IPC if a contract was entered into between a performer and a producer of an audiovisual work. This contract specifies separate compensation for each mode of exploitation, and entails the authorisation to fix, reproduce and communicate to the public the performance of the artist.

Any contract, by which a person ensures, against remuneration, the participation of a performer for the purpose of producing a performance, is presumed to be an employment contract if the exercise of the artist's activity does not require their registration in the commercial registry (Article L.7121-3 of the Labour Code). There is no need to establish the subordination of the artist for the application of the presumption of Article L. 7121-3 of the Labour Code. In addition, the contract does not have to be concluded directly with the artist, or the remuneration be paid directly by the entertainment entrepreneur for it to apply.[23]

Case law also mentions that the existence of an employment contract does not change the attribution of the intellectual property rights to the employee. The authorisation of the performer is thus required for each use of their performance. For example, the making of a phonogram is subject to authorisation by the performer[24].

The IPC distinguishes between cases where the remuneration is included in the salary of the performer, and where it is not included. Article L.7121-8 of the Labour Code states that the remuneration is not considered to be a salary if:

21. ECJ 9 November 2004 C-302/02 (*Nils Laurin Effing*); Cass Civ., 5 March 2009, N°07-19.734.
22. It complies with the definition found in the WIPO Performances and Phonograms Treaty.
23. Cass Soc 25 January 1990, No 86-10188.
24. Cass Civ.1, 6 March 2001, No 98-15502.

(i) The physical presence of the actor is not required to exploit the recording of his performance
(ii) This remuneration is not based upon the salary received for the production of his interpretation, execution of his performance, but is based upon the product of the sale or of the exploitation of the recording.

The Court of Cassation has stated, as a principle, that the remuneration allocated to the performer should be distinguished from the remuneration for their artistic performance, and a clause concerning global remuneration is not valid.[25]

If the remuneration is not mentioned in the contract or in an applicable collective agreement, for one or several ways of exploitation, its amount can be fixed by a specific agreement between the employee's and employer's professional representatives, according to Article L.212-5 IPC.

2.3 The Parties' Duty to Inform

There is no duty to inform the employer about the creation of an intellectual work made by an employee, unlike in patent law (see section 4 below). It should however be noted that there is a general duty of loyalty for both parties during the course of an employment relationship, which can be inferred from Article L.1222-1 of the French Labour Code: 'The employment contract shall be executed in good faith.'

Failure to comply with this obligation may be a cause for a civil suit. However, to our knowledge, there is no case law in which the employee has been held liable for a violation of a duty to inform their employer about the creation of an intellectual work.

Equally, there is no specific duty requiring the employer to inform the employee that they have transferred the rights assigned to them to a third party.

2.4 Employee Remuneration Right

Unlike patents, no distinction is made between a creation made during the course of or outside an employment relationship. In theory the calculation of the remuneration is the same as for any other transfer of right. Article L.122-7 IPC regulates the possibility to transfer rights for free. This should however not apply in the case of an employee's work, as it is necessary to demonstrate the (liberal) intent of the author.[26] This intent is difficult to demonstrate because the employer can exert hierarchical power over their employees. It seems that a payment should always be the rule in this case.

The first paragraph of Article L.131-4 IPC requires the payment of a proportional share of income to the benefit of the employee derived from the sale or exploitation of the work when rights to the work have been transferred (partially or wholly) by the employee. This is a public policy rule, which means that the parties cannot chose to exclude the provision in a contract. It is instituted with a view to protect the author, and

25. Cass Soc., 10 February 1998, No 95-43.510.
26. CA Paris 25 November 2005, No 04/02005.

is applied by the courts as such. This means that the courts can cancel a contract if they consider that the proportional remuneration is not sufficient.[27]

The second paragraph of Article L. 131-4 IPC lists the cases where a lump sum payment is admissible. This is in particular the case when: (1) the basis of calculation of the proportional share cannot be determined in practice, (2) the means to control the application of the proportional share are missing, (3) the cost of calculation and control would be out of proportion with the results to be reached, (4) the contribution of the author does not constitute one of the essential features of the intellectual creation of the work, or the use of the work is merely ancillary to the subject of the exploitation, (5) software is transferred. By virtue of Article L. 131-5 IPC the price can be modified in the case of a lump sum payment if it is lower than seven-twelfths of what would be a fair price, if the author requests it.

Provisions that do not respect Article L.131-4 IPC are invalid (relative nullity). It is thus advisable to mention the reasons why proportional remuneration is not possible in order to avoid invalidation of the clause.

3 EMPLOYEES' SEMICONDUCTOR CHIPS

3.1 The Legal Framework

By virtue of Article L.622-1 IPC, the topography of a semiconductor product is protected if it is the result of an intellectual effort of its creator, if it is not commonplace and if it is registered no more than two years after the first commercial exploitation of the topography, or no more than 15 years after its fixation or coding if it has never been used.

The protection is granted for ten years from the earliest of the following dates: (a) the end of the calendar year in which the topography is first exploited anywhere in the world or (b) the end of the calendar year in which the application for registration was filed in due form. If the topography was not commercially exploited anywhere in the world within a period of 15 years from its first fixation or encoding, any exclusive rights in existence shall come to an end (Article 7 of Directive 87/54/ EEC of 16 December 1986 on the legal protection of topographies of semiconductor products and Article L.622-2 IPC).

Article 10 of Directive 87/54/EEC mentions that '*the provisions of the Directive shall be without prejudice to legal provisions concerning patent and utility model rights*'. As a result, if the semiconductor is patentable, the legislation on employee's inventions will apply.

27. TGI Paris 30 November 1999, No 98/16766.

3.2 Ownership of Rights

According to Article L. 622-3 IPC, the right belongs to the creator or their successor in title.[28] The IPC does not mention any derogation from this rule when the topography has been created by an employee. It seems to us that the right should primarily belong to the employee, regardless of whether it has been created during the course of their work or not. It would therefore be advisable for the employer to conclude a contract for the transfer of rights with the employee-creator of the topography. To our knowledge this question has, however, not been controversial in practice.

If the topography were registered in violation of the rights of the creator or their successor in title, the employee can claim the benefit of the registration within five years from the publication of the registration application, by virtue of Article L.622-3 IPC.

3.3 The Parties' Duty to Inform

There is no specific duty to inform about topographies of semiconductor products.

3.4 Employee Remuneration Right

As the right belongs primarily to the employee, remuneration would arise from the transfer of rights or a licence. The transfer should be made in accordance with the requirements of Articles L.613-8 and L.613-9 IPC, referred to in Article L.622-7 IPC: it should be made in writing and it has to be registered to be enforceable against third parties.

4 EMPLOYEES' INVENTIONS

4.1 The Legal Framework

In principle the right to a patent belongs to the inventor or their successor in title as specified in Article L.611-6 IPC. However, most inventions are made by employees in the course of their duties. When the inventor is an employee, a specific regime is instituted in Article L.611-7 IPC[29]. This Article determines both the ownership of the rights to the invention and the type of compensation the employee may be entitled to receive.

---

28. Provided that the creator or their successor in title comply with the requirements concerning nationality, residency or the place of establishment as mentioned in Art L.622-2 IPC.
29. Art. L.611-7 IPC:
'Where the inventor is a salaried person, the right to the industrial property title, failing any contractual clause more favorable to the salaried person, shall be defined in accordance with the following provisions:
°*1.Inventions made by a salaried person in the execution of a work contract comprising an inventive mission corresponding to his effective functions or of studies and research which have*

With regard to employees' inventions, the regime applies to any patentable invention, whether patented or not. Under French law, an invention is protected when it is new, comprises an inventive step and is susceptible of industrial application (Article L.611-10(1) IPC). The rules regarding employees' inventions are public policy provisions, and a contract can only derogate from them if it is more favourable to the employee. It is necessary to define the scope of application of Article L. 611-7 IPC before a detailed analysis of these provisions can be made.

First, this regime is applicable to all employees irrespective of whether they are in the public or private sector, and it only applies to 'employees', i.e. company representatives,[30] interns[31] or external consultants[32] fall outside the scope of the regime. However, the term 'employee' is not defined in the French framework. Pursuant to case law, an employment agreement concerns the performance of a professional activity in subordination to another party in return for remuneration. French law does not require a written employment contract for permanent employment contracts.[33] If the inventor was an employee at the time he/she created the invention, even if he/she left the company when the company filed the patent application, the regime of employees' inventions is applicable.

---

been explicitly entrusted to him, shall belong to the employer. The employer informs the salaried person author of the invention, when an application for an industrial property right has been filed on such invention and when the title has been granted as the case may be. The conditions under which the salaried person who is the author of such an invention belonging to the employer shall enjoy additional remuneration shall be determined by the collective agreements, company agreements and individual employment contracts Where the employer is not subject to a branch collective agreement, any dispute relating to the additional remuneration shall be submitted to the commission of conciliation set up by Art L615-21 or by the First Instance Court.

°2. All other inventions shall belong to the salaried person. However, where an invention is made by a salaried person during the execution of his/her functions or in the field of activity of the company or by reason of knowledge or use of technologies or specific means of the company or of data acquired by the company, the employer shall be entitled, subject to the conditions and the time limits laid down by decree in the State Council (Conseil d'Etat) to have assigned to it the ownership or enjoyment of all or some of the rights in the patent protecting the employee's invention The salaried person shall be entitled to obtain a fair price which, failing agreement between the parties, shall be stipulated by the joint conciliation board set up by Art L615-21 or by the First Instance Court; these shall take into consideration all elements which may be supplied, in particular by the employer and by the employee, to compute the fair price as a function of both the initial contributions of either of them and the industrial and commercial utility of the invention.

°3. The salaried author of an invention shall inform his/her employer thereof and the latter shall confirm receipt in accordance with the terms and time limits laid down by regulation. The salaried person and the employer shall communicate to each other all relevant information concerning the invention. They shall refrain from making any disclosure which would compromise, in whole or in part, the exercise of the rights afforded under this Book. Any agreement between the salaried person and his/her employer concerning an invention made by the salaried person shall be recorded in writing, failing what it shall be invalid.

°4. The implementing rules for this article shall be laid down by decree of the State Council.

°5. This article shall also apply to servants of the State, of local authorities and of any other public legal person under the terms to be laid down by decree of the State Council.'

30. Cass Com. 21 June 1988, No 86-19166.
31. Cass Com. 25 April 2006, No 04-19482.
32. TGI Paris 18 April 2013, No 11/00543.
33. But a written contract is required for short-term employment contracts or part-time contracts.

Secondly, the legal entity to which this regime applies is the employer of the inventor – at the time the invention was made – whether or not the patent was applied for by another entity or was transferred.[34] It should however be verified that the branch of activity was transferred (i.e. merger or sale of a business unit). If there is a continuation of activity between the companies, the rules could apply to the new entity, even though it was not the employer of the inventor at the time of the creation of the invention. Indeed, Article L.1224-1 of the Labour Code mentions that:

> When a modification of the legal situation of the employer occurs, in particular by succession, sale, fusion, conversion of the fund, or the company creation of the business, all the ongoing employment contracts at the date of the modification subsist between the new employer and the staff of the company.

Finally, the rules concerning employees' inventions are applicable only when the employment contract is governed by French law, regardless of the nationality of the employee and employer. Article 8 of the 'Rome I' EC Regulation (No. 593/2008) dated 17 June 2008 defines the cases in which French Law governs an employment contract:

> 1. An individual employment contract shall be governed by the law chosen by the parties in accordance with Article 3 [Freedom of choice]. Such a choice of law may not, however, have the result of depriving the employee of the protection afforded to him by provisions that cannot be derogated from by agreement under the law that, in the absence of choice, would have been applicable pursuant to paragraphs 2, 3 and 4 of this Article.
> 2. To the extent that the law applicable to the individual employment contract has not been chosen by the parties, the contract shall be governed by the law of the country in which or, failing that, from which the employee habitually carries out his work in performance of the contract. The country where the work is habitually carried out shall not be deemed to have changed if he is temporarily employed in another country.
> 3. Where the law applicable cannot be determined pursuant to paragraph 2, the contract shall be governed by the law of the country where the place of business through which the employee was engaged is situated.
> 4. Where it appears from the circumstances as a whole that the contract is more closely connected with a country other than that indicated in paragraphs 2 or 3, the law of that other country shall apply.

4.2 Ownership of Rights

Article L. 611-7 IPC distinguishes between three categories of inventions, depending upon the extent of the employee's duties.

Any agreement between the inventor concerning the remuneration for an invention shall be executed in writing, otherwise it will be found invalid (Article

34. Cass Com. 9 December 2014, 13-16.559. See also, CA Paris 30 June 2015 No 13/10097 JurisData n°2015-030469 which confirmed the admissibility of the action brought by the employee (in order to obtain a supplementary remuneration) against the new owner of the patent – such decision is contrary to the main case law which decides that the supplementary remuneration can only be asked from the legal successor of the employer.

L.611-7(3) 3rd sentence IPC). In any case the inventor retains the moral right to be mentioned in the patent, or to oppose being mentioned (Article L.611-9 IPC), irrespective of the qualification of the invention.

4.2.1 *Inventions Made in the Course of the Employee's Normal Duties*

This type of invention (*invention de mission*) is made by the employee in the course of either an employment contract comprising an inventive duty which corresponds to their effective duties (general inventive duty), or of studies and research specifically assigned to the employee by the employer (specific inventive duty).

The general inventive duty specified in the contract shall not be vague and shall correspond to the normal duties of the employee; otherwise courts can classify the invention as falling outside the employee's employment duties.

Where the invention is made in the course of the employee's normal duties, it belongs to the employer *ab initio*, but the employee benefits from supplementary remuneration. The remuneration is compulsory by virtue of Law No. 90-1052 of 26 November 1990 (previously, supplementary remuneration was optional, and could depend on collective agreements which differed from one industrial sector to another).

Concerning public agents, inventions made in the course of their normal duties also belong to the public entity which employs them. However, the first paragraph of Article R.611-12 IPC provides that in the event that the public entity decides not to value such inventions, the employee can dispose of the corresponding economic rights, under conditions laid down in an agreement concluded with the public entity.

4.2.2 *Inventions Made Beyond the Employee's Normal Duties*

An invention made beyond the normal duties (*invention hors mission attribuable*) is an invention made by an employee either: (i) during the execution of their functions,[35] but where the invention does not fall into the previous category, or (ii) in the employer's field of activity, or (iii) by the use or knowledge of techniques or means specific to the employer, or data procured by the employer. This invention belongs to the employee but the employer has the right to request a compulsory assignment (or license) of the patent rights during a certain time limit, following the procedure mentioned in Articles R. 611- 33 IPC et seq.

Employers can decide whether or not they want the ownership of the invention to be assigned to them within four months starting from the receipt of the declaration of invention (Article R.611-7 IPC). Furthermore, after their decision and the signature of an act of transfer, the employer cannot change their mind and waive this right.[36]

In case of assignment to the employer, the employee is entitled to receive a 'fair amount' (see section 4.4.3 below).

35. This means during their working hours.
36. Cass Com., 18 February 2004, 02-17421.

4.2.3 Inventions Made Beyond Employment Duties

All other inventions fall into this category (*invention hors mission non-attribuable*). They belong to the employee who can exploit the invention freely. The employee cannot receive any financial remuneration from their employer for this invention, but they have the duty to inform their employer, as for any other invention. This regime is applicable to employees working in the private as well as public sector.

4.3 The Parties' Duty to Inform

Article L.611-7(3) IPC states:

> The salaried author of an invention shall inform his/her employer thereof and the latter shall confirm receipt in accordance with the terms and time limits laid down by regulation.
>
> The salaried person and the employer shall communicate to each other all relevant information concerning the invention. They shall refrain from making any disclosure which would compromise, in whole or in part, the exercise of the rights afforded under this Book.

4.3.1 The Procedure

Articles R.611-1 to 10 IPC provide the terms and time limits concerning the duty to inform. All communications or declarations of the parties should be made by registered letter with acknowledgement of receipt, or any other way in order to ensure proof of receipt by the other party.

Once the employee has made the invention, the employee should immediately declare it to the employer. In case of co-inventors, the declaration can be made by all or some of the inventors.

The declaration shall contain sufficient information for the employer to recognise the classification of the invention as an invention made in the course of duty or beyond it. The second paragraph of Article R.611-2 IPC provides that the information should refer to: (i) the subject of the invention and the application considered, (ii) the circumstances of the invention (instructions or guidelines received, experiences or company resources that were used, collaborations obtained) and (iii) the classification of the invention considered by the employee.

This declaration can also be provided by the French Patent and Trademark Office ('INPI') to which the employee (previously) sent the declaration, by virtue of Article R.611-9 IPC. However if the internal policy of the employer requires direct information, the employee cannot rely solely on the information provided through the INPI insofar as the duty to inform the employer is concerned.

In addition, if the employee considers that the classification allows the employer to exert their assignment right (invention outside the employee's normal duties), they shall include a description of the invention, presenting the problem which has been raised, possibly taking into account the prior state of the art, the solution given to the problem, and at least one example of achievement possibly accompanied by drawings (Article R.611-3 IPC). This description should also be added when the invention qualifies as an invention made beyond the employee's normal duties after the declaration by the employee.

Once the employer receives the declaration, the employer has two months to decide if they agree with the classification made by the employee, or to choose a category for the invention in a substantiated notice if the employee did not mention any. Within this timeframe, the employer can ask the inventor to complete the declaration, explaining the precise points at issue. If so, the two-month period starts anew when the declaration has been completed by the employee. If the employer does not answer within this time limit, they are presumed to agree with the classification chosen by the employee.

When the invention is considered as having been made outside the normal duties of the employee, unless otherwise agreed by the parties before the invention is made, the employer can claim the assignment of rights within four months, starting from the declaration of the inventor or from the completion of the declaration if supplementary information was requested.

In addition, Article L.611-7(3) and R.611-10 IPC state that both parties have a duty to abstain from any prejudicial disclosure of the invention as long as there is no agreement concerning its classification. If one of the parties decides to apply for a patent during this period, it should send a copy of the application to the other and use all the means offered by the legislative framework and regulation in order not to delay the publication of the application.

In case of litigation concerning the declaration or the classification of the invention, referral can be made to the Commission of Conciliation. This leads to a suspension of the time-limit until a final decision is rendered.

4.3.2 Failure to Comply With the Duty to Inform

Concerning the time-limits, even though the IPC states that the employee should inform the employer immediately, nothing is said about a belated declaration. According to case law[37] a late transmission has the sole consequence of delaying the starting point of the time-limits mentioned in Article R.611-6 and R.611-7 IPC. The employer can, however, claim damages if they have suffered damage as a result of this delay. This could, e.g., be the case if the invention was patented or disclosed in the meantime by a competitor.

As mentioned above, both parties have a duty of loyalty in the execution of an employment contract and its breach can justify the termination of the contract.

In the event that the employee is dismissed for not having informed the employer in due time, Article L.1234-9 of the Labour Code states that the employee has a right to

37. CA Reims 11 October 2010, No 09/02151. CA Bordeaux 13 March 2017, No16/00887.

compensation, unless the termination of the contract results from serious misconduct. Depending on the salary and experience of the employee, the compensation can be very high.

In order to determine if the lack or delay of information justifies the termination of the contract, case law takes into account the classification of the invention:

- if the invention is made as part of the employee's normal duties, the employer can generally end the employment relationship with the inventor with no gratification.
- if the invention has not been made as part of the employee's normal duties or beyond the employment relationship, case law generally considers that the absence of declaration does not justify the termination of the contract.[38]

4.3.3 *New Duty to Inform of the Filing and Granting of a Patent*

Article L.611-7 (1) IPC, as recently modified by Law n°2015-990 dated 6 August 2015, states that:

> The employer informs the salaried person author of the invention, when an application for an industrial property right has been filed on such invention and when the title has been granted as the case may be.

This new duty to inform bearing on the employer does not have a direct impact on the supplementary remuneration owed to the employee or its amount.

However, even though there is no sanction attached to the violation of this duty, it could have an effect on the statute of limitations within which the employee has to bring an action to claim his supplementary remuneration.

In some cases, the date on which the employee has been informed of the filing or the granting of the patent could be considered as the starting point of the time-limit if no other circumstance is likely to occur afterwards (e.g. in the case where the patent is not exploited). The scope of this new duty will certainly be assessed by case law.

4.4 Employee Remuneration Right

As mentioned above, the employee's compensation varies depending on the type of invention. Where the invention is made as part of the employee's normal duties, the employee is entitled to a supplementary remuneration (*rémuneration supplémentaire*), while they will obtain a fair amount (*juste prix*) in the case of an invention made outside the normal duties. In the case of an invention made outside the scope of the employment relationship, the employee will of course not receive any compensation from their employer, since they can freely exploit the invention.

38. Cass Soc. 21 September 2011 No 09-69.927; Cass Soc. 15 January 2015 No 13-14.811.

4.4.1 Supplementary Remuneration in the Private Sector

The right to a supplementary remuneration in case of an invention made as part of the employee's normal duties is a public order policy since 1990[39]. The right to compensation is compulsory, even when the invention is not exploited.

Any collective bargaining agreement or employment contract that would decide otherwise is invalid. Accordingly, an agreement that would provide for a payment, but condition it on specific facts, such as the grant of a patent, the exploitation of the invention within a certain time period, or the exceptional interest of the invention for the employer would not be valid.[40]

The IPC leaves the calculation of the supplementary remuneration to the parties, and does not provide any method for the employees in the private sector. The amount of supplementary remuneration can be ascertained by the following types of agreements, by virtue of Article L. 611-7 IPC:

- collective bargaining agreements (i.e. agreements governing all employment contracts in a particular industrial sector);
- company agreements (i.e. agreements negotiated between a company and the union partners, and applicable to all company employees);
- individual employment contracts.

The types of agreements are listed exhaustively, so a unilateral policy decided by the employer without negotiations with the union partners would be unenforceable against the employees.[41] If a company agreement is set up, case law considers that it '*has normative value, is applicable to everyone and regulates the current situations*'[42].

A survey conducted by the French Patent and Trademark Office in 2016 reveals that companies tend to grant several bonuses to the inventor according to the occurrence of an event (for instance: patent filing, first grant of the patent, first foreign filing, first licence granted).

According to this survey, for each event, the bonus paid by a sample of 97 French companies amounted to:

39. The previous version of the law stated that the inventor '*may enjoy*' supplementary remuneration. The 1990 Law changed this Article in providing that the inventor '*shall enjoy*' supplementary remuneration.
40. Cass Com. 12 February 2013, No 12-12.898: Art. 29 of the pharmaceutical industry collective bargaining agreement stated that the inventor is entitled to a supplementary remuneration only if the invention is '*of an exceptional interest to the company*', the Supreme Court dismissed this provision since it is contrary to Art. L. 611-7 IPC.
41. TGI Paris 7 November 2008 PIBD, No 889 III; TGI Paris 24 October 2014.
42. Cass Com. 22 February 2005, No 02-18.790.

Table 9.1

	minimum/ median (EUR)	maximum/ median (EUR)
Bonus	500	2,000
Combination of the remuneration (bonus + remuneration dependent on the exploitation)	400	15,000

If no collective bargaining or company agreement has been instituted concerning inventions, contractual freedom applies, and the remuneration is determined by the parties. If the amount of the remuneration offered unilaterally by the employer is too low, the courts can also decide to allocate to the inventor an additional amount of remuneration.

Article L.611-7 IPC states that if the employer is not subject to a branch collective bargaining agreement, any litigation relating to the supplementary remuneration is subject to the Commission of Conciliation (known as the 'CNIS'), the secretary of which is assumed by the French Patent and Trademark Office ('INPI'), or directly to the Court of First Instance[43] (see below).

In determining the amount of the supplementary remuneration, when no agreement has set up or a method for its calculation, the court takes into account the scope of research, the contribution of the inventor, the difficulties in developing the invention, the commercial and industrial use of the invention.[44]

Historically, the supplementary remuneration was based on the employees' monthly salary (usually one to six months). In 2000, the Court of Cassation stated that a court has no legal obligation to take into account the salary.[45] In this case, the Court upheld a decision awarding around EUR 600,000 to the inventor of a treatment for prostate cancer. Since this decision, the supplementary remuneration awarded by courts has ranged from a few thousand euros to several hundred thousand euros in very specific circumstances, depending on the importance of the invention.[46]

If the invention is not exploited, the supplementary remuneration generally amounts to several hundred euros, unless it is an important invention (for instance, in case of a defensive patent remuneration of EUR 20,000 was granted)[47].

In a few cases, in the absence of negotiations between the employer and the union partners or the employee, courts have analogously applied to employees of the

43. Paris First Instance Court has exclusive jurisdiction in first instance regarding patent litigation, including employee's invention proceedings.
44. CA Paris 30 January 2013, No 11/10036 approved in that regard by Cass Com. 9 December 2014, No 13-16.559.
45. Cass Com. 21 November 2000, No 98-11.900.
46. TGI Paris, 24 October 2014: a remuneration of EUR 280,000 is granted in view of a turnover of EUR 1.7 billion; TGI Paris, 19 May 2009: the Court awarded EUR 500,000 to the inventor for an invention that allowed a company to save around EUR 15 billion.
47. CA Paris 26 June 2013. CA Paris 30 October 2015 n°13/05568 awarded EUR 20,000 for five defensive patents.

private sector provisions applicable to public agents.[48] Indeed, for public agents a precise guideline for assessing the supplementary remuneration exists (see below 4.4.2).

The statute of limitation of the action for supplementary remuneration has been subject to debates and is crucial in many cases because proceedings concerning the remuneration for the invention are often initiated years after the invention was first developed.

Since a Law No.2008-561 dated 17 June 2008 Act regarding statutes of limitations, the action for the payment of the supplementary remuneration must be initiated within five years, by virtue of Article 2224 of the Civil Code

However, Article L.3245-1 of the Labour Code created by Law No. 2013-504 of 14 June 2013 mentions that '*the action in payment or in recovery of the salary*' is subject to a three-year limitation. This statute of limitation could apply with respect to the supplementary remuneration. The starting point of the statute of limitation is the moment when the inventor has the necessary elements for assessing the supplementary remuneration that they are entitled to receive.[49]

In case of a settlement agreement with an employee after a dispute, the supplementary remuneration should be expressly mentioned, so that the court can verify that it has been paid. If the formulation is too general, the court will consider that the settlement does not include the supplementary remuneration.[50]

For tax purposes the supplementary remuneration is considered as a salary.

4.4.2 Supplementary Remuneration in the Public Sector

The supplementary remuneration for agents of the public sector is subject to a specific regime. Like employees of the private sector, Article L.611-7 IPC is applicable to public servants, as mentioned explicitly in Article R.611-11 IPC. However, the supplementary remuneration of public servants is specifically determined in Article R.611-14-1 IPC: it consists of a bonus on the profits generated by the invention and a patent bonus. In case of co-inventors the payment is divided between them, in proportion of their contribution to the invention.

The bonus on the profits generated by the invention is paid annually, and is calculated on the basis of tax free profits generated by the invention each year for the entity, after deduction of all the direct costs borne by the public entity for the current year and the previous ones if they have not been already deduced before, and adjusted by a coefficient representing the contribution of the public servant to the invention. The bonus on the profits is equal to 50% of such amount, up to a threshold.[51] Beyond this threshold, the bonus equals 25% of the amount.

48. TGI Paris 9 March 2004 and 11 November 2008.
49. Cass Com. 22 February 2005, No 03-11.027; Cass Com. 12 June 2012, No 11-21.990. Cass. Soc. 26 January 2012, No 10-13825.
50. CA Bordeaux 24 February 2011, No 09/05759.
51. This threshold corresponds to a reference value of the annual salary of a public servant of a 'high category' (about EUR 68,660).

The patent bonus is a lump sum payment, paid in two parts. Its amount is decided by order of Secretary of State (currently EUR 3,000[52]). 20% of the sum is paid a year after the filing of an application for a patent, and the remaining 80% is paid at the execution of a licence agreement or an assignment agreement.

4.4.3 Fair Amount

Where the employee has made an invention outside their normal duties, they are entitled to a fair amount (*juste prix*), working as a form of remuneration. The fair amount is negotiated on an individual basis, and cannot be fixed in the employment contract. It should be calculated at the time where the employee exerts their attribution right. However, the rules for calculating this amount can be mentioned in the employment contract before the invention is made.

The parties are free to determine how the fair amount is calculated: it could be a lump sum or a percentage of the exploitation of the invention. It should be noted that even though in principle the employer does not have to patent the invention or to exploit it, case law considers that if the calculation of the fair amount is based on the turnover for the exploitation of the invention, the employer must exploit it.[53]

If the parties cannot reach any agreement, the fair amount will be determined by the Commission of Conciliation or by the Paris Court of First Instance (exclusive jurisdiction) that will take into consideration all the elements which may be supplied by the employer and the employee to determine the amount according to the initial contribution of both parties as well as the industrial and commercial use of the invention.

The action for payment of the fair amount must be initiated within five years, according to Article 2224 of the Civil Code. Since 2013, however, a shorter limitation period of two years may be applicable by virtue of Article L.1471-1 of the Labour Code, which states that '*any action concerning the execution or the termination of the employment contract is limited by two years starting from the day upon which the person bringing it knew, or ought to have known, the facts allowing him to exert his rights*'.

Yet, in a decision dated 18 December 2015, the First Instance Court of Paris held that: (i) the fair amount of a payment action could not be regarded as a salary debt since such fair amount does not compensate the work carried out by the employee, and that (ii) the statute of limitations should therefore be of five years[54].

The starting point of the limitation period to challenge the amount of the fair amount paid for the assignment of rights is the moment when the invention is assigned to the employer.

The fair amount should be assessed at the time when the employer requests the assignment of the invention (taking into account the exploitation prospects based upon

52. A first sum of EUR 600 is granted one year after the patent application and EUR 2,400 when a licence is granted.
53. Cass Com. 18 February 2004, No 02-17421.
54. TGI Paris 18 December 2015 No 14/08230, JurisData 2015-030479.

a business plan). Furthermore, in determining the price, the court can also take into account elements that follow from the assignment.

Recently, the Court of Cassation[55] held that '*if the fair amount is to be evaluated at the time when the employee exerts his attribution right, elements posterior to this day can be taken into account to confirm the appreciation of the perspectives of development of the invention*'.

For tax purposes, the fair amount is considered as a non-commercial benefit.

5 EMPLOYEES' UTILITY MODELS

On the European level, the Commission issued a proposal for a Directive on the protection of inventions by a utility model in 1996, but there was no harmonisation on that matter. The proposal did not contain any specific regulations concerning utility models made during the course of an employment relationship and the proposal was abandoned.

The IPC provides for the protection of a patentable invention by a certificate of utility (*certificat d'utilité*) in Article L.611-1 and L.611-2 IPC. The criteria are the same as for the grant of a patent, but there is no prior art search made by the INPI, and the time of protection is only six years from the filing date. In other words, the utility certificate is granted as filed.

The certificate of utility is subject to the specific regulation of employees' inventions discussed in section 4 above. In practice the certificate of utility is rarely used in France.

6 EMPLOYEES' DESIGN RIGHTS

6.1 The Legal Framework

Design Rights are protected by French law and the EU legislative framework. The French framework results from the transposition of EU Directive 98/71.

A French design can be granted for the appearance of a product (any industrial or artisanal object), or for parts of a product, characterised in particular by its lines, its outlines, its colours, shape, texture or materials. These characteristics can be those of the product itself or its ornamentation (Article L511-1 IPC).

The criteria for validity of design protection in France are:

(i) Novelty: no identical design should have been disclosed at the registration or priority date;
(ii) Individual character: a design must produce an overall impression on the informed observer (and not user) that differs from the overall impression produced on this observer by any other previous design;
(iii) The design must not be dictated solely by its technical function.

55. Cass Com. 9 July 2013, No 12-22.157.

Protection of designs is acquired through registration, as mentioned in Article L511-9 IPC.

Unlike the EU framework[56], French law does not provide protection for unregistered designs. Indeed, in accordance with the principle of the unity of art, a creation can be protected either by copyright (acquired without registration) and/or as a design. This is reaffirmed in Article L.513-2 IPC, in Article 96(2) of EU Regulation 6/2002, as well as in recital 8 and Article 17 of Directive 98/71. However, the criteria for the protection under those two IP rights are different, and they only overlap when the conditions for the two protections are fulfilled[57] (i.e. originality is not identical to individual character). Protection is often claimed on both design rights and copyright, or only on copyright, by employees but also by employers against third parties.

6.2 Ownership of Rights

In principle, the right belongs to the creator of the design (i.e. the designer) or their successor in title.

6.2.1 Ownership of a French Registered Design

Unlike EU Regulation 6/2002, French law does not include provisions concerning the ownership of a French design created by a designer in the course of an employment relationship. However, when the employer is the applicant of the design, they benefit from the presumption of Article L. 511-9 IPC[58] and will therefore not have to prove their ownership. Nonetheless, it is advisable for the employer to have the rights of their employees assigned to them, by including a specific clause in the employment contract. Such a clause should mention the right for the employer to register the creation as a design. Otherwise, in application of the unity of art theory, the employee could claim ownership of the registered designs on the basis of copyright.

6.2.2 Ownership of a Community Design and French Copyright

According to Article 14(3) of Regulation 6/2002:

> Where a design is developed by an employee in the execution of his duties or following the instructions given by his employer, the right to the Community design shall vest in the employer, unless otherwise agreed or specified under national law.

56. The unregistered Community Design is protected under certain conditions. Article 19(2) of Regulation 6/2002 of 12 December 2001 on Community designs specifies that: '*the contested use shall not be deemed to result from copying the protected design if it results from an independent work of creation by a designer who may be reasonably thought not to be familiar with the design made available to the public by the holder*'.
57. Cass Crim. 13 December 2011, No 10-80.623.
58. According to which '*The applicant for registration is, failing proof to the contrary, considered to be the beneficiary of this protection.*'

It seems that in such case there is no need for the transfer of right to the employer, who is considered the right holder *ab initio*.

The European Court of Justice has stated[59] that this rule, by virtue of being an exception, should be strictly interpreted and should thus not apply to designs that have been produced as a result of a commission, i.e. Article 14(3) is only applicable in case of an employment relationship. In the case of a commission contract, '*the right to the Community design vests in the designer, unless it has been assigned by way of contract to his successor in title*'. The rule applies '*unless otherwise agreed or specified under national law*'. The parties may therefore derogate from Article 14(3) contractually.

The question is whether this automatic transfer of design rights to the employer is contrary to French law. Indeed, French copyright law contains a rule (Article L.111-1 IPC) contrary to Article 14(3) since it provides that the mere existence of an employment contract does not imply any derogation to the ownership of intellectual property rights of the author. It has been debated whether this rule, specific to copyright, could also apply to registered designs because of to the unity of art. However, case law seems to consider that this rule is not applicable.[60]

In addition, Article 18 of the EU Regulation provides for the moral right of the designer to be cited as the designer. This is similar to the right of the inventor to be mentioned as the inventor in the patent application. However there is no corresponding provision under French design law.

6.3 The Parties' Duty to Inform

There is no specific duty to inform concerning a design created by an employee.

The previously mentioned general duty of loyalty of both parties in the execution of the employment relationship should also apply in this case.

6.4 Employee Remuneration Right

French design law does not provide for any specific remuneration. Unless otherwise stipulated by the contract, the employee is not being entitled to any remuneration for the registration of their work as a design.

As mentioned earlier, because of the unity of art theory, the design can also be protected by copyright, and the transfer of such a right should respect the conditions mentioned in the First Book of the IPC, and notably compensation (see section 2.4 above).

59. *FEIA v Cul de Sac Espacio Creativo SL and Acierta Product and Position SA* C-32/08 (2 July 2009).
60. TGI Paris 27 January 2011, No 09/15874.

7 EMPLOYEES' PLANT VARIETIES

Plant varieties can be protected in the European Union by a single title since the entry into force of Regulation 2100/94 of 27 July 1994 on Community plant variety rights. In 2014, France became the second European country behind the Netherlands for the registration of plant varieties.[61] The Regulation states in Article 11(4) that:

> ... if the breeder is an employee, the entitlement to the Community plant variety shall be determined in accordance with the national law applicable to the employment relationship in the context of which the variety was bred, or discovered and developed.

Plant varieties can also be protected under French law according to Articles L.623-1 to -44 IPC. The variety is protected if it is new (clearly different from all other varieties the existence of which, at the date of registration, is commonly known), homogeneous (sufficiently uniform in its relevant characteristics), and remains stable (identical to its initial definition after its reproduction). Protection is granted for 25 years from the grant, and for 30 years for specific plants listed in Article L.623-13 IPC.

Until 2011 there was no specific provision concerning employment relationships. Law No. 2011-1843 of 8 December 2011 concerning plant varieties certifies modified Article L. 623-24 IPC. This Article refers to the provisions concerning employees' inventions, which are now applicable to plant varieties as well (see section 4 above). Consequently, when a Community plant variety has been bred or discovered and developed by an employee *'in the context of the employment relationship'*, the same rules should apply.

8 INSTANCES FOR DISPUTES

8.1 The Commission of Conciliation

The National Commission on Employees' Inventions ('CNIS') is a panel seeking conciliation. It deals with cases regarding patentable inventions in application of Article L.611-7 IPC. The Commission sits at the INPI (or somewhere else in France if the circumstances require it).

A similar Commission was recently created for plant varieties. This Commission shall sit at the National Authority for Plant Varieties. Composition and operation of this Commission are dealt with in Article D.623-58-1 IPC implemented by Decree No. 2015-164 of 12 February 2015 which refers to the applicable rules concerning the CNIS.

The CNIS is presided over by a judicial magistrate, presently a judge from the Court of Cassation (Article L.615-21 IPC). The president is appointed for a renewable three-year period by ministerial order. The president is assisted by two assessors, one of which is chosen amongst the employees' unions, and the other one amongst the employers' unions. When the litigation pertains to an invention made by a public

61. http://www.cpvo.europa.eu/statistiques/AC_stats_October_2014.pdf.

servant, the assessors are chosen in a specific list comprising individuals listed upon proposition of the ministers and of organisations representative of the employees. A specific list is also established for inventions made by military agents (Article R.615-31 IPC). Once they are designated, each party can require that the assessors be changed for a serious and legitimate ground which is accepted by the president (Article R.615-17 IPC).

The CNIS has jurisdiction over all cases regarding the application of Article L. 611-7 IPC concerning the classification of the invention (invention made as part of the employee's normal duties, beyond normal duties and outside of the employment relationship) as well as to assess the amount of the financial remuneration (except if a branch collective bargaining agreement applies to the employment contract). However, it has no jurisdiction to judge inventorship or ownership claims (only the Paris Court of First Instance has jurisdiction over these claims).

Any party can institute proceedings before the CNIS, generally before doing it before the Court of First Instance, but also while a trial is ongoing. In such case, and if the CNIS has not already made a proposal of conciliation, the Court will stay the proceedings until the end of a six-month period. If the parties do not reach an agreement, only the proposal of the CNIS is communicated to the Court (Article R.615-27 and 28 IPC).

The request is signed by the plaintiff or its representative and indicates the name, first names, profession and address of the plaintiff and of the other parties, the subject matter of the litigation, the legal means and arguments of the plaintiff as well as all the exhibits that are useful. The request is notified to the other party which shall communicate its written observations on the merits (Article R.615-15 IPC).

The INPI communicates the elements in its possession that can be disclosed without prejudice of the rights of third parties (a report is made by the INPI notably regarding the patentability of the invention). A copy of this communication is sent to the parties.

The procedure is adversarial. The CNIS hears the parties and tries to reach conciliation. The parties can represent themselves or be represented. The president can order any measure of instruction and take note of a conciliation agreement between the parties or set another hearing.

As long as both parties are present at the hearing, there can be conciliation, total or partial. The record mentions the content of the agreement, as well as the contested points if the conciliation is not total. If the parties do not reach total agreement or if one of the parties is not present at the hearing, the CNIS establishes a proposal of conciliation within six months of the referral (but this time limit is not mandatory, and can be extended by the CNIS or with the agreement of the parties). The proposal is notified to the parties by the CNIS.

The proposal becomes final if no party challenges it before the Court of First Instance within a month of its notification (Article L.615-21 IPC). In the event proceedings were already ongoing before the Court of First Instance, the proposal may also be challenged within a month before this Court by filing written pleadings[62].

The agreement between the parties resulting from the proposal of the CNIS is made enforceable by decision of the president of the Paris Court of First Instance.

8.2 IP Courts

Civil litigation regarding intellectual property rights can only be brought before specific Courts of First Instance that have an exclusive jurisdiction. This means that even though the invention was made during the course of an employment relationship, the case will not be heard by the Labour Court, i.e. *'Conseil des Prud'hommes'* (except for the supplementary remuneration for the benefit of the employee if a branch bargaining agreement applies to the employment contract). As a general rule, the Court of First Instance issues a decision on the merits within 12 to 18 months from the summons for this type of matter.

Recourse to arbitration courts is possible, but it should be mentioned that it is very limited regarding employment relationships[63]. Indeed, the Court of Cassation has held that an arbitration clause in an employment contract cannot be enforced against employees since the employee's right to access the Labour Courts is deemed to be a public policy matter.

In respect of patents, utility certificates and supplementary protection certificates, Article L. 615-17 IPC and D. 211-6 of the Judicial Organisation Code (since the Decree No. 2009-1205 of 9 October 2009), hold that only the Paris Court of First Instance is competent. This is also applicable to topographies of semiconductor chips (Article L.622-7 IPC).

Specific Courts of First Instance also have exclusive jurisdiction over plant varieties litigation. The territorial jurisdiction is defined in an Annex to Article D.211-5 of the Judicial Organisation Code:

Table 9.2

Seat	Jurisdiction
Aix-en-Provence Court of Appeal	
Marseille	Jurisdiction of the Courts of Appeal of Aix-en-Provence, Bastia and Nîmes.
Bordeaux Court of Appeal	
Bordeaux	Jurisdiction of the Courts of Appeal of Agen, Bordeaux and Poitiers.
Colmar Court of Appeal	
Strasbourg	Jurisdiction of the Courts of Appeal of Colmar and Metz.
Douai Court of Appeal	

62. CA Paris, 15 November 2012, 11/22902.
63. Cass Soc., 30 November 2011, No 11-12.905; see also the recent Proposal for a Labour Arbitration Center.

Seat	Jurisdiction
Lille	Jurisdiction of the Courts of Appeal of Amiens and Douai.
Limoges Court of Appeal	
Limoges	Jurisdiction of the Courts of Appeal of Bourges, Limoges and Riom.
Lyon Court of Appeal	
Lyon	Jurisdiction of the Courts of Appeal of Chambéry, Grenoble and Lyon.
Nancy Court of Appeal	
Nancy	Jurisdiction of the Courts of Appeal of Angers, Caen and Rennes.
Paris Court of Appeal	
Paris	Jurisdiction of the Courts of Appeal of Basse-Terre, Cayenne, Fort-de-France, Nouméa, Orléans, Papeete, Paris, Reims, Rouen, Saint-Denis and Versailles, and of the Superior Tribunal of Appeal of Saint-Pierre
Rennes Court of Appeal	
Rennes	Jurisdiction of the Courts of Appeal of Angers, Caen and Rennes.
Toulouse Court of Appeal	
Toulouse	Jurisdiction of the Courts of Appeal of Montpellier, Pau and Toulouse.

As for French designs, Article L.521-3-1 IPC mentions the exclusive jurisdiction of specific Courts of First Instance. This is also applicable to copyright (Article L.331-1 IPC).

The territorially competent courts are listed in Annex VI of Article D. 211-6-1 of the Judicial Organisation Code:

Table 9.3

Seat	Jurisdiction
Bordeaux	Jurisdiction of the Courts of Appeal of Agen, Bordeaux, Limoges, Pau and Toulouse.
Lille	Jurisdiction of the Courts of Appeal of Amiens, Douai, Reims and Rouen.
Lyon	Jurisdiction of the Courts of Appeal of Chambéry, Grenoble, Lyon and Riom.
Marseille	Jurisdiction of the Courts of Appeal of Aix-en-Provence, Bastia, Montpellier and Nîmes.
Nanterre	Jurisdiction of the Court of Appeal of Versailles.
Nancy	Jurisdiction of the Courts of Appeal of Besançon, Dijon, Metz and Nancy.
Paris	Jurisdiction of the Courts of Appeal of Bourges, Paris, Orléans, Nouméa, Papeete, Saint-Denis and of the Superior Tribunal of Saint-Pierre.
Rennes	Jurisdiction of the Courts of Appeal of Angers, Caen, Poitiers and Rennes.
Strasbourg	Jurisdiction of the Court of Appeal of Colmar.

Seat	Jurisdiction
Fort-de-France	Jurisdiction of the Courts of Appeal of Basse-Terre, Cayenne and Fort-de-France.

As for Community designs, Article L.522-2 IPC and R.211-7 of the Judiciary Organisation Code provide for the exclusive jurisdiction of Paris Court of First Instance.

The decision of the Court of First Instance can be appealed before the Court of Appeal in law and in fact. Additionally, the decision of the Court of Appeal can be challenged in front of the Court of Cassation, but only in law.

8.3 Labour Court

The French Labour Courts ('*Conseil des Prud'hommes*') are composed of elected members, representative of employers and employees (Article L.1421-1 Labour Code).

They have exclusive jurisdiction over employment relationships in general (Article L.1411-4 Labour Code), and can therefore decide upon litigation pertaining to creations made by employees, but only if outside the exclusive jurisdiction of the Courts of First Instance. This is for instance the case when an employee was dismissed because they did not notify the employer of their invention, or because they refused to transfer their creation to their employer.

9 CONCLUDING REMARKS

After studying the different applicable provisions concerning employees' intellectual property Rights under French law, the lack of homogeneity can be highlighted. The multiplicity of provisions leads to uncertainty for practitioners and to conflicting situations that can transpire years after the creation of the protected work.

The present provisions do not seem to take into consideration the existing creation process where creators are often employees. Moreover, the absence of precise provisions concerning the remuneration assessment makes it highly unpredictable and leads to great variations.

The provisions of the IPC are quite old (1957 for copyright and 1978 for inventions). Due to the increasing number of cases, we hope that French law will be modified in the coming years to provide harmonised rules regarding the ownership and compensation of employees' creations.

BIBLIOGRAPHY

Table of Sources

M. Abello & J. Tassi, *Employees' creations* (Créations de salariés), No. 613 (LexisNexis 2017).

M. Abello & J. Tassi, *Litigation on employees' inventions* (Contentieux des inventions de salariés), (LexisNexis 2016).
M. Abello & J. Tassi, *The employee author of a software and fundamental rights: a priority preliminary ruling on constitutionality for the future?* (L'employé auteur de logiciel et les droits fondamentaux: une question prioritaire de constitutionnalité d'avenir ?), Revue Communication Commerce électronique No.4 (LexisNexis April 2011 Etude 8).
N. Bouche, *Community Protection of Plants Varieties* (Protection communautaire des obtentions végétales), JurisClasseur No. 1780 (LexisNexis 2014).
P. de Cande, *Concerning the ownership of a non-registered community design in employee/employer relationships: towards a more comprehensive application* (De la titularité du dessin ou modèle non communautaire non enregistré dans les relations employeur/employé: vers un application plus compréhensive de l'article 14-3 du règlement (CE) No. 6/2002, Revue Communication Commerce électronique No. 6 – June 2014, Etude 12).
J.-F. Debarnot, *Contractual practice: Concerning the remuneration of an interpreter hired for the realisation of an audiovisual work* (Pratique contractuelle - De la rémunération de l'artiste-interprète engagé pour la réalisation d'une œuvre audiovisuelle), Revue Communication Commerce électronique No. 2 (February 2015, 2).
J. Frantz et al., *Employees' creations and inventions Breaking the stereotypes* (Créations et inventions de salariés Rompre avec les schémas reçus), Colloque de l'IRPI, Droit des affaires, *Propriété Intellectuelle* No. 35 (LexisNexis 2010).
Proposal for a European Parliament and Council Directive approximating the legal arrangements for the protection of inventions by utility model 1996 Commission of the European Communities.
La rémunération des inventions de salariés *(The remuneration of employees' inventions)* - 2016, INPI - Observatoire de la Propriété Intellectuelle.
Statistics Meeting of the Administrative Council - 09/10/2014 Community Plant Variety Office.

Table of Legislation

Directive 2009/24/EC on the legal protection of computer programs.
Directive 98/71/EC on the Legal Protection of Designs.
Directive 87/54/EEC of 16 December 1986.
Regulation 6/2002/EC on Legal Community Design.
Rome I EC Regulation No. 593/2008.
Regulation 2100/94 of July 1994 on Community plant variety rights.
Civil Code.
Intellectual Property Code.
Judicial Organisation Code.
Labor's Code.
Law No. 2015-990 dated 6 August 2015
Law No. 2013-504 of the 14 June 2013.

Law No. 2011-1843 of 8 December 2011.
Law No.2008-561 dated 17 June 2008
Law No. 90-1052 of 26 November 1990.
Decree No. 2015-164 of 12 February 2015.
Decree No. 2009-1205 of 9 October 2009.

Table of Cases

ECJ 2 July 2009 C-32/08.
ECJ 9 November 2004 C-302/02
Cass. Com. 3 March 2015, No. 14-10.568.
Cass. Soc. 15 January 2015, No. 13-14.811.
Cass. Civ.1, 15 January 2015, No. 13-23.566.
Cass. Com. 9 December 2014, No. 13-16.559.
Cass. Com., 9 July 2013, No. 12-22.157.
Cass. Com., 12 February 2013, No. 12-12.898.
Cass. Com., 12 June 2012, No. 11-21.990.
Cass. Soc. 26 January 2012, No 10-13825.
Cass. Crim. 13 December 2011, No. 10-80.623.
Cass. Soc. 21 September 2011, No. 09-69927.
Cass. Soc. 18 January 2011, No. 09-41636.
Cass Civ., 5 March 2009 N°07-19.734
Cass. Civ. 1, 14 February 2008, No. 07-12176.
Cass. Soc. 30 November 2011, No. 11-12.905.
Cass. Civ.1, 21 November 2006, No. 05-19294.
Cass. Com. 25 April 2006, No. 04-19482.
Cass. Com. 22 February 2005 No. 03-11.027.
Cass. Com. 22 February 2005, No. 02-18.790.
Cass. Com. 18 February 2004, No. 02-17421.
Cass. Com. 18 February 2004, No. 02-17421.
Cass. Civ.1, 6 March 2001, No. 98-15502.
Cass. Com. 21 November 2000, No. 98-11.900.
Cass. Soc. 10 February 1998, No. 95-43.510.
Cass. Civ.1, 13 October 1993, No. 91-11241.
Cass. Soc. 25 January 1990, No. 86-10188.
Cass. Com. 21 June 1988, No. 86-19166.
Cass. Ass. Plen. 7 March 1986, No. 85-91465.
Cass. Civ.1, 12 April 1976, No. 74-12149.
CE 21 November 1972, Ofratem, N. 309.721.
CA Bordeaux 13 March 2017, No. 16/00887.
CA Lyon 10 June 2016, No. 15/01025
CA Paris 30 October 2015, No. 13/05568.
CA Paris 30 June 2015, No. 13/10097.
CA Rennes 1 July 2014, No. 12/06459.

CA Paris 10 November 2014, No. 14/00024.
CA Paris 30 January 2013, No. 11/0036.
CA Paris, 15 November 2012, No. 11/22902.
CA Bordeaux 24 February 2011, No. 09/05759.
CA Reims 11 October 2010, No. 09/02151.
CA Paris 25 November 2005, No. 04/02005.
TGI Paris, 18 December 2015, No. 14/08230.
TGI Paris 24 October 2014, No. 13/05568.
TGI Paris 18 April 2013, No. 11/00543.
TGI Paris 27 January 2011, No. 09/15874.
TGI Paris 7 November 2009, No. 2007/02765.
TGI Paris 19 May 2009, No. 2007/07300.
TGI Paris 10 November 2008, No. 07/06280.
TGI Paris 9 March 2004, No. 02/09023.
TGI Paris 30 November 1999, No. 98/16766.

CHAPTER 10

Germany

Jan Bernd Nordemann & Christian Czychowski[*]

1 BACKGROUND

The legal relationship between employers and employees in the area of intellectual property rights is in Germany not entirely regulated by specific intellectual property law provisions. While for some matters, such as the allocation of rights, special intellectual property law provisions exist,[1] others, such as the limitation of claims,[2] are regulated through general law provisions or remain subject to contractual agreements between the parties.[3] Even where specific intellectual property law provisions exist, there are no umbrella provisions covering all intellectual property rights. Rather, each type of intellectual property right attracts its own specific intellectual property law provisions governing the regulation solely of the respective right in employment relationships.

* The authors would like to express special thanks to Ms Greta Harnisch and Ms Stefanie Jehle, Berlin, for their helpful contributions to the chapter on Germany.

1. Cf. *infra* '2.2 Ownership of Rights and Termination' within the different subs.
2. Cf. *infra* '8. Instances for Disputes'.
3. Cf. Rother, *Group Report Germany for the AIPPI Working Committee No. 183: Employers' Rights to Intellectual Property* (in the German and English language), p. 10, available at: https://www.aippi.org/download/commitees/183/GR183germany.pdf (accessed 16 May 2017).

2 EMPLOYEES' COPYRIGHT

2.1 The Legal Framework

In Germany, protection of copyrightable works is regulated by the Act on Copyright and Related Rights ('UrhG').[4] According to the very first Section of the Act, the author of the work enjoys protection for their work in accordance with the legal provisions.[5] With regard to the definition of the term 'author', the core principle of the Act states that it is – invariably – the creator of a work who is considered its author – the so-called *authorship principle* of Section 7 UrhG. The authorship principle applies – without exception – to all works, including employee-created ones.

Protection under copyright arises automatically with the creation of a work. No formal requirement, such as the work's registration, exists. Neither is the work's publication a necessary precondition for its protection under the UrhG.[6]

With regard to employment relationships, in the absence of any European copyright directive, the German UrhG provides one core provision on the allocation of rights in employee-created works, namely Section 43 UrhG:

> The provisions of this [subchapter] shall also apply where the author has created the work in the fulfilment of obligations resulting from an employment or service relationship, unless otherwise provided in accordance with the terms or nature of the employment or service relationship.

Section 43 UrhG applies to all types of works, whereas Section 69b UrhG constitutes a special regulation for employee-created software. Provisions of the Act on Employees' Inventions ('ArbNErfG')[7] cannot be applied to copyright matters – neither directly nor by way of an analogy. Copyright matters not regulated by the UrhG or other applicable statutory provisions may be freely negotiated in individual contractual arrangements between the parties.

2.2 Ownership of Rights

Because of the authorship principle, all rights originally belong to the employee as the work's author, irrespective of whether or not creation took place during the course of

4. *Act on Copyright and Related Rights of 9 September 1965 as last amended by Art. 8 of the Act of 1 October 2013 ('Urhebergesetz', 'UrhG')*, Official English version available at: http://www.gesetze-im-internet.de/englisch_urhg/index.html (accessed 16 May 2017).
5. Cf. s. 1 UrhG.
6. Axel Nordemann, *Urheberrecht: Kommentar zum Urheberrechtsgesetz, zum Verlagsgesetz und zum Urheberrechtswahrnehmungsgesetz (Authors law: Commentary on the Act on Copyright and Related Rights, on the German Publishing Act and the Copyright Administration Act)*, Introduction UrhG, marginal n. 17 (Fromm/Nordemann, 11th ed., Kohlhammer, 2014).
7. For details on the provisions of the Act on Employees' Inventions cf. *infra* '4. Employees' Inventions'.

employment.[8] Therefore, Section 7 UrhG precludes legal constructions, according to which rights in employee-created works initially vest in the employer. Under German copyright law, authors' rights may only originate in natural persons creating the work. They cannot initially vest in non-natural persons having commissioned the work.

What is more, authors' rights as such are not assignable according to Section 29 (1) UrhG.[9] Thus, a change in ownership of authors' rights is not possible. With regard to employment relationships therefore, presuming a transfer of ownership of rights in employee-created works in favour of the employer does not constitute a possible solution. Rather, the employee is always considered the work's author and the original rights owner.

Although ownership as such cannot be transferred, a transfer of economic rights in the works remains possible. This is where Section 43 UrhG steps in: for works created on behalf and for the account of the employer, with no deviating contractual regulations existing, Section 43 UrhG presumes a transfer of the employee's economic rights in favour of the employer.[10] Although at first sight its wording[11] implies the exact opposite, Section 43 UrhG generally awards the employer all economic rights in employee-created works.[12]

2.2.1 Transfer of Economic Rights in Employees' Works

For the presumption of Section 43 UrhG to apply, first, the author must be employed by way of either a private employment contract in the sense of Section 611 of the German Civil Code[13] ('BGB')[14] or a public sector employment contract within the meaning of Sections 3(1) and 8(1) of the Act on the Regulation of the Officials' Status in the Federal States ('BeamtStG').[15] Secondly, the created work must be a product of the employee's work obligations. Regarding the scope of these work obligations, it is not authoritative whether or not the employment contract expressly commits the employee to creative operations. Nor is the potential use of the employer's resources for the work's creation,

8. Dreier, *Urheberrechtsgesetz, Urheberwahrnehmungsgesetz, Kunsturhebergesetz: Kommentar (Act on Copyright and Related Rights, Copyright Administration Act, Act on Copyright for Works of Art: Commentary)* § 43 UrhG marginal n. 1 (Dreier/Schulze, 5th ed., C.H. Beck, 2015).
9. The preclusion of transferring ownership of author's rights as such, as codified in s. 29(1) UrhG, is an inevitable result of the monistic theory, upon which German author law is based.
10. Cf. Dreier, *supra* n. 8, § 43 UrhG marginal n. 1.
11. Cf. *supra* '2. Employees' Copyright' section 2.1. 'Employees' Copyright: The Legal Framework'.
12. Lindhorst, *Beck'scher Online-Kommentar: Urheberrecht (Beck Online-Commentary: Copyright)*, § 43 UrhG marginal n. 2 (Möhring/Nicolini, 4th ed., C.H. Beck Munich, 2014); Axel Nordemann, *supra* n. 6, § 43 UrhG marginal ns. 1, 27.
13. *German Civil Code* ('Bürgerlicher Gesetzbuch', 'BGB'), available in the English language at: http://www.gesetze-im-internet.de/englisch_bgb/index.html (accessed 16 May 2017).
14. Axel Nordemann, *supra* n. 6, § 43 UrhG marginal ns. 9 et seq.
15. Cf. Dreier, *supra* n. 8, § 43 UrhG marginal n. 7; Axel Nordemann, *supra* n. 6, § 43 UrhG marginal n. 11; Rojahn, *Urheberrecht Kommentar (Copyright Law: Commentary)*, § 43 UrhG marginal ns. 10, 19 et seq. (Schicker/Loewenheim, 5th ed., C.H. Beck Munich, 2017); Wandtke, *Praxiskommentar zum Urheberrecht (Copyright Law: Commentary)*, § 43 marginal n. 13 (Wandtke and Bullinger, 4th ed., C.H. Beck Munich, 2014); Axel Nordemann, *Handbuch des Urheberrechts (Manual on Copyright Law)*, § 63 marginal n. 10 (Loewenheim, 2nd ed., C.H. Beck Munich, 2010).

the time or place creation took place. In particular, works that have been created by the employee at home or at night may come within the provision's scope of application.[16] Ultimately the inner will of the employee not to act on behalf and for the account of the employer is irrelevant.

The one relevant criterion is, whether the work has been created within the scope of the employee's duties or field of work, which requires an inner connection between the employee's contractual work obligations and the creative process.[17] Additionally, from a temporal angle, the work must have been created during the validity of the employment.

2.2.2 Inner Connection: Private Sector

Regarding the creative fields of industry within the private sector, two scenarios should be distinguished: for employees with a creative function, such as photographers, journalists, designers, architects, etc., the inner connection between their employment obligations and their creation may be reasonably assumed in general. Whereas in the case of employees not entrusted with creative tasks, such as stage workers, telephone operators, or IT technicians, who yet become creatively active, an inner connection will normally be missing, although it might be established in individual cases nonetheless.[18]

For the non-creative private sector the existence of an inner connection between the employee's work obligations and the creative process is more difficult to establish and depends on the precise task the employee has been entrusted with. Does this task comprise elements of creativity wherefore the employee is obliged to become creatively active? Does their obligation to become creatively active correspond with the process of creation in question? In this case an inner connection must be affirmed and a transfer of economic rights presumed according to Section 43 UrhG.

16. *Museumsführer, 3 U 3053/96*, ZUM 1999, 656, 657 (Higher Regional Court (Oberlandesgericht, 'OLG') Nuremberg, 18 February 1997); *Softwareentwickler im Arbeitsverhältnis, 16 O 136/96*, ZUM 1998, 167 (Kammergericht, 'KG') 3 July 1996); Wandtke, *Reform des Arbeitnehmerurheberrechts? (Reforming Employees' Copyright?)*, GRUR 1999, 390, 391 et seq. (1999); Axel Nordemann, *supra* n. 15, § 63 marginal n. 14; Rojahn, *supra* n. 15, § 43 UrhG marginal n. 21; Wandtke, *supra* n. 15, § 43 UrhG marginal n. 20.
17. *Wetterführungspläne, X ZR 72/98*, GRUR 2001, 155, 157 (Federal Court of Justice (Bundesgerichtshof, 'BGH') 24 October 2000); *TESY-M2/Schutzfähigkeit eines Computerprogramms – Vergütungsansprüche eines Beamten der Deutschen Bundespost, 29 U 2437/97*, ZUM-RD 2000, 8, 12 (OLG Munich, 25 November 1999); *Softwareentwickler im Arbeitsverhältnis, supra* n. 16, 167; *POLDOK/Urheber- und Nutzungsrechte eines Hochschullehrers, 5 U 2189/93*, ZUM-RD 1997, 175, 178 et seq. (KG, 6 September 1994); Axel Nordemann, *supra* n. 6, § 43 UrhG marginal n. 14; Wandtke, *supra* n. 16, 392; Axel Nordemann, *supra* n. 15, § 63 marginal n. 14.
18. Cf. Axel Nordemann, *supra* n. 6, § 43 UrhG marginal n. 15.

2.2.3 *Inner Connection: Public Sector*

When it comes to public sector employment contracts, a differentiation must be made between university matters and the remaining public sector. While officials employed in the public sector generally come within the scope of Section 43 UrhG, this is not true for matters involving universities.[19]

Outside the context of universities, it is the official's function and their duties that will determine whether an inner connection to the creative process exists. With regard to defining the official's public duties, public law may clarify these in some cases. In other cases though, where legal provisions are lacking, the tasks usually performed by the official will be determinant.

For scientific and artistic works of university professors and academic staff on the other hand, an inner connection is generally denied as works are presumed to originate from the academic's private sphere. This presumption is based on the assumption that university employees do not engage in researching, publishing and other creative activities because they are obligated to do so by their public duties, but on their own authority and responsibility.[20] This is even true for works that have been created during hours of public service and by means of the university's human and material resources. Only if the university employee's duties comprise an obligation of operating creatively, are the economic rights in such works transferred to the university. In all other cases though, rights in works by university employees remain with the latter.

2.2.4 *Scope of Transfer*

Once the requirements for a transfer pursuant to Section 43 UrhG are met, the precise extent of such transfer has to be determined. The scope of transfer, i.e. which economic rights have been transferred and to which extent, may vary. It may be the case, that the scope is expressly defined by specific contractual provisions. Otherwise, if the employment contract is silent on this subject, the scope has to be determined according to the 'Specific Purpose Principle' of Section 31(5) UrhG:

> If the types of exploitation have not been specifically designated when an exploitation right (economic right) was granted, the types of use to which the right extends shall be determined in accordance with the purpose envisaged by both parties to the contract.

According to the Specific Purpose Principle therefore, the scope of transfer is confined to those rights reasonably necessary to achieve the purpose for which the

19. Cf. Axel Nordemann, *supra* n. 6, § 43 UrhG marginal ns. 17, 21 et seqq.; Rojahn, *supra* n. 15, § 43 UrhG marginal ns. 126 et seq.
20. *Grabungsmaterialien, I ZR 244/88*, GRUR 1991, 523, 525 (BGH, 27 September 1990); Heermann, *Der Schutzumfang von Sprachwerken der Wissenschaft und die urheberrechtliche Stellung von Hochschulangehörigen (Scope of Protection for Literary Scientific Works and Academic Staff's Status with Regard to Copyright)*, GRUR 1999, 468, 472, 475 (1999); Axel Nordemann, *supra* n. 6, § 43 UrhG marginal n. 21; Wandtke, *supra* n. 15, § 43 UrhG marginal n. 40; Axel Nordemann, *supra* n. 15, § 63 marginal n. 21.

rights are transferred.[21] With regard to employment relationships, it is the object of the employer's business that constitutes the specific purpose. Therefore, the employer will receive only those rights necessary for the work's exploitation within business operations in order to achieve the object of the business. The object of the business will not only determine the type of rights transferred, but also the rights' characteristics in a territorial and temporal manner (Section 31(1) sentence 2 UrhG).

2.2.5 Transfer of Rights Beyond the Object of the Business

Employers should keep in mind that all rights beyond the specific purpose, i.e. the object of the business, have to be acquired by means of an individual contract with the employee. Contractual clauses precisely listing the rights to be transferred as well as the latter's scope in terms of content, territory and time, are essential.[22] Generally speaking, the UrhG does not contain any formal requirements as to the contract by which rights are transferred. Only insofar as the contract concerns rights in respect of unknown types of exploitation does it have to be in writing in order to be valid (Section 31a(1) UrhG).[23] According to Section 31a(4) UrhG this formal requirement is indispensable. Whether or not this also applies unrestrictedly to employment contracts is hotly debated. According to Section 2(1) of the Act on Documentation of Employment Contracts (NachwG) the employer is obliged to provide the essential terms of contract in writing within one month after beginning of employment. In order to fulfil this duty imposed by the NachwG, as well as for reasons of proof, employers are therefore well advised to lay down the essential terms of contract, including provisions on the transfer of rights, in writing.

2.2.6 Moral Rights

Although moral rights are not transferable as such, certain restrictions in this regard must be made for employment relationships. The employer compensates the employee for works to be created by the employee and to be later on exploited by the employer on their own authority and responsibility. Due to the very purpose of the employment relationship it is therefore self-evident that the employee cannot maintain and enforce their moral rights to the fullest extent.[24] Thus, it needs to be determined, which moral

21. *Pauschale Rechtseinräumung, I ZR 215/93*, GRUR 1996, 121, 122 (BGH, 27 September 1995); *Comic-Übersetzungen I, I ZR 189/95*, GRUR 1998, 680, 682 (BGH, 20 January 1998); *Honorarbedingungen Freie Journalisten, I ZR 73/10*, BeckRS 2012, 15227, marginal n. 15 (BGH, 31 May 2012) ; Rojahn, *supra* n. 15, § 43 UrhG marginal n. 48.
22. Jan Bernd Nordemann, *Urheberrecht: Kommentar zum Urheberrechtsgesetz, zum Verlagsgesetz und zum Urheberrechtswahrnehmungsgesetz (Authors law: Commentary on the Act on Copyright and Related Rights, on the German Publishing Act and the Copyright Administration Act)*, § 31 UrhG marginal n. 178a (Fromm/Nordemann, 11th ed., Kohlhammer, 2014).
23. For further details, cf. Jan Bernd Nordemann, *supra* n. 22, § 31a UrhG marginal n. 18.
24. Cf. *Schadensersatz im Wege der Lizenzanalogie wegen unberechtigter Verwendung von Filmbeschreibungen, 28 O 250/09*, ZUM 2010, 369, 370 (Regional Court (Landgericht, 'LG') Cologne, 23 September 2009); Dreier, *supra* n. 8, § 43 UrhG marginal n. 34; Axel Nordemann, *supra* n. 6,

rights remain with the employee, which moral rights are to be exercised by the employer, and which moral rights are restricted or waived altogether.

Usually, the parties will implicitly agree that the right of publication[25] is to be solely exercised by the employer. The employer bears the entire entrepreneurial and financial risk by pre-financing the work's creation, which is why it must be up to them to decide whether and how the work shall be published.[26]

With regard to the author's paternity right codified in Section 13 UrhG a distinction must be made. As a starting point, the author's paternity right, i.e. the right to be identified as the work's author,[27] is an inalienable right, therefore in principle, its assertion cannot be transferred to the employer.[28] As an exception to the rule however, the employee's waiver of their paternity right will be legitimate where the specific nature of the type of employment requires it.[29] The employee shall be permitted to contractually commit themselves not to enforce their paternity right – and thereby waive their right to some extent. The most common example is the ghost-writer's contractual waiver of their right to be identified as the work's author,[30] but also architects may agree to waive their paternity right.[31]

Contrary to the author's paternity right, the right to determine whether the work shall bear any designation of authorship at all[32] is commonly considered to be fully waivable.[33] However, the employee has to expressly waive their right of designation. Otherwise, the employer's obligation to have the work bear the designation of the employee's authorship depends on the industry's customs in that regard and has to be established on an individual case-by-case basis.[34]

Generally, a licensee's economic rights – in particular the adaptation right – are furthermore limited by the author's right of integrity codified in Section 14 UrhG. The right of integrity entitles the author to object to derogatory treatment. Thus, where the adaptation of a work by a licensee amounts to derogatory treatment, the author may stop the licensee from adapting the work. The very basic core of this moral right cannot

§ 43 UrhG marginal n. 52; Wandtke, *supra* n. 15, § 43 UrhG marginal n. 84; Axel Nordemann, *supra* n. 15, § 63 marginal n. 58; Rojahn, *supra* n. 15, § 43 UrhG marginal ns. 73 et seq.

25. S. 12 UrhG.
26. *Schadensersatz im Wege der Lizenzanalogie wegen unberechtigter Verwendung von Filmbeschreibungen*, cf. *supra* n. 24; Dreier, *supra* n. 8, § 43 UrhG marginal n. 35; Axel Nordemann, *supra* n. 6, § 43 UrhG marginal n. 53; Rojahn, *supra* n. 15, § 43 UrhG marginal ns. 73 et seqq.; Wandtke, *supra* n. 15, § 43 UrhG marginal n. 87; Axel Nordemann, *supra* n. 15, § 63 marginal n. 59.
27. Cf. s. 13 Sentence 1 UrhG.
28. Axel Nordemann, *supra* n. 6, § 43 UrhG marginal n. 54.
29. Dietz/Peukert, *Urheberrecht Kommentar (Copyright Law: Commentary)*, § 13 UrhG marginal ns. 26 et seq. (Schicker/Loewenheim, 5th ed., C.H. Beck Munich, 2017); Rojahn, *supra* n. 15, § 43 UrhG marginal n. 77; Wandtke, *supra* n. 15, § 43 UrhG marginal n. 88; Bayreuther, *Münchener Handbuch zum Arbeitsrecht (Munich Manual on Employment Law)*, § 91 marginal n. 16 (Richardi/Wlotzke/Wißmann/Oetker, 3rd ed., C.H. Beck Munich, 2009).
30. Cf. Groh, *Mit fremden Federn: Zur Wirksamkeit von Ghostwritervereinbarungen (Validity of Ghost-writer Agreements)*, GRUR 2012, 870(2012); Rojahn, *supra* n. 15, § 43 UrhG marginal n. 77.
31. *Namensnennungsrecht des Architekten, I ZR 3/92*, GRUR 1995, 671 (BGH, 16 July 1994).
32. Cf. s. 13 Sentence 2 UrhG.
33. Axel Nordemann, *supra* n. 6, § 43 UrhG marginal n. 54.
34. Axel Nordemann, *supra* n. 6, § 43 UrhG marginal n. 54 with further references.

be subject to a contractual waiver.[35] Apart from the moral right's core however, contractual restrictions are very well possible and common in employment relationships. It remains a question of the industry's customs if an implied waiver of the integrity right may be read into the employment contract, but in many cases this will be permissible. With regard to buildings, a conflict of interests often arises between the builder invoking their right of adaptation and the architect invoking their right of integrity. In this case, contractual arrangements as to restrictions on the right of integrity determining concurrently the limits of the adaptation right are essential, although such restrictions on the right of integrity must be precisely defined and limited.[36]

Through the restrictions imposed on the employee's moral rights, the employer is allowed to exercise their acquired economic rights in a more extensive manner. However, the legitimacy and limits of such restrictions and waiver clauses are highly controversial.[37] Yet it is commonly agreed upon, that the very core of the employee's moral rights, where important fundamental personal interests are affected, remains inalienable and cannot be further restricted or waived.

2.2.7 Free Works: Employee's Duty to Offer

Works that have been created by the employee but do not constitute a product of the employee's work obligations amount to so-called *free works*.[38] These works have not been created in exercise of the employee's employment obligations, i.e. their creation process lacks an inner connection to the employee's work obligations. Consequently, free works do not come within the scope of Section 43 UrhG, therefore rights in them generally remain with the employee. Thus, for the employer to acquire rights in free works, an individual contractual arrangement with the employee is required. In this regard, it is hotly debated whether or not the employee is obligated to offer to their employer a transfer of economic rights in free works.[39]

For free works that do not show any connection to the employer's business and are therefore neither exploitable within the employer's business nor capable of

35. *Das zweite Mal, I ZR 32/69*, GRUR 1971, 269, 271 (BGH, 27 November 1970); Honschek, *Der Schutz des Urhebers vor Änderungen und Entstellungen durch den Eigentümer (Protecting the Author against Alterations and Derogatory Treatment by the Owner)*, GRUR 2007, 944, 945 (2007); Dreier, *supra* n. 8, § 43 UrhG marginal n. 37; Axel Nordemann, *supra* n. 6, § 43 UrhG marginal n. 55; Rojahn, *supra* n. 15, § 43 UrhG marginal n. 83; Bayreuther, *supra* n. 29, § 91 marginal n. 17; Wandtke, *supra* n. 15, § 43 UrhG marginal n. 99; Axel Nordemann, *supra* n. 15, § 63 marginal n. 62.
36. *Architektenvertrag, 13 U 12/85*, BauR 1986, 601 (OLG Celle, 9 October 1985); *Unendliche Geschichte, U 2114/85*, GRUR 1986, 460, 463 (OLG Munich, 1 August 1985).
37. Cf. Groh, *supra* n. 30, 870 et seqq.
38. Wandtke, *supra* n. 15, § 43 UrhG marginal n. 22.
39. Dreier, *supra* n. 8, § 43 UrhG marginal ns. 23-29; Axel Nordemann, *supra* n. 6, § 43 UrhG marginal ns. 23 et seq.; Wandtke, *supra* n. 15, § 43 UrhG marginal ns. 30 et seqq.; Wiebe, *Recht der elektronischen Medien: Kommentar (Law of Electronic Media: Commentary)*, § 43 UrhG marginal ns. 5 et seq. (Spindler/Schuster, 2nd ed., C.H. Beck Munich, 2011); Rojahn, *supra* n. 15, § 43 UrhG marginal ns. 100 et seq.

competing with it[40] a duty to offer is commonly denied. On the other hand though, no consensus has yet been reached in respect of 'voluntary works', i.e. free works that are either exploitable within or capable of competing with the employer's business.[41] Some support a duty to offer for voluntary works created by means of the employer's resources or during working hours,[42] while others deny an obligation to offer altogether.[43] The question is yet to be decided by the courts.

With regard to works by university employees though, it is commonly agreed that such employees are not obliged to offer to license economic rights in their works to the university. By reason of the Freedom of Sciences as constitutionally guaranteed by Article 5(3) Basic Law for the Federal Republic of Germany[44] (GG) such duty does not arise even if creation took place using the university's resources or during working hours.[45]

2.2.8 Termination

The Berne Convention for the Protection of Literary and Artistic Works[46] (Berne Convention) obliges its members to provide for a term of protection of at least 50 years *post mortem auctoris* (*p.m.a.*), i.e. 50 years after the author's death.[47] Germany however, is bound by the European Copyright Duration Directive,[48] which foresees a minimum and maximum duration of protection of 70 years p.m.a.[49] Member States have to concur with this requirement and may neither allow a term of protection below or above this threshold. Consequently, protection of copyright-protected works under German law – which is automatically effective with the work's creation – lasts until 70

40. So-called *spare time works* according to Axel Nordemann, *supra* n. 6, § 43 UrhG marginal ns. 23 et seq.
41. Axel Nordemann, *supra* n. 6, § 43 UrhG marginal ns. 23 et seq.
42. Axel Nordemann, *supra* n. 6, § 43 UrhG marginal ns. 25 et seq.
43. Wandtke, *supra* n. 15, § 43 UrhG marginal ns. 34 et seqq.
44. Cf. Art. 5(3) *Basic Law for the Federal Republic of Germany ('Grundgesetz', 'GG')*, available in the English language at: http://www.gesetze-im-internet.de/englisch_gg/index.html (accessed 16 May 2017).
45. Heermann, *supra* n. 20, 472; Marcus Hirschfelder, *Open Access – Zweitveröffentlichungsrecht und Anbietungspflicht als europarechtlich unzulässige Schrankenregelungen?(Open Access – Are a Second Publication Right and a Duty to Offer Limitations on Copyright Contrary to European Law?)*, MMR 2009, 444, 446 et seq. (2009); Axel Nordemann, *supra* n. 6, § 43 UrhG marginal n. 21; Rojahn, *supra* n. 15, § 43 UrhG marginal n. 131; Wandtke, *supra* n. 15, § 43 UrhG marginal n. 32; Axel Nordemann, *supra* n. 15, § 63 marginal n. 21; for an obligation to offer based on the employee's duty of good faith cf.: *Grabungsmaterialien*, *supra* n. 20, 249; *Wegfall des Hochschullehrersprivileg bei Diensterfindungen, 2 U 19/05*, GRUR-RR 2006, 178, 181 (OLG Braunschweig, 10 November 2005); *Rechte an einem im Dienstverhältnis entwickelten Datenverarbeitungsprogramm, 7 O 15354/91*, ZUM 1997, 659, 655 (LG Munich, 16 January 1997).
46. *Berne Convention for the Protection of Literary and Artistic Works ('Berne Convention')*, available at: http://www.wipo.int/treaties/en/text.jsp?file_id=283698#P127_22000 (accessed 16 May 2017).
47. Cf. Art. 7(1) Berne Convention.
48. *Council Directive 93/98/EEC of 29 October 1993 harmonising the term of protection of copyright and certain related rights (Copyright Duration Directive)*, available at: http://eur-lex.europa.eu/legal-content/EN/TXT/PDF/?uri=CELEX:31993L0098&from=en (accessed 16 May 2017).
49. Cf. Art. 1 Copyright Duration Directive.

years p.m.a. (Section 64 UrhG).[50] Apart from trademarks - which can be infinitively renewed - copyright therefore provides the longest term of protection in Germany. Contrary to patents, utility models, registered designs or plant breeders' rights, no renewal or maintenance fees are due in order to benefit from the entire term of protection. On the other hand, no additional extension of the term of protection is possible.

Whereas Section 64 UrhG refers to the temporal restriction of authors' rights as such, the temporal scope of the transferred economic rights is determined through contractual arrangements by the parties themselves. With regard to employment relationships, it is hotly debated whether or not the employer's economic rights cease to exist with the end of the employee's employment.[51] According to German jurisprudence and the vast majority of legal literature, the employer's economic rights are not temporarily restricted per se, therefore they do not expire with the end of employment.[52] Others though, assume that economic rights in employees' works are merely transferred for the time of employment and cease to exist with the latter's end.[53] They consider an additional contractual arrangement necessary in order for the employer to maintain economic rights in the works after the end of employment. The correct approach though, is to take the object of the employer's business into consideration.[54] As mentioned earlier, it is the object of the business that determines the scope of transfer and thereby the temporal scope of the transferred economic rights.[55] Thus, the employer's economic rights do not necessarily cease to exist with the end of employment. Rather, the employer may maintain their rights in the employee's work depending on the object of the business and the employee's specific obligations resulting from the employment relationship.[56] In respect of rights in computer programs, the specific provision of Section 69b UrhG applies, therefore the employer indisputably acquires a perpetual right to use the software.[57]

2.2.9 Specific Provision on Employees' Software

In case of employee-created software, the UrhG provides a specific provision, namely Section 69b UrhG, comprehensively regulating the matter of rights allocation in

50. Axel Nordemann, *supra* n. 6, § 64 UrhG marginal n. 1.
51. Wilhelm Nordemann/Axel Nordemann, *Die zeitliche Dauer der Nutzungsrechtseinräumung im Arbeitsverhältnis (Duration of Economic Rights Granted within Employment Relationships)*, Festschrift für Artur-Axel Wandtke zum 70. Geburtstag, 141-147 (2013); Wandtke, *supra* n. 15, § 43 UrhG marginal n. 76.
52. *Statikprogramme, 3 AZR 371/81*, GRUR 1984, 429, 432 (Federal Labour Court (Bundesarbeitsgericht, 'BAG'), 13 September 1983); Dreier, *supra* n. 8, § 43 UrhG marginal n. 20; Axel Nordemann, *supra* n. 6, § 43 UrhG marginal n. 32; Wiebe, *supra* n. 39, § 43 UrhG marginal n. 11.
53. Wandtke, *supra* n. 15, § 43 UrhG marginal n. 76 with further references; Wandtke, *supra* n. 16, 394; dissenting: Dreier, *supra* n. 8, § 43 UrhG marginal n. 20.
54. Axel Nordemann, *supra* n. 6, § 43 UrhG marginal ns. 31 et seq.
55. Axel Nordemann, *supra* n. 6, § 43 UrhG marginal n. 56.
56. Dreier, *supra* n. 8, § 43 UrhG marginal n. 20; Axel Nordemann, *supra* n. 6, § 43 UrhG marginal n. 32; Wiebe, *supra* n. 39, § 43 UrhG marginal n. 11.
57. Cf. *infra* '2. Employees' Copyright' section 2.2.9 'Ownership of Rights and Termination: Specific Provision on Employees' Software'.

employment relationships. Pursuant to Section 69b UrhG, the application of the Specific Purpose Principle is precluded, therefore the scope of transfer is not confined to those rights reasonably necessary to achieve the object of the employer's business.[58] Rather, the employer acquires a comprehensive, unlimited and perpetual right to use the software regardless of the object of the business or any other operational or business necessity. The transferred right of use includes the right of adaptation as well as the right to further transfer or sublicense the acquired right. The employee therefore does not maintain any economic rights themselves.[59]

2.3 The Parties' Duty to Inform

The UrhG is silent on the question of whether or not the employee is obligated to inform the employer about newly created works. A statutory obligation to inform does therefore not exist. However, such duty is implied in Section 43 UrhG: Section 43 UrhG allocates all economic rights in the employee's work necessary to achieve the object of the business to the employer. It is a matter of logic that in order to benefit from these rights, the employer must gain knowledge of the existence of the work in question. Protection through copyright arises automatically without any registration of the work, through which the employer could become aware of the work's existence. Hence, the employer depends on a notification by the employee. Thus, in order to attach practical value to the employer's rights under Section 43 UrhG, the employee is obligated to inform the employer about any new work that comes within the scope of said Section. In regard of free works, the above[60] applies.

2.4 Employee Remuneration Right

Specific provisions on the employee's right to equitable remuneration – such as Section 9 ArbNErfG[61] for inventions – do not exist within the UrhG. Applying Section 9 ArbNErfG by way of analogy is no alternative, as an analogy cannot be drawn.[62] However, the UrhG does comprise provisions on contractual and statutory remuneration rights of the author in general, regardless of whether or not the author happens to be an employee. These provisions are likewise applicable to employment relationships although certain restrictions must be made.

58. Cf. *supra* '2. Employees' Copyright' section 2.2.4 'Ownership of Rights and Termination: Scope of Transfer'.
59. Axel Nordemann, *supra* n. 6, § 43 UrhG marginal n. 56.
60. Cf. *supra* '2. Employees' Copyright' section 2.2.7 'Ownership of Rights and Termination: Free Works'.
61. Cf. *infra* '4. Employees' Inventions' section 4.4 'Employee Remuneration Right'.
62. Axel Nordemann, *supra* n. 6, § 43 UrhG marginal n. 64; Rojahn, *supra* n. 15, § 43 UrhG marginal n. 64.

2.4.1 Contractual Remuneration Right

Sections 32 et seqq. UrhG contain legal provisions on contractually conferred remuneration rights. They have no origin in European copyright law. Whether or not the rights conferred to the author by Sections 32 et seqq. UrhG constitute statutory or contractual remuneration rights remains unclear.[63] In any case, Sections 32 et seqq. UrhG restrict the parties' freedom of contract in terms of remuneration clauses. Only collective agreements are not affected by these restrictions.[64]

Before turning to the individual provisions of the UrhG, one basic distinction between a freelance author and an employee author must be emphasised: while freelance authors highly depend on an equitable remuneration in return for the transfer of economic rights in their works in order to make a living, authors employed by way of a private or public sector employment contract are already ensured the possibility to earn a living through their basic salary.[65] The social function, which the provisions of the UrhG on the author's right to equitable remuneration fulfil – i.e. providing a basis for authors to make a living from their creative undertakings – fades into the background. The employee's salary is considered to include both – remuneration for the employee's general services and equitable remuneration for the transfer of economic rights.[66] Therefore, the transfer of economic rights under Section 43 UrhG does not generally entitle the employee to further additional remuneration.[67]

Nonetheless, the employee author may rely – just as any other author – on the provisions of Sections 32 and 32a UrhG,[68] both of which are not applicable if collective agreements exist.[69] Section 132 UrhG further regulates the applicability of Sections 32 and 32a UrhG to older cases. Accordingly, Section 32a UrhG only applies to cases, the facts of which have arisen after 28 March 2002.[70] Section 32 UrhG applies to contracts concluded after 30 June 2002, but also to contracts concluded between 1 June 2001 and 30 June 2002 provided that the transferred economic rights have been exercised after 30 June 2002.[71]

Section 32 UrhG entitles the author to equitable remuneration against their contractual partner in return for the transfer of economic rights. Thereby the author

63. Czychowski, *Urheberrecht: Kommentar zum Urheberrechtsgesetz, zum Verlagsgesetz und zum Urheberrechtswahrnehmungsgesetz (Authors law: Commentary on the Act on Copyright and Related Rights, on the German Publishing Act and the Copyright Administration Act)*, § 32 UrhG marginal n. 16 (Fromm/Nordemann, 11th ed., Kohlhammer, 2014).
64. Cf. ss. 32(4) and 32a(4) UrhG.
65. Axel Nordemann, *supra* n. 6, § 43 UrhG marginal ns. 3, 58.
66. *Zwischen Marx und Rothschild, 3 U 77/75*, GRUR 1977, 556, 558 (OLG Hamburg, 22 January 1976); Czychowski, *supra* n. 63, § 32 UrhG marginal n. 28; Rojahn, *supra* n. 15, § 43 UrhG marginal n. 64.
67. *Wetterführungspläne II, X ZR 72/98*, GRUR 2002, 149, 151 (BGH, 23 October 2001); Dreier, *supra* n. 8, § 43 UrhG marginal n. 30; Axel Nordemann, *supra* n. 6, § 43 UrhG marginal n. 58; Rojahn, *supra* n. 15, § 43 UrhG marginal n. 64; Axel Nordemann, *supra* n. 15, § 63 marginal n. 64; dissenting: Wandtke, *supra* n. 15, § 43 UrhG marginal ns. 134 et seqq. with further references.
68. Axel Nordemann, *supra* n. 6, § 43 UrhG marginal ns. 59 et seq.
69. Cf. ss. 32(4) and 32a(4) UrhG.
70. Cf. s. 132(3) sentence 2 UrhG.
71. Cf. s. 132(3) sentence 3 UrhG.

shall have a fair share in profits generated from the exploitation of their work.[72] As mentioned earlier, with regard to employees, it is generally assumed that part of the basic salary already constitutes equitable remuneration for the rights' transfer. However, this assumption has to be verified on an individual case-by-case basis. Whether or not remuneration is equitable is to be determined according to Section 32(2) UrhG.[73] In case collective agreements exist, the equitable remuneration has to be taken from such Rules. Otherwise, it is up to the courts to determine, what is equitable. Provided that the employee's general salary does not amount to equitable remuneration for the transfer of rights, the employee may invoke Section 32(1) sentence 3 UrhG: they may demand the employer's consent to a modification of the original contract in such a way that the author is granted equitable remuneration. This scenario is particularly common in short employment relationships, in which the employee is not guaranteed a long-term possibility to earn a living.[74] The author cannot waive their rights conferred by Section 32 neither in advance nor afterwards.[75] Contractual agreements deviating from Section 32(1) and (2) to the detriment of the author may not be invoked by the other party (Section 32(3) sentence 1 UrhG).[76] The provisions of Section 32(1) and (2) UrhG shall apply even if they are circumvented by other arrangements (Section 32(3) sentence 2 UrhG). Whether or not the author may assign their rights conferred by Section 32 UrhG is hotly debated.[77] While the author's right to an alteration of the original contract is considered unassignable, the author shall be allowed to assign their right to payment as such, provided that they are awarded equitable consideration in return.[78]

In case of extraordinarily and unforeseeably successful exploitation, the employee author may additionally rely on Section 32a UrhG, the so-called *bestseller provision*. If a noticeable disproportion exists between the employee's original remuneration and the profits generated from the work's exploitation, the employee is accorded the right to further participate in the benefits generated from the work's exploitation.[79] Pursuant to Section 32a(1) the other party – i.e. the employer – 'shall be obliged, at the author's request, to consent to a modification of the agreement which grants the author further equitable participation appropriate to the circumstances'. While Section 32a(1) UrhG awards the author a right to further compensation against their contractual partner, i.e. the employer, Section 32a(2) confers on the author a claim to further participation against a third party. Section 32a(2) only applies in cases where the author's contractual partner has sublicensed or assigned their economic

72. Axel Nordemann, *supra* n. 6, § 43 UrhG marginal ns. 3, 58.
73. For more details cf. Czychowski, *supra* n. 63, § 32 UrhG Nos. 33 et seqq.
74. Axel Nordemann, *supra* n. 6, § 43 UrhG marginal n. 59.
75. Czychowski, *supra* n. 63, § 32 UrhG Nos. 114 et seqq.; Schulze, *Urheberrechtsgesetz, Urheberwahrnehmungsgesetz, Kunsturhebergesetz: Kommentar (Act on Copyright and Related Rights, Copyright Administration Act, Act on Copyright for Works of Art: Commentary)* § 32 UrhG marginal n. 77 (Dreier/Schulze, 5th ed., C.H. Beck, 2015).
76. Schulze, *supra* n. 75, § 32 UrhG marginal n. 75 with further references.
77. Czychowski, *supra* n. 63, § 32 UrhG No. 22; Schulze, *supra* n. 75, § 32 UrhG marginal n. 76.
78. Czychowski, *supra* n. 63, § 32 UrhG No. 22.
79. Axel Nordemann, *supra* n. 6, § 43 UrhG marginal n. 60; for further details cf. Czychowski, *supra* n. 63, § 32a UrhG Nos. 1 et seqq.

rights, assertion of which then led to an extraordinary and unforeseeable success. In this case the author may invoke their right to further compensation directly against the third party.[80] As with Section 32, the author is not permitted to waive their right conferred by Section 32a in advance.[81] If the employed author's remuneration has been determined in accordance with a collective agreement and explicitly provides for a further equitable participation in cases under Section 32a(1), the author is not entitled to further participation (Section 32a(4) UrhG).

The general assumption, that the employee's salary includes equitable remuneration for the transfer of economic rights, refers only to those works that come within the scope of Section 43 UrhG and only to the transfer of those economic rights that are necessary to achieve the object of the employer's business. As mentioned above, the acquisition of further economic rights beyond the object of the business requires a contractual agreement between the employer and employee, which precisely and expressly stipulates the additional economic rights to be transferred.[82] An extra contractual agreement between the parties is furthermore needed, if the employer wishes to acquire economic rights in free works by the employee.[83] Such agreements will usually comprise clauses on the issue of remuneration. The transfer of additional economic rights or of rights in free works therefore entitles the employee to further remuneration, unless otherwise contractually agreed upon by the parties.[84]

The employee-created work may well be exploited after the employment relationship has ended and the employee is no longer paid a general salary (see 2.2.8). However, such subsequent exploitation does not necessarily entitle the employee to additional remuneration. Often, potential profits generated through post-employment exploitation are taken into account when determining the salary the employee is paid during their employment. In individual cases though, additional compensation might be awarded under the preconditions of Section 32 or Section 32a UrhG.[85]

The exploitation of works created prior to the employment relationship on the other hand, generally entitles the employee to separate remuneration. Only in individual cases, in which the basic salary includes remuneration for the already existing works' exploitation, is the employee denied further compensation.[86]

From March 1, 2017, certain additional provisions where introduced by the German legislator to further protect the author. They should have a limited effect also for employed authors.

The author is granted two separate new claims for information. According to the new Section 32d UrhG the author may once a year request from his contracting party

80. For further details cf. Czychowski, *supra* n. 63, § 32a UrhG Nos. 28 et seqq.
81. Cf. s. 32a(3) sentence 1 UrhG.
82. Cf. *supra* '2. Employees' Copyright' section 2.2.4 'Ownership of Rights and Termination: Scope of Transfer'.
83. Cf. *supra* '2. Employees' Copyright' section 2.2.7 'Ownership of Rights and Termination: Free Works'.
84. Dreier, *supra* n. 8, § 43 UrhG marginal n. 33; Axel Nordemann, *supra* n. 6, § 43 UrhG marginal n. 61.
85. Dreier, *supra* n. 8, § 43 UrhG marginal n. 32; Axel Nordemann, *supra* n. 6, § 43 UrhG marginal n. 62.
86. Axel Nordemann, *supra* n. 6, § 43 UrhG marginal n. 62.

information and accountability in respect of the extent of the use of the work and the proceeds and benefits derived therefrom on the basis of information which is generally available in the ordinary course of business activities. This is ruled out if the author has made only a secondary contribution to a work, product or service. The new Section 32e UrhG grants a claim for information and accountability in a licence chain. Where the author's contracting partner has transferred the right of use or granted further rights of use, the author may also demand relevant information and accountability from those third parties. Such information claims are meant to help the author to identify, if and to what extent the author is entitled to an additional remuneration pursuant Sections 32 and 32a. Consequently, the new Sections 32d and 32e should only apply to employed authors, in case they come under the provisions of Sections 32 and 32a. This is in particular likely for authors in short term employments. For other employed authors – generally speaking – this is not the case (see above). In any case, employed software programmers are out of the scope of Sections 32d and 32e by law (Section 69a (5) UrhG).

The new Section 40a UrhG provides the author with a right to other exploitation after ten years in the case of flat-rate remuneration. That means, where the author has granted an exclusive right of use against payment of flat-rate remuneration he shall nevertheless be entitled to exploit the work in another manner after the expiry of ten years. The first owner's right of use shall continue as a non-exclusive right of use for the remainder of the period for which it was granted. Probably, the new Section 40a UrhG does not apply to employees, as they are not remunerated 'flat-rate' (but e.g. monthly). Again for short term employments, Section 40a may be relevant.

2.4.2 *Statutory Remuneration Rights*

Beside the employee's contractual right to an equitable remuneration, statutory remuneration rights are provided for by the UrhG from which the employee benefits as the work's author.

First of all, the UrhG prescribes payment of equitable remuneration in connection with certain economic rights.

2.4.3 *Remuneration for Rental and Lending*

Section 27 UrhG comprises provisions on remuneration for rental and lending implementing Articles 5 and 6 of the European Rental Directive.[87]

Section 27(1) UrhG foresees the lessor's obligation to pay an equitable remuneration to the author, where the latter has granted rental rights (Section 17 UrhG) in respect of a video or audio recording to the producer of the audio recording or of a film. Thereby, Section 27(1) UrhG implements Article 5(1) of the European Rental Directive.

87. Directive 2006/115/EC of the European Parliament and of the Council of 12 December 2006 on rental right and lending right and on certain rights related to copyright in the field of intellectual property ('Rental Directive'), available at: http://eur-lex.europa.eu/legal-content/EN/TXT/HTML/?uri=CELEX:32006L0115&from=en (accessed 16 May 2017).

As the Directive further prescribes,[88] the claim for remuneration may not be waived and may be assigned in advance only to a collecting society (Section 27(1) sentences 2 and 3 UrhG).

Section 27(2) UrhG refers to the exhaustion of the author's lending right pursuant to Section 17(2) UrhG. Where the original work or copies of it have been brought to the market by sale with the consent of the person entitled to distribute them within the territory of the European Union or another state party to the Agreement on the European Economic Area, their dissemination shall be permissible, except by means of rental, according to Section 17(2) UrhG. Based on Article 6(1) of the Rental Directive, Section 27(2) UrhG foresees compensation for this exhaustion of the author's lending right: Where originals or copies of a work whose dissemination is permissible according to Section 17(2) UrhG are lent, the author shall be paid an equitable remuneration, provided that the lending is conducted through a publicly accessible institution.

According to Section 27(3) UrhG, claims for remuneration pursuant to this Section may only be asserted through collecting societies.

2.4.4 Cable Retransmission Right

Section 27 UrhG has functioned as a role model for another provision on equitable remuneration for the transfer of economic rights, namely Section 20b(2) UrhG, which implements Articles 8 and 9 European Cable and Satellite Directive 93/83.[89] Section 20b(2) UrhG states the following:

> Where the author has granted the right of cable retransmission to a broadcasting organisation or to the producer of an audio recording or film, the cable operator shall nevertheless pay the author an equitable remuneration for the cable retransmission. The claim for remuneration may not be waived. It may be assigned in advance only to a collecting society and may be asserted only by a collecting society. This provision shall not conflict with the collective agreements, works agreements and joint remuneration agreement of broadcasting organisations insofar as the author is thereby granted equitable remuneration for each cable retransmission.

88. Cf. Art. 5(2) and (3) Rental Directive.
89. Council Directive 93/83/EEC of September 1993 on the coordination of certain rules concerning copyright and rights related to copyright applicable to satellite broadcasting and cable retransmission, available at: http://eur-lex.europa.eu/legal-content/EN/TXT/ PDF/?uri = CELEX:31993L0083&from = EN (accessed 16 May 2017).

2.4.5 *Resale Right*

Another provision on equitable remuneration introduced to implement European standards is Section 26 UrhG. In accordance with the Resale Right Directive,[90] Section 26 UrhG introduced the author's resale right thereby providing another statutory right to remuneration, as contained in Section 26(1) sentence 1 UrhG:

> If the original of an artistic work or of a photographic work is resold and if an art dealer or an auctioneer is involved as purchaser, vendor or intermediary, the vendor shall pay to the author a share of the selling price.

However, the vendor's obligation only arises where the selling price does not amount to less than EUR 400 (Section 26(1) sentence 4 UrhG). As clarified by Section 26(8) the provisions of Section 26 UrhG shall not apply to architectural works and works of applied arts. As with the aforementioned remuneration rights, the resale right shall be inalienable and may not be waived in advance (Section 26(3) UrhG). Section 26(4) and (5) provide the author with a right to certain information necessary to assert their resale right. These information claims however, may only be asserted through a collection society (Section 26(6) UrhG).

2.4.6 *Compensation for Limitations on Authors' Rights*

Chapter 4 of the UrhG contains – in accordance with Article 5 European Copyright Directive 2001/29[91] – limitations on the author's rights in their work. For some of these limitations, the author is awarded a statutory right to equitable remuneration.

Section 45a UrhG declares the reproduction and distribution of a work permissible for non-commercial purposes exclusively for persons whose access to the work is considerably impeded by a disability, if such reproduction and distribution facilitates access to the work. As compensation for this limitation of the author's reproduction and distribution right, Section 45a(2) UrhG prescribes the payment of equitable remuneration. Claims pursuant to this Section may only be asserted through collecting societies.

Once a work has been published, its reproduction, distribution and making available shall be permissible for limited parts of the work incorporated in collections for religious, school or instructional use (Section 46 UrhG). In return, the author is accorded a right to equitable remuneration (Section 46(3) UrhG).

90. Directive 2001/84/EC of the European Parliament and of the Council of 27 September 2001 on the resale right for the benefit of the author of an original work of art, available at: http://eur-lex.europa.eu/LexUriServ/LexUriServ.do?uri=CELEX:32001L0084:EN:HTML (accessed 16 May 2017).
91. Directive 2001/29/EC of the European Parliament and of the Council of 22 May 2001 on the harmonisation of certain aspects of copyright and related rights in the information society, available at: http://eur-lex.europa.eu/legal-content/EN/TXT/HTML/?uri=CELEX:32001L0029&from=EN (accessed 16 May 2017).

Section 47(2) UrhG requires payment of an equitable remuneration for school broadcasts unless the broadcasted recordings are deleted within one year after the transmission of the school broadcast.

The reproduction and distribution of individual broadcast commentaries, individual newspaper articles and illustrations published in connection with the former shall be permissible for the purpose of communicating them to the public provided that they concern current political, economic or religious issues and do not contain a statement reserving rights. The author is to be paid an equitable remuneration for such reproduction, distribution and communication to the public unless only short extracts of several commentaries or articles in the form of an overview are concerned (Section 49(1) UrhG). Claims may only be asserted through collecting societies.

The communication and making available of a work is declared permissible in certain situations, all of which trigger the author's right to an equitable remuneration, Sections 52 (communication to the public for non-profit-making purposes, in religious services or at religious celebrations), 52a (making works available to the public for instruction and research), and 52b UrhG (communication of works at terminals in public libraries, museums and archives). Claims pursuant to Sections 52a and 52b may only be asserted through collecting societies (cf. Sections 52a(4) and 52b UrhG).

Section 54 UrhG further awards the author the right to an equitable remuneration where the nature of a work makes it probable that it will be reproduced for private and other personal use pursuant to Section 53 UrhG. In this regard, Section 53a UrhG provides for another right to equitable remuneration in case of an order for dispatch of copies. Claims under Section 53a UrhG may only be asserted through collecting societies. Sections 54a et seqq. contain further provisions on the issue of statutory remuneration.

For the greater part, claims under the provisions on equitable remuneration may only be asserted through collecting societies. This leads to the conclusion that the author's statutory remuneration rights may not be assigned to someone other than a collecting society.[92] In regard of employment relationships therefore, the employee cannot assign their statutory remuneration rights to the employer. For most of the statutory remuneration rights, the employee is furthermore not permitted to waive them in advance.[93] However, in case the employer is a publisher, they may nevertheless be entitled to a share in the remuneration of the author. Section 27a Verwertungsgesellschaftsgesetz (VGG-Law on Collection Societies) requires consent by the author after publication or after notifying the work to the collecting society.

92. Cf. Axel Nordemann, *supra* n. 6, § 43 UrhG marginal n. 63.
93. Cf. e.g. s. 20b(2) UrhG, s. 26(3) UrhG; cf. also Axel Nordemann, *supra* n. 6, § 43 UrhG marginal n. 63.

3 EMPLOYEES' SEMICONDUCTOR CHIPS

3.1 The Legal Framework

The German Semiconductor Chip Protection Act ('HalblSchG') actually originates from the US Semiconductor Chip Protection Act of 1984. Pursuant to the principle of material reciprocity, foreign nationals could only benefit from the protection of the US Semiconductor Chip Protection Act if in their country of origin the same protection was accorded to US nationals. Therefore, in 1986, only two years after the US Act had been passed, Directive 87/54/ECC[94] ('Semiconductor Directive') was adopted in order to ensure semiconductor chip protection in the US to all European nationals. In Germany, the Directive was implemented through the HalblSchG in 1987.[95]

Despite the semiconductor industry's economic importance, protection of semiconductor chips has been of only little practical significance. So far, there is no jurisprudence on the issue of semiconductor chip rights in Germany. In the US, the UK and Australia only one case has been decided. Likewise, the number of applications for registration[96] and actual registrations has been comparatively small over the years. By the end of 2007, only 109 semiconductor chip rights registrations were still in force.[97]

As the HalblSchG constitutes the German implementation of the Semiconductor Directive, its provisions have to be applied and interpreted in the light of the Directive, its provisions and objectives. Section 1 HalblSchG defines the subject matter, which is not the microelectronic semiconductor product itself, the microchip, but its three-dimensional pattern.[98] According to Section 1(1) and (2) HalblSchG, topographies, i.e. three-dimensional patterns of microelectronic semiconductor products, enjoy protection insofar as they constitute the result of their creator's own intellectual effort and are not commonplace. Protection furthermore extends to independently exploitable pieces of topographies, such as gate-arrays,[99] as well as to images used to produce

94. Council Directive 87/54/ECC of 16 December 1986 on the legal protection of topographies of semiconductor products, available at: http://eur-lex.europa.eu/legal-content/EN/TXT/?uri=CELEX:31987L0054 (accessed 16 May 2017).
95. Falck, *Computerrechts-Handbuch (Computer Law Handbook)*, s. 1, Ch. 5, Topographieschutz marginal ns. 2 et seq. (Kilian/Heussen, 32th ed., C.H. Beck Munich, 2013); Marly, *Praxishandbuch Softwarerecht: Rechtsschutz und Vertragsgestaltung (Handbook Software Law: Legal Protection and Contract Design)*, marginal n. 590 (6th ed., C.H. Beck, 2014); Koch, *Der Halbleiterschutz nach nationalem, internationalem und europäischem Recht*, in: *Rechtsschutz und Verwertung von Computerprogrammen (Semiconductor Chip Protection under National, International and European Law*, in: *Legal Protection and Commercialisation of Computer Programmes)*, marginal ns. 10 et seqq. (Lehmann, 2nd ed., Verlag Dr. Otto Schmidt Köln, 1993); Kraßer, *Patentrecht (Patent Law)*, s.1, Ch. 1, § 2, II, a) Halbleiterschutz marginal n. 1 (6th ed., C.H. Beck Munich, 2009).
96. In 2006 and 2007 only two applications per year were made; cf. Marly, *supra* n. 95, marginal n. 589.
97. Cf. Marly, *supra* n. 95 marginal n. 589; cf. also Falck, *supra* n. 95, s. 1, Ch. 5, Topographieschutz marginal n. 4.
98. Falck, *supra* n. 95, s. 1, Ch. 5, Topographieschutz marginal n. 6.
99. Falck, *supra* n. 95, s. 1, Ch. 5, Topographieschutz marginal n. 7.

topographies, in particular so-called *mask works*.[100] Although the German definitions differ slightly from the wording of the Semiconductor Directive, these regulations meet the European objectives of Articles 1 and 2 Semiconductor Directive.

Concerning employees' semiconductor chips, the HalblSchG provides merely one core provision, namely Section 2(2) HalblSchG, which concerns the ownership of rights to employee-created semiconductor chips. Otherwise, no specific regulations are provided. In particular, the provisions of the ArbNErfG are not applicable to employees' semiconductor chips.[101] Therefore, despite the issue of rights ownerships, all remaining aspects of the employer-employee relationship in regard to semiconductor chips have to be regulated by contract between the parties.

3.2 Ownership of Rights

3.2.1 Allocation of Rights to Employee-Created Topographies

Article 3(1) Semiconductor Directive states, 'the right to protection shall apply in favour of persons who are the creators of topographies of semiconductor products'. Accordingly, Section 2(1) HalblSchG stipulates that it is the creator of a topography who is entitled to its protection. With regard to employment relationships, Section 2(2) HalblSchG states the following:

> Where a topography has been created in the course of an employment or commissioned by another party, the right to protection shall vest in the employer or the party to the contract by whom the topography has been commissioned.

Thus, under German law, rights to employee-created topographies are allocated to the employer. This regulation stems from Article 3(2) Semiconductor Directive. Nevertheless, the Directive does not mandatorily command for a transfer of rights in favour of the employer. Thus, Member States remain free to choose whether rights in employee-created topographies belong to the employee or the employer. The regulation under the HalblSchG resembles the corresponding provisions in the UrhG, the ArbNErfG, and the DesignG. Rights to employee-created topographies belong to the employer – even though it remains unclear whether rights directly vest in the employer or are transferred to them by law[102] – unless deviating contractual agreements exist.

For Section 2(2) HalblSchG to apply, the topography must have been created in the course of employment. Due to the lack of case law in the field of semiconductor chip protection, there is no guidance by the courts as to how this requirement is to be interpreted. Even literature remains silent on this matter. Basic guidelines may be deduced from the interpretation of parallel provisions for other intellectual property rights (Section 4(2) ArbNErfG, Section 43 UrhG, Section 7(2) DesignG). Foremost,

100. Cf. Kraßer, *supra* n. 95, s. 1, Ch. 1, § 2, II, a) Halbleiterschutz marginal n. 2; Marly, *supra* n. 95, marginal n. 589.
101. Keukenschrijver, *Patentgesetz Kommentar (Patent Act: Commentary)*, Halbleiterschutzgesetz, s.1, § 2 marginal n. 9 (Busse, 7th ed., De Gruyter Recht Berlin, 2013).
102. Keukenschrijver, *supra* n. 101, Halbleiterschutzgesetz, s.1, § 2 marginal n. 9 with further references.

however, any potential interpretation of Section 2(2) HalblSchG has to be in conformity with European law, i.e. the Semiconductor Directive.

Section 2(2) to (5) further limit the group of people entitled to protection under the HalblSchG. Only natural persons who are nationals of a Member State of the European Union or who have their habitual residence on the territory of a Member State, companies or other legal persons, which have a real and effective industrial or commercial establishment on the territory of a Member State, shall benefit from the protection under the HalblSchG (Section 2(3) HalblSchG). According to Section 2(4) HalblSchG, the right to protection shall also apply in favour of the persons referred to in Section 2(3) who first commercially exploit within a Member State a topography which has not yet been exploited commercially anywhere in the world, and who have been exclusively authorised to exploit commercially the topography throughout the Community by the person entitled to dispose of it. Those who do not come within the scope of Section 2(3) or (4) are entitled to protection only if they fulfil the requirements of Section 2(6) HalblSchG.[103] These limitations also apply to employees and employers who must fulfil the outlined preconditions to benefit from protection under the HalblSchG.

3.2.2 *Termination*

Section 5(1) HalblSchG defines the prerequisites for protection to become effective. A topography enjoys protection from the day of its first non-confidential commercial exploitation provided that an application for registration in due form is filed with the German Patent and Trademark Office within two years after that day. Otherwise, protection is effective with the application for the topography's registration provided that it has not yet been commercially exploited.[104]

According to Section 5(2) HalblSchG the term of protection amounts to ten years. This time limit begins to run at the end of the legal year in which one of the prerequisites of Section 5(1) HalblSchG has been met.[105]

A topography that has not been commercially exploited, or for which no application for registration has been filed, within 15 years after it was recorded for the first time is no longer eligible for protection (Section 5(4) HalblSchG).

3.3 The Parties' Duty to Inform

The HalblSchG does not contain a statutory duty for the employee to inform the employer about new topographies. However, in order to ensure the employer's right to

103. Other persons are only entitled to protection if they benefit from the principle of national treatment by virtue of an international treaty or European law, or if their country of origin provides protection of topographies for German nationals as well (material reciprocity), cf. s. 2(6) HalblSchG.
104. Cf. Falck, *supra* n. 95, s. 1, Ch. 5, Topographieschutz marginal n. 15; Marly, *supra* n. 95, marginal n. 591; Kraßer, *supra* n. 95, s. 1, Ch. 1, § 2, II, a) Halbleiterschutz marginal n. 4.
105. Cf. Marly, *supra* n. 95, marginal n. 592; Falck, *supra* n. 95, s. 1, Ch. 5, Topographieschutz marginal n. 15; Kraßer, *supra* n. 95, s. 1, Ch. 1, § 2, II, a) Halbleiterschutz marginal n. 4.

the topography conferred to them by Section 2(2) HalblSchG, such an obligation has to be implied.[106] Thus, as soon as an employee's topography comes within the scope of Section 2(2) HalblSchG, the employee is obliged to inform the employer about said topography.

3.4 Employee Remuneration Right

Although the HalblSchG confers on the employer a right to employees' topographies created in the course of employment, it does not, on the other hand, provide for a statutory right to equitable remuneration for the employee. The lack of such a right to additional remuneration might originate from the universal employment law principle, according to which the employee's salary is supposed to constitute consideration for all work results achieved by the employee on behalf and for the account of the employer.[107] Thus, the employee's salary already includes compensation for potentially created topographies. Deviating contractual arrangements, i.e. a right to additional remuneration, may nevertheless be entered into by the parties.

4 EMPLOYEES' INVENTIONS

New inventions based on an inventive step and industrially applicable[108] can be protected through either patents or utility models. However, patents and utility models - and the protection conferred through them - show a number of differences: while according to Section 9 No. 1 and 2 of the 1981 German Patent Act ('PatG') patentable inventions comprise both products and processes, only products - not processes - are capable of being applied for as a utility model, Section 2 No. 3 German Act on Utility Models ('GebrMG'). Before a patent is registered, all substantial and formal requirements will be checked by the German Patent and Trademark Office, whereas for a utility model's registration only formal requirements have to be fulfilled. This difference in the registration process leads to the following consequences: having a utility model registered turns out to be easier, faster and cost-saving. While the patent registration process may take two and more years and cost about EUR 450, a utility model is often registered for EUR 40 within two to five months. On the other hand, utility models are under a higher risk of being attacked and annulled as their substantial requirements are not checked prior to their registration. Moreover, the maximum term of protection is ten years for utility models (Section 23 GebrMG), while it can last up to 20 years for patents (Section 16 PatG).

Inventors should keep the differences regarding preconditions, scope and duration of protection in mind when deciding whether to file an application for a patent or

106. For an implied obligation to notify the employer about new service designs, cf. Eichmann, *Geschmacksmustergesetz (Registered Designs Act)*, § 7 GeschmMG marginal n. 23 (Eichmann/ Vogel von Falckenstein, 4th ed., C.H. Beck Munich, 2010).
107. For service designs, cf. Eichmann, *supra* n. 106, § 7 GeschmMG marginal n. 25.
108. Cf. s. 1 *German Patent Act ('Patentgesetz', 'PatG')*; s. 1 *German Act on Utility Models('Gebrauchsmustergesetz', 'GebrMG')*.

a utility model. With regard to employment relationships, Section 13(1) ArbNErfG generally obliges the employer to file a patent application. Only if protection through a utility model appears to be more appropriate, should the employer opt for this alternative.

4.1 The Legal Framework

German law does not provide in general for a great amount of specific intellectual property law provisions regulating the relationship between employers and employees in matters of intellectual property rights. However, the ArbNErfG elaborately regulates the legal relationship between employers and employees regarding technical inventions with regard to both patents and utility models. It also includes provisions dealing with the issue of suggestions by the employee for improvements, such as technical improvements within the employer's company, which are not patentable or capable of being applied for as a utility model. Where the employment relationship concerns patentable inventions, the PatG provides further provisions Although deviating contractual agreements are legitimate in general, no agreement between the employer and the employee may change the provisions of the ArbNErfG to the detriment of the employee prior to the notification of the invention (Section 22 ArbNErfG).

4.2 Ownership of Rights

4.2.1 Allocation of Rights in Employees' Inventions

As with the authorship principle of Section 7 UrhG, the inventor principle stipulates that all rights in an invention originally belong to the inventor (Section 6 PatG).[109] This applies to employee-made inventions as well; the rights therefore initially belong to the employee.

However, in case of so-called *service inventions* within the meaning of Section 4(2) ArbNErfG, the employer has an original appropriation right, according to Section 6(1) ArbNErfG:

> The employer may claim a service invention by declaration to the employee.

Service inventions refer to inventions either made by an employee within business operations (Section 4(2) No. 1 ArbNErfG) or inventions building upon the experience or results achieved by the employee in connection with the employment (Section 4(2) No. 2 ArbNErfG). For such inventions, Section 6(1) ArbNErfG entitles the employer to claim the invention thereby acquiring all rights in it originally belonging to the employee.

Before the ArbNErfG was amended in 2009, it also entitled the employer, by virtue of Section 7 ArbNErfG, to claim a mere non-exclusive right to use the invention

109. Cf. also Mes, *Patentgesetz, Gebrauchsmustergesetz: Kommentar (Patent Act, Act on Utility Models, Commentary)*, § 6 PatG marginal n. 1 (Mes, 3rd ed., C.H. Beck Munich, 2001).

within business operations instead, so-called *claim with restriction*. Although this right was abolished in 2009, it still applies to inventions made before 30 September 2009.[110] For inventions made after this date, the employer may either claim the invention without restriction or decide not to exercise their appropriation right at all.

In order to claim the invention, in principle, a declaration by the employer is required according to Section 6(1) ArbNErfG. The declaration must be given to the employee within a certain time period after the employee has notified the employer of the invention.[111] However, in the absence of such declaration, the invention is deemed to be claimed if the employer does not release it to the employee within four months, according to Section 6(2) ArbNErfG. This arrangement benefits the employer insofar as they can make use of the invention immediately without prior agreements with their employees on the allocation of rights in the invention.[112] In practice, the scenario of Section 6(2) ArbNErfG constitutes by far the more frequent manner in which inventions are claimed by employers.

If, on the other hand, the employer releases the invention to the employee by means of a written declaration, the invention is considered free by virtue of Section 8 ArbNErfG.[113] This also applies to inventions within the meaning of Section 4(3) ArbNErfG, i.e. inventions that have been made outside business operations and do not build upon the experience or results achieved in connection with the employment.[114] Free inventions remain with the employee as their inventor. However, Section 19 ArbNErfG obliges the employee to offer the employer a non-exclusive right of use in free inventions during their employment, provided that the invention comes within the scope of the employer's business. The employee may subject the transfer of such right of use to certain terms and conditions, which, however, must be fair and reasonable.[115]

4.2.2 Employers' Duty of Registration

If the employer chooses to claim a service invention, they do not only acquire all rights in it, but incur a number of duties imposed on them by the ArbNErfG as well.

Most importantly, the employer is not only entitled to file an application for an inland patent registration, but is also obliged to do so pursuant to Section 13(1) sentence 1 ArbNErfG. Only when the protection of the invention as a utility model seems more appropriate, may the employer apply for the registration of the invention as a utility model instead, by virtue of Section 13(1) sentence 2 ArbNErfG. Failure to

---

110. Mes, *supra* n. 109, § 6 PatG marginal ns. 31 et seqq.; Schoenen, *Münchener Anwaltshandbuch: Gewerblicher Rechtsschutz (Munich Lawyer's Manual on the Protection of Industrial Property)*, § 43 Arbeitnehmererfindungsrecht marginal n. 13 (Hasselblatt, 4th ed., C.H. Beck Munich, 2012).
111. Czychowski/Langfinger, *German Law on Employees' Inventions Regarding University Employees*, Les Nouvelles, December 2010, 221, 221 (2010); Rother, *supra* n. 3, p. 11.
112. Bayreuther, *Neue Spielregeln im Arbeitnehmererfinderrecht (New Rules of the Game Concerning Law on Employees' Inventions)*, NZA 2009, 1123, 1123 (2009).
113. Cf. Schoenen, *supra* n. 110, § 43 Arbeitnehmererfindungsrecht marginal n. 18; Rother, *supra* n. 3, p. 11.
114. Czychowski/Langfinger, *supra* n. 111, 221.
115. Schoenen, *supra* n. 110, *§ 43* Arbeitnehmererfindungsrecht marginal n. 5.

comply with this duty entitles the employee to have the patent registered in the employer's name and at the employer's expense (Section 13(3) ArbNErfG).

Section 14(1) ArbNErfG entitles the employer to file a patent application abroad. As with the above, the employer is obliged to release the invention to the employee with regard to countries in which they do not want to pursue a patent registration. Thereby the employee becomes entitled to file a patent application in their own name in countries for which the invention has been released (Section 14(2) ArbNErfG).

According to Section 15(1) ArbNErfG the employer shall provide the employee with copies of the application documents; they shall inform the employee about the course of proceedings and allow them to look into the respective correspondence. The employee on the other hand, is obliged to support the employer and provide all necessary statements as respects the patent registration (Section 15(2) ArbNErfG).

However, Sections 14 and 15 ArbNErfG do not constitute mandatory law and are often ruled out by the parties by means of standard regulations.

4.2.3 *Ownership of Rights in University Employees' Inventions*

While the aforementioned regulations apply nowadays to all inventions made by employees, this has not always been the case for inventions made by university employees. Before 2002, the Act on Employees Inventions provided in Section 42(1) ArbNErfG the so-called *professors' privilege*[116] for inventions by university employees. According to this provision, inventions made by university staff within their scientific function were always considered as free inventions. Hence, such inventions were allocated to the inventor themself and not to the university as the inventor's employer. As compensation, universities were given the right to claim an equitable interest in the benefits generated through the invention's exploitation by virtue of Section 42(2) ArbNErfG. However, this right only arose if the invention had been based on investments made by the university.[117]

Since the amendment of Section 42 ArbNErfG, the legal treatment of inventions made by university staff has been more or less inverted. The professor's privilege in its former form does not exist anymore. Rather, the general rules of Section 42(1) and Sections 4-7 ArbNErfG nowadays apply likewise to inventions by scientific university employees. In consequence, inventions made by scientific employees at a university in their scientific function qualify as service inventions within the meaning of Section 4(2) ArbNErfG. If the university chooses to enforce its right to claim the invention pursuant to Section 6 ArbNErfG, the invention will be allocated to the university instead of the respective employee-inventor. However, in order to comply with the constitutionally guaranteed Freedom of Sciences (Article 5(3) GG) the amendment introduced a number of scientific privileges into Section 42 Nos 1-4 ArbNErfG. These include the right of positive (No. 1) and negative (No. 2) publication, the right to use the invention in connection with ongoing scientific research and teaching (No. 3), and the right to

116. Czychowski/Langfinger, *supra* n. 111, 221.
117. Czychowski/Langfinger, *supra* n. 111, 221.

claim an interest of 30% of the benefits arising from the invention's exploitation (No. 4).[118] Such benefits within the meaning of Section 42 No. 4 ArbNErfG comprise all monetary advantages including saved costs.[119] The personal scope of Section 42 ArbNErfG now includes 'all employees at universities or at a university of applied science regardless of whether they are employed as scientific, technical or other personnel and regardless of whether they are formally employed by the state or the university itself'.[120] The privileges of Section 42 No. 1–3 on the other hand, are confined to scientific employees, whereby guest lecturers, PhD students and regular students are excluded.[121]

Along with the new rights, universities now face the duties imposed on employers by the ArbNErfG. Of particular importance is the university's duty to file a patent application 'within a reasonable period of time',[122] i.e. without undue delay, according to Section 13 ArbNErfG. This duty causes an increase in organisation and costs, for which it was feared that universities might attempt to circumvent their newly introduced duties. However, this risk is constrained by Section 22 ArbNErfG, proscribing the university employee from waiving their right to have a patent application filed for their invention prior to the communication of the invention to the employer. Even afterwards, the right can only be waived by means of an individual agreement with the university. The amended Act is supposed to increase opportunities for universities to exploit inventions by their employees in order to strengthen their financial position. The administration of patents through the university shall further lead to an increase in the transfer of knowledge and technology and thereby guarantee an effective flow of information within universities.[123]

4.2.4 Termination

Patents enjoy protection for at least three years. The maximum term of protection through a patent is 20 years from the day following the day of registration (Section 16 PatG). In order to benefit from the entire term of protection, beginning from the third year after the invention's registration, an annual fee has to be paid (Section 17 PatG). Failure to pay the annual maintenance fee leads to the patent's termination (Section 20(1) No. 2 PatG). Prior to the general expiration date, protection ends if the patent is revoked (Section 21 PatG) or annulled (Section 22 PatG).

Furthermore, the right holder may choose to waive their right to the patent. This scenario generally leads to the end of protection, pursuant to Section 20(1) No. 1 PatG. In employment relationships, however, Section 16 ArbNErfG applies. Pursuant to Section 16(1) ArbNErfG, the employer may choose to no longer pursue a patent application or to waive their right to the patent. They are then obliged to notify the

118. Czychowski/Langfinger, *supra* n. 111, 222.
119. Cf. *Genveränderungen X ZR 59/12*, NZA 2013, 509 (BGH, 5 February 2013).
120. Czychowski/Langfinger, *supra* n. 111, 222.
121. Czychowski/Langfinger, *supra* n. 111, 222.
122. Cf. s. 13 *Act on Employees' Inventions* (*'Arbeitnehmererfindergesetz', 'ArbNErfG'*).
123. Czychowski/Langfinger, *supra* n. 111, 222.

employee and, if requested, to transfer the patent to the employee at the latter's own expense. However, the employer is entitled to maintain a non-exclusive right to use the invention subject to the payment of reasonable remuneration by virtue of Section 16(3) ArbNErfG. If the employee does not respond to the employer's notification within three months, the employer may give up the patent altogether by virtue of 16(2) ArbNErfG.

The termination of employment has no effect on the acquired inventions and rights.

4.3 The Parties' Duty to Inform

In order to ensure the employer's ability to exercise their appropriate right to service inventions (Section 6 ArbNErfG), the employee is obliged to notify the employer about such inventions by virtue of Section 5(1) ArbNErfG.[124] Once an invention has been made, notice has to be given in writing without undue delay by virtue of Section 5(1) ArbNErfG. The notification shall state the technical problem, its solution and describe the process of invention. It shall include all records necessary for comprehension of the invention and indicate business instructions and guidelines, experience or results achieved in connection with the employment that have been used in the process of invention as well as other employees involved. Original achievements by the inventor personally shall then be marked accordingly (Section 5(2) ArbNErfG). The employee's obligation to notify the employer pursuant to Section 5 ArbNErfG only applies to service inventions within the meaning of Section 4(2) ArbNErfG.

Free inventions, however, have to be communicated to the employer as well, according to Section 18 ArbNErfG. This obligation of communication is a necessary pre-condition to ensure the employer's right to acquire a non-exclusive right of use in free inventions pursuant to Section 19 ArbNErfG. Only free inventions that are evidently not exploitable within the employer's business – and accordingly do not come within the scope of Section 19 ArbNErfG – are exempted from the obligation of communication by virtue of Section 18(3) ArbNErfG. The communication has to be in writing and shall include all information necessary for the employer to determine whether or not the invention comes within the definition of a free invention within the meaning of Section 4(3) ArbNErfG.[125]

4.4 Employee Remuneration Right

The ArbNErfG provides the employee with a right to equitable compensation against their employer provided that the employer has claimed a service invention (Section 9 ArbNErfG).[126] It shall enable the employee to take a share in all economic advantages

124. Czychowski/Langfinger, *supra* n. 111, 221.
125. Schoenen, *supra* n. 110, *§ 43* Arbeitnehmererfindungsrecht marginal n. 5.
126. Cf. also Czychowski/Langfinger, *supra* n. 111, 221; Rother, *supra* n. 3, p. 13.

their employer enjoys on the basis of the invention.[127] The right exists for as long as the patent is valid, provides the employer a monopoly, and is used by the employer.

The amount of compensation is to be agreed upon by the employer and employee within a reasonable period of time after the employer has claimed the invention (Section 12(1) ArbNErfG). If the parties fail to reach such an agreement, the amount of compensation is to be determined and paid by the employer no later than three months after the patent has been granted (Section 12(3) ArbNErfG). Determination of the amount of compensation is not only the employer's prerogative but also their duty, the fulfilment of which the employee can claim before the courts.[128] Moreover, the employee may object to the employer's decision in case of disagreement, within two months (Section 12(4) ArbNErfG). The employee has a right to information from their employer concerning the type and extent of the use of the invention.[129] This right allows the employee to obtain a certain degree of transparency with regard to the process of calculation. If circumstances that have been authoritative for the calculation of the compensation change substantially, both parties can oblige the other one to agree to a new assessment of compensation (Section 12(6) ArbNErfG). A right to a refund is, on the other hand, not provided for by the law.

With regard to the amount of compensation, the ArbNErfG in connection with the 'Guidelines on the compensation for employees inventions in private employment' by the Federal Minister of Labour provide guidelines as to the process of calculation (Section 11 ArbNErfG). According to the guidelines, the amount of compensation is to be determined by two parameters: the inventive value and the so-called *share factor*.

4.4.1 The Inventive Value

The inventive value represents the economic exploitability of the invention, which is the most authoritative element in the assessment of the amount of compensation.[130] The inventive value can be determined in three different ways: by licence analogy, on the basis of its quantifiable business use or by mere estimation.[131]

The most significant calculation method in practice is calculation by licence analogy. It is especially applied if the invention is used in commercial transactions.[132] The value of the invention is determined by a fictitious licence fee, which the employer would have had to pay to a third party licensor in order to use the invention.[133]

127. *Spulkopf, X ZR 6/96*, GRUR 1998, 686 (BGH, 13 November 1997); *Copolyester II, X ZR 132/95*, GRUR 1998, 689 (BGH, 13 November 1997); Schoenen, *supra* n. 110, *§ 43* Arbeitnehmererfindungsrecht marginal n. 21.
128. *Cromegal, I ZR 28/61*, GRUR 1963, 135 (BGH, 28 June 1962); Schoenen, *supra* n. 110, *§ 43* Arbeitnehmererfindungsrecht marginal n. 23.
129. *Drahtinjektionseinrichtung, X ZR 215/00*, GRUR 2002, 609 (BGH, 6 February 2002); *Abgestuftes Getriebe, X ZR 127/99*, GRUR 2002, 801 (BGH, 16 April 2002); Schoenen, *supra* n. 110, *§ 43* Arbeitnehmererfindungsrecht marginal n. 26.
130. Schoenen, *supra* n. 110, *§ 43* Arbeitnehmererfindungsrecht marginal ns. 34 et seq.
131. Cf. Rother, *supra* n. 3, p. 14.
132. Cf. Schoenen, *supra* n. 110, *§ 43* Arbeitnehmererfindungsrecht marginal n. 36; Rother, *supra* n. 3, p. 14.
133. Cf. Rother, *supra* n. 3, p. 14.

Calculation pursuant to quantifiable business use depends on the advantage the invention creates for the employer. The costs the employer incurred as well as a reasonable employer profit are to be deducted from the profits generated through the invention.[134] The difference between these factors constitute the business use and thereby the inventive value. This calculation method is employed when no sales business is undertaken but the invention is used internally within the process of production allowing cost savings.[135]

Calculation by way of estimation is conducted by analogy with a purchase agreement. The sum that would have had to be paid as consideration for the acquisition of the invention from a third party inventor constitutes the inventive value. This calculation method is the most appropriate for inventions that are yet to be used.[136]

4.4.2 *The Share Factor*

From the determined inventive value the share of the business operation in the creation of the invention has to be deducted.[137] This deduction takes into account that it is not a free invention that is compensated but a service invention, i.e. an invention that has been made in connection with the employee's work for which the employee receives a regular salary. The share factor is expressed in percentage points and ranges from 8-25% at an average of 16.5% in practice. According to the guidelines it is calculated from three factors: 'Who presented the problem for the invention? How did the inventor come to the solution? What actual position did the employee have at the time of the invention within the business operation of the employer?'[138]

With regard to the first factor, the following applies: the higher the employee's initiative and involvement in the development of the problem to be solved and in the indication of operational deficiencies and needs, the higher their share in the creation of the service invention and the smaller the share factor to be deducted. The guidelines contain a list of scenarios each involving a different level of initiative by the employee to what a certain share factor is then respectively assigned.[139] The second factor considers whether or not, and to which extent, the solution - and therefore the invention - has been based on work experience and results achieved through the employment, the resources provided by the employer that have been used in the process of invention, etc.[140] Finally the employee's actual position in the business and the tasks they are entrusted with are taken into account. Based on the employee's position and their salary, could the employer expect them to contribute to technical developments and inventive solutions? What insight did the employee have because of

134. Cf. Rother, *supra* n. 3, p. 14.
135. Cf. Schoenen, *supra* n. 110, *§ 43* Arbeitnehmererfindungsrecht marginal n. 37; Rother, *supra* n. 3, p. 14.
136. Schoenen, *supra* n. 110, *§ 43* Arbeitnehmererfindungsrecht marginal n. 38.
137. Cf. Rother, *supra* n. 3, p. 14.
138. Cf. Rother, *supra* n. 3, p. 14.; cf. also Schoenen, *supra* n. 110, *§ 43* Arbeitnehmererfindungsrecht marginal n. 41.
139. Cf. Schoenen, *supra* n. 110, *§ 43* Arbeitnehmererfindungsrecht marginal ns. 42 et seq.
140. Cf. Schoenen, *supra* n. 110, *§ 43* Arbeitnehmererfindungsrecht marginal ns. 46 et seqq.

their position within the company? The guidelines once again provide a list of categories of employees that are assigned a different share factor according the employee's position and tasks.[141]

Taking into account all three factors, the share factor is determined and deducted from the inventive value. The difference remaining constitutes the equitable compensation the employee is entitled to.[142]

4.4.3 Remuneration in Practice

Contractual agreements to waive the inventor's rights of Sections 14 (2), 16 (1) and 13 ArbNErfG are very popular. The amount which is paid by the employer must not be inequitable. Otherwise the agreement is not valid. A comparison between the amount to be paid and the concrete economic advantages the inventor will miss has to be drawn. Recently case law has changed insofar in Germany. Each case must be analysed separately. To fulfil the requirements of Sections 14 and 16 ArbNErfG instead of buying out the inventor's rights, means much more administrative work for the employer. Usually remuneration between EUR 100 and EUR 500 per each invention is paid to buy out the rights of Sections 14 and 16 ArbNErfG. If additionally the right of Section 13 ArbNErfG is bought out and possibly also claims for information, the average remuneration is slightly higher. Another option is a flat-rate remuneration. One aim can be to reduce administrative work for the employer. Another aim can be to agree on a specific amount to be fair. Such provisions must be carefully drafted with regard to the right of Section 12 (6) ArbNErfG. One option can be to predict the overall turnover including specific reference figures. The compensation in whole corresponds with the predicted overall turnover. Every four or five years the model has to be checked if still adequate and fair. If this is not the case anymore, the inventor can receive an additional remuneration.[143]

5 EMPLOYEES' UTILITY MODELS

5.1 The Legal Framework

As mentioned earlier, the legal relationship between employees and employers regarding technical inventions - including utility models - is prevailingly regulated by the ArbNErfG.[144] Contractual agreements may complement these statutory regulations but must not deviate from them to the detriment of the employee prior to the communication of the invention (Section 22 ArbNErfG). In addition, the GebrMG contains general provisions on inventions capable of being applied for as a utility model.

141. Cf. Schoenen, *supra* n. 110, *§ 43* Arbeitnehmererfindungsrecht marginal ns. 52 et seqq.
142. Cf. for details: Schoenen, *supra* n. 110, *§ 43* Arbeitnehmererfindungsrecht marginal ns. 68 et seqq.
143. Trimborn, Mitt. 2015, 116-122.
144. Cf. *supra* '4. Employees' Inventions' section 4.1 'The Legal Framework'.

5.2 Ownership of Rights

5.2.1 Allocation of Rights in Employee's Inventions

Section 6 ArbNErfG confers on the employer an original appropriation right, which entitles them to claim inventions made by the employee in connection with their employment, i.e. so-called *service inventions within the meaning of Section 4(2) ArbNErfG*. According to Section 2 ArbNErfG the term 'invention' refers to inventions that are either patentable or capable of being applied for as a utility model. Once the employer has claimed an invention, they are generally obligated to submit an application for the grant of a patent (Section 13(1) sentence 2 ArbNErfG). Only if – following a reasonable assessment of the invention's likely exploitation – protection of the invention as a utility model seems more appropriate, the employer may choose to apply for the registration of the invention as a utility model instead (Section 13(1) sentence 2 ArbNErfG).

Claiming the invention pursuant to Section 6 ArbNErfG requires a declaration by the employer, to be given to the employee within a certain time period after communication of the invention.[145] In the absence of a declaration, the invention is deemed to be claimed if it is not released to the employee within four months after the notification (Section 6(2) ArbNErfG). Following that, the employer can start using the invention right away without any prior agreements with their employees on the allocation of rights.[146]

If the employer releases the invention by means of a written declaration, the invention is considered free (Section 8 ArbNErfG).[147] According to Section 4(3) ArbNErfG free inventions furthermore include those inventions that have been made outside business operations and do not build upon the experience or results achieved in connection with the employment.[148] With regard to rights in free inventions the aforementioned applies also in the context of utility models.[149]

5.2.2 Ownership of Rights in Inventions by University Employees

For inventions made by academic staff the general regulations on rights allocation apply.[150] However, professors and other academic staff benefit from the newly introduced scientific privileges contained in Section 42 Nos. 1-4 ArbNErfG. These include the right of positive (No. 1) and negative (No. 2) publication, the right to use

145. Cf. Czychowski/Langfinger, *supra* n. 111, 221; Rother, *supra* n. 3, p. 11.
146. Cf. Bayreuther, *supra* n. 112, 1123.
147. Schoenen, *supra* n. 110, *§ 43* Arbeitnehmererfindungsrecht marginal n. 18.
148. Czychowski/Langfinger, *supra* n. 111, 221.
149. Cf. *supra* '4. Employees' Inventions' section 4.2.1 'Ownership of Rights and Termination: Allocation of Rights in Employees' Inventions'.
150. Cf. *supra* '4. Employees' Inventions' section 4.2.3 'Ownership of Rights and Termination: Ownership of Rights in University Employees' Inventions'.

the invention in connection with ongoing scientific research and teaching (No. 3) and the right to claim an interest of 30% of the benefits arising from the exploitation of the invention (No. 4).[151]

5.2.3 Termination

Duration and termination of protection under the GebrMG is comprehensively regulated by Section 23 GebrMG. According to Section 23(1) GebrMG, protection is effective on the day of registration. The minimum term of protection is three years, beginning at the end of the month in which the utility model has been registered. To maintain protection, from the fourth year on, an annual fee has to be paid (Section 23(2) GebrMG). In this way, the term of protection can be extended up to ten years. Failure to pay the annual maintenance fee, on the other hand, leads to the expiration of the utility model (Section 23(3) No. 2 GebrMG). Moreover, the utility model ceases to exist with a respective waiver by the right holder (Section 23(3) No. 1 GebrMG). In case of service inventions, Section 16 ArbNErfG applies accordingly in this regard. Before waving the right to the utility model for good, the employer has to notify the employee and transfer their rights if requested.[152]

5.3 The Parties' Duty to Inform

For the employer's ability to exercise their appropriation right to service inventions, the employee is obliged to notify the employer about such inventions (Section 5(1) ArbNErfG).[153] Notice has to be given in writing without undue delay after an invention has been made (Section 5(1) ArbNErfG). Section 5(2) ArbNErfG states the precise requirements with regard to the contents of the notification. The obligation of notification pursuant to Section 5 ArbNErfG only applies to service inventions.

Nonetheless, free inventions have to be communicated to the employer as well, by virtue of Section 18 ArbNErfG. This is a necessary pre-condition to ensure the employer's right to acquire a non-exclusive economic right in the free invention pursuant to Section 19 ArbNErfG. Only free inventions that are evidently not exploitable within the employer's business are exempted from the obligation of communication (Section 18(3) ArbNErfG). The communication must be in writing and shall include all information necessary to determine whether the invention qualifies as a free invention within the meaning of Section 4(3) ArbNErfG.[154]

151. Czychowski/Langfinger, *supra* n. 111, 222.
152. Cf. *supra* '4. Employees' Inventions' section 4.2.4 'Ownership of Rights and Termination: Termination'.
153. Czychowski/Langfinger, *supra* n. 111, 221.
154. Schoenen, *supra* n. 110, *§ 43* Arbeitnehmererfindungsrecht marginal n. 5.

5.4 Employee Remuneration Right

Section 9 ArbNErfG, which confers on the employee a right to equitable compensation from their employer, also applies to inventions that are protected as utility models. Therefore the remarks on the employee's remuneration right with regard to patentable inventions apply likewise in the context of inventions protected as utility models.[155]

6 EMPLOYEES' DESIGN RIGHTS

6.1 The Legal Framework

Until 2014, the so-called *Geschmacksmustergesetz*, i.e. Registered Designs Act of 1876, regulated protection through design rights in Germany and implemented the European Design Directive 98/71.[156] On 1 January 2014, the Geschmacksmustergesetz was replaced by the Designgesetz (DesignG). The most evident change through the newly introduced DesignG has been the change in terminology for the purpose of adapting it to the language used internationally.[157]

The core provision regarding employee-developed designs, Section 7(2) DesignG, shows strong parallels with Section 69b UrhG.[158] The latter provision applies to employee software protected under copyright and is not directly applicable to employees' designs. However, due to the provisions' similarity, a number of considerations and analysis made with regard to Section 69b UrhG can be taken into account in the context of employees' designs.[159]

6.2 Ownership of Rights

6.2.1 General Allocation of Rights to Designs

Section 7(2) DesignG regulates the allocation of rights to designs in employment relationships. The provision replicates Article 14(3) of the European Regulation on Community designs.[160] Instead of a derivative acquisition of rights that was initially foreseen by the GeschmMG, Section 7(2) DesignG now prescribes an initial acquisition of rights by the employer: where a design has been developed by an employee in the

155. Cf. *supra* '4. Employees' Inventions' section 4.4 'Employee Remuneration Right'.
156. *Directive 98/71/EC of the European Parliament and of the Council of 13 October 1998 on the legal protection of designs*, available at: http://eur-lex.europa.eu/legal-content/EN/ TXT/HTML/?uri=CELEX:31998L0071&qid=1428516903311&from=EN (accessed 16 May 2017).
157. Cf. Schicker/Haug, *Grundzüge des Designgesetzes (Main features of the Registered Designs Act)*, NJW 2014, 726, 727 (2014).
158. Cf. *supra* '2. Employees' Copyright' section 2.2.9 'Specific Provision on Employees' Software'.
159. Eichmann, *supra* n. 106, § 7 GeschmMG marginal n. 17.
160. *Council Regulation (EC) No. 6/2002 of 12 December 2001 on Community designs*, available at: http://eur-lex.europa.eu/LexUriServ/LexUriServ.do?uri=OJ:L:2002:003:0001:0024:en:PDF(accessed 16 May 2017).

execution of their duties or following the instructions given by their employer, the right to the registered design shall vest in the employer, unless otherwise specified in the contract.

The provision contains significant parallels to Section 4 ArbNErfG and differentiates likewise between service and free designs.[161] Only service designs, i.e. those that have been developed in the execution of the employee's duties or following the instructions given by the employer, come within the provision's scope of application.

For the determination of the employee's duties, the actual sphere of obligations with which the employee has been entrusted at the time of the design's completion is authoritative. Although the obligations contractually agreed upon are to be taken into account as well, the employee's duties can comprise a wider range of obligations than originally enumerated in the contract of employment.[162] An employer providing the employee with resources for the design's creation may indicate a service design, although this is not necessarily the sole inevitable consequence.[163] General references in the contract of employment that encourage the employee to pay special attention to potential business improvements do not suffice as reasonable grounds to qualify a design as work-related. Nor does the employee's general duty of good faith.[164]

Where a design has not been created in the execution of the employee's duties, it might still qualify as a service design if created following instructions by the employer. Instructions are specific individual work assignments given by the employer by virtue of their managerial authority that are characteristically accomplished with the design's completion.[165] Instructions may go beyond the employee's general duties or exceed the employer's managerial authority. Such exceeding instructions nonetheless lead to service designs if created in execution of them.[166]

Service designs must be distinguished from free designs, the right to which vests in the employee and can only be acquired by the employer by means of a contract with the employee.[167] Such individual contracts not only contain provisions on the transfer of rights, but also regulations concerning the employee's right to respective compensation.[168] Often, arrangements on free designs are merely implied, based on the conclusive conduct by the employer and employee.[169] In this case precise provisions on

161. Cf. Eichmann, *supra* n. 106, § 7 GeschmMG marginal n. 20.
162. *Gummielastische Masse, X ZR 130/93*, NJW-RR 1995, 696, 698 (BGH, 17 January 1995); cf also Eichmann, *supra* n. 106, § 7 GeschmMG marginal n. 20.
163. Eichmann, *supra* n. 106, § 7 GeschmMG marginal n. 20; for s. 69b UrhG cf. Grützmacher, *Praxiskommentar zum Urheberrecht (Copyright Law: Commentary)* § 69b marginal n. 7. (Wandtke/Bullinger, 4th ed,. C.H. Beck Munich, 2014).
164. Eichmann, *supra* n. 106, § 7 GeschmMG marginal n. 20.
165. Eichmann, *supra* n. 106, § 7 GeschmMG marginal n. 20.
166. Loewenheim, *Urheber in Arbeits- und Dienstverhältnissen (Authors in Private and Public Sector Employment Relationships)*, § 69b UrhG marginal n. 6 (Schricker/Loewenheim, 5th ed., C.H. Beck Munich, 2017).
167. Eichmann, *supra* n. 106, § 7 GeschmMG marginal ns. 20 et seq.
168. Cf. *infra* '6. Employees' Design Rights' section 6.4 'Employee Remuneration Right: Service Design'.
169. Eichmann, *supra* n. 106, § 7 GeschmMG marginal n. 27.

the scope of rights transferred are usually lacking, therefore it is assumed that rights are only transferred to such an extent that is necessary for the design's exploitation within business operations.[170]

Where a design has not been developed within the material and temporal scope of the employment relationship, it is considered to be free, as are designs that are not causally determined by the employee's duties or instructions by the employer.[171] Contrary to Section 4(2) No. 2 ArbNErfG, the DesignG does not consider designs as service designs that are solely based on experience gained by the employee through the employment but are otherwise not work-related.[172] The usage of business experience for the design's creation may constitute a violation of the employee's duties of good faith or a potential non-compete clause. Nonetheless, these designs qualify as free designs.[173] On the other hand, however, designs developed outside working hours or the place of work do not necessarily amount to free designs. If their development took place in the execution of the employee's duties or instructions given by the employer, they rather constitute business-related designs.

The authoritative point of time to determine whether designs fulfil the requirements of Section 7(2) DesignG is the finalisation of the design.[174] If the preconditions are then given, the right to the design vests in the employer not with notification of the design but already with the design's completion by the employer.[175] The acquisition of rights is comprehensive; the Specific Purpose Principle that applies to the transfer of copyrights is not applicable in the context of designs. The employer acquires the exclusive economic right in the design, which entitles them to decide unilaterally if, when, for how long and to what extent the design shall be exploited. On the other hand, the employer does not incur an obligation to exploit the design. The employer may license or assign this right of exploitation. The right depends neither on a prior registration nor on the continuity of the employment relationship. Only in designs created before the start of employment is no economic right conferred on the employer.[176]

With the acquisition of the right to the design, the employer does not incur an obligation to register the design in general. Whether the service design shall be registered at all, as well as the countries the design shall be registered in, the type of registration and the duration of protection, are all decisions to be made solely by the employer.[177] Only under certain specific circumstances may an obligation to register exceptionally arise. This is the case where a special contractual regulation exists or where the employee shows a special vulnerability. The latter may arise out of a special interest of the employee in the right to be named as the designer. In these cases, the employer shall register the design within a reasonable period of time. With the expiry

170. Eichmann, *supra* n. 106, § 7 GeschmMG marginal n. 27.
171. Eichmann, *supra* n. 106, § 7 GeschmMG marginal n. 21.
172. For § 69b UrhG cf. Grützmacher, *supra* n. 163, § 69b UrhG marginal n. 31.
173. Eichmann, *supra* n. 106, § 7 GeschmMG marginal ns. 21 et seq.
174. Eichmann, *supra* n. 106, § 7 GeschmMG marginal n. 22.
175. Eichmann, *supra* n. 106, § 7 GeschmMG marginal ns. 17, 24.
176. Eichmann, *supra* n. 106, § 7 GeschmMG marginal n. 24.
177. Eichmann, *supra* n. 106, § 7 GeschmMG marginal n. 24.

of said period, considering the legal rationale of Section 13(3) ArbNErfG, the employee shall be entitled to register the design in their own name and on their own account. The employer may subsequently still demand the transfer of rights in the design in return for a compensation of the employee's expenses.[178]

6.2.2 *Allocation of Rights to Designs in the Public Sector*

The concept of allocation of rights pursuant to Section 7(2) DesignG applies likewise to designs developed by employees in the public sector. Nonetheless, designs by university employees are generally considered to be free. This follows from considerations similar to those taken into account in the context of employee software protected under copyright. The creation of computer programs is generally not considered to be part of the university employee's public duties. Thus, software created by university employees under the Freedom of Sciences guaranteed by Article 5(3) GG, amounts to free software, to which Section 69b UrhG does not apply. Only in rare cases, where the creation of software explicitly falls within the university employees' contractual duties, does the computer program come within the scope of 69b UrhG.[179,180] The same applies to designs created by university employees. Unless the creation of designs is explicitly listed as the university employee's duty, they amount to free designs, rights to which remain with the employee. If a university employee's design is considered a service design though, rights to it are transferred to the university. In this case, it should be noted that a provision similar to Section 42 ArbNErfG[181] providing special scientific privileges does not exist within the DesignG.[182]

6.2.3 *Termination*

Protection is effective with the design's registration with the German Patent and Trade Mark Office.[183] The registered design lasts up to 25 years subject to the payment of renewal fees to be paid every five years (Section 28(1) DesignG). Twenty-five years after the date of registration the right holder's monopoly ceases to exist and the design falls into the public domain (Section 27 DesignG).[184]

178. Eichmann, *supra* n. 106, § 7 GeschmMG marginal n. 24.
179. Cf. *supra* '2. Employees' Copyright' section 2.2.9 'Specific Provision on Employees' Software'.
180. Cf. Czychowski, *supra* n. 63, § 69b UrhG marginal n. 10; Grützmacher, *supra* n. 163, § 69b UrhG marginal n. 15; Loewenheim, *supra* n. 166, § 69b marginal n. 6.
181. Cf. *supra* '4. Employees' Inventions' section 4.2.3 'Ownership of Rights in University Employees' Inventions'.
182. Cf. also Eichmann, *supra* n. 106, § 7 GeschmMG marginal n. 21.
183. Cf. Schicker/Haug, *supra* n. 157, 727.
184. Cf. Schicker/Haug, *supra* n. 157, 727; Kraßer, *supra* n. 95, s. 1, Ch. 1, § 2, II, b) Geschmacksmusterschutz marginal n. 3.

6.3 The Parties' Duty to Inform

Section 7(2) DesignG contains an implied obligation on the employee to notify the employer of the service design. This duty to inform is a mandatory precondition to ensure the original acquisition of rights in service designs by the employer pursuant to Section 7(2) DesignG. It can also be deduced from the employee's general obligation of good faith. Notification shall be given without undue delay after the design's completion.[185]

Free designs, on the other hand, do not need to be communicated to the employer. This is true for as long as there are no doubts concerning the qualification of the design as a free design. Otherwise, the employee is obliged to provide all necessary information for the employer to be able to determine whether or not the design comes within the scope of Section 7(2) DesignG. Generally speaking, the employee does not incur an obligation to offer the right to a free design to their employer. However, there are some exceptions to this rule. First, in situations that are – with a view to the interests involved – similar to the situation dealt with in Section 19(1) ArbNErfG,[186] the obligation to offer a free design may occur. For this duty to arise, the employee must intend to exploit their free design during the employment relationship. Additionally, the free design must show a connection to the employer's field of work. The second exception concerns designs based on work experience but nonetheless considered free designs,[187] which have to be offered to the employer occasionally owing to the employee's obligation of good faith or resulting from the violation of a non-compete clause.[188]

If the employer is already aware of the design, notification is unnecessary. Formal requirements regarding the notification do not exist. Failure to notify the employer of a service design infringes the employer's right to registration and their economic right in the design, for which the employer may claim damages against their employee. Registration of the design in the employee's own name in breach of Section 7(2) DesignG entitles the employer to claims under Section 9(1) DesignG: the employer may either claim a transfer of the registered design or the employee's consent to the registration's deletion.[189]

6.4 Employee Remuneration Right

6.4.1 Service Designs

Neither the DesignG itself nor the official explanatory memorandum to the DesignG contains provisions on the matter of remuneration. However, it is commonly agreed

185. Eichmann, *supra* n. 106, § 7 GeschmMG marginal n. 23.
186. Cf. *supra* '4. Employees' Inventions' section 4.3 'The Parties' Duty to Inform'.
187. Cf. *supra* '6. Employees' Design Rights' section 6.2.1 'Ownership of Rights and Termination: General Allocation of Rights to Designs'.
188. Eichmann, *supra* n. 106, § 7 GeschmMG marginal n. 23.
189. Eichmann, *supra* n. 106, § 7 GeschmMG marginal n. 23.

that, generally speaking, the employer's acquisition of rights to service designs pursuant to Section 7(2) DesignG does not entitle the employee to further separate remuneration.[190] This finding is deduced from the universal employment law principle that the employee's salary constitutes consideration for all work results achieved by the employee on behalf and for the account of the employer, which includes service designs within the meaning of Section 7(2) DesignG.[191] The special remuneration obligation provided by the ArbNErfG cannot be applied to service designs - neither directly nor as an analogy.[192] The compensation for the transfer of design rights therefore remains reserved to individual contract regulations.[193]

The absence of a right to equitable remuneration has been called into question nowadays. The protective effects of registered designs resemble those of utility models; the latter conferring on the employee a statutory right to equitable remuneration pursuant to Section 9 ArbNErfG. With both intellectual property rights providing a similar level of protection for the right holder, it could be argued that their transfer should lead to a right to equitable remuneration in both cases.[194] Employers may therefore be well advised to at least consider an extra remuneration as an incentive for employees to engage in the development of service designs.

Even where the employee provides service designs that go beyond their contractually owed design performance they are not necessarily entitled to extra remuneration. Only if such designs are also copyright protected and thus come within the mandatory scope of Section 32(1) sentence 3 UrhG or Section 32a UrhG may the employee claim additional remuneration.[195] In case of copyright protection authors may also profit from statutory remuneration rights.[196]

According to Section 7(2) DesignG, the parties remain free to enter into deviating contractual arrangements. These may concern the allocation of rights in service designs, the right to register a service design, and also the issue of compensation. In this regard, the parties may agree upon current compensation during the time of the design's exploitation or on a lump sum as compensation for the transferred rights in service designs.[197]

190. Eichmann, *supra* n. 106, § 7 GeschmMG marginal n. 25.
191. Eichmann, *supra* n. 106, § 7 GeschmMG marginal n. 25; cf. also for ss. 69b and 32 UrhG Dreier, *supra* n. 8, § 69b UrhG marginal n. 10.
192. *Wetterführungspläne II*, *supra* n. 67, marginal n. 16; cf. also Eichmann, *supra* n. 106, § 7 GeschmMG marginal n. 25; Rother, *supra* n. 3, p. 14.
193. Rother, *supra* n. 3, p. 14.
194. Eichmann, *supra* n. 106, § 7 GeschmMG marginal n. 25.
195. Eichmann, *supra* n. 106, § 7 GeschmMG marginal n. 26; for further details on s. 32 and s. 32a UrhG cf. *supra* '2. Employees' Copyright' section 2.4.1 'Employee Remuneration Right: Contractual Remuneration Right'.
196. For further details cf. *supra* '2. Employees' Copyright' section 2.4.2 'Employee Remuneration Right: Statutory Remuneration Right'.
197. Eichmann, *supra* n. 106, § 7 GeschmMG marginal n. 27.

6.4.2 *Free Designs*

Individual contractual regulations are particularly important when it comes to free designs. As free designs do not fall within the scope of Section 7(2) DesignG, rights remain with the employee. The employer may therefore acquire rights in free designs only by means of an individual contract with the employee. Naturally, such agreements will contain provisions on extra remuneration as compensation for the rights' transfer.[198] As mentioned earlier, these individual arrangements on free designs are often not a matter of expressed contractual clauses but rather implied contracts on the basis of conclusive conduct by the parties.[199]

7 EMPLOYEES' PLANT BREEDER'S RIGHTS

7.1 The Legal Framework

Protection for breeders and new plant varieties in Germany is regulated by the Plant Breeders' Rights Act (SortSchG). Breeders enjoy protection for plant varieties that are distinct, uniform, stable and new (Section 1(1) SortSchG). Protection does not arise automatically but is effective with the registration of the plant variety's denomination with the Federal Plant Varieties Office (Bundessortenamt) (Section 1(1) No. 5 SortSchG). Plant Breeders' Rights granted under the SortSchG are limited in territory and provide protection only within the country's national borders. Besides national protection, breeders may benefit from protection at the European level through the Community plant variety rights, established by Regulation No. 2100/94 on Community plant variety rights (Community Regulation).[200]

The SortSchG does not contain any specific provisions on the relationship between employers and employees concerning rights to plant varieties. What is more, the provisions of the ArbNErfG[201] are not applicable in this context.[202] Therefore, the regulation of the relationship, including all arising questions (e.g. allocation of rights, equitable remuneration) remains reserved to individual contractual arrangements.

7.2 Ownership of Rights

The person who bred, or discovered and developed the variety, or their successor in title, shall be entitled to the plant variety right (Section 8(1) SortSchG). There are no exceptions from this general rule for employment relationships, therefore rights in a plant variety bred or discovered by an employee remain with the latter. For the

198. Eichmann, *supra* n. 106, § 7 GeschmMG marginal n. 27.
199. Eichmann, *supra* n. 106, § 7 GeschmMG marginal n. 27.
200. *Council Regulation (EC) No. 2100/94 of 27 July 1994 on Community plant variety rights*, available at: http://eur-lex.europa.eu/legal-content/EN/TXT/PDF/?uri=CELEX:31994R2100&from=EN (accessed 16 May 2017).
201. Cf. *supra* '4. Employees' Inventions'.
202. Rother, *supra* n. 3, p. 14.

employer to acquire rights in an employee's plant variety a corresponding contractual arrangement with the employee is therefore necessary. The parties may choose to include such an arrangement within the contract of employment. The employee might then be obliged to transfer rights in new plant varieties or at least offer a respective transfer to the employer on certain individually negotiated terms. Most probably, the parties will agree that this obligation only arises if a certain connection exists between the plant variety's breeding or discovery and the employee's work obligations. If such explicit contractual arrangements on the allocation of employees' plant variety rights are missing, one could consider an analogy with the aforementioned provisions for the other types of intellectual property rights.[203] In this case, an implied obligation for the employee to transfer their plant variety rights to the employer could be assumed for as far as an inner connection with the employee's work obligations exist.[204] However, no jurisprudence exists in this regard, therefore employers are well advised to include explicit clauses on the allocation of rights in employees' plant variety in the contract of employment.

Unlike copyright, plant variety rights are fully assignable, according to Section 11(1) SortSchG. The employee may therefore choose to either assign their rights (Section 11(1) SortSchG) or license them (Section 11(2) SortSchG) to the employer.

7.2.1 Assignment

If the employee chooses to assign their rights to the employer, the latter will acquire all rights in and to the plant variety. This transfer technically includes the employee's moral rights. Only the hard core of the employee's moral rights, i.e. those that affect significant personal interests, remains with the original right holder.[205]

The employer acquires not only the right to use the new plant variety but most importantly the right to the plant variety as such, i.e. the right to register the plant variety with the Register of Plant Varieties. They thereby become solely entitled to the exclusive rights listed in Section 10 SortSchG. Outside of employment relationships, the most common means of transferring plant variety rights is licensing them to a third party.[206] In employment relationships, however, the employer has a strong interest in acquiring the right to the plant variety as such, including the right of registration. Therefore, a comprehensive rights transfer by way of an assignment will be the more common and frequent means of transfer between employees and employers.

---

203. Cf. *supra* '2.2 Ownership of Rights and Termination' within the different subs. – for s. 43 UrhG cf. *supra* '2. Employees' Copyright', for s. 2(2) HalblSchG cf. *supra* '3. Employees' Semiconductor Chips', for s. 6 ArbNErfG cf. *supra* '4. Employees' Inventions' and '5. Employees' Utility Models', for s. 7(2) DesignG cf. *supra* '6. Employees' Design Rights'.
204. For further details on the requirement of an inner connection in the context of employees' copyright cf. *supra* '2. Employees' Copyright' section 2.2.1 'Ownership of Rights and Termination: Transfer of Economic Rights in Employees' Works'.
205. Leßmann/Würtenberger, *Deutsches und europäisches Sortenschutzrecht (German and European Plant Breeders' Rights)*, § 4 marginal n. 8 (2nd ed., Nomos, 2009).
206. Cf. *infra* '7. Employees' Plant Breeders' Rights' section 7.2.2 'Ownership of Rights and Termination: Licence'.

The contract between the employer and the employee may not include precise provisions on the scope of transfer. In this case, the universal intellectual property law principle of transfers being limited to their specific purpose applies.[207] Rights are transferred only to the extent necessary to achieve the purpose pursued by the parties, which – in this context – will mostly be the exploitation of the plant variety within the object of the employer's business. The contract may also be silent on whether rights are entirely transferred or merely licensed to the employer. Because of the employer's strong interest in the right of registration (cf. *supra*), and the importance of registration for the exploitation of the plant variety within the object of the business, an assignment – as the more common means of transfer within employment relationships – may be well assumed.

Sections 398 et seqq. BGB contain general provisions on assignments. Unlike the Community Regulation,[208] German law does not impose any formal requirements on the transfer of plant breeders' rights by assignment. Additionally, no formal disclosure is mandatory. The registration of change in ownership with the Register of Plant Varieties is of mere declarative effect – in contrast to the Community Regulation regime, where a formal disclosure act is a mandatory precondition of the validity of the assignment.[209] However, also under German law, the person originally registered as the rights owner remains formally entitled and obligated for as long as the change in ownership is not registered with the Register of Plant Varieties.[210]

7.2.2 *Licence*

Outside employment relationships, in practice, the most important means of transferring plant variety rights is licensing them to a third party.[211] However, a rights transfer by means of licensing is not precluded within employment relationships, even though it is rarely to be found. The right to make use of the plant variety to the extent contractually agreed upon is conferred on the licensee, i.e. the employer. The right to the plant variety as such on the other hand, remains with the licensor, i.e. the employee.

The importance of licences in the context of plant breeders' rights originates in the right holder's inability to economically exploit the plant variety and enforce their exclusive rights to the fullest extent. Costs and efforts necessary for such a comprehensive exploitation and rights' enforcement would far outweigh the right holder's potential benefits gained from it.[212] This is even more true where an employee is the plant variety rights holder. Therefore, the employee might be well advised to license

207. Leßmann/Würtenberger, *supra* n. 205, § 4 marginal n. 11.
208. Cf. Art. 23(2) Community Regulation requiring the contract to be in writing and signed by both parties; cf. also Leßmann/Würtenberger, *supra* n. 205, § 4 marginal n. 13.
209. Cf. Art. 23(4) Community Regulation.
210. Leßmann/Würtenberger, *supra* n. 205, § 4 marginal n. 14.
211. Leßmann/Würtenberger, *supra* n. 205, § 4 marginal n. 23; for employment relationship cf. *supra* 7. Employees' Plant Breeders' Rights' section 7.2.1 'Ownership of Rights and Termination: Assignment'.
212. Leßmann/Würtenberger, *supra* n. 205, § 4 marginal n. 23.

their plant variety right to the employer, thereby benefiting from the latter's business relations, their professional know-how and expertise, and, most importantly, from the employer's market position. In addition, the employee maintains the right to the plant variety as such and thus remains able to object to any potential abuse of the plant variety or the propagating material.

7.2.3 Termination

The duration of plant breeders' rights in Germany is regulated by Section 13 SortSchG, which is based on Article 19 of the International Convention for the Protection of New Varieties and Plants ('UPOV Convention'). According to Article 19 UPOV Convention, plant varieties have to be protected for 20 years in general, while a number of varieties even enjoy protection for at least 25 years. Members of the Convention may only deviate from this minimum duration period if they provide longer protection.[213] Accordingly, in Germany, the general protection period lasts for 25 years, according to Section 13 SortSchG. Similarly, Community plant variety rights are protected for 25 years in general.[214] A limited number of plant varieties are protected in Germany and under the CPVR regime for up to 30 years.[215] However, for plant breeders to benefit from the potential duration of 25 or 30 years, an annual maintenance fee has to be paid.[216]

Reasons for termination of plant breeders' rights other than the expiry of the protection period are listed in Section 31 SortSchG. This list is exhaustive. For the Community plant variety rights, reasons for termination can be found accordingly in Articles 19(3), 20, and 21 Community Regulation. The plant breeder's right ceases to exist with the plant breeder's waiver (Section 31(1) SortSchG), the right's annulment (Section 31(2) SortSchG), or when the right is revoked (Section 31(3) and (4) SortSchG).[217]

7.3 The Parties' Duty to Inform

As the SortSchG does not confer on the employer an original appropriation right to employees' plant varieties, the employee is under no statutory obligation to notify the employer. However, the employee might be bound to inform the employer by contract. The contract of employment may expressly obligate the employee to notify the employer of any new plant variety bred or discovered in connection with their employment. However, even if the contract is silent on that matter, a duty to inform may be implied. This will be the case if the contract of employment contains an arrangement on the transfer of rights in new employees' plant varieties to the benefit

213. Leßmann/Würtenberger, *supra* n. 205, § 6 marginal n. 1.
214. Leßmann/Würtenberger, *supra* n. 205, § 6 marginal n. 3.
215. Leßmann/Würtenberger, *supra* n. 205, § 6 marginal n. 6.
216. Leßmann/Würtenberger, *supra* n. 205, § 6 marginal n. 7.
217. Leßmann/Würtenberger, *supra* n. 205, § 6 marginal n. 11.

of the employer.[218] To ensure the employer's contractually conferred rights, the employee is bound to notify them about plant varieties that are subject to these rights.

7.4 Employee Remuneration Right

A statutory right to equitable remuneration for employee plant breeders does not exist. This is because of the lack of statutory provisions on the transfer of rights in employees' plant varieties. As mentioned previously, the transfer of rights in plant varieties requires an individual agreement between the employer and the employee. Such contractual agreement will naturally contain provisions on the employee's right to equitable compensation in return for the transfer of rights.

8 INSTANCES FOR DISPUTES

Legal disputes concerning non-technical protected rights, such as copyright, occur less often than disputes regarding employee inventions. They often deal with the question of whether a performance capable of copyright protection was produced during or outside the obligations arising from the employment contract.[219]

The UrhG does not foresee a prior conciliation stage for copyright litigation matters. According to Section 104 sentence 1 UrhG, copyright litigation matters, i.e. all legal disputes by means of which a claim is asserted on account of a legal relationship regulated under the UrhG, generally fall within the jurisdiction of the Ordinary Courts (Ordentliche Gerichte).[220] However, for 'copyright litigation matters resulting from employment or service relationships which have as their object only claims for payment of an agreed remuneration, recourse to the Labour Courts and the Administrative Courts shall remain unaffected' (Section 104 sentence 2 UrhG).

Section 105(1) UrhG entitles the federal state governments to 'assign ... copyright litigation matters for which the Regional Court [Landgericht] is competent as the Court of First Instance or as appeal court to one of the several regional courts competent within a district where this serves the administration of justice'. Likewise, the federal state governments may assign copyright litigation matters within the jurisdiction of the Local Courts (Amtsgericht) to one of several local courts within a district where this serves the administration of justice (Section 105(2) UrhG). Section 105 UrhG takes into consideration that only courts dealing with copyright litigation matters on a regular basis gain sufficient expertise and knowledge in this specific field of law in order to be able to reasonably decide on copyright disputes. Thus, Section 105 UrhG entitles the federal state governments to demand copyright claims to be brought before such specifically designated courts with broad expertise and knowledge in copyright litigation matters, whereby high efficiency in legal proceedings is ensured.[221] Except for

218. Cf. *supra* '7. Employees' Plant Breeders' Rights' section 7.2 'Ownership of Rights and Termination'.
219. Rother, *supra* n. 3, p. 15.
220. Axel Nordemann, *supra* n. 6, § 43 UrhG marginal n. 65. Rother, *supra* n. 3, p. 12.
221. Jan Bernd Nordemann, *supra* n. 22, § 105 UrhG marginal ns. 2 et seqq.

two federal states, all federal state governments have made use of this authorisation and declared specific courts competent for copyright litigation matters.[222]

Claims must be brought forward within the respective legal limitation period. With regard to disputes concerning copyright issues between employers and employees, no special limitations exist, therefore the general provisions on limitation periods, Sections 194 et seqq. BGB, apply. The limitation period commences with knowledge of the creation of the claim as of the end of the calendar year. From a subjective point of view, knowledge includes all circumstances providing the basis for the claim. In the absence of subjective knowledge, the absolute limitation period of ten years from the date of the creation the claim applies (Section 199(4) BGB).[223] A potential termination of employment does not affect the lawsuit between the employer and the employee.[224]

With regard to employees' semiconductor chips, the HalblSchG does not envisage any special prior conciliation stage. Section 11(2) HalblSchG declares Section 27 GebrMG applicable to semiconductor chips matters accordingly. Thus, all semiconductor chip litigation matters, i.e. all legal disputes by means of which a claim is asserted on account of a legal relationship regulated under the HalblSchG, come within the jurisdiction of the Regional Courts. With regard to limitation periods, the general provisions of Sections 194 et seqq. BGB apply.

The number of legal disputes concerning ownership of inventions between employees and employers is rather great. Disputes often concern the reasonable amount of the inventor compensation. Preliminary questions are whether the employee is the sole inventor or only a co-inventor and whether the object, which the employer uses, is based on the service invention.[225]

For disputes between employees and employers concerning patentable inventions – including those that concern inventions capable of being applied for as a utility model – the ArbNErfG envisages a special prior conciliation stage (Sections 28 et seqq. ArbNErfG). Before reverting to the ordinary courts, the parties are obliged to attempt reconciliation in a proceeding before the Arbitration Board at the German Patent and Trademark Office. According to Section 28 ArbNErfG this obligation only exists insofar as the employee is still employed with the employer. The Arbitration Board is an independent public institution, not a court, therefore it cannot pronounce a binding judgment. However, it can make a non-binding settlement suggestion. In case the parties do not agree upon this settlement suggestion, the arbitration proceeding is declared terminated and each party is free to revert to the ordinary courts. The jurisdiction of the Arbitration Board is confined to compensation claims of the employee and the calculation of such compensation. Questions of substantive ownership of the patentable invention cannot be decided by the Arbitration Board but – as a civil law issue – must be brought before a court of law.[226]

222. Cf. Jan Bernd Nordemann, *supra* n. 22, § 105 marginal n. 3.
223. Cf. also Rother, *supra* n. 3, p. 15.
224. Rother, *supra* n. 3, pp. 12,13.
225. Cf. Rother, *supra* n. 3, p. 15.
226. Cf. Rother, *supra* n. 3, p. 12.

After the arbitration proceeding the parties may bring their claim before a competent court. Legal disputes concerning employees' inventions fall within the exclusive jurisdiction of the courts generally competent in case of patent litigation (Section 39(1) ArbNErfG in connection with Section 143 PatG). Hence, the competent courts possess high professional knowledge and expertise in respect of patent disputes ensuring high efficiency in terms of court proceedings.[227]

Proceedings in the first instance can take from six to twelve months. Where the issue of compensation is concerned, actions are generally brought before the courts in stages. First, the employee will demand information relating to the scope of the invention's use by the employer, whereas at the second stage, once the necessary information has been provided, the court will determine a reasonable compensation (Section 38 ArbNErfG).[228]

Like all types of actions, actions for the attribution of an intellectual property right must be filed within a certain limitation period. For claims that pursue the transfer of a patent, specific limitation periods exist. According to Section 8 PatG such claims must be filed within two years after the patent was granted, provided that the patentee did not acquire the patent in bad faith. Otherwise, the general provisions on limitation of civil law claims, Section 194 et seqq. BGB, apply.[229]

There is no special prior conciliation stage envisaged for employer-employee disputes on designs. According to Section 52 DesignG design litigation matters, i.e. all legal disputes by means of which a claim is asserted on account of a legal relationship regulated under the DesignG, generally fall within the jurisdiction of the Regional Courts, except for claims concerning the invalidity of the design. Otherwise, the general procedural law provisions apply. Claims concerning employee-created designs are subject to the general limitations on civil claims pursuant to Sections 194 et seqq. BGB.[230]

Section 38(1) SortSchG determines the Regional Courts to be competent for all plant variety right litigation matters, i.e. all legal disputes by means of which a claim is asserted on account of a legal relationship regulated under the SortSchG. Neither a prior conciliation stage nor specific limitation provisions exist. Rather, according to Section 37 f SortSchG, the general limitations on civil claims pursuant to Sections 194 et seqq. BGB apply.

227. Cf. Rother, *supra* n. 3, p. 15.
228. Cf. Rother, *supra* n. 3, p. 15.
229. Cf. Rother, *supra* n. 3, pp. 12 et seq.
230. Rother, *supra* n. 3, pp. 12 et seq.

BIBLIOGRAPHY

Table of Sources

Bayreuther, *Münchener Handbuch zum Arbeitsrecht (Munich Manual on Employment Law)*, (Richardi, Wlotzke, Wißmann & Oetker, 3rd ed., C.H.Beck Munich 2009).

Bayreuther, *Neue Spielregeln im Arbeitnehmererfinderrecht (New Rules of the Game Concerning Law on Employees' Inventions)*, NZA 2009, 1123 (2009).

Czychowski, *Urheberrecht: Kommentar zum Urheberrechtsgesetz, zum Verlagsgesetz und zum Urheberrechtswahrnehmungsgesetz (Authors law: Commentary on the Act on Copyright and Related Rights, on the German Publishing Act and the Copyright Administration Act)*, (Fromm/Nordemann, 11th ed., Kohlhammer 2014).

Czychowski/Langfinger, *German Law on Employees' Inventions Regarding University Employees*, Les Nouvelles, December 2010, 221 (2010).

Dietz/Peukert, *Urheberrecht Kommentar (Copyright Law: Commentary)*, (Schicker/ Loewenheim, 5th ed., C.H.Beck Munich 2017).

Dreier, *Urheberrechtsgesetz, Urheberwahrnehmungsgesetz, Kunsturhebergesetz: Kommentar (Act on Copyright and Related Rights, Copyright Administration Act, Act on Copyright for Works of Art: Commentary)*, (Dreier/Schulze, 5th ed., C.H.Beck 2015).

Eichmann, *Geschmacksmustergesetz (Registered Designs Act)* (Eichmann/Vogel von Falckenstein, 4th ed., C.H.Beck Munich, 2010).

Erhardt, *Praxiskommentar zum Urheberrecht (Copyright Law: Commentary)*, (Wandtke/Bullinger, 4th ed., C.H.Beck Munich 2014).

Falck, von, *Computerrechts-Handbuch (Computer Law Handbook)*, (Kilian/ Heussen, 32th ed., C.H.Beck Munich 2013).

Groh, *Mit fremden Federn: Zur Wirksamkeit von Ghostwritervereinbarungen (Validity of Ghost-writer Agreements)*, GRUR 2012, 870 (2012).

Grützmacher, *Praxiskommentar zum Urheberrecht (Copyright Law: Commentary)* (Wandtke/Bullinger, 4th ed., C.H.Beck Munich 2014).

Heermann, *Der Schutzumfang von Sprachwerken der Wissenschaft und die urheberrechtliche Stellung von Hochschulangehörigen (Scope of Protection for Literary Scientific Works and Academic Staff's Status with Regard to Copyright)*, GRUR 1999, 468 (1999).

Hirschfelder, *Open Access – Zweitveröffentlichungsrecht und Anbietungspflicht als europarechtlich unzulässige Schrankenregelungen? (Open Access – Are a Second Publication Right and a Duty to Offer Limitations on Copyright Contrary to European Law?)*, MMR 2009, 444 (2009).

Honschek, *Der Schutz des Urhebers vor Änderungen und Entstellungen durch den Eigentümer (Protecting the Author against Alterations and Derogatory Treatment by the Owner)*, GRUR 2007, 944 (2007).

Keukenschrijver, *Patentgesetz Kommentar (Patent Act: Commentary)*, (Busse, 7th ed., De Gruyter Recht Berlin 2013).

Koch, *Der Halbleiterschutz nach nationalem, internationalem und europäischem Recht*, in: *Rechtsschutz und Verwertung von Computerprogrammen (Semiconductor Chip Protection under National, International and European Law, in: Legal Protection and Commercialisation of Computer Programmes)*, (Lehmann, 2nd ed., Verlag Dr. Otto Schmidt Köln 1993).

Kraßer, *Patentrecht (Patent law)*, (6th ed., C.H.Beck Munich 2009).

Leßmann/Würtenberger, *Deutsches und europäisches Sortenschutzrecht (German and European Plant Breeders' Rights)*, (2nd ed., Nomos 2009).

Lindhorst, *Beck'scher Online-Kommentar: Urheberrecht (Beck Online-Commentary: Copyright)*, (Möhring/Nicolini, 4th ed., C.H.Beck Munich 2014).

Loewenheim, *Urheber in Arbeits- und Dienstverhältnissen (Authors in Private and Public Sector Employment Relationships)*, (Schricker/Loewenheim, 5th ed., C.H.Beck Munich 2017).

Marly, *Praxishandbuch Softwarerecht: Rechtsschutz und Vertragsgestaltung (Handbook Software Law: Legal Protection and Contract Design)*, (6th ed., C.H.Beck 2014).

Mes, *Patentgesetz, Gebrauchsmustergesetz: Kommentar (Patent Act, Act on Utility Models, Commentary)*, (Mes, 3rd ed., C H.Beck Munich 2001).

Nordemann, Axel, *Urheberrecht: Kommentar zum Urheberrechtsgesetz, zum Verlagsgesetz und zum Urheberrechtswahrnehmungsgesetz (Authors law: Commentary on the Act on Copyright and Related Rights, on the German Publishing Act and the Copyright Administration Act)*. (Fromm/Nordemann, 11th ed., Kohlhammer 2014).

Nordemann, Axel, *Handbuch des Urheberrechts (Manual on Copyright Law)*, (Loewenheim, 2nd ed., C.H.Beck Munich 2010).

Nordemann, Jan Bernd, *Urheberrecht: Kommentar zum Urheberrechtsgesetz, zum Verlagsgesetz und zum Urheberrechtswahrnehmungsgesetz (Authors law: Commentary on the Act on Copyright and Related Rights, on the German Publishing Act and the Copyright Administration Act)* (Fromm/Nordemann, 11th ed., Kohlhammer 2014).

Nordemann, Wilhelm/Nordemann, Axel, *Die zeitliche Dauer der Nutzungsrechtseinräumung im Arbeitsverhältnis (Duration of Economic Rights Granted within Employment Relationships)*, Festschrift für Artur-Axel Wandtke zum 70. Geburtstag, 141 (2013).

Rojahn, *Urheberrecht Kommentar (Copyright Law: Commentary)*, (Schicker/Loewenheim, 5th ed., C.H.Beck Munich 2017)

Rother, *Group Report Germany for the AIPPI Working Committee No. 183: Employers' Rights to Intellectual Property* (in the German and English language), available at https://www.aippi.org/download/commitees/183/GR183germany.pdf (accessed 16May 2017).

Schoenen, *Münchener Anwaltshandbuch: Gewerblicher Rechtsschutz (Munich*

Lawyer's Manual on the Protection of Industrial Property), (Hasselblatt, 4th ed., C.H.Beck Munich 2012).

Schulze, *Urheberrechtsgesetz, Urheberwahrnehmungsgesetz, Kunsturheb ergesetz: Kommentar (Act on Copyright and Related Rights, Copyright Administration Act, Act on Copyright for Works of Art: Commentary)*, (Dreier/ Schulze, 5th ed., C.H. Beck, 2015)

Trimborn, *Aktuelle Entwicklungen im Arbeitnehmererfindungsrecht*, Mitt. 2015, 116

Wandtke, *Reform des Arbeitnehmerurheberrechts? (Reforming Employees' Copyright?)*, GRUR 1999, 390 (1999).

Wandtke, *Praxiskommentar zum Urheberrecht (Copyright Law: Commentary)*, (Wandtke/Bullinger, 4th ed., C.H.Beck Munich 2014).

Wandtke, *Zum Vergütungsanspruch des Urhebers im Arbeitsverhältnis (Employee Remuneration Right in Employment Relationships)*, GRUR 1992, 139 (1992).

Wiebe, *Recht der elektronischen Medien: Kommentar (Law of Electronic Media: Commentary)*, (Spindler/Schuster, 2nd ed., C.H.Beck Munich 2011).

Table of Legislation

Act on Copyright and Related Rights of 9 September 1965 as last amended by Article 8 of the Act of 1 October 2013 (Urhebergesetz, UrhG), Official English version available at: http://www.gesetze-im-internet.de/englisch_urhg/ index.html (accessed 16 May 2017).

Basic Law for the Federal Republic of Germany (Grundgesetz, GG), available in the English language at: http://www.gesetze-im-internet.de/englisch_gg/ index.html (accessed 16 May 2017).

Berne Convention for the Protection of Literary and Artistic Works (Berne Convention), available at: http://www.wipo.int/treaties/en/ip/berne/ (accessed 16 May 2017).

Council Regulation (EC) No. 2100/94 of 27 July 1994 on Community plant variety rights, available at: http://eur-lex.europa.eu/legal-content/EN/TXT/HTML/? uri = CELEX:31994R2100&from = EN (accessed 16 May 2017).

Council Regulation (EC) No. 6/2002 of 12 December 2001 on Community designs, available at: http://eur-lex.europa.eu/LexUriServ/LexUriServ.do?uri = OJ:L:2 002:003:0001:0024:en:PDF (accessed 16 May 2017).

Directive 87/54/ECC of 16 December 1986 on the legal protection of topographies of semiconductor products, available at: http://eur-lex.europa.eu/legal-content/ EN/TXT/?uri = CELEX:31987L0054 (accessed 16 May 2017).

Directive 93/83/EEC of September 1993 on the coordination of certain rules concerning copyright and rights related to copyright applicable to satel lite broadcasting and cable retransmission, available at: http://eur-lex.eur opa.eu/legal-content/EN/ TXT/PDF/?uri = CELEX:31993L0083&from = EN (accessed 16 May 2017).

Directive 93/98/EEC of 29 October 1993 harmonizing the term of protection of copyright and certain related rights (Copyright Duration Directive), available at: http://eur-lex.europa.eu/legal-content/EN/TXT/PDF/?uri = CELEX:31993 L0098&from = en (accessed 16 May 2017).
Directive 98/71/EC of the European Parliament and of the Council of 13 October 1998 on the legal protection of designs, available at: http://eur-lex.europa.eu /legal-content/EN/TXT/HTML/?uri = CELEX:31998L0071&qid = 1428516903 311&from = EN (accessed 16 May 2017).
Directive 2001/29/EC of the European Parliament and of the Council of 22 May 2001 on the harmonisation of certain aspects of copyright and related rights in the information society, available at: http://eur-lex.europa.eu/legal-content/ EN/ TXT/PDF/?uri = CELEX:32001L0029&from = EN (accessed 16 May 2017).
Directive 2001/84/EC of the European Parliament and of the Council of 27 September 2001 on the resale right for the benefit of the author of an original work of art, available at: http://eur-lex.europa.eu/legal-content/EN/TXT/?uri = celex:32001L0084 (accessed 16 May 2017).
Directive 2006/115/EC of the European Parliament and of the Council of 12 December 2006 on rental right and lending right and on certain rights related to copyright in the field of intellectual property (Rental Directive), available at: http://eur-lex.europa.eu/legal-content/EN/TXT/PDF/?uri = CELEX:32006L0 115&from = EN (accessed 16 May 2017).
German Civil Code ('Bürgerlicher Gesetzbuch', 'BGB'), available in the English language at: http://www.gesetze-im-internet.de/englisch_bgb/index.html (accessed 16 May 2017).

Table of Cases

Cromegal, I ZR 28/61, GRUR 1963, 135 (BGH, 28 June 1962).
Das zweite Mal, I ZR 32/69, GRUR 1971, 269 (BGH, 27 November 1970).
Zwischen Marx und Rothschild, 3 U 77/75, GRUR 1977, 556 (OLG Hamburg, 22 January 1976).
Statikprogramme, 3 AZR 371/81, GRUR 1984, 429 (Federal Labour Court (Bundesarbeitsgericht, BAG), 13 September 1983).
Unendliche Geschichte, U 2114/85, GRUR 1986, 460 (OLG Munich, 1 August 1985).
Architektenvertrag, 13 U 12/85, BauR 1986, 601 (OLG Celle, 9 October 1985).
Grabungsmaterialien, I ZR 244/88, GRUR 1991, 523 (BGH, 27 September 1990).
Namensnennungsrecht des Architekten, I ZR 3/92, GRUR 1995, 671 (BGH, 16 July 1994).
POLDOK/Urheber- und Nutzungsrechte eines Hochschullehrers, 5 U 2189/93, ZUM-RD 1997, 175 (KG, 6 September 1994).
Gummielastische Masse, X ZR 130/93, NJW-RR 1995, 696 (BGH, 17 January 1995).

Pauschale Rechtseinräumung, I ZR 215/93, GRUR 1996, 121 (BGH, 27 September 1995).

Softwareentwickler im Arbeitsverhältnis, 16 O 136/96, ZUM 1998, 167 (Kammergericht (KG) 3 July 1996).

Rechte an einem im Dienstverhältnis entwickelten Datenverarbeitungsprogramm, 7 O 15354/91, ZUM 1997, 659 (LG Munich, 16 January 1997).

Museumsführer, 3 U 3053/96, ZUM 1999, 656 (Higher Regional Court (Oberlandesgericht, OLG) Nuremberg, 18 February 1997).

Spulkopf, X ZR 6/96, GRUR 1998, 686 (BGH, 13 November 1997).

Copolyester II, X ZR 132/95, GRUR 1998, 689 (BGH, 13 November 1997).

Comic-Übersetzungen I, I ZR 189/95, GRUR 1998, 680 (BGH, 20 January 1998).

TESY-M2/Schutzfähigkeit eines Computerprogramms – Vergütungsansprüche eines Beamten der Deutschen Bundespost, 29 U 2437/97, ZUM-RD 2000, 8 (OLG Munich, 25 November 1999).

Wetterführungspläne, X ZR 72/98, GRUR 2001, 155 (Federal Court of Justice (Bundesgerichtshof, BGH) 24 October 2000).

Wetterführungspläne II, X ZR 72/98, GRUR 2002, 149 (BGH, 23 October 2001).

Drahtinjektionseinrichtung, X ZR 215/00, GRUR 2002, 609 (BGH, 6 February 2002).

Abgestuftes Getriebe, X ZR 127/99, GRUR 2002, 801 (BGH, 16 April 2002).

Wegfall des Hochschullehrersprivileg bei Diensterfindungen, 2 U 19/05, GRUR-RR 2006, 178 (OLG Braunschweig, 10 November 2005).

Filmbeschreibungen, 28 O 250/09, ZUM 2010, 369 (Regional Court (Landgericht, LG) Cologne, 23 September 2009).

CHAPTER 11
Hungary

Gusztáv Bacher

1 BACKGROUND

The introduction of a market economy – parallel to the political changes in 1989-1990 – required the start of a modern legislative framework in Hungary. Such modern legislation had to have its roots in Hungary's legal tradition relating to the laws concerning the protection of intellectual property. The Act II of 1969 on patent protection (effective before 1996) contained detailed rules on employee's inventions, and comprehensive jurisprudence emerged on this issue based on significant number of cases.

The IP laws in Hungary are in compliance with the international treaties and EU law. This overview focuses on the state of employees' intellectual property rights.

2 EMPLOYEES' COPYRIGHT

2.1 The Legal Framework

Hungary is a member of the EU and a member of all copyright/neighbouring rights conventions such as the Berne Convention (accession date ('a.d.') 14 February 1922), the Phonograms Convention (a.d. 28 May 1975), the Rome Convention (a.d. 10 February 1995), the WIPO Convention (a.d. 26 April 1970), the WIPO Copyright Treaty (a.d. 6 March 2002), the WIPO Performances and Phonograms Treaty (a.d. 20 May 2002) and the TRIPS Agreement/the WTO (a.d. 1 January 1995). As an EU Member

State, Hungary has implemented all EU Copyright Directives in addition to one Commission Recommendation (in relation to the compulsory licensing of the use of orphan works).[1]

In the Hungarian legal regime copyright protects all literary, scientific and artistic creations and it is regulated by the Act LXXVI of 1999 on copyright ('Copyright Act'). According to Article 4 of the Copyright Act the copyright belongs to the person who has created the work (the author). The legal protection begins at the moment the work is created by the power of law, meaning that no further steps are necessary from the author to achieve protection. The copyright (author's right) consists of two elements: moral rights and economic rights.

Moral rights are listed in Chapter II of the Copyright Act. These include the right of publication (including the right to withdraw such authorisation); the author's right to indicate their name and authorship in connection with the work; and the right to protect the work's integrity. The moral rights of the author are inalienable, they cannot be assigned to another person in any way and they cannot be waived.

Economic rights are listed in Chapter III of the Copyright Act. These include, but are not limited to, the right of reproduction; the right of distribution; the right of public performance; the right to communicate the work to the public; and the right of adaptation. In principle, these rights cannot be assigned or transferred, except in certain cases specified in the Copyright Act: works prepared within the framework of employment relationship, software, and works ordered for advertising. Further, the economic rights pass by inheritance, and authors are free to dispose of them for on their death.

The Copyright Act contains special provisions with regard to the economic rights of employees' copyright.

A work qualifies as a *service work* if the creation of the work was the author's obligation within the scope of the responsibilities of their employment. Service works may be created by employees in labour relationships, or by public service employees or public employees, by professional soldiers or professional law enforcement officers on active duty, or by the employed members of cooperatives. Unless agreed otherwise, the employer, as the legal successor to the author, obtains economic rights over the service

1. Council Directive 91/250/EEC of 14 May 1991 on the legal protection of computer programs, Council Directive 93/83/EEC of 27 September 1993 on the coordination of certain rules concerning copyright and rights related to copyright applicable to satellite broadcasting and cable retransmission, Directive 96/9/EC of the European Parliament and of the Council of 11 March 1996 on the legal protection of databases, Directive 2001/29/EC of the European Parliament and of the Council of 22 May 2001 on the harmonisation of certain aspects of copyright and related rights in the information society, Directive 2001/84/EC of the European Parliament and of the Council of 27 September 2001 on the resale right for the benefit of the author of an original work of art, Directive 2004/48/EC of the European Parliament and of the Council of 29 April 2004 on the enforcement of intellectual property rights, Directive 2006/115/EC of the European Parliament and of the Council of 12 December 2006 on rental rights and lending rights and on certain rights related to copyright in the field of intellectual property, Directive 2006/116/EC of the European Parliament and of the Council of 12 December 2006 on the term of protection of copyright and certain related rights, Commission Recommendation 2006/585/EC of 24 August 2006 on the digitisation and online accessibility of cultural material and digital preservation, Art. 6(a) and (c).

work on the condition that the work is delivered by the employee to the employer. In the case of legal succession of the employer, the economic rights obtained in service works devolve to the employers' legal successor (Article 30 of the Copyright Act).

Motion picture works shall not be subject to the general provisions pertaining to works made for hire (Article 65(5)).

2.2 Ownership of Rights

In the case that the above-mentioned conditions are met and the work qualifies as a work created in an employment relationship under Article 30 of the Copyright Act, the result is a complete transfer of the economic rights of the work without limitation between the author and the employer (unless the parties agreed otherwise).

Though theoretically the employer acquires the economic rights to the work, this transfer of rights through the operation of law (*cessio legis*) has effects on the author's moral rights as well. The author's right of publication and their right to withdraw such authorisation cease to exist by the operation of law (Article 30 of the Copyright Act) at the moment the work is handed over to the employer. The author's name will be indicated but they can opt out and remain anonymous by means of a legal statement. The author's moral right to protect the integrity of the work is narrowed down in an employment relationship as well, as the interest of the employer as an investor is at the forefront.[2] The author's name must be deleted if the employer alters the work by taking advantage of its rights as employer and the author does not agree with the changes.

In order for the *cessio legis* to take effect, the creation of the work needs to fall into the author's obligation in the scope of their employment. This obligation is judged based on objective criteria, meaning that the author's motivation at the time of the work's creation is not taken into account. When determining if a work was created within the author's obligation in the scope of their employment, the employment contract and the job description serve as sources of guidance. In case of doubt whether the creation of the work falls within the scope of the employee's duty, the burden of proof lies on the employer. It is usual to accept bylaws on intellectual property in workplaces where employees frequently create intellectual property (or include relevant provisions in the collective agreement). The author can naturally create works during their employment – even with the tools of the employer – that will not qualify as a work created as an obligation, meaning that economic rights will not be acquired by the employer by *cessio legis*.

2.3 The Parties' Duty to Inform

Unlike the Patent Act, there are no express provisions on the parties' duty to inform in the Copyright Act. However the provisions of Act I of 2012 on the Labour Code are applicable in the case of employment relationships. According to Article 6 of the Labour Code, the parties shall act in a manner which is consistent with the principle of

2. *Commentary on the Copyright Act* (Complex Kiadó 2014).

good faith and fairness, they shall be required to cooperate with one another, and they shall not engage in any conduct which breaches the rights or legitimate interests of the other party. Thus with regard to the obligation to cooperate in good faith and in fairness, an author in an employment relationship is considered to be obliged to provide all information to their employer about works created under the obligation of an employment relationship.

2.4 Employee Remuneration Right

While the employer acquires the economic rights of the work, unless agreed otherwise, the authors of service works are entitled to an equitable (fair) remuneration if the employer grants a licence for use of a service work or assigns the economic rights in the service work to a third person (Article 30(4) of the Copyright Act). If the same work is licensed several times by the employer, the employee is entitled to remuneration on the basis of the continuous income of the employer. The author can enforce their claim against the employer only. The author can expressly revoke their right to remuneration in a written statement (an employer can create a collective agreement that stipulates that the employees have no additional right to remuneration). In sum, the remuneration right of the employee author is based on a permissive set of rules.

Furthermore, the provisions on remuneration do not apply to software (Article 58(4)), or databases (Article 61(3)) created by an author as work made for hire. There is no place for the author's extra remuneration in cases where the author's work achieves extraordinary commercial success. The reason for this is that the remuneration should not be 'proportional' (e.g. in a best-seller clause) but 'equitable/fair' according to the Copyright Act (Article 16(5) in the Copyright Act).

3 EMPLOYEES' SEMICONDUCTOR CHIPS

3.1 The Legal Framework

Act XXXIX of 1991 on the legal protection of topographies of microelectronic semiconductor products ('Semiconductor Products Act') regulates the rights regarding the protection that can be granted to a creator of topography of a microelectronic semiconductor product. The Act implements EU Council Directive 87/54/EEC.[3] The Semiconductor Products Act provides that the general terms of the Patent Act apply to any issues which are not regulated specifically.

3.2 Ownership of Rights

According to Article 4 of the Semiconductor Products Act, where the creator of topographies is under an obligation due to their employment to create topographies (a

3. Council Directive 87/54/EEC of 16 December 1986 on the legal protection of topographies of semiconductor products.

so-called *service topography*), the right to protection of the topography shall generally belong to the employer. However, in these cases the creator of service topography will be entitled to remuneration as per their service contract. These issues are discussed in more detail with regard to the Patent Act.[4]

On the other hand, where an employee develops topography without an incumbent duty to do so arising from the employment relationship, the right to protection from the outset belongs to the employee as the creator of the topography, as per Article 2 of the Semiconductor Products Act. However, the employer is entitled *ex lege* to utilise the invention as per the Patent Act, although this right of utilisation is not exclusive.[5]

In any other matters concerning the moral rights of the creator of the topography, the right to protection and the service topography, the provisions of the Patent Act on the moral rights of the inventor, the right to a patent and service inventions shall apply *mutatis mutandis*.

3.3 The Parties' Duty to Inform

The Semiconductor Products Act does not contain any specific terms with regard to the parties' duty to inform. Rather, this is regulated by the Patent Act. (See section 4 below.)

3.4 Employee Remuneration Right

Neither does the Semiconductor Products Act contain any specific terms with regard to employee remuneration rights. This is also regulated by the Patents Act.

4 EMPLOYEES' INVENTIONS

4.1 The Legal Framework

The Patent Act differentiates between employees' inventions and service inventions.

4.2 Ownership of Rights

Service inventions refer to the inventions of persons on whom it is incumbent to develop solutions in the domain of the invention pursuant to the employment relationship. The patent claim belongs from the outset to the employer, as the legal successor of the inventor.[6] An inventor is the person who has created an invention and has the right to be indicated on the patent documents as such.

4. See section 4 below.
5. *Ibid.*
6. This provision explains the difference between the approaches of the continental and Anglo-Saxon patent legal systems. The right to claim to a patent in our system lies originally with the

On the other hand an *employee invention* is the invention of a person who develops an invention without an incumbent duty to do so arising from the employment relationship. The patent claim belongs from the outset to the inventor; the employer, however, is entitled *ex lege* to utilise the invention. The employer's right of utilisation is not exclusive and they may not grant a licence to exploit the invention.

4.3 The Parties' Duty to Inform

According to Article 11 of the Patent Act, the inventor is required to report the service invention to the employer immediately following its creation. The employer has 90 days following receipt of such a notification to make a declaration signalling their intention to claim the service invention, or to state their intentions concerning the exploitation of the employee invention. If there is no declaration in the prescribed time period, the employee-inventor may freely dispose of their rights in the service invention. The same is true if the employer consents to such an outcome. Furthermore, the right to a patent for an employee invention will belong to the inventor and will not be subject to the employer's right of exploitation if the employer gives their consent thereto or fails to make the required declaration or statement within 90 days.

In the event the employer chooses to claim the service invention, they are obliged to file a patent application within a reasonable time following receipt of notification regarding the existence of the invention. The Patent Act imposes an additional duty on the employer to act with the level of care generally expected in acquiring a patent. Moreover, if the employer omits or commits an act that would result in the rejection of the patent application, the employer must offer a free assignment of the patent claim to the inventor (although this will not apply where the inventor has already received fair remuneration).

If the employer considers the invention to be utilised as a trade secret, they will not file a patent application or withdraw it if it has already been submitted and has not been publicised. Under these circumstances, the employer must acknowledge that, in the absence of its trade secret characterisation, the service invention would otherwise be eligible for patent protection. The employee must be notified of this decision and in the case of a dispute, the burden of proof is on the employer to demonstrate that the invention was ineligible for patent protection.

4.4 Employee Remuneration Right

The fundamental principle of remuneration for the service invention is that the employee has achieved a significant result by creating the invention thanks to which

creator, i.e. the inventor, even if this person is employed and the employer may, pursuant to the service contract, be the legal successor of the economic rights only. In other words, this system does not acknowledge the original right ownership of the employer or the 'master' of creations (inventions and works) made for hire or within a labour relationship. From this approach stems the statutory right of the service inventor to be remunerated beyond their wages if the legal successor, i.e. the employer, utilises the service invention.

the employer gets into a monopolistic position and the employee must receive a share of the results arising from the exploitation of the patent. Consequently, the inventor's claim for remuneration is a legal relationship of a civil law nature between the inventor and the employee based on the statute which is independent from the employment relationship, and this relationship may continue in existence even after the termination of the employment relationship and alteration in the person of the contracting parties can also occur (i.e. succession in case of the death of the inventor, legal succession of the employer).

4.4.1 *Service Inventions*

There are two possibilities for the remuneration of inventors: (i) payment through a licence analogy or (ii) a lump sum amount: the parties may conclude a contract of remuneration in which a remuneration of a fixed amount is stipulated with respect to the inventions of the inventor to be created or exploited in the future (invention remuneration contract aiming at risk sharing). The inventor's right to remuneration arises upon the utilisation of the invention. As outlined in greater detail below, the concept of utilisation is broadly interpreted. The Patent Act includes a number of provisions in connection with the statutory claim of remuneration that is due to inventors.

Unless a remuneration agreement is concluded between the inventor and the employer, the following rules apply:[7]

Time span of the remuneration claim: the general rule provides for the presumed entire term of patent protection, i.e. 20 years (plus the term of the supplemental patent certificate in the case of medicines), even if the patent is terminated earlier through any act or omission of the employer.

Types of utilisation by the employer on the basis of which the inventor employee is entitled to remuneration: production, use of process, and/or licensing, and/or assignment of the patent, and/or any omission of utilisation in order to create or maintain an advantageous market position.

The inventor is entitled to remuneration separately in respect of each and every type of utilisation including the granting of any licences or assignments without consideration.

The remuneration claim may be enforced against the employer. In the case of licensing or assignment of the patent, the licensee or assignee may assume the obligation to pay remuneration. Under Article 6:203 of the new Civil Code, which is the background code behind the Patent Act, not only is the inventor's approval required for the transfer of the obligation, but such transfer qualifies as a genuine three-party agreement among the employer (original obligor), the new obligor and the employee (beneficiary). In the absence of a remuneration agreement, the claim may always be enforced against the employer (in case of multiple employers, against the employer utilising the service invention).

7. Art. 13 of the Patent Act.

The remuneration of the inventor must be governed by a remuneration agreement.

As mentioned above, the application of the statutory provisions to determine the amount of the inventor's claim in case of a dispute can be avoided through the use of well-drafted invention regulations (i.e. by-laws) and remuneration agreements. The invention regulations (i.e. by-laws) can be regarded as the employer's general contractual terms on the basis of which the employer makes the offer to the employee to conclude a remuneration agreement.

4.4.1.1 Remuneration on the Basis of the Licence Analogy

In the event the employer utilises the service invention, the remuneration provided to the employee shall be proportionate to the fee that would be paid by the employer as a licensee pursuant to a patent licensing agreement, in view of the licensing practice in the given technological field and in accordance with the subject matter of the invention. In licensing practice, in case of exclusive licences, the royalty rates based on the net turnover varies widely depending on the field of industry (in general between 0.5% and 5%[8], but in certain industries it can be higher as well).

If the service invention becomes the subject matter of a licence or an assignment, the royalty paid must be proportional to the proceeds of a licence or assignment, or to the transfer or economic benefit/advantage resulting from a free licence or free assignment.

Remuneration will be deemed proportional where due regard is made to the employer's contribution to the creation of the invention as well as to the obligations of the inventor in the employment relationship.[9] Accordingly, the remuneration shall be proportionate, on the one hand:

(i) to the extent the creation reflects the employer's research, plans, expenses arising from the research;
(ii) based on the employment contract, to the extent it was the employer's obligation to participate in the development of the creation of the invention; and
(iii) to the extent the invention reflects the initiative steps and the significant activity of the employee.

Further guidance in the form of jurisprudence is yet to come with respect to proportionality as well as licence analogy. Guidance can be found in the practice of the Body of Experts on Industrial Property established at the Hungarian Intellectual Property Office (abbreviation in Hungarian: ISZT).The opinions are published unless the party requesting the opinion asks for a confidential treatment. In professional questions arising in industrial property legal disputes, the courts and other authorities

8. The royalty rates in licence agreements usually constitute a trade secret. The above data are based on the judgment of the Metropolitan Court 3.P.21.730/2008/124., and Expert Opinion of ISZT No. 6/2012.
9. Art. 13(7)-13(9) of the Patent Act.

may request an expert opinion of the Body of Experts. Further, on request, the Body of Experts may, give an expert opinion in industrial property questions out of court as well. The Body of Experts shall give an expert opinion also on matters concerning the patentability of an invention kept secret and remuneration due to the inventor of a service or employee invention.

In case no licensing practice exists in a certain field of technology, the remuneration can be determined as follows: (i) for the period established for the actual case (approx. 3-5 years), (ii) based on the net profit realised through the utilisation, and (iii) applying in general royalty rate of 10%, however, several additional factors for raising and/or reducing the rate should be taken into account. However, if the calculation method cannot be based on the net profit but the net sale revenue, in general, 1 % royalty rate shall be applicable.[10]

4.4.1.2 Lump Sum Payments

As previously mentioned, the Patent Act allows for possible departures from the statutory remuneration provisions. In fact, the Patent Act expressly allows for lump sum (risk sharing) remuneration agreements.

As noted above, two possible avenues are allowed: (i) a remuneration of a fixed amount paid per invention and if necessary, determined specifically by the type of utilisation (own exploitation, licence, assignment); or (ii) a lump sum replacing all remuneration payable for all types of utilisation of all future inventions of the employee (risk sharing).

Companies often address remuneration issues by payment of a monthly fixed amount to the employee as remuneration for any potential service inventions. The employee can benefit from this payment even if no patentable solution is invented. Although feasible under the express terms of the Patent Act, this option could be fragile. A remuneration agreement qualifies as a civil law contract despite the fact that it is concluded in the context of an employment relationship, and this is reinforced by the Patent Act itself. The employer shall face the risk that as the enforcement of a claim by the inventor before the courts given a substantial change in circumstances can be more easily verified in the case of a general lump sum (risk sharing) agreement designed to satisfy all future claims. Further, there is no published case law on whether the inventor may challenge the agreement based on the legal cause of significant/remarkable disparity between the services rendered and the consideration paid for services, of error or of deceit.

Companies often address remuneration issues by the way of internal invention regulations (i.e. by-laws). Such invention regulations are in fact regarded as general terms of the (employment) contract to which all rules laid down by the Hungarian Civil Code are applicable. However, in the event a workers' council or other similar representative organisation duly accepts such a regulation, it qualifies as an individual

10. The above calculation method is set forth in the Expert Opinion of ISZT No. 7/2008.

agreement binding upon the employee pursuant to the employment contract, although under the Labour Code, the workers' council only has a right of consultation.

To sum up: there are two lawful ways to introduce or apply lump-sum (risk-sharing) remunerations:

(a) a remuneration of a fixed amount to be paid per invention and per inventor and, if necessary, broken down (specified) by utilisation types (own exploitation, licence, assignment); or
(b) a lump sum replacing all remunerations payable for all utilisations of all future inventions.

In both cases the due time for payment may precede (and this is typically the case) the commencement of actual utilisation, since the amount of the lump sum does not depend on the revenue or profit to be gained through utilisation. That is the expression of the risk-sharing feature as compared to an inventor's remuneration to be defined in proportion to the revenue/profit generated by the utilisation. Therefore, the due time of the payment of a lump-sum inventor's remuneration can be set as early as the filing of the patent application or the acceptance of the invention by the employer.

4.4.2 Employee Inventions

According to Article 14 of the Patent Act, remuneration for the right to exploit an employee invention shall be paid by the employer or, in the case of more than one employer and in the absence of an agreement to the contrary, by the employer exploiting the invention. The remuneration of the inventor will be governed by a contract concluded with the employer. In addition, the amount of the remuneration for the right to exploit an employee invention has to be equal to that which would be payable by the employer for a licence, on the basis of a patent licence agreement, taking into account the licensing conditions in the technical field of the subject matter of the invention.

5 EMPLOYEES' UTILITY MODELS

5.1 The Legal Framework

Act XXXVIII of 1991 on the protection of utility models ('Utility Models Act') regulates the protection that can be granted to an applicant for their utility model.[11] In connection with several issues, the Utility Models Act provides that the rules of the Patent Act apply *mutatis mutandis*.[12]

11. Act XXXVIII of 1991 on the Protection of Utility Models, Arts 1-4.
12. With respect to succession in title, to the pledging of rights deriving from utility model and from utility model protection, as well as to licence contracts, the provisions of the Patent Act shall apply *mutatis mutandis*. With respect to compulsory licences of protected utility models and the limitations and exhaustion of utility model protection, the common provisions of the Patent Act

5.2 Ownership of Rights

Under Hungarian law, the inventor of the utility model is the person who has created the utility model and their successor in title. The right to utility model protection shall belong to the inventor or their successor in title.

The provisions on service and employee inventions in the Patent Act[13] shall apply *mutatis mutandis* to utility models created by persons working in employment or in public service, by persons in service relationship or by members of a co-operative working within the framework of a legal relationship of an employment nature.

In matters concerning the moral rights of the inventor of the utility model and the right to utility model protection, the provisions of the Patent Act on the moral rights of the inventor and the right to a patent shall apply *mutatis mutandis*.

5.3 The Parties' Duty to Inform

As mentioned in section 5.2 above, the parties' duty to inform is governed by the provisions on service and employee inventions in the Patent Act, which are discussed in section 4.

5.4 Employee Remuneration Right

As mentioned in section 5.2 above, employees' remuneration rights are governed by the provisions on service and employee inventions in the Patent Act, which are discussed in section 4.

6 EMPLOYEES' DESIGN RIGHTS

6.1 The Legal Framework

Act XLVIII of 2001 on the legal protection of designs ('Design Protection Act') regulates the protection that can be granted to an applicant for any designs which are new and have individual character. The Design Protection Act implements EU Directive 98/71/EC.

6.2 Ownership of Rights

The Design Protection Act mirrors the Patent Act by drawing a distinction between a *service design* and an *employee design*.

on compulsory licences, and the provisions on limitations and exhaustion of patent protection thereof shall apply *mutatis mutandis*. With respect to joint right to a utility model protection and joint utility model protection, the provisions of the Patent Act on joint right to a patent and joint patent shall apply *mutatis mutandis*.

13. See section 4 above.

A *service design* is a design created by a person who, by reason of their employment, is under the obligation to develop solutions in the field of the design. In this case, the right to design protection for a service design will belong to the employer as the successor in title of the designer.

An *employee design* is a design created by a person who, without being under an obligation by reason of their employment, creates a design the exploitation of which falls within the field of business of their employer. In this case, the right to design protection for an employee design shall belong to the designer, but the employer shall be entitled to exploit the design. However, the employer's right of exploitation shall be non-exclusive and the employer may not grant a licence to exploit the design.

In any other matters concerning service and employee designs and the remuneration of the designer, the provisions of the Patent Act on service and employee inventions shall apply *mutatis mutandis*, with the following exceptions:[14]

(a) provisions on keeping a service invention secret and on the exploitation of service inventions as solutions subject to trade secrets shall not apply to service designs;
(b) in the case of service designs, failure to pay the maintenance fee shall mean failure to renew design protection and expiration shall mean the expiration of the term of design protection after which renewal is no longer possible;
(c) remuneration shall only be due to the designer for the exploitation of a service design that is covered by the design protection.[15]

In the case of designs created by persons working in public service, by persons in service relationships or by members of a co-operative working within the framework of a legal relationship of an employment nature, the above rules also apply *mutatis mutandis*.

6.3 The Parties' Duty to Inform

As mentioned in section 6.2 above, the parties' duty to inform is governed by the provisions on service and employee inventions in the Patent Act, which are discussed in section 4.

6.4 Employee Remuneration Right

As mentioned in section 6.2 above, employees' remuneration rights are governed by the provisions on service and employee inventions in the Patent Act, which are discussed in section 4.

14. Please see section 4.
15. Art. 14(5) of the Design Protection Act.

7 EMPLOYEES' PLANT VARIETIES

7.1 The Legal Framework

In Hungary, the Patent Act is the key piece of legislation in respect of plant variety protection, in particular Chapter XIII entitled Plant Varieties and the Protection of Plant Varieties.

7.2 Ownership of Rights

Matters relating to service and employee plant varieties and to the remuneration of the breeder are covered by the section of the Patent Act which deals with service and employee inventions (Articles 9-17).

7.3 The Parties' Duty to Inform

Please see section 4 above.

7.4 Employee Remuneration Right

Please see section 4 above.

8 INSTANCES FOR DISPUTES

In general, disputes arising in connection with IP rights, including disputes in connection with employees' intellectual property rights, fall under the exclusive jurisdiction of the County Courts (*törvényszék*) which are general courts, meaning that employees' intellectual property cases do not belong to the jurisdiction of the Labour Courts.

In Hungary, there is not a significant level of disputes in connection with employers' ownership claims. Disputes tend to occur in cases where it does not obviously result from the nature of the employment relationship that the elaboration of a given solution in the domain of the invention is the employee's obligation. This may arise, e.g., where an employee develops an invention or solution in connection with and during the course of employment that falls within the sphere of activities of the employer. These disputes however, can be minimised through the consistent use of appropriate contractual practices.

Numerous court cases have tackled issues such as whether an invention constitutes a service or employee invention (i.e., whether it falls within the scope of the employee's obligations under his or her employment contract). Legal disputes can arise relating to the proportionality of the inventor's remuneration; and consequences of the employer's failure to provide written declarations and responses to the employee's reporting of the invention within the timeframe set by law.

9 CONCLUDING REMARKS

The employee cannot be obliged under the employment relationship to create an invention that will (actually) satisfy the requirements of patentability. However, through their invention they may create a monopoly in favour of the employer. Otherwise, the employer can obtain an advantageous market position on the basis of a creative invention only if the employer acquires such creation at a fair market price. For this reason, payment to the inventor is justified. The basis of the payment may be the customary licence fee in the given field. In our view, remuneration provided for creations encourages developments of such intellectual creations.

Clearly drafted employment agreements and internal regulations governing remuneration issues are vital in the area of service inventions. While jurisprudence covering service inventions and remuneration agreements is still forthcoming in Hungary, consistently drafted employment and remuneration agreements, including invention by-laws, will ensure successful employment relationships and a competitive business.

BIBLIOGRAPHY

Table of Sources

Bacher Vilmos, *Jogalkalmazási kérdések a szolgálati és alkalmazotti találmányok új törvényi szabályozása alapján*, 5/96:13-22, Iparjogvédelmi és Szerzoi Jogi Szemle.

Bacher Vilmos, *Jogalkalmazási kérdések a szolgálati és alkalmazotti találmányok új törvényi szabályozása alapján*; 6/96:3-7 Gazdaság és Jog.

Szalma Borbála: *Know-how – a szolgálati találmány 'üzleti titka'*; 2015/1, Iparjogvédelmi és Szerzoi Jogi Szemle.

Szalma József, *A szolgálati és alkalmazotti találmány a magyar jogban,/* 2006:77-84 Magyar Jog 2.

Dr. Bogsch Attila, dr. Boytha György, dr. Faludi Gábor, dr. Grad, Gyenge Anikó, dr. Gyori Erzsébet, dr. Gyertyánfy Péter, dr. Kabai Eszter, dr. Kricsfalvi Anita, dr. Szinger András, dr. Tarr Péter, dr. Tomori Pál & dr. Tóth Péter Benjamin, *Commentary on the Copyright Act* (Complex Kiadó 2014).

Vilmos Bacher, Gusztáv Bacher, István Gödölle, Katalin Marmarosi & András Szecskay, *The AIPPI Report Q183 on Employers' rights to intellectual property.*

Dr. Legeza Dénes – A munkaviszonyban alkotott művek joggyakorlata napjainkban; in: Iparjogvédelmi és szerzői jogi szemle, 11. (121.) évf. 3. sz. / 2016, p. 100-128.

Table of Legislation

The Labour Code Act of 2012, I.
The Civil Code Act of 2013, V.
The Amendments to Certain Laws of Industrial Property and Copyright Act of 1994, VII.
The Protection of Inventions by Patents Act of 1995, Part V XXXIII.
The Protection of Utility Models Act of 1991, XXXVIII,
The Protection of Trade Marks and Geographical Indications Act of 1997, XI.
The Copyright Act of 1999, LXXVI.
Certain aspects of electronic commerce services and information society services (E-Commerce Act) of 2001, No. CVIII.
The Code of Civil Procedure Act of 2016, No. CXXX. The Civil Code Act of 1959, No. IV (Chapter VII: Inherent Rights and Rights related to Intellectual Products).

Table of Cases

EBH 2003. 948.
BH 1996. 249.
BH 1993. 546.
BH 1992. 168.
Metropolitan Court 3.P.21.730/2008/124.
ISZT - 7/2008
ISZT - 6/2012

CHAPTER 12

Italy

Andrea Ottolia

1 BACKGROUND

Italian law does not provide a unified regulation of employees' creations, but rather an extensive set of different rules.[1] Some provisions – not specifically regulating this subject but relevant to it – can be found in the Constitution, in the Civil Code and in labour law. Other provisions, specifically referring to employees' creations, are contained in the rules concerning each individual intellectual property right: while the most detailed and complex regulation is provided for patentable inventions by Articles 64 and 65 of the Italian Legislative Decree no. 30 dated 10 February 2005 – setting out the Italian Industrial Property Code, other rules refer to topographies for semiconductors (Article 89 of the Industrial Property Code), industrial and ornamental models (Article 38 of the Industrial Property Code), plant varieties (Article 101 of the Industrial Property Code), software and database rights (Article 12*bis* of Law 633/41, the 'Copyright Act'), industrial design (Article 12*ter* of the Copyright Act), simple photographs (Article 88 of the Copyright Act); no regulation is provided for copyrighted works in general.

Before discussing the details of the specific rules, we should step back and consider that, with the sole exception of university researchers' creations (regulated under Article 65 of the Industrial Property Code), a common principle of these laws is that (at least) when the employee is *hired to create*, the employer holds title on such creations.[2] This scheme has been commonly perceived as the result of a conflict between the employee, who should hold title for being the author, and the employer

1. A systematic approach can be undertaken by interpretation, see Ubertazzi, *L'appartenenza dei risultati creativi dei dipendenti*, in *Aida* 516 (2010).
2. This is not, however, a common rule as in some cases the employer holds title even if the employee is not 'hired to create', as is the case for software and databases, see section 5.2.

who should hold title for the (economic) reason that they incur the risk of the investment and should not bear the transaction costs (and employee's 'holdup rights'[3]) of an *ex post* negotiation. The solution given by the law to these conflicting interests often leaves a sense of 'expropriation' to justify the interpretation of the norm as exceptional and to reshape certain norms in favour of the employee.[4]

I suggest a more balanced reading of the interplay between the different interests involved. First, the Italian Constitution protects labour under Article 35[5] and assures the employee's right to an adequate remuneration under Article 36:[6] it does not, however, imply that employees should hold title on their creations.[7] Secondly, due to the cumulative nature of creative work and its growing complexity, the archetype of the solitary inventor is outdated: the employee typically invents and creates because they benefit from the employer's organisation and vision, they share knowledge with its members and focus their inventive efforts on practical application. The employer should not be regarded as a mere economical counterpart of the author but as an essential driver of the complex machinery of innovation. From this perspective I would draw some preliminary conclusions: (i) innovative enterprises should not only be protected by the principle of freedom of enterprise, under Article 41[8] of the Constitution, but should also be regarded as a key player in the production of culture, which is pursued as a major interest of the Republic according to Article 9;[9] (ii) based on Article

3. See Merges, *The Law and Economics of Employee Inventions*, 13 *Harv. J. L. and Tech*, 12 (1999-2000); Milgrom and Roberts, *Economics, Organization and Management*, 599 (1992).
4. This affects the way some commentators and courts interpret the difference between service and business inventions under Art. 64 (see section 4.2.1); it also justifies the rules on universities' researchers inventions under Art. 65 (see section 4.5).
5. Art. 35 of the Constitution provides that 'The Republic protects work in all its forms and practices. It provides for the training and professional advancement of workers. It promotes and encourages international agreements and organisations, which have the aim of establishing and regulating labour rights. It recognises the freedom to emigrate, subject to the obligations set out by law in the general interest, and protects Italian workers abroad'.
6. Art. 36 of the Constitution provides that 'Workers have the right to a remuneration commensurate to the quantity and quality of their work and in all cases to an adequate remuneration ensuring them and their families a free and dignified existence. Maximum daily working hours are established by law. Workers have the right to a weekly rest day and paid annual holidays. They cannot waive this right'.
7. It has been argued that the fact that employer hold rights over their creations in the first place, is a necessary corollary of constitutional norms protecting Culture (under Art. 9 of the Constitution) and labour (under Art. 35 of the Constitution), see Oppo, *Creazione intellettuale, creazione industriale e diritti di utilizzazione economica*, in *Riv. dir. civ.* 1, 13 (1969).
8. Art. 41 of the Constitution provides that 'Private-sector economic initiative is freely exercised. It cannot be conducted in conflict with social usefulness or in such a manner that could damage safety, liberty and human dignity. The law shall provide for appropriate programs and controls so that public and private-sector economic activity may be oriented and coordinated for social purposes.'
9. Art. 9 of the Constitution provides that 'The Republic promotes the development of culture and of scientific and technical research. It safeguards natural landscape and the historical and artistic heritage of the Nation.' The Protection of Culture, under Art. 9 of the Italian Constitution, has been interpreted as the foundation of the recognition of IPRs (see the Const. Court 20 March 1978, No. 20 in II *Riv. dir. ind.*, 3 ss. (1978) where the court declared the exclusion of pharmaceutical patents a violation of Art. 9) and as a preference of exclusive rights provided by intellectual property compared to public funding of research, in this sense cfr. Galli, *Diritti di proprietà intellettuale e remunerazione degli investimenti*, in *Aida* 78-79 (2005). The norm does not

9, innovative enterprises may be afforded peculiarly favourable rules (e.g. to reduce transaction costs[10] or facilitate capital collection[11]) without violating the equality principle under Article 3 of the Constitution;[12] (iii) the norm providing a full assignment of economic rights to the employer should not be regarded as exceptional.[13] The courts should therefore be more cautious before unduly restricting the application of the rule or over-expanding the scope of employee-favouring norms.[14]

In this more balanced scenario, the weaker negotiating position of the employee remains, however, undisputable and open to either regulation or market solutions.

Regulation intervenes by providing some specific rules in favour of employees. Labour law guarantees the greatest possible validity to employment contracts and their effects: it does so by not requiring contractual formalities, by allowing employment contracts to be formed by '*facta concludentia*', by excluding the possibility that the violation of administrative requirements (such as the reporting of new employment contracts to the Ministry of Labour) may affect contract validity. Labour law also provides a system of collective contracts that are negotiated and signed by Unions and which also bind those who are not members of them.[15] Finally, a general principle of labour law states that default rules established by regulation can be contractually derogated from only in favour of the employee[16] (i.e. the so-called ban of '*derogatio in*

discriminate between different types of creations (either creative or scientific), and should imply not to do so in substantive law of intellectual property, see Ubertazzi, cited, 518. Another legal foundation to IPRs is provided by Art. 42 of the Constitution which protects property: 'Property is publicly or privately owned. Economic assets belong to the State, to entities or to private persons. Private property is recognised and guaranteed by the law, which prescribes the ways it is acquired, enjoyed and its limitations so as to ensure its social function and make it accessible to all. Private property may, in the cases provided for by the law and with provisions for compensation, be expropriated for reasons of general interest. The law establishes the regulations and limits of legitimate and testamentary inheritance and the rights of the State in matters of inheritance.'

10. In general, transaction costs benefit from economies of scale and so favour larger companies. This justifies regulations meant to reduce such costs for SMEs, see Ferrarini and Ottolia, *Corporate Disclosure as a Transaction Cost: The Case of SMEs*, in 9(4) *European Review of Contract Law* 1-24 (2013); in the present work I argue that – also on the constitutional basis of Art. 9 – such transaction cost reduction should benefit cultural and innovative enterprises.
11. In this sense I would read the adoption of Italian crowdfunding regulation that is meant to benefit innovative start-up companies, see on the point Ottolia, *L'equity crowdfunding tra incentivi al reperimento di capitale di rischio per start up innovative e responsabilità*, in *Diritto della banca e del mercato finanziario* 43 (2014).
12. Art. 3 of the Constitution says 'All citizens have equal social dignity and are equal before the law, without distinction of sex, race, language, religion, political opinion, personal and social conditions.'
13. See Ubertazzi, *cited*.
14. E.g. in case of application of the 'equitable remuneration' under Art. 64(2).
15. E.g. see Cass. Civ. Sez. Lav. 14 April 1980, n. 2430; Cass. Civ. Sez. Lav. 16 October 1980, n. 5576; Cass. Civ. Sez. Lav. 9 January 1981, n. 190. Empirical evidence of Italian collective contracts shows, however, that too little attention is paid to the issue of employees' creation: a better use of this instrument could help to reduce uncertainty in the application of certain default rules. See Sarti, *Appunti sulle invenzioni dei dipendenti*, in Giur. ann. dir. ind., 1731 (1984).
16. This same principle applies to the relationship between collective contracts and individual contracts, see Vercellone, *Le invenzioni dei dipendenti* 67 (Milano 1961).

peius') so that employers cannot require the former to pre-assign title on occasional inventions,[17] or renounce an equitable reward.[18]

The market may also play a significant role. Economic studies[19] have shown that the parties may decide to provide specific benefits to the employee-creator and the latter may leave the company either to work for a different employer or to create a new venture. However, the real availability of such market-based solutions is highly dependent on the legal constraints on employee circulation and on the existence of a flourishing environment for innovative businesses.

2 EMPLOYEES' COPYRIGHT

2.1 The Legal Framework

Italian copyright is regulated by the Copyright Act, which does not contain a general provision on employees' copyright and leaves the subject entirely open to interpretation. Specific rules concern databases, software and simple photographs.[20]

Such a legislative gap can be filled by an analogous application of the rules concerning employees' inventions under which the employer holds title to the invention insofar as the invention is created in the execution of the employment.[21] Such method of interpretation requires considering first whether the norm is to be qualified as exceptional or not;[22] only in the latter case can it be applied by analogy to scenarios that, while different to the ones envisaged by the norm, share something in common with it that goes to the *reason of the norm*, its *ratio iurs*. [23] These conditions seem fulfilled in the present case because: (i) the patent provision under Article 64 is not to be considered exceptional, but rather as the result of the application of a general rule under which all the result of the employee's labour belongs to the employer; (ii) furthermore, copyright and patents seem to share a similar rationale. Applying a different rule to copyright creations cannot be justified by an argument relating to the 'sacredness of the person' and assumed cultural primacy of non-technical creations.[24] Indeed, creative works are no longer to be considered default rules. See Sarti, *Appunti sulle invenzioni dei dipendenti*, in *Giur. ann. dir. ind.*, 1731 (1984) the result of an individual flash of genius, but are rather produced under organised team activities of

17. See Art. 64(3) of the Industrial Property Code, discussed in para. 2.8.
18. See Art. 64(2) of the Industrial Property Code, in para. 2.2.
19. Merges, *cited*.
20. See Art. 12*bis* of the Copyright Act (databases and software) and Art. 88 of the Copyright Act (photographs).
21. See Art. 64(1) and first sentence of Art. 64.2 of the Italian Industrial Property Code.
22. Art. 14 of the preliminary norms of Italian Civil Code, so-called *Preleggi*.
23. According to the definition stated in Bobbio, *L'analogia e il diritto penale*, in *Riv. pen.*, 108 (1938). Indeed the definition is reminiscent of the traditional meaning of *Ubi eadem ratio, ibi eadem dispositio*.
24. The traditional perception of an ontological difference between technical and non-technical creations remains in the fact that while the Ministry of Culture is competent for copyright law, the Ministry of Industry is competent for inventions and trademarks. In my opinion, the need for homogeneousness and coherence of such convergent intellectual creations would justify unified governmental competence.

the cultural enterprise which, in practice, are not so different from the organisation of the inventive process at inventive enterprises. Even the difference between copyright-protected creations and inventions has been reduced because of the rise of technical creations protected by copyright law in some circumstances. In this sense a difference in treatment may contravene Article 3 of the Constitution that protects equality between citizens. Finally Article 9 that protects culture regardless of the technical or non-technical nature of creations.

A critical point is to understand is whether the employer holds title in works created by the employee undertaken as a consequence of the employment, or only where the employee is specifically hired to create. The first solution seems preferable. In patent law the employer holds title in both cases, however (at least under the interpretation of Article 64(2) supported here[25]) when the employee is not hired to invent they enjoy an equitable remuneration right for patentable innovations. The right to the reward seems exceptional and therefore not applicable by analogy to employees' copyright:[26] it is specifically conditioned to the unexpected merit to have provided to the employer not just an intellectual creation but that particular kind of creation that fulfils the patent requirements. This interpretation also fits with rules established for computer programs and databases by Article 12-*bis* of the Copyright Act which provides that the employer holds title in their employee s creations made in the execution of their duties (or upon the employer's instructions), regardless of whether the employee was hired to create,[27] and does not recognise any additional remuneration in the latter case. The same rule governs employees' simple photographs which, when created in the execution of the employment, belong to the employer (Article 88 of the Copyright Act).

2.2 Ownership of Rights

The assignment of copyright to the employer is to be considered 'derivative' (and not original) for a number of reasons:[28] (i) Article 2576 of the Civil Code and Article 6 of Copyright Act establish a link between the (original) acquisition of an IP right and the creation of the work; (ii) the assignment is based on a title that is the employment

25. Indeed the criteria under which Art. 64(1) and (2) are to be applied are still subject to debate. The issue is analysed further in section 4.2.
26. Neither the principle of equitable remuneration (Art. 64(2) second sentence) nor the other clauses under Art. 64(4)-(6) should be applicable to copyright works. Finally as for patents, these rules should apply to employment relationships and parasubordinate work while not to independent contractors for which the general rules of purpose of the contract would apply.
27. A similar argument has been put forth with reference to the European Software Directive, see Walter-Von Lewinski, *European Copyright Law* 118 (2010), who stresses that the European norm includes cases where the employee is not engaged for the purpose of creating a copyright work. It is interesting however to note that unlike the Italian norm, Art. 2(3) of the Software Directive requires Member States to confer on employers not necessarily the ownership, but merely the economic rights related to the computer program created by the employee in the execution of their duties.
28. Oppo, *Creazione intellettuale, creazione industriale e diritti di utilizzazione economica*, in I *Riv. dir. civ.* 1 (1969); the acquisition would not be derivative but original according to Vercellone, *Le invenzioni dei dipendenti* 75 ss. (Milan 1961).

contract;[29] (iii) no clause in Italian law refers to the employer as the author of the work;[30] (iv) by this interpretation original ownership of economic rights and moral rights converge on the same individual (i.e. the author[31]).

The derivative nature of the assignment has some consequences. Such an assignment would be void when the title (i.e. the employment contract) is void. This outcome would, however, be virtually nullified by Article 2126 of the Civil Code which provides that when the employment contract is invalid its effects remain valid and so the IPRs assigned by virtue of an invalid contract remain fully owned by the employer. However, the application of Article 2126 of the Civil Code to employees' creations has been subject to debate. If one considers the norm as protection provided in favour of the employee, a teleological interpretation would avoid the application of Article 2126 to guarantee the effects deriving from Article 64 which are mostly in favour of the employer. It has been noted that such teleological interpretation contrasts with the literal meaning of the clause, which does not discriminate between positive and negative effects in relation to the employee.[32] I believe that in this case the literal and teleological meaning converge, as the rule under Article 2126 is rather meant to avoid a possible problem where the potential invalidity of the contract will have disruptive consequences on the activities undertaken up to that moment. The clause is therefore applicable to employees' creations.

The derivative nature of the assignment is however open to the risk that the employee is deprived of the right before transferring it to the employer.[33]

Assignment of copyright under Article 64, however, while derivative, is not an effect of a contractual transfer to the employer[34] but rather the direct effect determined by the law,[35] therefore:

(i) we should not look at the purpose of the contract[36] in order to identify which rights are in fact transferred, since the natural effect provided by the

29. Oppo, *cited,* 10; Ubertazzi, *Profili soggettivi del brevetto, cited,* 28.
30. In North American law, the legislator specifically solved the matter of regulating copyright in 1999 by specifying that employers are the authors of the works created by their employees, thereby probably implementing acquisition by original title.
31. Ubertazzi, *L'appartenenza dei risultati creativi dei dipendenti, cited,* 528 ss.
32. Sena, *I diritti sulle invenzioni e sui modelli industriali* 197 (Milano 1990).
33. Some authors have stressed that the direct effect would impede the employee in disposing of their rights in breach of the transfer of rights toward the employer, see Auteri, *Fattispecie costitutiva e soggetti,* in Aa.Vv., *Diritto Industriale* 529 (Torino, 2005).
34. The derivative nature of an assignment does not necessarily imply a legal relationship between the parties or a transfer of rights between the two, see Pugliatti, *Voce 'Acquisto del diritto',* in I *Enciclopedia del Diritto* 510 ss. (Milan 1958). These are cases of derivative acquisition without a voluntary transfer such as succession in title based on legal schemes (succession *mortis causa* regulated without testament).
35. For similar reasoning, see Ubertazzi, *I diritti d'autore e connessi, cited,* 35 ss., Bertani, *Impresa culturale e diritti esclusivi, cited* 435 ss.; Cogo, *I contratti di diritto di autore nell'era digitale* 224 ss. (Torino 2010); Galli, *sub art. 12 ter l.a.,* in Ubertazzi, *Commentario breve al diritto della concorrenza,* 1384 in case law see the Court of Florence 18 September 2006, in *Aida* 2008, 1210; Court of Rome 23 January 1993, in *Aida* 665 (1994); since it is a direct effect it is not necessary a further act of consent to the transfer, see Cass. 329/1979, Cass. 2276/1979).
36. Circulation of IPRs is governed by the principle that – absent clear contractual statements – the parties transfer those, among the several economic rights, that are strictly necessary to pursue

employment contract is – where not otherwise agreed – the full assignment of all the economic rights to the employer.[37] Some courts have, however, applied the principle of contractual purpose in cases involving employees' creations, e.g. to limit the employer's right to use a television report made by its own reporter in its commercials,[38] or has prohibited an editor from exploiting comic strips for merchandising purposes created by an employee;[39]

(ii) the assignment has effect from the moment when the creation is made: there is no need to further accept it in order to undertake a full transfer of rights;[40]

(iii) this interpretation also affects the proof of the transfer. Indeed, under Article 110 of the Copyright Act the transfer of economic rights must be proved in writing. This condition should not apply to employees' creations where the assignment is not undertaken by a transfer but is effected by the law.[41] Furthermore, the norm is exceptional as it contrasts the general principle of

---

the purpose of the contract. The principle derives from the rule – based on Art. 19 of Copyright Act – that economic rights are independent and the exercise of one of them does not exclude the exclusive exercise of each of the others, and by Art. 119 of Copyright Act according to which the transfer of rights does not imply the transfer of other rights not necessarily dependent from the right transferred (see Auteri, *Contratti traslativi del diritto di autore e principio di indipendenza delle facoltà di utilizzazione*, in II *Riv. dir. ind.* 105-131 (1963)); the clause is considered applicable to neighbour rights (Sarti, *Diritti esclusivi e circolazione dei beni* 395 (Milano 1996); Bertani, *Impresa culturale e diritti esclusivi* 367 ss. (Milano 2000)) and also to the other IPRs.

It is debatable whether the norm is meant to protect the author (who is the ontologically weak party in the contract) from being deprived by counterparts that typically try to collect any possible right in the work (Oppo, *cited* 1 ss. (1969), De Sanctis, *Il carattere creativo dell'opera dell'ingegno* 167 (Milano 1971); in case law see the Court of Milan of 17 May 2002, Aida 2003; Court of Milan of 21 November 1991, Aida 1992, 818). I would rather support the idea that such clause is meant to incentivise the greatest possible economic exploitation of IPRs so favouring the fractioning of the rights and favouring the circulation of the work through different successors in title and different markets. If this perspective is correct, then the clause may also (i) regulate not just the contract between an author and their successor in title but also a contract between an author's successor in title and the third party; (ii) it should allow the transfer of all rights when expressly and clearly stated by the parties. Indeed, if the reason is a better exploitation, it should allow the parties to undertake a full transfer of rights when transaction costs render it more convenient.

37. Musso, *Recenti sviluppi normativi sulle invenzioni 'universitarie' (con alcune osservazioni sul regime delle altre creazioni immateriali)*, in *St. on. Vanzetti* (Milano 2004), who distinguishes between software, design and database (where there is an assignment of all rights) from other copyright in order to identify which rights are transferred we should look at the theory of the purpose of the contract would apply; see Court of Milan 17 May 2002 Aida 2003; against the theory of the total assignment of rights see De Sanctis, *Il carattere creativo dell'opera dell'ingegno, cited* 167; Oppo, *Creazione intellettuale, creazione industriale e diritti di utilizzazione economica, cited.*
38. Court of Milan 21 November 1991 in *Aida* 1992.
39. Court of Milan 17 May 2002 in *Aida* 2003.
40. In this sense with reference to copyrights see Carosone, *Riflessioni in tema di opera dell'ingegno create in rapporto di lavoro subordinato,* in *IDA* 219 (1979); in case law, Court of Florence 18 September 2006, in *Aida* 2008, 1210; contra De Sanctis, *I soggetti del diritto d'autore,* Milano 85 ss. (2000).
41. Art. 110 would not be applicable to employment contracts according to Court of Rome 18 March 2002, commented by Bertani in *Aida* 785 (2003); Cogo, *I contratti di diritto di autore nell'era digitale, cited,* 223; contra Oppo, *cited,* 37; Ristuccia and Zeno-Zencovich, *Prime notazioni sulla legge a protezione del software,* in *Dir. inf.* 243 (1994).

freedom of forms that governs contract law under Article 1325 of Civil Code[42] and should not be applied by analogy.[43]

In the field of employees' creations the mechanism of direct assignment of all economic rights to the employer is significantly preferable to that performed under the 'purpose doctrine':

(i) It reduces the typical problems of having multiple owners on the same asset[44] or on different rights over the asset. Often the (transaction) costs of reassembling the rights can swallow all gains:[45] certainly right holders can create institutions to solve this problem, but it is highly unlikely that creative employees can take part or benefit from such private initiatives.
(ii) Direct assignment avoids a typical holdup[46] problem which the employer may face where their successor in title wishes, e.g., to use a certain specific right not covered by the purpose of the contract. In this case the employer or their successor in title should negotiate for the 'last parcel' of economic rights left to the employee.[47]
(iii) Furthermore, since corporate creations often take the form of team production, it can be difficult to ascertain who does what because the creative process cannot per se be separated into parts. If this is possible, then it can be difficult to determine the contribution and the value of each component.[48]

The described mechanism of 'derivative and direct' assignment should be recognised for all creations performed in execution of an employment contract,[49] but not outside, such as where creations are undertaken by independent contractors.[50]

42. On the freedom of form in Italian contract law see Roppo, *Il contratto* 218 (Milano 2011); Castronovo *Principi di diritto europeo dei contratti* 151 (Milano 2001).
43. Galli, Sub art. 12*ter* l.a., in Ubertazzi, *Commentario breve al diritto della concorrenza,* 1384.
44. Heller, *The Boundaries of Private Property,* 108 *Yale L. J.* 1163 (1999).
45. Merges, *cited,* 4.
46. See Calabresi and Melamed, *Property Rules, Liability Rules, and Inalienability: One View of the Cathedral,* 85 *Harv. L. Rev.* 1089, 1106-07 (1972).
47. Often holdup problems in IP law rise when the 'last parcel' is a key component of the asset owned by the investing party, see Merges, *cited* 13.
48. Alchian and Demsetz, *Production, Information Costs, and Economic Organization,* 62 *Am. Econ. Rev.* 777 (1972).
49. It has to be noted that Art. 88 of the Copyright Act concerning simple photographs performed in execution of an employment refers to the rights which are necessary to the purpose of the contract. This could be regarded as an exception to the mechanism of complete and automatic assignment described above. However, there is common consensus in favour of a systematic reading of the norm so to apply to these works the same direct and complete assignment of rights provided for other employees' intellectual creations.
50. De Sanctis and Fabiani, *I contratti di diritto di autore* 88 (Milan 2007).

2.2.1 Employees Moral Rights

When copyright is assigned to the employer, the employee continues to hold title on moral rights.[51] Under Article 20 of the Copyright Act moral rights on creative works give the author the right to be recognised as such and the right to oppose to any modification or damage to the work that would create a prejudice to their honour and reputation. Furthermore the author of an anonymous work has the right to show up at any time and declare themself the author. Moral rights are then recognised for other kinds of IPRs: under Article 62 of the Industrial Property Code the author of an invention has the right to be recognised and indicated as the author; a moral right of paternity is then expressly recognised for Plant Varieties by Article 110 of the Industrial Property Code, while a paternity right to topographies is assumed by prevailing interpretation.[52]

The right to paternity obliges the employer to designate the employee as the author of the creative work (with regard to inventions, the author must be designated as such when filing an application according to Article 119 of the Industrial Property Code). However the employer is not obliged to name the employee in all commercial or non-commercial communications regarding the creative work.[53] The extent of such an obligation should be interpreted by taking into account the purpose of communication (e.g. citing the author may seem reasonable in a scientific communication[54] but not in a commercial one) and the existence of technical reasons that make such indication possible or adequate.[55]

The author's right to oppose illegitimate modifications of the work can be enforced by the author against any third party. However, it is likely that the employer will be interested in modifying the work, either for mere preferences, commercial reasons, or to serve an objective need. Article 20(2) of the Copyright Act contains an express limitation to the moral right which permits modification of an architectural work which is necessary to fulfil an objective need of the structure. This rare example of limiting a moral right could apply in cases where the modification is necessary to fulfil a requirement that is necessary to allow normal exploitation of the work.

---

51. De Sanctis, *Il carattere creativo dell'opera dell'ingegno* 167. The legal foundation of moral rights is Art. 2 of the Constitution says 'The Republic recognises and guarantees the inviolable rights of the person, as an individual and in the social groups where human personality is expressed. The Republic expects that the fundamental duties of political, economic and social solidarity be fulfilled.'
52. Art. 168.2 of the Industrial Property Code recognises the right of the author to be included during the registration of the topography. The right to paternity is supported in Musso, *La tutela giuridica delle topografie dei prodotti a semiconduttori*, in I *Riv. dir. ind.* 51 (1990).
53. Santini, *I diritti della personalità nel diritto industriale* 86 (Padova 1959); Court of Rome 25-1-1951, I *Foro Italiano* 505 (1951).
54. Ubertazzi, *Profili soggettivi del brevetto, cited,* 214.
55. Ammendola, *Il film pubblicitario,* in *Aida* 56 ss. (1994); with regard to design see Guizzardi, *La tutela d'autore del disegno industriale. Incentivi all'innovazione e regime circolatorio* 88 (Milano 2005).

Finally, the employee's moral rights cannot affect the decision about the way to exploit the work and so – for business inventions – it cannot limit the employer's discretionary decision whether to patent the invention or to keep it secret[56] or even disclose it to the rest of the world.

2.3 The Parties' Duty to Inform

There are no provisions governing the parties' duty to inform.

2.4 Employee Remuneration Right

There are no provisions governing the employee's remuneration right regarding copyright.

3 EMPLOYEES' SEMICONDUCTOR CHIPS

3.1 The Legal Framework

Article 89(2) of the Industrial Property Code concerning topographies of semiconductor products refers to the rules stated in Article 64 of the Industrial Property Code. They do not refer to Article 65. This partial extension: (i) confirms that the rules provided for university employees bear an exceptional nature[57] and that (ii) all the rules contained in Article 64 will be applicable to these kinds of IPRs: and consequently not only the different regimes referring to service, business and occasional inventions will apply, but also all the rules stated in Article 64(4)-(6).

3.2 Ownership of Rights

Solely concerning topographies, Article 89(3) extends the rule that 'employer holds title' to working relations different from employment. This typically refers to independent contractors. This specification not only means that all the rights belong to the buyer/employer, but that such assignment of rights should be considered – such as the mechanism under Article 64 – a direct effect of the law. Consequently, the assignment of rights will be direct and total and not limited to the sole scope of the contract.

Article 89(3) is a very interesting legal mechanism because it carries the advantages of direct assignment of IPRs, but its extension to other IPRs should be made by the legislature and not by courts because of the apparently exceptional nature of the clause.

In the event of contracts with independent creators for the development of complex products the parties will have to pay attention to this further default

56. Ubertazzi, *Profili soggettivi del brevetto, cited,* 221.
57. Ubertazzi, *Le invenzioni dei ricercatori universitari,* in *Contr. e impr./Europa* 1054 (2003).

asymmetry in the regimes for different components, i.e. topographies on the one side, and the software housed in the chips on the other.

3.3 The Parties' Duty to Inform

There are no provisions governing the parties' duty to inform.

3.4 Employee Remuneration Right

There are no provisions governing the employees' remuneration right regarding semiconductor chips.

4 EMPLOYEES' INVENTIONS

4.1 The Legal Framework

Article 64 of the Industrial Property Code regulates the allocation of rights concerning employees' inventions. It gives three different cases under which an invention can be created by an employee.

4.2 Ownership of Rights

Article 64(1)-(2) refers to '*invenzioni di servizio*' (i.e. service inventions) and '*invenzioni di azienda*' (i.e. business inventions) respectively, both of which are regulated under a general principle of labour law according to which all the results of the employees' labour belong to the employer, and, in the event of a business invention, this is only mitigated by the recognition of an equitable remuneration to the employee. Article 64(1)-(2) provides:

(1) where the invention is created in execution of a contractual obligation, where the inventive activity is part of such obligation and compensated for that purpose, all exploitation rights to the invention belong to the employer except only the right to be identified as author of the invention which shall belong to the employee.

(2) if there is no compensation for inventive activity and the invention is created in the execution or upon the performance of a contract of employment, all exploitation rights to the invention shall be owned by the employer, provided that the employee shall have the right to be identified as author of the invention, and, as far as the employer or his successors in title obtain a patent or use the invention as a trade secret, the right to be compensated by an equitable reward to be determined according to the importance of the

> invention, the tasks carried out by the employee, the employee's compensation, the contribution that the employee received from the employer's organisation.[58]

Under Article 64(3) of the Industrial Property Code, inventions that do not fall within the definition of service inventions and business inventions, and which are therefore not created in the execution of the employment, belong to the employee. This rule applies the traditional IP law principle that exploitation rights over a certain intellectual property belong to the author. It is, however, balanced by a rule in favour of the employer: if the invention belongs to the employer's field of activity, the latter enjoys an option right either to acquire the invention, to obtain a licence on it, or to file a patent application in foreign countries. The fee or the price is determined by subtracting from the market price an amount corresponding to the contribution that the inventor has received from the employer while creating the invention.

Since the right in favour of the employer is discretionary and not a pre-emption, it can be exercised regardless of the employee's intention to use or dispose of the invention. The employer exercises their option right by submitting to the inventor a declaration within three months after receiving the notification that the patent application has been filed.

Article 64(6) of the Industrial Property Code establishes that any invention for which an application is filed within one year of the end of the employment is presumed to have been developed during the execution of the employment, either as service invention or as a business invention. The rule provides an effective tool to solve some of the conflicts that may arise between the employer and employee but also between subsequent employers of the same inventor.[59]

Some consider this a refutable presumption,[60] others irrefutable.[61] I would support the first interpretation in view of the fact that nothing in the text seems to support a different interpretation and that the refutable presumption would be more coherent with the need to preserve inventors' rights in a time of great mobility in the labour market, particularly when innovative businesses are concerned. Even in this scenario, the burden of proof lies with the employee.

---

58. With regard to the calculation of the equitable reward see the recent Court of Cassation 9-4-14; the previous version of the norm (contained in Art. 23(2) of the Patent Law) only referred to the 'importance of the invention' but courts have traditionally applied the existing formula, see Mansani, *La determinazione dell'equo premio spettante al dipendente inventore secondo la 'formula tedesca'*, in *Contr. e Impr.* 720 ss. (1993).
59. Franzosi, *L'art. 26 l. brev. inv.: presunzione assoluta o relativa?* in *Studi in onore di R. Franceschelli* 305 ss. (Giuffrè, Milan, 1983).
60. Franzosi, *cited*, 301; Ubertazzi, *Profili soggettivi*, *cited*, 574-575.
61. Auletta, *Dei diritti sulle opere dell'ingegno e sulle invenzioni industriali*, in Commentario al Codice Civile Bologna-Roma 1947, 304; Vercellone, *Le invenzioni dei dipendenti*, 54.

4.2.1 *Distinguishing Between Service and Business Inventions*

Most of the debate concerning the interpretation of Article 64 of the Industrial Property Code focuses on the criteria applied to distinguish between service and business inventions. The choice affects the scope of application of the equitable reward mechanism and is therefore a particularly sensitive issue.

Labour courts have consistently reduced the scope of application of Article 64(1) by requiring that in order to qualify as a service invention, the inventive activity must be included in the purpose of the contract and in the employment tasks – i.e. the employee is 'hired to invent' and receives a separate and specific remuneration.[62] The criterion seems to originate from an underlying policy consideration designed to widen the economic rights of employees.[63] A different view does not consider remuneration a significant element of distinction: for service inventions the employee is hired to invent, while for business inventions the invention is merely created in execution of the employment. This view has limited support in case law,[64] but finds wider consensus among scholars.[65]

I would agree with the latter: the very nature of the equitable reward is not mere remuneration[66] but rather compensation for the unexpected and unforeseen advantage gained by the employer. It is not clear why the mere lack of a specific remuneration – that would just involve a matter of *quantum* of payment and adequateness[67] – would give rise to a reward. Indeed, if an employee is *hired to invent* – notwithstanding the absence of specific remuneration – the inventive activity is not an unforeseen result but the effect of a contractual obligation. Furthermore, the first interpretation seems to introduce a different treatment among employees 'hired to invent', which might even be in violation of the equality principle protected by Article 3 of the Constitution.

Another interpretative issue affecting the scope of application of Article 64(2) is whether an invention created by an employee not *hired to invent* but *hired to research* can be qualified as a service invention.[68] Supporters of a positive answer stress the overlap between the concept of inventing and researching. Others argue that the two

62. Cass. 6 November 2000, No. 14439, in *Giur. ann. dir. ind.*, 1999, No. 3883; Cass. 19 July 2003, No. 11305, therein, 2003, No. 4485; Cass. 24 January 2006, No. 1285, therein, 2006, No. 4941; see also Locatelli, *Invenzione del dipendente*, in *Contr. e impr.* 740 (1989); Ascarelli, *Teoria della concorrenza*, 598 ss. The (unwritten) requirement of a specific remuneration has been sometimes applied with a very rigorous standard, see Cass. 2 April, 1990, No. 2426, in *Giur. ann. dir. ind.* 1990, No. 2470.
63. Galli, *Le invenzioni dei dipendenti nel progetto di codice*, in Ubertazzi, *Il codice della proprietà industriale* 38 (Giuffrè 2004).
64. Cass. 2732 del 92; Cass. 6157 del 85 Cass. 329 del 79.
65. Di Cataldo, *I brevetti per invenzione e per modello*, in *Commentario* (Schlesinger) 374 ss. (Giuffrè 2012); Ubertazzi, *Profili Soggettivi del brevetto*, 15 ss.
66. Libertini, *I centri di ricerca e le invenzioni dei dipendenti nel codice della proprietà industriale*, in I *Riv. dir. ind.* 49 ss. (2006).
67. Cass. Civ. Sez. lav. 6-3-92 n. 2732.
68. According to Cass. 10851/1997 the employment task should not consist in the mere activity which preludes (and helps) the invention but in the invention activity; see also Cass. ci. Sez. lav. 23-5-91 No. 5803, Cass. 2732/1992.

activities differ substantially since an obligation to research does not imply undertaking inventive steps that remain an unpredictable outcome.[69]

I believe that the choice should not lie in an abstract consideration of the semantic difference between research and invention and should not receive a 'one fits all' solution. We should rather take into account the kind of research and the field of activity,[70] since in certain areas the inventive process is characterised by quasi-standardised research and (marginal) inventions are created by the use of complex software and routine research. Clear examples of such methodology are biotechnology and bioinformatics,[71] but the same can be said in other areas of highly dimensioned research centres in R&D offices.

4.2.2 The Causal Nexus Between the Employment Relation and the Invention

Under both Article 64(1) and (2) the invention must be carried out in execution of the employment and not merely coincide with it.[72] The point is relevant as it draws the boundary of the scope of application of the two categories and consequently affects the definition of a third case, the so-called occasional inventions.

The point is subject to debate. However, the dematerialisation of the tools of work and the possibility of working at a distance from the place of employment or outside the working hours should lead to rather substantial approach: we should look at what functionally derives from the execution of the employment regardless of the time or the place of the inventive activities. However some courts recognise that inventions undertaken outside office hours and during holidays are *per se* outside the application of Article 64(1)-(2).[73]

4.2.3 Defining 'Occasional Inventions'

A critical point concerns the definition of 'occasional inventions'. Courts have stressed that in order to be 'occasional' under Article 64(3), the invention must be undertaken outside the employment activity, on the free and independent initiative of the inventor, and must not have been envisaged even if the inventor has taken advantage of the premises and services of the business.[74] Insofar as such a definition is based on negative conditions (i.e. that the invention is not to be included under Article 64(1)-(2)), I would regard any possible invention not related to the employment as

69. Libertini, *cited*, 56; *contra*, Di Cataldo, *cited*, 188; Galli, *Problemi in tema di invenzioni dei dipendenti*, in I *Riv. dir. ind.* 19 ss. 24 (1997).
70. Some courts have tried to pursue a similar line of reasoning see Cass. 23 May 1991, No. 5803, therein, 1991, No. 2602; Cass. 5 November 1997, No. 19851, therein, 1997, No. 3579; see also App. Bologna, 28 December 1984, therein, 1984, No. 1823.
71. Ottolia, *Dati genetici (ed altre materie prime) nella innovazione bioinformatica*, in Arezzo, Falce and Olivieri, *Confini e fronti della innovazione biotecnologica per la proprietà intellettuale* (Giuffrè, 2014).
72. Di Cataldo, *cited*, 187; Ubertazzi, *Profili soggettivi del brevetto, cited*, 21.
73. See however Di Cataldo, *cited*, who suggests that office hours should not be relevant under Art. 64(1) but should be under Art. 64(2).
74. See Cass. 55/2276; 59/3380; 64/2517.

occasional, regardless of the fact that it was developed during business hours, or by using the premises and instruments of the employer.

When applying the concept of the employer's field of activity – the definition of which is a key condition for the applicability of the option right – one could adopt what I would refer to as an objective standard by referring to the activity defined under the corporate purpose of the company. The advantage of this standard is that it assures more predictable outcomes and reduces litigation. It has been supported by some legal scholars.[75] It is not rare to find reference to this kind of standard in case law concerning unfair competition and confusion between business names.[76] A different interpretative approach is to consider the actual activity undertaken by the company; some scholars suggest that such empirical standard may be even more specific and refer, at least in complex organisations, to the activity specifically undertaken in the branch of the business where the inventor is employed,[77] or ultimately depend on the foreseeable field of activity the company might be engaged in in the immediate future, depending on its potential and existing resources.[78]

The choice between objective and empirical standard should not be based on a literal interpretation or dogmatic arguments but rather on the function of the clause. If the reason for the option right afforded to the employer is to guarantee the fulfilment of the fiduciary duties that employees have towards the employer, then the objective standard – which is more employer-oriented – would apply. But I would be cautious to expand fiduciary duties to the extent to include an inventive activity that has not been undertaken in the execution of the employment. The reason for the option right is rather due to: (i) the benefit that the inventor gets from working in an organised business environment[79] and (ii) the fact that the employer is more likely to fruitfully exploit the invention. For these reasons I would adopt the empirical standard and even consider the specific branch where the employee is working

A critical theme for discussion is what the employee can and cannot do after the creation of an (occasional) invention and before the employer has exercised the option right. It has been argued that the creator of an occasional invention should be considered bound by an implied obligation not to take any action that is capable of undermining the validity of the future patent (e.g. not to disclose the invention before the employer has exercised the option right[80]), neither may they use the invention or sell it to third parties until the employer has decided whether to exercise the option right or not.

It is important to note that the option right under Article 64(3) is an exceptional benefit provided to the employer – radically different from other employers' rights

75. Bonelli, *Privativa per invenzione industriale*, in *Nuoviss. Dig. it.* 940 (Torino 1966).
76. See Cass. Civ. 23 November 1993, n.11570; Cass. Civ. 8 May 2009, n. 10587, in *Riv. dir. ind.* 573 (2009).
77. Di Cataldo, *cited*, 190, *contra* Rinaldi, *Le invenzioni industriali e gli altri prodotti dell'ingegno dei dipendenti e dei ricercatori universitari nel nuovo Codice della proprietà industriale*, in *Riv. dir. ind.* 437 (2005).
78. With reference to the previous norms this reasoning was adopted in Cass. 59/3380.
79. Galli, *Le invenzioni dei dipendenti nel progetto di Codice della proprietà industriale*, in *Quaderni di Aida* – in *Codice della proprietà industriale* 339 (Padova 2005).
80. Di Cataldo, *cited*, 238; Greco-Vercellone, *Le invenzioni e i modelli industriali* 243 (Utet 1969).

stated in Article 64(1)-(2) – since it is completely independent of the employment relationship. The application of the norm should therefore adhere to the strict conditions provided by the clause: the employer's rights arise only after the employee files the patent application and are not related to the secret invention, i.e. the invention that is kept secret before the filing. Furthermore, when the legislature has decided to include the secret invention in employers' rights it has stated it clearly as in Article 64(2). In this sense until the employee files the patent application, they should be able to freely decide whether to patent it, to keep it secret or even disclose it to the world. Going beyond such conditions would be an arbitrary reduction of the inventor's right to self-determination. Nevertheless, once the employee files the patent application, they will be bound by implied obligations that are necessary to preserve the employer's rights: the employee cannot sell the invention or do anything that can diminish the value of the invention (including disclosure during the secrecy period); the employee should inform the employer directly[81] and within such reasonable time to allow them to submit patent applications abroad.[82]

In the event of sale of the invention the employer may sue a third party that would be considered not entitled to file the patent application and would have the remedies against a person who files a patent application without title.[83] In any case the employee will face the consequences deriving from the breach of its loyalty obligations, including disciplinary actions, firing for just cause[84] and damages.

4.2.4 Article 64 Beyond the Employment Relationship

Several doubts have emerged concerning the application of Article 64 not only to employment relations but also to a broader set of working relations, such as so-called parasubordinate work where elements of the employment relation coexist with those of an independent contractor.[85]

It has been argued that the principle that the employer holds title is exceptional insofar as there is a common principle that intellectual creations belong to their authors in the first place: this would impede the analogical application of Article 64 to parasubordinate work. However, such argument is not decisive because in the assignment set forth by Article 64(1) the IPRs do belong to the employee in the first place but are then assigned to the employer upon the direct effect of the contract. On

81. Mansani, *cited*, 529, who stresses that when directly informed the employer will have the chance not only to exercise the option but also to intervene during the examination process.
82. In case of unjustified delay in communicating to the employer the submission of the patent application, the latter may institute proceedings for damages, see Mansani, *cited*, 530.
83. Galli, *Problemi in tema di invenzioni dei dipendenti*, in *Riv. dir. ind.* 19 (1997).
84. Not communicating the filing of the patent in order to give the employer the chance to exercise their option right is a valid reason for firing for just cause, see Cass. 1980/18077.
85. The use of the term 'employer' implies that such a clause refers to all private and public employees with the sole exception of university researches' inventions which are governed by Art. 65 of the Industrial Property Code.

the other hand, the assignment of the rights to the employer is the application of the general principle under which all results of the employees' activity belong to the employer.

Based on the ontological weakness of the employee, the law further provides a series of cogent norms meant to protect them, such as the equitable reward under Article 64(2) and the general principle of labour law, according to which contracts may not put the employee in a worse position than the law requires (the so-called ban '*derogatio in peius*').

I believe that both sets of rules should apply to parasubordinate work: the former by analogy, the latter as an application of a general principle. Indeed, I consider in parasubordinate work the same ratio of weakness in the working party as in traditional employment contracts: parasubordinate work is becoming a major scheme in Italy particularly for young workers who – while not technically employed – differ substantially from real independent contractors, as they are typically dedicated to one employer, perform their job personally and rely upon their employer's organisation. For lack of such additional protection, employers would easily be able to impose contracts where equitable reward is excluded, or significantly reduced, or where the customer/employer holds the title on occasional inventions.

The hypothesis envisaged here faces strong counter-arguments:

(i) It contrasts with the prevailing jurisprudence of the labour courts which traditionally excluded the application of labour law principles outside the formal boundary of the employment contract.[86] However, a possible shift may arise from the new competence given to ordinary courts.[87] Ordinary courts may establish that there are prevailing interests in homogeneous interpretation between these working relations beyond the qualification of employment.
(ii) In one case (for the so-called *collaborazioni coordinate e continuative a progetto*) the law extended Article 64 of the Industrial Property Code to parasubordinate work.[88] For supporters of the negative thesis this shows that when the law wants to extend the employees' inventions rules to parasubordinate work it says so clearly (the so-called principle of '*ubi voluit dixit*').

Italian law provides specific rules on family businesses by providing default protection to the interests of those who work together with a relative who owns an individual enterprise. The default rule includes economic rights and governance rules meant to guarantee a certain amount of participation while preserving the individual character of the business and not implying the creation of a corporation.[89]

86. Cass. 2-3-95, n. 2426, *Foro It.* 1995, I, 3515; Cass. 9-3-89, n. 1245, *Foro It.* 1989, 1480.
87. Pellicani, in *Aida* 2005, *cited*, 44-45.
88. Actually Art. 65 of Italian Legislative Decree 276/2003 does it indirectly by stating that the parasubordinate worker has not only the moral right to be recognised as author but carries also all rights and obligations as 'regulated by special laws': intellectual property rules under Art. 64 should be included in such special laws.
89. The rights of the relatives are to be considered to credit the individual enterprise and not the shareholders of a corporation.

Given that the existence of a family business does not imply the existence of a contract of employment between the parties, Article 64 should not apply. The ownership of intellectual creations produced during such relations should follow the civil law rules and should be entirely regulated by contract.[90]

4.2.5 University Researchers' Inventions

Traditionally innovations produced by university research have not been kept secret but have rather been disclosed due to the need for academics to publish the results of their works to obtain public recognition and prestige. The focus of university research has gradually shifted from basic to applied research and has converged with the interests of the commercial exploitation of innovation; the need to attract private funds and provide financial support to the universities has progressively led academia to pursue the route of intellectual property protection, especially using patents.

Article 65(1) of the Industrial Property Code states that university researchers who are employed by a university (this includes 'ordinary professors' 'associate professors' and so-called 'structured researchers') – or by public administration institutions pursuing research as their institutional focus – hold title to their inventions. The model of such IPR assignment is the opposite of that provided by Article 64. Even if the researcher holds title, Article 65(2) gives the university the power to determine the amount of royalties to be required for licences of the invention subscribed by the researcher: in the lack of any such determination, the university has the right to keep 30% of such income. In any case, the inventor has the right to keep not less than 50% of the income deriving from the exploitation of the inventions.

These rules do not apply where the research has been financed by private entities. In such cases the assignment regime should be regulated by Article 64.

4.2.5.1 Exceptional Nature of the Norm and Scope of Application

The principle that a researcher holds the title is open to critique: (i) the rule seems to be in clear violation of the principle of equality under Article 3 of the Constitution, since it provides a clear asymmetry with researchers employed in the private sector;[91] (ii) the norm violates the principle in labour law of avoiding disparities between public employees, since it provides rules that are more favourable towards researchers operating in technical fields; (iii) it seems to be based on the abstract and ideological assumption that IPR ownership represents a valid incentive for researchers to pursue

90. Spada, *Impresa familiare e invenzione del familiare*, in *Riv. dir. ind.* 30 (1986), who stresses the constitutional principle that creations should belong to authors.
91. For the detailed consideration of possible unconstitutionality of Art. 65 of the Industrial Property Code, see Ubertazzi, *Le invenzioni dei ricercatori universitari*, in AA. VV., in *St. on. Vanzetti* 1730 ss. (Giuffrè 2004). See also Granieri, *Invenzioni accademiche e modelli di appartenenza*, in *La gestione della proprietà intellettuale nella ricerca universitaria*; Granieri, (a cura di), *Il Mulino* 21-22 (Milano, 2010); Granieri, *Circolazione (mancata) dei modelli e ricerca delle soluzioni migliori. Il trasferimento tecnologico dal mondo universitario all'industria e la nuova disciplina delle invenzioni d'azienda*, in I *Riv. dir. ind.* 63 (2002).

qualitative inventions. However, it does not take into account the fact that researchers are not equipped with financial and organisational instruments to pursue the patenting and licensing process which is a key point for the commercial exploitation of IPRs.

In any case, as the norm is in complete contrast with the principle that the employer holds the title provided in patents – but also as provided in other kinds of IPRs – it should be considered exceptional and not susceptible to analogous application. For lack of an express norm referring to it – such as in the case of biotechnology patents and utility models – this should not be applied to other kinds of intellectual property, nor to working relationships other than employed researchers. I would not include parasubordinate researchers, such as those who benefit from fellowship research contracts, or other internal figures who are not employed by a university, such as PhD students, or contract-based professors who have no employment relationship but work merely as independent contractors. These researchers' creations will be assigned by general contractual law rules and under the principle of the 'purpose of the contract'.[92]

4.2.5.2 *Contract-Based Solutions*

The intrinsic defects of the system provided for university researchers inevitably lead to attempts to contract around the default rules, putting the IPRs back in the hands of the university. University regulations that are issued by the individual university and assure a certain level of autonomy of each structure can be used to undertake such reshaping.[93] The fact that these derogations are legitimate is however debatable since employees' rights cannot be modified in a way that puts employees in a worse position.

On the one hand, the rules of Article 65 merely provide a system of assignment and may not *per se* prevent the inventor, as patent holder, from negotiating and returning the rights to the university. However, I would consider such regulation void if they require university employees to automatically reassign their IPRs. I would consider valid only those university regulations that provide a particular mechanism meant to allow and facilitate researchers transferring their rights to the university as a free and optional decision.

4.3 The Parties' Duty to Inform

There are no provisions governing the parties' duty to inform.

92. Ubertazzi, *Le invenzioni dei ricercatori universitari, cited.*
93. Ubertazzi, *I diritti d'autore e connessi* 43-44 (Milan, 2003).

4.4 Employee Remuneration Right

4.4.1 The Equitable Reward under Article 64.2

Since 2010 an equitable reward for an employee in the case of business inventions is payable when the employer – or their successors in title – either obtains a patent or uses the invention as a trade secret.

Indeed, the secret referred in Article 64(2) should be defined under two conditions. First, it should fulfil the requirements laid down by Articles 98-99 of the Industrial Property Code[94] and the employer should be considered as having decided to keep it secret only in the technical meaning defined thereof. Secondly, not all trade secrets should afford the employee a right to equitable reward, but only those which are potentially patentable.[95] This reading would appear to be coherent both with the literal wording of the Article that implies patent and secrecy as a valuable alternative, and with the need to limit the reward to cases that provide a significant advantage to the employer by having satisfied the relevant standard of inventiveness.

The reason of including inventions protected as secret is meant to avoid the right to remuneration depending on an employer's discretionary choice between patenting the employee's invention, and to preserve the constitutional principle of protection of labour.[96]

However, even the existing system does not avoid asymmetries: in case of patenting, the title can be subsequently declared void[97] and according to Article 77 of the Industrial Property Code the employer does not have any obligation to pay further remuneration to the employer.[98] In case of trade secrets, this hypothesis does not exist. However, I would think that the employee should give the remuneration back to the employer once it is ascertained that the invention lacked elements of patentability or secrecy. Furthermore, limiting the equitable remuneration to patents, instead of including inventions kept secret, carries certain advantages: (i) it would reduce uncertainty about the level of inventiveness that justifies the remuneration; (ii) it would encourage better inventive activity;[99] (iii) it would incentivise employees to focus their efforts on imperative solutions, since the remuneration would be conditional on the employer's decision to file a patent application and bear any

94. Under Art. 98 of the Industrial Property Code: (i) the knowledge must be new, (ii) bear economic value for the holder, (iii) be kept secret by proper means by the holder.
95. I believe, in this sense, we have Mansani, cited, 527.
96. Galli, *Codice della Proprietà industriale*: *la riforma* 76 (2010); on the point Pellicani, *Tutela del lavoro e della proprietà intellettuale*, in *Aida* 42 ss. (2005). Case law has tried to solve the problem by stating that where the employer does not patent the invention, the employee could ask to do so, and eventually file the patent application v. Cass. Civ. Chambers. emp. 6-12-2002 No. 17398; it has been proposed that when the invention has not been patented for a reason attributable to the employer, the employee will nevertheless hold a right to equitable remuneration, see Libertini, *cited*, 57. I agree with the theory that the employer is free to decide whether to patent or not, see Ubertazzi, *cited*.
97. This happens however when the judgment becomes final, see Cass. 5-6-2000, 7484, in *Gadi* 1241 (2000).
98. While payments already provided to the inventor remain valid.
99. Cfr. Di Cataldo, *cited*.

corresponding costs; (iv) it would create a significant incentive to shift the high inventive attitude of Italian enterprises from a model of personal (and secret) know-how to a model of impersonal exclusive rights provided by intellectual property.

4.4.2 *Equitable Remuneration and Transfer of the Invention*

Some doubts have arisen about who is obliged to pay the equitable reward if the invention is transferred. Sometimes only the patent, the pending patent or the invention kept as a trade secret is sold, other times the sale involves the entire business unit where the invention was created.

In the past, some courts have ruled that the payment should not necessarily be made by the employer but by their successor in title who obtains the patent.[100] This interpretation would appear to be based on the assumption of a necessary connection between the obligation to pay (or the right to receive) the reward, on the one side, and the issuance of the patent[101] on the other. Establishing such a link is arbitrary since the norm creates the obligation on the sole employer who originally gets the advantage of the innovation, regardless of the following circulation or protection of the IP right.[102]

Sometimes the patent is transferred because of the transfer of the business unit to which it relates. In these cases the obligation to pay the remuneration should follow the norm of the Civil Code for which the debts of the business unit are also borne by the buyer as far as they are identified in the accounting records. However, according to Article 2112 of Civil Code, when debts relate to labour (such as, I would say, the right to the equitable remuneration) both the seller and the buyer are jointly responsible[103] even if they do not result from the accounting records.

5 EMPLOYEES' UTILITY MODELS

Italian law contains a series of norms regulating employees' creations beyond patents which have been adopted with different instruments at various times. This explains the existence of some differences in the discipline. Such differences should be abandoned in favour of a unified system regulating ownership and transfer of rights. A homogenous approach should be preferred since it reduces transaction costs (that are particularly high when the creation of a single product involves different ownership systems of its components) and would avoid different treatments between employees involved in different types of inventive activity. A higher level of homogeneous may be pursued by interpretation; however this finds a fundamental limitation in the rule that impedes analogical application of exceptional norms.

100. Court of Milan 6-12-99, *Gadi* 4111/1.
101. Cass. 5 June 2000, *Gadi* 4068/1.
102. Indeed trade secret right is not to be regarded an exclusive right.
103. Galli, *Problemi in tema di invenzioni dei dipendenti*, in I *Riv. dir. ind.* 19 ss. (1997).

Solely for utility models[104] - regulated by Article 86 of the Industrial Property Code - the law provides a complete application of the laws of patentable inventions contained in Section IV and so of the norms contained in Articles 64-65. The reason for such peculiar treatment is that there is a clear proximity between the two forms of IP, namely that both consist of a solution to a technical problem. Upon consideration of such proximity, Article 84 of the Industrial Property Code allows so-called alternative patenting, i.e. the possibility to file a patent application together with an application for a utility model and then choose the best, or the available, protection.

6 EMPLOYEES' DESIGN RIGHTS

Designs and models created by employees are not regulated by a referral to Article 64 since Article 38(2) simply states that as far as these creations are made in the execution of their duties, the employer holds the title, and the inventor holds the authorship. The reason seems to be that for this kind of IPRs, such as copyright, the assignment to the employer is built on the mere causal relationship between the employment and creative activity with no need to further specify that the creator is hired to create.

Nothing is said about independent contractors whose circulation of rights would depend on the general rules for IPRs' circulation: it would require a specific acceptance of the work, the rights transferred would depend on the particular purpose of the contract.

In the design field it seems reasonable to infer that the typical purpose of the contract would make the buyer/employer hold the title regarding production and commercialisation of the product incorporating the design.[105]

7 EMPLOYEES' PLANT VARIETIES

As in the case of employees' semiconductor-chips, the Industrial Property Code states (under Article 111(2)) that employees' plant varieties are regulated by the rules stated in Article 64. The missed referral to Article 65 confirms that the rules provided for university employees are exceptional and that all the rules contained in Article 64 will be applicable to these kinds of IPRs: and so not only the different regimes referring to service, business and occasional inventions, but also all the rules stated in Article 64(4)-(6).

Furthermore, and unlike semiconductors, the norm expressly recognises the author's moral right to paternity of the plant variety (under Article 111(1)). These asymmetries confirm that little attention has been paid by the legislature to the

104. It is worth noting that for biotechnology patents, Art. 81*bis* of the Industrial Property Code states that norms under section IV are in general applicable as far as not expressly derogated by section IV*bis*. But this seems to be an obvious specification since these fall within the category of patentable inventions.

105. See Court of Justice of the EU, 2 July 2009, C-32/08, commented by Sanna, in *Aida* 1314 (2010).

clearness of the wording, and indicate that the reading of the entire body of law concerning employees' intellectual creations should not be based on a mere literal interpretation.

8 INSTANCES FOR DISPUTES

Since its entry into force in 2005 the Industrial Property Code has removed (under Article 134(3)) cases concerning employees' inventions from the competence of labour courts – that would have otherwise had 'natural' competence because of clause 409 – and has recognised the competence of the IP Specialised Sections which decide according to the rules provided by Legislative Decree 5/2003 on the corporate process. This reform seems to have given preference to the speciality of intellectual property over labour law.[106]

The choice would appear to be also based on a policy consideration meant to stress the importance not only of the employee but also the better incentive for circulation of intellectual property rights. In this sense it is consistent with the more balanced vision described in paragraph 1.[107]

Such competence is however limited to deciding whether the employees' rights exist (i.e. the *an debeatur*) and do not concern the amount (i.e. the *quantum*), which in turn falls under the competence of a board of arbitrators.

9 CONCLUDING REMARKS

Legislative solutions to the issues involved in employee's creations may be integrated by market solutions. These typically consist of two voluntary solutions that may be regarded as private-based compensation to the rule that the employer holds the title to employees' creations.

Firstly, where the employee has an inventive attitude they may be able to insert in the employment contract a further compensation to their creative activity, beyond that provided by default rules.[108] Secondly, the creative employee may decide to leave the employer to create a new start-up or work for a new employer, by retaining creations that have not yet reached a sufficient level of specificity[109] and so are not yet to be attributable to their former employer.

The effectiveness of market-based solutions depends on the context. The law significantly restricts the possibility for an employee to extract knowledge during the

106. Galli and Gambino, *Codice commentato della proprietà industriale e intellettuale* (Utet 2011).
107. For a different view, see Pellicani, *Tutela del lavoro e tutela della proprietà industriale. Per una lettura costituzionalmente orientata della disciplina delle invenzioni del lavoratore*, in *Aida* 44 ss. (2005), who stresses that these rules concern the analysis of issues that could not have been removed from the competence of labour courts and enumerates arguments of unconstitutionality.
108. Merges, *Contracting into Liability Rules: Intellectual Property Rights and Collective Rights Organizations*, 84 *Cal. L. Rev.* 1293, 1361 (1996).
109. On the role of these alternative tools see Merges, *The Law and Economics of Employee Inventions*, 13 *Harv. J. L. and Tech.* 3 (2000).

employment relationship and use it outside the employer's company - under the loyalty principle of Article 2105 of the Civil Code the employee is not only banned from competing with the employer, but also from disclosing information on the organisation and methods of business' production, or act in a way prejudicial to it. If the employee discloses information to third parties it may also amount to a violation of trade secrets protected under Articles 98-99 of the Industrial Property Code.

After termination of employment, the employee is free to use knowledge developed and absorbed from the employer with significant limitations: (i) if the knowledge developed with the previous employer has already reached a sufficient level of innovativeness, it may have been already directly assigned to the former employer; (ii) if the knowledge is a trade secret protected under Articles 98-99 of the Industrial Property Code[110] the employer cannot use it or sell it without violating such rights; (iii) Article 64(6) of the Industrial Property Code contains a rebuttable presumption that all inventions filed for patent application within a year from the end of the employment relation are intended to belong to the employer; (iv) the insertion of non-competition clauses regarding the post-employment period are conditioned to strict requirements and therefore they are not always easy to be agreed upon by the parties: under Article 2125 of the Civil Code they are to be in writing, limited to the object, in time, and territory, and have a maximum duration of five years.

BIBLIOGRAPHY

Table of Sources

Ammendola, *Il film pubblicitario,* in *Aida,* 43 (1994).

Ascarelli, *Teoria della concorrenza e dei beni immateriali* (Milano 1960).

Auletta, *Dei diritti sulle opere dell'ingegno e sulle invenzioni industriali,* in Commentario al Codice Civile Bologna-Roma 1947.

Auteri, *Fattispecie costitutiva e soggetti,* in Aa.Vv., *Diritto Industriale,* 529 (Torino 2005).

Alchian & Demsetz, *Production, Information Costs, and Economic Organization,* 62 *Am. Econ. Rev.* 777 (1972).

Auteri, *Contratti traslativi del diritto di autore e principio di indipendenza delle facoltà di utilizzazione,* in *Riv. dir. ind.,* 105-131 (1963).

Auteri, *I contratti traslativi del diritto di autore nel diritto internazionale privato,* in *Riv. dir. ind.,* 288 (1965).

Bertani, *Impresa culturale e diritti esclusivi* (Milano 2000).

Bertani, *Diritto d'autore europeo* (Torino 2011).

Bonelli, *Privativa per invenzione industriale,* in *Nuoviss. Dig. it.,* 940 (Torino 1966).

Bobbio, *L'analogia nella logica del diritto* (Torino 1938).

Calabresi & Melamed, *Property Rules, Liability Rules, and Inalienability: One View of the Cathedral,* 85 *Harv. L. Rev.* 1089, 1106-1107 (1972).

110. Court of Milan sez. penale, 19 July 2011.

Castronovo *Principi di diritto europeo dei contratti* (Milano 2001).

Cogo, *I contratti di diritto di autore nell'era digitale* (Torino 2010).

Carosone, *Riflessioni in tema di opera dell'ingegno create in rapporto di lavoro subordinato*, in *IDA*, 219 (1979).

Carosone, *L'opera dell'ingegno creata nel rapporto di lavoro autonomo e subordinato* (Milano 1999).

De Sanctis & Fabiani, *I contratti di diritto di autore* (Milan 2007).

De Sanctis, *Il carattere creativo dell'opera dell'ingegno* (Milano 1971).

Di Cataldo, *I brevetti per invenzione e per modello*, in *Commentario* Schlesinger (Giuffrè 2012).

De Sanctis, *I soggetti del diritto d'autore* (Milano 2000).

Ferrarini & Ottolia, *Corporate Disclosure as a Transaction Cost: The Case of SMEs*, in European Review of Contract Law, 9(4), pp. 1-24 (2013).

Franzosi, *L'art. 26 l. brev. inv.: presunzione assoluta o relativa?* in *Studi in onore di R. Franceschelli*, 301 (Giuffrè, Milan 1983).

Greco-Vercellone, *Le invenzioni e i modelli industriali* (Torino 1969).

Galli, *Codice della Proprietà industriale: la riforma 2010* (Milano 2010).

Galli, *Problemi in tema di invenzioni dei dipendenti*, in *Riv. dir. ind.*, 19 (1997).

Galli & Gambino, *Codice commentato della proprietà industriale e intellettuale* (Torino 2011).

Galli, *Le invenzioni dei dipendenti nel progetto di Codice della proprietà industriale*, in *Quaderni di Aida* – in Codice della proprietà industriale (Milano 2004).

Galli, *Diritti di proprietà intellettuale e remunerazione degli investimenti*, in *Aida*, 68 (2005).

Granieri, *Circolazione (mancata) dei modelli e ricerca delle soluzioni migliori. Il trasferimento tecnologico dal mondo universitario all'industria e la nuova disciplina delle invenzioni d'azienda*, in *Riv. dir. ind.*, I, p. 53 (2002).0

Guizzardi, *La tutela d'autore del disegno industriale. Incentivi all'innovazione e regime circolatorio* (Milano 2005).

Heller, *The Boundaries of Private Property*, 108 *Yale L. J.* 1163 (1999).

Libertini, *I centri di ricerca e le invenzioni dei dipendenti nel codice della proprietà industriale*, in *Riv. dir. ind.*, 49 (2006).

Locatelli, *Invenzione del dipendente*, in *Contr. e impr.*, 740 (1989).

Mansani, *La determinazione dell'equo premio spettante al dipendente inventore secondo la 'formula tedesca'*, in *Contr. e Impr.*, 720 (1993).

Merges, *Contracting into Liability Rules: Intellectual Property Rights and Collective Rights Organizations*, 84 *Cal. L. Rev.* 1293, 1361 (1996).

Merges, *The Law and Economics of Employee Inventions*, 13 *Harv. J. L. & Tech.* 12 (2000).

Milgrom & Roberts, *Economics, Organization and Management*, 599 (1992).

Musso, *Recenti sviluppi normativi sulle invenzioni 'universitarie' (con alcune osservazioni sul regime delle altre creazioni immateriali)*, in *St. on. Vanzetti* (Milano 2004).

Musso, *La tutela giuridica delle topografie dei prodotti a semiconduttori*, in *Riv. dir. ind.*, 51 (1990).

Oppo, *Creazione intellettuale, creazione industriale e diritti di utilizzazione economica*, in *Riv. dir. civ.*, 1 (1969).

Ottolia, *Dati genetici (ed altre materie prime) nella innovazione bioinformatica*, in Arezzo, Falce & Olivieri, *Confini e fronti della innovazione biotecnologica per la proprietà intellettuale* (Giuffrè 2014).

Ottolia, *L'equity crowdfunding tra incentivi al reperimento di capitale di rischio per start up innovative e responsabilità*, in *Diritto della banca e del mercato finanziario*, 43 (2014).

Pellicani, *Tutela del lavoro e tutela della proprietà industriale. Per una lettura costituzionalmente orientata della disciplina delle invenzioni del lavoratore*, in *Aida*, 44 (2005).

Pugliatti, *Voce 'Acquisto del diritto'*, in Enciclopedia del Diritto, I, 510 (Milan 1958).

Rinaldi, *Le invenzioni industriali e gli altri prodotti dell'ingegno dei dipendenti e dei ricercatori universitari nel nuovo Codice della proprietà industriale*, in *Riv. dir. ind.*, 437 (2005).

Ristuccia & Zeno-Zencovich, *Prime notazioni sulla legge a protezione del software*, in *Dir. inf.*, 243 (1994).

Roppo, *Il contratto* (Milano 2011).

Santini, *I diritti della personalità nel diritto industriale*, 86 (Padova 1959).

Santoro Passerelli, *Interpretazione – Opera dell'ingegno – Contratto di lavoro*, in *Riv. dir. civ.*, 186 (1940).

Sarti, *Appunti sulle invenzioni dei dipendenti*, in *Giur. ann. dir. ind.*, 1731 (1984).

Sena, *I diritti sulle invenzioni e sui modelli industriali*, 197 (Milano 1990).

Sarti, *Diritti esclusivi e circolazione dei beni*, 395 (Milano 1996).

Spada, *Impresa familiare e invenzione del familiare*, in *Riv. dir. ind.*, 30 (1986).

Ubertazzi, *I diritti d'autore e connessi* (Milano 2000).

Ubertazzi, *Profili soggettivi del brevetto* (Milano 1985).

Ubertazzi, *Le invenzioni dei ricercatori universitari*, in *Contr. e impr./Europa*, 1109 (2003).

Ubertazzi, *L'appartenenza dei risultati creativi dei dipendenti*, in *Aida*, 516 (2010).

Vercellone, *Le invenzioni dei dipendenti* (Milano 1961).

Walter-Von Lewinski, *European copyright Law: A Commentary* (Oxford 2010).

Table of Legislation

The Constitution
The Copyright Act.
The Italian Industrial Property Code
Italian Civil Code

Table of Cases

Cass. 55/2276.
Cass. 59/3380.
Cass. 64/2517.
App. Bologna, no. 1823 (28 December 1984).
Cass. n. 1245, *Foro It* (1989).
Cass. ci. Sez. lav. no. 5803 (1991).
Cass. no. 5803 (23 May 1991).
Cass. Civ. Sez. lav. n. 2732 (1992).
Cass. Civ. n.11570 (23 November1993).
Cass. n. 2426, *Foro It.* (1995).
Cass. 10851 (1997).
Cass. no. 19351 (5 November 1997).
Court of Milan, *Gadi* 4111/1 (1999).
Cass. 7484, in *Gadi* (2000).
Cass. *Gadi* 4068/1 (5 June 2000).
Cass. Civ. n. 10587 (8 May 2009).
Court of Justice of the EU, C-32/08 (2 July 2009).

CHAPTER 13
Latvia

Ilmārs Šatovs

1 BACKGROUND

In Latvia the Employment Law does not regulate intellectual property ownership issues. All the ownership issues are regulated by separate intellectual property laws.

Since 1991, when Latvia restored its independence from the Soviet Union (in which ownership of intellectual property rights was not supported for many decades), the intellectual property laws of Latvia were adopted and revised several times in order to provide more efficient applicability or in order to comply with the laws of the EU. Hence, the legal regulation of employees' intellectual property rights is comparatively new and does not have a long lasting legal history or tradition.

The creator is the initial owner of intellectual property rights. No other person is considered as the initial owner of intellectual property rights by law. In most cases intellectual property rights may be transferred by the employment agreement to the employer, if the employer and the employee have included in the employment agreement such special transfer of rights agreement. Only in few cases when the employment agreement is concluded and specific circumstances are met, all the economic rights to the intellectual property are transferred to the employer on the basis of law, unless otherwise provided in the employment agreement.

2 EMPLOYEES' COPYRIGHT

2.1 The Legal Framework

Copyright is protected by the Copyright Law adopted on 6 April 2000. The Copyright Law is one of the most discussed legislative acts of all the intellectual property laws in Latvia as it provides controversial legal regulation and several legal norms are still not compliant with EU directives. The Copyright Law governs all the issues related to copyright, including the legal regulation of ownership of works created in the course of employment.

2.2 Ownership of Rights

2.2.1 General Rule

Article 12(1) of the Copyright Law provides that if the author has created a work while performing their duties arising out of an employment relationship, the moral and economic rights to the work shall belong to the author. The initial owner of copyright is always the author regardless of the existence of an employment relationship.

Article 12(1) of the Copyright Law provides that the economic rights of the author may be transferred to the employer by an employment agreement in the course of the employment relationship or at any time by a copyright assignment agreement. Article 14 of the Copyright Law provides that the author's moral rights cannot be transferred.

2.2.2 Employees' Computer Programs

The only exception to the said general rule of ownership is copyright in computer programs. Article 12(2) of the Copyright Law provides that if a computer program has been created by the employee while performing a work assignment, all economic rights to the computer program so created shall belong to the employer, unless specified otherwise by the agreement. In this case the law at the outset provides automatic transfer of copyright from the employee to the employer. However, the economic rights in the computer program are transferred only if the employee has performed a work assignment in the course of the employment, but not during the employment relationship in general. The employer has to document all the assignments to the employee with due care in order to obtain the economic rights in the course of the employment relationship.

Article 12(2) of the Copyright Law is inconsistent with Article 2(3) of the Directive 2009/24/EC on the Legal Protection of Computer Programs which provides that where a computer program is created by the employee in the execution of their duties or following the instructions given by their employer, the employer shall exclusively be entitled to exercise all economic rights in the program so created, unless otherwise provided by the agreement. Article 2(3) of the Computer Program Directive is not fully implemented in Article 12(2) of the Copyright Law and Article 12(2) of the Copyright Law excludes computer programs which are created by the employee in the execution of their employment duties in general.

2.2.3 Neighbouring Rights

According to Article 47(1) of the Copyright Law the initial owners of the neighbouring rights are performers, phonogram producers and film producers. The said initial holders of neighbouring rights may be natural persons.[1] The law does not provide any specific legal provisions for transfer of neighbouring rights in the course of employment. Transfer of neighbouring rights to the employer should be provided in the employment agreement, otherwise the neighbouring rights remain with their initial owner.

2.3 The Parties' Duty to Inform

The Copyright Law does not provide any duty of the employee to inform the employer about creation of a work.

2.4 Employee Remuneration Right

There are no specific legal norms that stipulate the employee remuneration right in respect of copyright. Remuneration for creation of works and transfer of economic rights is deemed to be received for performing duties under the employment agreement, unless otherwise provided by the agreement. The remuneration may be all-inclusive and covered only by salary.

3 EMPLOYEES' SEMICONDUCTOR CHIPS

3.1 The Legal Framework

The Law on Protection of Topographies of Semiconductor Products was adopted on 31 March 1998. The said law governs all the issues related to semiconductor chips, including the legal regulation of ownership in the course of employment. The Law on Protection of Topographies of Semiconductor Products is compliant with Directive 87/54/EC on the legal protection of topographies for semiconductor products, as the Directive leaves it to the Member States to regulate the presumption of ownership of rights under the employment agreement.

3.2 Ownership of Rights

Article 3(1) of the Law on Protection of Topographies of Semiconductor Products provides that the exclusive right to a topography belongs to its creator. Article 3(2) of the said law provides that the employer can obtain an exclusive right to a topography if the topography has been created while performing duties under the employment agreement which provides for creation of topographies and transfer of rights is provided by the individual or collective employment agreement.

1. Art. 1(6), (8) and (10) of the Copyright Law.

3.3 The Parties' Duty to Inform

The Law on Protection of Topographies of Semiconductor Products does not provide any duty to inform.

3.4 Employee Remuneration Right

The remuneration for creation of semiconductor chips is deemed to be received for performing duties under the employment agreement, unless provided otherwise. The remuneration may be all inclusive and covered only by salary.

4 EMPLOYEES' INVENTIONS

4.1 The Legal Framework

The Patent Law was adopted on 27 February 2007. The Patent Law governs the general issues related to patentable inventions, including the legal regulation of ownership of a right to a patent application or a patent in the course of employment. The Law on Scientific Activity of 14 April 2005 governs specific issues related to ownership of inventions of the employees of scientific institutions.

4.2 Ownership of Rights

4.2.1 General Rule

Article 15(1) of the Patent Law provides a general rule that the employer has the right to a patent if the invention in relation to which the patent application has been filed has been created by the employee whose work duties include either inventor activity or research, design, construction or preparation of technological development.

Article 15(1) has a gap in legal regulation since its text does not expressly provide that the employer has a right to a patent application which may also be an object of ownership within the meaning of the Patent Law. However, if Article 15(1) provides that the employer has a right to a patent, then it should be considered all the more so that the employer has a right to a patent application, as a granted patent is the result of a successful patent application. This interpretation is supported by other legal norms of the Patent Law, e.g., Article 13 of the Patent Law provides rules for filing of a claim in court to transfer a patent application or a patent when a patent application has been filed by a person who did not possess such rights or a patent has been granted to a person who did not possess a right to file a patent application.

Recently the court had to decide such a claim in a civil case where the inventor who was a shareholder and an employee in a company brought a civil action against the company seeking transfer of the rights to the patent granted in Latvia and the international patent application. He claimed that the company did not possess rights to the invention, therefore the patent was registered and the international patent application was filed by a person who did not possess the rights to file a patent application in accordance with the Patent Law. The company contended that the time for bringing such action in the court had lapsed and that the inventor had himself transferred the right to the invention and the right to file a patent application to the company. The employment agreement did not provide for a transfer of the inventor's rights to the invention or rights to the patent application or patent. The inventor had signed a power of attorney as the member of the board to file the Latvian patent application in the name of the company. The inventor had also signed invention assignment agreements in the countries that required express intent of the inventor to transfer the invention to the applicant. The Regional Court of Riga[2] and the Supreme Court[3] held that there was a lapse of time for filing of the claim to transfer the said rights and, notwithstanding that, the inventor had himself transferred the rights to the company, which was evidenced by the power of attorney having been signed in the capacity of a member of the board of the company, and by transferring the rights to the invention in countries which so required under national law. It is evident from this case that the employer may obtain a right to a patent application in the course of employment under the Patent Law.

Furthermore, Article 15(2) of the Patent Law provides that if the duties of the employee do not comprise either inventor activity or research, design, construction or preparation of technological development, but are nonetheless related to the field of activity of the employer, the right to a patent belongs to the inventor. The employer in this case has the right to utilise the invention under a non-exclusive licence without the right to grant a sublicence to third parties.

The Patent Law also grants the inventor moral rights which cannot be transferred.

4.2.2 Inventions of Scientific Personnel

The Law on Scientific Activity provides distinct and specific legal regulation of intellectual property ownership issues for scientific personnel. The Law on Scientific Activity regulates intellectual property ownership issues between scientific personnel and registered private and public scientific institutions. In accordance with the Law on Scientific Activity, scientific personnel consist of scientists, scientific technical personnel and service personnel.

Article 8(3) of the Law on Scientific Activity provides that all the rights to the intellectual property that has been created by a scientist during their scientific activity is owned by the scientist, unless otherwise provided by the agreement.

The exception to this rule concerns scientific activity which is financed by the state. Article 39 of the Law on Scientific Activity provides that the state owns economic rights to the invention, if the invention has been created as a result of research entirely funded by the state in accordance with the by-laws of the institution which is in charge

2. Judgment of the Regional Court of Riga of 5 May 2010, case No. C04304709.
3. Judgment of the Supreme Court of 5 May 2010, case No. C04304709.

of the research. Additionally, Article 39 of the Law on Scientific Activity provides that if the invention is created by the employees of the state scientific institution during the research financed by the state and whose duties under the employment agreement include inventor activity or research, design, construction or preparation of technological development, all the economic rights to the invention are owned by the state scientific institution. If the employees of the state scientific institution create an invention in the state scientific institution during the research financed by the third party, ownership of the invention is determined by the agreement according to which the research was carried out.

4.3 The Parties' Duty to Inform

Article 15(3) of the Patent Law provides a duty of the employee who has created the invention to inform the employer in writing immediately and imposes an obligation on the employee to file the documents allowing assessment of the invention. Article 15(4) of the Patent Law provides that if the employer waives the right to the invention or within a time limit of three months does not inform the employee regarding the intention to utilise this right, the right to the invention is automatically transferred to the employee. Similar legal norms are provided under Article 39 (1) of the Law on Scientific Activity in respect to the inventors employed by the state scientific institutions.

4.4 Employee Remuneration Right

The Patent Law leaves freedom for the employer and the employee to agree on additional remuneration for the creation and utilisation of the inventions in the individual or collective employment agreement.

However, Article 39 of the Law on Scientific Activity provides that the inventor has a right to fair remuneration from the income generated by the state scientific institution from exploitation of the rights to the invention.

5 EMPLOYEES' UTILITY MODELS

Latvian law does not protect utility models.

6 EMPLOYEES' DESIGN RIGHTS

6.1 The Legal Framework

The Design Law of 17 November 2004 governs all the issues related to ownership of design rights. The Copyright Law also provides legal regulation in respect to design works created in the course of employment. Additionally, Regulation 6/2002/EC on Community Designs regulates the employees' rights to the Community design rights.

6.2 Ownership of Rights

Article 10(1) of the Design Law provides that the designer is the initial owner of design rights under the Design Law. Article 10(3) of the Design Law provides that the designer is the owner of the rights to the design that has been created in the performance of a work assignment, unless provided otherwise by the agreement with the employer. The Design Law also grants the designer inalienable moral rights to a name and authorship which cannot be transferred.

The Copyright Law provides simultaneous protection for design works. Under the Copyright Law the author is the initial owner of copyright. In case of design works the economic rights of the author are maintained by the author in the course of employment, unless the employment agreement provides otherwise.

However, the legal regulation of ownership stipulated by the Design Law and the Copyright Law is distinct from the framework provided by Regulation 6/2002/ EC on Community Designs.

Article 14(3) of the Regulation provides that where a design is developed by the employee in the execution of their duties or following the instructions given by their employer, the right to the Community design vests in the employer, unless otherwise agreed or specified under national law.

Latvian Design Law specifies that the design rights may only be transferred to the employer, if the design has been created in the performance of a work assignment and the employment agreement provides transfer of design rights. Therefore in Latvia the owner of the rights to a Community design is the designer, unless the employment agreement provides otherwise.

6.3 The Parties' Duty to Inform

The Design Law does not provide any duty to inform.

6.4 Employee Remuneration Right

The remuneration for creation of designs is deemed to be received for performing duties under the employment agreement, unless provided otherwise. The remuneration may be all-inclusive and covered only by the salary.

7 EMPLOYEES' PLANT VARIETIES

7.1 The Legal Framework

The Plant Varieties Protection Law of 17 November 2005 provides general legal regulation for ownership of the plant varieties created in the course of employment. The Law on Scientific Activity regulates the ownership issues in relation to plant varieties that have been created by the scientific institutions. The Council Regulation (EC) No. 2100/94 on Community Plant Variety Rights is applicable to Community plant varieties.

7.2 Ownership of Rights

Article 8(1) of the Plant Varieties Protection Law provides that a natural person is considered as the initial owner of the breeder's economic rights to a plant variety. After conclusion of the employment agreement regarding breeding or isolation of the particular variety, the employer is considered as the owner of the economic rights to a plant variety. The same is applicable to a Community plant variety right under Article 11(4) of the Regulation on Community Plant Variety Rights as transfer of rights under the employment agreement to a plant variety is assessed according to national legislation.

The Plant Varieties Protection Law grants the natural person who has bred or isolated a plant variety moral rights which cannot be transferred.

The Law on Scientific Activity which is applicable to scientific activity of private and public scientific institutions provides distinct and specific legal regulation of plant variety ownership for scientific personnel. The legal regulation of ownership under the employment agreement for plant varieties and inventions is identical (see Chapter 4.2.2).

7.3 The Parties' Duty to Inform

The Plant Varieties Protection Law does not provide a duty to inform. Only Article 39(1) of the Law on Scientific Activity in respect of breeders employed by the state scientific institutions provides a duty of the employee who has created the plant variety to inform the employer in writing immediately. The Law on Scientific Activity provides that if the employer waives the right to the plant variety or within a time limit of three months does not inform the employee regarding the intention to utilise this right, the right to the plant variety is automatically transferred to the employee.

7.4 Employee Remuneration Right

The Plant Varieties Protection Law provides the employee a moral right to apply for an equitable remuneration for the utilisation of a variety in cases provided for in the agreement entered into with the employer. This declaratory legal norm does not create any additional right to remuneration for the employee other than provided in the employment agreement.

However, Article 39 of the Law on Scientific Activity provides that the breeder has a right to fair remuneration from the income generated by the state scientific institution from exploitation of the rights to the plant variety.

8 INSTANCES FOR DISPUTES

In Latvia all the disputes related to ownership of intellectual property rights are reviewed by civil courts.

The Court of First Instance for copyright- and plant variety-related matters is the District Court which is composed of one judge. The court of appeal is the Regional Court which is composed of three judges. The judgment of the Regional Court may be also reviewed by the Supreme Court Civil Cases Department (Cassation court) which is composed of chambers in which the cases are reviewed in writing by three judges. In exceptional cases the case may be reviewed by a full court of seven judges in oral proceedings.

All matters related to industrial property matters (e.g., patents, designs, semi-conductor chips) are reviewed only by Riga Vidzeme District Court as the Court of First Instance. The judgments of the Court of First Instance may be appealed to Riga Regional Court. The judgment of the Riga Regional Court may be also reviewed by the Supreme Court Civil Cases Department.

9 CONCLUDING REMARKS

The intellectual property laws of Latvia are new and there is still no guidance as to interpretation of legal norms related to intellectual property ownership. The disputes related to ownership of intellectual property rights in the course of employment are rare and usually settled out of court due to high litigation costs and unpredictability of court judgments.

Latvian intellectual property laws provide overlapping protection to several kinds of intellectual property.

The design may be simultaneously protected by the Design Law, the Copyright Law or by the Regulation 6/2002/EC on Community Designs. Plant variety is protected by the Plant Varieties Protection Law, however, biological material used to breed the plant variety may be protected by the Patent Law as a biotechnological invention. A computer program may be protected by the Copyright Law as a literary work and also may be a part of a computer-implemented invention under the Patent Law while performing a technical function.

The overlapping and interconnection of intellectual property rights allow development of a suitable strategy for intellectual property protection. At the same time it may also create confusion as to the scope of the exclusive rights to the intellectual property and may cause obstacles for the use of intellectual property.

The employer should always consider and include in the employment agreement transfer of the key intellectual property rights in order to safeguard their investment.

BIBLIOGRAPHY

Table of Legislation

The Law on Protection of Topographies of Semiconductor Products of 31 March 1998.
Copyright Law adopted of 6 April 2000.
Design Law of 17 November 2004.
Plant Varieties Protection Law of 17 November 2005.
Law on Scientific Activity of 14 April 2005.
Patent Law 27 February 2007.
Directive 87/54/EC on the legal protection of topographies for semiconductor products.
Directive 2009/24/EC on the Legal Protection of Computer Programs.
Regulation (EC) No. 2100/94 on Community plant variety rights.
Regulation 6/2002/EC on Community Designs.

Table of Cases

Judgment of the Regional Court of Riga of 5 May 2010, case No. C04304709.
Judgment of the Supreme Court of 5 May 2010, case No. C04304709.

Chapter 14
Lithuania

Edita Ivanauskiene

1 BACKGROUND

In Lithuania the employment law contains neither rules nor principles governing the employer's rights to the products of the employee's work. The employer's property rights to the employee's work results are governed by other legal rules, and, if a product of work is covered by an intellectual property right or it may be protected through industrial property registration procedures, the question of entitlement to these intellectual property rights is governed by intellectual property laws. As per Lithuanian intellectual property laws the legal grounds to acquire intellectual property rights are as follows: the law, the agreement and the inheritance. However, the rules on the employer's right to intellectual property created by the employee depends on the object of intellectual property as in Lithuania legal protection of nearly every intellectual property object is governed by a separate law and the employee-employer situation in each law is governed slightly differently.

In copyright law the general rule is that the creator (who is a natural person) is the first owner of non-economic (moral) and economic rights. Non-economic rights always stay with the author and the performer, irrespective of whether they have the status of an employee. The employer may only become a derivative owner of economic rights.

The Lithuanian industrial property laws apply different rules of ownership to an industrial property object created by the employee. According to the Lithuanian industrial property laws a creator is the first owner of non-economic rights irrespective of whether they have the status of an employee, while the ownership of exclusive economic rights to an industrial property object depends on the status of the creator. The rights to the industrial property object created by the employee is presumed to belong to the employer, except the moral rights, provided certain conditions are met.

However, despite different approaches in Lithuanian copyright and industrial property law, the employer's right to control intellectual property may be obtained through the employment relationship provided that certain conditions are met, and a separate agreement governing the transfer of employees' intellectual property rights is not needed. Such ownership rules to employees' intellectual property apply only if the employer and the employee do not agree on different terms in the agreement, i.e. the applicable laws do not forbid them to agree in the employment or other agreement otherwise than the law prescribes. However, it is very rare that employees dare to negotiate with the employer and change the rules of ownership of intellectual property rights prescribed by the law through individual employment, collective or other type of agreements.

2 EMPLOYEES' COPYRIGHT

2.1 The Legal Framework

In Lithuania the Law on Copyright and Related Rights ('Copyright Law')[1] provides that 'the author shall be a natural person who has created a work' setting the rule from which it may be implied that only natural persons are the creators and primary owners of copyright in the work.

Section VI of Chapter 2 (Articles 38-50) of the Copyright Law governs general rules on transfer of copyright. Article 38 of the Copyright Law provides that only authors' economic rights may be transferred. They may be transferred by an agreement, by succession or by other procedure prescribed by law. Article 60 of the Copyright Law governs transfer of related rights, to which transfer similar principles apply that apply to works protected by copyright.

The Copyright Law also sets specific rules concerning employees' copyright (Articles 9-10) according to which an employee-creator has the same rights to their work as other non-employed creators and an employer may only be a derivative owner of copyright. Article 60(6) of the Copyright Law sets specific rules concerning employees' performer rights according to which an employer may only be a derivative owner of performer rights.

2.2 Ownership of Rights

As mentioned before, an employer may only be a derivative owner of copyright. However, Section VI of Chapter 2 of the Copyright Law, governing general rules on transfer of copyright and stipulating certain rules of interpretation concerning copyright agreements, does not apply to transfer of employees' copyright to the employer.

For the transfer of employees' copyright to the employer special rules set in Articles 9-10 of the Copyright Law apply. These contain a rebuttable presumption of

1. Law on Copyright and Related Rights of the Republic of Lithuania of 18 May 1999 No. VIII-1185 (O.J. 1999, No. 50-1598).

transfer of employee's copyright on works to the employer. According to Articles 9-10 the transfer of employees' copyright to the employer is only effective through the employment relationship when an employee enters into the individual employment contract with an employer. A separate agreement, according to which an employee transfer their copyright to the employer, or express provisions effecting transfer of rights in the employment agreement, are not required.

However, the presumption of the transfer of employee's copyright to the employer is limited in time. Article 9(2) of the Copyright Law enables the employer to hold the employee's economic rights to works, other than a computer program, 'for the period of five years, unless otherwise provided for by an agreement'. Therefore, in contrast to what is generally applicable in copyright law, pursuant to Article 9 of the Copyright Law all the economic rights of the employed author subsisting in the work are transferred to the employer entirely for the period of five years. Copyright in computer programs created by employees, on the other hand, is transferred to the employer for the entire period of copyright protection subsisting in the work.

The presumption of transfer of economic rights to the work does not apply if the work has been commissioned. In such case general rules on transfer of copyright applies (Articles 38-50) which require a copyright agreement in writing to be concluded and be in compliance with other requirements set in the Copyright Law. The main condition which must be satisfied is that the copyright agreement must indicate at least the exclusive rights which are transferred or licensed under the copyright agreement.[2]

According to Chapter 3 of the Copyright Law related rights are protected according to similar principles that apply to works protected by copyright. Performers, phonogram producers, producers of audiovisual works and broadcasting organisations have rights of limited duration to their performance or their related rights objects, and most of the copyright transfer provisions apply *mutatis mutandis* to the transfer of related rights (Article 60).

In Lithuania, there is no statutory exemption for teachers with respect to copyrights, i.e. the above-mentioned rules of ownership of copyrights set in the Copyright Law apply.

The ownership to state university teacher's copyright is specifically governed by the Law on Higher Education and Research,[3] but these provisions are mostly non-mandatory. The issues of acquiring and transfer of intellectual activity results, including copyright, from the employee state university teacher to the state universities or research institutions are to be governed 'in accordance with the procedure laid down by the law and/or contracts'.[4] That means that the rules set in the Copyright Law apply also for state university teachers' copyrighted works and the copyright should be owned by the universities for the period of five years, unless otherwise agreed in the employment agreement between the state university teacher and the state university.

2. Alfonsas Vileita, *Lietuvos Respublikos autorių teisių ir gretutinių teisių įstatymo komentaras, Lietuvos Respublikos administracinių teisės pažeidimų kodekso 214(10) straipsnio komentaras*, Lietuvos rašytojų sąjungos leidykla, Vilnius, 2000, 104.
3. Law on Higher Education and Research of the Republic of Lithuania (O.J. 2009, No. 54-2140, No. 61, No. 101).
4. *Ibid.*

In contrast to the Copyright Law, the Law on Higher Education and Research sets the lowest limit of amount of remuneration which must be paid to state university employee researchers and doctoral students for the commercial use of their copyrighted works, which is: 'Not less than one-third of the profit gained for commercial use of the results of intellectual activities achieved by employees of a state higher education institution when carrying out research while fulfilling the job functions and by persons studying in doctoral studies when carrying out research must be allocated to an author (co-authors), unless otherwise provided for in an employment contract or another contract concluded between the state higher education institution and the employee (person studying in doctoral studies) (Article 82(3)).' However the law also gives the possibility for the parties to agree otherwise in the agreement. Therefore, it seems that state university researchers and doctoral students may be deprived of the right to receive remuneration by way of individual contract in Lithuania.

2.2.1 Employees' Copyright Economic and Moral Rights

As mentioned before, special provisions on employees' copyright on a work apply for the transfer of economic rights from employed authors to the employer. According to Article 9 of the Copyright Law 'an author's economic rights in a work, other than a computer program, created by an employee in the execution of his duties or fulfilment of work functions shall be transferred to the employer for the period of five years, unless otherwise provided for by an agreement'.

Following Lithuanian jurisprudence, two conditions should be met for the transfer of copyright from employed authors to the employer to be valid:

- the employment relationship between the employer and the employee must exist;
- copyright-protected works must be created 'in the execution of service duties[5] or fulfilment of work functions'.[6]

Pursuant to Article 32 and 34 of the Labour Code[7] the employment relationship exists when: (i) there is an employment agreement signed containing the work function (i.e. duties), the salary and the workplace; (ii) the work function is subordinate to the employer and executed for the employer's benefit.[8]

If the employment relationship exists, then it must be checked whether the work has been created in execution of the work function. The work function of the employee

5. It should be also applied to civil servants and statutory officers. See Jūratė Ūsonienė, Autorių teisių perdavimo ypatybės, Teisinės informacijos centras, Vilnius, 2008, 96.
6. Ūsonienė, *supra* n. 5, at 91-107; Vytautas Mizaras, *Autorių teisė*, Vol. I, Justitia, Vilnius, 2008, 261-265.
7. Labour Code of the Republic of Lithuania of 14 September 2016 (RLA, 19 September 2016, No. 23709)
8. Para. 2 of Art. 32 of the Labour Code clarifies that 'Subordination to the employer means the performance of a work function where the employer has the right to control or direct both the whole work process and its part and the employee obeys the employer's instructions or the procedures in place at the workplace.'

should be indicated in the employment agreement itself, may be with reference to staff duty instructions or to other internal documents of the employer, be in the collective agreements where such function may be specified in more detail. The employee must be acquainted with these internal documents of the employer where his work function is specified in more detail. A direct relationship must exist between the work function and the work which should be determined according to whether an employer could require or order an employee to carry out certain work based on the contents of employment agreement.[9]

The employee's copyright on the work created in the execution of work functions is acquired by the employer from the moment the employee transfers the work to the employer.[10] Non-transfer of the employee's created work to the employer would be considered as non-fulfilment of the employee's duties.[11]

If the work created by the employee has no direct relation to their work functions, the employer does not obtain any rights to the work although it may be created using the employer's technical means, information resources and/or at the premises of the employer.[12] However, the burden of proof to overturn the presumption lies with the employee: the employee will have to prove that creation of the work was not within their duties or work functions.[13]

The presumption is overturned if the employment agreement or any other agreement signed between the employee and the employer indicates that the economic rights to the works created 'in the execution of employees duties or fulfilment of work functions' stays with the author-employee or contains other provisions diminishing the scope of transfer.

As mentioned before under the Copyright Law the transfer of employee's economic rights to the employer is limited in time: the employer owns the employee's economic rights for only five years. In practice in order to avoid this time limitation employers are advised to put clauses into the employment agreement which provide for the transfer of employee's economic rights to the employer for the whole period of validity of these rights.

However, neither the law, nor case law give a clear answer as to whether the presumption of transfer of all the exclusive economic rights to the employer also covers the transfer of the right to fair compensation for statutory limitations of these rights: such as the right to remuneration for private copying, reprographic reproduction or for the lending of works. The right to remuneration for resale of original art works or

9. Vileita, *supra* n. 2, at 49; Usonienė, *supra* n. 5, at 98-99.
10. Vileita, *supra* n. 2, at 49.
11. *Ibid.*
12. *B. V.-F. v. Museum*, Fodio, Ministry of Culture, Decision of the Lithuanian Appeal Court of 2 January 2007 in civ. case No. 2A-98/2007 where the court rebutted one of the defence arguments that the defendants did not infringe the copyright on the work as the collection was created in the employment relationship and using the employer's collected information resources and technical means, and confirmed the finding of the lower court that a former museum employee had no duty to create a collection under her employment agreement concluded with the museum and, therefore, found that the museum and the publisher infringed copyright to her collection by publishing it without signing a copyright agreement with her.
13. Usoniene, *supra* n. 5, at 100.

original manuscripts, the right to remuneration for rental of audiovisual works, the right to remuneration for rental of phonograms which contain recording of works are set in the law as not transferrable to third parties, therefore as per legal doctrine the employer should have no title to these remuneration rights.[14] If statutory provisions on the presumption on transfer are not acceptable by the employee or the employer, they may generally agree otherwise on acceptable terms in the employment agreement or in a separate agreement. Then the general provisions on copyright transfer in Section VI of Chapter 2 of the Copyright Law are applicable to this agreement.

Author's moral rights and performer's moral rights may not be transferred (Articles 14 and 52). Authors and performers retain their moral rights to their works even after the transfer of economic rights to their employers.

2.2.2 Employee's Computer Programs

For the transfer of copyright on computer programs created by employees the same principles apply as to transfer of copyright on works created by other creative employees, except that Article 10(2) of the Copyright Law[15] which repeats the rule enshrined in Article 2(3) of the Software Directive,[16] provides for a different presumption: the employee's economic rights to computer programs are vested with the employer entirely without limitation in time irrespective of the five-year time period. Article 10 of the Copyright Law transfers the employer copyright (exclusive economic rights) subsisting in employee's computer programs in its entirety for the whole protection period of economic rights, unless provided otherwise in the agreement, except the moral rights.

Under Article 14 of the Copyright Law the moral rights of an author of a computer program are not transferrable, but they may not be used in a manner which unreasonably prejudices the employer's economic rights subsisting in these computer programs, including the right to carry out adaptation, alteration and distribution of these works at the employer's discretion, with the exception of the cases when such actions would be prejudicial to the author's honour or reputation.

2.2.3 Producers' Rights to Audiovisual Works

The Copyright Law sets the presumption of transfer of copyright on audiovisual works[17] (except copyright on musical works specifically created for an audiovisual

14. Ramūnas Birštonas et al., *Intelektinės nuosavybės teisė*, Registrų centras, Vilnius, 2010, 120.
15. The owner of an author's economic rights in a computer program created by an employee in the execution of their duties or fulfilment of work functions shall be the employer, unless otherwise provided by an agreement (Art. 10(2) of the Copyright Law).
16. Council Directive 91/250/EEC of 14 May 1991 on the legal protection of computer programs, *OJ L 122 (17 May 1991), 42-46.*
17. Art. 2(4) of the Copyright law defines 'Audiovisual work' as 'a cinematographic work or any other work created by means of cinematography that comprises a series of related images expressing a motion, with or without accompanying sounds, recorded (fixed) in an audiovisual communication medium'.

work or included in an audiovisual work) to the producer based on the agreement for the creation (production) of an audiovisual work (Article 11) and the presumption of the transfer of the performers' rights, except the right to fix the performance, based on the agreement of fixation of audiovisual recordings (Article 53(4)). An audiovisual work producer which may be 'a natural or legal person on the initiative and responsibility of which an audiovisual work is being made' (Article 2(3)) is a derivative owner of these rights. The presumption of transfer of copyright on the pre-existing works to the audiovisual work producer also applies to authors who have given their authorisation to adapt or incorporate their works in an audiovisual work. Therefore, for the presumption on the transfer of economic rights of co-authors of audiovisual works and the previously mentioned performer economic rights to audiovisual fixation to be valid, it is sufficient to conclude an agreement with the authors on the creation of audiovisual works or an agreement on adaptation or incorporation of their works into an audiovisual work, and with the performers – an agreement of fixation of an audiovisual recording.

An employment agreement is not required to be concluded, but possible, in which case the presumption of transfer of employee's copyright to the employer will apply.[18]

If statutory provisions on the presumption on transfer are not acceptable to the parties, they may generally agree otherwise on acceptable terms in the agreement. Then the general provisions on copyright transfer in Section VI of Chapter 2 of the Copyright Law are applicable to this agreement.

As mentioned before, the presumption of ownership of copyright on audiovisual work is not applicable to musical works which are incorporated in the audiovisual work, therefore, the producer should specifically address the issue of copyright transfer or licence to the music in the copyright agreement or the agreement for the creation (production) of an audiovisual work. The general provisions on copyright transfer in Section VI of Chapter 2 of the Copyright Law would be applicable to this agreement.

2.2.4 *Related Rights of Performers and Producers*

According to Article 2(11) of the Copyright Law related rights are accorded to the following: (i) performance of a work, including direct (live) performance and its sound or visual fixation; (ii) a phonogram (a sound recording); (iii) the first fixation of an audiovisual work (film) (audiovisual fixation of related images); (iv) radio and/or television broadcast by a broadcasting organisation.

In Lithuania, a performer has performer's economic and moral rights. The same type of rebuttable presumption of transfer of a performer's rights to the employer applies as in case of author-employee copyright transfer to the employer. A performer's economic rights are transferred to the employer for a period of five years, if a performer performs a work 'in the execution of his duties or following the instructions given by his employer', unless otherwise provided for by an agreement (Article 60(6)).

18. For more details please see Section 2.2.1.

Following the provisions of Articles 2(9), 2(12), 54 and 57 of the Copyright Law, producers of sound recordings and first fixation of images and the original holders of related rights are natural or legal persons on the initiative and responsibility of whom the first fixation of sound recording or of related images was made.

However, if employed authors and performers are considered to be the initial holders of producer rights, the rights are transferred to the employer producer if the work, sound recording or audiovisual recording was produced by an employee as part of their duties or work functions or as a result of instructions given by the employer (the latter applies only to the performer). As mentioned before, the transfer of the economic rights to the employer would be valid only for five years.

2.2.5 *Sui Generis* Database Producers Rights

Following Article 61 of the Copyright Law the original holder of *sui generis* rights to databases may be any person 'who shows that he has made a substantial qualitative and/or quantitative (intellectual, financial, organisational) investment in obtaining, arrangement, verification and presentation of the contents of that database'. The main criterion for deciding whether the *sui generis* rights to database becomes vested in the employer is a certain type of investment in certain types of activities which are generally of non-creative nature. Therefore, generally employees do not have any initial *sui generis* rights to such databases, and the initial owner of the *sui generis* database rights is the database producer.

Nevertheless, the database may contain works, sound recordings, audiovisual recordings and other protected subject matter, created by the employees, to which initial ownership is vested with the employees. In such a case the employee's rights are transferred to the employer-producer if the protected subject matter was created by an employee as part of their duties or work functions or as a result of instructions given by the employer (the latter applies only to the performer). As mentioned before, the transfer of their economic rights to the employer would be valid only for five years, except the rights to computer programs.

2.2.6 University Researchers' Copyright

Issues pertaining to the ownership of university researchers' and teachers' copyrights and university researchers' and teachers' rights and duties in connection therewith are specifically addressed in Article 82 of the Law on Higher Education and Research, but it appears that the provisions of this Article should only be applied to works created by state university researchers and teachers and research institute employees.[19] The provisions of Article 82 governing ownership of state university researchers' copyrights are of a general nature: 'Issues pertaining to the transfer of economic rights to a higher

19. Art. 11(3) of the Law on Higher Education and Research: 'The provisions of Art. 82 of this Law apply *mutatis mutandis* to the rights of a research institute and its staff to the results of their intellectual activities.'

education institution or conferring of such rights for use shall be regulated in accordance with the procedure laid down by the law and/or contracts.'

The Copyright Law does not contain any provisions specifically addressing university researchers' and teachers' copyrights, therefore general rules on transfer of copyright shall apply. The works created by university researchers and teachers shall belong to the university for a period of five years from the date of transfer of the work to the employer if they are created during the employment relationship with the university and in the execution of duties and work functions.

The Law on Higher Education and Research expressly governs an employee researcher's or teacher's duty to inform the employer *state* university about the created copyrighted work, and lays down special rules on remuneration for the commercial use of intellectual property (including copyright) of the researchers and doctoral students of a state university[20] and a provision that the conditions of management, use and disposal of intellectual property created in cooperation with other institutions must be stated in the agreement concluded between the state university and other institution.

2.3 The Employee's Duty to Give Notice

An express positive duty to give notice to an employer of the creation of intellectual property is present in the Law on Patents[21] (Article 11(2)), which is applicable to patentable inventions, in the Law on Designs[22] (Article 13(2)) which is applicable to registrable designs), in the Law on Higher Education and Research which is applicable to intellectual property created by an employee of a state higher education institution and of a research institution[23] through the use of its experience or technologies and/or equipment and/or in the process of a research with the said person performing his job functions during the research (Article 82(2)). The Law on Copyright does not contain such duty, therefore, it is advisable to include expressly such duty in the agreements concluded with the authors.

2.4 Employee Remuneration Right

Neither the Copyright Law, nor the Labour Code sets any special rules regulating employees' rights to remuneration, therefore, general rules applicable to remuneration

20. Art. 82(3) of the Law on Higher Education and Research: '*Not less than one third of the profit gained through commercial use of the intellectual activity results* achieved by employees of a state higher education institution in the process of a research with the employees performing their job functions during the research and by persons studying in doctoral studies in the process of a research must be allocated to the author (co-authors), unless otherwise provided for in an employment contract or a contract concluded between the state higher education institution and the employee (person studying in doctoral studies).' [emphasis added].
21. Law on Patents of the Republic of Lithuania of 18 January No. I-372 (O.J., 28 January 1994, No. 8-120).
22. Law on Designs of the Republic of Lithuania of 7 November 2002 No. IX-1181 (O.J., 22 November 2002, No. 112-4980).
23. In Lithuania there are two types of higher education institutions: universities and colleges (Art. 6(1) of the Law on Higher Education and Research).

under the employment agreement apply. According to the Labour Code the parties are free to decide on the amount of salary, however, it must be higher than the amount of minimum salary set by the government (currently EUR 380 per month, EUR 2.32/ hour)[24] as the minimum salary may be paid for the unqualified work only.[25]

Special rules on the amount of remuneration for the use of employee intellectual property results are only set by the Law on Higher Education and Research, but these rules apply only to employees of a state higher education institution and a research institution. According to Article 82(3) of this Law '*not less than one third of the profit gained through commercial use of the intellectual activity results* achieved by employees of a state higher education institution *in the process of a research with the employees performing their job functions during the research and by persons studying in doctoral studies in the process of a research* must be allocated to the author (co-authors), unless otherwise provided for in an employment contract or a contract concluded between the state higher education institution and the employee (person studying in doctoral studies).' (Emphasis added)

If a separate copyright agreement is concluded with an employee regarding creation of works, the creation of which is not covered by the employment agreement, then general rules governing the remuneration under the copyright agreement would apply. Under Article 48 of the Copyright Law it is up to the parties to determine the amount of remuneration under the copyright agreement except in certain cases provided in the Copyright Law (e.g. amount of remuneration for private copying, reprographic reproduction). The remuneration may be calculated as a certain percentage of the revenue obtained by the user for each mode of use of the work, as a lump sum or in any other way specified by the agreement.

The parties to an agreement may provide for an advance payment of the whole or part of remuneration. However, if the amount of remuneration is calculated as a certain percentage of the revenue obtained for each mode of use of the work, the author or any other owner of copyright shall have the right to receive information on the extent of use of the work, agreements concluded by the user and their revenues obtained from the use of the work.

3 EMPLOYEES' SEMICONDUCTOR CHIPS

3.1 The Legal Framework

In Lithuania the protection of topographies for semiconductors is governed by the Law on the Legal Protection of Topographies of Semiconductor Products enacted in 1998.[26] The provision on the employer's right to the employee's topographies mirrors Article 2(a) of Directive 87/54/EC on the Legal Protection of Topographies for Semiconductor

24. Government Decree on Increase of Minimum Salary of 22 June 2016 No. 644 (RLA, 27 June 2016, No. 17577).
25. Art. 141(2) of the Labour Code (RLA, 19 September 2016, No. 23709).
26. Law on the Legal Protection of Topographies of Semiconductor Products of 16 June 1998 No. VIII-791 (O.J., 01 July 1998, No. 59-1655), in force from 1 July 1998.

Products[27] despite the fact that Article 2(a) of the Directive is not mandatory. Following this provision, similarly to employees' copyright on computer programs, the presumption applies that exclusive rights to the topography are vested in its entirety with the employer if the topography is created under the employment agreement unless otherwise provided in the agreement.

3.2 Ownership of Rights

Following Article 4 of the Law on the Legal Protection of Topographies of Semiconductor Products the initial owner of exclusive rights to a topography is a natural person who has created the topography. The author of the topography is also vested with the moral rights: the right of authorship, the right to the author's name and the right to the inviolability of the topography.

As in the case of employees' copyright on a computer program, a rebuttable presumption is provided for in this Law that the exclusive rights to the employee's topography belong to the employer if it has been developed in the course of the employee's duties under the employment agreement, unless otherwise provided in the agreement (Article 4(3)).

If a topography has been created under a contract other than a contract of employment, the exclusive rights to the topography shall belong to that party to the contract which commissioned the topography, unless the contract provides to the contrary.

The exclusive rights to the topography come into existence on the following date, whichever is earlier:

- on the date when the topography is first commercially exploited anywhere in the world or
- on the date when an application for registration of the topography has been filed at the State Patent Bureau.

The exclusive rights to the topography are valid for ten years from the following date, whichever is earlier: from the end of the calendar year of the first commercial exploitation in the world, or from the end of the calendar year of its registration.

Authors may transfer their exclusive rights to the topography to other persons only if it is registered with the State Patent Bureau. The application must be filed within two years of the first commercial exploitation of the topography anywhere in the world. Exclusive rights to the topography may be transferred to another person under a written agreement which is only valid if it is entered into the Topographies Register of the Republic of Lithuania. Exclusive rights to the topography are also subject to inheritance. Unlike for authors, the Law on the Legal Protection of Topographies of Semiconductor Products does not require the employer or the commissioner to register the topography with the Topographies Register of the Republic of Lithuania for having

27. Council Directive 87/54/EC of 16 December 1986 on the legal protection of topographies of semiconductor products, O.J. L24, 27 January 1987, 36.

the right to transfer the exclusive rights to the topography. However, as the agreement on transfer is valid only if it is registered with the Register, the topography must be registered to enable such agreement to be entered into the Register.

No application to register a topography has yet been submitted to the State Patent Bureau in Lithuania. No case law exists on the matter in Lithuania.

3.3 The Parties' Duty to Inform

An express positive duty to give notice to an employer of the creation of intellectual property is present only in the Law on Patents, in the Law on Designs, both of which are not applicable to topographies, and in the Law on Higher Education and Research which is applicable to intellectual property, including topographies created by an employee of a state higher education institution or of a research institution which is created through the use of its experience or technologies and/or equipment and/or in the process of a research with the said person performing their job functions during the research. The technicalities of this duty are detailed by the order of the specific institution in question.

3.4 Employee Remuneration Right

Special rules governing employees' right to remuneration for the creation and use of the topography, except topographies created by an employee of a state higher education institution or of a research institution, do not exist. Therefore, it is up to the employee and the employer to decide on the technicalities of remuneration in the agreement.

In the case of topographies created by employees of a state higher education institution or of a research institution performing their job functions or by persons studying in doctoral studies during their research, the Law on Higher Education and Research obligates state higher education institutions and research institutions to pay them remuneration 'not less than one third of the profit gained through commercial use of the intellectual activity results achieved' by them, however it applies only if not agreed otherwise in the agreement that was entered into between the specific institution and the employee (Article 82(3) of the Law on Higher Education and Research).

4 EMPLOYEES' INVENTIONS

4.1 The Legal Framework

The Law on Patents[28] is the main legal act governing employees' inventions in Lithuania. Employees' inventions, provided certain conditions are met, are called 'service inventions' and special rules on ownership and remuneration apply.

28. Law on Patents of the Republic of Lithuania of 18 January 1994 No. I-372 (O.J., 28 January 1994, No. 8-120), in force from 1 February 1994.

Following Article 11 of the Law on Patents a service invention is an invention made:

- during the execution of an employment contract which provides for inventing activities;
- on a concrete assignment, in the course of designing, constructing, scientific research or the creation of technologies;
- when using the experience accumulated by the employer or its technologies and equipment.

The Law is not clear with regard to whether all three conditions or only one must be met to qualify the invention as a service invention. Neither legal doctrine nor case law has explained the issue yet. Exclusive rights to the service invention are vested with the employer, for which reason it is very important to properly regulate this issue in the individual collective agreements or internal legal acts in order to avoid the uncertainty of who actually owns the right to the patentable invention.

Article 11 of the Law lists the employer's and employee's rights and obligations with regard to service inventions. The Law applies to any service invention regardless of whether the employee is employed in the private or public sector.

4.2 Ownership of Rights

4.2.1 Employees' Inventions

As mentioned before, the Law on Patents governs the rules of ownership of rights to patents. These rules are mostly non-mandatory, and, therefore, individual employment agreements may govern the ownership of rights to patents. The employer has a duty to pay remuneration to the employee for the service invention, but this duty may be easily avoided by paying the employee an agreed increased salary for making inventions.

The law establishes that an inventor is always a natural person who created an invention, however, they are not always the primary holder of the right to a patent. Unless otherwise agreed in the agreement between the employer and the employee, the law establishes the presumption that in case of a service invention the right to a patent belongs to the employer and a patent application should be filed on behalf of the employer (Articles 10(1-3), 11(3)).

The employer must use the employee's right to a patent, otherwise it 'returns' to the employee-inventor. If the employer does not inform the inventor about the intention to make use of this right within four months, the right shall pass over to the inventor (Article 11(2)). The same rule applies if the employer renounces the right to a patent: the right to a patent passes over to the inventor (Article 11(3)).

The employer and the employee may agree that the right to receive a patent and a right to a patent belongs to the employee (Article 11(8)).

Different rules apply if an invention is created under a civil agreement concluded with the inventor and the commissioner. In such a case it is up to the parties to agree on the right to a patent: 'If an invention is made at an enterprise, institution or organisation which carries out scientific research, design, construction works and other works of creative character under contract with a client who finances an appropriate work, then the right to a patent for the invention is established by such contract' (Article 10(1)).

The employee-inventor is also provided with certain moral rights: their name must be indicated in the patent application and in the patent, except if they file a request to the State Patent Bureau requiring not to publish their name.

4.2.2 University Researchers' Inventions

The main legal acts governing the ownership of university researchers' inventions in Lithuania are the Law on Higher Education and Research and the Law on Patents. However, the provisions of the Law on Higher Education and Research governing the ownership aspects of university researchers' inventions are applicable only to the state university researchers and are mostly of general nature and non-mandatory. The issues of acquiring intellectual activity results, including inventions, from the employee state university researchers and transferring them to the universities are referenced as to be governed 'in accordance with the procedure laid down by the law and/or contracts'.

In this respect, the main legal act governing the ownership of employee university researchers as with other employees is the Law on Patents in Lithuania. That means that the rules set in the Law on Patents apply also for employee university researchers' inventions and the patent rights should be owned by universities unless otherwise agreed in the individual employment agreements, collective agreements or internal university documents (e.g. policies) to which the employee university researcher agreed.

4.3 The Parties' Duty to Inform

The Law on Patents also expressly contains an employee's duty to immediately inform an employer in writing about the created invention (Article 11(2)). This duty is also established in the Law on Higher Education and Research (Article 82(2)), except that this law requires a notification to be made in accordance with the procedure laid down by the state higher education institutions or research institutions.

Both the employer and the employee have the duty to keep secret the essence of an invention until the filing of a patent application. The employer has to inform the employee whether they intend to use the right to a patent within four months, counted from the date of receipt of the notice (although the Law does not expressly indicate the date from which four months start), otherwise the employer loses their right to a patent.

4.4 Employee Remuneration Right

The Law on Patents also contains an employee-inventor's mandatory right to remuneration which is formulated as an employer's duty to pay remuneration to an employee after the grant of a patent, except: (i) if they have been or are paid an agreed increased salary for making inventions or (ii) if they agreed in the agreement that the right to acquire a patent, the right to a patent application and the right to a patent belongs to an employee-inventor.

The amount and payment conditions of remuneration should be established in a remuneration agreement signed by the employer and the inventor. This agreement must be signed within six months following the granting of a patent or within one year of the beginning of the use of an invention, if the beginning of the use of the invention is prior to the granting of a patent. If no agreement exists between the parties, the amount of remuneration shall be established by the court.

The rights and duties of the employer and employee, related to service inventions, with the exception of the payment of remuneration, shall expire when the patent becomes null and void, or one year after the termination of labour relations.

4.4.1 *Remuneration*

As per Article 11(4) of the Law on Patents, the only factor that is important in determining the amount of remuneration for the employee remuneration is *the economical value or other value* which may be derived from the use of the invention.

Unlike the Law on Patents, the Law on Higher Education and Research sets the lowest limit of amount of remuneration which must be paid to state university employee researchers and doctoral students for the commercial use of their patentable inventions, which is: 'not less than one third of the profit gained through commercial use of the intellectual activity results achieved' (Article 82(3)), however the law also sets the possibility for the parties to agree otherwise in the agreement. Therefore, it seems that state university researchers may be deprived of the right to receive remuneration by way of agreement.

Case law concerning the establishment of the amount of remuneration or on the issue of remuneration arising out of a remuneration agreement which is signed between an employer and an employee-inventor regarding a service invention is non-existent in Lithuania. However, two Supreme Court cases[29] exist concerning a dispute between the inventors and the company regarding non-payment of remuneration for creation of the patented invention used in the company. The parties to the agreement regarding the payment of remuneration agreed that each inventor will in equal parts receive remuneration equal to 25% of net profit received for the use of the invention through the whole period of validity of the patent. The parties also agreed how the net profit arising out of the use of this invention would be calculated. Extensive expertise was needed to determine whether the company in fact used the invention and the economic value generated by the invention. After almost seven years of litigation, the inventor in the first case finally received a final judgement according to which remuneration in the amount of LTL 399,355.57 (~ EUR 115,661) was awarded

29. *M.B. v. Orlen Lietuva*, Decision of 1 March 2010 of the Supreme Court of Lithuania in civ. case No. 3K-3-66/2010; Decision of 2 December 2016 of the Supreme Court of Lithuania in civ. case No. 3K-3-496-378/2016

for the use of the patented invention; nearly four years needed for other seven inventors to receive a final judgment for the sum ranging between EUR 771,433.91 and EUR 871,237.55 for the use of the invention in the second case.

5 EMPLOYEES' UTILITY MODELS

Lithuanian law does not protect utility models.

6 EMPLOYEES' DESIGN RIGHTS

6.1 The Legal Framework

Employers' rights to employees' designs are specifically regulated by the Law on Designs.[30] As with the right to a patent the employer is also presumed to be a primary holder of the right to the service design unless otherwise agreed in the agreement between an employer and an employee (Article 12(3)(3)). The conditions as regards when an industrial design is considered a service design are clearer than in respect of patents as the Law on Designs expressly states that it is sufficient to satisfy one of the three conditions indicated in Article 13(1) of the Law.

The provisions on the duty to inform and on the right to remuneration are the same as in respect of patentable inventions created by the employees, except the following:

- a designer is not entitled to an increased salary or additional remuneration if they must create designs under the employment agreement;
- there is no provision on return of the right to a design to the designer if an employer does not intend to use the right to a design (Articles 12-13 of the Law on Designs).

A designer has also moral rights: their name must be indicated in an application and a certificate of registration of the design unless they file a request with the Lithuanian Patent Bureau requiring not to indicate their name. Further, a design may not be altered without a designer's consent, unless otherwise provided for in the agreement.

6.2 Ownership of Rights

Article 12(3)(3) of the Law on Designs states that an exclusive right to a design belongs to the employer or the commissioner, if the design is a service design and the

30. Law on Designs of the Republic of Lithuania of 7 November 2002 No. IX-1181 (O.J., 2002, No. 112-4980)

agreement does not provide otherwise. Pursuant to Article 13(1) of the Law a design shall be deemed to be a service design, if it has been developed under at least one of the following conditions:

- during the execution of an employment contract which provides for the development of a design;
- following concrete instructions (on order) under the contract in the course of design, construction, scientific research or creation of technologies;
- when making use of the experience accumulated by an enterprise, agency or organisation, or its technologies and equipment.

Although under Article 12(3)(3) of the Law in case of a service design the right to a design should belong to a commissioner unless the contract states otherwise, Article 13(7) of the Law provides that the commissioner's exclusive right to a design must be established in the contract between the designer and the commissioner who finances design works. Nevertheless, the Law further states that if the commissioner withdraws the right to the design or fails to inform the designer within four months about their intention to exercise such right, the right to the design shall be transferred to the designer. It seems that if the contract between the commissioner and the designer does not state otherwise the presumption that the commissioner owns the right to the service design should apply.

When a design is also protected by copyright as a work of applied art, transfer of rights will be governed by the rules indicated in the Copyright Law mentioned above.

When a design is protectable as a Community design, Article 14(3) of the Regulation on Community Designs [31] will apply according to which the right to the Community design is vested with the employer 'where a design is developed by an employee in the execution of his duties or following the instructions given by his employer', unless otherwise agreed or specified under national law.

All the mentioned laws governing design protection for Lithuania set the presumption that all rights to the design developed by an employee vest in the employer if the design has been developed in the employment relationship and in execution of their duties, work functions or following the instructions given by their employer, except that under the Copyright Law the transfer of the economic rights to a copyrighted design is restricted to five years. Under the Law on Designs the employer would also be presumed to own the right to the employee's design if the design has been developed by the employee when making use of the experience accumulated by the employer or its technologies and equipment even if the employment agreement does not contain any duty to create designs.

31. Council Regulation (EC) No. 6/2002 of 12 December 2001 on Community designs, O.J. L3, 05 January 2002, 1-24.

6.3 The Parties' Duty to Inform

Under the Law on Designs the employee who developed a service design must immediately inform the employer in writing. Until the filing of an application, the employer and the employee may not disclose the design.

6.4 Employee Remuneration Right

Article 13(3) of the Law provides for an employer's duty to pay remuneration upon the registration of an employee's design in the name of the employer, however this duty does not apply if 'the employment contract provides for the development of a design' (Article 13(4)). That means that an employer's duty to pay remuneration in case of a service design applies only when an employee has no duty under the employment agreement to develop a design and has created a design when making use of the experience accumulated by an employer or its technologies and equipment or under a separate civil contract following the instructions of the commissioner.

Further, similarly to the Law on Patents, Article 13(3) of the Law provides that 'The amount of such remuneration shall depend on the economic value of the design and any benefit derived by the employer from the utilisation of this design. *The conditions of remuneration and the payment thereof shall be established in a remuneration agreement which is signed by the employer and the designer*, and which is concluded not later than within six months following the issuing of a certificate of registration of the design or within one year of the beginning of the use of the design, if the beginning of the use of the design is prior to the issuing of the certificate of registration of the design. *If the employer fails to fulfil the contractual requirements or the said contract is not concluded within the aforementioned time limits*, the designer shall be entitled *to appeal to the court to be recognised as a design holder of a service design*.' (Emphasis added) Further, as with the Law on Patents, the Law on Designs requires the court to establish the amount of remuneration if the parties fail to conclude the agreement on remuneration.

Another provision borrowed from the Law on Patents is that the rights and duties of the employer and the employee, related to service designs, with the exception of the payment of remuneration, shall expire when the registration of the design is declared invalid or after one year of the termination of labour relations. It seems that following that provision the employer should pay the employee remuneration for the use of the design even if it has been invalidated or has expired.

7 EMPLOYEES' PLANT VARIETIES

7.1 The Legal Framework

The Law on the Protection of Plant Varieties[32] regulates the legal protection and use of plant varieties in Lithuania. The Law does not specifically regulate the legal relationship concerning employers' rights to employee breeders' rights except that it provides for the definition of 'breeders' (Article 2(10)), which includes legal persons, and general provisions on transfer of ownership in Article 10 of the Law. The absence of a specific provision in the Law has not, however, given rise to any disputes as mostly only scientific research institutes (legal persons) working in the plant industry register plant varieties in Lithuania and their employees breed, discover or develop plant varieties under employment agreements.

7.2 Ownership of Rights

As mentioned before Article 2(10) of the Law provides for the definition of breeders which are defined as follows: (i) one or several natural persons who bred or discovered and developed a new plant variety; (ii) a legal person *whose one or several employees, working under an employment contract or fulfilling an assigned task*, bred or discovered and developed a new plant variety; (iii) branches set up in the Republic of Lithuania by enterprises established in Member States of the European Union and other countries of the European Economic Area, which bred or discovered and developed a new plant variety; (iv) a successor in title who has acquired property rights to the variety from the persons indicated in subparagraphs (i)-(iii) above. Following the definition of 'breeder', an employer will be presumed to hold the right to the new plant variety bred or developed by the employee if the employee bred or discovered and developed the new plant variety under the employment agreement or under their assignment. This conclusion is strengthened by Article 9(1) of the Law which entitles 'the breeder' to file an application for the grant of a plant variety right. Therefore, the employment relationship between an employer and an employee covering breeding or development of new plant varieties is sufficient that potential rights to new plant varieties bred by an employee be vested with the employer in their entirety (exclusive right).

However, if an employer is a natural person, they are not considered a breeder, and therefore, the employment agreement between a natural person employer and a breeder employee should expressly mention transfer of the rights of employee to any new plant varieties to the employer.

One more peculiarity of the Law is that a Lithuanian branch of an EEA enterprise which has no legal person status in Lithuania can also be considered the primary owner of the rights to new plant varieties.

32. Law on Protection of Plant Varieties of 22 November 2001 No. IX-618 (O.J., 2001, No. 104-3701).

As regards the Community plant rights the employer will be entitled to employee breeder's Community plant rights. According to Article 11(4) of the Regulation on the Community Plant Variety Rights[33] 'if the breeder is an employee, the entitlement to the Community plant variety right shall be determined in accordance with the national law applicable to the employment relationship in the context of which the variety was bred, or discovered and developed'. Following that, the existence of the Community plant variety right in the employment relationship is assessed according to national legal provisions.

In the context of university researchers' plant varieties, the employee's research results would be owned by universities, unless otherwise agreed in the employment or collective agreements, or internal state university documents (policies).[34]

7.3 The Parties' Duty to Inform

An express positive duty to give notice of a new variety does not exist in the Law. However, this duty is established in the Law on Higher Education and Research (Article 82(2)) requiring teachers and/or researchers of state university or research institution to notify in accordance with the procedure laid down by the state universities or research institutions.

7.4 Employee Remuneration Right

Special rules regulating employee-breeder's remuneration right are not set by the Law. Consequently if, and to what extent, breeders are entitled to remuneration depends generally on the agreement that has been entered into by the employer and the employee, except breeders employed by state higher education institutions or research institutions who are entitled to '*not less than one third of the profit gained through commercial use* of the intellectual activity results achieved by employees of a state higher education institution in the process of a research with the employees performing their job functions during the research and by persons studying in doctoral studies in the process of a research must be allocated to the author (co-authors), unless otherwise provided for in an employment contract or a contract concluded between the state higher education institution and the employee (person studying in doctoral studies).' (Emphasis added)[35]

8 INSTANCES FOR DISPUTES

Disputes between the employer and the employee regarding the rights and duties enshrined in labour laws, other normative legal acts, the employment or collective

33. Council Regulation (EC) No. 2100/94 of 27 July 1994 on Community plant variety rights, O. J. L 227, 01 September. 1994, 0001 - 0030.
34. For more details see Section 4.
35. Art. 82(3) of the Law on Higher Education and Research.

agreements are considered individual labour disputes[36] which should be dealt by the Labour Dispute Commissions as the first instance unless otherwise provided in the Labour Code and other laws.[37] Labor disputes regarding the rights may be heard in commercial arbitration in accordance with the Law on Commercial Arbitration of the Republic of Lithuania, if that is agreed upon by the parties to the labor dispute after the dispute has arisen.[38]

The intellectual property laws governing protection of intellectual property rights and the Civil Procedure Code provide that the disputes stemming from the enforcement of intellectual property rights should be dealt by the court in the first instance. As per the Law on Patents and the Law on Designs the courts also have exclusive jurisdiction to determine the amount of remuneration if the employer fails to conclude with an employee an agreement on remuneration as required by these laws.

If a dispute arises between an employer and an employee in relation to the employee's economic copyright or related rights and the right to the plant variety, the general jurisdictional rules apply and the dispute should be dealt by the District Courts according to the employer's place of residence as the first instance court,[39] with the County Courts as the second instance court[40] and the Supreme Court as the last instance court.[41] If moral rights or both economic and moral rights have been infringed, the County Court has jurisdiction over the whole dispute with the Lithuanian Appeal Court as the second instance court and the Supreme Court as the last instance court.[42]

If the dispute arises between an employer and an employee in relation to the enforcement of the right to the patent, design or topography, then such dispute is subject to the jurisdiction of the Vilnius County Court as the first instance court by virtue of the provisions of the Law on Patents,[43] the Law on Designs,[44] the Law on the

36. According to Art. 213(3) of the Labour Code (RLA, 19 September 2016, No. 23709) an individual labour dispute is a disagreement between the employee or other participants in the employment relationship, on the one hand, and the employer, on the other hand, arising from the conclusion, modification, performance or termination of an employment contract, as well as the non-performance or inappropriate performance of labor laws in the employment relationship between the employee and the employer.
37. Art. 220(1) and 215 of the Labour Code (RLA, 19 September 2016, No. 23709).
38. Art. 216(2) of the Labour Code (RLA, 19 September 2016, No. 23709)
39. Art. 26(1) and 29 of the Civil Procedure Code of the Republic of Lithuania (the CPC) of 28 February 2002 No. IX-743(O.J., 2002, No. 36-1340).
40. Art. 301(3) of the CPC.
41. Art. 340(2) of the CPC.
42. Para. 2 of Art. 27 and Para. 4 of Art. 33 of the CPC.
43. Art. 51(1)(2) of the Patent Law: Disputes regarding the patenting and the use of inventions shall be settled by the Vilnius County Court, in respect of disputes regarding: decisions of the Appeals Division of the State Patent Bureau; assignment of a patent application or the patent ownership to a different person; invalidation of a granted patent in full or in part; infringement of a patent application for which temporary protection is granted; infringement of a granted patent; declaration of non-infringement of a patent; granting, revocation and change of conditions of the licences referred to in Arts. 48-49 of this Law; revocation of a patent.
44. Art. 46(1)(2) of the Law on Designs: Vilnius County Court shall deal with disputes relating to: 1) decisions of the Appeals Division; 2) invalidation of registration of a design; 3) enforcement of the rights of the design holder; 4) a Community design under the provisions of Art. 79-92 of the Regulation on Community designs.

Legal Protection of Topographies of Semiconductor Products[45] and the Civil Procedure Code,[46] with the Appeal Court of Lithuania as the second instance[47] and the Supreme Court as the last instance.[48]

Collective labor disputes are heard by a Labour Commission in the first place, with the assistance of a mediator if needed and by the Labour Arbitration formed at the territorial division of the State Labor Inspectorate, in the territory of which the employer or employer's organisation has its seat.[49] No collective labour disputes regarding issues of intellectual property rights have ever been heard in Lithuania.

9 CONCLUDING REMARKS

As described above, under the Lithuanian intellectual property laws the employers are presumed to own the exclusive economic rights to intellectual property objects created by the employees, if certain conditions set in the laws are met. This rule is not absolute, however. The employer and the employee may agree in an individual employment agreement, collective agreements or internal policies on different rules of transfer of intellectual property rights to the objects created by the employee.

Employers are advised to regulate this issue in detail in the employment or collective agreements or by the internal policies of the employer for the following reasons:

First of all, the conditions and the scope of transfer of the employee's rights to the employer prescribed by the law differ from one law to another, for which reason the applicable intellectual property regulatory framework may not satisfy their needs.

Secondly, the employee's duty to inform about the creation of intellectual property is not governed by all intellectual property laws, for which reason it is advisable to regulate this duty by the agreements or internal policies of the employer.

Thirdly, the product created by the employee might be protected by several types of intellectual property rights. If the employer would rely on the presumption of transfer of the employee's intellectual property rights to the employer, they might get into the situation whereby they do not own all the rights needed to freely exploit the product. The employer should evaluate the scope of protection needed to exploit their products without restrictions and secure additional protection needed through the employment contracts with the employees, the collective agreements or their internal policies.

45. Art. 20(1)(2) of the Law on the Legal Protection of Topographies of Semiconductor Products: disputes relating to the decisions of the Appeals Division of the State Patent Bureau and other disputes arising after the issuance of the certificate - by the Vilnius County Court.
46. Art. 28 and 33(4) of the CPC.
47. Art. 301(4) of the CPC.
48. Art. 340(2) of the CPC.
49. Art. 236-243 of the Labour Code.

BIBLIOGRAPHY

Table of Sources

Alfonsas Vileita, *Lietuvos Respublikos autorių teisių ir gretutinių teisių įstatymo komentaras, Lietuvos Respublikos administracinių teisės pažeidimų kodekso 214(10) straipsnio komentaras* (Lietuvos rašytojų sąjungos leidykla, Vilnius 2000)

Jūratė Usonienė, *Autorių teisių perdavimo ypatybės* (Teisinės informacijos centras, Vilnius 2008)

Ramūnas Birštonas et al., *Intelektinės nuosavybės teisė* (Registrų centras, Vilnius 2010)

Vytautas Mizaras, *Autorių teisė*, Vol. I, Justitia (Vilnius 2008)

Table of Legislation

Law on Copyright and Related Rights of the Republic of Lithuania of 18 May 1999 No. VIII-1185 (O.J.1999), No. 50-1598

Law on Protection of Plant Varieties of the Republic of Lithuania of 22 November 2001 No. IX-618 (O.J., 2001), No. 104-3701

Law on Higher Education and Research of the Republic of Lithuania (O.J. 2009), No. 54-2140, No. 61, No. 101

Labour Code of the Republic of Lithuania of 14 September 2016 No. XII-2603 (RLA, 19 September 2016, No. 23709

Law on Patents of the Republic of Lithuania of 18 January No. I-372 (O.J., 28 January 1994), No. 8-120

Law on Designs of the Republic of Lithuania of 7 November 2002 No. IX-1181 (O.J., 22 November 2002), No.112-4980

The Government Decree on Increase of Minimum Salary of 19 December 2012 No. 1543 (O.J., 22 December 2012), No. 152-7772

Law on the Legal Protection of Topographies of Semiconductor Products of 16 June 1998 No. VIII-791 (O.J., 01 July 1998), No. 59-1655

Civil Procedure Code of the Republic of Lithuania of 28 February 2002 No. IX-743 (O.J., 2002), No. 36-1340

Council Directive 91/250/EEC of 14 May 1991 on the legal protection of computer programs (O.J.L122, 17 May 1991), 42-46

Council Directive 87/54/EC of 16 December 1986 on the legal protection of topographies of semiconductor products (O.J.L24, 27 January 1987), 36

Council Regulation (EC) No. 6/2002 of 12 December 2001 on Community designs, O.J. L3 (05 January 2002), 1-24

Council Regulation (EC) No. 2100/94 of 27 July 1994 on the Community plant variety rights, O. J. L227 (01 September 1994), 1-30

Table of Cases

B. V.-F. v. Museum, Fodio, Ministry of Culture, Decision of the Lithuanian Appeal Court of 2 January 2007 in civ. case No. 2A-98/2007

M.B. v. Orlen Lietuva, Decision of 1 March 2010 of the Supreme Court of Lithuania in civ. case No. 3K-3-66/2010

Inventors v. Orlen Lietuva, Decision of 2 December 2016 of the Supreme Court of Lithuania in civ. case No. 3K-3-496-378/2016

CHAPTER 15

The Netherlands

*Rogier de Vrey & Willem Hoorneman**

1 BACKGROUND

Rules on the allocation of intellectual property rights generated in the course of employment are spread over various Dutch intellectual property acts. Below, we will provide an overview of these rules categorised by intellectual property right. The relevant wording from the legislation is cited at the beginning of each section.

2 EMPLOYEES' COPYRIGHT

2.1 The Legal Framework

As regards employees' copyright the relevant provision is Article 7 of the Dutch Copyright Act. Further, the allocation of neighbouring rights is provided for in Article 3 of the Neighbouring Rights Act ('NRA'). The allocation of database rights is provided for in Articles 1 and 1a of the Databases (Legal Protection) Act ('Database Act').

2.2 Ownership of Rights

2.2.1 Copyright

The Dutch Copyright Act is based on the principle of fictitious authorship. Not the actual creator, the employee, but the employer is deemed to own the copyrights of the copyrightable works according to Article 7 Copyright Act:

* We express our gratitude to Ms E. Besterveld for her assistance in the preparation of this chapter.

> Where labour carried out by an employee consists in the making of certain literary, scientific or artistic works, the employer shall be deemed the author thereof, unless otherwise agreed between the parties.

The term employee not only refers to an employee who has entered into a private employment contract, but also to the public officer appointed under a civil service appointment in the service of the government. Flexible workers such as freelance designers or journalists are considered employees as long as they have worked at least three months with an employer. Article 7 Copyright Act however does not apply to works generated by freelancers if they can prove the absence of an authoritative relationship with the employer. It is therefore advisable to enter into an agreement with a freelancer and stipulate who will own any resulting copyrights.

Temporary workers are also covered by Article 7 Copyright Act. As the employee is (formally) employed by an agency but (factually) works for the hiring enterprise, it is not always clear who owns the copyright: the employment agency or the hirer. Given this uncertainty, it is therefore important that the hirer makes arrangements in advance with the agency about the transfer of rights if necessary.

Not all works produced during employment fall under the provisions of Article 7 Copyright Act. The wording of this Article refers to employment activities that 'consist in the manufacture of certain works of literature, science or art'. If such activities are not part of the duties of the employee, Article 7 Copyright Act will not be applicable. However, any works created by the employee outside of their regular activities, but on the basis of an explicit, albeit incidental, instruction of the employer, will fall within the ambit of Article 7 Copyright Act.

2.2.1.1 Employment and Software

The Software Directive[1] is implemented in the Copyright Act. Articles 2 and 3 of this Directive detail the persons entitled to copyright protection. Article 2(3) Software Directive refers to software created 'by an employee in the performance of his duties'. This article has not been implemented in the Copyright Act; the current Article 7 Copyright Act has different language and mentions the making of 'certain works'. In practice judges in cases involving a technical work such as software will be inclined to award the claims of the employer (and will pay less attention to the fulfilment of the requirement of a 'certain work').

2.2.1.2 Directory Law

Article 7 Copyright Act is directory law. The employer and employee are free to make arrangements to the contrary, even orally. However, this is rarely the case, except for collective agreements that often contain specific provisions concerning intellectual property rights.

1. Directive 2009/24/EC on the legal protection of computer programs.

2.2.1.3 Moral Rights

As suggested, any works created by the employee in the line of their general or specific duties are vested with the employer. Article 7 Copyright Act not only attributes the exploitation rights with the employee, but also considers the employer as the 'creator' of the works. However, in most cases, the moral rights stipulated in Article 25 (1) Copyright Act[2] will remain with the 'factual' creator, the employee.

To a certain extent such moral rights can be waived by the employee. However, the Copyright Act only allows for the waiver of the right of attribution and the right of the author to oppose the name of the work or changes in the work, according to Article 25(3) Copyright Act. The original author will always be able to oppose distortion, mutilation or other adverse interference with the work and the change of its name.

Article 25 Copyright Act is generally assumed to apply to works where the employer is considered the 'creator' (author) and therefore the copyright holder under Article 7 Copyright Act, even though the works are actually created by the employee. There is still uncertainty as to who owns the moral rights in case of such employer-owned works. This will greatly depend on the circumstances of the case, but the employer will have to take into account that a court will likely attribute some or all of the moral rights to the employee. As a result, an employer may be limited in exploiting its copyrights.

2.2.2 Neighbouring Rights

The allocation of neighbouring rights is provided for in Article 3 of NRA:

> An employer shall be entitled to exploit the rights of a performer ..., where this has been agreed between the parties or derives from the nature of the employment contract concluded between them, from custom or from the requirements of reasonableness and fairness. Unless it has been agreed otherwise or arises otherwise from the nature of the contract, from custom or from the requirements of reasonableness and fairness, the employer shall pay the performer or his assignee an equitable remuneration for each form of exploitation of the latter's rights. The employer shall respect the rights of the performer as referred to in article 5.

---

2. Art. 25(1) Copyright Act:

> Even after assignment of his copyright, the author of a work has the following rights:
>
> a) the right to oppose the communication to the public of the work without acknowledgement of his name or other indication as author, unless such opposition would be unreasonable;
> b) the right to oppose the communication to the public of the work under a name other than his own, and any alteration in the name of the work or the indication of the author, insofar as it appears on or in the work or has been communicated to the public in connection with the work;
> c) the right to oppose any other alteration of the work, unless the nature of the alteration is such that opposition would be unreasonable;
> d) the right to oppose any distortion, mutilation or other impairment of the work that could be prejudicial to the name or reputation of the author or to his dignity as such.

Article 3 NRA aims to provide a framework for the pursuit of claims in case the (audiovisual) performances are carried out by the performer during their employment. Employment means – in any case – an employee who has entered into a contract with a private employer as well as the public officer appointed by official appointment in the service of the government. This Article does not apply to an agreement for services, nor does it apply to (works created by) film actors. The Copyright Act contains specific provisions for such cases.[3]

The employer has no legitimate claim to (the ownership of) any neighbouring right of its contracted performers. The intended purpose of the statutory provisions is that parties make contractual arrangements. In the absence of any agreement the employer can derive their authorisation to exploit the neighbouring right from the employment contract, good practices and the principle of reasonableness and fairness. In that case, important factors are the existing practices between the parties, the size of the group of employees (the bigger the group, the more obvious it is for the employer to exercise such rights), and how independent the majority of workers are acting versus the employer (if the employees are depending on the employer, it is more obvious that the employer exercises such rights).

2.2.2.1 *The Remuneration of the Employee*

In case any claims are awarded to the employer, the employee is in principle entitled to a fair compensation. However, this provision is directory law. The nature of the contract, good practice and the requirements of reasonableness and fairness may lead to a different result. The fact that the exploitation of the neighbouring right may generate (substantial) revenue, will often be an important factor in this respect.

2.2.2.2 *Moral Rights*

The moral rights are contained in Article 5 NRA.[4] These are the right to oppose to a different or absence of any attribution, the right to oppose any alteration in the performance and the right to oppose any distortion, mutilation or other modification of the performance. These moral rights are vested with the employee.

---

3. Arts. 45a-45g Copyright Act.
4. Art. 5 NRA: 'A performer shall be entitled, even after he has assigned the right referred to in article 2: a) to oppose the communication to the public of a performance without acknowledgement of his name or other designation as a performer, unless such opposition would be unreasonable; b) to oppose the communication to the public of a performance under a name, other than his own, and any alteration in the way in which he is designated, insofar as such name or designation is mentioned or communicated to the public in connection with the performance; c) to oppose any other alteration to the performance, unless the nature of the alteration is such that opposition would be unreasonable; d) to oppose any distortion, mutilation or other impairment of the performance that could be prejudicial to his name or reputation or his dignity as a performer.'

2.2.3 Database Rights

Database rights are regulated in the Database Act. The allocation of database rights is provided for in Article 1 and 1a of the Database Act:

> Article 1
> ... producer of a database: the person who bears the risk of the investment for creating the database ...
> Article 1a
> In the absence of proof to the contrary, the producer of a database is the person indicated in the database as such, or in the absence of such an indication, the one who was communicated as producer when the database was communicated to the public.

These Articles provide that the person bearing the risk of investment is entitled to the database right. The wording 'take the initiative' from the Database Directive (preamble 41) is not included in the Dutch Database Act, because the legislature wanted to express that the database right is primarily intended to protect the investment in the database. However, according to the parliamentary history these words were nevertheless considered relevant. It was pointed out, e.g., that the provider of venture capital in a company that produces a database should not be granted any database right because it did not directly invest in the database and furthermore did not take the initiative for the investment.[5] The Database Act (and Database Directive) do not provide any details on whether, in general, the party who actually manufactures the database or the commissioning party is to be considered the party carrying the risk of the investment. This will probably be determined case by case on the basis of their contractual relationship.[6]

As the party bearing the risk of investment is entitled to the database right, there is no legal issue regarding allocation of database rights in an employment context, as the employee will (almost) never be the party bearing the risk of investment.

2.3 The Parties' Duty to Inform

There is no statutory duty to give an employer notice of the creation of copyrighted works, neighbouring rights or database rights.

2.4 Employee Remuneration Right

If the employee is hired to create works protected by copyright, it is generally assumed that this has been taken into account in determining their salary. In case of an (ad hoc) instruction to create a work that is not part of the normal duties of the employee, the

5. NnV, Kamerstukken II 1998/99, 26 108, No. 3, p. 5.
6. See e.g. District Court (KG) Groningen, 28 June 2010, IEPT 20100628 (Donorkind Amsterdam/ Donorkind Groningen); District Court Utrecht 16 May 2012, IEPT 20120516 (Schending geheimhoudingslicht) and District Court Rotterdam 21 December 2011, IEPT 20120508 (EuroPsyche/ Fides).

agreed salary is more likely not to compensate for the creation of the work. The question is whether the employee is entitled to additional compensation if the work they have created turns out to be a blockbuster. The Copyright Act provides no rules for this.

In a patent case before the Dutch Supreme Court,[7] a general rule was formulated that in principle, the agreed salary constitutes a fee for *all* the agreed activities. That means that if the employee has as a (general or incidental) task to make such works, in principle no additional compensation can be claimed if the work turns out to be a blockbuster, even if the salary does not contain a specific component as compensation for the loss of the intellectual property. In exceptional cases, an employee may bring an action based on the requirements of reasonableness and fairness or the legal concept of good employment practices and may be thus entitled to an additional fee.

Finally, it is worth mentioning that a bill is pending concerning author's contract law.[8] This bill aims at strengthening the contractual position of authors and performers in relation to the parties commercialising their copyrighted works. The bill gives the maker the right to equitable remuneration for the exploitation of their work. Furthermore, the creator may claim additional compensation if the originally agreed fees are materially disproportionate to the proceeds of sale of the works (bestseller clause). Although this bill is in principle not applicable to the relationship between the employer and employee, it is possible that this scheme can be applied by analogy.

3 EMPLOYEES' SEMICONDUCTOR CHIPS

3.1 The Legal Framework

The Legal Protection of Original Topographies of Semiconductor Products Act ('Chips Act') introduced a separate intellectual property right for topographies of chips. Chips are three-dimensional as the electronic components are located in multiple layers of silicon on top of each other. Before starting with the production of a chip one needs to make a design. In the design, the position of the various electronic interconnections is indicated. This results in a three-dimensional design, called topography. The chips right is similar to copyright. Article 3 of the Chips Act contains a provision for chips designed during employment. This Article states:

> In case the employment activities performed in the service of another exist in the creation of topographies, [the employer] shall, unless otherwise agreed, be considered the creator of the topography.

7. Supreme Court 1 March 2002, ECLI:NL:HR:2002:AD7342 (TNO/Ter Meulen).
8. Wet Auteurscontractenrecht, Kamerstukken 33 308.

3.2 Ownership of Rights

As is the case with the Copyright Act, Article 3 of the Chips Act provides that the employer is considered the creator of the topography if it is designed during employment and if it belongs to the work of the employee to design topographies. The legislature has explicitly stated that this Article needs to be interpreted in accordance with the case law on Article 7 Copyright Act. The Chips Act is very rarely applied; relevant case law has not occurred to date.

3.3 The Parties' Duty to Inform

The Act does not require employees who create topographies to inform their employer.

3.4 Employees' Remuneration Right

The Act does not give employees a right to remuneration.

4 EMPLOYEES' INVENTIONS

4.1 The Legal Framework

The allocation of patent rights is provided for in Article 12 of the Patent Act 1995:

> (1) If the invention for which a patent application has been filed has been made by a person employed in the service of another party, the employee shall be entitled to the patent unless the nature of the service entails the use of the employee's special knowledge for the purposes of making inventions of the same kind as that to which the patent application relates, in which case the employer shall be entitled to the patent.
> (2) If the invention for which a patent application has been filed has been made by a person who performs services for another party in the context of a training course, the party for whom the services are performed shall be entitled to the patent unless the invention has no connection with the subject of the services.
> (3) If the invention has been made by a person carrying out research in the service of a university, college or research establishment, the university, college or research establishment in question shall be entitled to the patent.
> (4) For the purposes of Article 4(3) and (4), an application filed by a party who does not have a right to a patent shall be disregarded in respect of the subject matter of an application filed by the employer referred to in the last sentence of the first paragraph or by a party providing the opportunity to perform the services referred to in the second paragraph.
> (5) The provisions contained in paragraphs (1), (2) and (3) may be derogated from by written agreement.

(6) In the event that the inventor cannot be deemed to have been compensated in the salary he earns or the pecuniary allowance he receives or in any extra remuneration he receives for not having been granted a patent, the party who is entitled to the patent on the basis of paragraphs (1), (2) or (3) will be obliged to grant him equitable remuneration related to the pecuniary importance of the invention and the circumstances under which it was made. Any right of action on the part of the inventor in accordance with this paragraph shall lapse after the expiry of three years from the date of the grant of the patent.

(7) Any stipulation that is in derogation from the provisions contained in the sixth paragraph shall be void.

4.2 Ownership of Rights

Unlike most other intellectual property laws, the Dutch Patent Act provides that the employee is entitled to the patent, unless the nature of the employment entails the use of the employee's special knowledge for the purposes of making inventions of the same kind as that to which the patent application relates. Consequently, the employer bears the burden of proving these facts.

According to Article 75(1)(e) Patent Act, an employee may claim that the patent be invalidated by a court in case the patent was granted to the employer and the employer is not entitled to the patent. The employee can also opt for an action under Article 78 Patent Act claiming entitlement to a patent in whole, in part or in joint ownership if they believe to be (jointly) entitled to the patent. This claim lapses upon the expiry of a period of five years from the date on which the patent was granted or, if in case of a European patent, from the date on which the notification of the grant of the European patent was published, unless the employer had applied for the patent in bad faith as referred to in Article 78(7) Patent Act.

The term employment means, in relation to Article 12 of the Patent Act, an employee who has entered into a contract with a private employer as well as the public officer appointed by official appointment in the service of the government. Over the years, the concept of 'employment' has been extended. Article 12 of the Patent Act is also considered applicable to employees on secondment. Whether temporary workers also fall within the ambit of this Article is uncertain. In case of an assignment agreement or contract for work Article 12 is not applicable (meaning that the actual inventor may claim the patent).

As indicated above, Article 12(1) of the Patent Act stipulates that employers only have a claim in relation to employees whose job it is to invent. If making inventions is not part of the job of the employee, the employer will have no claim to (patentable) inventions. What matters, in this context, are the actual working activities of the employee, not the formal job description formulated in their contract. It has been established in case law that a one-time explicit instruction to do research work can also fulfil the requirements of Article 12(1) of the Patent Act. It is of no importance whether the invention is made at the workplace or at home. The performance of work-related activities at home can also belong to one's professional (work) responsibilities.

As far as interns and trainees are concerned, a company will have a claim under Article 12(2) Patent Act to any inventions made by a person who works for the company in the context of a training program, unless the invention has no connection with the subject of these services. If the invention has been made by a person carrying out research in the service of a university, college or research establishment, such entity shall be entitled to the patent according to Article 12(3) Patent Act.

Article 12(1)-(3) of the Patent Act are directory law and allow the parties to make alternate arrangements according to Article 12(5) Patent Act. This can, other than is the case for copyright law, only be effectuated by entering into a written agreement. As previously mentioned in section, many collective agreements include provisions concerning the allocation of intellectual property rights. The arrangements in collective agreements often have a broader scope than the equivalent provisions in the Patent Act.

Further, the patent applicant is obliged under Article 14(1) Patent Act to mention the name and residence of the inventor, unless the inventor has declared not to appreciate such a statement. The European Patent Convention ('EPC') also expressly provides that the patent applicant must state the name of the inventor (Article 81 EPC). Pursuant to Article 14(2) Patent Act this is mandatory law as far as the invention is made during employment.

4.3 The Parties' Duty to Inform

There is no statutory duty to give an employer notice of the creation of inventions.

4.4 Employee Remuneration Right

Under Article 12(6) Patent Act, those persons who have by virtue of paragraphs 1, 2 or 3 no rightful claim to the patent, have a right to an equitable remuneration. This clause is mandatory law. The Supreme Court has held that this clause only applies to inventions that fall under the scope of paragraphs 1, 2 and 3 of Article 12 Patent Act.[9] For inventions that fall outside of this scope, parties can make alternate arrangements including assignment by the employee of (its claims to) such (future) inventions. Such assignment is valid provided the assignment is not void or voidable, e.g., in case of undue influence by the employer.

In two cases before the Supreme Court, the issue of what is an equitable remuneration was addressed.[10] The Supreme Court ruled in both cases that the contractually agreed salary in principle is deemed to entail a fee for not being awarded a patent. Article 12(6) Patent Act is therefore an exceptional provision. If an equitable

9. Supreme Court 30 June 1950, NJ 1952, 35 (Perquin/Perira).
10. Supreme Court 27 May 1994, NJ 1994, 136(Hupkens/Van Girneken) and 1 March 2002, ECLI:NL:HR:2002:AD7342 (TNO/Ter Meulen).

remuneration is paid to the employee, this fee is considered a bonus. The remuneration may not be calculated on the basis of the financial value neither may it be calculated on the basis of an assumed licence agreement between the employee and the employer, although the court is allowed to take these amounts into account when determining an equitable remuneration. The Supreme Court thus opted for an employment law approach to the concept of equitable remuneration.

5 EMPLOYEES' UTILITY MODELS

Dutch law does not protect utility models.

6 EMPLOYEES' DESIGN RIGHTS

6.1 The Legal Framework

There is no specific legislation regarding designs in the Netherlands as such. The Convention on Intellectual Property (the BCIP) is applicable in the Benelux to the protection of designs. The BCIP contains two provisions with regard to the rights of employers and employees, Article 3.8 and 3.29 BCIP:

> Article 3.8 (Rights of employers and commissioning parties)
> 1. If a design has been developed by an employee in the execution of his duties, the employer shall, unless otherwise agreed, be deemed to be the designer.
> 2. If a design has been created on commission, the commissioning party shall, unless specified otherwise, be deemed to be the designer, provided that the commission was given with a view to commercial or industrial use of the product in which the design is incorporated.
>
> Article 3.29 (Copyright of employers and commissioning parties)
> Where a design is created in the circumstances referred to in Article 3.8, the copyright relating to the design shall belong to the party deemed to be the designer, in accordance with the provisions of that Article.

6.2 Ownership of Rights

The rights to a design created by an employee are as a general rule vested with their employer. An employee is in any case an employee who has entered into a contract with a private employer as well as the public officer appointed by official appointment in the service of the government. However, if the design is not created by the employee 'in the execution of his duties', e.g., in the absence of a clear link between the design on the one hand and the job of the employee-designer on the other, the employee remains entitled to the design.

Pursuant to Article 3.8(2) BCIP, a party who has commissioned a designer to create a design, will under certain circumstances be deemed the owner of the design right. This is, e.g., the case if a freelance designer creates a standard design for their client who intends to manufacture such a design on an industrial scale. The employer or commissioning party that is considered the designer, will have the right to apply for a design right. Article 3.8 is also applicable to unregistered designs.[11]

Article 3.8 BCIP is directory law and allows for parties to make alternate arrangements. The application is not obliged (contrary to patent law) to mention the name of the actual designer in the application.

6.2.1 *Ownership of Rights: Designs and Copyright Law*

To prevent third parties being faced with different right holders, the legislature has opted for a regulation ensuring that the rights to a design and copyright related to a certain object are under one control. For the employee and commissioned designer, Article 3:29 BCIP stipulates that not only the design right, but also the copyright is vested with the employer or commissioning party.[12] The Supreme Court has held[13] that it is not necessary for the application of Article 3:29 BCIP (in conjunction with Article 3.8 BCIP) that the design meets the requirements for design protection under Article 3.1(1) BCIP.[14] However, for these provisions to be applicable, the design has to meet the requirements for copyright protection under the Copyright Act. It is assumed that after expiration of the design right (after 25 years), or if the registration is cancelled, the related copyright will vest with the (possible) earlier owner of the design.

6.3 The Parties' Duty to Inform

The BCIP does not require employees who create designs to inform their employer.

6.4 Employees' Remuneration Right

The BCIP does not give employees a right to remuneration

7 EMPLOYEES' PLANT VARIETIES

7.1 The Legal Framework

Plant breeder's rights are regulated in the Seeds and Planting Materials Act. The allocation of plant breeder's rights is provided for in Article 51 of the Seeds and Planting Materials Act:

> 1. If the breeder has bred or discovered and developed a variety in the course of employment or under a contract for the provision of services for the

11. Benelux Court of Justice, 22 June 2007, A 2006/3/13 (Electrolux).
12. The Copyright Act contains no similar provision.
13. Supreme Court 25 October 2013, ECLI:NL:HR:2013:1036 (S&S/Esschert).
14. Art. 3.1. BCIP: '1. A design shall be protected to the extent that it is new and has individual character. 2. The appearance of the whole or a part of a product shall be regarded as a design.'

benefit of another, for payment other than wages, which employment or agreement entails that the breeder performs breeding or development work with respect to the cultivated species to which the variety belongs, the entitlement to the plant breeder's right shall notwithstanding Article 50 accrue to the employer, the commissioning party or the predecessors in title of the said employer or party.

2. In the event referred to in paragraph (1) the breeder shall be entitled to a fair remuneration, unless such remuneration may be deemed to be included in the wages received by him or the benefits enjoyed by him.
3. Any stipulation in derogation from paragraph (2) shall be null and void.

7.2 Ownership of Rights

The rules with regard to plant breeder's rights are similar to the rules regarding patents and we refer to our comments under section 5. The employer or the commissioning party is entitled to the plant breeder's rights plant as far as the breeder:

- is employed as an employee with the employer on the basis of a private employment;
- is appointed as a public official through an official appointment or acts as contractor;
- is employed by the employer as an intern, trainee or volunteer without pay.

In both these cases the employer will only acquire the right to a new variety if the employee has created the variety in the course of their work or training and if these activities are related to their duties. The duties must include the cultivation of cultivated species to which the variety belongs. If this is not the case, the breeder (employee, intern, etc.) will be entitled to the plant breeder's rights.

Article 51 of the Seeds and Planting Materials Act is directory law and allows the parties to make alternate arrangements, e.g., stipulating that the employer or the commissioning party may claim the plant breeder's rights for all varieties, belonging to any cultivated species that will be grown by the employee or contractor. These alternative arrangements can be included in applicable collective agreements.

7.3 The Parties' Duty to Inform

The Act does not require employees who create plant varieties to inform their employer.

7.4 Employee Remuneration Right

If the employer or commissioning party is granted a plant breeder's right created by an employee or contractor, the actual breeder of the variety is entitled to equitable remuneration. The salary and benefits involved (travel expenses, compensation for expenses for visiting foreign conferences, etc.) need to be taken into account when establishing the equitable remuneration. Article 51(2) Seeds and Planting Materials Act does not specify which criteria should be used in determining the equitable remuneration. It is likely that the two Supreme Court patent cases identified above regarding equitable remuneration can be applied by analogy. The Supreme Court ruled in those judgments[15] that the wage contractually agreed in principle entails a fee for loss of patent rights. The Supreme Court considered that equitable remuneration is only awarded in exceptional circumstances. Such compensation will take the form of a bonus. See also § 4.4 (Patent Law).

8 INSTANCES FOR DISPUTES

Regarding employees design rights and copyright according to Article 4.6 BCIP, the territorial jurisdiction of the court is determined by the residence of the defendant or by the place where the contractual obligation has originated. The claimant may choose which of these options to use as the venue. Parties can arrange for a different forum (either Brussels, the Hague or Luxembourg) through a (specific) contractual arrangement.

As regards employees' inventions, the determination of the remuneration of an employee is according to Article 83(2) Patent Act a claim relating to an employment contract. Consequently such a request must usually be submitted to the Dutch Subdistrict court. The claim will expire three years after the grant of the patent according to Article 12(6) Patent Act.

Should the plant breeder's right be granted to the wrong person, e.g., the employee or contractor instead of the employer or commissioning party, Article 76 Seeds and Planting Materials Act enables the person who has full or partial entitlement to the plant breeder's right the option to claim the right in its entirety or partly. The Hague District Court has sole jurisdiction for such claims. The period of limitation for such an action is five years after registration of the plant breeder's right.

9 CONCLUDING REMARKS

As shown above the Dutch rules concerning allocation of intellectual property rights for employees and employers provide a blurry image. The Dutch legislature unfortunately has no true coherent vision on the allocation of intellectual property rights in employment relations. Another striking point is that the scope of the concept of 'employment' is quite extensive, also covering 'flexi-workers' such as freelancers and journalists and as a result employers and commissioning parties can often claim the intellectual property rights in relation to these flexible workers.

15. *Supra* note 10.

BIBLIOGRAPHY

Table of Sources

Intellectuele Eigendom, Tekst & Commentaar (4th ed., Kluwer 2013).
Kort begrip van het Intellectuele Eigendomsrecht (red. Ch. Gielen, 11th ed., Kluwer 2014).
A.A. Quaedvlieg, *SDU Commentaar Arbeidsrecht* (digital), SDU, Auteurswet artikel 7.
A.A. Quaedvlieg, *SDU Commentaar Arbeidsrecht* (digital), SDU, Rijksoctrooiwet 1995 artikel 12.
Wet Auteurscontractenrecht, Kamerstukken 33 308.

Table of Legislation

The Dutch Copyright Act.
The Neighbouring Rights Act.
The Database Act.
The Chips Act.
The Patent Act 1995.
The European Patent Convention.
The Convention on Intellectual Property.
The Seeds and Planting Materials Act.
Directive 2009/24/EC on the legal protection of computer programs.

Table of Cases

Perquin/Perfra, Supreme Court (30 June 1950), NJ 1952, 35.
Hupkens/Van Ginneken, Supreme Court (27 May 1994), NJ 1994, 136.
TNO/Ter Meulen, Supreme Court (1 March 2002), ECLI:NL:HR:2002:AD7342.
Electrolux, Benelux Court of Justice (22 June 2007), A 2006/3/13.
Donorkind Amsterdam/Donorkind Groningen, Zie District Court (KG) Groningen (28 June 2010), IEPT 20100628.
EuroPsyche/Fides, District Court Rotterdam 21 December 2011, IEPT 20120508.
Schending geheimhoudingslicht, District Court Utrecht (16 May 2012), IEPT 20120516.
S&S/Esschert, Supreme Court (25 October 2013), ECLI:NL:HR:2013:1036.

CHAPTER 16

Poland

Kacper Szkalej[*]

1 BACKGROUND

Polish law provides general rules according to which rights to employees' intellectual property generated in the fulfilment of the employment contract belong to the employer. The scope of rights acquired by the employer, the time in which the transfer of rights occurs or other rules governing ownership of employees' generated intellectual property may differ, however, with respect to the type of intellectual property in question, specific employment contract provisions agreed by the parties, or specific provisions provided by legal instruments.

The following sections describe general as well as specific legal rules governing the rights to specific intellectual property objects generated by employees in the fulfilment of their obligations arising from the labour contract. Most of these rules are *ius dispositivum*, however some of them generate obligations of the *ius cogens* character.

Among specific legal rules which must be taken into consideration in the context of employees' intellectual property, are rules governing intellectual property generated by research employees employed at higher education institutions (*uczelnie wyższe*), the Polish Academy of Sciences (*Polska Akademia Nauk, hereinafter PAN*), the National Centre for Research and Development, and research and development institutes (*instytuty badawczo-rozwojowe*). Amendments to a number of legal instruments governing higher education institutions and research institutes, which entered into force on 1 October 2014,[1] brought significant change to the manner in which universities, polytechnics, and other higher education institutions or research

* This chapter is based upon work prepared by Mateusz M. Wozniak.

1. Act of 11 July 2014 amending the Act – Law on Higher Education, and certain other Acts, OJ 2014 pos. 1198.

institutes manage or dispose of rights to intellectual property and other research and development results generated by research staff in the fulfilment of their employment contracts.

The amendments essentially oblige the above-mentioned institutions to adapt their internal regulations governing the management, disposal or commercialisation of research and development results, and regulations governing the use of the institutions' research infrastructure; to adapt the institutions' infrastructure management and transfer rights to results (TTOs and academic entrepreneurship incubators); to introduce provisions providing basic rules on permitted ways of commercialisation, and provisions providing so-called limited 'affranchisement'[2] of research empoloyees (*uwłaszczenie pracowników naukowych*).[3]

These rules have further been completed by a number of amendments in 2016.[4]

According to the framework the higher education institutions, as well as PAN research institutes, will have three months to inform the research employee whether (or not) the institution will commercialise the research or development results generated by the employee.[5] If the institution decides not to commercialise the R&D results or if the time limit expires, it is obliged to submit to the employee within 30 days an offer to acquire from the institution the rights to such R&D the employee created at the institution.[6] The provisions also limit the maximum price for which the research employee will be able to purchase the rights to the R&D results from the institution to not more than 5% of the average wage for work for the previous year as published by the Central Statistical Office (GUS).[7] Furthermore the Act also defines a minimal contribution percentage of the research employee or institution of the profits obtained from the R&D results' which are the subject of commercialisation.[8]

2. The limited affranchisement of research employees is limited only to innovations, designs, utility models, topography of semiconductor chips and new plant varieties, as well as to results of development work.
3. According to the government's propositions, all results generated by the research employee will become their property and the higher education institution should be remunerated for providing the research infrastructure and services in the form of a contribution percentage of the profits obtained by the research employee form the commercialisation of the results. Such solution was broadly criticised by higher education institutions and research institutes during social consultations and parliamentary sessions.
4. Act of 4 November 2016 amending certain Acts defining the conditions for conducting commercial innovate activities, OJ 2016 pos. 1933 (*hereinafter* 2016 Amendment Act).
5. Act of 27 July 2005 – Law on Higher Education, Art. 86e(1) (*hereinafter* Higher Education Act) and Act of 30 April 2010 on the Polish Academy of Sciences (*hereinafter* Polish Academy of Sciences Act), Art. 94c(1) respectively. The wording of both provisions was amended by the 2016 Amendment Act, Arts. 6(3)(a) and 8(1)(a) respectively.
6. Higher Education Act, Art. 86e(2) and Polish Academy of Sciences Act, Art. 94c(2) respectively.
7. *Id.*
8. In case of commercialisation of the R&D results by the higher education or research institution, the employee is entitled to not less than 50% of the value of funds gained by the higher education or research institution, reduced by not more than 25% of direct costs associated with commercialisation incurred by the higher education or research institution (Art. 86f(1) Higher Education Act). In case of commercialisation of the R&D results by the employee, the higher education or research institution is entitled to not less than 25% of the value of funds gained by the employee, reduced by not more than 25% of direct costs associated with commercialisation incurred by the employee (Art. 86f(2) Higher Education Act).

The amendments are, however, not without criticism as many issues are still ambiguous, particularly the scope of certain definitions or even rights and obligations of the parties taking part in the process of generating and commercialising R&D results.[9]

It is also worth mentioning that Polish law does not really oblige in any direct way an employee to inform their employer about the creation of intellectual property (with the exception of employees in higher education). The duty to inform the employer must be contained in the internal regulations or individual employment contracts.

2 EMPLOYEES' COPYRIGHT

2.1 The Legal Framework

Provisions concerning the employer-employee relationship with regard to copyright are contained in the Act of 4 February 1994 on Copyright and Neighbouring Rights (*hereinafter* Copyright Act).[10] General provisions concerning the relationship between employers and employees – in particular rights and duties of the parties to the employment contract – are provided in the Act of 26 June 1974 the Labour Code.[11]

The Polish Copyright Act establishes rules for the transfer and ownership of rights to works, and to objects protected by neighbouring rights created by the employee in the performance of the employment agreement. The Copyright Act differentiates between three types of works to which the specific employer-employee rules are applied:

- employee's work (*utwór pracowniczy*) – the work created by the employee in the performance of their obligations arising from the employment relationship (Articles 12-13 of the Copyright Act),
- scientific work (*utwór naukowy*) – the work created by the employee of a scientific institution (*instytucja naukowa*) in the performance of the obligations arising from the employment relationship (Article 14 of the Copyright Act),
- computer program (*program komputerowy*) – the work, which is a computer program, created by the employee in the performance of the obligations arising from the employment relationship (Article 74(3) of the Copyright Act).[12]

9. For example, the Act does not provide any sanctions for higher education or research institutions for actually not commercialising the R&D results, despite declaring such an intention, the result of which appears to be the prevention of the research employee from purchasing the rights to the generated results.
10. OJ 1994, no 24, pos. 83 as amended.
11. OJ 1974, No. 24, pos. 141 as amended.
12. The Copyright Act does not provide any definition of 'computer program', however Art.74(1) provides that a computer program is subject to protection granted to literary works, 'if provisions of Chapter 7 of the Copyright Act do not state otherwise'.

As with neighbouring rights, the Copyright Act distinguishes artistic performances.[13] The remaining objects protected by neighbouring rights, i.e. phonograms, videograms, broadcasts (*nadanie programu*), publication or other lawful public communication of previously unpublished works (*pierwsze wydanie*) or critical and scientific publication (*wydanie krytyczne i naukowe*), are not specifically regulated in the context of the employer-employee relationship.

2.2 Ownership of Rights

The general rule is that by reason of the existence of an employment contract, the employer becomes the owner of the economic rights to a work created by the employee when the work is accepted by the employer (*przyjęcie utworu*) in fulfilment of the duties arising from the employment relationship.[14]

This general rule may be changed in almost every scope by the parties by way of a clause in the employment contract. It is therefore up to the parties to decide whether they want to regulate ownership under general rules arising from the Copyright Act, or under rules formed as a result of negotiation.

2.2.1 Employees' Works and Scientific Works

According to Article 12(1) Copyright Act the employer, whose employee created the work in the performance of their obligations arising from the employment relationship (*utwór pracowniczy*), acquires, at the time the work is accepted by the employer, the economic rights to such work, to the extent resulting from the purpose of the individual employment agreement and the concerted intention of the parties. The general rule is applied automatically, unless the individual employment contract or the provisions of the law state otherwise. What is more, the ownership of the object on which the work is embedded or recorded - e.g., paper, CD, USB pendrive[15] - at the time the work is accepted by the employer, is transferred to the employer (Article 12(3)). Article 12 does not however apply if the author is the sole shareholder and board member (director) of a one person company with which they do not have an employment contract *specifically* envisaging creative work.[16]

The employer's acquired ownership of the economic rights to the employee's work may be limited in time, unless the parties agree otherwise in the individual employment contract. According to Article 12(2) Copyright Act, if the employer does not commence the dissemination (exploitation) of the work within a period of two

13. Art. 92 of the Copyright Act states that, among other provisions of the Act, Art. 12 governing rules on employee's works are applied to employee-employer relationships where the object is an artistic performance.
14. Copyright Act, Art. 12(1).
15. If the employee does not want to transfer to the employer ownership of an object on which the work is recorded, e.g. personal CD, USB pendrive, the parties shall agree on that as otherwise the general rules on copyright will apply.
16. Supreme Court judgment of 25 March 2011 IV CSK 504/10 (graphic design of sonographic instrument case).

years after accepting the employee's work (where the work was intended for dissemination by virtue of the employment contract) the author of the work is entitled to impose a specific deadline by which the work shall be disseminated. Failing such dissemination by the imposed deadline, the economic rights, including the object on which the work is recorded, will revert back to the author (employee). The parties may however determine otherwise, including a different time limit than that provided in Article12(2).

By virtue of Article 13 Copyright Act, if the employer does not notify the employee within a period of six months after the delivery of the work of the non-acceptance of the work, or of alterations that should be made to the work, the work shall be deemed accepted without any reservation. The parties to the employment contract may determine a different period of acceptance notification.

In relation to scientific works, unless the employment contract states otherwise, a scientific institution is entitled to a priority right to publish the scientific work (*utwór naukowy*) of an employee who created the work in performance of their employment duties (Article 14(1) Copyright Act). The right to priority expires if, within a period of six months counted form the date of the delivery of the work the scientific institution does not conclude a publishing contract with the author (employee) of the work, or, within a period of two years from the date of the acceptance of the work, the work is not published. What is more, according to Article 14(2) Copyright Act, the scientific institution is entitled to use the scientific material included in the scientific work and make this work accessible to third parties, if it results from the agreed purpose of the work or was agreed in the employment contract.

The above-mentioned rules state clearly that the ownership of all the economic rights[17] to employees' works (*utwór pracowniczy*) or scientific works (*utwór naukowy*) are transferred from the employee to the employer as a result of acceptance of such work by the employer (unless the employment contract states otherwise). At the time of creating the work, the sole and primary owner of the economic rights is its author, and the employer – in the above-mentioned circumstances – becomes the derivative (or secondary) owner of such rights. The transfer of rights is 'automatic' and based on the provisions of the law; hence, specific provisions need not be contained in the employment contract.

Moral rights are not transferable and belong to the author of the work. However these general principles do not apply to works which are computer programs.[18]

17. The Copyright Act (Art. 17) grants the author the sole right to use and to dispose the work in every field of exploitation (*pola eksploatacji*) and the sole right to remuneration for the use of the work by third parties, with exceptions or limitations explicitly specified by the Act (not only the Copyright Act but any legal Act in Polish law).
18. Provisions referring to computer programs in the Polish Copyright Act are listed in Ch. 7 of the Act.

2.2.2 Research Results Financed by the National Centre for Research and Development

Rights to research results financed by the National Center for Research and Development (*Narodowe Centrum Badań i Rozwoju*) are governed by separate legislation.

In particular, by virtue of Art. 32(3) National Centre for Research and Development Act[19], economic rights subsisting in results[20] originating from research and development projects carried out 'for the benefit of national security or the protection of public order' financed by the Centre belongs to the State Treasury (which may be represented by the entity indicated in the financing agreement).

2.2.3 Computer Programs

Article 74(1) of the Copyright Act states that computer programs are protected by the Copyright Act in the same way as literary works, unless the provisions of Chapter 7 of the Act state otherwise. One such provision - Article 74(3) - establishes specific rules for ownership of the economic rights to computer programs, whereby such rights to computer programs created by the employee in the fulfilment of the obligations arising from the employment relationship belong to the employer, unless the employment contract states otherwise. This means that the employer is the sole and primary owner of all economic rights to the computer program from the outset. There is no transfer of economic rights from the employee to the employer. If the employee - the author of the computer program - wishes to become the owner (sole or co-owner) of such rights, the employment contract must provide this explicitly.

Moral rights to the computer program, since they cannot be transferred as such, belong to the employee. However the scope of moral rights is somewhat modified, compared to other works. In particular, the moral right to consent to any alterations of the work[21] is transferred to the scope of economic rights. According to the disposition of Article 74(4) of the Copyright Act, the economic rights to the computer program, save for the provisions of Article 75(2) and (3) of the Act, entail:

(1) 'permanent or temporary reproduction of the computer program in whole or in part by any means and in any form; where it is necessary to reproduce a computer program for its loading, displaying, running, transmitting and storing, consent of the rightholder shall be required for such acts;
(2) translation, adaptation, arrangement or any other modification of the computer program, without prejudice to the rights of the person that made such modifications; and

19. Act of 30 April 2010 on the National Centre for Research and Development, OJ 2010 nr 96 pos. 616. The wording of the provision was inserted by Art. 7(1) 2016 Amendment Act.
20. And rights to inventions, utility models and industrial designs.
21. Art. 16(3) Copyright Act provides for the inviolability of the contents and form of the work (as a moral right).

(3) dissemination of the computer program or a copy thereof, including lending *[użyczenie]* and leasing *[najem]*.'[22]

2.3 The Parties' Duty to Inform

The Copyright Act does not contain any explicit rules on a duty to inform of the creation of a copyright-protected work. Such duties may arise however from the employment regulations or employment contract.

However, the employer has an implied duty to accept the work created by the employee. Conversely, therefore, the employee must submit the work for acceptance. As mentioned in section 2.2.1 above, by virtue of Article 13 Copyright Act, such acceptance is deemed to have occurred where the employer does not notify the employee, within a period of six months after the delivery of the work, of the non-acceptance of the work, or of alterations that should be made to the work. The parties to the individual employment contract may determine a different period of acceptance notification. The importance of this duty is attached to the moment from which the employer acquire the economic rights to a work created by the employee (save for computer programs).

In relation to research employees employed by higher education or research institutions, although the Copyright Act establishes an implied duty, the question whether the rules on commercialisation of research results introduced by the 2014 amendment to the Higher Education Act (which does contain such a duty in Article 86e) is left unanswered by the Act. Moreover, whereas Article 86e Higher Education Act requires an employee to inform the employer (higher education institution) about the 'research results, development results as well as know-how related to these results', the position is exacerbated by the wording in Article 86d, by virtue of which the results of research which become industrial property or results from development work, created by an employee of a public higher education institution in the execution of their duties 'are governed by Arts 86e-86h'. The basis for the ambiguity is whether *governed by Article 86e* is meant to reduce the applicability of that provision to Article 86d, or merely ensure that the different forms of (industrial) property fall within the ambit of Article 86e, non-exclusively.

Seeing in particular that the Act in general, and Article 86e specifically, does not preclude the incorporation of works protected by copyright into its ambit, that know-how (whose expression might be protected by copyright) is expressly included, and that the Act does not as such define *research results*, it would appear that Article 86e might very well apply to works protected by copyright. Additionally, Article 86i

22. Authors' translation. Art. 74(4) of the Copyright Act provides an exhaustive catalogue of the economic rights to computer programs. Implementing any other kind of economic rights or field of explcitation, especially fields of exploitation listed in Art. 50 of the Copyright Act, in a contract concerning a computer program will be ineffective.

expressly makes it clear that matters which are left outside the scope of the Act are governed by *inter alia* the Copyright Act, thereby making the Higher Education Act take precedence over the overlapping issues.

2.4 Employee Remuneration Right

By virtue of Article 17 of the Copyright Act, if the Act does not provide otherwise, the author of the work is solely entitled to remuneration for the use of the work by third parties. However, the legal situation for the author-employee has to be analysed with specific legal circumstances in mind. First of all, as discussed above, the creation of works by employees is accompanied by the transfer of economic rights to the employer on the basis of the employment contract. The employee is, generally speaking, remunerated by the employer for their work, which can, and often does, take the form of the creation of works protected by copyright.

According to Article 43 of the Copyright Act, concluding a royalty-free transfer of rights or granting a royalty-free licence must be explicitly expressed in the contract. Otherwise, the author is entitled to remuneration. If the contract does not specify the amount of the author's remuneration, such amount is defined to reflect the scope of the concluded transfer or granted licence and benefits[23] arising from the use of the work. The position of the author is also strengthened by the rule specified in Article 44 Copyright Act, which entitles the author to claim an appropriate increase of the remuneration by the court, in case of gross disproportion between remuneration received and the benefits obtained by the acquirer of the rights or the licence.

If the contract does not determine otherwise, the author is entitled to separate remuneration for using the work in each separate field of exploitation *(pole eksploatacji)*.[24] If the author's remuneration is contingent upon the amount of income received from the use of the work, the author has the right to obtain information and access to documentation necessary to determine the amount of this remuneration.[25]

If the author's remuneration is defined as a percentage from the sale/licence price of the copies of the work, and the price is increased, the author is entitled to be remunerated an agreed percentage of the increased price. Unilateral price reduction before the end of one year from the start of the dissemination of the work does not affect the amount of remuneration. However the parties may extend this term.[26]

In practice most employment contracts concluded in Poland with authors-employees do not separate different types of remuneration included in the lump-sum salary (e.g. remuneration for labour and remuneration for the transfer of copyright). A well-defined employment contract – in the interest of both parties – should provide such information in its provisions, so that in case of any dispute the value of the object

23. Art. 43 Copyright Act is using the term 'benefit' *(korzysc)* and not 'profit' *(zysk)*. However in most cases, benefits obtained from using the work have a pecuniary character and can therefore include profits.
24. Copyright Act, Art. 45.
25. *Ibid.*, Art. 47.
26. *Ibid.*, Art. 48.

is already defined; this will most likely avoid the prolongation of proceedings and save unnecessary costs. Of course in some cases determining a fixed amount of remuneration for the transfer of rights can be impeded by the character of the workplace, the frequency of the creation of works or other factors attributable to the employment relationship.

Another common phenomenon is the failure to include in employment contracts any direct provisions referring to copyright or neighbouring rights. Most of the parties in employment contracts leave their rights and obligations to be determined by the general rules of the Copyright Act or other acts in Polish law, despite the fact that the parties may agree differently than provided by law in specified circumstances. Such situations arise mostly because of the very low awareness in matters concerning intellectual property and intellectual property law.

Also a common problem concerns both parties of the employment contract making incorrect assumptions that the employer always becomes the owner of copyright on every work created by the employee. Parties often forget that the employer is entitled, by virtue of the employment contract, only to those works which were generated by the employee in the fulfilment of the obligations arising from the (individual) employment relationship. If the employee creates works beyond the confines of their contractual obligations or regular work place, they are entitled to additional remuneration for transferring their rights to the employer.[27] The transfer of such rights must always be confirmed in a direct manner and in written form, otherwise it will be considered null and void.

The employee rules on remuneration concerning transfer or commercialisation of R&D results created by research employees employed at higher education or research institutions are discussed above in section 1, however regard ought to be had to the caveat made in section 2.3 above. Furthermore by virtue of Article 19(1) Regulation of the Minister of Science and Higher Education of 2 December 2016 concerning the conditions for remunerating work and provision of benefits relating to work for employees of public higher education institutions, OJ 2016 pos. 2063, as a result of 'increasing the duties of, or entrusting additional tasks to, the employee, or because of the *specific character of the work* or the conditions for its fulfillment [emphasis added]' the employee may be awarded special benefits. Such special benefit cannot however normally be higher than 80% of the base salary (Article 19(3)).

3 EMPLOYEES' SEMICONDUCTOR CHIPS

3.1 The Legal Framework

The topography of semiconductor chips (*topografia układu scalonego*) is protected by a registrable right (*prawo z rejestracji*) granted by the Patent Office of the Republic of Poland. The object of protection is the topography of the semiconductor chip and not

27. Such transfer shall also be accompanied by concluding a separate contract between the employer and employee.

the semiconductor chip itself. The registered topography right is regulated by Act of 30 June 2000 on Industrial Property Law ('Industrial Property Act').[28] The rules governing the employer-employee relationship in relation to the creation of intellectual property are common for all the industrial property rights. General provisions concerning the employer-employee relationship, in particular rights and duties of the parties to an employment contract, are provided in the Act of 26 June 1974, the Labour Code.[29]

By virtue of Article 196 Industrial Property Act:

(1) any solution consisting of a three-dimensional arrangement of the elements, however expressed, at least one of which is an active element, and of all or some interconnections in an integrated circuit, shall be considered as a topography of an integrated circuit;
(2) an integrated circuit shall mean any three-dimensional product having one or more layers, composed of elements of semiconducting material forming a continuous layer and of conducting interconnections and insulating spaces, inseparably interconnected, intended to perform electronic functions.

The Patent Office grants legal protection for topographies of semiconductor chips, if the topography is original[30] and has not been openly used for commercial purposes for more than two years before the application to the Patent Office,[31] nor has been created and recorded in any form 15 years before such application.[32] The rights will however not be granted, if the topography results from the function of the integrated circuit in which it is applied.[33]

3.2 Ownership of Rights

By virtue of Article 200 in connection with Article 11(3) Industrial Property Act, unless the employment agreement states otherwise, the person or entity entitled to obtain the registered topography right is the employer if the topography of the semiconductor chip is created by an employee in the fulfilment of their obligations arising from the employment relationship. If the employment agreement explicitly provides that the employee shall remain entitled to the right, any transfer of such right or the transfer of the topography for the purpose of exploitation from the employee to the employer (or other person or entity) needs to be in writing, otherwise it will be considered null and void. The Industrial Property Law Act enables the parties to conclude a gratuitous transfer.

28. OJ 2001, No. 49, pos. 508 as amended.
29. OJ 1974 No. 24, pos. 141 as amended.
30. Industrial Property Act, Art. 197(1). By virtue of Art. 198(1), a topography is deemed original if its creation is the result of the creator's intellectual work and the topography is not generally known at the moment of creation.
31. *Ibid.*, Art. 197(3).
32. *Ibid.*, Art. 197(4).
33. *Ibid.*, Art. 199.

When, by virtue of the employment contract, the employee (creator) is the sole owner of the topography, in case of transfer of rights by the employee to an entrepreneur *(przedsiębiorca)* for the purpose of exploitation, when the topography is submitted to the entrepreneur, the transfer of rights is concluded when the topography is accepted for use by the entrepreneur and if the creator of the topography is notified about the acceptance of the topography within one month (unless the parties have determined another period of time in the written transfer agreement).[34]

The registered topography right granted under the provisions of the Industrial Property Act provides for the exclusive use of the registered topography for commercial or professional purposes, on the whole territory of the Republic of Poland.[35] The right protects the holder from unlawful reproduction of the whole or part of the protected topography, except for reproduction of the non-original parts of the topography (i.e. such that do not satisfy the originality requirements), from unlawful import, sale or other introduction on the market of a copy of the protected topography, integrated circuits manufactured with the use of such copy or of the product embodying such integrated circuits. 'Reproduction' of the topography is interpreted as its reproduction in an integrated circuit on the basis of the pattern, documentation or analysis. [36]

The protection of the registered topography lapses after ten years from the end of the calendar year in which the topography or the semiconductor chip containing such topography was introduced to trading, or from the end of the calendar year in which the topography was registered at the Patent Office, whichever comes earlier. The right also expires after 15 years from the time when the topography is made and recorded, if this time lapses earlier than the period of protection counted from the registration of the topography, if the topography has not been exploited for commercial purposes during that time. [37]

3.3 The Parties' Duty to Inform

The Industrial Property Act does not contain any explicit rules on a duty to inform of the creation of the topography.

In relation to research employees employed by higher education or research institutions refer to section 1 above.

3.4 Employee Remuneration Right

Under the Industrial Property Act remuneration model the creator of the topography receives a right of remuneration for the use of the topography by an entrepreneur *(przedsiębiorca*, any person, legal person or entity which pursues business activities).[38] In the context of the employee-employer relationship, unless the employment

34. *Ibid.*, Art. 200 in connection with Arts 12, 20 and 21.
35. *Ibid.*, Art. 211.
36. *Ibid.*, Art. 212.
37. *Ibid*, Arts 220 and 221(2).
38. *Ibid.*, Art. 201 together with Arts 22-23.

agreement provisions state explicitly that the salary for fulfilling the employment obligations includes remuneration for the employer using a topography created by the employee, the employee is entitled to obtain additional remuneration besides the salary.

The amount of the additional remuneration can be established by the parties either in the employment agreement or any other additional agreement. If the parties do not establish such amount, the Industrial Property Act provides a framework in Arts 22(2)-(4) for the establishment of the amount of such additional remuneration. The sum is established in fair proportion to the benefits derived by the employer from the exploitation of the topography, and by taking into consideration circumstances in which the topography was created, in particular the scope of the employer's contribution/assistance in developing the topography and the scope of employment duties of the employee.

Unless the employment or other agreement states otherwise, remuneration is paid in full, not later than two months after the first benefit from the topography was obtained; or in part, not later than two months after the lapse of each year from obtaining such benefits, but no longer than for five years.[39] The above-mentioned remuneration should be increased if the benefits derived by the employer turn out to be significantly higher than the expected benefits that were used as a basis to calculate the paid out remuneration.[40]

In relation to research employees employed by higher education or research institutions refer to section 1 above.

Furthermore by virtue of Article 19(1) Regulation of the Minister of Science and Higher Education of 2 December 2016 concerning the conditions for remunerating work and provision of benefits relating to work for employees of public higher education institutions, OJ 2016 pos. 2063, as a result of 'increasing the duties of, or entrusting additional tasks to, the employee, or because of the *specific character of the work* or the conditions for its fulfillment [emphasis added]' the employee may be awarded special benefits. Such special benefit cannot however normally be higher than 80% of the base salary (Article 19(3)).

4 EMPLOYEES' INVENTIONS

4.1 The Legal Framework

In comparison to some jurisdictions, there is no special statute which specifically regulates employee's rights to inventions created in the course of their employment; rather these rules are contained in the Industrial Property Act and concern inventions

39. *Ibid.*, Art. 22(4). Provision inserted by Art.1(2)(b) of the Act of 24 July 2015 amending the Industrial Property Act and certain other Acts, OJ 2015 pos. 1266.
40. *Ibid.*, Art. 23.

which become the object of a patent application.[41] For this reason, any rights that an employer receives by virtue of the express provisions of the law concern the right to file a patent application (and by extension the rights attached thereto). There are no provisions in the Act that expressly regulate ownership of an invention which does not become the object of a patent application. The ownership of such an invention must be recognised under general provisions concerning the employer-employee relationship, in particular rights and duties of the parties to the employment contract, which are provided in the Labour Code.

Poland is a signatory of the European Patent Convention and for an invention to receive patent protection it must satisfy the requirements of novelty, inventive step and be capable of industrial application.[42]

4.2 Ownership of Rights

4.2.1 The Right to File a Patent Application and Transfer of Rights

The general rule for the ownership of a patent is that the creator of the invention is entitled to obtain a patent for the invention. However, one exception to this rule states that in case the invention is created in the fulfilment of obligations arising from employment relationship, the right to obtain the patent for such invention belongs to the employer, unless the parties to the employment agreement have agreed otherwise.[43] If the employment agreement explicitly determines that the employee remains entitled to obtain a patent for the invention, any transfer of such right or the transfer of the invention for the purpose of its exploitation by the employer (or any other entrepreneur) must be in writing, otherwise it will be considered null and void.[44]

The Industrial Property Act permits gratuitous transfer (Article 20). If the invention is transferred for the purpose of its exploitation to the employer (or any other entrepreneur), the rights to file a patent application becomes effective on the day the invention was reported to the employer (or any other entrepreneur) in writing if the following two conditions are satisfied:

- if the invention is accepted by the employer (or any other entrepreneur); and
- if the creator of the invention is notified about the acceptance within one month (unless the parties agree on a different period in the written transfer agreement).[45]

---

41. With the exception of rights to inventions whose creation results from research financed by the National Centre for Research and Development which is governed by separate legislation. See section 2.2.2 above.
42. *Ibid.*, Art. 24.
43. *Ibid.*, Arts 11(1) and (3).
44. *Ibid.*, Art. 12. The same rule is applied to the transfer of the already granted patent. The transfer of the patent (not the right to obtain the patent) is effective in relation to third parties at the moment of making the entry about this transfer to the patent registry provided by the Patent Office. *See* Arts 67(2)-(3) Industrial Property Act.
45. *Ibid.*, Arts 20-21.

This is also applicable to utility models and designs.

An employer, who is not the creator of the invention, filing a patent application at the Patent Office to obtain protection must specify the creator of the invention and provide the legal basis for the right to obtain the patent.[46]

4.2.2 Employer's Right to Exploit the Invention

By virtue of Article 11(5) of the Industrial Property Act where the invention is created with assistance *(pomoc)* provided by an entrepreneur *(przesiębiorca*, e.g. employer, under circumstances mentioned below), the entrepreneur has the right to exploit the invention for their own purposes. The parties to the contract for granting assistance to the creator may establish in the contract that the entrepreneur granting such assistance is entitled in whole or in part to the right to obtain a patent for the invention.[47]

The above-mentioned rule may be applied in the employer-employee relationship but outside the employment contract and its fulfilment, e.g., when the employee creates the invention outside working hours, outside their employment obligations, using the employer's materials, laboratories or funds to pay for testing the invention. In such circumstances - if the parties have not concluded any separate contract - the employer is entitled by law, in the form of a licence (irrespective of whether the employer knew or consented to such extra activity), to use the invention, but only for their own purposes. This must be interpreted in such a way that the employer is not entitled to disseminate or commercialise such invention without the employee's (inventor's) consent. What is more, the employee is fully entitled to the rights to the invention and to obtain the protection for such invention, if the parties did not agree otherwise. On many occasions the employer using the invention on the basis of the above-mentioned licence might also have difficulties to use it in full scope. According to the Industrial Property Act the employee-inventor is not obliged to transfer any know-how or trade secrets associated with the invention if the invention was created outside the employment relationship. In the absence of such information the employer (entrepreneur) may in some circumstances even be (technically) unable to exploit the invention.

4.3 The Parties' Duty to Inform

The Industrial Property Act does not state any explicit rules on duties to inform in the context of inventions made in the course of fulfilling obligations arising from the employment relationship. Such duties shall arise directly from the employment contract or employment regulations. However, as mentioned above, where rights are transferred to the employer (if, by operation of an agreement, they belong to the employee from the outset rather than the employer), the employer has an obligation to inform the employee of the acceptance of the invention (which is one of the conditions

46. *Ibid.*, Art. 32.
47. The rule is also applied for utility models and designs.

necessary for the receipt of the right to file a patent application). Likewise, the employee must 'inform' the employer of the invention, however as this procedure is attached to the discretionary transfer of the invention (Article 20 of the Act provides that the creator of an invention *may* transfer their rights), this will occur automatically when the employee proposes to transfer their rights.

For rules on the duty to inform about inventions being R&D results created by research employees employed at a higher education or research institution refer to section 1 above.

4.4 Employee Remuneration Right

The general rule of the Industrial Property Act, stated in Article 22(1), grants the creator of the invention (as well as the utility model or design) the right to obtain remuneration for exploiting the invention by an entrepreneur (e.g. employer on the basis of Art. 11(3) or (5)), both on the basis of an employment agreement and a transfer of the invention under Article 21.[48]

In the context of the employee-employer relationship, unless the employment agreement provisions state explicitly that the salary for fulfilling the obligations arising from the employment agreement includes remuneration for using the invention by the employer, the employee is entitled to obtain additional remuneration to the salary. The Supreme Court has however held that the right to receive such remuneration arises from the moment in time when the employer actually makes use of the invention and receives, as a result, specific benefits from such use,[49] rather than, e.g., from the moment of creation of a *patentable* invention by the employee or the granting of a patent.

The amount of the additional remuneration for the employee can be established by the parties either in the employment agreement or any other additional agreement. In case when parties do not establish such amount, the Industrial Property Act provides a framework for the establishment of the amount of such additional remuneration. The sum is established in fair proportion to the benefits derived by the employer from the exploitation of the invention, and by taking into consideration circumstances in which the invention was created, in particular the scope of the employer's contribution/ assistance in developing the invention and the scope of employment duties of the employee.

Unless the employment or other agreement states otherwise, remuneration is either paid in full, in which case payment is made within two months after the first benefits from the invention were obtained; or in part, in which case payment is made for the first part as above, and for the remaining parts within two months after the lapse of each year from obtaining such benefits, but no longer than for five years.[50] The

48. Art. 22(1) in conjunction with Arts 20-21 of the Industrial Property Law Act.
49. Supreme Court judgment of 8 July 2015 Case II PK 168/15 (applicability of transitional provisions re remuneration right), p.8. This has subsequently been confirmed in Supreme Court judgment of 11 August 2016 Case II PK 246/14.
50. *Ibid.*, Art. 22(4).

above-mentioned remuneration should be increased if the benefits derived by the employer turn out to be significantly higher than the expected benefits that were used as a basis to calculate the paid out remuneration.[51]

For rules on remuneration concerning transfer or commercialisation of inventions being R&D results created by research employees at a higher education or research institutions, refer to section 1 above.

Furthermore by virtue of Article 19(1) Regulation of the Minister of Science and Higher Education of 2 December 2016 concerning the conditions for remunerating work and provision of benefits relating to work for employees of public higher education institutions, OJ 2016 pos. 2063, as a result of 'increasing the duties of, or entrusting additional tasks to, the employee, or because of the *specific character of the work* or the conditions for its fulfillment [emphasis added]' the employee may be awarded special benefits. Such special benefit cannot however normally be higher than 80% of the base salary (Article 19(3)).

5 EMPLOYEES' UTILITY MODELS

5.1 The Legal Framework

Rules concerning ownership of rights to utility models created in employment relationships are provided by the Industrial Property Act.[52] General provisions concerning the employer-employee relationship, in particular rights and duties of the parties to the employment contract, are provided in the Labour Code.

According to the Industrial Property Act, a utility model *(wzór użytkowy)* is a new and useful solution of a technical character affecting the shape, construction or durable assembly of an object. The utility model is regarded as a useful solution, if by means of that solution a practical effect is attainable, expedient in the process of manufacturing or exploitation of a product.[53]

5.2 Ownership of Rights

The Patent Office grants protection rights for utility models. Such protection rights entitle exclusive use of the utility model for commercial or professional purposes, on the whole territory of the Republic of Poland. The subjective scope of the protection right is determined by the claims included in the description. The time limit for protection is ten years from the date of filing the application to the Patent Office.[54]

The provisions that apply to inventions also (expressly) apply to utility models. See section 4.2 above.

51. *Ibid.*, Art. 23.
52. With the exception of rights to utility models whose creation results from research financed by the National Centre for Research and Development which is governed by separate legislation. See section 2.2.2 above.
53. Art. 94 of the Industrial Property Law Act.
54. See Arts 95-96 of the Industrial Property Law Act.

5.3 The Parties' Duty to Inform

The Industrial Property Act does not contain any explicit rules on duties to inform in the context of utility models created in the course of fulfilling obligations arising from the employment relationship. Such duties shall arise directly from the employment contract or employment regulations.

For rules on the duty to inform about utility models being R&D results created by research employees employed at higher education or research institution, refer to section 1 above.

5.4 Employee Remuneration Right

The provisions that apply to inventions also (expressly) apply to utility models. See section 4.4 above.

6 EMPLOYEES' DESIGN RIGHTS

6.1 The Legal Framework

Rules concerning ownership rights to designs created by employees in the fulfilment of the employment agreement are regulated by either the Copyright Act or the Industrial Property Act.[55] General provisions concerning the employer-employee relationship, in particular rights and duties of the parties to the employment contract, are provided in the Labour Code.

The Copyright Act protects *inter alia* 'industrial design works' (*utwór wzornictwa przemysłowego*). If the design satisfies the statutory requirements for originality, the rules discussed in the context of copyright in section 2 above are applicable to such designs. Otherwise, the question of ownership will be governed by the Industrial Property Act, on the assumption that the design satisfies the relevant requirements for the subsistence of rights to a design, i.e. is an industrial design (*wzór przemysłowy*).[56]

6.2 Ownership of Rights

The ownership of rights to the design must be considered in the context of two legal frameworks – either the Copyright Act or the Industrial Property Act concerning registered designs.

55. With the exception of rights to designs whose creation results from research financed by the National Centre for Research and Development which is governed by separate legislation. See section 2.2.2 above.
56. To be an industrial design it must be new and possess an individual character (Art. 102 Industrial Property Act).

6.2.1 Copyright in Designs

See section 2 above for a discussion of ownership of rights to employees' designs created in the course of duty.

6.2.2 Industrial Property Rights in Designs

The Patent Office grants exclusive rights once the industrial design is registered, which allows the holder to exploit the design for commercial or professional purposes in the whole territory of the Republic of Poland.

The same provisions that apply to inventions also (expressly) apply to industrial designs, for which reason see section 4.2 above.

6.3 The Parties' Duty to Inform

The Industrial Property Act does not contain any explicit rules on duties to inform in the context of designs created in the course of fulfilling obligations arising from the employment relationship. Such duties shall arise directly from the employment contract or employment regulations.

For rules on the duty to inform about designs being R&D results created by research employees employed at a higher education or research institutions, refer to section 1.1 above.

6.4 Employee Remuneration Right

Depending on the type of design, different legal frameworks will be applicable. If the design falls under copyright protection, the framework established by the Copyright Act will be applicable, for which reason see section 2.4 above. If, on the other hand, the design falls under the scope of the Industrial Property Act, the same rules will apply as apply to inventions, for which reason see section 4.4 above.

7 EMPLOYEES' PLANT VARIETIES

7.1 The Legal Framework

Rules concerning ownership of breeders' rights to plant varieties bred by the employee in the fulfilment of the employment agreement are contained in the Act of 26 June 2003 on the Legal Protection of Plant Varieties ('Plant Varieties Act').[57] General provisions concerning the employer-employee relationship, in particular rights and duties of the parties to the employment contract, are provided in the Labour Code.

57. OJ 2003, no 137, pos. 1300 as amended.

According to Article 4(1) Plant Varieties Act, a breeders' right to a plant variety *(odmiana roslin)* relates to a variety which is distinct, uniform, stable and new, and its denomination complies with the requirements listed in Article 9(1) of the Act. The entity entitled to grant a breeders' right to a variety is the director of the Research Centre for Cultivar Testing *(Centralny Osrodek Badania Odmian Roslin Uprawnych,* COBORU).

7.2 Ownership of Rights

By virtue of Article 4(1) Plant Varieties Act, the person entitled to obtain the exclusive right is the breeder, who by virtue of Article 2(1) point 7 is defined as a person who:

(a) had bred or discovered and developed a variety; or
(b) is or was the employer of the person referred to in (a) or has concluded an agreement under which another party to the agreement has bred or discovered and developed a variety; or
(c) is successor in title of persons referred to in (a) and (b).

According to the rules provided by the Plant Varieties Act, when a plant variety is bred and developed by the employee in the course of fulfilling the obligations arising from the employment relationship, the sole right to obtain the exclusive right for such plant variety belongs to the employer, unless the parties have agreed otherwise.[58]

The employer and employee may freely determine their relationship regarding the right to obtain and hold the exclusive right, e.g., they may determine that the employee is solely, or together with the employer, entitled to obtain and/or use the exclusive right, or even transfer such rights to a third party. The Plant Varieties Act grants the parties to the employment agreement absolute freedom to determine ownership of rights.

There are also no obstacles to transferring the exclusive right from the employer to the employee, after obtaining such a right by the employer. Such agreement needs to be concluded in writing.[59]

Contrary to provisions concerning inventions and designs, under the rules of the Plant Varieties Act the employer (or any other person or entity) is not obliged to specify in the application the actual breeder of the plant variety.

7.3 The Parties' Duty to Inform

The Plant Varieties Act does not expressly contain any duty to inform either party of the breeding of a new plant variety, or of any benefits derived by the employer from

58. Plant Varieties Act, Art. 2(2) in conjunction with Art. 2(1) point 7.
59. *Ibid.*, Art. 4(5).

commercial use of the protected plant variety. For this reason, the employer and employee may freely establish such duties in the employment agreement or regulations.

For rules on the duty to inform about plant varieties being R&D results created by research employees employed at a higher education or research institution, refer to section 1 above.

7.4 Employee Remuneration Right

According to the disposition of Article 4a Plant Varieties Act, the employee, who actually bred or discovered and developed the plant variety in the course of fulfilling their obligations arising from the employment agreement has the right to remuneration for any commercial use of the variety, unless the employment agreement states otherwise. The Act does not provide any model for calculation of the level or amount of such remuneration, nor any rules concerning the timing of payments.

As this gives room for significant uncertainty, the parties ought to determine the specific and explicit rules concerning the payment of remuneration by the employer to the employee in the agreement, while having regard to provisions enabling the employee to access the employer's books concerning benefits derived from the commercial use of the plant variety.

For rules on remuneration concerning transfer or commercialisation of plant varieties being R&D results created by research employees employed at a higher education or research institution, refer to section 1 above.

Furthermore by virtue of Article 19(1) Regulation of the Minister of Science and Higher Education of 2 December 2016 concerning the conditions for remunerating work and provision of benefits relating to work for employees of public higher education institutions, OJ 2016 pos. 2063, as a result of 'increasing the duties of, or entrusting additional tasks to, the employee, or because of the *specific character of the work* or the conditions for its fulfillment [emphasis added]' the employee may be awarded special benefits. Such special benefit cannot however normally be higher than 80% of the base salary (Article 19(3)).

8 INSTANCES FOR DISPUTES

Disputes in the field of intellectual property in the context of the fulfilment of obligations arising from an employment relationship (civil law relationship) must be treated as cases falling under the scope of labour law disputes *(sprawy z zakresu prawa pracy)*.[60] Such disputes may be resolved amicably between the parties, in writing, or in civil proceedings before a competent court.

60. *See generally* Art. 476(1) of the Act of 17 November 1964 – Civil Procedure Code as amended and Arts 294-295 Industrial Property Act. The Supreme Court has indicated that disputes relating to additional remuneration provided for the use of an invention created by an employee fall under the scope of labour law disputes – Supreme Court Resolution of 16 March 2016 III PZP 8/15 (a Supreme Court Resolution such as this one is equivalent to a legal rule).

The court of first instance having jurisdiction over disputes based on claims in civil law arising from an employment relationship combined with intellectual property rights is the Labour Court *(sąd pracy)*, which is a separate organisational unit of the Regional Court *(sąd okręgowy)*.[61] The court of second instance for such disputes is the Appellate Court *(sąd apelacyjny)* territorially competent for the given courts of first instance. In events provided for by law, a dispute may further be appealed to the Supreme Court *(Sąd Najwyższy)*.

Disputes concerning labour law are brought before the court that has territorial jurisdiction over the place of the defendant's domicile, or before the court in the region of which the work is, was or was to be performed, or before the court in the region of which the seat of the employer is located.[62]

According to the rule of general territorial jurisdiction of the court, actions shall be brought before courts of first instance in the region of which the defendant has their residence (natural persons) or their seat (legal entities).[63] If the defendant who is a natural person has no permanent place of residence on the territory of the Republic of Poland, general jurisdiction is designated to the place of their temporary residence in the Republic of Poland, and if such place is unknown or is not on the territory of the Republic of Poland the defendant's last place of permanent place of residence in the Republic of Poland.[64] If the rules on general jurisdiction provide for the jurisdiction of several courts, the plaintiff has the right to bring the action to the court of their choice.[65] If on the basis of the circumstances of the case general rules of territorial jurisdiction of courts do not enable the establishment of a competent court, the competent court is determined by the Supreme Court.[66]

In some circumstances, when facts which give rise to the dispute concerning employees' intellectual property occur after the termination of the employment relationship, there might be a difficulty in establishing jurisdiction, i.e. deciding whether to initiate proceedings before the Labour Court or the District or Regional Court. In such cases the competent Court of First Instance will always be the Regional Court, which should *ex officio* decide whether it is competent for the action or not. The defendant has the right to bring objections to the established competence of the court.

9 CONCLUDING REMARKS

Awareness of legal rules in the field of employees' intellectual property in Poland is rather low, however it is continuously increasing. It should also be pointed out that such awareness is rather higher among businesses in the private sector, many of which are employers in the IP industry, than among employees. The level of awareness is also related to the size of a company – the smaller the company, the lower the awareness of

61. Art. 262 § 1 Labour Code together with Art. 17 point 2 Civil Procedure Code.
62. Civil Procedure Code, Art. 461(1).
63. *Ibid.*, Art. 27(1).
64. *Ibid.*, Art. 28.
65. *Ibid.*, Art. 43(1).
66. *Ibid.*, Art. 45.

these legal rules. In many cases employers believe that all intellectual property created or generated by employees belongs to the employer, and employees, on the other hand, believe that they can use generated or created intellectual property beyond the workplace.

Nonetheless, awareness of legal rules in this field is rising due to the rising number of research and development projects, mainly financed by public or EU funds, as well as because of broad public awareness campaigns on rules concerning intellectual property which were provided and conducted in Poland during the EU 2007-2013 financial framework.

Disputes arising in connection with employees' intellectual property are – because of this low public awareness – rather common, however most of them are settled in the form of mutual agreements outside of court. If the case is brought before a court, the express provisions are in most cases sufficient to issue a judgement. Problems in connection with employees' intellectual property really concern the establishment of the competent court, especially when exploitation or disposal of rights occurs after the termination of employment.

New rules on the management and disposal of rights related to R&D results created or generated by research employees, especially in franchise-related research co-operation, will require members participating in R&D projects, as well as members and parties of the R&D results commercialisation process, to adopt a new approach for preparing adequate rules and contracts for the management and disposal of rights. This will certainty contribute to increased awareness in connection with ownership, transfer and terms of use of intellectual property; both for parties representing the education and research sector, as well as for parties representing the business sector and implementing the R&D results. However, such phenomena will also be very positive for building higher awareness of the rules governing employees' and employers' intellectual property rights generally.

BIBLIOGRAPHY

Table of Legislation

Act of 17 November 1964 –Civil Procedure Code, OJ 1964 nr 43 pos. 296, as amended

Act of 26 June 1974 – the Labour Code, OJ 1974 nr 24 pos. 141 as amended

Act of 4 February 1994 on Copyright and Rights Neighbouring Copyright, OJ 1994 nr 24 pos. 83 as amended (Copyright Act)

Act of 30 June 2000 – Industrial Property Law, OJ 2001 nr 49 pos. 58, as amended (Industrial Property Act)

Act of 26 June 2003 on Legal Protection of Plant Varieties, OJ 2003 nr 137 pos. 1300, as amended (Plant Varieties Act)

Act of 27 July 2005 – Law on Higher Education, OJ 2005 nr 164 pos. 1365, as amended (Higher Education Act)

Act of 30 April 2010 on the National Centre for Research and Development, OJ 2010 nr 96 pos. 616
Act of 30 April 2010 on the Polish Academy of Sciences, OJ 2010 nr 96 pos. 619 (Polish Academy of Sciences Act)
Act of 11 July 2014 amending the Higher Education Act and certain other Acts, OJ 2014 pos. 1198
Act of 24 July 2015 amending the Industrial Property Act and certain other Acts, OJ 2015 pos. 1266
Supreme Court Resolution of 16 March 2016 No III PZP 8/15
Act of 4 November 2016 amending certain Acts defining the conditions for conducting commercial innovate activities, OJ 2016 pos. 1933 (2016 Amendment Act)
Regulation of the Minister of Science and Higher Education of 2 December 2016 concerning the conditions for remunerating work and provision of benefits relating to work for employees of public higher education institutions, OJ 2016 pos. 2063

Table of Cases

Supreme Court judgement of 25 March 2011 Case IV CSK 504/10 (graphic design of sonographic instrument case)
Supreme Court judgement of 8 July 2015 Case II PK 168/15 (applicability of transitional provisions re remuneration right)
Supreme Court judgement of 11 August 2016 Case II PK 246/14 (applicability of transitional provisions re remuneration right)

CHAPTER 17
Portugal

Lígia Gutierrez Setúbal

1 BACKGROUND

Portugal, as a civil law country, has its legal framework for intellectual property law established in various codes. Of course, court decisions and legal opinions, although not being binding, are always welcome and very much helpful to properly interpret the law. However, (and unfortunately), the specific theme of employees' rights to their intellectual property, either being inventions, copyright or design, is not highly developed either by the courts or authors in Portugal. Hence, at least for now, we may only resort to the law and what it establishes.

The legal framework below first of all emphasises the parties' freedom to agree about ownership and remuneration rights to the author/inventor which is, as a rule, considered in the majority of employment contracts that may involve the employee's inventive activity. If the rules are not established, then the law provides the solution.

The nature of the rights determines, of course, different approaches: whenever the parties have failed to agree upon rules concerning *copyright* ownership in the employment contract, the law establishes a presumption that the copyright belongs to the intellectual creator of the work; for *patent* ownership, the presumption goes in favour of the employer.

There are no rules without exceptions and the Portuguese law also establishes all the conditions and specificities for the applicability of the regime which, as will be shown, are of course different for each intellectual property right.

The *iter* starts with copyright.

2 EMPLOYEES' COPYRIGHT

2.1 The Legal Framework

The legal framework for copyright ownership is determined by Chapter II, Section II of the Portuguese Copyright and Neighbouring Rights Code ('CNC')[1].

As concerns the employee's rights to their intellectual creations, the most relevant provision is Article 14 CNC, which establishes the applicable rules concerning copyright ownership when the work was made in the course of an employment contract. Article 15 also plays a relevant role in this matter, as will be explained in the sections below.

It should be noted that even in the case that the employee's ownership is vested in the employer, the intellectual creator always keeps the moral rights to their work.[2]

2.2 Ownership of Rights

2.2.1 General Rules

The first rule on copyright ownership is provided by Article 11 of the CNC and states that copyright ownership is vested in the intellectual creator of the work, unless otherwise expressly stated. Article 12 then clarifies that copyright is recognised regardless of registration of the work, its deposit or any other formality.

Exceptions to the author's ownership of their works are determined by Article 14(1) CNC, which establishes that if the work is made within the course of an employment contract, copyright ownership is assessed in accordance with what was established in said contract. Hence, the parties may agree that the employer has *ab initio* ownership of the works created by the employee during the course of the employment contract.

Whenever the parties have failed to agree upon rules concerning copyright ownership in the employment contract, there is a legal presumption that the copyright belongs to the intellectual creator of the work (Article 14(2) CNC). It should be noted that this is a *iuris et de iure* presumption, which means that it cannot be rebutted by the employee. However, the following provision establishes that if the name of the intellectual creator of the work is not mentioned on the work itself, or in the place intended for this purpose, it is legally presumed that the copyright belongs to the employer/the legal person that commissioned the work for hire (Article 14(3) CNC).[3]

Article 15 CNC sets forth the limits for the use of copyrighted works. First, Article 15(1) CNC states that in the cases of Articles 13 and 14 CNC, when the copyright

1. The CNC was approved by Decree-Law No. 63/85, of 14 March 1985.
2. As in many other jurisdictions, Portugal also adopted a two-fold protection system, dividing economic rights (the patrimonial part) and moral rights (the personal part) of copyright.
3. In accordance with Oliveira Ascensão, in *Direito Civil. Direito de Autor e Direitos Conexos*, Coimbra, 1992, 143, this last presumption is more concrete than that of Art. 14(2) CNC. Hence, in case of doubt, this should prevail over the presumption of Art. 14(2) CNC.

belongs to the intellectual creator, the work can only be used for the purposes specified in the contract executed between the parties. Hence, even if the copyright ownership is of the employer, the latter cannot introduce changes to the work without specific and delimited consent of the creator (Article 15(2) CNC).

Finally, Article 15(3) CNC establishes that the creator of the work cannot make use of their work in a way that may jeopardise the purpose intended by the employer for such work. This is a negative delimitation of the powers of the author/intellectual creator of the work,[4] which goes along with the employee's responsibilities emerging from the employment contract, as established in Article 128(1)(b), (c), (f), (g), (e) and (h) of the Portuguese Labour Code ('LC').[5] [6]

2.2.2 Exceptions

2.2.2.1 Journalistic Works

The exception to the rule of Article 14(1) CNC above relates to journalistic works. In this particular case, the Portuguese legislator ruled that if the work is made during the course of the employment contract which entails the identification of its author, the copyright of the work belongs to the intellectual creator (Article 174(1) CNC).

In this vein, it was established that unless authorised by the company that owns the newspaper or similar publication, the author of the work can only separately publish their work three months after the first publication (Article 174(2) CNC).

Another special feature of journalistic works is mentioned in Article 174(4) CNC: if the work is neither signed nor contains any identification of its author, copyright will vest in the company that owns the newspaper or the publication in which said work is included. In this case, the author may only separately publish their work with the authorisation of the mentioned company.

2.2.2.2 Photographic Works

Despite not being specifically mentioned in Article 14 CNC, photographic works also have a specific regime enshrined in Article 165 CNC. Essentially, the author of a photographic work has the exclusive right to reproduce, broadcast and control the first sale of their work (Article 165(1) CNC) [with the restrictions concerning exhibitions, reproduction of portraits and without prejudice to the copyright of the work that is being reproduced in what concerns photography of plastic arts works]. In accordance with Article 165(2) CNC, if the photograph has been taken within the scope of an employment contract, it is presumed that the copyright belongs to the employer.

4. *Ibid.*
5. The LC was approved by Law No. 7/2009, of 12 February.
6. Pedro Romano Martinez, *Direito do Trabalho*, 5th Edition, Almedina, 2010, 552-558.

2.2.3 *Computer Programs*

The legal framework for employees' copyright to computer programs is determined by Article 3 of the Decree-law that establishes the Legal Protection of Computer Programs ('PCP') which transposes into the Portuguese framework the Council Directive of 14 May 1991 on the legal protection of computer programs (91/250/EEC) [7] and provides that the authorship and ownership of computer programs are governed by the provisions of the CNC (Article 3(1) PCP). Hence, the parties may agree that the employer has *ab initio* the right to become the copyright owner of the works created by the employee during the course of the employment contract. If the computer program has been made in accordance with specific instructions given by the one who benefits from the activity, the rights belong to such person/company, unless otherwise established or it arises from the purpose of the contract (Article 3(3) PCP). It shall be noted that this may only occur if those instructions are part of the employer's management powers and are related to the employee's activity or, *in limine*, if such instructions may be included in the *ius variandi* of the employer.[8]

Moreover, Article 3(2) contains a presumption that a computer program developed in a company is a collective work.

2.3 The Parties' Duty to Inform

There are no provisions regarding this issue in the CNC. However, the duty to inform may be provided for by the parties in the employment contract.

2.4 Employee Remuneration Right

Article 14(4) CNC states that even when the ownership of the economic rights belongs to the employer, the intellectual creator is still entitled to a special remuneration:

(a) when the intellectual creation clearly exceeds the task assigned to its creator;
(b) when the work is used for additional purposes (or additional advantages are created) neither included nor envisaged in the compensation remuneration agreed with the employer.

This provision clearly states that the special remuneration has to be paid *in addition* to an adequate remuneration for the benefit of the intellectual creator of the work.

Article 165(3) CNC concerning *photographic works* includes a specific rule concerning the author's remuneration. It establishes that the one who reproduces the photograph for commercial purposes must pay the author an equitable remuneration.

7. Decree-Law No. 252/94, of 20 October 1994.
8. Pedro Romano Martinez, in *Direito do Trabalho*, 5th Edition, Almedina, 2010, 552-558, and Lopes Rocha, *Contratos de Licença de Utilização e Contratos de Encomenda de 'Software', Num Mundo Novo do Direito de Autor*, T. II. Lisboa, 1994, 695.

Moreover, Article 3(4) PCP rules that the special remuneration mentioned in Article 14(4) is also applicable for *computer programs*.

3 EMPLOYEES' SEMICONDUCTOR CHIPS

3.1 The Legal Framework

The legal framework for employees' semiconductor chips is determined by Articles 156, 157 and 158 of the Portuguese Industrial Property Code ('IPC')[9] which essentially provide that topographies of semiconductor chips created by an employee within the execution of an employment contract shall be governed by Articles 58, 59 and 60 IPC, which is the legal regime established for inventions and is further discussed below.

3.2 Ownership of Rights

For further information please see section 4.2 below.

3.3 The Parties' Duty to Inform

For further information please see section 4.3 below.

3.4 Employee Remuneration Right

For further information please see section 4.4 below.

4 EMPLOYEES' INVENTIONS

4.1 The Legal Framework

Portuguese employment contracts include, in some circumstances, rules concerning the employee's inventive activity and the ownership of the rights arising from those inventions. The relevant legal framework is provided by the IPC, notably Articles 58 and 59, which respectively establish general and special rules on patent ownership. Article 60 IPC also plays an important role in this matter, as will be explained in the sections below.

The general rule on patent ownership provided by Article 58(1) IPC is that the right to obtain a patent is originally vested in the inventor or their successors. Article 58(2) of the same provision then clarifies the rules concerning co-ownership stating that if there are two or more inventors, anyone of them has the right to file for a patent application for the benefit of all. Article 59 IPC establishes special rules concerning ownership of inventions, notably where the inventor is an employee and their

9. The IPC was approved by Decree-Law No. 36/2003 of 5 March.

employer claims the rights to the invention based on the employment contract executed between the parties. As will be further explained, in this case the right to the patent will, in principle, be vested in the employer and the inventor is entitled to a monetary remuneration in return. However, it should be highlighted that, *unless otherwise agreed* between the parties, paragraphs (1), (2), (4) and (5) of Article 59 IPC mentioned below and the remarks that will be made concerning said provisions also apply, with the necessary adaptations, *to inventions for hire*_(Article 59(7) IPC).

Moreover, unless otherwise set forth by means of a special law that overrules the IPC, all the content of Article 59 IPC and respective observations below are also applicable to the *Portuguese State and public entities* and to their relationships with civil servants and service providers working for the state at any title (Article 59(8) IPC). Due to this provision, Portuguese universities draft their own Regulations for the Protection of Intellectual Property in which ownership and exploitation regime of the so-called *university inventions* is defined. Those regulations must be interpreted in accordance with Article 59(8) IPC.[10]

4.2 Ownership of Rights

The Portuguese Labour Code states that the employer must provide the employee with mandatory information regarding, for instance, the activity rendered by the employee, their remuneration and their working hours. However, the rules referring to the ownership of rights of employee's inventions are not among such information, which means that the employer is not obliged to provide the employee with said information.

In light of the above and regarding the ownership of rights, the employer and the employee are free to determine whether the rules concerning such ownership shall or shall not be included in the content of the employment contract. The IPC has specific rules in this regard and distinguishes two different situations: (i) the employment contract includes a clause regarding the employee's inventive activity;[11] and (ii) the employment contract lacks a provision regarding the employee's inventions, but the invention is linked to the employer's activity.

In accordance with Article 59(1) IPC provided (a) that the employment contract covers the 'inventive activity' and (b) the invention is made in the course of said employment relationship the employer is the owner of the right to the patent of the employee's invention. In this case, if the employment contract does not provide for a specific remuneration due to the inventor for their inventive activity, the IPC surpasses the absence of said provision stating that they are entitled to a remuneration determined in accordance with the importance of the invention (Article 59(2) IPC).

To sum up, it is generally understood that the employer has the right to obtain the patent *ab initio* for that invention and become its owner.

10. António Campinos and Luís Couto Gonçalves in *Código da Propriedade Industrial Anotado*, Almedina, 2010, 222.
11. By 'covering inventive activity' should be understood an employment contract that generally covers the employee's ability to achieve an invention not a specific provision concerning the ownership of the invention.

Irrespective of the conditions set forth in Article 59(1), when the employment contract is silent regarding the employee's inventive activity, but the invention is linked to the employer's activity, the law refers to a type of pre-emptive right to the patent or a right to choose whether the employer wishes to obtain the patent or not.

Article 59(3)(a) IPC provides that the employer has the right to choose between:

(a) obtaining the patent and becoming the patent owner; or
(b) keeping the right to exclusively exploit the invention; or
(c) acquiring the patent; or
(d) having the ability to apply for or to acquire a foreign patent.

In the first case, i.e., *obtaining the patent and becoming the patent owner* the law uses the same criteria established in Article 59(2) IPC and provides expressly that the employer shall pay the inventor a monetary remuneration according to the importance of the invention. In the remaining cases, the wording of the law is quite ambiguous. In fact, the provision only refers to a monetary remuneration according to the importance of the invention when the employer chooses to obtain the patent and become its owner, but no mention is made regarding the remaining cases. However, it is believed that when the employer decides to keep the right to exclusively exploit the invention, the referred provision should be interpreted as to establish such compensation to the employee in this case as well.

When the employer decides *to keep the right to exclusively exploit the invention*, it is understood that the parties may also execute an exclusive licensing agreement, within which the employee pays the inventor (and patent holder) a price (as a rule, royalties) for the exploitation of the invention. Once again, the law is not clear and to date there is no case law providing guidance in this respect.

As will be further explained in section 4.3, the inventor shall inform the employer of their invention within three months after the invention is achieved (Article 59(3)(b) IPC). This notification then triggers a three-month term within which the employee can exercise the option to apply for the patent or claim the right to its exclusive exploitation (Article 59(3)(e) IPC). If the employer does not comply with this deadline, the company loses its pre-emptive right and the inventor will own all the rights to their invention (Article 59(4) IPC).

Another interesting feature of the Portuguese system regarding employees' rights to their intellectual property relates to the cases when employees change their workplace (Article 59(5) IPC).

If a patent application is filed by the inventor (or even by the new company where they work) within the first year after having changed workplace, it is legally presumed that the right to the patent belongs to the first employer. This is, however, a *iuris tantum* presumption, which means that it may be rebutted by the employee or by the new employer if they provide evidence that the invention was made afterwards and has nothing to do with the inventor's prior work for the first employer.[12] Moreover, the

12. António Campinos and Luís Couto Gonçalves, *Código da Propriedade Industrial Anotado*, Almedina, 2010, 221.

statute renders void and unenforceable any provision to the extent where the employee waives their rights in advance (Article 59(9) IPC).

Insofar as ownership is concerned, a final remark to highlight the content of Article 60 IPC should be made. If the patent is not requested in the name of the inventor, they have the right to be mentioned as inventor both in the patent application and in the title of the patent, unless the inventor requests, in writing, not to be mentioned as such in any publications resulting from the application (Article 60(1) and (2) IPC).

4.3 The Parties' Duty to Inform

As mentioned, in accordance with Article 59(3)(b) IPC, the inventor shall inform the employer of their invention within three months after the date on which the invention is deemed to have been completed. If during such period the inventor/employee has applied for a patent, the deadline to inform the employer about the invention and the patent application is one month as of the filing of the patent application before the Portuguese Industrial Property Institute (*Instituto Nacional da Propriedade Industrial*, 'INPI'[13]) (Article 59(3)(c) IPC).

Although the provision simply mentions *patent applications* and narrows the duty to inform to those filed before the INPI, it is believed that this understanding applies also to European patent applications filed before the European Patent Office in accordance with the rules established in the European Patent Convention ('EPC')[14] and also to international patent applications filed directly with the World Intellectual Property Organization ('WIPO') within the terms of the Patent Cooperation Treaty ('PCT').[15] Failing to comply with these obligations may result in the inventor's civil and labour liability (Article 59(3)(d) IPC).

4.4 Employee Remuneration Right

With regard to the remuneration right, three situations where the inventor/employee shall be paid a monetary remuneration for their invention must be distinguished:

13. The Institute website is http://www.marcasepatentes.pt/ and is available both in the Portuguese and English language.
14. Convention on the Grant of European Patents), of 5 October 1973, as revised by the Act revising Art. 63 EPC, of 17 December 1991, and the Act revising the EPC, of 29 November 2000, available at http://www.epo.org/law-practice/legal-texts/html/epc/2013/e/ma1.html
15. Patent Cooperation Treaty, done at Washington on 19 June 1970, amended on 28 September 1979, modified on 3 February 1984, available at http://www.wipo.int/pct/en/texts/Articles/atoc.htm.

4.4.1 *Employment Contract With a Clause Establishing the Employee's Monetary Remuneration for Their Inventive Activity*

When the parties envisage that the company's activity involves the likelihood the employee will come up with an invention during the course of the employment contract, they usually draft clauses in the employment contract to reflect that possibility.

The parties can therefore agree on a monetary remuneration for the inventor's inventive activity during the course of the employment contract *in exchange for the ownership of their inventions*. The remuneration is the result of the parties' free will to agree, in advance, the terms of their relationship, which means the inventive activity that is set out and rewarded from the outset by the contract itself irrespective of the importance of the invention (Article 59(1) IPC and Article 59(2) *a contrario*).

The downside of the agreement on the employee's remuneration for their inventions on the employment contract is that the parties cannot foresee what the invention will be or its economic value. Therefore, the remuneration agreed between the parties may be considered, subsequently, to be inequitable, either because the remuneration agreed was overstated (damaging the employer's interests) or underestimated (damaging the employee-inventor's interests).[16] However, this is a risk that both parties must agree in advance to assume.

4.4.2 *Employment Contract Without a Clause Establishing the Employee's Monetary Remuneration for Their Inventive Activity*

In this second case, the parties have also executed an employment contract covering the employees' inventive activity in the course of the employment relationship, but the contract does not provide for a specific remuneration to be paid to the inventor for their inventions.

As mentioned above, the IPC surpasses the contract's omission by stating that the employee is entitled to a compensation determined in accordance with the importance of the invention (Article 59(2) IPC). The difficulty of this criterion is to determine how to measure and calculate the *importance of the invention*, as neither the law provides guidance nor is there available case law clarifying how to determine the amount of such compensation.

As already mentioned above, it is believed that the importance of the invention is related to its economic value, which can only be determined after launching the invention on the market and subsequently assessing its success. This way, the foregoing risks of an inequitable remuneration may also occur in this case.

Unlike the case where the remuneration is agreed before the invention is made, Article 59(6) IPC states that in case the parties do not reach an agreement regarding the

16. Id. António Campinos and Luís Couto Gonçalves, *Código da Propriedade Industrial Anotado*, Almedina, 2010, 218-219. According to these Professors, if the invention was not 'tested' on the market the evaluation should be based, at least, considering its potential market power.

amount of the remuneration due to the employee, the dispute shall be solved by arbitration proceedings (mandatory arbitration).

4.4.3 *When the Employee's Inventive Activity Is Not Covered by the Employment Contract*

When the employment contract is absent of the employee's inventive activity, but the invention is linked to the employer's activity, the employee has the right to choose between the right to obtain the patent and become the patent owner, to keep the right to exclusively exploit the invention, to acquire the patent or to apply for, or acquire, a foreign patent (Article 59(3)(a) IPC).

As highlighted above, despite only referring to a monetary remuneration according to the importance of the invention when the employer chooses to obtain the patent and become its owner, it is commonly accepted that this remuneration is also due when the employer decides to keep the right to exclusively exploit the invention.

The law then provides that the employer shall choose this option within three months from the receipt of the notification of the invention by the inventor (Article 59(3)(e) IPC).[17]

The outcome established in Article 59(4) IPC is the following: if the employer fails to pay the due monetary remuneration within the three-month period established by law, it loses the right to the patent to the inventor/employee. In consequence, the inventor-employee also loses their right to demand the monetary remuneration from the employer.

Some authors also maintain that the consequences established in Article 59(4) IPC and the three-month deadline of Article 59(3)(e) shall also apply to the cases where the parties have executed an employment contract covering the employee's inventive activity in the course of the employment relationship, but the contract does not provide for a specific remuneration to be paid to the inventor for their invention, i.e. the cases of Article 59(2) above. [18] Although it seems to be the most reasonable approach, this opinion is not undisputed as the literal element of the mentioned provisions only specifically refers to Article 59(3)(a) and its particular conditions. Also in this case, Article 59(6) IPC establishes that any dispute concerning the amount of the employee's compensation will necessarily be solved by mandatory arbitration. However the statute has no provision about the cut-off date of the three-month period for payment and if this deadline should or should not be suspended until the arbitration proceedings solve the dispute.

17. *Ibid.*, 219.
18. *Ibid.*, 221.

5 EMPLOYEES' UTILITY MODELS

5.1 The Legal Framework

The legal framework for employees' utility models is determined by Articles 121, 122 and 123 IPC and provides that utility models created by an employee within the execution of an employment contract shall be governed by Articles 58, 59 and 60 IPC, i.e., the same legal regime concerning inventions.

5.2 Ownership of Rights

For further information please see section 4.2 above.

5.3 The Parties' Duty to Inform

For further information please see section 4.3 above.

5.4 Employee Remuneration Right

For further information please see section 4.4 above.

6 EMPLOYEES' DESIGN RIGHTS

6.1 The Legal Framework

The legal framework for employees' design rights is determined by Articles 181, 182 and 183 IPC and provides that designs created by an employee within the execution of an employment contract shall be governed by Articles 58, 59 and 60 IPC, i.e. the same legal regime concerning inventions.

It should be highlighted that Article 182 IPC, although referring to Article 59 IPC, also mentions that this provision should be applicable 'without prejudice of the provisions regarding copyright'.

A horizontal characteristic of these rights is the overlap of the applicable rules. Hence, if an employee creates an industrial design that entails an intellectual creation triggering copyright, the rules of the IPC and the CNC should be applied together without competing with each other.

6.2 Ownership of Rights

For further information please see sections 2.2 and 4.2. above.

6.3 The Parties' Duty to Inform

For further information please see sections 2.3 and 4.3 above.

6.4 Employee Remuneration Right

For further information please see sections 2.4 and 4.4 above.

7 EMPLOYEES' PLANT VARIETIES

7.1 The Legal Framework

The legal framework for plant varieties is determined by Decree-Law No. 213/90 of 28 June, which establishes the legal regime of breeders' rights to plant varieties, and by the Ministerial Order No. 940/90 of 4 October, approved by the Ministry of Agriculture, Fisheries and Food,[19] laying down the Regulation on the Protection of New Varieties of Plants ('Plant Varieties Regulation').[20]

The first aspect that should be emphasised is that the IPC specifically excludes plant varieties from patent protection (Article 53(3)(b) IPC).[21]

According to Article 5(2) Plant Varieties Regulation:

> For the purpose of this Regulation, 'variety' shall be taken to mean a plant grouping within a single botanical taxon of the lowest known rank, which grouping, irrespective of whether the conditions for the grant of a plant variety right are fully met, can be:
>
> - defined by the expression of the characteristics that results from a given genotype or combination of genotypes;
> - distinguished from any other plant grouping by the expression of at least one of the said characteristics; and
> - considered as a unit with regard to its suitability for being propagated unchanged.

Therefore, plants that are not stable cannot be considered as plant varieties and, thus, might be patentable. The rules for patentable inventions mentioned above shall apply in such case. For the remaining cases, the application for granting the right of the breeder of a plant variety and registration for a variety denomination are made with the Directorate General of Food and Veterinary – Seeds and Varieties Division[22] which is controlled by the Ministry of Agriculture, Forests and Rural Development.

---

19. Currently Ministry of Agriculture, Forests and Rural Development.
20. Council Regulation (EC) No. 2100/94, of 27 July 1994 on Community plant variety rights.
21. Art. 53 IPC reflects Art. 53 of the CPE and 27 of The Agreement on Trade Related Aspects of Intellectual Property Rights ('TRIPS').
22. The law originally created a National Centre for the Register of Protected Varieties.

The law does not provide for any provision regarding the employee's rights to plants varieties and there is no available case law regarding ownership, remuneration or the employee's duty to inform in case of dispute or in the absence of specific clauses in the employment contract.

7.2 Ownership of Rights

In accordance with Article 8 Plant Varieties Regulation, the breeder's right to a plant variety may be required by its breeder, their successor in case of death or a third party, who is entitled to make this request by means of a contract (which naturally includes employment contracts), provided that the applicant is:

(a) of Portuguese nationality;
(b) a foreign citizen living in Portugal;
(c) a legal person with registered office in Portugal;
(d) a national of a Member State of the International Union for the Protection of New Varieties of Plants ('UPOV') any natural or legal person living or having its registered office in one of these states, provided that the state concerned grants protection to the genus or species to which the variety subject of the application belongs; or
(e) a national of another Member State of the UPOV or a natural or legal person living or having its registered office in the state in which Portuguese citizens, foreign citizens living in Portugal and legal persons with registered office in Portugal, enjoy the same protection as granted to their nationals with regard to the varieties of the genus or species applied for.

Moreover, Article 24 of the Plant Varieties Regulation states that breeders' rights may also be transferred by succession or by contract. As there is no exclusionary provision, it is believed that the reference made to 'contracts' shall be broadly understood, thus, including employment contracts. As the law is completely silent concerning the ownership of the breeder's right to the plant varieties, this may be included in the employment contract, in accordance with the will of the parties.

Problems may arise whenever the employment contract fails to provide a solution concerning the ownership of a new plant variety. As mentioned above, there is still no available case law determining what the applicable criteria should be.

7.3 The Parties' Duty to Inform

There are no provisions regarding the parties' duty to inform in Decree-Law No. 213/90 of 28 June nor in the Plant Varieties' Regulation. The remarks regarding the definition of these aspects in the employment contract and respective issues made above are also applicable in this case.

7.4 Employee Remuneration Right

There are no provisions regarding the employee remuneration right in Decree-Law No. 213/90 of 28 June nor in the Plant Varieties' Regulation. The remarks regarding the definition of these aspects in the employment contract and respective issues made above are also applicable in this case.

8 INSTANCES FOR DISPUTES

8.1 Arbitration

First, it should be emphasised that the Portuguese system tends to give preference to alternative dispute resolution systems to solve intellectual property disputes.

In fact, there are several cases in which the law expressly states that the parties should resort to Arbitral Courts to solve their disputes.[23]

8.1.1 Mandatory Arbitration

As mentioned above, Article 59(6) IPC establishes that any dispute concerning the amount of the employee's remuneration for their inventions will necessarily be solved by mandatory arbitration. The provision is also applicable to utility models, designs and semiconductor chips.

8.1.2 Voluntary Arbitration

Unless otherwise is set forth, parties may also decide and establish in the employment contract that any disputes concerning employee's industrial property should be solved by arbitration. These voluntary arbitration proceedings may take place at ARBITRARE, an institutionalised Arbitration Centre, competent to resolve disputes between private parties (and also between private parties and the competent registration bodies) regarding industrial property, domain names, trademarks and corporate names.[24]

ARBITRARE also allows mediation as an optional phase, in which the parties try to reach an amicable settlement with the intervention of a neutral and qualified third party (the mediator).

23. Despite falling outside the scope of this paper, it should be noted that the most paradigmatic case was the enactment of Law No. 62/2011 of 12 December, which established mandatory arbitration for the disputes that involve reference and generic medicines.
24. The ARBITRARE website is https://www.arbitrare.pt/.

8.2 Labour Courts

In accordance with the Portuguese Law on the Organisation of the Judiciary System ('LOJS'),[25] the jurisdiction of Labour Courts includes claims arising from employment relationships and claims arising from disputes between parties in an employment relationship, or between one of those parties and a third party, whenever arising from relationships linked to the employment relationship by a complementary or dependent bond, and such claim is cumulated with another one included in the jurisdiction of the Labour Courts (Article 126 LOJS).

Whenever the above-mentioned claims arise from a services agreement (and not from an employment relationship) the jurisdiction of Labour Courts is, as a rule, excluded.

Nonetheless Labour Courts have jurisdiction over claims which arise from employment relationships and are directly related to such relationships. For instance, according to Article 59(3) IPC, if an employee makes an invention that is included or linked to its activity and does not inform the employer of such invention within the legal periods of time, they shall be liable before the employer for not having complied with their labour duties; therefore, failure to comply with labour duties and its disciplinary consequences (namely termination for cause) have to be analysed and decided by Labour Courts.

8.3 Intellectual Property Court

The necessity of an Intellectual Property Court was only recently recognised in Portugal. In operation since 2012, the jurisdiction of the Intellectual Property Court includes claims where the cause of action arises from copyright and industrial property rights (Article 111 LOJS). Hence, it is understood that claims related to the ownership of those rights and the relevant remuneration, even though arising from employment relationships, should in principle be submitted to the Intellectual Property Court (instead of the Labour Courts). It should be noted that there is no available case law on this matter, so no relevant guideline may be drawn regarding this issue.

9 CONCLUDING REMARKS

The Portuguese system respects, beyond any specific regulation, the employer and the employee's free will to determine whether the rules concerning ownership and the remuneration due to the employee shall be included or not in the content of the employment contract. Those particular aspects have to be negotiated and agreed between the parties and the respective conclusions included in the employment contract. However, if the parties fail to agree, Portuguese law establishes a general principle of the employer's ownership for the Intellectual Property created by the

25. Law 62/2013 of 26 August 2013.

employee within the course of an employment contract, provided that an adequate remuneration (and, in certain cases, a special remuneration) is paid to the inventor/ intellectual creator of the work.

For dispute resolution, it is worth noting the absence of available case law concerning employee's intellectual property rights; thus, no relevant guideline may be drawn regarding some aspects of the law, which therefore remain unanswered.

BIBLIOGRAPHY

Table of Sources

Oliveira Ascensão, *Direito Civil. Direito de Autor e Direitos Conexos* (Coimbra 1992).

Pedro Romano Martinez, *Direito do Trabalho* (5th Edition, Almedina 2010).

Lopes Rocha, *Contratos de Licença de Utilização e Contratos de Encomenda de 'Software', Num Mundo Novo do Direito de Autor*, T. II. (Lisboa 1994).

António Campinos e Luís Couto Gonçalves, *Código da Propriedade Industrial Anotado* (Almedina 2010).

Table of Legislation

Copyright and Neighbouring Rights Code approved by Decree-Law no. 63/85, of 14 March.

Labour Code approved by Law no. 7/2009, of 12 February.

Legal Regime on the Protection of Computer Programs approved by Decree-Law no. 252/94, of 20 October.

Industrial Property Code approved by Decree-Law no. 36/2003, of 5 March.

Legal Regime of breeders' rights for plant varieties approved by Decree-Law no. 213/90, of 28 June.

Council Regulation (EC) no. 2100/94, of 27 July on Community plant variety rights approved by the Ministerial Order no. 940/90, of 4 October.

The Agreement on Trade Related Aspects of Intellectual Property Rights (TRIPS).

Law on the Organisation of the Judiciary System approved by Law no. 62/2013 of 26 August.

CHAPTER 18
Slovenia

Elizabeta Zirnstein & Špelca Mežnar

1 BACKGROUND

The Constitution of the Republic of Slovenia establishes in Article 60 the protection of rights concerning innovation activities, in which the protection of copyrights and other rights related to works of art, science, research and inventor's activities is provided. The constitutional provision is regulated more precisely in different legislative acts.

The Slovene legislation regarding creative works done in the course of employment is mostly non-mandatory and sets out only minimal standards of protection of employee and employer rights. The legal provisions governing ownership of IP rights in Slovenia are found in a number of separate statutes. The legal relationship between the employer and the employee regarding technical innovations is regulated in the Act on Employment Relationship Inventions. The Copyright and Related Rights Act contains provisions on the legal relationship between the employer and the employee regarding copyright, models and designs. Detailed regulations of topographies created by employees are contained in the Act on the Protection of the Topography of Semiconductor Circuits. The relationship between the employer and the employee regarding other innovations, e.g., technical improvements, suggestions and similar are not regulated by positive law, however it can be regulated contractually, in the form of collective agreements, employment agreements and other agreements.

2 EMPLOYEES' COPYRIGHT

2.1 The Legal Framework

Copyright works created in the course of employment are governed by the Copyright and Related Rights Act ('Copyright Act').[1] Articles 101 and 102 of the Copyright Act, which specifically regulate the question of exclusive assignment of the economic and other rights of authors to the employer, represent a general rule that applies to all employment relationships (i.e. both in the public and private sector). A different special provision is provided for certain categories of copyright-protected works, such as databases, collective works and computer programs created in the course of employment (Article 102 and Article 112 Copyright Act), and for performances in the course of employment (Article 126 Copyright Act).

2.2 Ownership of Rights

When a copyright work is created by an employee: (i) in the execution of their duties or (ii) following the instructions given by their employer (copyright work created in the course of employment), it shall be deemed that the economic rights and other rights of the author to such work are exclusively assigned to the employer for a period of ten years from the completion of the work, unless otherwise provided by contract (Article 101 Copyright Act). This means that the economic and other rights of the author to copyright work are *ex lege* transferred to the employer for at least the first ten years.[2] After ten years, copyright is automatically returned to its full original position,[3] which means that the transferred rights are reverted to the employee, while the employer can claim a new exclusive assignment of these rights, in exchange for adequate remuneration (Article 101/II of Copyright Act). Irrespective of the exclusive assignment of the author's economic and other rights to the employer, an employee retains the exclusive right to use a work, created in the course of employment, as part of their collected works (Article 102 Copyright Act).

The exclusive assignment of economic and other rights in a designated copyright work means that the employer may, in accordance with Article 74(2) of the Copyright Act, exclude the author (employee) as well as any other third person from the use of this work. The employee is therefore not entitled to use the work themself (with the exception of use in the context of collected works).[4]

1. *Zakon o avtorski in drugih pravicah ('ZASP')*, Official Gazette of Republic of Slovenia, No. 16/07, 68/08, 110/13.
2. Zirnstein, *Učinkovitost pravne ureditve imetništva pravic iz inovacij, ustvarjenih v delovnem razmerju*, III(2) LeXonomica 188 (2011).
3. Trampuž/Oman/Zupančič, *Zakon o avtorski in sorodnih pravicah ZASP s komentarjem*, 232 (1997) *('Trampuž ')*.
4. Zirnstein, *Pravna ureditev inovativnosti v delovnem razmerju - med pravičnostjo in učinkovitostjo*, 119 (2012) ('Zirnstein').

2.2.1 Non-obligatory Nature of the Statutory Provision

The statutory exclusive assignment of the work for ten years is not mandatory and applies only in the absence of an agreement between the parties providing otherwise.[5] The author and their employer are entitled to a different regulation of the ownership of economic rights. They may decide on a different period or the extent of the transfer, etc. Usually, such agreement is included in the contract of employment, but may also be governed in a special contract. The parties enjoy complete autonomy with regard to the content of such contracts, with the exception of moral rights, which always remain with the author (employee).[6] The legal regime is thus only applicable if the parties do not regulate these issues themselves.

2.2.2 Exceptions

As already stated in section 2.1 above, the Slovenian Copyright Act provides different, or more specific, regulation for certain categories of copyright-protected works. In the case of databases, collective works and computer programs that are created in the course of employment, the economic rights and other rights of the author are assigned exclusively and without limitations to the employer, unless otherwise provided by contract (Article 102 Copyright Act).

Thus in the case of a computer program created by an employee, the employer obtains more rights than they would for other categories of works under Article 101 of the Copyright Act. Unless the parties have agreed otherwise, the employer is entitled to exclusive commercial exploitation for the entire duration of the economic rights (70 years p.m.a., rather than merely ten years after the completion of the work), while the employee (the author of a computer program) has no right to adequate remuneration after the expiration of the first ten years.[7]

An exception is also provided for performances in the course of employment, for which the Copyright Act in Article 126 stipulates that the relationships with respect to such performance are governed by a collective agreement or other contract.

2.3 The Parties' Duty to Inform

The assignment of economic and other copyrights occurs automatically (*ex lege*). Neither party has a duty to inform the other party.

5. Decisions of the Supreme Court reference No. I Up 211/2002(5 July 2005) and I Up 584/2002 (7 June 2005).
6. Zirnstein, p. 119.
7. Zirnstein, p. 120.

2.4 Employee Remuneration Right

If the economic and other rights are assigned to the employer in accordance with Article 101 of the Copyright Act, the author does not have a right to additional remuneration. The underlying reason is that the employee is already rewarded for their work with a salary and other benefits provided by the employment contract.[8] The Copyright Act does not envisage any special monetary remuneration for the author (apart from the regular salary), unless the parties agree differently. Because of the non-obligatory nature of Article 101 of the Copyright Act, the parties certainly have the right to conclude an agreement providing otherwise regarding the employee's remuneration right.

2.4.1 Compulsory Licence

Since there is a risk that the employee would enforce a new exclusive assignment of rights to the employer with disproportionate conditions after the expiry of the ten-year period, the Copyright Act establishes a compulsory licence scheme in favour of the employer. The compulsory licence gives the employer a right to demand the assignment of rights from the author after the expiry of the first ten-year period, but only in exchange for adequate remuneration.[9] The amount (or method of calculation) of remuneration is not provided in the Copyright Act. Remuneration is therefore determined primarily by the agreement between the parties, taking into account the usual fees for a particular category of works, the scope and duration of exploitation, and other circumstances (Article 81(1) Copyright Act). Where the profit derived from the exploitation of the work is manifestly disproportionate to the agreed amount or determined remuneration, the author may demand that the contract be revised so that a more equitable share of the revenues is provided for them (Article 81(2) Copyright Act). This right cannot be waived.

3 EMPLOYEES' SEMICONDUCTOR CHIPS

3.1 The Legal Framework

The ownership of rights to semiconductor chips (topographies) created in the course of employment is regulated by the Protection of Topographies of Integrated Circuits Act.[10] Slovenian legislation follows to some extent Article 3 of Directive 87/54/EEC of 16 December 1986 on the Legal Protection of Topographies of Semiconductor Products.[11]

8. Zirnstein, p. 118.
9. Trampuž, p. 232.
10. *Zakon o varstvu topografije polprevodniških vezij ('ZVTPPV')*, Official Gazette of Republic of Slovenia, No. 21/95, 96/02, 7/03, 60/06, 81/06.
11. Official Journal L 024, 27/01/1987, P. 0036 - 0040.

3.2 Ownership of Rights

The right to topographies of semiconductor products created in the course of employment is granted to the creator's employer, unless the terms of employment provide otherwise (Article 3(2) Protection of Topographies of Integrated Circuits Act). The employer's right commences with the date of filing the application for registration of the topography or with the date when the respective topography was first commercially exploited anywhere in the world (Article 7(2) Protection of Topographies of Integrated Circuits Act). The exclusive rights will cease after ten years. The ten-year period is calculated from the end of the calendar year in which the application for registration was filed or from the end of the calendar year in which the topography was first commercially exploited (Article 7(3) Protection of Topographies of Integrated Circuits Act).

3.3 The Parties' Duty to Inform

The Protection of Topographies of Integrated Circuits Act regulates employees' semiconductor chips very basically merely in two articles. Apart from Article 3(2), where the rights are granted to the employer (originally), Article 17(2) requires logical (reasonable) application of the provisions of the law governing the acquisition and protection of intellectual property rights (in compliance with the Industrial Property Act). In the Industrial Property Act the duty to inform is not specifically regulated.

3.4 Employee Remuneration Right

The employer has the *original right* to protection of topographies created in the employment relationship. In contrast to the provisions on employee inventions, the employee who created a topography of a semiconductor chip does not have a remuneration right.[12]

4 EMPLOYEES' INVENTIONS

4.1 The Legal Framework

The first records of legislation regarding innovation in Slovenia date back to 1922, when the Act on Protection of Industrial Property came into force. The Act provided that the owner of all the rights to the invention is the employee, unless otherwise stipulated by the employment contract or by specific work regulations. After the Second World War that principle changed. In line with the socialist principles of the new Yugoslav Republic, the results of innovation processes belonged to society and

12. In addition to this, the Protection of Topographies of Integrated Circuits Act does not provide for the employee a right to be mentioned as the creator (author) of the topography.

innovation is regarded as a common patrimony.[13] The patent system was preserved only to a limited extent and mainly because of commercial activities with foreign countries. The state had a pre-emption right for all innovation.[14] Besides that, the patent proceedings outside Yugoslavia were not possible without the approval of the competent national authority (usually the Ministry for Foreign Affairs). Twenty years later new legislation in this field was adopted, in line with the socioeconomic and political changes that occurred after 1960. The Act on Patents and Technical Improvements[15] had a whole chapter dedicated to employees' inventions. The conflict between labour law principles and principles of intellectual property law was solved in favour of employers, but employees gained the right to a fair remuneration.

Currently employees' inventions are regulated by the Employees' Inventions Act, in force since 1995.[16] In the beginning the Act regulated the ownership of all (employees') industrial property rights and was originally titled the Act on Employees' Industrial Property Rights.[17] The official title and its contents were changed in 2006, when the legislator decided to only regulate employees' inventions. Technical improvements, designs and other results of employees' creative endeavour were left out of the scope of the Act.[18]

The Act covers employees and employers in the private and public sector. In defining whether an innovation was created as a result of the employment relationship, the decisive factor is the definition of employment relationship as it is understood in labour law.

The Act is only applicable to inventions which are eligible for patent or utility model protection (Article 2). Proposals for technical improvement, registered designs, topographies, plant varieties, etc., do not fall under the Employees' Inventions Act.

4.2 Ownership of Rights

Inventions that are made (whether in a company or other organisation) by an employee during the period of employment are subject to the Employees' Inventions Act. The Act distinguishes two types of employees' inventions: 'service inventions'[19] (bound, job-related, non-free inventions) and 'free inventions'[20] (not job-related, independent

13. The Act on Innovation and Technical Improvements was enacted on 1 December 1948 by the new Yugoslav socialist government; the ownership of rights was regulated in Art. 12.
14. The Act on Innovation and Technical Improvements, Art. 19.
15. *Zakon o patentih in tehničnih izboljšavah,* Official Gazette of Republic of Slovenia, No. 44/1960.
16. *Zakon o izumih iz delovnega razmerja ('ZIDR'),* Official Gazette of Republic of Slovenia, No. 45/1995, 96/2002, 139/2006.
17. *Zakon o pravicah industrijske lastnine iz delovnega razmerja ('ZPILDR').*
18. The reason for that was the opinion of the legislator that it is not appropriate to regulate so different results of employees creative endeavour (innovations, new designs, technical improvements etc.) on the same principles. Only employees' inventions, as the most important result of innovative thinking, should be regulated by the law (explanatory text to the Proposal on Employees' Innovations Act, published in the Reporter of the Slovenian Parliament, No. 120/2006, p. 1).
19. 'Službeni izumi'.
20. 'Prosti izumi'.

inventions).[21] Service inventions are those which have been made during the period of employment and may be: (i) *direct service inventions* which have arisen from the obligatory task or activity of the employee or at the employer's explicit request, or on the basis of a special contract concluded between the employer and the employee, or (ii) *indirect service inventions* which are substantially based on the employees' experiences gained at their workplace, or result from the assets made available to them by the employer. All other inventions made in the course of employment are free inventions.

The Act recognises the employee as the inventor and therefore as the original holder of the rights to the invention.[22] This does not, however, mean that the employer is without rights. The rights arising from employees' inventions are transferred to the employer under specific circumstances by a one-sided declaration. Those circumstances relate to the question of whether the invention is a service invention or a free invention. For service inventions, the employer has the possibility of transferring rights, which is discussed further below.

The legislative framework discussed in this section is applicable to employees' inventions in the private and public sector. However there are special rules regarding inventions made by employees (academic and research staff) at public universities and public research institutions (Article 21 of the Employees' Inventions Act).

Inventions made by university staff at public universities, as well as by researchers in public research institutions, can be claimed by the university (the employer) if:

(i) the university/research institution provides the organisational infrastructure required for an efficient exploitation of such inventions;
(ii) the university/research institution governs the rules of procedure regarding the claiming of such inventions in the manner suited to the needs of scientific research work and publication of scientific achievements;
(iii) the university/research institution determines in the rules of procedure the shares which, in the case of the exploitation of the invention, are appointed to the university/institution, to the unit of the university/institution employing the inventor, and to the inventor, whereby the inventor's share shall not be less than 20% of the gross royalty acquired by the institution from the exploitation of the invention. If the mentioned conditions are not fulfilled, the inventions created by academic and research staff at public universities/ research institutions are deemed to be free inventions.

4.2.1 Reporting of Inventions

The Employees' Inventions Act imposes an obligation on employees to inform the employer without delay about the created invention (Article 5 Employees' Inventions Act). This is further discussed in section 4.3 below. On the basis of such notification

21. Employees' Inventions Act, Art. 3.
22. Mežnar, *Izumi v delovnem razmerju – med delovnim pravom in intelektualno lastnino*, 9 Delavci in delodajalci 52-53 (2009).

(the so-called *report of innovation*) the employer can decide whether they agree with the employee's categorisation of the invention (service invention/free invention) and whether they want to claim the right of a reported service invention. The most important legal consequence of the report of innovation is the initiation of a three-month period, within which the employer should claim their rights (Article 6(2) Employees' Inventions Act). Within this period the employee is not allowed to use the invention (Article 8(4) Employees' Inventions Act).

4.2.2 Claiming Rights

The employer can claim the right to a reported service invention in an *unrestricted* or *restricted* manner (Article 6(1) Employees' Inventions Act).With a *restricted claim* the employer obtains a non-exclusive right to use the invention. Employers will usually opt for *unrestricted claims*, as all rights to the service invention are transferred to the employer. As a consequence, the employer has the right to apply for a patent and gain all the rights to that patent (Article 8(2) Employees' Inventions Act). In both cases, the employer has, however, some duties to fulfil to the benefit of the employee. First, the employee has the right to be mentioned (named) as inventor; they also have the right to a compensation (section 4.4 below concerning the remuneration right). Secondly, the Employees' Inventions Act prescribes that the employer must without delay apply for a patent for a service invention which has been claimed unrestrictively (Article 10(1) Employees' Inventions Act).[23] He also has to bear all the costs deriving from that. If the employer fails to do so, the employee may themself file an application on behalf of and at the expense of the employer (Article 10(2) Employees' Inventions Act). The employer must submit to the employee a copy of the application and the attachments thereto simultaneously with the filing of the patent application. Moreover the employer must inform the employee of the course of the proceedings (Article 11(1) Employees' Inventions Act). If, during the application assessment process, the employer loses interest in gaining the patent, they have to offer it back to the employee (Article 10(2) Employees' Inventions Act). The Act also prescribes a duty to keep the invention secret, for both parties, as long as necessary (Article 7 Employees' Inventions Act). Thirdly, in the case of an unrestricted claim, it can happen that the utilisation of the service invention is rendered unjustly difficult by the employer's rights. In such case the employee may demand that the employer either claims the right to the service invention without restriction, or releases it to the employee (Article 8(3)/III Employees' Inventions Act).

23. The employer is not obligated to submit a patent application only if: (i) the employee agrees to non-submission of the patent application; (ii) if the service invention has become free; (iii) if the legitimate interests of the company require that a reported service invention should not be made known (Arts. 10 and 14 Employees' Inventions Act).

4.2.3 *Released Service Inventions*

If the employer does not claim the right to a service invention within three months from receipt of the proper innovation report, the service invention becomes free and the employee can use it freely. A service invention also becomes free in the case of employer's restricted claim, if the utilisation of that invention by the employee is rendered unjustly difficult by the employer's right of use and the employer does not issue an unrestricted claim within two months. The service invention also becomes free if the employer releases it in writing (Article 9 Employees' Inventions Act).

4.2.4 *Free Inventions*

The employee who has made a free invention during the period of employment must also notify the employer of that invention. The notification must be in writing and must contain sufficient information for the employer to be able to decide whether the invention is in fact free (Article 19(1) Employees' Inventions Act). The employer can agree to or challenge the status of the invention within two months from the receipt of the notification. If the employer does not dispute, within two months, that the invention is not free, the employer can no longer claim that the invention is a service invention (Article 19(3) Employees' Inventions Act). However, no obligation is imposed on the employee to notify the employer of the free invention when the invention is clearly not within the employer's field of activity (Article 19(5) Employees' Inventions Act).

In the case of a free invention, the employee must first offer it to the employer before starting to exploit it. The employer has a pre-emption right to licence a non-exclusive right for the use of the invention from the employee within two months from the receipt of the offer. If the employer wishes to accept the offer, but disagrees with the terms, they may bring proceedings before the Conciliation Board within two months of receipt of the offer (Article 19(4) Employees' Inventions Act).

4.2.5 *Insolvency Proceedings*

If, after an unrestricted claim to the employee's invention has been submitted, insolvency proceedings are instituted against the employer, the employee has a pre-emption right to the invention (Article 14(1) Employees' Inventions Act). If remuneration has not been paid out yet, the employee has a higher priority than other insolvency creditors, as the Act provides expressly that the amount of remuneration has to be regarded as a cost during insolvency proceedings (the regime of proportionate payment is not applied; the employee has the right to the whole amount of remuneration) (Article 14(2) Employees' Inventions Act).

4.3 The Parties' Duty to Inform

The Employees' Inventions Act imposes an obligation on the employee to inform the employer about the created invention without delay. Notification of the invention must be in writing (Article 5 Employees' Inventions Act), although electronic notification will also be considered appropriate (by virtue of Article 13 Electronic Commerce and Electronic Signature Act).[24] The employee must clearly indicate that the notification constitutes a *report of an invention* (Article 5(1) Employees' Inventions Act). The invention report should contain: (i) a detailed description of the invention, i.e. its technical functions and its solution, as well as the drawings necessary to make the invention understandable, (ii) the data on how the invention was made, and the statement whether the invention is a direct or an indirect service invention and (iii) data on the co-workers involved in the creation of the invention, with the description of their contributions (Article 5(3) Employees' Inventions Act). If the notification of an invention does not contain all the aforementioned information it will still be considered correct if the employer does not declare within one month that the report requires complementation (Article 5(4) Employees' Inventions Act). However, the employer has a duty to confirm, in writing, the receipt of the employee's notification of the service invention (Article 5(2) Employees' Inventions Act).

The employer must also, within three months, notify the employee whether they agree with the classification of the invention (Article 6(2) Employees' Inventions Act). Within this period the employee is not allowed to use the invention.

4.4 Employee Remuneration Right

To compensate the employee for the transfer of rights to the employer, the employee has right to a reasonable remuneration (Article 15 and 16 Employees' Inventions Act). The employee's remuneration claim arises as soon as they receive the employer's unrestricted claim (Article 15(1) Employees' Inventions Act) or, in the case of a restricted claim, as soon as the employer starts exploiting the invention (Article 16(1) Employees' Inventions Act). Decisive factors for calculating the amount of remuneration are mainly: (i) the economic exploitability of the service invention, (ii) the duties and position of the employee in the company[25] and (iii) the involvement of the company in the creation of the service invention (Article 15(2) Employees' Inventions Act). The type and concrete amount of remuneration should be established by agreement between the employer and the employee within three months from the employer's unrestricted or restricted claim (Article 18(1) Employees' Inventions Act). If several employees were involved in the service invention, remuneration must be established for each of them separately (Article 18(2) Employees' Inventions Act). The

24. In Slovenian: *Zakon o elektronskem poslovanju in elektronskem podpisu ('ZEPEP')*. Official Gazette of Republic of Slovenia, No. 57/2000.
25. On the basis of this factor, the amount of remuneration will be higher in the case of indirect service inventions.

employer must notify all the involved employees of the total amount of remuneration and the shares that go to each individual inventor (Article 18(2) Employees' Inventions Act).

If no agreement concerning remuneration is reached between the employer and the employee within the three months after the employer has claimed the right to a service invention, then the employer must fix the remuneration with a substantiated written declaration (Article 18(3) Employees' Inventions Act). The employer has to pay the amount of remuneration in accordance with that declaration within three months from the grant of the patent or, in the case of a restricted claim, within three months from the day when the employer started exploiting the invention (Article 18(3) Employees' Inventions Act). The employee has the right to contest the employer's decision regarding the amount of remuneration within one month of receipt of that decision. The dispute is heard by the Conciliation Board. If the employee does not contest it, then the remuneration is considered to be determined by agreement (Article 18(3) Employees' Inventions Act). When several employees are involved in the service invention, every one of them has the right to contest the amount of the remuneration (Article 18(5) Employees' Inventions Act). If the remuneration is contested or otherwise disputed, the employer and the employee must, before going to court, attempt to reach an amicable agreement before the Conciliation Board (Article 19(3) Employees' Inventions Act).

If the circumstances under which the remuneration was determined change significantly, then both parties can require each other to revise the remuneration. Nevertheless, the employer cannot demand the return of remuneration already paid out (Article 18(6) Employees' Inventions Act).

For the concrete assessment of the amount of remuneration, a special Regulation has been issued.[26] The Regulation prescribes the minimum amount of remuneration, which is the amount of at least two last salaries of the employee (for direct service invention), or at least five last salaries of the employee (for indirect service invention; Article 8 of the Regulation). The remuneration for indirect service inventions is normally a lump sum. For direct service inventions, the amount of remuneration is calculated on the basis of the value of the invention, the value of the employee's duties, and the employers' contribution to the invention (Article 4 of the Regulation). If the amount of remuneration cannot be calculated on this basis, then a licence analogy should be used. In any case, the employer and the employee can always agree to determine the amount of the remuneration on some other basis (Article 5 of the Regulation). The guidelines for the determination of the amount of remuneration are used only in cases when they are more favourable for the employee in comparison to the employment agreement or collective agreement (Article 17(2) Employees' Inventions Act).

26. Regulation regarding Remuneration for Employees Inventions (*Pravilnik o nagradah za izume iz delovnega razmerja*, Official Gazette of Republic of Slovenia, No. 65/2007).

4.4.1 *Arbitration*

As mentioned before, in the case of any disagreement regarding the amount of the employees' remuneration, the employer and the employee must, before going to court, go through an arbitration process before the Conciliation Board (Article 19(3) Employees' Inventions Act). The Conciliation Board is not a court, however its decisions, if an agreement is reached, are binding on the parties.

The composition of the Conciliation Board is regulated in Article 26 of the Employees' Inventions Act. The Board is set up at the Slovene Intellectual Property Office and consists of a chairman, their substitute and four assessors. The Chairman and their substitute are appointed by the minister responsible for industrial property matters, for a period of three years. Both are required to possess qualifications for justiceship. The two assessors are appointed for every dispute separately by the President of the Intellectual Property Office and are chosen among experts in the technical field to which the employees' invention relates. The other two assessors are appointed by the parties in the conciliation proceeding (one by the employer and one by the employee). The purpose of the arbitration process is to bring about an amicable agreement between the employer and the employee in order to avoid costly disputes before the regular courts.

5 EMPLOYEES' UTILITY MODELS

Slovenia does not have a registration system for utility models. The Industrial Property Act provides for a specific type of protection called a *short term patent* which,[27] to some extent, corresponds to the utility model protection as it is known in other countries. The maximum term of a short-term patent is ten years from the day of filling the application. The Employees' Inventions Act applies also to the short-term patent.

6 EMPLOYEES' DESIGN RIGHTS

The protection of designs/models in general is regulated by the Industrial Property Act. However, neither the Industrial Property Act nor any other Act regulates the relationship between an employee and their employer regarding the design of a product. If the employees' creation is protectable by copyright, then the provisions of the Act on Copyright and Related Rights apply. Those provisions are discussed in detail in section 2.2 above. However it should be stressed out that the Act on Copyright and Related Rights applies only if the new design of a product was made while executing employee's duties with the employer or following the employer's instructions (Article 101). By virtue of the Copyright Act, and as discussed in section 2.2, the employer will not have to consider contractual assignment of copyright for the first ten years, and the creative employee will not be entitled to any special remuneration (in addition to the

27. *Zakon o industrijski lastnini ('ZIL-1')*, Official Gazette of Republic of Slovenia, No. 5/2006, 100/2013.

salary). On the expiration of the term of ten years the economic rights revert to the actual author – the employee; however the employer has a pre-emption right whereupon they may claim a new exclusive assignment of the rights. In such case the employer will be required to pay an adequate remuneration.

7 EMPLOYEES' PLANT VARIETIES

7.1 The Legal Framework

New plant varieties created in the course of employment are governed by the Protection of New Varieties of Plants Act.[28]

7.2 Ownership of Rights

The Act stipulates that the person entitled to plant variety protection and to the plant breeder's rights is the breeder of the variety or their legal successor. Where the breeder is an employee of a legal person, mutual rights and obligations should be settled by a contract and the entitlement to variety protection and plant breeder's rights should be determined by this contract. If this is not the case, the provisions of the regulation on rights deriving from the employment relationship concerning patents (Employees' Inventions Act) are to be applied analogously (Article 12(3) Employees' Inventions Act).

The Act does not provide for the type of contract to be used to determine the rights and obligations of the employer and the employee regarding the variety protection and plant breeders' rights. This article should therefore be interpreted broadly, in the sense that all kinds of contracts can be used, e.g., employment contracts, service contracts, collective agreements. However, the Protection of New Varieties of Plants Act requires all contracts for the transfer of plant breeder's rights to be in writing, otherwise they are null and void (Article 40(3)).

7.3 The Parties' Duty to Inform

When there is no contract regarding the entitlement to variety protection and plant breeder's rights, the provisions of the Employees' Inventions Act are to be applied analogously. This means that the employee has the obligation to inform the employer in writing about the new plant variety that was bred (Article 5 Employees' Inventions Act). Consequently, the employer is entitled to claim rights by submitting an unlimited or limited claim. Given that Council Regulation (EC) No. 2100/94 of 27 July 1994 on Community Plant Variety Rights[29] prohibits cumulative protection (Article 92(1)), the employer should also inform the employee which type of protection will be sought (Community plant variety rights or national plant variety rights). The relevance of this

28. *Zakon o varstvu novih sort rastlin*, Official Gazette of Republic of Slovenia, No. 113/2006.
29. Official Journal L 227, 1.9.1994, pp. 1-30.

information is connected to the fact that the employee's remuneration will be higher in cases where the broader Community right is sought.

7.4 Employee Remuneration Right

As stated above, when there is no contractual regulation between the employer and the employee regarding plant varieties, legislation concerning rights to inventions in employment relationships is to be applied analogously. For the reason that the law in this field (the Employees' Inventions Act) clearly obliges the employer to always remunerate the employee for inventions, the employee is entitled to a reasonable remuneration also for new plant varieties. The criteria for the amount of this remuneration are set up in the Employees' Inventions Act and in the Rules on Compensation for Work-Related Inventions.

8 INSTANCES FOR DISPUTES

The regulation of jurisdiction for disputes between an employee and an employer with regard to copyright-protected works, inventions, semiconductor chips, utility models and design rights as well as plant varieties created in the course of employment is fragmented. For this reason several different laws must be taken into account.

8.1 Copyright-Protected Works

Disputes relating to copyright-protected works created in the course of employment are normally subject to the exclusive jurisdiction of the District Court in Ljubljana. However, because of inconsistencies in the legislation, it may be the case that the Labour and Social Court in Ljubljana is also competent.[30]

The question of 'subject matter jurisdiction' of the court, in relation to copyright works created in the course of employment, has been dealt with by the Higher Labour and Social Court.[31] The plaintiff (an employed performer) did not base his claim on compensation for the transfer of economic and other rights or the provisions of the Copyright Act, but instead relied on the contract of employment and the employer's general act, both based in labour law. The Court concluded that the dispute between the parties arose from the employment relationship, which made the dispute fall under the competence of a Labour Court.

8.2 Inventions and Utility Models

According to Article 33 of the Employees' Inventions Act, the Labour Court has the subject matter jurisdiction over disputes concerning industrial property rights arising

30. Zirnstein, p. 143.
31. Decision of the Higher Labour and Social Court, reference No. Pdp 1348/2008 (21 May 2009).

from employment (i.e. when an employee creates an invention which has received patent protection, a new design, short-term patent). The same provision is found in Article 8(2) of the Labour and Social Courts Act[32] which provides that such disputes shall be within the exclusive jurisdiction of the Labour and Social Court in Ljubljana. The reason for exclusive jurisdiction of the Labour and Social Court is supposed to be the specialisation of judges and therefore resulting in a higher quality of dispute resolution.[33]

Pursuant to Article 32 of the Employees' Inventions Act the judicial protection of rights in certain cases is only granted upon prior (and unsuccessful) conciliation proceedings. Such proceedings are conducted before the Conciliation Board and begin upon written request, which should contain a brief description of the issue and its subject matter. A substantiated proposal for a mutual settlement of the dispute is then formed by the Conciliation Board. Within a one-month period, either party may object to the proposal in writing. The proposal for a compromise solution is adopted when and if the parties accept the proposed settlement in writing. Judicial protection is available directly (without the prior institution of conciliation proceedings) only if the employee's employment relationship was terminated and a claim for an interlocutory order is filed.[34]

Conciliation as a procedural precondition for filing a claim has been introduced only for disputes over inventions arising from an employment relationship. Other laws governing the rights and obligations of the employee and the employer in respect of innovations arising from employment are not familiar with such requirement.[35]

8.3 Plant Varieties and Semiconductor Chips

Disputes involving plant varieties and semiconductor chips are not covered by the Labour and Social Courts Act. Neither the Protection of New Varieties of Plants Act nor the Protection of Topographies of Integrated Circuits Act contain provisions on subject matter court jurisdiction.

According to the general rule on jurisdiction in Article 101 of the Courts Act,[36] such disputes fall within the subject matter jurisdiction of District Courts of general jurisdiction. However, a general delegation of the District Court in Ljubljana has been introduced. This means that the District Court in Ljubljana has the exclusive territorial jurisdiction over first instance disputes on plant varieties and semiconductor chips (Article 103(2) Courts Act).

32. *Zakon o delovnih in socialnih sodiščih ('ZDSS-1')*, Official Gazette of Republic of Slovenia, No. 2/04, 10/04 - popr., 45/08 - ZArbit, 45/08 - ZPP-D, 47/10 - odl. US in 43/12 - odl. US.
33. Zirnstein, p. 142.
34. For specified conciliation proceedings see Arts 24-31 of the Employees' Inventions Act
35. Zirnstein, p. 148.
36. *Zakon o sodiščih (ZS)*, Official Gazette of Republic of Slovenia, No. 94/07 – uradno prečiščeno besedilo, 45/08, 96/09, 86/10 – ZJNepS, 33/11, 75/12 – ZSPDSLS-A in 63/13.

9 CONCLUDING REMARKS

The analysis of Slovenian legislation on employee intellectual property rights shows that in most cases the Slovenian legislator favours an *ex lege* transfer of the rights to the employer. The reason for such regulation is the presumption that the employer will protect the rights more efficiently than the employee (has more knowledge and more resources for this), but will also exploit them more efficiently.

In our opinion, the allocation of rights between the employee and employer deriving from employees' intellectual creative endeavour in Slovenia is comparable to other legislations in Europe. There are some minor inconsistencies (ten years *ex lege* transfer for the copyright, unclear legislation regarding disputes, possible problems in practice when the law provides for the judicial application), but it can generally be concluded that the legislation in this field is appropriate and consistent.

BIBLIOGRAPHY

Table of Sources

Mežnar, *Pravno varstvo izumov na univerzah*, 3-4 Podjetje in delo (2001).

Mežnar, *Izumi v delovnem razmerju – med delovnim pravom in intelektualno lastnino*, 9 Delavci in delodajalci (2009).

Trampuž, *Zakon o avtorski in sorodnih pravicah ZASP s komentarjem* (1997).

Zirnstein, *Pravna ureditev inovativnosti v delovnem razmerju – med pravičnostjo in učinkovitostjo* (2012).

Zirnstein, *Učinkovitost pravne ureditve imetništva pravic iz inovacij, ustvarjenih v delovnem razmerju*, III(2) LeXonomica (2011).

Table of Legislation

Zakon o avtorski in drugih pravicah ('ZASP'), Official Gazette of Republic of Slovenia, no. 16/07, 68/08, 110/13.

Zakon o delovnih in socilanih sodiščih ('ZDSS-1'), Official Gazette of Republic of Slovenia, no. 2/04, 10/04 - popr., 45/08 - ZArbit, 45/08 - ZPP-D, 47/10 - odl. US in 43/12 - odl. US.

Zakon o sodišc?ih (ZS), Offcial Gazette of Republic of Slovenia, no. 94/07 - uradno prec?išc?eno besedilo, 45/08, 96/09, 86/10 - ZJNepS, 33/11, 75/12 - ZSPDSLS-A in 63/13.

The Council Regulation (EC) No. 2100/94 of 27 July 1994 on Community plant variety rights. Official Journal L 227, 1.9.1994, pp. 1-30.

Directive 87/54/EEC of 16 December 1986 on the legal protection of topographies of semiconductor products. Official Journal L 024, 27.1.1987, pp. 36-40.

Zakon o varstvu topografije polprevodniških vezij ('ZVTPPV'), Official Gazette of Republic of Slovenia, No. 21/95, 96/02, 7/03, 60/06, 81/06.

Zakon o patentih in tehničnih izboljšavah, Official Gazette of Republic of Slovenia, No. 44/1960.
Zakon o izumih iz delovnega razmerja ('ZPILDR'), Official Gazette of Republic of Slovenia, No. 45/1995, 96/2002, 139/2006.
Zakon o elektronskem poslovanju in elektronskem podpisu ('ZEPEP'). Official Gazette of Republic of Slovenia, No. 57/2000.
Zakon o industrijski lastnini ('ZIL-1'). Official Gazette of Republic of Slovenia, No. 5/2006, 100/2013.
Zakon o varstvu novih sort rastlin. Official Gazette of Republic of Slovenia, No. 113/2006.

Table of Cases

Decision of the Supreme Court reference no. I Up 211/2002 (5 July 2005).
Decision of the Supreme Court reference no. I Up 584/2002 (7 June 2005).
Decision of the Higher Labour and Social Court reference no. Pdp 1348/2008 (21 May 2009).

CHAPTER 19
Spain

Luis-Alfonso Durán & Laura Nadal

1 BACKGROUND

In the last few years two specific factors have resulted in an increased concern and attention regarding employees' inventions:

- Research and development of new inventions is becoming increasingly expensive and, therefore, new inventions are mostly developed within companies that can assume such a huge investment. Given the high risk and investment required to start a new development project, companies tend to collaborate between each other, with public research bodies and with universities to establish a framework in which risk and investment is shared.
- In recent years, Spanish companies have increasingly realised and become aware of the value of intellectual property as an important part of their assets.

Hence, it is of utmost importance to clearly establish ownership of intellectual and industrial property creations made by employees.

2 EMPLOYEES' COPYRIGHT

2.1 The Legal Framework

The legal framework for employees' copyright is determined by Articles 8, 51 and 97 of the Royal Legislative Decree 1/1996 of 12 April 1996 approving the new text of the Intellectual Property Law.[1]

1. Royal Legislative Decree 1/1996 approving the new text of the Intellectual Property Law, Art. 8, 51 and 97 (12 Apr. 1996), BOE No. 96 of 22 Apr. 1996.

2.2 Ownership of Rights

First, it is important to know whether the rights generated can be considered a collective work under the Royal Legislative Decree 1/1996, that is, if the work has been done under the initiative and coordination of a legal or natural person that is in charge of editing and publishing the work and that the work has been done by several persons, whose personal contribution is merged in a single creation, making it impossible to assign to any of them separately and specifically a right on the particular work done and the creation (copyright).

If the work can be considered a collective work then, unless otherwise agreed, the ownership corresponds to the natural or legal person that edits, publishes and discloses the copyright-protected work under its name, that is, the employer.

If the work cannot be considered a collective work then, ownership and, therefore, assignment of the rights to the work generated by virtue of an employment relationship will be governed by what is established in the corresponding contract or agreement, which must be in writing.

In case the contract does not include provisions regarding ownership and assignment of the created work, then it is to be presumed that the rights to a work or the exploitation rights are exclusively assigned, with the scope required to exercise its customary activity, to the employer at the time of the delivery of the work done by virtue of the employment relationship.

It is important to note that the above provision is included in Article 51 of the Royal Legislative Decree 1/1996 and the Article is entitled 'transmission of rights of the salaried author'. Nevertheless, courts have interpreted Article 51 broadly, not only including salaried authors (with an employment contract), but also including other types of agreements and service contracts which explicitly or implicitly require the transfer of rights mentioned in Article 51 to the employer to enable them to perform their business.[2]

In addition, for computer programs similar provisions apply. If the computer program is considered a collective work, the ownership corresponds to the natural or legal person that edits, publishes and discloses the computer program under its name, that is, the employer. For other instances, when an employee creates a computer program when performing in the course of employment or under the direction of the employer, ownership of the right to work the computer program belongs exclusively to the employer, unless otherwise agreed.

Moral rights belong to the author and cannot be waived or transferred.

2.3 The Parties' Duty to Inform

There are no provisions regarding this issue in the Royal Legislative Decree 1/1996.

2. *NEUTRIN S.L. v Mr. Gerardo, Mr. Claudio and Mr. Victor Manuel*, 199/2007, Court of Second Instance (civil), section 15, Barcelona.

2.4 Employee Remuneration Right

According to what is stated in section 2.2 above, it is to be understood that the right of the employee to receive remuneration or compensation for the creation of the work and the amount of said remuneration is to be established in the employment or service contract, that is, the parties must negotiate and agree the applicable compensation.

If the contract does not include provisions relating to compensation or extra remuneration, it should be understood that the employee is not entitled to the benefit.

3 EMPLOYEES' SEMICONDUCTOR CHIPS

3.1 The Legal Framework

The legal framework for employees' semiconductor chips is determined by Article 3(2)(a) of Law No. 11/1988 of 3 May 1988 on the Legal Protection of Topographies of Semiconductor Products,[3] which determines that topographies of semiconductor chips created by an employee during the term of their contract or work relationship with an employer are governed by Articles 15-21 of the Spanish Patent Law No. 24/2015 of 24 July 2015.[4]

3.2 Ownership of Rights

For further information see section 4.2 below.

3.3 The Parties' Duty to Inform

For further information refer to section 4.3 below.

3.4 Employee Remuneration Right

For further information see section 4.4 below.

4 EMPLOYEES' INVENTIONS

4.1 The Legal Framework

The legal framework for employees' inventions is primarily determined by Articles 15-21 and 133-135 of the Spanish Patent Law No.24/2015.[5]

3. Law No. 11/1988 on the legal protection of topographies of semiconductor products, Art. 3.2 a) (3 May 1988), BOE No. 108 of 5 May 1988.
4. Law No. 24/2015 on Patents, Art. 15 to 21, 133 to 135 and 150 (24 July 2015), BOE No. 177 of 24 July 2015.
5. Law No. 24/2015 on Patents, 2015, *supra* n. 4.

Royal Decree RD 55/2002 establishes a corresponding legal framework for certain public research bodies.[6]

For the purpose of providing a more comprehensive view of the legislative framework concerning employee's inventions, the subsections below will be divided or explained in terms of three categories of employers: universities, public research bodies, as well as private and public companies (which includes private and public companies other than universities and public research bodies; as well as civil servants, employees and workers of the State, Autonomous Communities, provinces, municipalities and other public bodies).

4.2 Ownership of Rights

4.2.1 Private and Public Companies

According to Article 10 of the Patent Law No. 24/2015, the right to a patent shall belong to the inventor or to his/her successors in title and it shall be transferable by any of the means recognised in the Law. In case that an invention has been made jointly by a number of persons, the right to obtain a patent shall belong to them jointly.

Regarding ownership of the rights to an invention we can distinguish three situations:[7]

(1) Where the invention is made by an independent inventor, the invention belongs to the inventor.
(2) When the inventor is an employee two different situations can be distinguished:
    (a) invention made by an employee during the term of his/her contract, employment relationship or service; and the result of a research that is explicitly or implicitly the object of his/her contract, shall belong to the employer. The inventor could only be entitled to a supplementary remuneration if the personal contribution of the inventor and the importance of the invention to the company have obviously exceeded the content of the contract of employment;
    (b) invention made by an employee, not contemplated in paragraph i), related to its professional activity and the knowledge gained within the company have had a decisive influence thereon or they have used means provided by the company, the employer shall have the right to claim ownership of the invention or to reserve the right to use it. Where the

6. Royal Decree 55/2002 on the work and assignment of inventions made in public research bodies, in accordance with the provisions of Art. 20 of Law 24/2015 of 24 July (18 Jan. 2002), BOE No. 26 of 30 Jan. 2002.

7. Carlos Fenández-Nóvoa, José Manuel Otero Lastres and Manuel Botana Agra, *Manual de la Propiedad Industrial*, 141-143 (Marcial Pons, 1st ed., Ediciones Jurídicas y Sociales, S.A., 2009) and José Manuel Otero Lastres, Alberto Casado Cerviño, et al., *Compendio Práctico sobre la Protección de la Propiedad Industrial. Una Visión Renovada en España, Europa y el Mundo*, 119 (Thompson Reuters, 1st ed., Lex Nova, December 2012).

employer claims ownership of an invention or reserves a right to its use, the employee shall have the right to equitable financial remuneration fixed in relation to the invention's industrial and commercial importance and taking into account the value of the means or knowledge provided by the company and the contribution made by the employee himself/herself.

(3) When an invention is made in the frame of a service contract between two parties, the rights to any invention derived from the contract will belong to what is established in the contract.

4.2.2 *Universities*

Ownership of inventions has a similar structure to that described above. Hence, inventions made by a professor as a result of their research responsibilities at a university and which are within the scope of their teaching and research functions belong to the university.[8] Notwithstanding this, the university may assign ownership of an invention to the professor who is its creator. In such case the university may keep a non-exclusive, non-transferable and free licence to use the invention.[9]

The foregoing does not exclude the possibility that a professor can hold or sign agreements or contracts with public or private entities for the development of, e.g., technical or scientific works. In those situations the contract or agreement should specify which of the parties is the owner of the invention generated as a result of the contract or agreement.

4.2.3 *Public Research Bodies*

Public research bodies have a similar ownership structure as that described for universities.[10] Therefore, inventions made by research staff as a result of their research responsibilities at a public research body, and which are within the scope of their functions, belong to the public research body. The public research body may assign ownership of an invention to the researcher who is its creator. In such case, the public research body may keep a non-exclusive, non-transferable and free licence to use the invention.

Also, as in the case of universities, researchers from public research bodies can hold or sign agreements or contracts with public or private entities for the development of, e.g., technical or scientific works. In those situations the contract or agreement should specify which of the parties is the owner of an invention generated as a result of the contract or agreement.

It is important to note that Royal Decree 55/2002 defined the Spanish Patent Law No. 24/2015 regarding public research bodies, which include the following: the Higher

8. *UNIVERSIDAD POLITECNICA DE VALENCIA* Ruth 401/2011, Court of Second Instance (civil), section 9, Valencia.
9. Fenández-Nóvoa, *supra* n. 7, at 143.
10. Fenández-Nóvoa, *supra* n. 7, at 143.

Council for Scientific Research ('CSIC'), the Centre for Energy, Environmental and Technological Research ('CIEMAT'), the Geological and Mining Institute of Spain ('IGME'), the National Institute for Aerospace Technology 'Esteban Terradas' ('INTA'), the National Institute of Agricultural and Food Research Technology ('INIA'), the Spanish Institute of Oceanography ('IEO'), the Health Institute Carlos III, the Astrophysics Institute of the Canary Islands, the Channel of Hydrodynamic Experiences El Pardo, the Spanish Centre of Metrology and the Centre for Studies and Experimentation on Public Works.[11]

In the case of these public research bodies, as for other public research bodies, inventions made by research staff as a result of the performance of activities in the specific area of their duties belong to the respective public research body. If the public research body decides not to keep ownership of the invention, the research staff inventor is entitled to file a patent application. The public research body may, however, keep a non-exclusive, non-transferable and free licence to use the invention.

As in the case of universities and the other public research bodies, in the public research bodies mentioned in the Royal Decree 55/2002, when research staff makes an invention as a result of an agreement between the public research body and public or private entities, the contract or agreement must specify ownership and benefit sharing.

For each of the three categories of employers discussed above, the inventor always keeps the moral right over the invention to appear or be mentioned as the inventor thereof.

4.3 The Parties' Duty to Inform

4.3.1 Private and Public Companies

Regarding inventions on request and service inventions, items 1 and 2 of section 4.2.1 above, the employee must inform the employer by a written notification of the invention, including all the data and reports required to evaluate the invention. The employer must decide within three months from the date they are informed, whether they are interested in the invention and, therefore, whether to exercise its rights regarding ownership or right to use the invention. Failure to fulfil the duty to inform causes the employee to lose their rights recognised under employee's inventions legal framework (Articles 15-21) in the Spanish Patent Law.

4.3.2 University Researchers

When a professor at university creates an invention as a result of their research responsibilities, they must notify the university immediately. The university must notify the author of the invention of his/her intention to maintain his/her rights over the invention within a period of three months.

11. Fenández-Nóvoa, *supra* n. 7, at 143.

4.3.3 *Public Research Bodies*

The same provisions and requirements relevant for universities also apply in the case of public research bodies. However, as stated above, Royal Decree RD 55/2002 further defines the duty to inform of the research staff working at a public research centre mentioned above. Such research staff, when creating an invention, must notify the Director or President of the public research body. Such notification must be in written form and must be accompanied by all relevant data and reports allowing the centre to properly evaluate the invention. The public research body must notify the research inventor within three months from the day after the receipt of the notification from the researcher whether it decides to exercise its rights to the invention and, hence, whether it wants to keep ownership of the invention. In the latter case the research centre must file a patent application within one month.

Failure to fulfil the duty to inform causes the relevant research staff to lose their rights to the invention.

It is worth mentioning that the relevant research staff cannot publish the results relating to the invention during the above-mentioned three-month term or, if applicable, until the corresponding patent application has been filed.

4.4 Employee Remuneration Right

4.4.1 *Private and Public Companies*

In the case of inventions on request (item 1 of section 4.2.1 above), the employee is not entitled to an additional remuneration (other than what is already contemplated in their contract)[12] for the performance of the invention, except where both their personal contribution to the invention and the importance of the invention to the company obviously go beyond the explicit or implicit terms of their contract or work relationship.

With regard to service inventions, whenever the employer decides to claim ownership of the invention or to reserve a right to use the invention, the employee is entitled to an equitable economic compensation, which amount is fixed taking into account the invention's industrial and commercial importance, the value of the means or knowledge provided by the company and the contribution made by the employee himself.

4.4.2 *Universities*

If the university decides to keep ownership of the invention, the professor inventor has the right to a share of proceeds that the university obtains from using or assigning the invention. The terms and amount depend on the university and are defined by the

12. *Mr. Victorino and Mr. Carlos Manuel v AMBIESYS, S.L.*, 551/2011, Court of Second Instance (civil), section 15, Barcelona.

statutes of each university. However, the university may assign the ownership of the invention to the professor who is its inventor. In such case, the university is entitled to:

- keep a non-exclusive, non-transferable and free licence to use the invention; and
- when the professor obtains proceeds from using (including assigning and licensing) the invention, the university has the right to a share of the proceeds as determined by the statutes of the university.

4.4.3 *Public Research Bodies*

The same provisions and requirements stated for universities apply in the case of public research bodies. However, as stated earlier, Royal Decree RD 55/2002 further defines remuneration rights of research staff of the research centres listed earlier.

If any of these research centres decides to retain ownership of an invention, the benefits obtained by the research centre in connection with using (including assigning and licensing) the invention are distributed as follows:

- one-third for the public research body;
- one-third for the inventor(s) (research staff); and
- the remaining one-third is distributed according to the criteria established by the board of a given institution, taking into account the importance and significance of the invention (patent), the benefits that can be generated in the future, and the participation or collaboration of staff different from the research staff stated in the point above (inventors).

As mentioned above, the public research body can decide not to keep ownership of the invention. In such case the research staff inventors are entitled to file a patent application. The public research body will in return keep a non-exclusive, non-transferable and free licence to use the invention and 20% of the benefits derived from using (including assigning and licensing) the invention when a patent application is filed by the research staff inventor(s).

5 EMPLOYEES' UTILITY MODELS

5.1 The Legal Framework

The legal framework of employees' utility models is determined by Articles 15 to 21, 133 to 135 and 150 of the Spanish Patent Law No. 24/2015.[13]

13. Law No. 24/2015 on Patents, 2015, *supra* n. 4.

5.2 Ownership of Rights

As established by Article 150 of the Spanish Patent Law No. 24/2015 given the fact that there are no specific provisions in this law that regulate employees' utility models, the legal framework and the provisions applicable in the case of (patentable) inventions also apply in the case of utility models. Therefore, for further information see section 4.2 above.

5.3 The Parties' Duty to Inform

The provisions regulating this matter in the context of (patentable) inventions also apply for utility models. Hence, for further information see section 4.3 above.

5.4 Employee Remuneration Right

The provisions regulating this matter in the context of (patentable) inventions also apply for utility models. Hence, for further information see section 4.4 above.

6 EMPLOYEES' DESIGN RIGHTS

6.1 The Legal Framework

The legal framework of employees' design rights is determined by Article 15 of Law No. 20/2003 of 7 July 2003 on the legal protection of industrial designs.[14]

6.2 Ownership of Rights

When an industrial design has been created by an employee performing their work obligations or under the instructions of the employer or as a request under a service agreement or relationship, then the employer or contracting party has the right to register or file the industrial design unless otherwise provided in the contract.

However, the author of a design (e.g., the employee) has the right to be mentioned as such in the application, in the registry, and in the publication.

6.3 The Parties' Duty to Inform

There are no provisions in the law regarding this issue in the context of design rights.

14. Law No. 20/2003 on legal protection of industrial designs, Art. 15 (7 Jul. 2003), BOE No. 162 of 8 Jul. 2003.

6.4 Employee Remuneration Right

Spanish design legislation does not contain any provision regarding remuneration rights or compensation for employees. Therefore, it is to be understood that it is up to the parties to agree or negotiate such remuneration or compensation.

7 EMPLOYEES' PLANT VARIETIES

7.1 The Legal Framework

The legal framework of employees' plant varieties is determined by Article 10.5 of Law No. 3/2000 of 7 January 2000 establishing the legal framework for the protection of plant varieties.[15] This Article includes a reference to by Articles 15 to 21 of the Spanish Patent Law No. 24/2015.[16]

7.2 Ownership of Rights

According to Article 10.5 of Law No. 3/2000 ownership of a plant variety obtained by an employee is determined according to the rules applicable to the employment or service relationship. This means that the employment or service contract or general rules governing them should establish ownership of employees' plant varieties. However, where such contracts do not determine the ownership of the plant variety obtained by the employee the provisions of Articles 15-21 of the Spanish Patent Law No. 24/2015 are applicable. Therefore, for further information see section 4.2 above.

7.3 The Parties' Duty to Inform

As stated above, first, this feature would be governed by rules applicable to the employment or service relationship. In case they fail to include specific provisions regarding this matter Articles 15-21of the Spanish Patent Law No. 24/2015 are applicable. Hence, for further information refer to section 4.3 above.

7.4 Employee Remuneration Right

Like the subsections above, at first instance rules applicable to the employment or service relationship, and if such contracts or rules do not include provisions regulating or clarifying this matter, Articles 15-21 of the Spanish Patent Law No. 24/2015 would be applicable. Similarly to the above, refer to section 4.4.

---

15. Law No. 3/2000 establishing the legal framework for the protection of plant varieties, Art. 10.5 (7 Jan. 2000), BOE No. 8 of 10 Jan. 2000.
16. Law No. 24/2015 on Patents, 2015, *supra* n. 4.

8 INSTANCES FOR DISPUTES

Only in the case of inventions (patents) and utility models does Spanish law establish a special procedure regarding disputes in the field of employees' rights. In those instances, before instituting proceedings in court, the dispute must be brought for conciliation before the Registry of Industrial Property. For the purpose of such conciliation, a Commission is set up which is composed of an expert from the Registry of Industrial Property acting as the chairman of the Commission and who is appointed by the Director of the Registry. The Commission further consists of an expert appointed by the employees of the company of the inventor and another expert appointed by the employer. In case the inventor is an employee of a public administration body, the Commission will be chaired by an expert from the Registry of Industrial Property appointed by the Director and will include those members as established by Royal Decree in accordance with the legislation on civil servants.

A proposal of agreement must be issued by the Conciliation Commission within two months of the filing of the request of conciliation. Then, the parties must state within 15 days whether they agree (or not) with the proposal (silence by any of the parties is interpreted as acceptance of the proposal). If any of the parties does not agree with the proposal issued by the Conciliation Commission, the Director of the Industrial Property Registry will issue a certificate noting that fact. Subsequently the dispute can be submitted to the court and the different instances normally available can be used.

For the other intellectual property rights mentioned in this chapter, that is, copyright, semiconductor chips, design rights and plant varieties, the legal framework does not provide for special instances to which disputes should be directed and, therefore, said disputes can be submitted directly to the court and the different instances normally available can be used.

9 CONCLUDING REMARKS

As can be readily ascertained from this chapter, Spanish law establishes as a principle that ownership of rights, or at least the right to exclusivity, to intellectual property created by an employee while performing their job duties belongs to the employer. In addition, as a general rule, compensation and extra remuneration should be negotiated and agreed between the employer and the employee and reflected or included in the employment or service contract in advance. Finally it can be noted that the provisions in Spanish law related to intellectual property rights set a quite homogenous legal framework for the different modalities discussed in this chapter.

BIBLIOGRAPHY

Table of Sources

Carlos Fenández-Nóvoa, José Manuel Otero Lastres & Manuel Botana Agra, *Manual de la Propiedad Industrial*, 141-143 (Marcial Pons, 1st ed., Ediciones Jurídicas y sociales, S.A. 2009).

José Manuel Otero Lastres, Alberto Casado Cerviño, et.al., *Compendio Práctico sobre la Protección de la Propiedad Industrial. Una Visión Renovada en España, Europa y el Mundo*, 119 (Thompson Reuters, 1st ed., Lex Nova, December 2012).

Mónica del Corral Alarcón, Antonio Castán Perez-Gómez, Javier Fernández Lasquetty, Isidro José García Egea, Iván L. Sempere Massa & Jordi Romaní Lluch, *Report Q183 Employers' Rights to intellectual property in Spain*.

Table of Legislation

Law No. 24/2015 on Patents, Article 15 to 21 and 133 to 135 (24 July 2015), BOE No. 177 of 24 July 2015.

Royal Legislative Decree 1/1996 approving the new text of the Intellectual Property Law, Article 8, 51, and 97 (12 April 1996), BOE No. 96 of 22 April 1996.

Law No. 11/1988 on the legal protection of topographies of semiconductor products, Article 3.2 a) (3 May 1988), BOE No. 108 of 5 May 1988.

Law No. 11/1986 on Patents, Article 15 to 20, 140 to 142, and 154 (20 March 1986), BOE No. 73 of 26 March 1986.

Royal Decree 55/2002 on the work and assignment of inventions made in public research bodies, in accordance with the provisions of Article 20 of Law 11/1986 of 20 March on Patents (18 January 2002), BOE No. 26 of 30 January 2002.

Law No. 20/2003 on legal protection of industrial designs, Article 15 (7 July 2003), BOE No. 162 of 8 July 2003.

Law No. 3/2000 establishing the legal framework for the protection of plant varieties, Article 10.5 (7 January 2000), BOE No. 8 of 10 January 2000.

Table of Cases

NEUTRIN S.L. v Mr. Gerardo, Mr. Claudio and Mr. Victor Manuel, 199/2007, Court of Second Instance (civil), section 15, Barcelona.

Mr. Alexander v NATYSAL SA, 213/2009, Court of Second Instance (civil), section 28, Madrid.

Mr. Victorino and Mr. Carlos Manuel v AMBIESYS, S.L., 351/2011, Court of Second Instance (civil), section 15, Barcelona.

UNIVERSIDAD POLITECNICA DE VALENCIA Ruth, 401/2011, Court of Second Instance (civil), section 9, Valencia.

CHAPTER 20

Sweden

Sanna Wolk

1 BACKGROUND

In Swedish employment law the employer is central, and the products of the employee's work belong to the employer insofar as they are created in connection with the employee's work. The employee is compensated for their efforts by means of a salary and other employment benefits. It is a general principle of law that the fruits of an employee's work belong to the employer, and it is the contract of employment which constitutes the basis of the employer's rights. The employment relationship, and therefore also the employment contract, are not static, however, but undergo constant change. This is why the employer, in their capacity as a supervisor and work coordinator, determines the employee's factual duties, the way and the place for the performance of work as well as the working times.

However, if an intellectual property right covers the product of work, or if it can be protected by intellectual property registration procedures, the situation is governed by other legal rules. The main rule in Swedish intellectual property law is that the creator is central, irrespective of whether they have the status of an employee. The applicable fundamental principle is that any person who has actually created the work covered by proprietary rights, or which can be protected by intellectual property law, becomes the holder of that right. This means that a proprietary right emerges in relation to the person who has created the product protected under intellectual property law. An employee-creator has thus the same right to their intellectual property as other non-employed creators. This fundamental principle is well established in the field of Swedish intellectual property law. Since the right arises in the first place in relation to employees, the employer may only receive a derivative right from the employee in order to exploit the product and obtain protection under intellectual property law. The employer's right to control intellectual property cannot thus be obtained merely through the employment relationship, but must be acquired by means of a separate agreement. Such transfer of rights may be effected under applicable law either, by means of an agreement, or, in some cases, based on implied terms and conditions of employment.[1]

2 EMPLOYEES' COPYRIGHT

2.1 The Legal Framework

In Sweden the Act on Copyright in Literary and Artistic Works (1960:729)[2] stipulates that a person who has created a literary or artistic work shall have copyright in that work (Section 1). In the Act there are no general rules concerning employees' copyright. The Act only has an exclusionary provision relating to employees' computer programs which can be found in Section 40a of the Copyright Act.[3] However the provisions of Section 40a are not intended to function as guidelines for protected works other than computer programs. Such guidelines are instead found in general legal principles.

2.2 Ownership of Rights

As mentioned above, an employer may only obtain a *derivative* right from an employee in order to exploit the work. For the transfer of copyright in an employee-employer relationship it is the collective agreement and/or the individual employment contract that is decisive, i.e. whether the employment contract entails express or implied provisions effecting transfer of rights, as well as what the scope of such transfer is. Copyright in computer programs created by employees is regulated specifically by Section 40a of the Copyright Act.

Chapter 3 of the Copyright Act contains general rules on transfer of copyright. Section 27 provides that copyright may be transferred entirely or partially to another natural or legal person, subject to the limitations following from Section 3 regulating moral rights. According to Section 3 the author, even when employed, may only waive the right in relation to uses which are limited as to their character and scope. Section 28 of the Act stipulates certain rules of interpretation concerning transfer agreements. Furthermore, in the absence of an agreement to the contrary, the person to whom a copyright has been transferred may not transfer the copyright further to third parties (Section 28).

In Chapter 5 of the Copyright Act neighbouring rights to copyright are protected according to similar principles that apply to works protected by copyright. These rights are said to be adjacent to, or originate from, copyright law. Sections 45–49 of the Act protect performing artists (actors, musicians, singers, etc.), producers of phonographic

1. See further Sanna Wolk, *Arbetstagares immaterialrätter* (Norstedts Juridik, Stockholm, Sweden 2008).
2. Lag (1960:729) om upphovsrätt till litterära och konstnärliga verk.
3. Which implements Art. 2(3) of the Directive 2009/24/EC on the legal protection of computer programs.

recordings and film, radio and television organisations, producers of catalogues (databases) and photographers. Performing artists, photographers and producers have a right of limited duration to their performance and works, and some of the copyright provisions apply *mutatis mutandis* to the neighbouring rights.

In Sweden there is no statutory exemption for teachers with respect to copyrights. Such a right has however been created by custom.

2.2.1 Employees' Copyright and Personal Neighbouring Rights

For the transfer of copyright from employed authors to the employer the general provisions on copyright transfer in Chapter 3 of the Copyright Act are applicable, subject to a limitation in Section 3 concerning moral rights Furthermore, performing artists (Section 45) and photographers (Section 49a) are the original holders of such rights (personal neighbouring rights). These rights are linked to individual persons, which is why they are in a close relationship to copyright ('personal neighbouring rights'). This is reflected, *inter alia,* in the fact that the provisions refer to Section 3 regarding authors' rights and to Chapter 3 of the Copyright Act. The preparatory work preceding the enactment of the Copyright Act stipulates in general terms that the transfer of personal neighbouring rights resembles transfer of rights protected by copyright, which is why equivalent legislative provisions apply to the transfer of rights of an employed performing artist or photographer.[4] This indicates that the same general legal principles, which apply to the use of employed copyright holders' works, should also apply to performing artists and photographers.

A transfer of employees' copyright can be made by means of a collective agreement or individual employment contracts; in cases where nothing has been expressly agreed a transfer could apply implicitly in certain cases by virtue of the nature of the employment relationship. In such cases guidance should be sought from the nature of the employment relationship as well as the general terms and customary practices applicable in the sector. However, in the absence of express provisions concerning transfer of copyright in an employment relationship and given there are few court decisions in the area, a general principle has evolved in legal literature stipulating that if there is no express agreement, the employment contract should imply that some of the copyright holder's rights in a given work will pass on to the employer. The transfer of rights depends in each individual case on whether there is a connection between the employment and the creation of the work, and the strength of such connection. The determining criterion concerning the use of employees' works found in the literature has come to be referred as *the rule of thumb.*[5] In accordance with this rule an employer may, as a result of the employment contract, use for their regular

4. See Swedish Government Official Reports 1956:25 p. 382 et seq. and Government Bill 1960:17 p. 242 (performing artists), as well as Government Bill 1993/94:109 p. 21 et seq. (photographers).
5. Many definitions of the rule of thumb have been provided by different authors, all implying principally the same, however. The rule of thumb has also been used in the context of legislation, see, *inter alia,* Swedish Government Official Reports 1980:42 p. 114, Government official reports 1985:51 p. 110, Government Bill 1988/89:85 p. 2, Government Bill 1992/93:48 p. 116 and Government Bill 1996/97:111 p. 32 et seq.

business activities such works that have been created by an employee as part of their duties, or as a result of instructions given to the employee by the employer. The scope of such use is determined on the basis of the point in time at which the work was created. The employer's regular business activities are deemed to be the actual activities that are conducted by the business. In practice the scope of exploitation of the employee's work by the employer will be determined by the extent to which the employer requires the work to conduct their business activities normally. If the employer conducts temporarily other operations, the use of the employee's copyright will not come into question as a rule. Furthermore, any changes to the work that must be made in order to meet the purpose for which the work was created in course of the employment are permissible.

The rule of thumb was confirmed by the Labour Court in case 2002, No. 87. In this case, the employee requested through the Swedish Armed Forces Union that the Labour Court should declare that the musicians' neighbouring rights under Section 45 of the Copyright Act (rights to control recording, broadcasting and/or transmission as well as making their performances available to the public) had not passed on to the employer. The employer, the state, claimed through the Swedish Armed Forces Union that the performing musicians' rights had passed on to the employer. The Court found that the employer did not show that the neighbouring rights of the musicians to the video recording had passed on to the employer by means of the individual employment contracts or some other agreement. The employee's claim was thus granted and the decision was unanimous.

The Labour Court case 2002 No. 87 is crucial because it was the first case in which the rule of thumb was tried by a higher instance. The Court also emphasised in the reasons for the decision that in the case of uncertainty about rights transfer the employer must prove that transfer has actually taken place. This is consistent with the notion that the employee is the original holder of copyright, which follows from the fundamental principle in copyright law. Further, the decision shows that when there is no special agreement, the rule of thumb governing the transfer of rights must be considered as a guiding principle when determining transfer of a personal neighbouring right or a copyrighted work. The Court has further emphasised that transfer of rights must rely on the express or implied content of the contract.

The Labour Court case 2002 No. 87 also shows that other uses, falling outside the employer's regular business activities, devolve, in principle, on the employee, unless otherwise agreed upon. The employee's exclusive proprietary rights may, however, be limited by other principles, for instance the duty of loyalty applicable under employment law. Therefore, whereas uses falling outside the employer's regular activities may devolve upon the employee, the employee may not compete with the employer or hurt the employer's business in any way.

In another case the Court of Appeal for Western Sweden had established that an intern at a newspaper should be considered an employee and be encapsulated by the rule of thumb.[6] A person had been accepted for an internship at a city district

6. See Court of Appeal for Western Sweden, case FT 3542-07, 31 October 2008.

administration unit which was part of the Gothenburg Municipality. The intern had as their assignment the creation of editable material – articles and images – for publication in the unit's newspaper. One article and photograph was also published in another district administration periodical. The Court established that the intern was an employee under principles of employment law and that Gothenburg Municipality was not an employer, but each city district administration unit was an employer, for which reason using the intern's article and photograph in another periodical was deemed infringement of copyright.

2.2.2 *Employees' Computer Programs*

Section 40a of the Copyright Act stipulates that employers gain exclusive rights to computer programs or their preparatory material developed by employees. It must be pointed out that in contrast to what is generally applicable in copyright law, Section 40a of the Act transfers the copyright subsisting in the work entirely. This means, *inter alia*, that Sections 3, 27 and 28 of the Act are not applicable when copyright is vested in the employer pursuant to the provisions of Section 40a of the Act.

As regards the employer's right to employees' computer programs, the main criteria for deciding whether the work becomes vested in the employer are the employee's duties. This follows from the wording of Section 40a ('... as a part of his tasks or following instructions by the employer'). If one considers more closely the meaning of the expression 'as a part of the employee's tasks', the expression applies to employees whose duties consist in the development of computer programs. The expression covers duties in which computer programs may be developed pursuant to the employment contract, or following well-established practice, etc. Nothing indicates, however, that these activities have to constitute the employee's main or only duties.[7] The expression in Section 40a of the Copyright Act encompasses therefore not only persons who are employed solely for the purpose of developing computer programs or systems; rather, an employee's duties may include development of new computer programs or parts of programs, or it is sufficient that the employee participates in the development of existing programs.[8] The above activities do not have to constitute the main field of their employment – the employee may have other working duties, such as, e.g., work as a project leader or a systems administrator. The decisive factor with regard to the employer's possible right in accordance with the presumptive provision is that an employee's duties must include work that may result in a computer program protected by copyright. Further, under the presumptive provisions of the Copyright Act the employer is sometimes entitled to rights in computer programs which have not been developed as part of the employee's de facto duties. This follows from the expression 'or following instructions by the employer' in Section 40a of the

7. Cf. s. 3(1) of the Swedish Act on the Right to Employee's Inventions.
8. See also Lund District Court, case T 1216-10, 9 March 2010. A computer program (a search engine) was not made as part of the employee's duties. The decision was appealed by the employer but the Court of Appeal did not grant the appeal, see Scania and Blekinge Court of Appeal, decision Ö 611-10, 11 March 2010 and Labour Court decision B 19/10, 15 March 2010.

Act. This expression entails instructions concerning activities which lie outside the employee's regular working tasks, but which fall within the framework of their working duties. If an employer issues supplementary instructions concerning the execution of regular working tasks by the employee, those will frequently be covered by the expression 'as a part of the employee's tasks'. The statute does not specify the character of the employer's instructions. In my opinion the formulation 'following instructions by the employer' should be interpreted in such a way as to require that an employer shall have issued instructions to solve a certain problem. The employer need not have provided any guidelines for the solution of the problem, or specified in detail the way in which the problem should be solved.[9] The employer's instructions do not thus need to be detailed, insofar as the scope of the provision is concerned.

The basic condition for the application of Section 40a of the Copyright Act is that no other agreement has been entered into between the employer and the employee. Where the parties have concluded some other agreement, it will invalidate the presumption, and the scope of the employer's right will be instead determined by the parties' agreement. In many cases this does not necessarily lead to any difficulties, but in individual cases disputes may arise. For example, questions may arise concerning the scope of the employer's rights when the company is sold, or when the employer is declared insolvent.

In reality complications of this kind are most common in the area of copyright law, which is why agreements invalidating the presumption are discussed below in relation to the provisions of the Copyright Act. However, the description of the latter applies also to design law, except as regards aspects of a non-economic nature.

The provisions of Section 40a of the Copyright Act differ markedly from existing Swedish copyright law principles concerning works created in employment relations. Pursuant to the provision, copyright in a computer program created by an employee as part of their duties is transferred in its entirety to the employer. If there exists, on the other hand, an agreement concerning copyright transfer, the presumption becomes completely invalidated by that agreement. In this case, copyright will not automatically be transferred to the employer; instead, the employer's right to use the program shall be determined by the parties' agreement. One important consequence thereof is that an employer may have lesser rights with respect to their employees' computer programs when the transfer of rights is regulated by contract. This occurs when Sections 3 and 27-28 of the Act become applicable.

The explanatory statements to Section 40a of the Copyright Act do not provide any guidance regarding more detailed conditions applying to agreements that could invalidate the presumption. Since Section 40a of the Act is formulated in such a way that the creator has no express right to their computer program, it is usually necessary for both parties to wish to regulate the transfer of rights by agreement in advance. This is necessary because of the basic premise of Section 40a of the Act, stipulating that the employer is automatically entitled to an employee's computer programs. No action on the part of the employer is thus necessary or required. These conditions indicate that an

9. Cf. s. 3(1) of the Swedish Act on the Right to Employee's Inventions.

explicit agreement between the employer and the employee is necessary if the presumption is to be set aside. Such an agreement can be made in writing, e.g., by incorporating terms and conditions to this effect in the employment agreement, or by means of a collective agreement. If no such terms have been concluded, the presumption cannot normally be set aside by virtue of the nature of the employment relationship, since the default position favours the employer. Consequently, one cannot assume that the presumption has been set aside solely on the grounds that the employer has been passive. In an individual case special circumstances may however imply that the presumption may be considered to have been set aside.

Since the employer is presumed to be the copyright holder under Section 40a of the Copyright Act, questions arise concerning the burden of proof. As regards the allocation of the burden of proof between the employee who has created a copyrightable computer program and the employer, the burden has been shifted in relation to the rules generally applicable in copyright law. Thus, it is up to the employee to show that the parties have agreed otherwise, and they must also show evidence of the existence of an agreement invalidating the presumption. In Swedish law this is not a question of the burden of proof in any strict sense, however. The courts apply a relaxed standard of proof, which takes into account all the circumstances in order to determine what has really been agreed between the parties.

If the parties have concluded a rights transfer agreement, it will be the terms and conditions stipulated in this agreement that will be crucial for the interpretation of the employer's rights, and not the provisions of Section 40a of the Copyright Act. Questions concerning interpretation of such agreements are treated in accordance with the general rules of contract law and Section 3 of the Act. There is nothing that supports the supplementary application of the presumptive provisions in such a situation. That would be going too far.

In principle agreements in the field of copyright law are interpreted strictly and words are attributed their natural meaning. This means that when an agreement is unclear or vague as regards the scope of transfer of copyright, it is normally presumed that copyright will not be transferred entirely (specification principle). Since this interpretation principle of copyright law applies in its entirety when the presumption of Section 40a of the Copyright Act has been invalidated, the agreement can be interpreted in favour of the employee copyright holder. If, on the other hand, rights transfer is determined on the basis of the statutory provisions, this means derogation from the principle of contract interpretation applicable in copyright law.

2.2.3 Neighbouring Producers Rights

As regards producers' neighbouring rights in Sections 46-48 concerning producers of sound recordings and of images, and Section 49 on producers of catalogues and databases, the original holder of a producer right may be a natural or legal person. Nevertheless, if an employee is considered to be the initial holder of a producer right, the right is transferred to the employer if the recording or database was produced by an employee as part of their duties or as a result of instructions given by the employer.

2.2.4 *University Researchers' Copyright*

The Swedish Copyright Act is absent of specific provisions regulating the subsistence of researchers' and teachers' rights. Instead customary practice has resulted in a special exception whereupon researchers and teachers enjoy ownership of teaching materials and the products of their research, and extends over articles, dissertations, teaching books, etc.

In particular the 'teacher exemption'[10] provides that any copyright subsisting in teaching materials or products of research belongs to the teacher/researcher. The custom is based on a long tradition predating modern technology, with roots in academic freedom and creativity. Traditionally the employment conditions of researchers and teachers are characterised by a broad right to self-determination and autonomy in teaching and researching (arguably originating from the era of Classical Greece). The unique nature of the employment of academics can be viewed as being of benefit to both creativity and research. The teachers' exemption can be described as an unwritten legal rule constituting current state of the law regarding the teacher's educational materials. The exact scope of this rule is not uniform, however, as some universities may have developed their own local custom.

Educational institutions are however increasingly departing from this rule. For instance, several Swedish universities have as their policy that rights are to be vested in the university and not the teacher or researcher. One reason for this practice is the digitisation of educational environments which carries with it new ways to utilise copyright protected works, not only regarding university courses. Moreover, e-learning solutions have increased the interest that institutions of higher education share in their capacity as an employer of teachers and researchers, who, by virtue of their employment, produce valuable copyrightable works such as presentations, web-published compendiums, instruction videos, etc. Nonetheless, despite that such institutions by virtue of their employer status have the rights to lead and distribute work, it has historically been the case that teachers are given autonomy to teach in a manner that the teacher (and not their employers) has considered to be best, e.g., by choosing the course literature, scheduling teaching, choosing tuition and grading forms. This 'package' is not considered to be the product of work belonging to the institutions as it is strongly tied personally to the teacher and falls within the framework of the teachers exemption. Teaching in digital environments and the virtual teacher, however, place these issues at their extreme.[11]

2.3 The Employees' Duty to Give Notice

An express positive duty to give notice to an employer of the creation of intellectual property is only present in the Act (1949:345) on the Right to Employee's Inventions,

10. Lärarundantaget.
11. See further Carlson & Sanna Wolk, *Virtual teachers – A copyright paradox?*, in *Individualism and Collectiveness in Intellectual Property Law* 321 et seq. (Rosén ed., Edward Elgar, Cheltenham, UK 2012)

which is applicable only to patentable inventions.[12] Statutes regulating other intellectual property rights do not expressly contain such a duty, however an employee is by implication still required to give notice to their employer, despite the lack of express provisions, as a result of the loyalty principle in employment law. The general duty of the employee's loyalty towards the employer is an unwritten rule complimenting the employment contract and is a general principle of law.

2.4 Employee Remuneration Right

Special rules regulating employees' right to remuneration, other than for inventions, do not exist.[13] If and to what extent creators of other intellectual property are entitled to remuneration depends instead on any agreement that has been entered into by the employer and the employee.

3 EMPLOYEES' SEMICONDUCTOR CHIPS

3.1 The Legal Framework

The general principle in Section 40a Copyright Act is mirrored in the law regulating topographies for semiconductors. The provision on the employer's right to the employee's printed circuit board was enacted in 1995 in Section 3(2) of the Act (1992:1685) on the Protection of Topographies for Semiconductor Products.[14] The provision can be seen as implementing Directive 87/54/EC on the legal protection of topographies for semiconductor products, even though the presumption rule in the directive is discretionary. It follows from the preparatory works of the legislation that regulation is motivated because employees' development of circuit designs has much in common with the development of computer programs; it is not unusual that several people participate in such development, it is a capital intensive activity, and computer programs are often installed in circuit boards. Furthermore the enactment of the law did not provoke any discussions. [15]

3.2 Ownership of Rights

An employer may only obtain, as mentioned, a derivative right from an employee in order to exploit the chip right. However, from Section 3(2) of the Act on the Protection of Topographies for Semiconductor Products it follows that the employee's chip right belongs to the employer if it was developed in the course of the employee's duties. The basic condition for the application of Section 3(2) of the Act is that no other agreement has been entered into between the employer and the employee. Where the parties have

12. See below section 4.
13. See below section 4.
14. Lag (1992:1685) om skydd för kretsmönster för halvledarprodukter.
15. See Government Bill 1992/93:48 p. 51.

concluded another agreement, it will invalidate the presumption, and the scope of the employer's right will be determined instead by that agreement.

3.3 Employees' Duty to Give Notice

An express positive duty to give notice to an employer of the creation of intellectual property is only present in the Act on the Right to Employee's Inventions, which is applicable only to patentable inventions.[16] Statutes regulating other intellectual property rights do not expressly contain such a duty, however an employee is by implication still required to give notice to their employer, despite the lack of express provisions, as a result of the loyalty principle in employment law. The general duty of the employee's loyalty towards the employer is an unwritten rule complimenting the employment contract and is a general principle of law.

3.4 Employee Remuneration Right

Special rules regulating employees' right to remuneration, other than for inventions, do not exist.[17] If and to what extent creators of other intellectual property are entitled to remuneration depends instead on any agreement that has been entered into by the employer and the employee.

4 EMPLOYEES' INVENTIONS

4.1 The Legal Framework

The Act (1949:345) on the Right to Employee's Inventions[18] contains a detailed description of the parties' rights and obligations. The law applies to patentable inventions made by employees in the private as well as public sector. The Act is, however, not applicable to inventions made by university researchers (Section 1(2) of the Act).

In addition, since the Act is mostly non-mandatory, rights in employee inventions in the private sector are regulated by the Collective Agreement on the Right to Employee Inventions of 1995 or 2015.

Since 1 April 1995 there has been an agreement between the Confederation of Swedish Enterprise and the trade union (PTK) on the right to employees' inventions. However, the Collective Agreement was terminated by PTK on 10 December 2014 which concurrently asked for the negotiation of a new agreement. The reason for this was that PTK wanted to renegotiate the agreement to seek conditions that were more favourable to its members. The negotiations resulted in a new collective agreement, namely the 'Contract concerning the right to employees' inventions of Swedish

16. See below section 4.
17. See below section 4.
18. Lag (1949:345) om rätten till arbetstagares uppfinningar.

Enterprise and PTK,' the so-called Inventors' Agreement. As of 2 July 2015 the agreement is valid until further notice with a termination notice period of one year. Inventions notified to the employer from 1 December 2015 should be handled according to the agreement of 2015, whereas previously notified inventions are examined under the agreement of 1995.

The Inventors' Agreement of 2015 contains much of that of the agreement of 1995. This is the case regarding, e.g., the categorisation of inventions, employee notification, employer rights to employee inventions (so-called A-, B- and C-inventions), patent applications after terminated employments and the handling of disputes. The factors to be taken into account when calculating the employee's reasonable compensation for the right that the employer acquires are also the same. Nevertheless, a number of important changes have been added to the agreement, e.g., guidelines for flat-rate compensation to employees as well as that the provision of fair compensation is now mandatory. Also, the limitation period has been extended regarding when an action later to be brought before the tribunal. Furthermore, the employee's right to retry a decision made on compensation has been taken out of the agreement. Another important change in the agreement of 2015, compared to the agreement of 1995, concerns the limitation periods for initiating proceedings. The collective agreement now clarifies that proceedings, after local and central negotiations, must be initiated within ten years from the date the patent application was handed in. In the event that the employer has not submitted an application, the time is counted from the day on which the invention was reported to the employer. If the notification is not made within that time the claim cannot be brought. This is an important change because it normally takes a long time, somewhere between 5 and 15 years (but sometimes even up to 20 years), before the value of the invention can be established.

4.2 Ownership of Rights

4.2.1 Employees in the Private and Public Sectors

In Sweden, employee inventions are divided into three categories: *research inventions*, *mixed inventions* and *other inventions*. The category of a given invention depends on the strength of the link between the invention and the employment relationship. The primary requirement for all three categories, however, is that the utilisation of the invention must fall within the sphere of the business activity of the employer. 'Business activity' refers in this context to the company's industrial and R&D activities.

Under Section 3(1) of the Act on the Right to Employee's Inventions, an invention should be categorised as a *research invention* if the employee's duties mainly include research or invention activities, and where the invention has come into existence substantially as a result of these activities. The employer may acquire an exclusive right to such inventions.

Mixed inventions are inventions whose utilisation fall within the sphere of activity of the employer's business and have come into existence in connection with the employment other than that stated above (see Section 3(2) of the Act on the Right to Employee's Inventions). Regarding such inventions, the employer may acquire the right to exploit the invention in their business without hindrance from the employee (non-exclusive licence). The employer is also given precedence over anyone else to enter into an agreement with the employee in order to acquire a more comprehensive right than that just stated.

Finally, as regards *other inventions* in Section 3(3) of the Act on the Right to Employee's Inventions, which have not come into existence in connection with the employment, but which fall, nevertheless, within the sphere of the employer's business activity, the employer has a discretionary right to conclude an agreement on conditions stipulated by the employee.

4.2.2 University Researchers' Inventions

The Act on the Right to Employee's Inventions applies, as described above, to patentable inventions made by employees in their private or public employment. However, from Section 1(2) of the Act it follows that the Act is not applicable to inventions made by university researchers (*teacher exception*). Swedish universities are supposed to carry out educational activities, rather than to pursue commercial activities, and as a matter of principle teachers and researchers employed by universities own their inventions. This is prescribed by the provisions of the Higher Education Act (1992:1434).[19] Hence, the objectives of the Swedish universities do not include the pursuit of commercial activities by exploitation of their researchers' inventions.

The social and economic changes, as well as changes in electronic commerce, have however led to the fact that certain universities are becoming more and more interested in exercising influence over the application and exploitation of their researchers' inventions. It is now up to the Swedish government to consider whether the universities ought to take care of their researchers' patents by means of abolishing the teacher's exception and revising the Higher Education Act for the purpose of improving the universities' finances.

4.3 The Employees' Duty to Give Notice

The Act on the Right to Employee's Inventions contains provisions concerning an employee's duty to give notice (Sections 4–5). The employee-inventor is to 'immediately' inform the employer of an invention whose utilisation falls within the sphere of activity of the employer. The employer is to inform the employee whether they wish to acquire the right to the invention within four months from the date of receipt of the notice.

19. Högskolelag (1992:1434).

4.4 Employee Remuneration Right

The Act on the Right to Employee's Inventions is mostly non-mandatory, and individual contracts or collective agreements may govern the transfer of patent rights to inventions. However, the most important employee rights are mandatory. In Sweden, employee inventors have a mandatory right to remuneration which serves as compensation for the rights transferred to the employer (see Section 6). This means that agreements which limit the employee's right entered into before the creation of the invention are invalid. If the parties, on the other hand, agree after the creation of the invention that some, or none, remuneration is payable, such agreement will be respected.

An employee inventor has to raise any claim for compensation under Section 6 of the Act on the Right to Employee's Inventions within ten years from the day the employee informed the employer about a developed invention.[20]

4.4.1 Mandatory Reasonable Remuneration

Section 6 of the Act on the Right to Employee's Inventions imposes that remuneration is to be *reasonable*. The concept of reasonableness has no fixed boundaries, which is why there are no uniform criteria specifying what is to be understood as reasonable remuneration. Nevertheless, Section 6 puts forth certain factors that may influence the determination of remuneration. The first refers to the *value of the invention* (and in relevant cases, to the scope of patent claims); the second to the *terms of employment* (the weaker the connection between the scope of employment and the invention, the stronger the right to remuneration); and the third to the *scope of the rights acquired* by the employer. In addition to these, a number of other factors may be considered when determining compensation. As regards research inventions, however, employee-inventors are only entitled to remuneration if the value of the invention exceeds that which can be considered as reasonable for the employee to accomplish, having regard to the inventor's salary or other benefits and advantages of employment.

In the private sector, the Collective Agreement on the Right to Employee Inventions of 1995 and 2015 contains a recommendation for a compensation scheme. Calculating the reasonable compensation to the employed inventor is based on the same requirements in the 2015 agreement as those of the agreement of 1995, regarding both A-inventions (invention made in the performance of the employee's duties/ special assignments) and B-inventions (invention which is not an A- invention). In the 1995 agreement SEK 1,700–2,550 should be paid when the employer was notified of the invention and an additional SEK 2,800–22,500 when the patent is granted in Sweden. The 2015 agreement has, however, introduced a new comment on the compensation to be paid for the A- and B-inventions. The comment expresses that an employee who has made an A-invention is only entitled to a special compensation in addition to salary if

20. See s. 2 of Act (1981:130) on Limitation, and clause 6 of the Collective Agreement on the Right to Employee Inventions of 1995.

the value of the rights to the invention that the employer has taken over exceeds what the employee- in terms of salary, flat-rate compensation and other employment benefits- has been assumed perform. This must also be related to the value of the invention. Furthermore, the commentary states that, if there are grounds for further compensation to be paid to an A- or B-invention, royalty payments may in some cases be more appropriate than a lump sum. Changes have also been made regarding the standard amount that is payable to the employed inventor. Regarding the flat rates, a comment in the 1995 agreement said that the compensation should be set in advance at flat rate. Furthermore, as mentioned, the comment stated that the compensation would preferably start at SEK 1,700–2,550 at the invention notification and additional SEK 2,800–22,500 if patent is applied for the invention and leads to a patent. However, these amounts were not indexed, which in practice meant that once relatively high amounts in later currency were perceived as low. In the current agreement, this has changed and the flat rate is linked to the base amount instead (as of 2017 it is estimated to SEK 44,500). According to the wording of the agreement a half a base amount should be paid – but in the event that the invention has significant value to the company should the sum should increase to an entire base amount.

It is not very common for the issue of remuneration to come under the scrutiny of the courts. In the only two cases from higher courts, both from the Swedish Labour Court, considering the reasonableness of the amount paid, the remuneration was set relatively low (SEK 100,000–300,000 for employer's licence right).[21] However, the two precedents from the Labour Court are over 25 years old, and more recent cases from lower courts have determined higher amounts as reasonable compensation, up to SEK 500,000.

A special source of law for matters relating to the Act on the Right to Employee's Inventions is composed of statements by the National Board for Employee Inventions. The National Board for Employee Inventions is an authority with the mandate to decide, through non-binding advisory opinions, disputes concerning the interpretation of the Act. As mentioned, a significant part of the private sector of the Swedish labour market is regulated by collective agreements, through the Collective Agreement on the Right to Employee Inventions of 1995. The Industrial Inventions Board issues arbitral awards, with binding effect on the parties, in questions concerning the application of the collective agreement.

During the 2000s, the National Board for Employee Inventions has issued opinions in several cases, which have all concerned the issue of reasonable remuneration in addition to salary according to Section 6 of the Act on the Right to Employee's Inventions. The Industrial Inventions Board, has also issued arbitral awards concerning reasonable remuneration according to the Collective Agreement on the Right to Employee Inventions. The Board delivers two or three decisions each year.

However, the decisions by the two Boards are not public, for which reason there is no case law explaining how they assess questions relating to remuneration.

21. See Labour Court case 1982 No. 21 and Labour Court case 1983 No. 19.

Remuneration ordered by the Boards to employee-inventors during the 2000s ranges between SEK 200,000 and SEK 6,000 million.

5 EMPLOYEES' UTILITY MODELS

Swedish law does not protect utility models.

6 EMPLOYEES' DESIGN RIGHTS

6.1 The Legal Framework

The Swedish Design Protection Act (1970:485)[22] does not contain any special provisions concerning employers' rights to employees' designs other than the general provisions on transfer of ownership in Sections 25–27 of the Act. The absence of provisions concerning this issue does not mean that there have been many conflicts in the area. Nevertheless, following the increasing importance of design rights as a form of protection, disputes concerning employees' rights in the designs developed by them may become more frequent in the future.

6.2 Ownership of Rights

In cases where there is no express agreement, the legal literature provides certain guidelines for the determination of the employer's rights. Regarding industrial designs the Act on the Right to Employee's Inventions may be consulted. Where a design protected by copyright for applied art is concerned, rights transfer shall be governed by the rule of thumb concerning copyright transfer mentioned above. This can be explained by the fact that splitting rights transfer cannot be considered as suitable with regard to one and the same object. Practical implications of the above have not been established, however, and the legal position is unclear.

However, considering the provisions of Council Regulation (EC) No. 6/2002 on Community designs, it is no longer desirable to rely on the provisions of the Act on the Right to Employee's Inventions or on the general principles of copyright law when determining employers' rights to employees' designs. Especially in cases where a design registered in Sweden may be simultaneously protected as an unregistered Community design. A reasonable recommendation would thus be that rights transfer ought to be inspired by the presumptive provisions of Article 14(3) of the Regulation. Additionally, the governmental enquiry into design protection concluded that the presumptive rule specified in the EC Regulation expresses something that already applies in Sweden.[23]

Article 14(3) of the Regulation on Community designs mirrors the general and fundamental principle in Swedish law concerning the employer's right to employees'

22. Mönsterskyddslag (1970:485).
23. See the Swedish Government Official Reports 2001:68 p. 227.

products of intellectual work; in accordance with the provisions of the Regulation an employer is also presumed to have an exclusive right to Community designs. Under Swedish industrial designs law this means that all rights to a design developed by an employee vest in the employer if the design has been developed by the employee in the execution of their duties or following instructions given by the employer.

It should be also mentioned that the Designs Protection Enquiry from 2001 proposed that Swedish law should expressly regulate the relationship between employers and employees. The proposal formulated the following presumptive provision: 'Unless otherwise agreed, the right to a design developed by an employee in the execution of his duties shall vest in the employer.' [24]

6.3 The Employee's Duty to Give Notice

An express positive duty to give notice to an employer of the creation of intellectual property is only present in the Act on the Right to Employee's Inventions, which is applicable only to patentable inventions.[25] Statutes regulating other intellectual property rights do not expressly contain such a duty, however an employee is by implication still required to give notice to their employer, despite the lack of express provisions, as a result of the loyalty principle in employment law. The general duty of the employee's loyalty towards the employer is an unwritten rule complementing the employment contract and is a general principle of law.

6.4 Employee Remuneration Right

Special rules regulating employees' right to remuneration, other than for inventions, do not exist.[26] If and to what extent creators of other intellectual property are entitled to remuneration depends instead on any agreement that has been entered into by the employer and the employee.

7 EMPLOYEES' PLANT VARIETIES

7.1 The Legal Framework

The question of the employer's right to the employee's plant varieties has been left to the parties to decide because the legal framework does not contain any particular rule regulating the legal relationship in question. The Act (1997:306) on the Protection of Plant Breeders' Rights[27] only contains general provisions on the acquisition of rights (see Chapter 7 Sections 2–3 of the Act). The absence of a specific provision in the law has not, however, given rise to any disputes because it is not common that employees

24. See the Swedish Government Official Reports 2001:68 p. 227.
25. See above section 4.
26. See above section 4.
27. Växtförädlarrättslag (1997:306).

claim a better right. There are few actors in the plant industry that register plant varieties, and the rights are usually regulated in agreements. However, absent any express clauses, the legal framework is less than clear in comparison to other rights where the starting point is defined by statutory provisions, despite these being dispositive.

7.2 Ownership of Rights

The potential rights of the employer regarding the employee's plant varieties are, like those concerning employees' inventions, not dependent on registration – it is sufficient that the right be registrable. It is accordingly the right to register a plant variety right through the Swedish national system that the employer would be provided with (registration right). The same is applicable for the Community plant variety right, since its existence is assessed according to national legislation when the employment relationship is concerned.[28]

The question of the employer's rights following breeding or development of new plant varieties has been discussed in the preparatory works of the Act on the Protection of Plant Breeders' Rights.[29] It is pointed out in the preparatory works that the exclusive rights belong in the first place to the employee, and any potential transfer of rights must consider general principles of employment law. It is further provided that there need not be an express agreement on rights transfer for the employer to assume ownership of a plant variety right. There is room for the argument that when a person is employed as a breeder, the employment agreement may, as a rule, be interpreted as containing an implied term of transfer of rights. It is however not possible to decisively answer whether rights are transferred generally, especially in the absence of a special agreement, and the question will have to be decided in every particular case. If, on the other hand, transfer of rights occurs, then the right that is vested in the employer is final and transferred in its entirety (exclusive right).

In the context of university researchers' plant varieties, the default point may, analogously to the exceptions in the Act on the Rights of Employee's Inventions, be that researchers who breed or develop protectable plant varieties are provided with the right to their research results. Another reason favouring this outcome is the fact that plant breeders' rights and patent law are increasingly overlapping as a result of new technology. It is for instance not uncommon that genetic inventions are also constituents of new plant varieties and there are specific rules on cross-licensing these two forms of protection.

28. See Art. 11(4) Council Regulation (EC) No. 2100/94 on Community plant variety rights
29. See Government Bill 1971:40 p. 36,73 and 90, Government Bill 1996/97:128 p. 75 et seq. and Ministry Memorandum 1995:69 p. 110 et seq.

7.3 The Employee's Duty to Give Notice

An express positive duty to give notice to an employer of the creation of intellectual property is only present in the Act on the Right to Employee's Inventions, which is applicable only to patentable inventions.[30] Statutes regulating other intellectual property rights do not expressly contain such a duty, however an employee is by implication still required to give notice to their employer, despite the lack of express provisions, as a result of the loyalty principle in employment law. The general duty of the employee's loyalty towards the employer is an unwritten rule complimenting the employment contract and is a general principle of law.

7.4 Employee Remuneration Right

Special rules regulating employees' right to remuneration, other than for inventions, do not exist.[31] If and to what extent creators of other intellectual property are entitled to remuneration depends instead on any agreement that has been entered into by the employer and the employee.

8 INSTANCES FOR DISPUTES

There are several different courts and boards that can rule on disputes between employees' and employers' duties and rights.

8.1 Disputes Concerning Employees' Copyright, Semiconductor Chips, Designs and Plant Varieties

A dispute relating to employees' intellectual property rights, with an exception for inventions, are in some disputes heard by the Labour Court and in others by a District Court. However, a decision by a District Court may be appealed to the Labour Court because a dispute on employees' intellectual property rights relates to the relationship between an employer and their employee, and is considered to be a labour dispute.[32]

8.2 Disputes Concerning Employees' Inventions

Disputes concerning the interpretation of the Act on the Rights of Employee's Inventions are settled by the Patent and Market Court,[33] which has exclusive jurisdiction by

30. See section 4 above.
31. See section 4 above.
32. See Scania and Blekinge Court of Appeal, decision Ö 611-10, 11 March 2010 and Labour Court decision B 19/10, 15 March 2010 (if an employer had a right to employees computer program under s. 40a of the Copyright Act).
33. Patent- och marknadsdomstolen.

virtue of the Patents Act (1967:837),[34] with the Patent and Market Court of Appeal[35] as the second instance and the Supreme Court[36] as final instance. There is additionally the National Board on Employees' Inventions,[37] which on the request of the parties or the court may provide advisory opinions in disputes concerning employees' inventions. Disputes arising out of the invention collective agreement are heard by an arbitral body – the Industrial Inventions Board.[38]

The National Board on Employee's Inventions, the members of which are appointed by the parties to the dispute, is an authority with the task to rule, through non-binding advisory opinions, on disputes that concern the application of the Act on the Right to Employee's Inventions (see Section 10 of the Act). Opinions by the National Board on Employee's Inventions may, e.g., concern the employer's right to acquire an employee's invention, the size of the remuneration or the category of invention that is applicable. The National Board on Employee's Inventions is composed of a chairman and six members with one deputy for each member. Two of the members and the chairman are appointed among persons who are independent of the employer's and the employee's interests. The chairman and one of these members are experienced judges, while the other member has special insight into, and experience in, patent rights and associated issues. In addition, two members are appointed to each of the parties as representatives.

Disputes regarding the application, and interpretation, of the Collective Agreement on the Right to Employee's Inventions are heard by the Industrial Inventions Board. The Industrial Inventions Board is an arbitral tribunal and cases before it are determined as arbitration awards with binding effect on the parties. The Industrial Inventions Board may, e.g., determine the category a certain invention belongs to or the size of the remuneration. The Industrial Inventions Board is made up of a chairman and two members, along with deputies. The chairman of the arbitral tribunal is appointed jointly by the employer and employee associations, while the parties to the dispute appoint one member each.

9 CONCLUDING REMARKS

9.1 Overlapping Intellectual Property Rights

Thus far provisions concerning each type of intellectual property and the general principles concerning proprietary rights appearing in an employment relationship have been described. Intellectual property can no longer be divided and segregated so simply. In today's world, where borders between different types of protection often diffuse and overlap, it is not uncommon that one and the same intellectual property product is covered by several intellectual property rights (accumulation).

34. Patentlagen (1967:837).
35. Patent- och marknadsöverdomstolen.
36. Högsta domstolen.
37. Statens nämnd för arbetstagares uppfinningar.
38. Industrins Uppfinnarnämnd.

It is often advantageous for a rights holder that the boundary between different proprietary rights is blurred, especially from the point of view of competition. Intellectual property becomes an important strategic asset for companies. Overlapping protection may also create more flexibility and opportunities for those whose works are copied, which can be explained, among other things, by the special character of intellectual property.

Accumulation does not normally entail any special difficulties regarding different types of intellectual property protection, where the various protective schemes are gathered together in one hand. The fact that sole proprietary rights show an increasing tendency to overlapping may have ramifications, however, for rights transfer in the employment relationship. Today, the necessary prerequisites for, and the scope of, employer's rights differ depending on the type of the applicable intellectual property regulatory framework.

The overlapping rights which come into play when rights are transferred in employment relationships concern copyright *vis-à-vis* design rights in the context of the design of products, and copyright and patent rights with respect to computer related inventions. The latter is relevant despite the fact that the object of protection may de facto not be the same in both frameworks. Problems of applicability may also be relevant in the context of overlapping intellectual property rights. The Patents Act excludes plant varieties from patentability, however biotechnological development has led to some form of integration of patent law and plant breeder's rights. A plant variety that is protected as a plant breeder's right may also contain a patent protected genetic invention, or a plant variety modified in such a way may be encapsulated by a process patent where a gene creates certain attractive or necessary characteristics. In Sweden there are however only a handful of entities which register plant varieties and the question of the employer's or employee's rights is in this context of minor practical relevance. Overlapping protection for printed circuit boards is probably also without particular practical importance, however it should be noted that by virtue of Section 10(2) of the Copyright Act copyright protection does not subsist in topographies for semiconductor products. Laws regulating topographies for semiconductor products do not, however, exclude other rights, e.g., patents or design protection.[39]

9.2 Interconnection Between Copyright Law and Patent Law

A special relationship has developed over the years between copyright law and patent law in the area of protection of computer programs. In special cases a computer program may contain and express a patentable invention (as regards its technical character), and it may also be embraced by copyright law (as regards its administrative character). This means that both the Act on the Right to Employee's Inventions and the presumptive provisions of the Copyright Act may become applicable. Trying to apply

39. See also Art.10 of the Directive on the Legal Protection of Topographies of Semiconductor Products (87/54/EEC). Cf. recital 7 of the Directive 98/71/EC on the Legal Protection of Designs and recital 31 of the Regulation 6/2002/EC on Community Designs.

the statutory provisions in such a case will result in a situation in which the separate proprietary rights will not be gathered together in one hand

This may result in the employer enjoying copyright protection in a computer program pursuant to the provisions of Section 40a of the Copyright Act, and the employee enjoying patent rights pursuant to provisions of Section 3(2) of the Act on the Right to Employee's Inventions. In such a case the employer's copyright may be restricted by the employee's patent rights. On the other hand, the employee's exercise of their patent right may be limited by the employer's copyright. It is true that there is a possibility to separate the literary side of the contribution from the technical side. However, the employer cannot use another code for the same result, as it would likely lead to an infringement of the employee's patent rights.

If the employee has not applied for a patent, they will hold no proprietary rights which could limit the employer's copyright. Furthermore, there will be no problems in the sphere governed by the collective agreement, since class A inventions cover both inventions which have come into existence as a result of research activities and those resulting from the employee's duties (Section 3(1)-(2) of the Act on the Right to Employee's Inventions).

9.3 Interconnection Between Designs Law and Copyright Law

Overlapping of rights may also create problems when a design is an object of applied art within the meaning of copyright legislation. Under Article 14(3) of the Community Design Regulation, and the general principles of Swedish design law, the employer has an exclusive right to those parts of an object which fall under design law. Under the copyright principles the employer will only have a restricted right of use however.

When a design is covered by protection which extends to works of applied art, various questions arise concerning its utilisation by the parties. For example, are the employee's copyright prerogatives restricted during the time when the employer enjoys design protection? What will apply when the employer's operations expand and new forms of utilisation become part of their activities? Will the employee's copyright extend over the employer's design rights? What happens when design protection ceases to apply? Will the employer have to obtain permission from the employee to be able to use the design in a new way or in other activities?

9.4 Concluding Remarks Concerning Overlapping Rights

There is no general answer to the question of how the parties' utilisation of their respective proprietary rights ought to be handled when an object is covered by several intellectual property rights. In each particular case a separate solution has to be found. This is something to think about when parties intend to conclude an agreement, which is the common ground for each of the various intellectual property rights.

BIBLIOGRAPHY

Table of Sources

Laura Carlson & Sanna Wolk, *Virtual teachers – A copyright paradox?*, in *Individualism and Collectiveness in Intellectual Property Law*, p 321 et. seq (Jan Rosén ed., Edward Elgar, Cheltenham, UK 2012).

Sanna Wolk, *Arbetstagares immaterialrätter* (Norstedts Juridik, Stockholm, Sweden 2008).

Official Reports Etc.

Government Official Reports 1956:25.
Government Official Reports 2001:68.
Government Bill 1960:17.

Table of Legislation

Act on Copyright in Literary and Artistic Works (1960:729).
Act (1997:306) on the Protection of Plant Breeders' Rights.
Design Protection Act (1970:485).
Patents Act (1967:837).
Directive 2009/24/EC on the legal protection of computer programs.
Directive 98/71/EC on the Legal Protection of Designs.
Regulation 6/2002/EC on Community Designs.

Table of Cases

Labour Court case 1982 no. 21.
Labour Court case 1983 no. 19.
Labour Court decision B 19/10, 15 March 2010.
Lund District Court, case T 1216-10, 9 March 2010.
Court of Appeal for Western Sweden, case FT 3542-07, 31 October 2008.
Scania and Blekinge Court of Appeal, decision Ö 611-10, 11 March 2010.

CHAPTER 21

Switzerland

Yaniv Benhamou & Cyrill Rieder[*]

1 BACKGROUND

In Switzerland, the legal framework for employees' intellectual property rights is provided by both intellectual property law as well as general rules of law.

Swiss intellectual property law is governed by the *creator principle*: the creator is placed in the foreground, independently of their status as an employee or independent creator, and only the person who has actually created the work becomes the *original holder* of the right.[1] An employee-creator thus has the same intellectual property right as other non-employed creators, and their employer may only receive a derivative right from the employee by means of an agreement in order to exploit the product and become a *derivative holder* (except where a specific legal transfer to the employer is provided by law, i.e. for service inventions and designs, and for computer programs).

Whether an agreement entails provisions affecting transfer of rights (express or implied), as well as what the scope of such transfer (by means of an assignment or a licence) is, will be interpreted according to the general *rules of contract law* (subjective and objective interpretation) and by the specific interpretation of rules of intellectual property law (in particular the *theory of the finality* for copyright).[2] This interpretation plays a crucial role in intellectual property law because it applies to all IPR and to all relationships, i.e. it applies not only to employment contracts but also to agency

* The authors wish to thank Prof. Jacques de Werra and Dr. Konrad Becker for their fruitful comments, as well as James Walker for his reading of the text.

1. For copyright, Ivan Cherpillod, *Commentaire romand de la propriété intellectuelle* (de Werra/Gilliéron ed., Basel 2013), LDA 6 N 5. For patents, Nathalie Tissot, *Commentaire romand de la propriété intellectuelle* (de Werra/Gilliéron ed., Basel 2013), LBI 3 N 3.
2. See section 2.1.

contracts and contracts for work and service.[3] It also plays a crucial role because it determines the scope of the transfer, i.e. whether the transfer will be a licence or an assignment, the latter being larger in scope as it allows the recipient the right to transfer to third parties. Moreover, in the case of an employer's bankruptcy, the right belongs to the employer's assets with an assignment, while it remains the employee's asset in the case of a licence.[4]

Agreements affecting the transfer of rights are treated differently by the various intellectual property regulations. For copyright, there are no special provisions, i.e. there is no legal transfer to the employer (except for computer programs). Thus, the creator-employee is always deemed to be the original holder of copyrights, and the employer can only acquire a derivative right by means of an agreement.[5] However, for inventions, designs, semiconductor chips, and plant varieties, there is a specific provision providing for a legal transfer to the employer. The employer is deemed to be the original holder of rights for works produced by the employee in the performance of contractual obligations, and the derivative holder of rights for works produced by the employee through the performance of their contractual obligations.[6]

Moreover, employment relationships under private law are regulated by the Code of Obligations (CO),[7] within the employment law section (Articles 319-362 CO), which outlines the respective duties of the parties. The main duties of the employer include paying the agreed or customary salary (Articles 319(1), 322(1) CO) and safeguarding the employee's health and personal rights (Article 328 CO). The main duties of the employee are to assume the tasks personally, according to the employer's instructions, to exercise due care and loyalty, and to hand over any work produced in the course of his/her contractual activities (Article 321b CO).[8]

However, employment law only applies to employment contracts, i.e. those contracts characterised by four criteria (a subordinate relationship, effort to be provided, against salary, and with duration, the first criterion being the most important), and does not apply to other contracts.[9] Thus, the legal transfer for service inventions and designs provided in employment law does not apply to other contracts, such as agency contracts or contracts for work and service (although legal transfers are

3. See Roland von Büren, *Der Übergang von Immaterialgüterrechten, Schweizerisches Immaterialgüter und Wettbewerbsrecht* vol. I/1, 269 (Basel/Geneva/Munich 2002); and Tissot, *supra* n. 1, LBI 3 N 22, who notes that the *theory of the finality* applies to 'all contracts regarding the creation and the transfer of rights related to intangible assets'.
4. Reto M. Hilty, Lizenzvertragsrecht: Systematisierung und Typisierung aus schutz- und schuldrechtlicher Sicht 754 et seq. (Zurich 2001).
5. See section 2.
6. See sections 4 and 6.
7. Federal Act on the Amendment of the Swiss Civil Code of 30 March 1911 (RS 220), http://www.admin.ch/opc/en/classified-compilation/19110009/201401010000/220.pdf (accessed 23 May 2014).
8. Regarding due care and loyalty, one can think of the obligation to keep confidentiality about the invention and to collaborate for the registration procedure, see Tissot, *supra* n. 1, LBI 3 N 36. Regarding hand-over the work produced, one can think of the obligation to deliver the source code programmed by the employee, see Jacques de Werra, *Commentaire romand de la propriété intellectuelle* (de Werra/Gilliéron ed., Basel 2013), LDA 17 N 6.
9. Ivan Cherpillod, *Protection des logiciels et des bases de données: la révision du droit d'auteur en Suisse*, in RSPI/SMI 49, 56 et seq. (1993).

sometimes provided elsewhere).[10] Moreover, employment law only applies to the private sector. The situation of employees in the public sector is principally regulated by the Federal Personnel Act[11] and cantonal and municipal regulations. In this area, each public institution (i.e. an institution that is financed by public resources) can regulate IPR through its own provisions (e.g. specific Regulations for IPR of researchers at universities). Thus, there are different solutions for employees of public institutions: some regulations provide that principles of private law, in particular employment law, apply *mutatis mutandis*; others provide that IPR belong to the creator-professor (or creator-researcher) or that they belong to the public institution (e.g. ETH Regulation, Act of the University of Geneva).[12]

2 EMPLOYEES' COPYRIGHT

2.1 The Legal Framework

For copyright, there are no special provisions regarding employment relationships. The Federal Act on Copyright (CopA)[13] determines who is the author, co-author or performer according to the Creator Principle, but provides no legal transfer to the employer (except for computer programs).

The creator-employee is always deemed to be the original holder of copyrights and the employer can only acquire a derivative right by means of an agreement. Thus, employee-employer relationships, and in particular agreements affecting the transfer of copyright, are governed by the general rules of contract law and by certain rules of interpretation of the CopA.

2.2 Ownership of Rights

2.2.1 Employees' Copyright and Performers' Rights

According to the CopA, the author is the natural person who has created the work (Article 6 CopA), while works created by several persons belong jointly to all

10. For instance the agency contract provides a legal transfer to the client (Art. 401 CO), but the client is the derivative holder, and the inventor the original holder (Tissot, *supra* n. 1, LBI 3 N 26).
11. Federal Personal Act of the Confederation of 24 March 2000 (RS 172.220.1)
12. See Art. 18 of the Federal Act on the Federal Institutes of Technology of 4 October 1991 (RS 414.110), https://www.admin.ch/opc/en/classified-compilation/19910256/201705010000/414.110.pdf (accessed 20 July 2017); Art. 15(1) of the cantonal Act of the University of Geneva of 13 June 2008(C 1 30), https://www.geneve.ch/legislation/rsg/f/s/rsg_C1_30.html (accessed 20 July 2017).
13. Federal Act on Copyright and Related Rights of 9 October 1992 (RS 231.1), http://www.admin.ch/ch/e/rs/2/231.1.en.pdf (accessed 23 May 2014). This CopA (last amendment in 2008) is under revision since 2012, which led to a draft for a revised CopA (11 December 2015) and the result of the consultation procedure (2 December 2016). See https://www.ige.ch/en/law-and-policy/national-ip-law/copyright-law/modernisation-of-copyright.html (accessed 20 July 2014).

co-authors (Article 7(1) CopA). This is known as the *creator principle*: only the creative contribution by an individual entitles that person to hold copyright (*original rightholder*).[14] Apart from computer programs, Swiss Law contains no instances where copyright is bestowed on legal entities from the start; only a transfer by contract can give the legal entity the whole or part of the copyright (*derivative rightholder*).[15]

The rights of performers follow the same principles: the performer is the natural person who performs or participates artistically in the performance of the copyrighted work (Article 33(1) CopA).[16] Thus, the rights of performers belong to the performer (*original rightholder*) and only a transfer by contract can give the employer the whole or part of those rights (*derivative rightholder*).[17]

For works made for hire by an employee, the employee is deemed to be an author as soon as he/she contributes to the work in a creative way. The sole criterion is that it is the creative contribution of the employee; the status of the employee being irrelevant (auxiliary, assistant, manager, director, professor, etc.).[18]

For instance, in the case of audio-visual works, where the legislator did not want a special provision entitling the producer to all the rights and depriving the other participants of their rights, Swiss practice distinguishes between main authors (art director, scriptwriter, scenario writer, composer of the music) and the ancillary contributors who are not entitled to any rights (camera man, cutter, hair stylist, dress and set designers).[19] However, an ancillary contributor may be considered to be a co-author if their contribution is creative and subject to copyright protection (e.g. the person in charge of visual effects or dress and set designers creating a copyrighted work),[20] or considered as a performer if their contribution participates artistically in the performance of the copyrighted work (e.g. a comedian). The use of pre-existing music or scenario for a film is in principle a derivative work, not a joint work. It is however considered to be a joint work when the music is specifically created for the film or the scenario is adapted during the filming and requires collaboration with the persons in charge of that pre-existing work.[21]

In the case of scientific works, one can also distinguish between authors or co-authors, who contribute to the work in a creative way (e.g. collaborator, assistant, professor, student), and the ancillary contributors (e.g. a collaborator who just collects data, does some experiments, writes summary reports, corrects mistakes or reads critically).[22] Furthermore, a person who merely gives starting ideas or general topics is

14. Cherpillod, *supra* n. 1, LDA 6 N 5 and LDA 7 N 6.
15. François Dessemontet, *Intellectual Property Law in Switzerland* 43 (Zurich 2012); Cherpillod, *supra* n. 1, LDA 6 N 5.
16. The provision is based on the international law, in particular on Art. 2(a) WPPT that gives the following definition of performers: 'actors, singers, musicians, dancers, and other persons who ... otherwise perform literary or artistic works or expressions of folklore'.
17. Catherine Mettraux Kauthen, *Commentaire romand de la propriété intellectuelle*, (de Werra/ Gilliéron ed., Basel 2013), LDA 33 N 11.
18. Cherpillod, *supra* n. 1, LDA 7 N 8-9.
19. Dessemontet, *supra* n. 15, 44.
20. Willi Egloff, *Urheberrecht und Urhebervertragsrecht in der audiovisuellen Produktion* 17 (sic! 1998); Cherpillod, *supra* n. 1, LDA 7 N 14.
21. Supreme Court, 699 (sic! 2003). Cherpillod, *supra* n. 1, LDA 7 N 15-16.
22. Cherpillod, *supra* n. 1, LDA 7 N 10.

not categorised as an author, because copyright protects the expression of the idea, not the idea itself. And only those persons who influence the concrete work are considered to be co-authors, e.g., the persons whose instructions influence the work in the structure, the choice of examples, and the structure of the reasoning can be deemed to be co-authors.[23]

Once the original holder has been determined, the next step is to determine whether any copyrights have been transferred to a derivative holder. According to Article 16(1) CopA, the following copyrights may be assignable to third parties: (in particular the rights listed in Article 10(2) CopA, i.e. the right of reproduction (lit. a), the right of distribution (lit. b), the right of execution (lit. c), the right of broadcasting (lit. d), the right of retransmission (lit. e), the right of diffusion (lit. f), in whole or in part).[24]

The transfer of copyright is governed by the general rules of contract law and by certain rules of interpretation within the CopA.

The transfer of copyright is governed by the general principle of *freedom of contract* and the *freedom of form*. [25] However, there are some restrictions. According to Article 27(2) CC, the privacy rights limit the transfer of right against the risk of excessive engagements, such as the transfer of the whole future works.[26] According to Article 19(2) CO, some moral rights may not be transferable, depending on certain factors, such as the nature of work (classic or industrial) and the circumstances of the work's creation (individual or joint work, independent creation or made for hire).[27]

As a result of freedom of contract, disputes frequently come up regarding the existence, the nature, or the scope of copyright transfer. This is particularly an issue when the agreement has been concluded orally. In these situations, the contract has to be interpreted.[28] The scope of assignment of copyrights is determined both on the basis of general rules of contract law and on the basis of the real intention of the involved parties (subjective interpretation); or, if the latter cannot be determined, on the basis of what the parties would have agreed to in good faith and based on the specific circumstances of the agreement (objective interpretation). If the objective interpretation is uncertain, then the scope of assignment of copyrights is determined on the basis of the specific rules of interpretation of copyright.[29]

The first rule is provided by Article 16(2) CopA which stipulates: 'The assignment of a right subsisting in the copyright does not include the assignment of other partial

23. Cherpillod, *supra* n. 1, LDA 7 N 6-7.
24. de Werra, *supra* n. 8, LDA 16 N 6.
25. Some authors plead for the introduction of the written form, see authors quoted by de Werra, *supra* n. 8, LDA 16 N 34.
26. de Werra, *supra* n. 8, LDA 16 N 13.
27. de Werra, *supra* n. 8, LDA 16 N 17-21: The jurisprudence and the traditional approach consider that moral rights are not transferable: Supreme Court, 526 c. 3.3 (sic! 2010). Denis Barrelet/Willi Egloff, *Le nouveau droit d'auteur* (Bern 2008), LDA 16 N 6. A more liberal approach is for the transfer of moral rights based on the commercialisation of personal rights, such as the right of image and of name: Hilty, *Urheberrecht* 161 et seq., 186 et seq. (Bern 2010); Christian Laux, *Vertragsauslegung im Urheberrecht* 10 et seq. (Berne 2003); François Dessemontet, *Le droit d'auteur* 604 (Lausanne 1999).
28. de Werra, *supra* n. 8, LDA 16 N 37.
29. de Werra, *supra* n. 8, LDA 16 N 38-39.

rights, unless such was agreed.' This rule works to help interpret the scope of the transfer of a right. For instance, an assignment of the right of reproduction (Article 10(2)(a) CopA) and of distribution (Article 10(2)(c) CopA) does not automatically include the right to modify (Article 11(1)(b) CopA).

The second rule is provided by Article 16(3) CopA which stipulates: 'The assignment of the ownership of a copy of a work does not include the right to exploit the copyright, even in the case of an original work.' Accordingly, the sole disposal of the copy of a work does not lead to a transfer of copyright, and thus the copyright and the copy of the work remain separate from each other.

These rules reflect a broader third rule, the so-called *theory of the finality* (*Zweckübertragungstheorie, théorie de la finalité*), under which the transfer of rights is limited to the purpose of the contract.[30] Parliament did not follow the provision proposed in the 1989 Bill according to which, in the absence of a covenant to the contrary, the copyright on works made for hire by an employee during their activities for the employer and in performance of their contractual obligations would be transferred to the employer. This provision was advocated as a restatement regarding employee-employer relationships and the prevailing legal theory of the finality.[31] However because the provision was not passed, the current state of the law reflects the *theory of finality*.

The *theory of the finality* is considered by the doctrine to be the first rule of copyright contract law[32] and has recently been expressly recognised by the judiciary.[33] This theory applies only if the objective interpretation of the agreement is too uncertain to be determined.[34] When it does apply it has a broader scope of application than the rule provided in Article 16(2) CopA, because it is not limited to the interpretation of the scope of transfer, but also applies to the existence, the nature (assignment or license), and the duration of transfer.[35] Furthermore, some authors are in favour of a further reaching approach and propose that this rule should apply not only to the field of copyright, but to all contracts regarding the creation and the transfer of rights related to intangible assets.[36] However, this extended approach is not currently the settled law, and the courts tend to apply the general rules of contract law in fields other than copyright, i.e. to determine the scope of assignment according to the subjective and the objective interpretation.

30. Supreme Court, 23 April 2013, 4A_643/2012, c. 3.1; de Werra, *supra* n. 8, LDA 16 N 40.
31. Dessemontet, *supra* n. 15, 45.
32. Dessemontet, *supra* n. 27, 598. Its application is, however contested and criticised, see Hilty, Urheberrecht, 239 et seq., 265 et seq.; Dessemontet, *supra* n. 15, 45: 'in Switzerland, the theory is not conceived as a limitation to the transfer of rights, but as its justification: the purpose of the contract may imply that a transfer has taken place'.
33. Supreme Court, 23 April 2013, 4A_643/2012, c. 3.1; Supreme Court, 8 May 2008, 4A_104/2008 c. 4.2; Supreme Court, 30 April 1997, 4C_442/1996, c. 4 bb. The theory was applicable for a long time without being explicitly quoted ATF 101 II 102, JdT 1976 I 525; Roland von Büren/Michael A. Meer, *Der Urheber*, in SIWR vol. II/1, 171 (von Büren/David ed., Basel 2006).
34. de Werra, *supra* n. 8, LDA 16 N 41.
35. de Werra, *supra* n. 8, LDA 16 N 42.
36. Tissot, *supra* n. 1, LBI 3 N 22; Roland von Büren, *Übergang*, p. 269.

According to the *theory of the finality*, if there is a doubt about the transfer of right, only the transfer of those rights necessary to obtain the purpose of the contract will be deemed to have taken place (*in dubio pro auctore*).[37] For example, courts have determined that an advertising agency keeps all copyrights; the client can only use it once.[38] However, it has been judged that a journalist who made reports for radio broadcasts of the Swiss public broadcasting organisation (SRG) without expressly addressing the issue of copyright, assigned his copyright to the SRG because the purpose of the contract was broadcasting, and re-broadcasting reports could not be achieved without assignment of copyrights.[39]

In the case of an employment relationship, the *theory of the finality* plays a crucial role because, unlike design or patent rights, there is no legal transfer of copyright to the employer (except for computer programs), and the employer can only acquire the copyrights to their employee's work by means of an employment contract or a separate agreement.[40] This usually leads to the conclusion that the employer can use the copyrighted work as usual and that the employee retains only moral rights, as well as residual rights to uses of the work that fall in the ordinary exploitation of the work by the employer. New uses of the work belong to the employee.

For works made by the employee during their activities and in performance of their contractual obligations (and in the absence of a covenant to the contrary), the copyrights are deemed to be transferred to the employer to the extent required by the employment contract, in particular the right to use and the right to modify the work.[41] Indeed, the employer supports the whole costs and economic risk of creation and must be able to use, continue, terminate or improve the works made by their employees.[42]

For works made by the employee during the performance of their contractual obligations, a licence or an assignment may be established since the *theory of the finality* does not automatically lead to the conclusion of the transfer of right.[43]

For works made by scientific researchers and within their contractual obligation, the professor (or research team leader) is deemed to have the rights on the research outcome of the collaborators;[44] they can use and modify the outcome as well as decide

37. Supreme Court, 23 April 2013, 4A_643/2012, c. 3.1; von Büren/Meer, p. 169; CR PI-de de Werra, *supra* n. 8, LDA 16 N 41.
38. Cantonal Court Basel, BJM 1995, p. 248
39. Supreme Court, 23 April 2013, 4A_643/2012, c. 3.1.
40. de Werra, *supra* n. 8, LDA 16 N 44. As an example of disputes regarding the existence, the nature, or the scope of copyright transfer in the case of an employment relationship, see the decision of the Bern Cantonal Court: HGer Berne, 17 juin 2015, HG 15 39 CAA, sic! 2016, 56: the Court indicated that there is no implied transfer of several employees, if the employer is not in a position to designate each co-author individually and their role in the creation.
41. District Court Zurich 227 c. 3-1 (sic! 2011); Manfred Rehbinder/Adriano Viganò, LDA 11 N 3; de Werra, *supra* n. 8, LDA 16 N 45.
42. Edgar Philippin, *Commentaire romand de la propriété intellectuelle* (de Werra/Gilliéron ed., Basel 2013), LDA 11 N 23.
43. District Court Zurich, 227, c. 3.1 (sic! 2011); de Werra, *supra* n. 8, LDA 16 N 46.
44. Martina Altenpohl, thesis, p. 265 et seq., who applies the rule even when the scientific collaborator is the sole author, with reference to Rehbinder, *Festschrift Hubmann*, p. 371.

whether it has to be published under their name and their responsibility.[45] For works made under an employment contract but out of the contractual obligation, an express assignment of right has to be made.[46]

However, two cases must be distinguished if a scientific collaborator has contributed to the work in a creative way.[47] When the scientific collaborator and their superior (generally a professor) intend to publish the work under both of their names and responsibilities, the work is a joint work governed by Article 7 CopA; the collaborator has the right to be named as co-author (paternity right) and to accept or not accept modifications of the work.[48]

When the work has to be published under the sole name and responsibility of the superior, the work provided by the employee can be considered a 'service work' (i.e. a work that has been produced by the employee in the course of their work and in performance of their contractual obligations and accordingly belongs to the employer, see Article 332 CO). The creative contribution of the collaborator is at the professor's disposal, even if it is protected by copyright, the collaborator has been hired to provide assistance to the superior and, according to the *theory of the finality*, the superior must have full disposal of the work made through the fulfilment of contractual obligations. Thus, the professor can modify and publish the work under their sole name and responsibility (in that case, there is an implied waiver of the employee's paternity right and of the remuneration right).[49]

A third case may apply for public institutions, since they are regulated by public regulations; some of them provide the obligation to mention the name of the persons who take part in the work subject to the publication (see e.g. Article 18 of the ETH-Federal Act: 'Scientific publications must list all those who have made a scientific contribution').[50] Finally, the more the collaborator takes part in the creation of work, the more likely they are to be deemed a co-author. Inversely, the smaller their contribution is, the more they are to be deemed an auxiliary, and it may be usual to consider that they transferred their copyrights and waived their moral rights.[51]

2.2.2 Employees' Computer Programs

For employees' computer programs, a special provision provides a legal transfer to the employer. Article 17 CopA states that an employer has the exclusive right to use software made by their employee within the framework of their activities for the

45. Cherpillod, *supra* n. 1, LDA 7 N 12.
46. de Werra, *supra* n. 8, LDA 16 N 44-46.
47. Cherpillod, *supra* n. 1, LDA 7 N 10.
48. Cherpillod, *supra* n. 1, LDA 7 N 11.
49. Cherpillod, *supra* n. 1, LDA 7 N 12.
50. Federal Act on the Federal Institutes of Technology ('FIT Act') (RS 414.110)., https://www.admin.ch/opc/en/classified-compilation/19910256/201501010000/414.110.pdf (accessed on 20 July 2017).
51. Cherpillod, *supra* n. 1, LDA 7 N 13.

employer and in conformity with their contractual obligations. The provision has been integrated in order to harmonise Swiss Law with EU Directive 91/24.[52]

Article 17 CopA, as a dispositive provision, only applies in the absence of an agreement to the contrary. Thus, the parties may exclude its application. In that case, the question of transfer of right will be interpreted according to the principles of contract law and the *theory of the finality*.[53]

Article 17 CopA refers to a 'computer program [that] has been created', i.e. to the copyrights on the computer program, not to other adjacent copyrighted works, such as the user's guide.[54]

Article 17 CopA only applies where a computer program has been created 'under an employment contract'.[55] Thus, Article 17 CopA shall neither apply when a computer program has been created by an independent programmer, nor when it has been created within a public relationship (in the latter case, one must refer to the public regulations).[56] Moreover, Article 17 CopA shall only apply to a computer program created during the employment relationship. Thus, the employee who leaves the company before achieving the computer program needs the agreement of their former employer to use the unachieved computer program.[57] The same applies to a computer program that has been created before the beginning of contract; the employee retains the right to that computer program and the employer needs a transfer of rights if they want to commercialise it.[58]

Finally, Article 17 CopA applies to computer programs that have been created 'in the course of discharging professional duties or fulfilling contractual obligations'. This formulation corresponds to Article 332 CO, which relates to inventions and designs produced by the employee. The first condition, *in the course of discharging professional duties*, requires a logical link between the creation of the computer program by the employee (or their participation in the creation) and their activity in the company.[59]

52. In particular Art. 2(3) of Directive 91/24, see de Werra, *supra* n. 8, 17 N 2.
53. Emil F. Neff/Mathias Arn, *Urheberrechtlicher Schutz der Software*, in SIWR vol. II/2, 278 (Basel/Frankfurt/München 1998); de Werra, *supra* n. 8, LDA 17 N 4.
54. de Werra, *supra* n. 8, LDA 17 N 7 ; V. Salvadé/N. Tissot, La réalisation d'un site web ou l'ouverture d'un compte par le travailleur. Qui est titulaire des droits ?, in: Jean-Philippe Dunand, Pascal Mahon (publishers), Schulthess 2014, 239.
55. Cherpillod, RSPI 1993, p. 56; Tissort/Salvadé, *supra* n. 54, 239, making a distinction between 'service softwares' (created in the course of work for the employer and in performance of a contractual obligation, and subject to CopA 17) and website (including its layout and designs, which is not subject to CopA 17).
56. de Werra, *supra* n. 8, LDA 17 N 8. However, see Stutz/Ambühl, *Rechte an Computerprogrammen, geschaffen im öffentlich-rechtlichen Arbeitsverhältnis – eine Schweizer Sonderlösung?*, GRUR Int 667 et seq. (2010), who consider that the European Directive 2009/24/EG does not differentiate between employment under private or public law and that Art. 17 CopA shall thus apply to public and private law.
57. Neff/Arn, *supra* n. 53, 281.
58. Neff/Arn, *supra* n. 53, 281; de Werra, *supra* n. 8, LDA 17 N 13.
59. Supreme Court, 382, 387 (sic! 1997); whether the computer program has been created at work or during working time is irrelevant. de Werra, *supra* n. 8, LDA 17 N 10.

The second condition, *fulfilling contractual obligations*, requires that the employee had an obligation via express or implied provisions to provide a creative activity according to their position within the company.[60]

In the employer's interest, it is advisable to adopt a contractual provision that precisely defines the scope of activity, in particular that the creation of computer program is covered under *professional duties and contractual obligations*. It is even more advisable to adopt a global transfer of rights related to computer programs created under the employment contract and excluding in that way the application of Article 17 CopA.[61]

There are discussions about the nature of the transfer, i.e. whether the provision entails an assignment of copyright or an exclusive licence.[62] According to a teleological and systematic interpretation, it provides a legal transfer to the employer and is thus similar to Article 332 CO in that it provides original ownership to the employer.

There are also discussions about the scope of the transfer, i.e. whether the provision refers only to the economic rights or whether it also covers moral rights as well. According to a teleological and systematic interpretation, it refers to all rights needed for the use of the computer program by the employer, in particular the right of modifying and adapting the computer program.

2.2.3 Neighbouring Producer's Rights

With regard to neighbouring rights of phonogram and audio-visual fixation producers, Article 36 CopA provides that:

> [a] producer of phonograms and audiovisual fixations has the exclusive right: a. to reproduce the fixations and to offer, transfer or otherwise distribute the reproductions; b. to make the fixations available through any kind of medium in such a way that persons may access them from a place and at a time individually chosen by them.

The original holder of a producer right may be the natural or legal person that provides services of a certain complexity in charge of the first and the whole fixation process.[63] When several natural persons or legal entities jointly provide these services, the CopA does not expressly contain a provision for joint work, such as Article 7 CopA. Thus, the general rules of co-ownership apply, in particular that each co-producer can

60. ATF 72 II 270, p. 273 et seq. Whether this obligation has been explicit or not is irrelevant; it all depends on whether parties could reasonably wait for such an activity. Thus, Art. 17 does not apply if an employee did not have such an obligation and created a computer program outside their contractual obligations in order to facilitate their work, such as a compatibility program, Briner, PJA 577 (1993).
61. For a template, see Urs Egli, *Softwareentwicklung im Arbeitsverhältnis*, ArbR 15 et seq., 32 (2007); de Werra, *supra* n. 8, LDA 17 N 12.
62. See Roland von Büren, *Der Werkbegriff*, in *Schweizerisches Immaterialgüter und Wettbewerbsrecht* ('SIWR') vol. II/1, 228 (Basel/Genf/München 2005); Barrelet/Egloff, *supra* n. 27, LDA 17 N 6.
63. Mettraux Kauthen, *supra* n. 17, LDA 36 N 12, who observes that this definition is similar to Art. 2(d) WPPT.

alienate and pledge their share (Article 646 (3) Civil Code). In practice, co-producers generally include a co-production contract that governs their relationships and, in the absence of a specific provision, the court shall decide in accordance with the rule that it would make as a legislator (Article 1(2) CC).[64]

Nevertheless, the general rules of contract law (subjective and objective interpretation) and the specific interpretation rules of intellectual property law (in particular the *theory of the finality*) also apply to performer rights. Thus, according to the *theory of the finality*, if an employee is considered to be the initial holder of a producer right, the right is transferred to the employer if the recording was produced as part of the employee's duties or as a result of instructions given by the employer.[65]

2.2.4 Neighbouring Broadcasting Rights

With regard to neighbouring broadcasting rights, Article 37 CopA provides that:

> [a] broadcasting organisation has the exclusive right ... c. to offer, transfer or otherwise distribute copies of the fixations of its broadcast; ... e. to make its broadcasts available through any kind of medium in such a way that persons may access them from a place and at a time individually chosen by them.

The original holder of a broadcasting right may be the legal entity that organises, plans and collects the whole program until its broadcast, independent of the control of the technical broadcasting process.[66]

In the case of an employment relationship, the employee will not be considered to be the original holder, since he/she is either a natural person or a legal entity that is not in charge of the whole program until broadcasting.

2.3 The Parties' Duty to Inform

An express provision providing a positive duty to give notice to an employer of the creation of intellectual property, other than for inventions and designs, does not exist (see sections 4 and 5). However an employee is by implication still required to give notice to their employer, despite the lack of express provisions, as a result of the loyalty principle in employment law (Article 321a CO).[67]

2.4 Employee's Remuneration Right

Special rules regulating an employee remuneration right, other than for inventions and designs, do not exist (see sections 4 and 5). If and to what extent creators of other

64. Mettraux Kauthen, *supra* n. 17, LDA 36 N 13. Another doctrine proposes referring to the provisions regarding the *société simple* Egloff, 543 (sic! 1999).
65. See section 2.2.1.
66. Mettraux Kauthen, *supra* n. 17, LDA 37 N 6.
67. See section 1 under employees' duties.

intellectual property are entitled to remuneration depends instead on any agreement that has been entered into by the employer and the employee.

In the absence of a covenant to the contrary, the employment relationship is interpreted according to the general rules of contract law (subjective and objective interpretation) and, if the latter are uncertain, according to the *theory of the finality*. The agreement and the salary provided are deemed to cover the employee's copyrights needed for the purpose of the contract, so that the employee is not entitled to a supplemental remuneration. In a recent case, relating to a freelance journalist who made reports for the Swiss public broadcasting organisation (SRG) based on an oral agreement that did not address copyrights, the Federal Court held that the agreement and the payment covered the assignment of the freelancer's copyright to the SRG because the purpose of the contract, which was to broadcast reports, could not be achieved without assignment of copyrights.[68]

For completeness, a doctrine, based on the general rule of contract law *clausula rebus sic stantibus*, is for an adaptation of the contract providing an additional remuneration's right, when a copyrighted work exceeds the projections made by the parties at the time of the assignment.[69]

3 EMPLOYEES' SEMICONDUCTOR CHIPS

3.1 The Legal Framework

The Federal Act on the Protection of Topographies of Semiconductor Products (Topographies Act, ToA)[70] protects three-dimensional designs of semiconductors provided they are not commonplace at the time of development (Article 1(1) ToA). For a design to be protected, a certain degree of intellectual work must be involved. However, neither an inventive step (as required for patents), nor individual character (as is necessary for copyright protection) is required.[71] Even a design composed of known elements is protectable if the specific combination is unusual (Article 1(2) ToA). The scope of protection is limited to the design and does not include the semiconductor's electronic function or coding.[72]

The term of protection begins with the creation of the design and ends ten years after its registration or first publication (Article 9(1) ToA). The right to register a design expires two years after its first distribution (Article 9(2) ToA).[73] The use of semiconductor designs may also be restricted by copyright or unfair competition law. However,

68. Supreme Court, 23 April 2014, 4A_643/2012.
69. For patents, see Adrian Staehelin/Frank Vischer, *Obligationenrecht*, V 2c, Der Arbeitsvertrag, Art. 319-362 OR, (3rd ed., Zurich 1996), CO 332, N 11; Supreme Court, 20 March 2009, 4A_595/2008 consid. 6 and the ref.
70. Swiss Federal Act on the Protection of Topographies of Semiconductor Products of 9 October 1992 (SR 231.2).
71. Swiss Federal Journal, BBl, 1989, p. 572 and 574. Comparison between German and US law: Gert Lück, *Halbleiter-Topographieschutz*, in SIWR vol. II/2, 363 et seq.
72. Swiss Federal Journal, *supra* n. 71, at 572.
73. Swiss Federal Journal, *supra* n. 71, at 577.

once the ten-year term of protection under the Topographies Act has expired use is also permitted under any other title.[74] Since 1993 only 19 designs have been registered.[75]

3.2 Ownership of Rights

The Topographies Act is guided by the idea of investment protection. Therefore, all rights are originally acquired by the person or entity who developed the design at its own expense and risk (Article 3 ToA). Thus, the original owner of an employee's design is virtually always the employer.[76]

3.3 The Parties' Duty to Inform

Swiss law does not stipulate an express duty of disclosure regarding employees' semiconductor designs. However, the employee must carry out the work assigned to him with due care and loyally safeguard the employer's legitimate interests (Article 321a Code of Obligations, CO).[77] Based on this general duty of loyalty, the employee is obliged to inform the employer immediately of newly created semiconductor designs.

3.4 Employee's Remuneration Right

Neither the Topographies Act nor Swiss labour law stipulates an obligation to provide compensation for the creation of a protectable semiconductor design.

4 EMPLOYEES' INVENTIONS

4.1 The Legal Framework

The Federal Act on Patents for Inventions (Patent Act, PatA)[78] contains no specific provisions on employee inventions, but accords the right to be granted a patent to the inventor, the successor in title, or a third party who owns the invention under another title (Article 3(1) PatA). Thus, in accordance with the creator principle,[79] the employee in question originally acquires all rights to the invention unless the law explicitly assigns these rights to the employer. As the original holder of all rights, the employee may be under a contractual obligation to transfer some or all rights to the invention to the employer based on the employment contract. However, in such cases the employer will acquire these rights only as a successor in title and provided that the employee did not transfer these rights to a third party beforehand.

74. Swiss Federal Journal, *supra* n. 71, at 508.
75. https://www.ige.ch/de/service/datenbanken/topographien-register.html.
76. Swiss Federal Journal, *supra* n. 71, at 574; Wolfgang Portmann, BSK, Art. 332 N 4.
77. Portmann, *supra* n. 76, at n. 12; Streiff/von Kaenel/Rudolph, *Arbeitvertrag*, Art. 321a n. 2.
78. Federal Act on Patents for Inventions of 25 June 1954 (SR 232.14).
79. See section 1.

4.2 Ownership of Rights

4.2.1 Employees in the Private Sector

The Code of Obligations (CO) distinguishes three categories of employee inventions: *service inventions, reservable inventions* and *other inventions* (Article 332 CO).

'*Service inventions*' are created in the course of work for the employer and in performance of a contractual obligation (Article 332(1) CO). Accordingly, service inventions may occur if the employee in question is expected to work on technical advances.[80] Express instructions to carry out development or research tasks is however not essential. Service inventions may also arise if the employee could reasonably have been expected to take the steps required in order to resolve technical problems at hand. Thus, the required contractual obligation may not only arise from explicit employer instructions or the contract of employment, but may also result from the employee's general responsibilities, education, salary and position within the firm.[81] Subject to such contractual obligation service inventions do not necessarily have been created during working hours or by using the employer's resources. Service inventions are fully owned by the employer who is regarded as the original holder of all rights.[82] Employees will only retain a (waivable) right to be named as creator (Article 5 PatA).[83]

'*Reservable inventions*' are created in the course of work but not in performance of a contractual obligation. Thus, reservable inventions may occur if the employee in question is neither obliged nor expected to make innovative advances.[84] What is required, however, is a material link both to the employer's field of activity and to the employee's range of tasks.[85] Subject to such material link reservable inventions do not necessarily have been created during working hours or by using the employer's resources.[86] Reservable inventions are fully owned by the employee in question who is the original holder of all rights. The employer may reserve in writing the option to

80. Supreme Court: BGE 6B_310/2014 E. 3.4.1; BGE 4A_691/2011 E. 3.1, BGE 100 IV 167 E. 1 and BGE 72 II 270 E. 4; Streiff/von Kaenel/Rudolph, *Arbeitsvertrag*, CO Art. 332 n. 7.
81. Supreme Court: BGE 6B_310/2014 E. 3.4.1; BGE 4A_691/2011 E. 3.1, BGE 2A.204/2006 E. 7.1 and BGE 100 IV 167 E. 1; Decision of the Cantonal Court of Nidwalden of 11.3.2008, E. 4.3, in: sic! 2010, p. 41 ff.; Decision of the Commercial Court of St. Gallen of 13.3.2001 E. 2, in: GVP 2001, No. 41; Peter Heinrich, PatG/EPÜ, Art. 3 N 9.
82. Decision of the Cantonal Court of Nidwalden of 11.3.2008, E. 4.3, in: sic! 2010, p. 41 ff.; Thierry Calame, *Die Berechtigung an der Erfindung*, in SIWR vol. IV, 195 et seq.; Heinrich, *supra* n. 81, at Art. 3 n. 9.
83. Decision of the Commercial Court of Zurich of 28.8.2007, E. 3.2, in: ZR 2008, p. 55 ff.; Decision of the Cantonal Court of Nidwalden of 11.3.2008, E. 4.3, in: sic! 2010, p. 41 ff.; Streiff/von Kaenel/Rudolph, *supra* n. 77, at n. 9; Heinrich, *supra* n. 81, at Art. 5 n. 1.
84. Decision of the Commercial Court of St. Gallen of 13.3.2001 E. 2, in: GVP 2001, No. 41; Heinrich, *supra* n. 81, at Art. 3 n. 13.
85. Supreme Court: BGE 4A_691/2011 E. 3.1 and BGE 72 II 270 E. 4; Calame, *supra* n. 82, at 195. Material link between an invention and employee's work is assumed if the invention incorporates the employer's knowledge: Decision of the Cantonal Court of Nidwalden of 11.3.2008, E. 4.3, in: sic! 2010, p. 41 ff.
86. Supreme Court: BGE 6B_310/2014 E. 3.4.1; BGE 4A_691/2011 E. 3.1 and BGE 72 II 270 E. 4; Decision of the Commercial Court of St. Gallen of 13.3.2001 E. 2, in: GVP 2001, No. 41; Streiff/von Kaenel/Rudolph, *supra* n. 80, at n. 10.

acquire such inventions as a successor in title (Article 332(2) CO).[87] The employee can agree to assigning such rights beforehand, effectively giving the employer sole discretion to acquire the rights to a reservable invention once it has been created.[88] Even if no written reservation of rights to a reservable invention was agreed by the parties, the employee may nevertheless be obliged to offer such invention to the employer based on the employee's general duty of loyalty (Article 321a CO).[89]

'Other inventions' are neither produced in the course of work nor in performance of contractual obligations, and thus have no connection to the employer's field of activity or the employee's work. The employee remains the original holder of all rights to such other inventions (Article 3(1) PatA).

4.2.2 Employees in the Public Sector

As there is no unified employment legislation for the public sector in Switzerland, different laws apply to the employment relationships of the various governmental bodies. Thus, the original holder of all rights to employee inventions under public law will be defined by the applicable federal, cantonal or municipal public labour law in question. Many of these regulations contain similar rules to the Code of Obligations or even expressly refer to its provisions.[90] If the relevant public labour regulation does not contain any specific provision, all rights to the invention will vest with the employee (Article 3(1) PatA) in accordance with the creator principle and the employer may acquire these rights as a successor in title.[91]

The legal regulation of the Swiss university sector is as fragmented as the public sector in general. However, in order to stimulate technology transfer between the public and the private research sectors, most universities have meanwhile adopted regulations assigning the rights to service and reservable inventions of academic researchers *ex lege* to the university.[92]

87. Decision of the Commercial Court of St. Gallen of 13.3.2001 E. 3, in: GVP 2001, No. 41; Streiff/von Kaenel/Rudolph, *supra* n. 80, at n. 11.
88. Supreme Court: 4A_442/2014 E. 4; BGE 127 III 461 E. 4a; BGE 57 II 304 E. 1; Calame, *supra* n. 82, at 199; Heinrich, *supra* n. 81, at Art. 3 N 13; Mondini/Bürge, *Zuordnung der Ergebnisse gemeinsamer Forschung und Entwicklung in der Praxis*, AJP 16 (2008); Mosimann/Graf, *Arbeitnehmererfindungen*, in *Schweizerisches und europäisches Patentrecht* 170 (Bertschinger/Münch/Geiser eds., Basel 2002); Streiff/von Kaenel/Rudolph, *supra* n. 80, at. n. 10. Different opinion: Münch/ Herzog, *Berechtigung an der Erfindung*, in *Schweizerisches und europäisches Patentrecht* 170 (Bertschinger/Münch/Geiser eds., Basel 2002).
89. Adrian Andermatt, *Das Recht an im Arbeitsverhältnis geschaffenen immaterialgüterrechtlich geschützten Erzeugnissen*, p. 298 et seq.; Calame, *supra* n. 82, at. 195; other opinion e.g.: Streiff/von Kaenel/Rudolph, *supra* n. 80, at n. 14.
90. For federal officials: Art. 6(2) Bundespersonalgesetz (SR. 172.220.1). For cantonal officials see e.g., Art. 4 Personalgesetz of the Canton of Basel or Art. 60 Personalgesetz of the Canton of Bern.
91. Supreme Court denies analogous application of Art. 332 CO: BGE 74 II 106 E. 4a; Calame, *supra* n. 82, at 195; Cyrill Rieder, *Eigentum an Hochschulerfindungen* 196 et seq., 245 et seq. (Bern 2010). Different view: Streiff/von Kaenel/Rudolph, *supra* n. 80, at n. 4. Also see section 1.
92. E.g., Art. 36 Bundesgesetz über die Eidgenössischen Technischen Hochschulen (SR 414.110); Art. 10a UniG of the Canton of Friborug; Art. 12a UniG of the Canton of Zurich; Art. 70 UniG of the Canton of Waadt; Art. 73 UniG of the Canton of Neuenburg; Art. 30 UniG of the Canton of

4.3 The Parties' Duty to Inform

Based on Article 332(3) CO private sector employees have to notify their employer in writing about the creation of a 'reservable invention'.[93] The employer then has to decide within six months whether to acquire the invention in question. In regards to 'service inventions'[94] the law does not expressly stipulate a similar duty to inform. However, the employee must at all times carry out the work assigned to him with due care and loyally safeguard the employer's legitimate interests (Article 321a CO). [95] Based on this general duty of loyalty, the employee is obliged to inform the employer immediately of any inventions created within the employer's field of activity.[96] Furthermore, the employee has to support in protecting the invention and sign all required documents.[97] In regards to public sector employees the applicable federal, cantonal or municipal public labour regulation has to be consulted.

4.4 Employee's Remuneration Right

Unless otherwise agreed, private sector employers acquire the rights to 'service inventions' without having to pay a financial compensation.[98] Some authors[99] argue that such compensation is due if the efforts made by the employee exceed what reasonably could be expected, or if the service invention generates unexpected revenue. However, Swiss court practice does not follow this minority opinion.[100] Financial compensation is however always due if the employer wishes to acquire a 'reservable invention'.[101] The amount depends on the economic value of the invention, the degree to which the employer contributed, any potential reliance on other staff and on the employer's facilities, the expenses incurred by the employee and their position in the firm (Article 332(4) CO). The obligation to pay for reservable inventions cannot be modified to the employee's disadvantage (Article 362 CO).[102]

Genf; Art. 4; Art. 60 Personalgesetz Canton of Bern and Art. 70 UniG of the Canton of Bern. Overview: Rieder, *supra* n. 91, at 187 ff.; Calame, *supra* n. 82, at 203 ff.

93. See section 4.2.1 under 'reservable invention'.
94. See section 4.2.1 under 'service invention'.
95. Supreme Court: 4A_688/2014 E. 3.3.4; Cantonal Court of Nidwalden, 41 et seq. (sic! 2010), E. 4.2; Wolfgang Portmann, BSK, 5th ed., CO Art. 321a N 12; Streiff/von Kaenel/Rudolph, *supra* n. 77, at Art. 321a N 2.
96. Wolfgang Portmann, *Die Arbeitnehmererfindung* 101 (Bern); Mosimann/Graf, *supra* n. 88, at 977; Andermatt, *supra* n. 89, at 298.
97. Supreme Court: 4A_688/2014 E. 3.3.4; Federal Patent Court: S2014_007 E. 4.3.
98. Calame, *supra* n. 82, at 197 f.; Portmann, *supra* n. 76, at 76 f.; Wyler/Martin, *Droit du travail* 375 (2nd ed.). See section 4.2.1 under 'service invention'.
99. Staehelin/Vischer, CO 332 N 11; Manfred Rehbinder, BK OR-331-355, CO Art. 332 N 8; Henneberger-Sudjana/Henneberger, *Gesonderte Honorierung von Diensterfindungen als Beitrag zur (besseren) wirtschaftlichen Entwicklung*, Diskussionspapier Nr. 128 der Reihe des Forschungsinstituts für Arbeit und Arbeitsrecht an der Universität St. Gallen.
100. Supreme Court: BGE 6B_310/2014 E. 3.4.3, implicitly confirming that compensation is due only for reservable inventions.
101. See section 4.2.1 under 'reservable invention'.
102. Andermatt, *supra* n. 89, at. 294 ff.; Calame, *supra* n. 82, at 200.

In regards to public sector employment relationships the applicable federal, cantonal or municipal public labour regulation has to be consulted.[103] Special remuneration provisions apply to university researchers who generally receive about a third of the licensing revenues generated by the invention.[104]

5 EMPLOYEES' UTILITY MODELS

Swiss law does not protect utility models.

6 EMPLOYEES' DESIGN RIGHTS

6.1 The Legal Framework

The Federal Act on the Protection of Designs (Design Act, DesA)[105] does not contain a specific provision for employee's designs, but assigns the right to a design to its creator,[106] the successor in title, or to a third party who owns the design under another title (Article 7(1) DesA). Thus, as in case of inventions, the employee originally acquires all rights to designs unless an explicit provision of the law assigns these rights to the employer. As the original holder of all rights, the employee may be under a contractual obligation to transfer some or all rights to the design to the employer based on the employment contract. However, in such cases the employer will acquire these rights only as a successor in title and provided that the employee did not transfer these rights to a third party beforehand.

6.2 Ownership of Rights

6.2.1 Employees in the Private Sector

As in case of inventions, the Code of Obligations distinguishes three categories of employee designs: *service designs, reservable designs* and *other designs*.[107] However, unlike inventions, designs may also be protected by copyright,[108] which the employer must acquire in addition to the design right.[109]

All rights to 'service designs', which are defined as designs created in the course of work and in performance of contractual obligations, originally vest in the employer

103. For the Canton of Zurich and Canton of Bern: Calame, *supra* n. 82, at 202 f.
104. E.g. Art. 6 Verordnung des ETH-Rates über die Immaterialgüter im ETH-Bereich und die Beteiligung an Unternehmungen (SR 414.172); Calame, *supra* n. 82, at 203 ff.
105. Federal Act on the Protection of Designs of 5 October 2001 (SR 232.12).
106. See *creator principle*, section 1.
107. Stutz/Beutler/Künzi, Designgesetz DesG, Art. 7 DesG, N 49 ff.; Peter Heinrich, *Schweizerisches Designgesetz/Haager Musterschutzabkommen*, DesG 7 N 11; Markus Wang, *Entstehung und Bestand des Designrechts*, in SIWR vol. VI, 151 et seq.; Streiff/von Kaenel/ Rudolph, *supra* n. 80, at n. 7 ff.
108. Wang, *supra* n. 107, at 153; Stutz/Beutler/Künzi, *supra* n. 107, at Art. 7 N 78.
109. See section 2.

(Article 332(1) CO).[110] 'Reservable designs',[111] which are created in the course of work but not in performance of contractual obligations, as well as other designs, basically belong to the employee. However, by written agreement the employer may reserve the right to acquire reservable designs as a successor in title (Article 332(2) CO). The employee can agree to assigning such rights beforehand, effectively giving the employer sole discretion to acquire the rights to a reservable design once it has been created.[112] Even if no written reservation has been agreed with regard to reservable designs, the employee may nevertheless be obliged to offer such design to the employer based on the general employee's duty of loyalty (Article 321a CO).[113]

6.2.2 Employees in the Public Sector

As there is no unified employment legislation for the public sector in Switzerland, different laws apply to the employment relationships of the various governmental bodies. Thus, the original holder of all rights to employee designs under public law will be defined by the applicable federal, cantonal or municipal public labour law in question. Many of these regulations contain similar rules to the Code of Obligations or even expressly refer to its provisions.[114] If the relevant public labour regulation does not contain any specific provision, all rights on the invention will however vest with the employee (Article 7(1) DesA) in accordance with the creator principle and the employer may acquire these rights as a successor in title.[115]

6.3 The Parties' Duty to Inform

Based on Article 332(3) CO private sector employees have to notify their employer in writing about the creation of a 'reservable design'. The employer then has to decide within six months whether to acquire the design in question. In regards to 'service designs' the law does not expressly stipulate a similar duty to inform. However, the employee must at all times carry out the work assigned to him with due care and loyally safeguard the employer's legitimate interests (Article 321a CO).[116] Based on this general duty of loyalty, the employee is obliged to inform the employer immediately of

110. Also see explanations in section 4.2.1 under 'service invention'.
111. Also see explanations in section 4.2.1 under 'reservable invention'.
112. Supreme Court: 4A_442/2014 E. 4; BGE 127 III 461 E. 4a; BGE 57 II 304 E. 1; Streiff/von Kaenel/Rudolph, *supra* n. 80, at n. 13; Stutz/Beutler/Künzi, *supra* n. 107, at 66; Wang, *supra* n. 107, at 152.
113. Andermatt, *supra* n. 89, p. 230; Wolfgang Portmann, BSK, 5th ed., CO Art. 332 N 13 f.; for other opinion see e.g.: Streiff/von Kaenel/Rudolph, *supra* n. 80, at 14.
114. With regard to federal officials: Art. 6 para. 2 Bundespersonalgesetz (SR. 172.220.1). For cantonal officials e.g., Art. 4 Personalgesetz of the Canton of Basel or Art. 60 Personalgesetz of the Canton of Bern.
115. Supreme Court denies analogous application of Art. 332 CO: BGE 74 II 106 E. 4a; Stutz/Beutler/Künzi, *supra* n. 107, at 77; Different view: Streiff/von Kaenel/Rudolph, *supra* n. 80, at 4.
116. Supreme Court: 4A_688/2014 E. 3.3.4; Cantonal Court of Nidwalden, 41 et seq. (sic! 2010), E. 4.2; Wolfgang Portmann, BSK, 5th ed., CO Art. 321a N 12; Streiff/von Kaenel/Rudolph, *supra* n. 77, at Art. 321a N 2.

any designs created within the employer's field of activity.[117] Furthermore, the employee has to support in protecting the design and sign all required documents.[118] In regards to public sector employees the applicable federal, cantonal or municipal public labour regulation has to be consulted.

6.4 Employee's Remuneration Right

Unless otherwise agreed, private sector employers acquire the rights to 'service designs' without having to pay a financial compensation.[119] Financial compensation is however always due if the employer wishes to acquire a 'reservable design'. The amount depends on the economic value of the design, the degree to which the employer contributed, any potential reliance on other staff and on the employer's facilities, the expenses incurred by the employee and their position in the firm (Article 332(4) CO). The obligation to pay for reservable designs cannot be modified to the employee's disadvantage (Article 362 CO).[120] In regards to public sector employment relationships the relevant federal, cantonal or municipal public labour regulation has to be consulted.

7 EMPLOYEES' PLANT VARIETIES

7.1 The Legal Framework

The Federal Act on the Protection of Plant Varieties (Plant Varieties Act, PVA)[121] implements the International Convention for the Protection of New Varieties of Plants ('UPOV Convention')[122] and other international treaties (Article 4 PVA). In Switzerland, a plant variety protected under the Plant Variety Act may not be additionally patented as an invention (prohibition of double protection).[123]

7.2 Ownership of Rights

The Plant Varieties Act assigns the right to protect a variety to the breeder or their successor in title and expressly refers to Article 332 Code of Obligations with regard to

117. Wolfgang Portmann, *Die Arbeitnehmererfindung* 101 (Bern); Mosimann/Graf, *supra* n. 88, at 977; Andermatt, *supra* n. 89, at 298. Regarding inventions: Supreme Court: 4A_688/2014 E. 3.3.4; Federal Patent Court: S2014_007 E. 4.3.
118. Wang, *supra* n. 107, at 152 f.; Portmann, *supra* n. 77, at Art. 321a OR N 12; Streiff/von Kaenel/Rudolph, *supra* n. 77, at CO Art. 321a N 2.
119. Swiss Supreme Court: BGE 4A_691/2011 E. 4.1; Stutz/Beutler/Künzi, *supra* n. 107, at 61; Heinrich, *supra* n. 107, at 18; different opinions: Manfred Rehbinder, BK, OR-331-355, CO Art. 332 N 8.
120. Streiff/von Kaenel/Rudolph, *supra* n. 77, at n. 11; Stutz/Beutler/Künzi, *supra* n. 107, at N 68; Heinrich, *supra* n. 81, at N 14.
121. Federal Act on the Protection of Plant Varieties of 20 March 1975 (SR 232.16).
122. Switzerland has been a member State of the UPOV since 10 July 1977 (SR 0.232.163). The Convention was adopted in Paris in 1961 and revised in 1972, 1978 and 1991.
123. Swiss Supreme Court: BGE 121 III 125.

employees' plant varieties (Article 9(1) PVA). Thus, as in the case of employee's inventions or designs, three categories of employees' plant varieties can be distinguished: *service plant varieties, reservable plant varieties* and *other plant varieties.*

All rights to 'service plant varieties' that have been created in the course of work and in performance of contractual obligations are originally acquired by the employer (Article 332(1) CO). 'Reservable plant varieties', which have been created in the course of work but not in performance of contractual obligations, and 'other plant varieties' belong to the employee. However, by written agreement the employer may reserve the right to acquire reservable plant varieties as a successor in title (Article 332(2) CO).[124] Even if no written reservation was agreed by the parties with regard to reservable plant varieties, the employee may nevertheless be obliged to offer such plant variety to the employer based on the employee's general duty of loyalty (Article 321a CO).[125]

As a general rule, Article 332 CO applies only to employment relationships under private law. In the public sector, the allocation of rights is usually defined by the applicable federal, cantonal or municipal public labour law in question.[126] The fact that the Plant Varieties Act expressly refers to Article 332 CO without distinguishing between private or public law suggests however, that this provision applies to all kinds of employee plant varieties including those created in the public sector. This legal opinion is supported by a statement of the Federal Council[127] but remains so far unconfirmed by the courts.

7.3 The Parties' Duty to Inform

Based on Article 332(3) CO all private and public sector employees have to notify their employer in writing about the creation of a 'reservable plant variety'. The employer then has to decide within six months whether to acquire the plant variety in question.

In regards to 'service plant varieties' the law does not expressly stipulate a similar general duty to inform. However, in the private sector employees must at all times carry out the work assigned to him with due care and loyally safeguard the employer's legitimate interests (Article 321a CO).[128] Based on this general duty of loyalty, the employee is obliged to inform the employer immediately of any plant variety created within the employer's field of activity.[129] Furthermore, the employee has to support in

124. Decision of the Commercial Court of St. Gallen of 13.3.2001 E. 3, GVP 2001, No. 41; Streiff/von Kaenel/Rudolph, *supra* n. 80, at n. 11.
125. Andermatt, *supra* n. 89, p. 298 ff.; Thierry Calame, *Die Berechtigung an der Erfindung* in SIWR vol. IV, 195; different opinions e.g.: Streiff/von Kaenel/Rudolph, *supra* n. 80, at n. 14.
126. Peter Heinrich, *Schweizerisches Designgesetz/Haager Musterschutzabkommen*, DesG 7 n. 24; Streiff/von Kaenel/Rudolph, *supra* n. 77, at n. 4.
127. Swiss Federal Gazette, BBl, 2004, p. 4182.
128. Cantonal Court of Nidwalden, sic! 2010, p. 41 f, E. 4.2; Wolfgang Portmann, BSK, 5th ed., CO Art. 321a, n 12; Streiff/von Kaenel/Rudolph, *supra* n. 77, at n. 2.
129. Cantonal Court of Nidwalden, 41 et seq. (sic! 2010), E. 4.2; Streiff/von Kaenel/Rudolph, *supra* n. 77, at n. 2.

protecting the plant variety and sign all required documents.[130] In regards to public sector employees the applicable federal, cantonal or municipal public labour regulation has to be consulted.

7.4 Employee's Remuneration Right

Unless otherwise agreed, the employer acquires the rights to 'service plant varieties' without having to pay a financial compensation.[131] Financial compensation is however always due if the employer wishes to acquire a 'reservable plant variety'. The amount depends on the economic value of the plant variety, the degree to which the employer contributed, any potential reliance on other staff and the employer's facilities, the expenses incurred by the employee and their position in the company (Article 332(4) CO). The obligation to pay for reservable plant variety cannot be modified to the employee's disadvantage (Article 362 CO).[132]

8 INSTANCES FOR DISPUTES

Since 1 January 2011, civil proceedings related to employee ownership are regulated nationally by the Civil Procedure Code ('CPC') and the Federal Act on the Federal Patent Court ('PatCA').[133] Disputes related to employees' IPR are heard either by a Labour Court as an action relating to employment law (Article 34(1) CPC), by the Cantonal Court as a dispute in connection with IPR (Article 5(1) a CPC), or by the Federal Patent Court as a dispute concerning employees' inventions (Article 26(2) PatCA). Disputes arising from public institutions are subject to Administrative Courts. The decisions of those Courts can be challenged before the Federal Supreme Court.

8.1 Dispute Concerning Employees' IPR

Disputes related to employees' IPR (with the exception of inventions) are heard by a Labour Court or by a Cantonal Court.

The Labour Court has jurisdiction for disputes related to employment law, such as termination of contract. The *territorial jurisdiction* is where the defendant has its domicile or registered office or where the employee normally carries out their work (Article 34 CPC). The *material jurisdiction* depends on the legal situation in each

130. Wang, *supra* n. 107, at 152 f.; Portmann, *supra* n. 77, at Art. 321a OR N 12; Streiff/von Kaenel/Rudolph, *supra* n. 77, at CO Art. 321a N 2. Regarding inventions: Supreme Court: 4A_688/2014 E. 3.3.4; Federal Patent Court: S2014_007 E. 4.3.
131. Supreme Court: BGE 4A_691/2011 E. 4.1; Wyler/Martin, *Droit du travail* 375 (2nd ed.); Wolfgang Portmann, *Die Arbeitnehmererfindung*, p. 76 f.; Other opinions: Manfred Rehbinder, BK, OR-331-355, CO Art. 332 n. 8.
132. Andermatt, *supra* n. 89, at 294 ff. Calame, *supra* n. 82, at 200.
133. Civil Procedure Code of 19 December 2008 ('CPC') (RS 272) https://www.admin.ch/opc/en/classified-compilation/20061121/201701010000/272.pdf (accessed 20 July 2017). Federal Act on the Federal Patent Court of 20 March 2009 (PatCA), https://www.admin.ch/opc/en/classified-compilation/20071763/201212010000/173.41.pdf (accessed 20 July 2017).

canton (e.g. Arbeitsgericht in Zurich, Tribunal des Prud'homme in Geneva). Decisions of the Labour Court may be objected first at a Cantonal Court (Article 308 and 319 CPC) then at the Federal Supreme Court (Article 75(1) LTF).[134]

The Cantonal Court has jurisdiction for disputes related to IPR, including disputes concerning the ownership of IPR (Article 5(1)a CPC). The *territorial jurisdiction* is where the service has to be provided, i.e. where the defendant has its domicile or registered office (Article 36). Nevertheless, the appellant could argue that the dispute concerns employment law rather than IPR and that only the Labour Court has jurisdiction rather than the Cantonal Court for IPR disputes. The *material jurisdiction* depends on each Cantonal Court (e.g. Handelsgericht in Zurich, Cour de Justice in Geneva).[135] Decisions of the Cantonal Court may be contested directly before the Federal Supreme Court (Article 75(2) LTF).

8.2 Disputes Concerning Employees' Inventions

Disputes concerning employees' inventions are heard by the Federal Patent Court (Article 1(1) PatCA; Article 76 LBI; Article 5(1)a CPC), since it has jurisdiction in civil actions with a connection to patents, in particular actions that concern the right to patents or their assignment (Article 26 (2) PatCA). Decisions of the Federal Patent Court may be challenged before the Federal Supreme Court (Article 75(1) LTF).

The Federal Patent Court, which began operations on 1 January 2012, is composed of judges with expert knowledge of patent law (Article 8(1) PatCA) and is situated in the canton of Saint-Gall.[136] For proceedings, provisions of the CPC apply *mutatis mutandis* (Article 27 PatCA), with some particularities, such as that the language of proceedings can be English with the consent of the parties and the Court (Article 36 PatCA).

8.3 Disputes Concerning Public Institutions

Disputes arising from public institutions are subject to Administrative Courts. For federal public institutions, the decisions can be challenged first before the Federal Administrative Court (Article 31 LTAF),[137] then before the Federal Supreme Court (Article 75 LTF). For cantonal public institutions, the decisions can be contested first before Administrative Cantonal Courts, then before the Federal Supreme Court (Article 75 LTF).

---

134. Federal Act on Supreme Court, https://www.admin.ch/opc/fr/classified-compilation/20010204/201706010000/173.110.pdf (accessed 20 July 2017).
135. See the list of Cantonal Courts: Michael Widmer/Timo Leis, *Zuständigkeit gemäss ZPO im Immaterialgüterrechtsprozess* 363, 374 (sic ! 2013).
136. Other special locations in other cantons are possible according to the circumstances (Arts. 6-7 PatCA).
137. Federal Act on Administrative Supreme Court, https://www.admin.ch/opc/fr/classified-compilation/20010206/201701010000/173.32.pdf (accessed 20 July 2017).

8.4 Competing Jurisdiction Between IPR and Labour Courts

Questions may arise concerning competing jurisdiction. A dispute regarding copyright ownership between the employer and the employee could be heard either by a Labour Court as an action relating to employment law (Article 34(1) CPC) or by the Cantonal Court (Article 5(1) a CPC) or the Federal Patent Court (Article 1(1) PatCA), as a dispute in connection with IPR.

While several authors give priority to the IPR Courts which may entertain all questions arising from the case, which is particularly true for a procedural economy,[138] the Federal Supreme Court seems to admit both jurisdictions without priority: in a recent case regarding an employee's inventions, after a cantonal decision rejected the request of an employee, the employee raised the question before the Cantonal Court as an appeal relating to employment law, and at the same time before the Federal Supreme Court as an objection in connection with IPR. Finally, the Cantonal Court ruled on the case in the capacity of an appellate court relating to employment law, while the Federal Supreme Court suspended the proceedings because the disputes could be considered as a dispute in connection with IPR.[139]

9 CONCLUDING REMARKS

9.1 Overlapping Intellectual Property Rights

Thus far, each type of intellectual property has been discussed separately. However, the boundaries between the different types of intellectual property rights can blur and sometimes even overlap. It is therefore possible that the product of an employee's work is protected by more than one type of intellectual property right (e.g. a design of a semiconductor may also enjoy copyright protection).

This 'belt and braces' protection does not generally lead to any specific legal difficulties. However, given that each type of intellectual property right is governed by its own rules, some intellectual property rights to a specific product of an employee's work may be transferred to the employer, whilst other rights remain with the employee. Accordingly, the employer may enjoy design protection (e.g. pursuant to Article 332(2) CO), while the employee enjoys copyright protection (e.g. in accordance with the creator principle[140]). In such cases, both the employer and the employee may be restricted by the other's intellectual property rights. However, given their duty of loyalty (Article 321a CO), the employee must not unreasonably impede the employer's exploitation of rights. Depending on the interests involved, the employee may therefore be obliged to offer their intellectual property right to the employer in return for a

138. Rainer Wey, *Kommentar ZPO* (Sutter-Somm/Hasenböhler/Leuenberger ed., 2nd ed., Zurich/Basel/Geneva 2013), 5 N 8; Widmer/Leis, sic! 2013, 363, 373 seq.; Christian Hilti and Sarah Hennenberger-Sudjana, *Kompetenzattraktion vor Bundespatentgericht in Fällen objektiver Klagenhäufung und/oder Anspruchsgrundlagenkonkurrenz?* 84 et seq. (sic ! 2013).
139. Supreme Court, 6 November 2012, 240, 242 (sic ! 2013).
140. See section 1.

reasonable payment. However, it would seem that these potentially conflicting intellectual property rights are of no practical relevance; the authors are unaware of any court decisions concerning such scenarios.

9.2 Discrepancy in IPRs

It appears that there is a discrepancy in employees' IPRs: for copyright no legal transfer to the employer is provided (except for computer programs), while for other intellectual property rights there is a legal transfer to the employer for works made in performance of contractual obligations. The difference is, however, more theoretical than factual because *the theory of finality*, as applied by the judiciary, leads to an objective interpretation rather than to a specific interpretation in copyright law and accordingly most of the time to the conclusion that the copyright of works made for hire is transferred to the employer. Thus, the result seems equal either with a legal transfer provided or with *the theory of the finality*.

Another discrepancy within IPRs is that *the theory of the finality*, under which the transfer of rights is limited to the purpose of the contract and which determines the existence, the nature, and the duration of transfer, applies to copyright, while the general rules of contract law (subjective and the objective interpretation) apply to other intellectual property rights. Such a discrepancy does not appear justified since other IPRs face the same challenges as copyright issues, in particular the need to determine the existence, the nature, and the duration of transfer. Moreover, there is no legal basis for such a determination (the *theory of the finality* has a broader scope of application than the rule provided in Article 16(2) CopA and arises from the jurisprudence), so that there is no reason to treat copyright differently from other IPRs and vice versa. Finally, it appears appropriate to have a transversal approach and to consider that this rule shall apply to all contracts regarding the creation and the transfer of rights related to intangible assets.[141]

9.3 Adaptation of the Contract Once Transfer Has Been Made

Once the transfer has been made, there shall be no additional compensation (either for service works by means of legal transfer or for copyright by means of an employment contract or a separate agreement, the employer acquires the work). However, the general rules of contract law allow the court to adapt the contract if the initial contractual obligations appear disproportionate due to a substantial change of circumstances. Thus, the court may have discretional power according to the *clausula rebus sic stantibus* theory to provide the employee with additional compensation based on

141. Tissot, *supra* n. 1, LBI 3 N 22.

the economic value of the work, if the transferred work leads to income that exceeds the parties' forecasts or when the efforts provided exceed that which could be reasonably expected.[142]

BIBLIOGRAPHY

Table of Sources

Adrian Andermatt, *Das Recht an im Arbeitsverhältnis geschaffenen immaterialgüter-rechtlich geschützten Erzeugnissen*, p. 298 et seq.

Adrian Staehelin & Frank Vischer, *Obligationenrecht*, V 2c, Der Arbeitsvertrag, CO 332, N 11, Article 319-362 OR (3rd ed., Zurich 1996).

Bertschinger, Münch & Geiser, *Schweizerisches und europäisches Patentrecht*, 170 (Basel 2002).

Catherine Mettraux Kauthen, *Commentaire romand de la propriété intellectuelle*, LDA 33 N 11 (Werra/Gilliéron eds., Basel 2013).

Christian Laux, *Vertragsauslegung im Urheberrecht*, 10 et seq. (Berne 2003).

Christian Hilti & Sarah Hennenberger-Sudjana, *Kompetenzattraktion vor Bundespatentgericht in Fällen objektiver Klagenhäufung und/oder Anspruchsgrundlagenkonkurrenz?*, 84 et seq. (sic !, 2013).

Cyrill Rieder, *Eigentum an Hochschulerfindungen*, 196 et seq., 245 et seq. (Bern 2010).

Denis Barrelet & Willi Egloff, *Le nouveau droit d'auteur*, LDA 16 N 6 (Bern 2008).

Edgar Philippin, *Commentaire romand de la propriété intellectuelle*, LDA 11 N 23 (eds: Werra/Gilliéron, Basel 2013).

Emil F. Neff & Mathias Arn, *Urheberrechtlicher Schutz der Software*, in SIWR vol. II/2 (Basel/Frankfurt/München 1998).

François Dessemontet, *Intellectual Property Law in Switzerland* 43 (Zurich 2012).

François Dessemontet, *Le droit d'auteur* 604 et seq. (Lausanne 1999).

Gert Lück, *Halbleiter-Topographieschutz*, in SIWR vol. II/2, 363 et seq.

Henneberger-Sudjana & Henneberger, *Gesonderte Honorierung von Diensterfindungen als Beitrag zur (besseren) wirtschaftlichen Entwicklung*, Diskussionspapier Nr. 128 der Reihe des Forschungsinstituts für Arbeit und Arbeitsrecht an der Universität St. Gallen.

Hilty, *Urheberrecht*, 161 et seq., 186 et seq. (Bern 2010).

Ivan Cherpillod, *Protection des logiciels et des bases de données: la révision du droit d'auteur en Suisse*, in RSPI/SMI, 49 (1993).

Jacques de Werra, *Commentaire romand de la propriété intellectuelle*, LDA 17 N 6 (Werra/Gilliéron eds., Basel 2013).

142. Staehelin/Vischer, *supra* n. 69, CO 332 N 11; see also Supreme Court of 20 March 2009, 4A_595/2008, consid. 6 and references.

Manfred Rehbinder & Adriano Viganò, LDA 11 N 3.
Manfred Rehbinder, *BK OR-331-355*, CO Art. 332 N 8.
Markus Wang, *Entstehung und Bestand des Designrechts*, in SIWR vol. VI, 151 et seq.
Martina Altenpohl, thesis, p. 265 et seq.
Michael Widmer & Timo Leis, *Zuständigkeit gemäss ZPO im Immaterialgüterrechtsprozess* (sic ! 2013).
Mondini & Bürge, *Zuordnung der Ergebnisse gemeinsamer Forschung und Entwicklung in der Praxis*, AJP 16 (2008)
Mosimann & Graf, *Arbeitnehmererfindungen*.
Münch & Herzog, *Berechtigung an der Erfindung*, in *Schweizerisches und europäisches Patentrecht* 170 (Bertschinger/Münch/Geiser eds., Basel 2002).
Nathalie Tissot, *Commentaire romand de la propriété intellectuelle* (de Werra/ Gilliéron eds., Basel 2013).
Peter Heinrich, *Schweizerisches Designgesetz/Haager Musterschutzabkommen*, DesG 7 N 11.
Rainer Wey, *Kommentar ZPO* (2nd ed., Zurich/Basel/Geneva 2013).
Reto M. Hilty, *Lizenzvertragsrecht: Systematisierung und Typisierung aus schutz- und schuldrechtlicher Sicht* 754 et seq. (Zurich 2001).
Roland von Büren, *Der Übergang von Immaterialgüterrechten, Schweizerisches Immaterialgüter und Wettbewerbsrecht* (Basel/Geneva/Munich 2002).
Roland von Büren & Michael A. Meer, *Der Urheber*, in SIWR vol. II/2 (von Büren/ David eds., Basel 2006).
Roland von Büren, *Der Werkbegriff*, in *Schweizerisches Immaterialgüter und Wettbewerbsrecht* ('SIWR') vol. II/1, 228 (Basel/Genf/München 2005).
V. Salvadé/N. Tissot, La réalisation d'un site web ou l'ouverture d'un compte par le travailleur. Qui est titulaire des droits ?, in: Jean-Philippe Dunand, Pascal Mahon (publishers), Schulthess 2014, 227 et seq.
Staehelin & Vischer, CO 332 N 11.
Streiff, von Kaenel & Rudolph, *Arbeitsvertrag, Praxiskommentar zu Art. 319-362 OR*, (7th ed.) CO Articles 321a and 332.
Stutz & Ambühl, *Rechte an Computerprogrammen, geschaffen im öffentlich-rechtlichen Arbeitsverhältnis – eine Schweizer Sonderlösung?*, 667 et seq. (GRUR Int 2010).
Stutz, Beutler & Künzi, *Designgesetz DesG*, Art. 7 DesG, N 49 ff.
Swiss Federal Gazette, BBl, 2004, p. 4182.
Thierry Calame, *Die Berechtigung an der Erfindung*, in SIWR vol. IV, p. 195 ff.
Urs Egli, *Softwareentwicklung im Arbeitsverhältnis* 15 et seq., 32 (ArbR 2007).
Widmer & Leis, (sic! 2013).
Willi Egloff, *Urheberrecht und Urhebervertragsrecht in der audiovisuellen Produktion* 17 (sic! 1998).
Wolfgang Portmann, *BSK*, CO Art. 321a N 12 (5th ed.).
Wolfgang Portmann, *Die Arbeitnehmererfindung*, 101 (Bern).
Wyler & Martin, *Droit du travail*, 375 (2nd ed.).

Table of Legislation

Cantonal Act of the University of Geneva of 13 June 2008 (C 1 30), https://www.geneve.ch/legislation/rsg/f/s/rsg_C1_30.html (accessed 20 July 2017).
Civil Procedure Code of 19 December 2008 ('CPC') (RS 272),
https://www.admin.ch/opc/en/classified-compilation/20061121/201701010000/272.pdf (accessed 20 July 2017).
Directive 91/24.
Federal Act on the Amendment of the Swiss Civil Code of 30 March 1911 (RS 220), https://www.admin.ch/opc/en/classified-compilation/19110009/201704010000/220.pdf (accessed 20 July 2017).
Federal Personal Act of the Confederation of 24 March 2000 (RS 172.220.1).
Federal Act on the Federal Institutes of Technology of 4 October 1991 (RS 414.110), https://www.admin.ch/opc/en/classified-compilation/19910256/201705010000/414.110.pdf (accessed 20 July 2017).
Federal Act on Copyright and Related Rights of 9 October 1992 (RS 231.1), http://www.admin.ch/ch/e/rs/2/231.1.en.pdf (accessed 20 July 2017).
Federal Act on Patents for Inventions of June 25, 1954 (SR 232.14).
Federal Act on the Protection of Designs of October 5, 2001 (SR 232.12).
Federal Act on the Protection of Plant Varieties of March 20, 1975 (SR 232.16).
Federal Act on the Federal Patent Court of 20 March 2009 (PatCA), https://www.admin.ch/opc/en/classified-compilation/20071763/201212010000/173.41.pdf (accessed 20 July 2017).
Federal Act on Supreme Court, https://www.admin.ch/opc/fr/classified-compilation/20010204/201706010000/173.110.pdf (accessed 20 July 2017).
Federal Act on Administrative Supreme Court, https://www.admin.ch/opc/fr/classified-compilation/20010206/201701010000/173 32.pdf (accessed 20 July 2017).
Swiss Federal Act on the Protection of Topographies of Semi-conductor Products of October 9, 1992 (SR 231.2).
Swiss Federal Journal, BBl, 1989, p. 572 and 574.
UPOV since July 10, 1977 (SR 0.232.163).

Table of Cases

Supreme Court: BGE 72 II 270 (1946).
Supreme Court: BGE 100 IV 167 (1974).
Supreme Court: BGE 121 III 125 (27 March 1995).
Cantonal Court Basel, BJM, p. 248 (1995).
Supreme Court, 4C_442/1996, c. 4 bb (30 April 1997).
Commercial Court of St. Gallen, E. 2, in: GVP 2001, No. 41 (13 March 2001).
Supreme Court: BGE 127 III 461 (30 May 2001).
Supreme Court: BGE 2A.204/2006 E. 7.1 (22 June 2007).
Commercial Court of Zurich, E. 3.2, in: ZR 2008, p. 55 ff (28 August 2007).

Cantonal Court of Nidwalden, E. 4.3, in: sic! 2010, p. 41 ff (11 March 2008).
Cantonal Court of Bern, HGer Berne, HG 15 39 CAA, sic! 2016 (17 June 2015)
Supreme Court, 4A_104/2008 c. 4.2 (8 May 2008).
Supreme Court, 4A_595/2008 (20 March 2009).
Cantonal Court of Nidwalden, sic! 2010, p. 41 f. (2010).
District Court Zurich, sic! 2011, p. 227 (2011).
Supreme Court: BGE 4A_691/2011 E. 3.1 (6 November 2012).
Supreme Court, 4A_643/2012 (23 April 2014).
Federal Patent Court: S2014_007 (30 October 2014).
Supreme Court: BGE 4A_442/2014 (7. January 2015).
Supreme Court: BGE 4A_688/2014 (15. April 2015).
Supreme Court: BGE 6B_310/2014 (23 November 2015).

CHAPTER 22
Turkey

Ekrem Soylu & Korcan Dericioğlu

1 BACKGROUND

According to Turkish labour law, the employee-employer relationship is subject to the principle of interpretation in favour of the employee. However, a balance of mutual interests is sought in creations during the *relationship* both under labour law, law of obligations and IP law. Once an employer has paid a certain sum in return for the performance of a certain service, they acquire the rights on the creations arising with relation to such service. In fact, the employer defines the scope of the work in question, provides the necessary technical facilities and concurrently takes the risk of a failure, and in turn, is the right holder over the creations developed within the framework of this relationship. On the other hand, the employee receives remuneration for their services in form of wages and/or other similar financial gains.

In this context, the employer is, as the rule of thumb in IP, regarded as the right holder over all creations developed within the scope of a *service relationship* on the condition that these creations hold relevance to the work undertaken in the same scope.

If the determined conditions have not been satisfied, the employee is in the position of a right holder. However, if the work has been undertaken with the equipment of the employer, an obligation of notification, as well as the employer's remuneration rights are invoked pursuant to the balance of interests.

In addition, specific legal arrangements are determined to ascertain that moral rights of the creator remain with themselves in any case. Indeed, the legislation stipulates that these rights are closely and strictly connected with the creator and that any pre-determined ownership of right defined in an economically based relationship may only remain within the framework of economic rights.

Furthermore, as copyright law mostly focuses on individual interests rather than industrial interests in contrast to other types of IP protection, the ownership is assigned in full to the author and the employer is only afforded a right to exercise.

Among the general rules, the new Industrial Property Code entered into force on 10 January 2017. Unlike previous Decree-Laws, IP Code no. 6569 is a unified code regulating rights concerning trademarks, patents, designs and geographical indications. The definition of 'employee' is provided in Article 2(d). An employee is a person who carries out work under a contractual relationship with a legal entity in which the specification of particular work on behalf of the employer is provided. There is no specific regulation with respect to the legal status of different employees contracted under the same group of companies (subsidiaries). Therefore, they are to be considered as independent entities.

2 EMPLOYEES' COPYRIGHT

2.1 The Legal Framework

In Turkey, copyright is addressed within the framework of Law No. 5,846 on Intellectual and Artistic Works. Chapter 1 of the above-mentioned Law covers the range of works and the criteria necessary to establish the subsistence of copyright, whereas Chapter 2 covers the definition of 'author'. Chapter 3 addresses the rights relating to authorship. Ownership of rights over works belonging to employees is covered within the scope of Article 18 in Chapter 3.

2.2 Ownership of Rights

According to Article 1(B) of the Law, a work is defined as:

> Any intellectual or artistic product bearing the characteristics of its author, which is deemed a scientific and literary or musical work or work of fine arts or cinematography.

The protection granted depends on the precondition that the work in question falls within the scope of this definition. The author of a work is, in turn, defined in Article 8 as:

> The author of a work is the person that has created it.

Article 18, governs the ownership of right over works created by employees. According to this article:

> The rights in works created by civil servants, employees and workers during the execution of their duties shall be exercised by the persons who employ or appoint them; provided that the contrary may not be deducted from a special contract between such persons or from the nature of the work. This rule shall also apply to the organs of legal persons.

Article 18 is a general provision that addresses the issue in such a manner as to cover all types of works including neighbouring rights, databases and computer programs. Moreover, it does not make any distinction between university researchers, public-sector employees or private-sector employees, but generally regulates the nature and criteria of the right and covers all subjects.

2.2.1 *Nature of the Right*

Combined consideration of Article 8 and Article 18 leads to the observation that the right holder over the work is the person who created it regardless of the conditions under which such work was created. In fact, Article 18 addresses the exercise of rights over a work but not the ownership of rights. In other words, the author of a work is the person who created it under any circumstance, whereas the authority to exercise the rights affiliated with the work is granted to the employer on the condition that such work has been created within the framework of an employee-employer relationship and that certain conditions have been satisfied.

The Law does not specify any clear provision to ascertain whether the rights to be utilised when such conditions have been fulfilled will be economic rights or moral rights. However, the prevailing opinion in the doctrine is that the rights to be exercised by the employer will be the economic rights, whereas moral rights may also be exercised to the extent that the exercise of the economic rights makes it possible.[1]

2.2.2 *Definition of Employee*

The Law does not include a definition of 'employee'. Nevertheless, the term is considered broad, to cover employees subject to labour law, law of obligations, press labour law and maritime labour law. Persons subject to a concrete business relationship are also regarded as employees.

The concept of 'civil servants', on the other hand, is to be construed as persons subject to the Civil Servants Law. 'Civil servant' refers to a person providing public services. The authority to exercise rights over works created by these persons rests with the administrative institutions that have employed/appointed them.

2.2.3 *Criteria Required for the Exercise of Rights by the Employer*

In order for the rights to a work to be exercised by the employer, the legislation deems it sufficient that such work has been created by the employee during the execution of their duties unless the contrary may not be deduced from the nature of the work. In this framework, where the work has been created by an employee within office hours, the authority to exercise rights over the work rests with the employer. However, if such

1. Firat Öztan, *Fikir ve Sanat Eserleri Hukuku* 551 (Ankara 2008); Funda Cinoğlu, *Eser Üzerinde Işverenin Hak Sahipliği*, 14 (Istanbul 2010); http://sites.khas.edu.tr/tez/FundaCinoglu_izinli.pdf (accessed 15 May 2014)

work bears no relevance to the duties undertaken by the employee, the employer may not hold the authority to exercise rights over the work. The Supreme Court has in this respect held[2] that the employer does not have any rights over a book on property law written by an attorney at law employed at the legal department of a media company, because the preparation of a book during the performance of their routine duties is not amongst the usual tasks of the profession of attorney.

Article 18 also incorporates a possibility whereby the employee and the employer may execute a private contract to establish the party/parties that will hold or exercise the rights in the work.

Works created by university researchers are also addressed within the framework of the same criteria. However in practice, the work created by a university researcher, even within office hours, is not considered to be part of their duties unless specifically requested by the university for a project or publication. Although there is no precedent case pertaining to this issue, numerous universities have published their own IP policies in order to eliminate any uncertainties. They emphasise therein that all rights arising from works created will rest with the university researcher with the exception of works requested directly by the university for a project or to be included subsequently in a publication of the university.

2.3 The Parties' Duty to Inform

The Law on Intellectual and Artistic Works does not include any specific provision for the duty to inform the employer. However, despite the lack of express provisions, within the scope of the loyalty principle the employee must inform the employer of any work created within the framework of an employment contract during the performance of their routine duties.

2.4 Employees' Remuneration Right

The Law on Intellectual and Artistic Works does not include any specific provision on the remuneration of employees. This matter only applies in the presence of a private contract executed between the parties with agreed conditions thereto.

3 EMPLOYEES' SEMICONDUCTOR CHIPS

3.1 The Legal Framework

In light of scientific and technological developments, integrated circuit topographies, previously protected by design law before specific legal instruments were passed, are granted protection under Code No. 5,147 on the Protection of Integrated Circuit Topographies, dated 22 April 2004. Chapter 3 of the Code addresses the ownership of

2. Supreme Court 11. Civil Chamber, 1988/5471E. 1989/384K., 30.01.1989.

rights, in which Article 8 specifically regulates ownership of rights in employment contracts, Article 9 regulates ownership of rights of members of universities and Article 10 regulates ownership of rights in other service contracts.

3.2 Ownership of Rights

In the case of integrated circuit topographies, the right holder is the designer or the designer's legal successors. If more than one person has designed an integrated circuit topography, the rights are held jointly by these persons. However, such joint ownership is not a mandatory requirement as the parties may agree differently by concluding a contract.

3.2.1 Ownership of Rights in Employment Contracts

The ownership of rights to integrated circuit topographies designed by civil servants, workers and employees during the execution of their duties, shall be vested in their employers. Nevertheless, this is not a mandatory requirement and Article 8 permits the parties to agree differently by concluding a contract.

The Code does not distinguish, in relation to ownership of rights, between the designs of topographies where this falls within the nature of one's work, and where it falls outside of it. Hence, designs created by an employee within office hours rests with the employer regardless of whether it is relevant to the duties undertaken by the employee. However, if the employee has designed an integrated circuit topography by using the information and equipment available at work but in conditions that are not required by the contract or are outside the nature of one's work, they may claim a certain amount of remuneration.

The Code does not include a definition of 'employee'. Nevertheless, as with copyright, the term 'employee' is considered broadly, to cover employees subject to labour law, law of obligations, press labour law and maritime labour law. Persons subject to a certain business relationship are also regarded as employees.

The concepts of civil servants, on the other hand, are to be construed as persons subject to the Civil Servants Law. The term 'civil servant' refers to a person providing public services. The authority to exercise rights over works created by such persons rests with the administrative institutions that have employed/appointed them.

3.2.2 Ownership of Rights of University Researchers

University researchers fall into a special category and are addressed in Article 9. According to this Article, rights over integrated circuit topographies designed by university members involved in scientific experiments and studies rest with the respective member. However, if a member of the university uses tools and equipment available at the university and uses funds available to the university, an obligation to notify the university, as well as a form of a remuneration/reimbursement right to the benefit of the university, will become applicable.

3.2.3 Ownership of Rights in Other Service Contracts

The right holder of integrated circuit topographies designed within the framework of other service contracts is the client. The establishment of such a case requires the service contract between the parties to be related to the design of the integrated circuit. However, this is not a mandatory requirement and according to Article 10, the parties may agree differently by concluding a contract.

3.3 The Parties' Duty to Inform

3.3.1 University Researchers

If a university has made available specific equipment and means for the research work that led to the design of the semiconductor chip topography, the researcher must give notification in writing of the creation of the design, specify the modality of exploitation and the amount of proceeds achieved if requested by the university.

3.3.2 Employees

The duty to inform within the scope of an employment relationship is not expressed through a specific provision. However, despite the lack of express provisions, within the scope of the loyalty principle the employee must inform the employer of any work created within the framework of an employment contract during the performance of their routine duties.

3.4 Employees' Remuneration Right

3.4.1 University Researchers

According to Article 9 of Code No. 5147, rights over integrated circuit topographies designed by university members involved in scientific experiments and studies rest with the respective member. However, if the university has made available specific equipment and means for the research that led to the design, the university may request a reasonable remuneration within three months following the receipt of the notification by the university researcher. This amount shall, however, not exceed the costs expended by the university.

3.4.2 Employees

If an employee has designed an integrated circuit topography by using the information and equipment available at work but in a condition that is not required by the contract, they may claim a certain amount of remuneration from the employer. The determination of the amount will be subject to a calculation that takes into account the economic value of the integrated circuit topography in question. If the parties cannot agree the amount of remuneration, the courts will decide it. In addition, the parties have the right to determine the amount by concluding a contract in advance.

4 EMPLOYEES' INVENTIONS

4.1 The Legal Framework

The new Industrial Property Code has an entire chapter (Part 5) dedicated to employee inventions. The chapter comprises definition of service inventions, duty to report, employers rights on the invention and compensation for a claim, first filing requirement for the employer, common provisions, mutual rights and duties, inventions of the professors and arise out of the government supported projects. Additionally, legalisation includes comprehensive guidelines wherein the calculation of the remuneration is explained.

4.2 Ownership of Rights

Pursuant to Article 109, only the inventor or their successor in title has right to a patent and the right is transferable. Article 115(2) of Chapter 5 grants the right to an employer to claim full or partial rights of the invention from the inventor. In case of claiming full rights, all rights on the invention is transferred to the employer at the date of the receipt of the claiming of the invention and employer is obliged to first file of a patent application regarding the subject matter of the invention disclosure before the Turkish Patent Office. However, after filing a patent application full or partial assignment of rights back to the employee is possible.). If the employer claims conditional rights, the invention will be a free invention but the employer will hold a right to use the invention.

After claiming full rights, the employer is under an obligation to file a patent application before the Turkish Patent Office. Exceptions to this obligation include the release of rights by the employer, rejection due to the prior art, agreement with the inventor not to file an application, and situations where the information contained in the invention is a trade secret of the employer. If the employer does not file an application with the Turkish Patent Office within a time limit defined by the employee, they have the right to file the application on behalf of the employer. In case of release of rights by the employer, the employee is free to file a patent application as an applicant. If release has been done after a first application filed by the employer, patent rights has to be assigned to the employee.

Two types of employee inventions are recognised: *free inventions* and *service inventions*. In order for an invention to be considered a service invention, at least one of the two following conditions must be fulfilled: the invention must have been created while performing an obligatory task assignment; or, the invention must be based, to a great extent, on the experience and activity of the company. All other inventions are considered free inventions and the right to file a patent belongs to the inventor, i.e., the employee.

According to the applicable legislation, university researchers' inventions are considered service inventions, and the right to file a patent application for an invention wherein the inventor is an academic member of a university (during scientific research) belongs to the university.

4.3 Parties' Duty to Inform

4.3.1 Invention Notification

Pursuant to Article 114, the employee is under duty to inform the employer in writing without any delay of the making of an invention. The form of the notification has two requirements. The first requirement concerns the formal completeness of the notification. The notification should include: share of contribution to the invention for each inventor. Secondly, the notification should include the employer's previous work on the subject matter and explain the invention, describing the technical problem and its solution.

It is advisable for the employer to provide a template document to the inventor, having regard to the employer's obligation to support the inventor in complying with all legal requirements to report an invention disclosure. Multiple employees are allowed to jointly submit combined information in one document.

4.3.2 Information after Notifying – Pending Potential Patent Application

Pursuant to Article 114(4), if any missing part or insufficient information is detected in the notification as submitted, the employer should inform the employee of the information to be completed within two months after the receipt of the incomplete notification. If the employer does not inform the employee within a two-month period, the disclosure is considered complete.

After receiving the invention notification, the employer has a duty to inform the employee of the receipt of the notification in writing without delay. It is advised that the confirmation document include date, name of the employee and at least a title of the invention to prevent any confusion in the future. Furthermore, the employer should claim full or partial rights in writing within four months from the date of receipt. All rights to the invention will be transferred to the employer on the date of receipt of the claim by the inventor. According to Article 20, if it is not sent in due time, the invention will be considered as a free invention and the employee will have the right to file a patent application. Other conditions for an invention to be considered a free invention are the release of rights to the invention and claiming partial rights to the invention.

4.3.3 *Information after Patent Application*

After filing an application, the employer should provide a copy of the application documents to the employee (Article 114(1)). Upon the request from the employee, the employer should provide information on the file and provide documents created during the prosecution of the file. The employer should provide supporting information against challenges to the patentability of the invention.

Pursuant to Article 118(2), if the employer decides to surrender patent protection before the payment of the remuneration, the employee should be informed regarding the decision. Upon the request of the employee, the employer should provide documents required for assignment of the patent application or patent to the employee. If the employee does not respond within three months from sending the notification of the decision, the employer is free to surrender the application and the rights arising from the application without any further obligation.

In case of filing a subsequent application, the same conditions are applicable without any specific time limit. This requirement is implicit in Article 116(6) under the term of enabling the employee to make a patent application for which the employer does not wish to file a patent. This clearly requires a notification to the inventor with respect to the subsequent filing decision. For example, sending a list of designated countries in case a regional application is indicated.

4.4 Employees' Remuneration

Article 115(6) introduces a right to claim compensation. After claiming full or partial rights to the invention, the employee has a right to request a reasonable amount to be paid. When calculating the compensation, the economic value of the invention, the title of the employee and the company's share of contribution to the invention must be taken into consideration. The amount of remuneration should be calculated separately for each inventor.

According to Article 115(7), after claiming the right, the employer cannot withhold payment of the remuneration based on the employer's internal evaluation on patentability. However, if the Turkish Patent Office or Court holds that the patentability requirements are not met, the employee cannot claim compensation.

The IP Code acknowledges freedom of contract between the employer and employee concerning the term and amount of the payment. In case of a disagreement on the payment amount, an arbitration board is the first address to evaluate the case.

The employee invention guidelines being part of the legalisation explain effective calculation methods of the compensation in case of a conflict between the employer and employee. Compensation is divided into two main parts namely, an incentive award and the remuneration itself in case of the use of the invention. According to Rule 7 of the guidelines, the incentive award payment should be paid until two months after claiming the invention and should be not less than one net minimum wage. In case of multiple inventors, total amount is shared by each inventor according to their share.

There are three main elements to be considered in order to calculate the remuneration amount. The first one is the economic value of the invention. It is defined by direct or indirect income from the invention. Third chapter of the guideline explain how to calculate income from an invention. Turkish Corporate Tax Law includes a well-known tax exemption regulation called *Patent Box*. According to Rule 12(1), if a company enjoys tax exemption, patent valuation submitted to the tax authorities is by default presumed to be equal to the economic value of the invention for remuneration calculation. Otherwise, the employer has to select a convenient calculation method from one of three alternatives, namely, *licence analogy*, *operational benefit* or *estimation*.

Licence analogy is suitable for industries with a common patent licence and sales agreements. If a similar free invention comparable to the service invention already exists, product unit value as defined is multiplied by number of items or weight. If there is no comparable invention, then contribution of the invention to the sales is calculated to determine the economic value.

The contribution method is suitable for inventions related to the cost saving such as new production methods or tools for manufacture of a known product. Positive difference between costs and incomes by use of the invention is the economic value of the invention. General accounting principles are used to calculate the value.

The last method is estimation which is suitable if any other method cannot be applied for the calculation. The method offers finding suitable similar free invention on the market and determine the buying price of the patent.

Remuneration applies even for patents not exploited by use but licence out the patent. Sale of the patent or cross- licence are also commercialisation methods which require calculation of the remuneration. Licence stacking is also considered under Rule 19 of the guideline. If a number of inventions are employed on the same product or production method, the total economic value is shared among the inventors of all the patents.

After calculation of the economic value, the title of the inventor and contribution of the company to the invention is considered by separately assigning each one of the co-inventors under corresponding groups. A maximum limit to the remuneration amount is also defined by the guideline. If the income from the invention is more than 150,000 times of the minimum wage, employer is not obliged to pay exceeding amount to the inventor. A matrix is presented by the guideline to facilitate calculation of the remuneration. Accordingly, 0.1% to 0.6% of the economic value of the invention has to be paid to the inventors according to their share of contribution respectively. Remuneration is paid to the each inventor annual basis unless it is agreed otherwise. The term of the remuneration is the lifetime of a patent.

5 EMPLOYEES' UTILITY MODELS

The same conditions for inventions are applicable to utility models.

6 EMPLOYEES' DESIGN RIGHTS

6.1 The Legal Framework

Designs fall under the protection of the new Industrial Property Code No. 6569, dated 10 January 2017. Article 70 of the mentioned Code relates to ownership of rights, while Article 73-74 addresses the employee rights.

6.2 Ownership of Rights

In the case of designs, as a rule of thumb, the designer or the designer's legal successors is the right holder. If more than one person is designer, the entitlement shall be vested jointly by these persons. However, such joint ownership is not mandatory and the parties may define differently through the execution of a contract.

6.2.1 Ownership of Rights in Employment Contracts

The ownership of rights to a design developed by civil servants, workers and employees during the execution of their duties, shall be vested in their employers unless otherwise understood from the nature of the work. Nevertheless, this is not mandatory and the Article incorporates that respective parties may define this differently through executing a contract.

The IP Code also establishes that if the employee has developed a design by using the information and equipment available at work but in a condition that is not required by the contract or as a nature of the work, they may claim a certain amount of remuneration.

6.2.2 Ownership of Rights of University Researchers

University researchers are not considered separately from the employment relationship. Rights over designs developed by university researchers involved in scientific experiments and studies rest with the university.

6.2.3 Ownership of Rights in Other Service Contracts

According to Article 73(4), ownership of rights of a design developed within the framework of other service contracts will be defined through the articles of contract.

6.3 The Parties' Duty to Inform

6.3.1 Duty to Inform for University Researchers

The duty to inform for university researchers is not expressed in any specific provision. However, despite the lack of express provisions, within the scope of the loyalty principle, if a university has made available specific equipment and means for the research work that led to the creation of the design, the researcher is obliged to notify the university.

6.3.2 Duty to Inform for Employees

The duty to inform within the scope of an employment relationship is not expressed through a specific provision. However, despite the lack of express provisions, within the scope of the loyalty principle, the employee must inform the employer of any work created within the framework of an employment contract during the performance of their routine duties.

6.4 Employees' Remuneration Right

6.4.1 For University Researchers

Rights over designs developed by university researchers involved in scientific experiments and studies rest with the university. However, the university has to pay a reasonable remuneration that should be at least 50% of the revenue made from the commercialisation of the design in question.

6.4.2 For Employees

If an employee has developed a design by using the information and equipment available at work but in a condition that is not required by the contract, the employee may claim a certain amount of remuneration from the employer. The determination of the amount will be subject to a calculation that takes into account the economic value of the design in question. If the parties cannot reach an agreement on the amount of remuneration, the courts will decide it. In addition, the parties have the right to determine the amount beforehand by way of a contract.

7 EMPLOYEES' PLANT VARIETIES

7.1 The Legal Framework

In Turkey, plant varieties are under protection of the Law No. 5,042 on the Protection of Plant Breeders' Rights for New Plant Varieties. Chapter 2 includes provisions relating to ownership of rights and its scope. Article 12 concerns ownership of rights in the case of an employment relationship and Article 13 concerns ownership of rights in other service contracts. In addition, a specific regulation is in force, which includes provisions for plant varieties bred or discovered and developed by employees working at public institutions.

7.2 Ownership of Rights

The right holder of a new plant variety is the breeder and their legal successor. If more than one person is breeder, the entitlement shall be vested jointly in these persons; however, the parties may agree differently by way of a contract as the law is not mandatory.

7.2.1 Ownership of Rights for Employees of Private Companies

The ownership of the right to a plant variety bred or discovered and developed by employees during the execution of their duties shall be vested in their employers unless otherwise understood from the nature of the work. Nevertheless, this is not mandatory and the parties may agree otherwise in a contract.

Law No. 5,042 also provides that if the employee has bred or developed new plant varieties by using the information and equipment available at work but in a condition that is not required by the contract or as a nature of the work, they may claim a certain amount of remuneration.

If the employee has bred or developed new plant varieties by themselves without any relevance to their contract, the rights over such plant varieties rest with the employee.

7.2.2 Ownership of Rights for Other Service Contracts

The ownership of a right to a plant variety bred or discovered and developed within the framework of other service contracts is the client. The establishment of such a case requires the service contract between the parties to be related to breeding or discovery and development of a plan variety. Nevertheless, this is not mandatory and parties may agree otherwise in a contract.

7.2.3 Ownership of Rights by Employees of Public Institutions

The ownership of rights to a plant variety bred or discovered and developed by employees of public institutions during the execution of their duties, shall be vested in the relevant institutions, unless otherwise understood from the nature of the work. The extent of, and benefits deriving from, breeder's rights applicable to employees of public institutions is determined by the Regulation on Employees of Public Institutions and

Agencies Benefiting From Breeder's Rights.[3] This Regulation also defines universities as public institutions. Breeding and development activities undertaken by members of universities are addressed within the scope of the provisions of the same Regulation.

7.3 The Parties' Duty to Inform

The duty to inform is not expressly provided in a specific provision. However, despite the lack of express provisions, the employee must inform the employer, within the scope of the loyalty principle, of any work created within the framework of the employment contract and during the performance of their routine duties.

7.4 Employees' Remuneration Right

7.4.1 For Employees of Private Companies

If an employee has bred or discovered and developed a plant variety by using the information and equipment available at work but in a condition that is not required by the contract, they may claim a certain amount of remuneration from the employer. The determination of the amount will be subject to a calculation that takes into account the economic value of the plant variety in question. If the parties cannot agree on the amount of remuneration, the courts shall decide it. In addition, the parties have the right to determine the amount beforehand by way of a contract.

7.4.2 For Employees of Public Institutions

The extent of, and benefits deriving from, breeder's rights applicable to employees of public institutions and university researchers is determined by the Regulation on Employees of Public Institutions and Agencies Benefiting From Breeder's Rights. The percentage, criteria and the amount to be paid are specified within the framework of this Regulation. Moreover, and by virtue of the Regulation, a part of the earnings generated resulting from the exploitation of breeder's rights is to be distributed to the employees that have bred or developed or contributed to the variety in question. Such sum shall, however, not exceed 50% of the earnings remaining after all mandatory deductions.

3. Regulation on Employees of Public Institutions and Agencies Benefiting From Breeder's Rights, Section 3 Arts 8-10 (30 April 2005).

8 INSTANCES FOR DISPUTES

The competence of courts to rule over legal disputes in Turkey is addressed in the Code of Civil Procedure. The jurisdiction of the court is related to the specific Code with which the dispute is concerned.[4]

Under the rule of thumb specified in the Code of Civil Procedure, the competent court is the court of the domicile of the defendant or the place where the act was committed.

For IP-related disputes, the competent court for the institution of legal proceedings by the right-holder against third parties is the court of the domicile of the plaintiff, of the place where the act was committed, or of the place where it produced an effect. However, the competent court for the institution of proceedings by third parties against the right-holder is the court of domicile of the defendant.

Jurisdiction of courts for IP-related disputes is specified in each Code. Specialised IP Courts exist in the provinces of Ankara, Istanbul and Izmir, and if proceedings are initiated in one of these three provinces, the courts mandated for these cases will be the Specialised IP Courts. For the remaining provinces, the Civil and Criminal Courts of First Instance are designated as the competent courts.

9 CONCLUDING REMARKS

9.1 General Remarks

IP legislation in Turkey incorporates specific provisions concerning employee's creations. The fundamental principle within the framework of these provisions mandates the employer as the right holder of creations developed within the scope of an employment relationship. Nevertheless, there are significant differences in the interests protected thereby in different cases. There is no publicly available Supreme Court decision involving matters pertaining to the establishment of rights and procedures in the employee-employer relationship were considered.

9.2 Cumulative Protection

Even though the scope and limitations of each IPR are clearly defined, creations may be subject to protection under more than one IPR.

The most obvious of such overlapping protection is the cumulative protection of design and copyright. Under the principle of cumulative protection, issues may arise by reason of the inclusion of designs in the scope of the protection offered by the Copyright Law and the inconsistency observed in the two legislative instruments in relation to ownership of right of employees. Even though there has not been any dispute or public court decision relating to this matter, the rules of thumb require the separation of rights to the greatest extent possible and if such separation is not possible, the relationship

4. Law on Civil Procedures, Section 1, Art. 1-19 (12.01.2011).

between the parties and the manner in which rights have been claimed on both sides will be assessed within the framework of the characteristics of each individual case and the principle of balance of interests.

Having regard to the practical implications of such issues, doctrine suggests the identification of the right holder and the individual assignment of rights in advance by way of a contract.[5]

BIBLIOGRAPHY

Table of Sources

Firat Öztan, *Fikir ve Sanat Eserleri Hukuku* (Ankara 2008).

Funda Cinoğlu, *Eser Üzerinde Işverenin Hak Sahipliği*, Istanbul 2010 http://sites.khas.edu.tr/tez/FundaCinoglu_izinli.pdf (accessed 15 May 2014).

Cahit Suluk, Ali Orhan, *Uygulamali Fikri Mülkiyet Hukuku* (Istanbul 2005).

Table of Legislation

Law on Intellectual and Artistic Works, (No. 5846; 05.12.1951).

Law on Protection of Integrated Circuit Topographies, (No. 5147; 22.04.2004).

Industrial Property Code (No. 6569; 10.01.2017).

Law on the Protection of Plant Breeders' Rights for New Plant Varieties (No. 5042; 15.01.2004).

Regulation on Employees of Public Institutions and Agencies Benefiting From the Breeder' Rights (30.04.2005).

Law on Civil Procedures, (No. 6100; 12.01.2011).

Table of Cases

Supreme Court 11. Civil Chamber, 1988/5471E. 1989/384K., 30.01.1989.

5. Cahit Suluk, Ali Orhan, *Uygulamali Fikri Mülkiyet Hukuku*, vol 3, 201 (Istanbul 2005).

CHAPTER 23
The United Kingdom

Belinda Isaac[*]

1 BACKGROUND

In the United Kingdom the allocation of ownership rights in relation to intellectual property ('IP') as between employers and employees is first and foremost governed by IP statutes which, generally speaking, provide for ownership by the employer of those works created in the course of employment but the parties have, with a number of limited exceptions, the freedom to negotiate and agree between themselves alternative allocations of ownership rights.

'In capitalist societies it is an assumption by now largely unremarked that the product of labour belongs to the owner of the business'[1] however, this was not always the case. Indeed, it was not until the twentieth century that this principle began to be applied in England in relation to IP rights created by an employee. The principle was first introduced by statute in Section 5 Copyright Act 1911 but it was not applied to patentable inventions until 1955 when Lord Simon declared that 'it is an implied term in the contract of service of any workman that what he produces by the strength of his arm or the skill of his hand or the exercise of his inventive faculty shall become the property of his employer'.[2] Prior to this, it had been necessary for the employer to show that it had the benefit of an express term in a contract if it was to succeed in a claim for ownership of any invention made by an employee in the course of employment even if it was made using the employer's materials and time.[3]

* I am grateful to Professor Lionel Bently for reading through and commenting on an earlier draft, however, the errors that remain are entirely mine.
1. W. Cornish *Intellectual Property* (1st ed., Sweet and Maxwell, London, 1981) p. 215.
2. *Patchett v Sterling* (1955) 72 RPC 50.
3. *Marshall and Naylor's Patent* (1900)17 RPC 553 at 555.

The presumption thus developed at common law that the employer would be the owner of any invention created by an employee in the course of employment but that the parties were free to contract as they wished and could by express agreement vary the operation of the presumption. Some employers however went too far, claiming ownership of inventions unrelated to their business and as a consequence, when patent law was reformed, specific provisions were included to make the common law principle a statutory default in specific situations and curtailing the power of employers to require wide-ranging transfers of rights.[4]

Although the provisions included in the Copyright Act 1911 made it clear that the employer would be the owner of copyright works developed during the course of employment, subject to any agreement with the employee to the contrary, unlike patent law, no provisions were included restricting the ability of employers to seek an assignment of broader rights from employees if they so wished.

In addition to the creation of the statutory presumption of ownership of IP created by employees, the twentieth century also witnessed the considerable proliferation of new forms of IP rights as well as the extension in breadth and duration of some already in place. The implementation of international treaties relating to IP and various directives, the product of European harmonisation,[5] has meant that the scope of copyright and related rights protection in the UK has increased, performer's rights extended, and new rights created such as the rental and lending rights and copyright related rights such as the artist's resale right, database rights, rights in semiconductor chips along with supranational rights relating to registered and unregistered Community designs. All these rights have added significantly to the portfolio of IP rights available to employees and employers in the twenty-first century. New forms of inalienable rights have also been introduced in to UK legislation with a view to protecting the interests of creators of copyright works (as well as performers): these include moral rights and, in certain limited circumstances, the right to remuneration distinguished from the underlying property right. The right to remuneration in respect of copyright works appears to be a European initiative originally developed in certain Member States in response to the resurgence of interest in demands for natural justice,[6] and subsequently implemented throughout the EU by specific EU legislation (see below). The result of such changes is that whilst, in general terms, it is true to say that the first owner of IP rights created in the course of employment by an employee usually belong to the employer, there are notable exceptions and limitations. There is, however, no standard rule as regards an employee's entitlement to remuneration beyond that of the employer's contractual obligations.

As will be seen below, whilst recent developments in UK copyright law have provided certain employees with a right to remuneration these have been left largely

4. Patents Act 1977, s. 42(2).
5. Notably the Berne Convention for the Protection of Literary and Artistic Works 1886 (1971 revision with 1979 amendments) and WIPO Performances and Phonograms Treaty 1996 also declared to be a Community Treaty by European Communities (Definition of Treaties) (WIPO Copyright Treaty and WIPO Performances and Phonograms Treaty) Order 2005 (SI 2005/3431).
6. In particular, the argument that an inventor or creator should not go unrewarded for the fruits of their intellectual endeavour.

undefined and their calculation left to the parties themselves to determine or to refer to the Copyright Tribunal if they cannot agree. Only in relation to patents is the law more specific as to when an employee might be entitled to supplementary compensation above that in the employee's contract. However, even here the criteria to be taken into account to assess entitlement and the calculation of quantum are opaque and the consequences unpredictable – a situation far from ideal given the potential cost to employees of bringing a claim for compensation.

As we turn now to consider the specifics of each IP right and how the interests of employees are affected, we will see that freedom of contract remains the guiding principle, subject to a few statutory exceptions, with the parties largely free to specify who should own any IP rights created by the employee in the course of employment. In the event that the employment contract is silent, a statutory framework now exists to ensure that there is a clearly defined default position.

2 EMPLOYEES' COPYRIGHT

2.1 The Legal Framework

In the UK ownership of the copyright in works created after 1 August 1989 is regulated by the Copyright, Designs and Patents Act 1988 (as amended) ('CDPA 1988').[7] Section 11 of the CDPA 1988, establishes the general rule that the author is the first owner of copyright in a work subject to certain exceptions the most notable of which, for the purposes of this text, concerns literary, dramatic, musical or artistic works or films made by employees in the course of their employment. In such situations it is the employer, not the author that is the first owner of copyright, subject to any agreement between the parties to the contrary.[8] The fact that the employer is deemed to be the first owner of copyright does not mean that the employee is left without any rights in their creation. Chapter IV of the CDPA 1988 introduced into UK law the concept of 'moral rights' which provide an author of a qualifying work[9] with a limited range of inalienable, non-pecuniary rights (discussed at 2.2.4 below), but the employee's ability to assert such rights is seriously restricted where exploitation occurs with the consent of their employer. The implementation in the UK of the EU Rental Right Directive has

7. Ownership of copyright in works created before this date continues to be governed by the rules in place at the time of creation as preserved in Schedule 1 para. 11(1) of the CDPA 1988.
8. Other notable exceptions to the general relate to Crown copyright and Parliamentary copyright (ss 163 and 165) where the first owner of copyright is, in the case of the former, the Crown if the work was made by Her Majesty or by a servant or officer of the Crown in the course of their duties, or in the case of Parliamentary copyright, where a work is made by an officer or employee of the House of Commons or House of Lords in the course of their duties (s.165(4)) where it belongs to either House as appropriate as the employer. Pursuant to s.168 of the CDPA 1988 copyright that does not qualify for protection by reference to the author or place of first publication may nevertheless receive protection if the international organisation is among those identified by an Order in Council as benefiting from protection.
9. Sections 77-89 CDPA 1988 set out the main provisions relating to moral rights which generally arise in relation to literary, dramatic musical and artistic works but the rights are subject to the limitations and qualifications mentioned in sections 78, 79, 81 and 82.

resulted in the creation of additional inalienable economic rights for authors in connection with the rental of original, and copies of, copyright works (discussed further below at 2.2.5 and 2.4).

It is important to note that the derogation from the general rule, that the author of a work is the first owner of any copyright, does not apply to all forms of copyright work.[10]

2.2 Ownership of Rights

As noted above, the general statutory rule is that copyright subsisting in works created by an employee during the course of their employment belongs automatically to the employer, making the employer the first owner of the copyright unless there is an agreement between the parties to the contrary. Unlike the Patents Act 1977,[11] the CDPA 1988 provides no statutory control over the terms of any agreement between an employee and employer limiting the scope of any transfer of copyright ownership and so, in theory, it is possible for the parties to agree that *all* copyright works created by the employee whilst employed (even those unrelated to their employment) will belong to the employer, if the parties so wish but such a contractual term may be resisted or rendered unenforceable on the grounds that it is in restraint of trade.[12]

No documentation is required to transfer or vest in the employer the copyright ownership to which it is entitled by statute since Section 91(1) of the 1988 Act provides that the copyright will vest in the employer automatically once the prospective work is created. A written agreement is, however, needed to transfer the rights in any work created outside the scope of employment. Such an agreement (which could form part of the employment contract) should, amongst other things, set out the commitment of the parties and be signed by the employee.[13]

In the absence of an agreement with the employee that ownership of all copyright works created during the period of their employment will belong to the employer, the critical question when determining ownership of any specific copyright material is whether or not the work concerned was created by the employee under a contract of employment, that is, a 'contract of service or apprenticeship'[14] as opposed to being provided on a self-employed basis (referred to as a 'contract for services'). This is a question of law and fact and must be determined by reference not just to the duties listed in the employee's contract or job description (if one exists) but also those duties that are routinely assigned to or performed by the employee, together with the surrounding circumstances including the nature of the relationship between the employer and the employee where the status of the employee is in doubt. Three

10. Different rules apply to Crown copyright and Parliamentary copyright and to the ownership of copyright in films soundtracks, sound recordings, and rights in performances. The latter are discussed further in the sections that follow.
11. See below at 4.2.
12. Discussed further below at 4.3.
13. In order to enforce the agreement it would need to be recorded in writing, supported by consideration and pre-date the making of the work concerned to satisfy s. 90(3) CDPA 1988.
14. CDPA 1988, s. 178.

elements need to be present for an employment contract to be found: (i) mutuality of obligation – by which the employee agrees to work and the employer agrees to provide work; (ii) an obligation on the employee to provide the service personally, and (iii) the employer must be in a position to exercise control over the manner in which the work is performed by the employee.[15] The ability of the 'employee' to substitute someone else to undertake the work they are contracted to do is considered inconsistent with a contract of service.[16] The fact that the copyright work is made during working hours with or without use of the employer's materials is not, of itself determinative. The key questions are: (a) whether the work was made as part of the employee's job or was merely ancillary to it and (b) the nature of the circumstances under which the work was produced.

Stephenson, Jordan & Harrison Limited v Macdonald & Evans [17] is a Court of Appeal decision, and the leading authority in England and Wales, on the relevant factors to be taken into account when determining whether a work was created during the course of employment. The case concerned the ownership of copyright in two books written by a former employee. One book included a number of public lectures given by the employee whilst working for his former employer as well as case studies based on actual situations encountered by the employee whilst working for the claimant company. Whilst the text of one book had been written before the employee's departure, the text of the second was written after he had left the claimant's business. In determining the ownership of the copyright in the two books the Master of the Rolls, Sir Raymond Evershed, considered the level of control exercised by the employer over the different kinds of work undertaken by the employee and the manner in which the work was undertaken, in particular whether the activities and materials produced could be said to have formed part of the employee's regular duties. It was held, on the facts, that the employee was the owner of the copyright in the written lectures on the grounds that these were considered an accessory to, and not a part of, the employee's work even though they were typed up by a secretary during office hours. In contrast, the copyright in the case study, which the employee had based on a report that he had prepared for a client (whilst an employee of the claimant), was considered part of the employee's work as a servant of the company under his contract of service and copyright in it therefore belonged to the company. Copyright in material written after the employee had left the claimant's employment was held to belong to the employee.

In *King v South African Weather Service*,[18] King, an employee of the South African Weather Service ('SAWS'), was employed to collect and analyse meteorological information. During the course of his employment King wrote several computer programs to help with the gathering of meteorological information and its subsequent

15. *Ready-Mixed Concrete (South East) v Minister of Pensions and National Insurance* [1968] 2 QB 497 approved in *Ultraframe v Fielding* [2004] RPC 24 (CA) ('*Ultraframe*').
16. *Ultraframe*.
17. [1952] 69 RPC 10 (CA) ('*Stephenson* case'). See also *Noah v Shuba* [1991] FSR 14.
18. *King v The South African Weather Service* [2009] FSR 6, Sup. C. of A. South Africa. Although the case concerned provisions of South Africa copyright law which were based on the UK Copyright Act of 1956, the issue of whether a work was made in the course of employment is the same as that under the 1988 Act. *Noah v Shuba* [1991] FSR 14.

analysis. After King left the employment of SAWS, the question arose as to the ownership of the copyright in the computer programs and whether they had been written by King in the course of his employment. On appeal, it was held that the programs had been written in the course of King's employment. Although there was no requirement in his contract to write any software, and writing software was not part of the normal duties of a meteorological technician, the programs had been written by King to assist him in performing his duties of collecting and analysing data on behalf of SAWS and prior to his departure from SAWS he had spent a significant proportion of his working day developing the programs. Accordingly, it was held that SAWS was the owner of the copyright in the programs.

2.2.1 Works Made by Directors

Directors of companies are not necessarily employees and so the copyright works they produce will not always belong to the company automatically. For copyright to vest in the company on its creation the director must have a contract of employment with the company and the work must be made in the course of the director's duties. If the director has a contract to provide services to the company but is otherwise independent, self-employed or divides their time between a number of different companies and there is no written agreement addressing the issue of copyright ownership there is a risk that all copyright in works produced will belong to the director personally. If there is no contract of employment but the director creates the work in the course of their duties as a director of the company for the benefit of the company the court may regard the company as the beneficial owner of the work and the director as the legal owner. The director may then be obliged by a court to transfer legal ownership of the copyright to the company and in the interim is obliged to hold the work on trust for the company.[19]

2.2.2 Special Situations

Notwithstanding the statutory provisions relating to first ownership of copyright in materials created by employees in the course of their employment outlined above, there may be special situations where the factual matrix is such that a different ownership arrangement is upheld by the courts, *Noah v Shuba* [20] was one such case. The claimant was a scientist employed by the government Public Health Laboratory Service ('PHLS') with particular responsibility for issues relating to the spreading of infections. During the course of his employment Dr. Noah wrote material with a view educating specific industry sectors on specific topics. In his own time, at evenings and weekends, Dr. Noah produced a document entitled a 'Guide to Hygienic Skin Piercing' which he sent to industry recipients and practitioners. He was not asked to write the Guide by his employer. One recipient reproduced a substantial portion of the Guide in

19. *Antocks Lairn Ltd v Bloohn Ltd* [1972] RPC 219.
20. [1991] FSR 14 ('*Noah v Shuba*').

its own publication without attribution alongside quotes attributed to Dr. Noah but which Dr. Noah had not said. As Dr. Noah's name appeared as the both copyright owner and author on the Guide itself it raised a presumption that Dr. Noah was in fact the copyright owner and this was not rebutted by the defendant. Mummery J likened the position of Dr. Noah to that of the accountant in the *Stephenson* case indicating that even if the defendant had been able to rebut the presumption he would have held that Dr. Noah was the copyright owner on the basis of the evidence which supported his finding that there was an implied term in his contract of employment that excluded the application of the statutory provisions in favour of the employer. The evidence to which Mummery J referred was to a long and established practice within the PHLS whereby employees were acknowledged as both the authors and copyright owners of materials such as the Guide and PHLS had never challenged this.

A further area that gives rise to claims that it is a special situation concerns materials created by academics in the course of their employment by academic institutions.[21] In particular, it is said that the ownership of copyright in lecture notes, teaching materials, published articles and research materials is the property of the academic as opposed to that of the institution employing them[22] on the basis: (1) that there is a long and established practice of allowing academics to enter into contracts in their own names with publishers of journals of their choosing publishing works in their own name and naming the academic as copyright owner (as was the case in *Noah v Shuba*); (2) that maintenance of academic freedom requires the academic to be free to determine the content and direction of their lectures, research notes and publications as opposed to directed by their employer and (3) it would be 'unjust' and 'defy common sense' if the lecture notes of a former academic employee could be used, adapted or modified by a subsequent lecturer by whom he or she was replaced especially given the considerable effort and intellectual investment made by the former employee and the absence of moral rights that might otherwise enable the parting academic to object to the actions of the subsequent employee. The application of such special circumstances, were they to apply, would generally be construed narrowly and so would be unlikely, in any event, to extend to all areas of work undertaken by academics such as email correspondence entered into by the academic relating to the business of the institution such as admissions policies and procedures, the day to day running of the institution, the employment and/or management of staff etc. all of which might still be regarded as being subject to the usual statutory default as would be the materials created by non-academic co-workers or support staff. Whilst the employers of such academics might argue that the roles of academics has changed since the *Stephenson* case was decided 50 years ago, that greater emphasis is now placed on the preparation of lecture notes and teaching materials produced by academics as universities and colleges have developed on-line and distance learning courses which require such materials and that, given the need to generate new sources

21. See J. Pila, *Who owns the intellectual property rights in academic work?*, 32 (12) EIPR 609-613 (2010).
22. *Ibid.*

of income for the institutions they have a financial interest in retaining ownership of copyright material, it remains something of a moot point.

If academe is regarded as a special situation for copyright purposes the provisions are likely to be construed narrowly and institutions may seek to counter the terms of any exception by incorporating express terms in the academic's contract of employment, insofar as it is able.[23] Some institutions have already sought to do this by developing IP policies that set specific boundaries with regard to, amongst other things, ownership of copyright allowing academics to retain ownership in externally published material but otherwise retaining copyright in all other materials produced by the academic in addition to other IP rights, subject to any statutory limitations,[24] but such policies, which tend to be incorporated into the contract of employment by reference, tend to vary from one institution to another are hard for individual academics to negotiate variations or exceptions and the practice of individual universities is not always consistent with the policies held.

2.2.3 *Performer's Rights*

Although for many years the rights granted to performers lagged behind those granted to authors and directors in other creative industries, changes to UK legislation as a consequence of the implementation of international conventions and EU directives combined with domestic lobbying has meant that, with the exception of the duration of rights and the depth of protection, performers now enjoy rights equivalent to those in other creative industries.

In contrast to the position concerning literary, artistic and musical works, the first owner of performer's property and non-property rights[25] is the performer. There are no provisions for employers or those commissioning performances to acquire ownership of the performer's rights automatically. Whilst a performer's property rights are fully assignable and transmissible,[26] the only way that someone other than the performer can become the first owner of the performer's property rights is by virtue of a prospective assignment by the performer.[27] In contrast, the performer's non-property rights may not be assigned or transferred except on the death of the performer.[28] Producers with exclusive recording agreements may also have rights equivalent to a performer's non-property rights.[29]

23. Although the academic, as author of a work, has the benefit of certain in alienable moral rights the author can agree in writing to waive these rights if they so choose (s. 87 CDPA 1988).
24. See e.g. s. 42(2)(b) Patents Act 1977 which provides that any term in an employment contract which seeks to diminish the rights of the employee in inventions or applications for a patent will be unenforceable.
25. These are the rights conferred on a performer by s. 182 (consent required for recording of live performances), s. 183 (infringement of performer's right by use of a recordings made without consent) and s. 184 (infringement of performer's rights importing, possessing or dealing with illicit recordings).
26. CDPA 1988, s. 191B.
27. CDPA 1988, s. 191C.
28. CDPA 1988, s. 192A.
29. CDPA 1988, s. 180(1)(b).

2.2.4 *Moral Rights*

An author who creates a work that qualifies for copyright protection in the UK as a literary, dramatic, musical or artistic work or the director of a copyright film is granted under the CDPA 1988 certain *moral rights.* [30] These rights include the right to be identified as the author or director of the work when the work is disseminated to the public (the right of attribution);[31] the right not to be named as the author of a work that a person did not create (the right to object to false attribution);[32] and the right to control the form in which the work appears (the right of integrity).[33] Equivalent rights are also granted to performers.[34] In order to enforce these moral rights, they need to be asserted at the appropriate time, that is, at the time of publication, performance, exhibition by the author or director of the work, as the case may be.

Moral rights are personal to the author, director or performer concerned and may not be assigned *inter vivos* although they may be transferred on the death of the holder. The rights may be waived in whole or in part.

The right to be identified as the author of the work, or the director of a film, does not apply in certain situations, notably, to anything done by or with the authority of the copyright owner, where the works were created by an employee in the course of their employment and copyright in the work originally vested in the employer.[35]

Where copyright in a literary, dramatic, musical or artistic work is created by an employee or by a director of a copyright film in the course of their employment, the right to object to derogatory treatment of the work does not apply unless the author or director was identified at the time of the relevant act, or has previously been identified, as the author or director.[36] Where the right does apply it is not infringed if there is a sufficient disclaimer.[37]

2.2.5 *Rental & Lending Rights*

Article 1 of the EU Rental Right Directive[38] required Member States to provide, a right to authorise or prohibit the rental and lending of originals and copies of copyright works, and other subject matter,[39] subject to a derogation under Article 6 in relation to the public lending of works, provided that authors obtain some form of remuneration for such lending.[40]

30. CDPA 1988, ss. 77-89.
31. CDPA 1988, s. 77.
32. CDPA 1988, s. 84.
33. CDPA 1988, s. 80.
34. The Performances (Moral Rights) Regulations 2006 SI 2006/18 however these rights only apply to performances given after 1 February 2006.
35. CDPA 1988, s. 79(3) (as amended by SI 2003/2498 Sch. 1 Part 1 para. 1. 18(1)).
36. CDPA 1988, s. 82(2).
37. *Ibid.*
38. Directive 2006/115/EC of 12 December 2006 on rental right and lending right and certain other related rights (codified version) ('Rental Right Directive').
39. Detailed in Art. 3(1).
40. Rental Right Directive, Art. 6(1).

In the UK the requirements of the Rental Right Directive were implemented by means of the Copyright and Related Right Regulations.[41]

If the author of the underlying copyright work or the performer is an employee, and the copyright work is made in the course of the author's employment, then the rental and lending rights in the underlying copyright work will belong to the author's employer although the right to receive remuneration will remain with the author or performer as discussed below.[42]

2.2.6 Artists' Resale Right

The artist's resale right, also known as the *droit de suite*, subsists in certain original works of art in which copyright subsists and was introduced into the UK in the Resale Right Regulations 2006[43] which implemented an EU harmonising Directive on the topic.[44] Although the right is not strictly a copyright, it is considered here because it only arises where there is an underlying copyright work. The existence of the right gives the author an entitlement to a royalty on the sale of certain original works subsequent to the first transfer of ownership by the author[45] subject to certain conditions[46] but irrespective of the ownership of the copyright in the work concerned.

The right belongs to the author of the qualifying work and is inalienable (whilst the author is alive) which means that it cannot be assigned, transferred, charged or waived.[47] Following the author's death, the right can be inherited by one or more natural persons or by certain specified charities. If the artist was acting under a contract of employment when the original work of art was made so that ownership of the work (together with the copyright in it) first belonged to the employer, then a first transfer of the work by the employer will not give rise to liability to make a payment of a royalty on subsequent sales. This is because the Directive only requires that the right arise on sales 'subsequent to the first transfer of the work *by the author*' (emphasis added), and in such cases there is no transfer by the author although one might make parallel arguments to those made below[48] in relation to the equitable right of remuneration namely that the 'first transfer' is that which vests the work of art in the employer.

An alternative approach would be for the author to buy back the work on the open market at a later date in which case any subsequent transfer by the author and further subsequent sales could then trigger payment of the royalty.[49]

41. SI 1996/2967.
42. Rental Right Directive, Art. 6(1).
43. SI 2006/346 as amended by the Artist's Resale Right (Amendment) Regulations 2009 (SI 2009/2792) ('Resale Right Regulations').
44. Directive 2001/84/EC dated 27 September 2001 on the Resale Right for the Benefit of the Author of an Original Work of Art.
45. Resale Right Regulations, Reg. 3(1).
46. Resale Right Regulations, Reg. 12.
47. Resale Right Regulations, Reg. 7 and 8.
48. See section 2.4.
49. See Laddie Prescott and Vitoria, *The Modern Law of Copyright and Designs*, para. 31.17 footnote 4.

2.2.7 *Database Rights*

The database right is a *sui generis* right that subsists in a database, that is, a collection of independent works, data or other materials arranged in a systematic or methodical way and individually accessible by electronic or other means[50] which by reason of the selection and arrangement of the content are protected by virtue of the Database Directive which was implemented in the UK by the Database Regulation.[51] The Database Regulation thus provides a form of protection for a database that is quite separate from any copyright that might exist in any of the individual works within the database. The right is discussed here for convenience only.

The first owner of the database right is the 'maker' of the database and the maker is said to be the person who takes the initiative in obtaining, verifying or presenting the contents of the database and who assumes the risk of investing in that process. Where the database is made by an employee in the course of employment then, unless there is an agreement between the parties to the contrary, the employer is regarded as the maker of the database and is thus the initial holder of the database right.[52]

2.3 The Parties' Duty to Inform

There is no statutory duty on the employee to inform their employer that a copyright work or a database has been created or that any related rights exist. However, express duties to inform the employer might be set out in the employee's contract of employment or in an employment manual or form part of a corporate IP policy which is incorporated into the contract of employment by reference.

In the absence of any express term, there will, in any event, be an implied general duty of loyalty and fidelity between employees and employers in addition to a duty of confidence in relation to any confidential information disclosed to the employee during the course of their employment. Thus if an employee, before leaving their employment, makes unauthorised copies of copyright works or copies an entire database, or a substantial proportion of it, and takes these with them to their new employer it could be actionable as an infringement of copyright and/ or breach of confidence, or infringement of the former employer's database rights depending on the nature of the documents concerned. To avoid any doubt as regards ownership of the copyright materials, consent to their copying or removal, or the copying and removal of any database in such a situation it is sensible to include in the terms of employment contractual provisions restricting the copying and removal of the company's property (or copies of it) from the premises on termination of the employment contract.

As the artists' resale right is not engaged until there has been a qualifying act of resale, there is no duty on the employee to inform their employer that a qualifying work has been created either at the time of creation of the work or subsequently.

50. Directive 96/9/EC of 11 March 1996 on the legal protection of databases ('Database Directive'), Art. 1(2).
51. Copyright and Rights in Databases Regulation 1997 SI 1997/3032 ('Database Regulations').
52. Database Regulations, Reg. 14(2).

In the UK it is compulsory for collecting societies to manage the artists' resale right on behalf of authors and so the prudent employee will need to identify which collecting society he/she wishes to engage, if it is not to be left to the Design and Artists Copyright Society, by default.

2.4 Employee Remuneration Right

With the exception of rental and lending rights, copyright based rights do not ordinarily give rise to employee rights of remuneration.

The right to receive equitable remuneration in the UK from the exercise of a rental right derives from the Copyright and Related Rights Regulations which implemented in the UK the Rental Right Directive[53] and resulted in the insertion of Sections 93B (relating to copyright in a film or sound recording) and 191G (relating to performer's rights) in the CDPA 1988.

The right to receive equitable remuneration arises where there is a transfer of the rental right.[54] The right is itself inalienable and cannot be waived by authors or performers, although the right may be assigned to a collecting society so that the rights can be enforced.[55] The purpose of the provision is to protect the income of authors and performers and so encourage further creative effort and the investment required for the production of films and sound recordings[56] but the Directive does not provide for remuneration as a consequence of all forms of exploitation, only those *where the author or performer has transferred* their right to exploit the work by the rental of phonograms or films which embody the author's/performer's contribution.

Where the author or performer is employed, the rental right will automatically belong to their employer. Does the right to equitable remuneration therefore arise? On one reading, the author of the copyright work or performance must have transferred or assigned the rental right in the copyright in the sound recording or film to the producer of a film or sound recording for the right to equitable remuneration to arise. If there has been no transfer or assignment of the rental right by the employee, there can be no payment of equitable remuneration. However, if the right to equitable remuneration is regarded as personal to the author (of the copyright work or the principal director of a film) or the performer, it may be argued that it is retained by the author or performer whatever might be the mechanism by which the right is vested in the third party. The fact that the UK law vests the rental right in the employer by operation of law, rather than by assignment or transfer, should not, is it submitted, make a material

53. Rental Right Directive, Art. 5. This states: 'Where an author or performer has transferred or assigned his rental right concerning a phonogram or an original or copy of a film to a phonogram or film producer, that author or performer shall retain the right to obtain an equitable remuneration for the rental.'
54. CDPA 1988, s. 93B (where the transfer is from the author of a sound recording or film to the producer).
55. CDPA 1988, s. 93B(2).
56. Rental Right Directive, recital 5 states that 'the creative and artistic work of authors and performers necessitates an adequate income as a basis for further creative and artistic work ...'.

difference.[57] However the salary received by the author or performer may be considered sufficient to satisfy the obligation to pay equitable remuneration.[58]

The amount payable by way of equitable remuneration is as agreed by, or on behalf of, the person by or to whom it is payable, subject to the determination of the Copyright Tribunal in the event that the parties are unable to agree the appropriate amount to be paid.[59] An agreement between the parties as to the amount payable by way of equitable remuneration has no effect insofar as it purports to exclude or restrict the author's or performer's rights to question the amount of equitable remuneration to be paid. In determining the amount that should be paid, account should take as to the importance of, and the contribution made by, the author or performer to the sound recording or film in question and any payment already received.

Under the Rental Right Directive, the author of a copyright work has the exclusive right to authorise or prohibit the lending of the original or copies of their work. If a Member State derogates from this right it must provide the author with remuneration for such lending. In the UK an author can receive payment out of public funds in respect of the lending of books by public libraries provided that the author registers their interest under the Public Lending Right Act 1979.[60] Payments made to authors are not large and are capped at a modest level[61]. If there is more than one author in respect of the work, the payment is shared. Having registered their right the lending right becomes the author's property and may be assigned freely. Although the public lending right originally only applied to authors and books, the scheme now extends to narrators and producers and to audio books and e-books downloaded from a public library[62].

An employee that has created a database during the course of their employment is not entitled to any remuneration from their employer beyond payment of their salary unless they have negotiated an agreement with the employer to the contrary.

An employee that has created a relevant work of art during the course of their employment is not entitled to any remuneration from their employer in respect of artist's resale rights beyond payment of their salary, unless they have negotiated an agreement with the employer to the contrary

57. Cf. recital 6, which assumes that most creators are self-employed ('These creative, artistic and entrepreneurial activities are, to a large extent, activities of self-employed persons').
58. See recitals 5-6 of the Rental Right Directive.
59. CDPA 1988, s. 93C.
60. This has been updated and enabled by the Public Lending Right Act 1982 (Commencement) Order 1982 SI 1982/719.
61. In the financial year 2013/2014 the maximum payout was capped at GBP 6,600 per annum.
62. Public Lending Right Scheme 1982 (Commencement of Variations) Order 2014 SI 2014/1457

3 EMPLOYEES' SEMICONDUCTOR CHIPS

3.1 The Legal Framework

In 1986 an EC Directive was adopted requiring Member States to provide protection for the topography of semiconductor chips.[63] In the UK protection is currently provided under provisions relating to design rights as set out in Sections 213–264 of the Copyright, Designs and Patents Act 1988 (as amended) ('CDPA 1988').[64] Unregistered design rights are said to subsist in a pattern fixed in or on, or intended to be fixed in or on, a layer of a semiconductor product or a layer of material used in the course of or for the purposes of making a semiconductor product or the arrangement of patterns fixed or intended to be fixed in or on the layers of a semiconductor product in relation to one another.[65] The topography must be 'original' and not 'commonplace' in order to qualify for protection, as with other designs.

Semiconductor chips are also, in principle, protectable as Registered Community Designs, national registered designs and Unregistered Community Design Right which are discussed further below.[66]

3.2 Ownership of Rights

Design rights in semiconductor topographies arise either because the designer is a qualifying person or the products qualify as a result of their first being marketed within certain qualifying countries.

For the purposes of semiconductor topographies, a qualifying person is either a citizen of, or subject of, or an individual habitually resident in, a qualifying country which includes the UK and other EU Member States and certain additional countries,[67] or is a body corporate which has a place of business or carries out substantial business activity in any qualifying country or in Gibraltar or the person falls within a special class of people identified in the Regulations.[68]

Where the design of the semiconductor topography was created by an employee in the course of their employment then, unless there is a an agreement to the contrary (or the design qualifies for protection as a result of first marketing), the design rights

63. Council Directive 87/54/EEC of 16 December 1986 on the legal protection of semiconductor products, OJ L24/36.
64. Protection was initially provided by the Semiconductor (Protection of Topography) Regulations 1987 which was subsequently replaced by the Design Right (Semiconductor Topographies) Regulations 1989 SI 1989/1100 (hereafter referred to as SI 1989/1100) which amended the Copyright, Designs and Patents Act 1988. The provisions were further modified by the Design Right (Semiconductor Topographies) (Amendment) Regulations 2006, SI 2006/1833 and the Design Right (Semiconductor Topographies) (Amendment) (No. 2) Regulations 2008, SI 2008/1434.
65. SI 1989/1100 reg 2.
66. See section 6 below.
67. SI 1989/1100 reg 4(2) as amended by the British Overseas Territories Act 202, s. 2(3).
68. Added by the Design Right (Semiconductor Topographies) (Amendment) Regulations 1991, SI1991/2237 as amended by SI 1993/2497 reg 2 Schedule 2. Countries listed include Australia, French overseas territories, Finland, Iceland, Japan, Norway, Switzerland and the USA.

will belong to the employer provided the employer qualifies for protection by reason of the fact that the employer carries on substantial business activities in a qualifying country.[69] Where the design qualifies for protection as a result of its place of first marketing the above rules do not apply.[70]

3.3 The Parties' Duty to Inform

The 1988 Act provides no positive obligation on the employee to notify the employer that a design right has been created. However, there might be express duties set out in the employee's contract of employment or in an employment manual or as part of a corporate IP policy of the employer incorporated into the contract of employment by reference, which obligate the employee to disclose, notify or deposit copies of new designs with the employer's legal department, or some other part of the business.

In the absence of any express terms, there will, in any event, be an implied general duty of loyalty and fidelity between the employee and the employer in addition to a duty of confidence in relation to any confidential information disclosed to the employee during the course of their employment or developed by the employee in the course of their work. Thus if an employee before leaving their employment made copies of any design document and took these with them to their new employer it might be actionable as an infringement of the former employer's design right and/or breach of confidence depending on the nature of the design documents concerned, the circumstances of the copying and whether they might be considered confidential. To avoid any doubt as regards ownership of the design documents in such a situation and whether they were produced with consent it is sensible to include in the terms of employment express contractual provisions restricting the copying and removal of the employer's property (including design documents) from the premises on or prior to termination of the employment contract and prohibit the subsequent use of the employer's designs without their prior express consent.

3.4 Employee Remuneration Right

An employee that has created design rights in relation to semiconductor topography designs during the course of their employment is not entitled to any remuneration from their employer beyond payment of their salary unless they have negotiated an agreement with the employer to the contrary.

69. CDPA 1988, s. 215(3).
70. CDPA 1988, s. 215(4).

4 EMPLOYEES' INVENTIONS

4.1 The Legal Framework

The ownership of inventions made by employees is governed solely by the Patents Act 1977 (as amended) and, in particular, Section 39(1) which applies to an employee: who is mainly employed in the UK at the time the invention is made;[71] or if the employer had a place of business in the UK, to which the employee was mainly attached when the invention was made and the employee was not mainly employed elsewhere;[72] or whose place of employment cannot be determined.

Thus an employee that spends a significant amount of their working time overseas may not be able to rely on these provisions unless their place of employment cannot be determined.[73]

Although some of the relevant statutory provisions can be supplemented by contract, many of them are mandatory and can only be altered by improving the position of the employee, not by depriving the employee of their rights or seeking to diminish them in any way.[74]

It should be noted that the ownership rights discussed below are in addition to, and quite distinct from, the employee inventor's right to be identified as the inventor in any patent for the invention in accordance with Section 13 Patents Act 1977.

4.2 Ownership of Rights

Section 39(1) Patents Act 1977 makes it clear that despite anything in any rule of law,[75] an invention made by an employee[76] shall, as between the employee and the employer, be taken to belong to the employer if:

(a) it was made in the course of the normal duties of the employee or in the course of duties specifically assigned to the employee, and in either case, the

71. Patents Act 1977, s. 43(2)(a).
72. Patents Act 1977, s. 43(2)(b).
73. There have been no cases since the enactment of this provision where this issue has been pertinent.
74. Patents Act 1977, s. 42(2).
75. In *Re Harris' Patent* [1985] R.P.C. 19 at p. 28 Falconer J held that common law provisions that operated before the implementation of s. 39(1) no longer apply although earlier authorities might provide guidance as to how the courts have assessed the duties of employees in particular circumstances. In CIPA Guide to the Patents Act at para. 39.07 the authors argue that the opening words of s. 39 renders precedents decided under the common law of no assistance for cases relating to s. 39(1)(a) and of only limited help in ss.(1)(b) cases but acknowledge that Jacob LJ comments in *LIFFE Administration v Pankava* [2007] RPC 30 ('*LIFFE v Pankava*') mentioned below that the common law may indeed re-appear where the employee's ownership rights clash with duties of confidentiality.
76. An employee is said to be a person who works or (where the employment has ceased) worked under a contract of employment or in employment under or for the purposes of a government department or a person who serves (or served) in the naval, military or air forces of the Crown – Patents Act 1977, s. 130.

circumstances were such that an invention might reasonably be expected to result from the performance of the employee's duties; or

(b) the invention was made in the course of the duties of the employee, and at the time of making the invention, due to the nature of the duties of the employee and their particular responsibilities arising due to the nature of their duties, the employee had a special obligation to further the interests of the employer's business.

In any other situation, an invention created by an employee shall, as between the employee and the employer, belong to the employee.[77]

The first step in determining whether a particular invention belongs to an employee or an employer is to consider the relationship between the employer and employee to ensure that a contract of service exists (as opposed to a contract for services)[78] and then to look at the duties of the employee. Whilst it is useful to consider the employee's job description as set out in the contract of employment (if one exists) to ascertain the nature of the employee's duties, the courts must also have regard to the actual duties of the employee.[79] In part this is to protect the employee since it would be relatively easy for an overzealous employer to overstate the scope of the employee's duties with a view to broadening the terms by which to claim ownership of a greater number of the employee's inventions. In part too, it reflects the practical reality that the scope of an employee's duties may change over the period of time that a person is employed, either by a process of subtle or incremental variation or by a reassignment of duties. In any event, the extra duties or different duties 'specifically assigned' can become the employee's normal duties over time.[80]

In *Re Greater Glasgow Health Board's Application* the employer sought to argue that a junior registrar's specially assigned duties not only included treating patients but also devising new ways to diagnose and treat patients. Whilst recognising that the doctor concerned developed an optical spacing device for an ophthalmoscope whilst at home revising for exams, the Greater Glasgow Health Board maintained that the invention was developed in the course of his employment. Jacob J (as he then was) disagreed, holding that the particular circumstances in which the invention was made had nothing to do with the registrar's duties at all, even though the invention itself could be used for the purposes of his employment.

Having considered the nature of the employee's duties the second step is to determine whether the 'circumstances ... were such that an invention might reasonably be expected to result from the carrying out of [the employee's] duties'. Jacob LJ described this phrase as comprising 'particularly incompetent wording' on the basis

77. Patents Act 1977, s. 39(2).
78. Employment Rights Act 1996 defines 'contract of employment' at s. 230(2) as a 'contract of service or of apprenticeship'. The same definition is used in s. 295 Trade Union and Labour Relations (Consolidation) Act 1992. See 2.2 above regarding the essential elements of a contract of service.
79. *LIFFE v Pinkava* [2007] RPC 30.
80. *LIFFE v Pinkava* [2007] RPC 30 at p698. See also Bank's Committee Report on the British Patent System 1970 Cmnd. 4407 at p.684 para. 39.

that the nature of a patentable invention was such that you simply could not *reasonably expect* an invention to result from research, even within a research department, no matter how hard the researchers might try. What you can do, he said, is *hope* that an invention results and this is what the legislator must have intended. In *LIFFE Administration v Pinkava*,[81] the Court of Appeal considered whether the particular attributes of the employee concerned were relevant to the question of whether an invention could reasonably be expected to result from the performance of the employee's duties. The majority indicated that the individual qualities of the employee (e.g. their creativity and inventiveness) were relevant when assessing the likelihood of an invention being made but dissenting, Jacob LJ disagreed, observing that this would mean that 'a thick employee would own his own inventions whereas a brilliant one would not!'. Instead, Jacob LJ paraphrased the requirements of this aspect of the statute by saying 'In the end one is asking whether the employee is employed to innovate and, if he is, what general sort of areas his innovation duties cover'.[82]

If the employee is not employed to innovate, or the invention actually made is outside the general areas covered by their duties, the invention will be the property of the employee[83] unless Section 39(1)(b) is satisfied, that is, that due to the nature of the employee's duties the employee had a special obligation to further the interests of the employer's business. This limb of the statute is most relevant where the employee concerned is a director or a member of the senior management of the employer since employees in this position are regarded as being under a general duty to further the interests of their employer. In such situations, unless the invention concerned is unrelated to the business of the employer, there is a greater likelihood that inventions made by members of the senior management team will likely to belong to the employer.

In *Harris Patent*, [84] one of the first cases to be decided under the Patents Act, the court had to consider whether the employee, who was a manager in the business, was sufficiently senior for this general obligation to apply. In determining whether it did apply the court considered among other things the status and responsibilities of the employee, and his salary level, noting that he had no power over recruitment, or even to agree holiday dates, that he did not attend board meetings even when his department might have been under discussion. Furthermore, on one occasion when he had suggested an improvement to a product he had been told to limit his endeavours to sales. For all these reasons the court held that he was not under a special obligation to further his employer's business interests within the meaning of Section 39(1)(b).

4.3 The Parties' Duty to Inform

There is no express provision in the Patents Act obliging an employee to disclose their invention to their employer. However, express provisions might, and often are,

81. [2007] RPC 30.
82. *LIFFE v Pinkava* at p. 698.
83. Patents Act 1977, s. 39(2).
84. [1985] R.P.C. 19.

included in an employees' contract of employment, in employment manuals and/or as part of a corporate IP policy of the employer incorporated into the contract of employment by reference.

Any express provisions in the contract of employment will be subject to Section 42(2) Patents Act which makes it clear that any term which seeks to diminish the rights of employees in relation to inventions of any description made by them after the date of the contract, or in relation to patents, or applications, for such inventions, will be unenforceable against the employee to the extent that the provisions seek to diminish the employee's rights in the invention or the patent application. Whether a requirement simply to disclose an invention would be sufficient to 'diminish the employee's right in the invention' is a question of fact. If the information is held in confidence by the employer it should not reduce or interfere with the employee's ability to seek patent protection but if the obligation goes beyond mere disclosure and, e.g., grants the employer an option over the invention, or if it constitutes a waiver of any compensation to which the employee might be entitled, it is likely to overstep the mark and render the provision unenforceable either as a consequence of Section 42(2) or because the provision is regarded as being in restraint of trade, especially if the employee is not employed on a full time basis.[85]

In the absence of any express obligations to disclose inventions, there still exists an implied duty that the employee will act in good faith, with loyalty towards their employers and with mutual trust and confidence. Whether such an implied duty extends as far as requiring an employee to disclose to their employer any invention they make during the course of their employment that might reasonably be expected to belong to their employer is a moot point.[86] In *Prout v British Gas plc* [87] it was held that the voluntary disclosure of an invention belonging to an employee as part of an employee suggestion scheme must be treated in confidence by the employer, especially where the invention was not taken up by the employer under the scheme and the rules of the scheme were silent on the issue of confidentiality.

In *LIFFE v Pinkava* LJ Jacob held that if an employee created an invention to which they felt they were entitled, using confidential information belonging to the employer but that due to the obligations of confidence owed by the employee he or she could not seek patent protection without breaching the obligations of confidence this would suggest that the invention rightly belonged to the employer under Section 39(1) Patents Act, 'Any other conclusion leads to the absurd result that an invention would belong to an employee and yet he would owe a duty of confidence to his employer. Parliament cannot have intended such a stalemate.'[88]

Whilst filing an application for a patent, or holding a patent to which the employee is entitled, does not amount to breach of fiduciary duty,[89] the secret

85. *Electrolux Ltd v Hudson* [1977] FSR 312.
86. See e.g. *British Syphon Co Ltd v Homewood* [1956] 1WLR 1190; *Ransom v Customer Systems Plc* [2012] IRLR 769; *Threlfall v ECD Insight Ltd* [2013] IRLR 185; *Kynixa Ltd v Hynes* [2008] EWHC 1495 (QB); *Tullet Prebon Plc v BGC Brokers* [2010] EWHC 484.
87. [1992] FSR 478.
88. *LIFFE v Pinkava* at para.100.
89. Patents Act 1977, s. 39(2).

patenting of an invention to which the employer is entitled would be considered a breach of fiduciary duty.[90] The exploitation of an invention by an employee in competition with their employer may well constitute a breach depending on the circumstances.[91] A higher standard of loyalty will be expected from an employee in a position of responsibility.[92]

If an employee makes an invention that ordinarily would belong to their employer but fails to disclose it to their employer before leaving their employment and instead takes the invention with them to a new employer, this would amount to a breach of the employee's fiduciary duty.[93]

4.4 Employee Remuneration Right

Where an invention created by an employee is identified as belonging to the employer in accordance with Section 39(1), the inventor may be entitled to statutory compensation, by virtue of Section 40 Patents Act 1977. To qualify for compensation, the employer must have obtained a granted patent for the invention, and following the making of an application by the employee to the court or the comptroller of patents in accordance with Section 40,[94] the court or comptroller must consider it just to award the employee compensation, taking into account, amongst other things, the size and nature of the employer's business and the fact that the invention, or the patent, (or the combination of both) is or was of outstanding benefit to the employer.[95]

Although the statute provides no guidance as to what amounts to an 'outstanding benefit', the statute does provide some guidance in Section 41(1) as to how the courts or comptroller should approach the determination of the level of compensation payable in that it should be such as to secure for the employee a fair share of the benefit[96] obtained by the employer, or that might reasonably be expected to be derived by the employer, from any of the following, taking into account all the circumstances:

(a) the invention in question;
(b) the patent for the invention; or
(c) the assignment, assignation or grant of any right in or under, an application for the patent, to a person connected with the employer.

Section 41(2) goes on to state that for the purposes of Section 41(1)(c) the court should assume that the benefit obtained from a transaction entered into with a person

90. *Yeda Research and Development v Rhone-Poulenc Rorer International Holdings Inc* [2007] UKHL 43.
91. *Helmet Integrated Systems Ltd v Tunnard* [2007] FSR (16) 437, 451 (CA).
92. *Ransom v Customer Systems Plc* [2012] IRLR 769.
93. *British Syphon Co Ltd v Homewood* [1956] 1WLR 1190.
94. In relation to the court, in accordance with Rule 63.12 of Part 63 of the Civil Procedure Rules and in relation to the comptroller in accordance with Rule 71 of the Patent Rules 2007.
95. Patents Act 1977, s. 40(1).
96. S. 43(7) Patents Act 1977 makes it clear that the only benefit that should be taken into account is that in money or money's worth.

connected with the employer, should be taken to be the amount which could reasonably be expected to have been derived had the transaction been one concluded at arm's length.

In summary, therefore, where a patent has been granted in respect of an invention created by an employee in the course of their employment (and where there was a reasonable expectation that the employee might invent something in the course of their duties), and that patent or the invention itself is of outstanding benefit to the employer, the employee is entitled to a fair share of that benefit if the court/comptroller considers it just, taking into account all the surrounding circumstances. There is thus no absolute or automatic right to an award; it remains at the discretion of the court or comptroller.

Given the vagueness of many of the terms included in Sections 40 and 41 – 'just', 'outstanding benefit', 'fair share' – coupled with the fact that these provisions have no precedent in earlier legislation or common law practice, it is not surprising that there has been, and to some extent there remains, uncertainty as to their meaning and scope with the consequence that only a handful of claims have come before the courts since 1977 and, at the time of writing, only one has so far proved successful![97]

When considering whether an employee should apply for compensation under Section 40 the first step to be determined is whether the employer has received an outstanding benefit from the employee's invention. The courts that have considered the claims made to date have all agreed that it is not advisable to redefine the word 'outstanding', but have confirmed that the benefit must be something special', 'more than substantial or good',[98] 'something out of the ordinary and not what one would expect to arise from the results of the duties that the employee is paid for'.[99]

When Section 40 was first enacted the onus was on the employee to show that the patent[100] itself (as opposed to the invention) was of outstanding benefit to the employer, but this changed following the implementation of the Patents Act 2004. Now the employee only has to show that either the patent or the invention, or both together, provided the benefit to the employer.

The assessment of the benefit received by the employer is a matter of fact. Whilst it may be difficult evidentially for the employee to obtain full financial details of contracts entered into by the employer prior to commencement of proceedings these ought, in the ordinary course of proceedings, to be disclosed by the employer and if they are not, the employee should promptly seek their disclosure.[101]

Provided the employee can show that, on the balance of probabilities, the employer did receive a significant benefit it is for the employer to show that the benefit was not outstanding in the circumstances and/or that it would not be just to make an award in favour of the employee. To this end the employer may seek to rely on *ex gratia*

97. *Kelly and Chiu v GE Healthcare Ltd* [2009] EWHC 181; [2009] RPC 12.
98. Aldous J in *Memco-Med Ltd's Patent* [1992] RPC 403 at p. 414.
99. *GEC Avionics Limited's Patent (Ellis' Application)* [1992] RPC 107 per Superintending Examiner, and affirmed by Aldous J in *Memco-Med Ltd's Patent* [1992] RPC 403 at p. 414.
100. In this context the reference to a patent includes foreign patents as well as UK patents, Patents Act 1977 s. 43(4).
101. Terrell on the Law of Patents (7th ed., Sweet and Maxwell, London, 2011) p.130.

payments or bonuses already made to the employee, promotions or shares given or the financial investment or costs incurred by the employer in making the original invention, developing it or making its subsequent exploitation possible thus minimising the apparent level of benefit received or increasing the level of further investment that the employer had been required to make in order to generate the success that the invention achieved thereby seeking to decrease its apparent profitability.

In *Kelly & Chiu v GE Healthcare*, [102] the employees successfully demonstrated that their employer (at the time a relatively small life sciences company) benefited enormously from the medical compound they invented on a very limited budget. Not only was the company given awards for technological achievement as a result of the development of the product but the revenues it generated enabled the company to negotiate corporate deals, acquire stakes in joint ventures and acquire other companies. According to the employees, the gross profit margin on the product was very high. In the words of the judge, Floyd J, 'The benefits went far beyond anything which one could normally expect to arise from the sort of work that the employees were doing.'[103]

Having concluded that the employer had received an outstanding benefit Floyd J had to consider whether it was just to award compensation to the two employees making the claim. Although the employer sought to argue that it would be unjust to reward the two inventors above other members of the research team and that all the risk was taken by the company, Floyd J was not persuaded that this was a sufficient reason not to make an award. Valuing the benefit of the patents[104] to the employer conservatively at GBP 50 million, the employees were awarded a total of GBP 1.5 million (i.e. 3% of the benefit). Although Floyd J recognised that a fair share of the benefit might, in principle, lay somewhere between nil and 33%, he felt that on the facts the employees share fell somewhere towards the bottom of this bracket. In reaching this conclusion the judge was strongly influenced by the terms of a licence agreement between the company and a consultant in which the consultant was to be paid a royalty in respect of the commercialisation of any invention made by the consultant (and owned by the employer).

In contrast to the *Kelly* case, the *Shanks* case[105] illustrates the fortunes of a research professor employed by a large international organisation who made a number of inventions that were patented by his employer, Unilever UK Central Resources Limited, assigned to a holding company Unilever Plc for GBP 200 who subsequently licensed them to third parties several years later. Although it was accepted that Unilever benefited to the sum of GBP 23 million as a result of the patented inventions[106], it was held that, given the size and nature of Unilever's business, its turnover,

102. *Kelly and Chiu v GE Healthcare Ltd* [2009] EWHC 181; [2009] RPC 12.
103. [2009] EWHC 181 (Pat) para. 148.
104. This case was decided prior to 2004 and so the relevant assessment related to the benefit provided to the employer by the patents rather than by the invention itself.
105. *Shanks v Unilver Plc and Others* [2014] EWHC 1647 (Pat) [2017] EWCA Civ 2 (CA).
106. At an earlier stage of the proceedings Unilever had sought to argue that pursuant to Section 41(2), the benefit it had received should be limited to the nominal consideration paid to the subsidiary that had employed the inventor by the group holding company to which the patents applications were transferred but this argument was rejected by the court at an interim hearing holding that the benefit in this case was the actual amount that the company received in licence

profits, licensing activities and general activities that the benefit provided by the patents fell short of being outstanding. This decision therefore will make it far harder for employee inventors from large and successful research organisations to make successful claims against their employers.[107] At first instance the Hearing Officer held that had the benefit been outstanding that a fair share would have been 5% (i.e. GBP 1.15 million) but on appeal Arnold J. criticised his finding, holding that it would not have been appropriate to award Professor Shanks more than was awarded (in percentage terms) to both of the litigants in the *Kelly* case given the efforts expended by the employees in *Kelly* to implement the invention compared to those of Professor Shanks and the transformative effect of the patents at issue in the *Kelly* case on the business of the employer compared to the situation within Unilever.

From these two cases it can be seen that to qualify for an award of compensation: (a) the financial benefit from the invention (or the patent, or both) to the employer must be considerable and (b) the larger the organisation the greater the size of the benefit must be for the employee to succeed. If the employer has invested significantly in the creation of the invention, the development of the technology and/or the marketing of the invention and if the employee was only one member of a large team then, based on current case law, the less likely it is that the benefit will be considered outstanding or an award considered just.

The opportunity for an employee inventor to seek compensation from their employer is not restricted to those situations where the invention is deemed to belong to the employer in accordance with Section 39(1), it is also possible for an employee to claim compensation under Section 40(2) in relation to an invention that the employee was deemed to own under Section 39(2) but which the inventor subsequently agreed to assign or license to their employer and the benefit derived by the employee from the contract of assignment is considered to be inadequate in relation to the benefit derived by the employer from the invention and the court considers it just to make an award of compensation. In determining the amount of such an award the court is required to take into account the same criteria set out in Section 41 that apply to an employee invention as discussed above. To date there have been no reported cases based on Section 40(2).

Under Section 40(3) of the Patents Act an employee is not entitled to make an application for compensation when at the time their invention the employee was a member of a trade union which had negotiated a relevant collective agreement which provided for the payment of compensation in respect of inventions of the same description to employees of the same description. A collective agreement is defined[108] as any agreement or arrangement made by or on behalf of a trade union and one or more employer or employers' associations relating to, amongst other things, conditions of employment. Few such collective agreements are understood to exist.

fees from the actual third parties licensees involved and not from some hypothetical entity. See Shanks v Unilever BL O/138/09 and subsequently [2010] RPC 11 and [2011] RPC 12.

107. *Ibid.* at para. 223.

108. Trade Union and Labour Relations (Consolidation) Act 1992, s. 178(1).

5 EMPLOYEES' UTILITY MODELS

A utility model is a form of intellectual property available in some countries to protect an invention similar in form to a patent but less stringent and offering a shorter term of protection. The UK does not provide any statutory form of protection for utility models of this kind.

6 EMPLOYEES' DESIGN RIGHTS

6.1 The Legal Framework

The legal protection of designs in the UK is complex because there are several different regimes that could apply to a design, some of these are home grown, others derive from EU legislation, several overlap in aspects of their application but they also contain marked differences. As a consequence, depending on the nature of the design at issue, it may be considered an artistic work (protectable by copyright under the CDPA 1988); a UK unregistered design (protectable within the UK under the CDPA 1988); a UK registered design (protectable under the Registered Designs Act 1949); a Community unregistered design or a Community registered design (both protectable under the Design Regulation[109]).

6.2 Ownership of Rights

Due to changes in UK legislation implemented in 2014[110] which altered the rights of commissioners of UK designs, the first owner of the right in a design is now the designer (that is, the creator of the design) not the commissioner, unless the design was developed by an employee in the course of their employment, in which case it shall belong to the employer (in the absence of any agreement between them to the contrary).[111]

Where a Community design is developed by an employee 'in the execution of his duties or following instructions given by his employer' the right to the Community design vests in the employer unless agreed or specified otherwise by national law.[112] This subtle difference in language compared to that relating to national UK design legislation suggests, on the face of it, that the employer's entitlement to ownership of Community designs is narrower than to their UK counterpart but it is submitted that in reality the differences are of no practical significance given that the tests applied by the court when determining what amounts to 'in the course of employment' for a national design will, in effect, require an assessment of the duties of the employee and the

109. Council Regulation (EC) No. 6/2002 ('Design Regulation').
110. The Intellectual Property Act 2014.
111. CDPA 1988, s. 215(3) (as amended); Registered Designs Act 1949, s. 2 (1B); Design Regulation, Art. 14(3).
112. Design Regulation, Art. 14(3).

directions given by the employer as required under the Design Regulation.[113] The results of the analysis are therefore likely to be the same.

Thus, the ownership rights to the various different forms of design protection now all reside in the same party irrespective of the nature of the design (that is, whether it is a functional design, decorative, two dimensional or three dimensional or a work of artistic craftsmanship) and irrespective of the nature of the right being relied on and whether it is a national right or one that is enforceable throughout the EU.

UK unregistered design right is a *sui generis* right introduced in 1989 to provide a limited form of protection for the shape or configuration of the whole or part of an article. Where copyright and design right both subsist in a design the rights do not overlap but offer alternative forms of protection.[114]

Some designs, notably graphic works, engravings, sculptures and works of artistic craftsmanship, may be protected by copyright (as well as registered designs) and as such the author of the artistic work may have the benefit of certain moral rights even if the artistic work was made in the course of employment.

6.3 The Parties' Duty to Inform

There is no statutory duty on an employee to inform an employer of the creation of a design whether it is protectable as an artistic work under copyright law or under the law relating to design rights registered or otherwise, a national or a Community right. There may however, be express duties to inform the employer of such designs in the employee's contract of employment or in any employment manual or corporate IP policy of the employer, incorporated into the contract of employment by reference.

In the absence of any express provisions, there will, in any event, be an implied general duty of loyalty and fidelity between the employee and employer in addition to a duty of confidence in relation to any confidential information disclosed to the employee during the course of their employment. Thus if an employee before leaving their employment creates a new design but fails to disclose it choosing instead to take it with them to their new employer it maybe actionable by the former employer as a breach of fiduciary duty and/or breach of confidence and/or copyright/design right infringement depending on the nature of the documents involved, the nature of the design concerned and whether there has been any reproduction of the design. To reduce the risk that design documents might be removed by employees when departing, it is sensible for employers to maintain a log of design documents and to include in the contract of employment express provisions restricting the copying and/or removal of the employer's design documents (or copies of them) from the employer's premises by employees on cessation of their employment or otherwise.

113. See *Stephenson, Jordan and Harrison Ltd v MacDonald and Evans* (1952) 69 RPC 10 for a discussion as to what may amount to 'in the course of employment'.

114. CDPA 1988, s. 236 states that 'it is not an infringement of design right in the design to do anything which is an infringement of the copyright in that work'.

6.4 Employee Remuneration Right

Neither the UK design right, nor the UK registered design nor their Community equivalents provide any form of economic rights for employee designers whose rights in their designs have been transferred to their employers.

Where an author of an original artistic work[115] has transferred their rental right concerning a sound recording or a film to the producer of the sound recording or film, the author retains the right to equitable remuneration under Section 93B CPDA 1998. If the author of the artistic work is an employee, and the copyright work is made in the course of the author's employment, then the rental rights in the underlying copyright work would automatically belong to the author's employer as discussed above.[116]

7 EMPLOYEES' PLANT VARIETIES

7.1 The Legal Framework

The protection of plant varieties in the UK is governed by the Plant Varieties Act 1997. This Act provides breeders with protection in the UK for new, distinct, uniform and stable varieties of plants and seeds that the breeder breeds, discovers or invents.[117]

In 1994 the European Community adopted a Community Plant Variety Regulation which made it possible to obtain a single Community-wide plant variety right.[118] Community Plant Variety Rights have uniform effect throughout the territory of the EU. Where a breeder obtains a Community plant variety right any national breeder's rights for the same variety are suspended and cannot be enforced for so long as the Community plant variety right remains effective.[119]

7.2 Ownership of Rights

Where the breeder is an employee and the new variety has been discovered, bred or invented in the UK during the course of their employment, then unless an agreement exists to the contrary, the employer (or their successor in title) will be the person entitled to the grant of national plant breeder's rights.[120]

Entitlement to a Community plant variety right follows the same principles of ownership as the UK national right and so if the breeder of the new variety is an employee the entitlement to the Community plant variety right is determined in

115. Which includes graphic works, photographs, sculptures or collages or works of artistic craftsmanship but not buildings and works of applied art as these are specifically excluded from the relevant provisions, see Art. 3(2) of Directive 2006/115/EC on rental and lending right (codified version).
116. See section 2.4 above.
117. Plant Varieties Act 1997, s. 4(2)-(3).
118. Council Regulation (EC) No. 2100/94 of 27 July1994 on Community plant variety rights ('CPVR Regulation').
119. CPVR Regulation, Art. 92(2).
120. Plant Varieties Act 1997, s. 4(4).

accordance with the national law applicable to the employee relationship which, in the UK, means that the employer would be entitled to the grant of the Community right subject to any agreement with the employee to the contrary.[121]

As with other IP rights there may be scope for argument as to whether any new variety was created by the employee in the course of employment, in which case the same principles discussed above would be applicable to determine the issue.

7.3 The Parties' Duty to Inform

There is no statutory obligation or duty on an employee to notify or inform an employer of the creation, invention or discovery of a new plant variety. However, express duties may be included in the employee's contract of employment or in any employment manual or corporate IP policy of the employer incorporated by reference into the contract of employment.

In the absence of any express provision, there will, in any event, be an implied general duty of loyalty and fidelity between the employee and employer in addition to a duty of confidence in relation to any confidential information disclosed to the employee during the course of their employment. Thus if an employee before leaving their employment creates, invents or discovers a new plant variety and instead of disclosing it to their employer they take it with them to their new employer it could be actionable as breach of fiduciary duty and/or a breach of confidence, depending on the facts at issue, in addition to any entitlement claim. To avoid any doubt as regards the ownership of such new plant variety rights in such situations, it is sensible for the employer to include in its standard terms of employment contractual provisions requiring employees to disclose new plant variety rights to the employer on creation and to restrict the employee's right to remove documents (electronic or otherwise) from the employer's premises (or copies of them) on termination of employment.

7.4 Employee Remuneration Right

There is no provision for employee remuneration in respect of the creation, invention or discovery of a new variety and the grant of either national or Community plant variety protection beyond the requirement in the contract of employment to pay the employee their usual salary and any contract term that might exist relating to the payment of a bonus.

121. CPVR Regulation, Art. 11(4).

8 INSTANCES FOR DISPUTES

Legal proceedings relating to intellectual property, whether between an employer and employee or more generally, can be brought in a variety of fora depending on the nature of the dispute, the value of the claim and whether there are any specific statutory provisions that might apply.

With the exception of claims made by: (a) employees for compensation in respect of inventions (discussed further below), (b) an employee concerning the right to be named as the inventor of an invention the subject of a patent application[122] and/ or the owner of a patent application[123] and (c) an author or a performer in a sound recording or film in respect of remuneration for the transfer of rental rights, for which statutory rules provide for an alternative forum,[124] all intellectual property disputes must be brought in the Patents Court (part of the Chancery Division of the High Court[125]). According to the value of the claim the case may be allocated to the High Court if the value of the claim is in excess of GBP 500,000 and/or if it involves complex issues that would require a trial of more than two days. Claims of between GBP 10,000 and GBP 500,000 are allocated to the Business and Property Courts ('BPC') (formerly the Intellectual Property and Enterprise Court ('IPEC')). Although BPC also has a small claims track for those claims with a value of GBP 10,000 or less, the court does not accept claims relating to registered or Community designs, patents or plant varieties.[126]

Whilst it may be difficult at the outset to put an accurate figure on the value of a claim it is acceptable to estimate the value on the basis of the potential loss to the claimant. The principal differences between the three courts concern procedural issues and the level of experience of the judge hearing the case, the estimated length and complexity of the final hearing, the level of damages that can be awarded and the maximum value of legal costs (if any) that can be recovered from the unsuccessful party. In the small claim track a simplified procedure applies, no legal representation is necessary, a decision can be made on the papers, the maximum financial award is GBP 10,000 and it is not possible to recover any costs but it is possible to obtain an injunction. By comparison, in the High Court there is no limit to the length of the trial, the level of damages that can be awarded or the level of costs that might be recovered. The BPC seeks to offer a balance between these two extremes with trials limited to two days, damages capped at GBP 500,000 and costs capped at GBP 50,000. It is possible, with the consent of the court, for the case to be transferred from one court to another, if appropriate.

122. Patents Act 1977, s. 13(1).
123. Patents Act 1977, s. 8 and 37.
124. These proceedings need to be brought before the Copyright Tribunal in accordance with s. 93C of the CDPA 1988.
125. If the employee is based in Scotland or Northern Ireland proceedings may be initiated in the Court of Session or the High Court in Northern Ireland as appropriate.
126. Civil Procedure Rules 63.2(1) and Practice Direction 63 para. 16.1.

8.1 Applications for Employee Inventor Compensation

An inventor seeking compensation under Section 40 of the Patents Act 1977 may issue a claim at the Patents Office or at the High Court at any time from the grant of the patent(s) at issue until 12 months after the date that the patent ceases to have effect.[127] The procedures for initiating a claim are set out in Rule 63.12 of Part 63 Practice Direction 13 of the Civil Procedure Rules for the High Court and Rule 71 of the Patent Rules 2007, for determination by the Comptroller of Patents.

When deciding whether to initiate a claim in the Patents Court or at the Patents Office it is worth bearing in mind not just the cost of the proceedings and any representation required but also the time taken by each tribunal to reach a final decision. Whilst there have been only a few such actions since 1977 when the provision was first introduced it is worth noting that the only successful claim so far was brought in the High Court and that a decision was made within two years of the claim having been issued, whereas the most recent claim brought before the Patents Office took seven years to reach a first instance decision in 2014. The case has since been appealed twice, and at the time of writing, is believed to be the subject of a pending appeal to the Supreme Court. Whilst some of the delay was down to the parties and how they chose to conduct the proceedings it is also true to say that proceedings before the comptroller do move more slowly than at the Patents Court where there is a strict timetable to which the parties much adhere.

It should also be born in mind that the person determining the case on behalf of the comptroller is not a qualified judge of the same calibre as those at the Patents Court although both are experienced in hearing patent related disputes. The main distinction between the two tribunals is the cost of the proceedings themselves but potential litigants would do well to take account of the value of an early decision, the advantages of an experienced judge and the fact that an appeal from the Patent Office to the High Court is not a rehearing which means that findings of fact made by the Patent Office continue to apply and only if there has been a misdirection in law by the Hearing Officer will the appeal court intervene.

9 CONCLUDING REMARKS

9.1 Overlapping Intellectual Property Rights

Due to the nature of intellectual property rights and the piecemeal manner in which they have evolved over the years to protect different aspects of intellectual creations it is not surprising that there are overlaps in protection in different areas. Such overlaps can be beneficial in providing owners with additional layers of protection against different challenges where the rights are in common ownership. Where however the rights are held separately this can prove problematic. To avoid such difficulties between the rights of the employee and those of the employer in relation to rights that

127. Rule 91 Patent Rules 2007.

arise during the course of employment, the legislation has included specific provisions in an effort to make the boundaries between the different intellectual property rights clear and indicating which rights will prevail.

9.2 Interaction between Moral Rights and Copyright

The CDPA 1988 makes it clear[128] that the moral rights of attribution and conferred on the employee author does not apply to anything done by or with the authority of the copyright owner where the copyright originally vested in the employer, thus delimiting the boundary between the two parties in favour of the employer. The right of integrity held by the employee is similarly restricted where the author was previously identified in or on the published copies of the work if there is a sufficient disclaimer by the employer. Both provisions, however, are subject to any waiver there might be of these and other moral rights in the employee's contract of employment.

9.3 Interaction between Copyright and Design Rights

Section 51 of the CDPA 1988 identifies the boundary between copyright works and designs making it clear that it is not an infringement of copyright in a design document or model recording or embodying a design for anything other than an artistic work or a typeface to make an article to the design or to copy an article made to the design. The effect of this provision is to make copyright in a design document secondary to the rights in the design itself and to prevent both copyright and design rights from being relied upon simultaneously.

In addition to the above provision, Section 236 of the CDPA 1988 makes the case that where copyright and design right both subsist in a work that includes a design that the design right is not infringed by any action that constitutes an infringement of the copyright in that work. Thus the possibility of infringing both the copyright and the design rights simultaneously by the same act is avoided.

The potential for a possible conflict between the rights of the employer and those of the employee in relation to copyright and design right ownership is unlikely given that the same rules of ownership apply to both forms of right.

9.4 Interaction between Copyright and Patents

Literary, dramatic, musical and artistic works and any aesthetic creation whatsoever are specifically excluded from the scope of an invention under the Patents Act 1977[129] as are computer programs and the presentation of information subject to the proviso that these things shall only be prevented from being treated as inventions to the extent that the patent or application for a patent relates to that thing as such. Thus potential overlaps between patentable subject matter and copyright material is reduced.

128. CDPA 1988, s. 79(3) and 82(2).
129. Patents Act 1977, s. 1(2).

However, documents, models, diagrams, etc. may have been created by an employee in the course of their employment and so the copyright in them would belong to their employer even though the invention itself does not. In recognition of this potential overlap between the rights of the employer and those of the employee, Section 39(3) Patents Act 1977 provides that in such a situation nothing done by the employee or by anyone on their behalf for the purposes of performing or working the invention shall be taken to infringe the copyright or design right in any document or model that the inventor's employer holds. Thus, the patent rights of the employee in the invention trump the copyright and design rights relating to the invention of the employer.

9.5 Concluding Remarks

To summarise, generally speaking the intellectual property rights created by an employee in the course of their employment belong to the employer subject to certain exceptions namely performer's rights and limitations, in the case of patent rights. Only in relation to certain rental and lending rights and patent rights is a creative employee entitled to remuneration as a consequence of the employee's creative actions. In the absence of a compelling justification for the exceptional treatment of rental and lending rights, it would be preferable to treat all authors of copyright works and those of related rights in the same manner for the sake of consistency.

BIBLIOGRAPHY

Table of Sources

Banks Committee Report on the British Patent System 1970 Cmnd. 4407.
Bently & Sherman, *Intellectual Property* (3rd ed., OUP, Oxford 2013).
CIPA Guide to the Patents Act 1977 (Sweet & Maxwell, London 2014).
W. Cornish, *Intellectual Property* (1st ed., Sweet & Maxwell, London 1981).
Gurry, *Confidential Information* (2nd ed., OUP, Oxford 2012).
Laddie Prescott & Vitoria, *The Modern Law of Copyright & Designs* (4th ed., LexisNexis, London 2011).
J. Phillips & M Hoolahan, *Employees' Inventions in the United Kingdom Law & Practice* (ESC Publishing Ltd, Oxford 1982).
J Pila, *Who owns the intellectual property rights in academic work?*, 32(12) E.I.P.R. 609-613 (2010).
Terrell, *Law of Patents* (17th ed., Sweet & Maxwell, London 2011).

Table of Legislation

Berne Convention for the Protection of Literary and Artistic Works 1886 (1971 revision with 1979 amendments).

WIPO Performances and Phonograms Treaty 1996.

Council Regulation (EC) No. 2100/94 of 27 July 1994 on Community plant variety rights.

Directive 96/9/EC of the European Parliament and of the Council of 11 March 1996 on the legal protection of databases.

Council Regulation (EC) No. 6/2006 of 12 December 2001 on Community designs.

Directive 2006/115/EC of the European Parliament and of the Council of 12 December 2006 on rental right and lending right and on certain rights related to copyright in the field of intellectual property (codified version).

Directive 2001/84/EC of the European Parliament and of the Council of 27 September 2001 on the resale right for the benefit of the author of an original work of art.

Council Directive 87/54/EEC of 16 December 1986 on the legal protection of topographies of semiconductor product or products.

Artist's Resale Right Regulations 2006.

Artist's Resale Right (Amendment) Regulations 2009 (SI 2009/2792).

British Overseas Territories Act 2002 SI 1989/1100.

Community Design Regulations 2005.

Copyright Act 1911.

Copyright and Related Rights Regulations 1997.

Copyright and Rights in Databases Regulation 1997 SI 1997/3032.

Copyright, Designs & Patents Act 1988.

Designs Act 1949.

Design Right (Semiconductor Topographies) (Amendments) (No. 2) Regulations 2008 SI 2008/1434.

Design Right (Semiconductor Topographies) (Amendments) Regulations 2006 SI 2006/1833.

Design Right (Semiconductor Topographies) (Amendments) Regulations 1991 SI 1991/2237.

Design Right (Semiconductor Topographies) Regulations 1989 SI 1989/1100.

Patents Act 2004.

Patents Act 1977.

Performances (Moral Rights) Regulations 2006 SI 2006/18.

Plant Varieties Act 1997.

Public Lending Right Act 1979.

Public Lending Right Act 1982 (Commencement) Order 1982 SI 1982/719.

Rental Right Regulations SI 1996/2967.

Semiconductor (Protection of Topography) Regulations 1987.

The Intellectual Property Act 2014.

Trade Union and Labour Relations (Consolidation) Act 1992.

Table of Cases

Antocks Lairn Ltd v Bloohn Ltd, RPC 219 (1972).
British Steel PLC's Patent, RPC 11 (1992).
British Syphon Co Ltd v Homewood, 1WLR 1190 (1956)
Electrolux Ltd v Hudson, FSR 312 (1977).
Kelly & Chiu v GE Healthcare Ltd, EWHC 181 (2009).
Kynixa Ltd v Hynes, EWHC 1495 (QB) (2008).
GEC Avionics Ltd's Patent, RPC 107 (1992).
Helmet Integrated Systems Ltd v Tunnard, FSR (16) 437, 451 (CA) (2007).
King v South African Weather Service, FSR 6, Sup. C. of A. South Africa (2009).
LIFFE Administration v Pinkava, RPC 30 (2007).
Marshall & Naylor's Patent, 17 RPC 553 (1990).
Memco-MedLtd's Patent, RPC 403 (1992).
Noah v Shuba, FSR 14 (1991).
Patchett v Sterling, 72 RPC 50 (1955).
Prout v British Gas plc, FSR 478 (1992).
Ransom v Customer Systems Plc, IRLR 769 (2012).
Ready-Mixed Concrete (South East) v Minister of Pensions and National Insurance, 2 QB 497 (1968).
Re Greater Glasgow Health Board's Application, RPC 207 (1996).
Re Harris' Patent, RPC 19 (1985).
Shanks v Unilever Plc & Othrs, EWHC 1647 (Pat) (2014).
Stephenson, Jordan & Harrison Limited v Macdonald and Evans, 69 RPC 10 (1952).
Threlfall v ECD Insight Ltd, IRLR 185 (2013).
Tullet Prebon Plc v BGC Brokers, EWHC 484 (2010).
Ultraframe v Fielding, RPC 24 (CA) (2004).
University of Southampton's Applications, EWHC 2107 (2004).
Yeda Research and Development v Rhone-Poulenc Rorer International Holdings Inc, UKHL 43 (2007).

B In Asia

CHAPTER 24
China

Ella Cheong & Li Luo

1 BACKGROUND

Before the enforcement of the current intellectual property laws in China, authors, designers and inventors received more spiritual awards than material rewards. Due to the then planned economy and de-privatisation policy, regulations[1] for the promotion of scientific and technological innovation and artistic creation were not well enforced.

The current Patent Law was introduced in 1984 and was amended three times in 1992, 2000 and 2008. The current Copyright Law was adopted in 1990, and was amended twice in 2001 and 2010. The current Regulations for Protection of New Plant Variety were promulgated in 1997 and revised in 2013. The Regulations on the Protection of Layout Designs of Integrated Circuits were passed in 2001. These laws, together with their implementing regulations provide some protection for employees for their employment-related intellectual property. Besides intellectual property laws, the Law on Promoting the Transformation of Scientific and Technological Achievements of 1996, Contract Law of 1999 and the Labour Contract Law of 2007 also contain some provisions concerning employees' intellectual property rights.

Recognising that innovation is the key to the development of the economy, the Chinese government proposed in the Outline of the National Mid- to Long-term Talent Development Planning (2010-2020) ('国家中长期入才发展规划纲要(2010-2020年)') 'to draft regulations on service technological achievement, to improve mechanisms of ownership of scientific and technological intellectual property and benefits sharing; to

1. E.g. Temporary Regulations for the Protection of Invention Right and Patent Right of 1950 ('保障发明权与专利权暂行条例'), Regulations on the Reward for Inventions of 1963 ('发明奖励条例'), Determination on Improving and Developing Publishing of 1950 ('关于改进和发展出版工作的决议'), Rules on Correcting the Phenomenon of Arbitrarily Reproducing Books of 1953 ('关于纠正任意翻印图书现象的规定').

protect the legitimate rights and interests of scientific and technological inventors; to clarify the rights and interests of service inventors, to promote the benefit ratio of the main inventors; to draft benefit sharing mechanism for employee inventors in movement'.[2] The State Intellectual Property Office ('SIPO') drafted the Regulations on the Service Invention ('职务发明条例' (草案)), which has been published and revised two times, and submitted to the State Council for review.[3]

2 EMPLOYEES' COPYRIGHT

2.1 The Legal Framework

The relevant legal framework is principally governed by statutory provisions under the Copyright Law and the Implementing Regulations of the Copyright Law. In particular, Article 16 of the Copyright Law stipulates the ownership of copyright in an employment work, and Articles 11 and 12 of the Implementing Regulations further set out the definitions involved in employment copyright and the licensing thereof.

2.2 Ownership of Rights

Under the Chinese Copyright Law, an employment work refers to the type of work performed by an employee for the purpose of fulfilling the work tasks assigned to them by a legal entity or an organisation (the employer). Such 'work tasks', as set out by Article 11 of the Implementing Regulations of the Copyright Law, refer to the duty that the employee should perform in such legal entity or organisation as their employer.

2.2.1 Copyright in the Employment Work Owned by the Employee

According to Article 16 of the Chinese Copyright Law, the copyright in the employment work shall belong to the employee who is the author of the employment work, whereas the employer has a priority right to use such work within its scope of business.

In addition to the priority right to use the employment work, the employer is also entitled to exclude any third party from using the employment work within a certain period of time. In particular, without the consent of the employer, the employee cannot license to a third party a right to use the employment work in the same manner the work is used by the employer within the first two years counting from the creation of the work,[4] which is the date that the work is handed to the employer by the employee.[5] The employee can license to a third party a right to use the work in the same manner that the work is used by the employer if consent to such licensing has been provided by

2. Ch. 4, subs. 10, the Outline of the National Mid- to Long-term Development Planning (2010-2020).
3. http://www.sipo.gov.cn/ztzl/ywzt/zwfmtlzl/tlcayj/201403/t20140331_925616.html.
4. Art. 16(1) of the Copyright Law.
5. Art. 12(2) of the Implementing Regulations of the Copyright Law.

the employer. The remuneration that the employee obtains from such licensing shall be distributed between the employer and the employee in agreed proportions.[6]

2.2.2 *Copyright in the Employment Work Owned by the Employer*

Article 16(2) of the Copyright Law also specifies works for which authorship remains with the employee, but where the remaining rights belong to the employer. Such works include:

(i) drawings of engineering designs, drawings of product designs, maps, computer software and other works, which are mainly created by virtue of the material and technical resources of the employer, and are also under the employer's responsibility; and
(ii) works in which the copyright shall belong to the employer as stipulated by specific laws and regulations, or agreements.

2.2.3 *Copyright in the Employment Work Co-owned by the Employer and Employee*

There were no laws to specify the ownership of copyright for works which were created before the implementation of the first PRC Copyright Law. For works that were created in that special historical period but were used and promoted by both the employer and employee in the following years, it is likely that courts would rule that both parties shall co-own the copyright. [7]

2.3 The Parties' Duty to Inform

There is no statutory duty to inform, neither for the employee nor the employer.

2.4 Employee Remuneration Right

For employment works in which the copyright shall belong to the employer, Article 16 of the Copyright Law sets out that the employer may reward the employee for their work done.

Other than that, the Copyright Law and its Implementing Regulations are both silent on how an employee should be remunerated for their copyright work.

6. Art. 12(1) of the Implementing Regulations of the Copyright Law.
7. *Shanghai Art Film Studio v Qu Jian Fang*, 上海知识产权法院(2015)沪知民终字第200号.

3 EMPLOYEES' SEMICONDUCTOR CHIPS

3.1 The Legal Framework

There is no special law or regulation concerning employees' semiconductor chips. The governing rules of this topic are scattered in Regulations on the Protection of Layout Designs of Integrated Circuits, and the Contract Law.

Current effective provisions with regard to employees' semiconductor chips are: Rule 9 of Regulations on the Protection of Layout Designs of Integrated Circuits, and Article 326 of the Contract Law. Among these provisions, the employment technological achievement prescribed in Article 326 of the Contract Law refers to the technical solutions which relate to the product, method, materials and improvements thereof by utilising knowledge of science and technology, information and experience, and not only comprise inventions, utility models, and design patents and patent applications, but also comprise know-how, computer programs, layout-designs of integrated circuits, plant varieties, etc.[8] Moreover, the employment technological achievement refers to an achievement which is accomplished in the course of performing the duties of an employee or by mainly using the employer's material and technical conditions. The right of use and the right of assignment of an employment technological achievement belong to the employer. The employer shall extract a certain percentage of the profit obtained from use or assignment of the achievement as reward or remuneration to the employee. Once the employer wishes to assign the achievement, the employee has the priority to become the assignee.

3.2 Ownership of Rights

According to the Regulations for the Protection of Layout Designs of Integrated Circuits, the creator of the layout-design of the semiconductor chip is the owner of the exclusive right unless otherwise prescribed.

For designs the creation of which is organised by the employer, created according to the employer's will, and where the responsibility of design is assumed by the employer, the employer is the creator and the owner of the exclusive right. If the design is, on the other hand, made by an individual, the individual is the creator and the owner of the exclusive right. That is to say, even though the individual is an employee, they still can own the exclusive right if they made the design and such design is not organised by the employer, created according to the employer's will and that the responsibility of the design is assumed by the employer.

Except for the special regulations for the chips, the layout-design of semiconductor chips is also governed by Article 326 of the Chinese Contract Law. The employer automatically has the right of use and right of assignment of an employee-created layout-design of semiconductor chips if the achievement is accomplished in the course

8. Interpretation of the Supreme People's Court concerning Some Issues on Application of Law for the Trial of Cases on Disputes over Technology Contracts (法释[2004]20).

of performing the duties of an employee or made mainly by using the employer's material and technology. However, if the employee and the employer have any agreement with regard to the right of layout-design of semiconductor chips during the employment period or after the employment, the court shall follow such agreement.

3.2.1 *Layout Design Made in Execution of Employee's Duty*

According to judicial interpretation, the employee's duty comprises:

(i) performing the employee's own duty and other duties of research and development ('R&D') assigned by the employer;
(ii) within one year from the employee's retirement, resignation or from termination of employment from the employer, wherein the invention-creation relates to their own duty or other task entrusted to them by their previous employer unless the laws or administrative regulations prescribe otherwise.[9]

If the employee makes the layout-design in execution of the employee's duty in one of the above-mentioned manners, then the layout-design is an employment layout-design, and the employer is the right owner. The employee has priority to become the assignee if the employer decides to assign it.

3.2.2 *Layout Design Made Mainly by Using Employer's Materials and Technology*

Also according to judicial interpretations, material and technology comprises money, devices, equipment, raw materials, undisclosed information and documentations, etc.[10]

Layout-designs made mainly by using such employers' conditions comprise the scenarios that the employee uses all or a major part of such conditions during the R&D procedure to make the designs and such conditions are substantively influential for obtaining the designs; and the substantive contents of the layout-designs are made on the basis of the employer's undisclosed technical achievement or known technical achievement. If the employee makes the layout-design mainly by using employer's materials and technology, then the layout-design is an employment layout-design, and the employer is the owner of the right. The employee also has priority to become the assignee if the employer decides to assign it.[11] However, if the employee and employer have an agreement of paying special fees for such conditions, or the employee uses the employer's conditions for testing or approving after layout-design is finalised, the layout-design is not regarded as made mainly by using employer's materials and

9. Art. 2 of Interpretation of the Supreme People's Court concerning Some Issues on Application of Law for the Trial of Cases on Disputes over Technology Contracts (法释[2004]20).
10. Art. 3 of Interpretation of the Supreme People's Court concerning Some Issues on Application of Law for the Trial of Cases on Disputes over Technology Contracts (法释[2004]20).
11. Arts. 3-4 of Interpretation of the Supreme People's Court concerning Some Issues on Application of Law for the Trial of Cases on Disputes over Technology Contracts (法释[2004]20).

technology, and therefore is not an employment layout-design. The creator, i.e. the employee, is the owner of the exclusive rights to such layout-design.

3.3 The Parties' Duty to Inform

There exists no statutory duty for any of the parties to inform each other with regard to the existence of the layout-design or profit gained from the design.

3.4 Employee Remuneration Right

According to Article 326 of the Chinese Contract Law and judicial interpretation, the employer shall extract a certain percentage of the profit obtained from use or assignment of the layout-design of semiconductor chips as reward or remuneration to the employee.

4 EMPLOYEES' INVENTIONS

Under the current rules and regulations, the right of use and the right of assignment of an employment technological achievement belong to the employer. The employer shall extract a certain percentage of the profit obtained from use or assignment of the achievement as reward or remuneration to the employee. Shanghai Higher People's Court issued Guidance[12] in June 2013 with regard to the reward and remuneration to the employee. The Draft Regulations of Service Invention cover the parties' duty to inform, i.e., the employee's duty to inform the employer of the existence of the employment and employer's duty to inform the employee of the remuneration to be obtained.

4.1 The Legal Framework

There is no special law or regulation concerning employees' inventions. The governing rules of this topic are scattered within the Patent Law and the Contract Law.

Current effective provisions with regard to employees' inventions are: Articles 6-7, 16 of the Patent Law, Articles 12, 76-78 of the Implementing Regulations of the Patent Law, and Article 326 of the Contract Law. Among these provisions, the employment technological achievements prescribed in Article 326 of the Contract Law refer to the technical solutions which relate to product, method, materials and improvement thereof by utilising knowledge of science and technology, information and experience, and not only comprise invention, utility model, and design patents and patent applications, but also comprise know-how, computer programs, layout-designs

12. Guidance on Trial of Disputes over Rewards and Remunerations for Inventors or Designers of Employment Inventions or Creations by Shanghai Higher People's Court, June 2013.

of integrated circuit, plant varieties, etc.[13] According to this provision, employment technological achievements refer to those which are accomplished in the course of performing the duties of an employee or by mainly using the employer's material and technology. The right of use and the right of assignment of an employment technological achievement belong to the employer. The employer shall extract a certain percentage of the profit obtained from use or assignment of the achievement as reward or remuneration to the employee. Once the employer decides to assign the achievement, the employee has the priority to become the assignee.

4.2 Ownership of Rights

As a general principle, ownership of an invention belongs to the employer if the invention is a service invention made by the employee. The inventor or designer of a non-service invention enjoys the patent application right and owns the patent. Non-service inventions are those which are not service inventions. There is also an exemption that makes it possible for the employee to become a patentee of a service invention if special conditions are met.

An invention made by the employee becomes a service invention under the Patent Law if the employee executes their employment tasks, or the employee makes the invention mainly by using the materials and technical means of the employer. The employer has the right to decide whether to apply for patent protection for the service invention and becomes the patentee if a patent is granted.

4.2.1 Inventions Made in Execution of Employment Tasks

The employee has no possibility of becoming a patentee if the invention was made executing employment tasks. Such inventions are made by an employee:

(i) in the course of performing their own duty; for example, an employee working in the R&D department of a company whose duty is to handle research, development, design etc.;
(ii) in execution of any task, other than their own duty, which was entrusted to them by the employer; for example, some written R&D duty assigned by the employer to an employee whose duty is not research;
(iii) within one year from the employee's retirement, resignation or from termination of their employment from the employer, wherein the invention relates to their own duty or other task entrusted to them by their previous employer.

13. Art. 1 of Interpretation of the Supreme People's Court concerning Some Issues on Application of Law for the Trial of Cases on Disputes over Technology Contracts (法释[2004]20).

4.2.2 Inventions Made Mainly by Using Employer's Materials and Technical Means

For inventions made mainly by using the employer's materials and technical means, employees can agree with employers on who has the right to apply for patent protection and who will own the patent. If there is no such agreement, the employer has the right to apply for patent protection and own the patent.

The employer's 'materials and technical means' refer to the entity's money, equipment, spare parts, raw materials or technical materials which are not disclosed to the public, etc.[14] According to Article 5 of the Brief of the Intellectual Property Meeting of Courts with regard to Disputes on Technical Contracts,[15] if the employee only uses the technical information which is disclosed to the public or commonly known to the person skilled in the art, the invention will not be regarded as being made mainly by using the materials and technical means of the employer. Thus, the technical materials which are regarded as employer's materials and technical means exclude prior art and common knowledge. Under such circumstances, since the employee can enter into a contract with the employer on who becomes the owner of the patent, it is possible for the employee to become the sole patentee or co-owner of the patent together with the employer.

4.3 The Parties' Duty to Inform

No laws prescribe employers' and employees' duty to inform, except the Regulating Rules on the Protection of University Intellectual Property of 1998, which was promulgated by the Ministry of Education. Therefore, only principle investigators, who are university employees or university-affiliated enterprise employees,[16] should promptly suggest their employers file a patent application and provide relevant materials for service inventions.[17]

4.4 Employee Remuneration Right

Generally, the employer shall reward the employee once the patent is granted; and shall provide the employee a reasonable remuneration after the patent right is exploited and depending on the obtained proceeds.[18]

The employee can enter into a contract with the employer to agree on the manner and amount of the reward and remuneration. Such contract can be in the form of a labour contract or company regulation. If the contract is silent on these issues, the

14. Art. 3 of Interpretation of the Supreme People's Court concerning Some Issues on Application of Law for the Trial of Cases on Disputes over Technology Contracts (法释[2004]20).
15. Art. 5 of Notification Concerning the Printing and Distribution of the Brief of Intellectual Property Meeting of Courts with regard to Disputes on Technical Contract from the Conference on National Intellectual Property Courts Trail Work (法[2001]84) 19 June 2001.
16. Art. 2 of the Regulating Rules on the Protection of University Intellectual Property of 1998.
17. Art. 18 of the Regulating Rules on the Protection of University Intellectual Property of 1998.
18. Art. 16 of the Patent Law and Art. 326 of the Contract Law.

Implementing Regulation of the Patent Law provides a statutory reward and remuneration, i.e. the employer shall award a prize of RMB 3,000 to the employee for an invention patent, RMB 1,000 to the employee for a utility model and a design patent within three months from the date of announcement of the grant of patent. In case there is neither such agreement nor internal regulations on the manner and sum of the reward and remuneration to the employee for the service invention, the employer shall annually draw a percentage of not less than 2% from the profit from exploitation of the invention during the period of the patent for the employee or award a lump sum of money to the employee once and for all by taking reference to the percentage. If the patent is licensed to others, the employer shall draw no less than 10% from the licence fee it receives and award it to the employee as remuneration.[19]

According to the Guidance issued by the Shanghai Higher People's Court in June 2013, in principle, the agreement between the employer and employee regarding the reward and remuneration should be regarded as reasonable if it is based on the nature of the R&D in the industry, the purpose of the patent application, and the characteristics of the patent exploitation. The court can adjust the amount if the agreed amount is extremely low and apparently unreasonable. Since the current effective governing rules do not mention the reward and remuneration issue of assignment of the patent, the Guidance clearly mentions that the manner and amount of the remuneration for any assignment of patent right shall refer to the above-mentioned statutory standard.

5 EMPLOYEES' UTILITY MODELS

Employee's utility models share the same legal structure as inventions.

6 EMPLOYEES' DESIGN RIGHTS

Designs can be patented as well as copyrighted in China. The designer can be granted a design patent if the design meets the requirements of the Patent Law. According to Chinese Patent Law, designs mean, with respect to a product, new designs of the shape, pattern, or the combination thereof, or the combination of the colour with shape and pattern, which are rich in an aesthetic appeal and are fit for industrial application. Although works of applied art are not explicitly included in the Copyright Law, it is widely recognised that they are copyrightable.[20] Hence, some designs which demonstrate originality fall within the scope of copyright protection.

For these reasons the employees' rights for designs vary depending on whether their designs are patented or not. For patented designs, the employee-designers have the same rights and duties as employee-inventors, except that they might receive less remuneration and reward than the latter when there are no contracts in this regard between the employees and the employers. From the profit of the exploitation of the

19. Art. 78 of the Implementing Regulations of the Patent Law.
20. See *Intelego AG v Kegao (Tianjin) Toys Company, Ltd*, 北京市第一中级人民法院(1999)一中知初字第132号, (2002)高民终字第279号; OKBABY Company *v* Cixi Jiabao Company, Ltd, 北京市第二中级人民法院(2008)二中民初字第12293号.

designs during the period of the patent, employers are legally obligated to draw no less than 0.2% for employee-designers, whereas no less than 10% for employee-inventors, or award a lump sum of money once and for all by taking reference to the percentage.[21] For unpatented but copyrighted designs, employee-designers' rights and duties are the same as those of employee-authors.

7 EMPLOYEES' PLANT VARIETIES

7.1 The Legal Framework

Plant varieties are not patentable under Chinese law.[22] New plant varieties are mainly protected by the Regulation for the Protection of New Varieties of 1997[23] and its two implementing rules were passed by the Ministry of Agriculture and the Bureau of Forestry. Articles 326, 327 and 328 of the Contract Law are also applicable to service breeding of new plant varieties.

7.2 Ownership of Rights

Article 7 of the Regulation for the Protection of New Varieties of 1997 stipulates the ownership of the plant variety right. For service breeding, the employer has the right to apply for new plant variety and owns the variety right once the application is approved. No negotiation for the ownership between employees and employers is allowed. Service breeding refers to breeding accomplished under one of the following conditions:

- (i) in the course of performing the duties;
- (ii) in executing tasks assigned by the employer other than the employee's regular duties;
- (iii) within three years from the date the employee left the employer, wherein the breeding relates to their own duty or other task assigned to them by their previous employer;
- (iv) by mainly using the employer's funds, equipment, experimental field, employer's unpublished breeding and technical information.[24]

According to Article 328 of the Chinese Contract Law, the employee-breeder has the right to have their name shown on the relevant certificates.

21. Art. 78 of the Implementing Regulations of the Patent Law.
22. Art. 25(4) of the Chinese Patent Law.
23. The current Regulations for the Protection of New Varieties is amended in 2013.
24. Art. 7 of the Implementing Rules to the Regulations for the Protection of New Varieties (Ministry of Agriculture) and Art. 5 of the Implementing Rules to the Regulations for the Protection of New Varieties (Bureau of Forestry).

7.3 The Parties' Duty to Inform

No legal instruments prescribe a duty to inform for either of the parties.

7.4 Employee Remuneration Right

The Regulation for the Protection of New Plant Varieties does not mention the employees' remuneration right. It merely rules that the People's Government or relevant departments above county level should award breeders for their new plant varieties which are related to state interests or public interests and have great application value.[25]

Without a contract on employees' remuneration right, employees can request remuneration according Article 326 of the Contract Law, which rules that the employer shall provide the employee a reasonable remuneration by drawing a certain percentage from the profit made (from the exploitation and assignment of the new plant variety).

8 INSTANCES FOR DISPUTES

Three routes are available to resolve disputes between employees and employers regarding intellectual property: civil litigation, arbitration and administrative mediation.

8.1 Litigation

Either party can file a lawsuit at the court within two years from the date they knew or should have known their right was infringed.[26] According to the relevant High Court announcements and interpretations, for proceedings in the first instance the competent courts are usually the Intermediate People's Courts of capital cities of each province, autonomous regions or municipality cities directly under the central government and some Intermediate and District Courts designated by the Supreme People's Court.[27] However, in practice, most of the copyright disputes would fall into the jurisdiction of the Basic People's Court of each city. Beijing, Shanghai and Guangzhou Intellectual Property Courts usually would not hear the first instance of copyright disputes.

25. Art. 4 of the Regulations for the Protection of New Plant Varieties.
26. Art. 135 in connection with Art. 137 of the General Principles of the Civil Law.
27. See Art. 2 of the Several Rules concerning the Applicable Law in Trial of Cases of Patent Disputes by the Supreme People's Court "最高人民法院关于审理专利纠纷案件适用法律问题的若干规定" 法释 (2001) 21号; Art. 3 of Announcement on Several Issues concerning the Intensive Implementation of the Copyright Law of the People's Republic of China by the People's Supreme Court ("最高人民法院关于深入贯彻执行"中华人民共和国著作权法"几个问题的通知", (法发 (1993) 44号); Art. 43 of the Explanation on Several Issues Concerning the Applicable Law in Trail of Cases of Technology Contracts Disputes by the Supreme People's Court ("最高人民法院关于审理技术合同纠纷案件适用法律若干问题的解释" 法释[2004]20号").

If neither party is satisfied with the decision of the Court of First Instance, they can appeal to the court at the next higher level.[28] The decision of the Court of Second Instance is final.[29] However, if the party insists that the decision of the Court of Second Instance is erroneous, they can file a retrial petition to a higher court.[30]

8.2 Arbitration

Employees and employers can apply for arbitration before an Arbitration Commission when they have an arbitration agreement.[31] Unlike in civil proceedings, where the parties can only file a lawsuit before courts with jurisdiction over the case, the Arbitration Commission can be chosen by the parties.[32] The arbitration award is final. No appeal or judicial review is available.[33]

Employees and employers can also choose to file an application for arbitration before a Local Labour Dispute Arbitration Commission if the ownership and the reward and remuneration of the intellectual property are written in the employment agreement, even if there is no arbitration agreement signed.[34] The parties should file the arbitration application within one year from the date they knew or should have known their rights have been infringed.[35] Employees who are not satisfied with the arbitration award can initiate proceedings before the competent court within 15 days upon the receipt of the statement of award.

8.3 Mediation

For copyright[36] and patent[37] related disputes, employees and employers can also call upon the relevant copyright administrative bodies and patent administrative bodies for mediation. Either party can file a lawsuit on the same disputes within two years from the day when the mediation application was filed, if they are not satisfied with the settlement.[38]

9 CONCLUDING REMARKS

At legislative level, employees' intellectual property rights are well recognised. Employees can primarily enforce their rights through contracts. However, in most cases employees are not in the position to negotiate with their employers. When there is no

28. Art. 164 of the Civil Procedure Law.
29. Art. 175 of the Civil Procedure Law.
30. Art. 179 of the Civil Procedure Law.
31. Art. 4 of the Arbitration Law.
32. Art. 6 of the Arbitration Law.
33. Art. 9 of the Arbitration Law.
34. Art. 2(5) Of the Law on Mediation and Arbitration of Labour Disputes.
35. Art. 27 of the Law on Mediation and Arbitration of Labour Disputes.
36. Art. 55 of the Copyright Law.
37. Art. 85 of the Implementing Regulations of the Patent Law.
38. Art. 140 of the General Principles of the Civil Law.

contract, employees can claim their rights according to the relevant law. There is no unified reward and remuneration system for employee authors, designers, service inventors and service breeders. Relevant provisions are scattered in various legal acts at different levels. Generally there is no duty to inform for both the employees and their employers.

BIBLIOGRAPHY

Table of Sources

Interpretation of the Supreme People's Court concerning Some Issues on Application of Law for the Trial of Cases on Disputes over Technology Contracts (法释(2004) 20).

Notification Concerning the Printing and Distribution of the Brief of Intellectual Property Meeting of Courts with regard to Disputes on Technical Contract from the Conference on National Intellectual Property Courts Trial Work (法(2001) 84) 19 June 2001).

Guidance on Trial of Disputes over Rewards and Remunerations for Inventors or Designers of employment invention or creation by Shanghai Higher People's Court (June 2013).

Table of Legislation

Arbitration Law.
Civil Procedure Law.
Chinese Patent Law.
Contract Law.
Copyright Law.
Determination on Improving and Developing Publishing of 1950 ('关于改进和发展出版工作的决议').
General Principles of the Civil Law.
Implementing Regulations of the Patent Law.
Implementing Regulations of the Copyright Law.
Implementing Rules to the Regulations for the Protection of New Varieties.
Law on Mediation and Arbitration of Labour Disputes.
Regulations on the Reward for Inventions of 1963 ('发明奖励条例').
Regulations for the Protection of New Plant Varieties (amended in 2013).
Regulating Rules on the Protection of University Intellectual Property of 1998.
Rules on Correcting the Phenomenon of Arbitrarily Reproducing Books of 1953 ('关于纠正任意翻印图书现象的规定').
Several Rules concerning the Applicable Law in Trial of Cases of Patent Disputes by the People's Supreme Court.

Temporary Regulations for the Protection of Invention Right and Patent Right of 1950 ('保障发明权与专利权暂行条例').

Table of Cases

Intelego AG v Kegao (Tianjin) Toys Company Ltd, 北京市第一中级人民法院（1999）一中知初字第132号, (2002) 高民终字第279号.

OKBABY Company v Cixi Jiabao Company Ltd, 北京市第二中级人民法院（2008）二中民初字第12293号.

Shanghai Art Film Studio v Qu Jian Fang, 上海知识产权法院(2015) 沪知民终字第200号.

CHAPTER 25
Hong Kong

Ella Cheong

1 BACKGROUND

Hong Kong has been a special administrative region of People's Republic of China since the transfer of sovereignty over Hong Kong from the United Kingdom to China in 1997. Under 'one country, two systems' and as provided by the Basic Law (mini-constitution for Hong Kong), the legal system in Hong Kong is separate and distinct from that in China and the laws previously in force in Hong Kong (that is, the common law, rules of equity, ordinances, subordinate legislation and customary law) are maintained, save for any that contravene the Basic Law, and subject to subsequent amendment by the Hong Kong legislature.

As Hong Kong has inherited the legal system prior to the transfer of sovereignty, it is not uncommon to find statute laws in Hong Kong making references to their counterparts in the UK laws and the courts in Hong Kong may refer to the precedents of other common law jurisdictions under the common law system which continues in Hong Kong.

2 EMPLOYEES' COPYRIGHT

2.1 The Legal Framework

The relevant legal framework is principally governed by statutory provisions stipulated under the Copyright Ordinance (Cap. 528). In particular, Section 14 of the Copyright Ordinance contains provisions concerning employee works.

2.2 Ownership of Rights

2.2.1 Private Sector

According to Section 14 of the Copyright Ordinance, in general, an employer is the first owner of any copyright in the work, be it a literary, dramatic, musical or artistic work, or a film, made by an employee in the course of their employment subject to the following exceptions:

(i) the first exception is where the parties contract to the contrary by an agreement; and
(ii) the second exception is where such work is exploited by their employer or by someone else with the employer's permission in a way that could not reasonably have been contemplated by the employer and the employee at the time of making the work unless there is any agreement to the contrary.

With respect to the second exception, Section 14 further stipulates that the employer shall pay an award to the employee in respect of such exploitation at such amount as agreed between the employer and the employee or, in case of failing to reach an agreement, as determined by the Copyright Tribunal.

According to Section 198(1) of the Copyright Ordinance, 'employee', 'employer' and 'employment' refer to employment under a contract of service or of apprenticeship.

2.2.2 Public Sector

Section 14 of the Copyright Ordinance does not apply to government copyright or Legislative Council copyright.[1]

Government copyright is governed under Section 182 of the Copyright Ordinance which stipulates that where a work is made by an officer of the government in the course of their duties, the government is the first owner of any copyright in the work.

Legislative Council copyright is governed under Section 184 of the Copyright Ordinance which stipulates that where a work is made by or under the direction or control of the Legislative Council, the Legislative Council is the first owner of any copyright in the work.

As Section 14 does not apply to government copyright under Section 182 and the Legislative Council copyright under Section 184, if the copyright work by a government employee unexpectedly becomes exceptionally popular and generates unprecedented revenues to the government, the government employee does not have any right to seek award therefor.

1. Section 16, Copyright Ordinance (Cap. 528).

2.2.3 *Case Law*

Some of the recent instances of disputes in relation to employees' copyright in Hong Kong can be summarised as follows:

(i) *Emagist Entertainment Ltd v Nether Games (Hong Kong) Ltd and Others*, HCA1659/ 2012; [2013] 1 HKLRD 898

The plaintiff was an online game developer and game provider company, while the defendant company was set up by former personnel of the plaintiff. The said personnel brought with them the source codes of the game and associated databases of the plaintiff to the defendant company. The plaintiff sought interlocutory injunctions to enjoin the defendants to deliver up to the plaintiff the source codes of the game and associated databases as well as to restrain the defendants from damaging or disposing of the same.

One of the key issues was whether the copyright work was created in the course of employment and the plaintiff's counsel relied heavily on Section 14 of the Copyright Ordinance to argue that the plaintiff was the owner of the copyright work in question. It was argued by the defendants that they were not under the employment of the plaintiff but were instead the creators of the game and licensed the use of the copyright work to the plaintiff.

Having considered all the relevant factors, including:

- (i) whether there was a serious question to be tried,
- (ii) whether the defendant would be adequately compensated in damages should it turn out that the interlocutory injunctions were improperly granted,
- (iii) the strength of the plaintiff's claim to copyright of the game as the employer as compared to the defendants' bare denials of their being employed, and
- (iv) whether it was just and convenient for the court to grant the relief or remedy when it does so at the interlocutory stage,

the court granted the injunctions sought.

(ii) *Total Lubricants Hong Kong Ltd and Others v Christophe De La Cropte De Chanterac and Others*, HCA1694/2008

One of the causes of action by the plaintiffs was copyright infringement by their ex-employees, the defendants. To constitute a full and complete plea of a reasonable cause of action for copyright infringement, the pleadings must contain the following elements:

- (i) the title of the plaintiff to sue, that is, as owner or exclusive licensee;
- (ii) the subsistence of copyright in the work, which must be identified with precision;
- (iii) the infringement by the defendant; and
- (iv) the relief claimed.

The plaintiffs failed to give an account of the work being copied by the defendants in their pleadings. In addition, regarding the employees' work during the course of their employment, the plaintiffs failed to provide the requisite particulars on fulfilment of the qualification requirement under Section 178 of the Copyright Ordinance, i.e. whether any of the authors was at the material time:

(i) an individual domiciled or resident or having a right of abode in Hong Kong or elsewhere; or
(ii) a body incorporated under the law of any country, territory or area.

Consequently, the court held that the plaintiffs failed to plead and prove the subsistence of copyright in their statement of claim. Neither did the proposed amendment cure the defect. For the foregoing reasons, the plaintiffs' claims for copyright infringement were struck out and the proposed amendment was disallowed.

(iii) *L & T (Asia Pacific) Limited and Others v Wong Hong Lai Trading As Chong Yip Concrete Mixer Engineering (A Firm) and Others*, HCA 4204/2002

The plaintiffs sued for, amongst other things, copyright infringement by the defendants. The subject of infringement was the drawings of concrete mixer drums which were the products sold and distributed by the plaintiffs.

As the drawings were found to be created by an employee of the plaintiffs, the court held that the plaintiffs owned the copyright under Section 14 of the Copyright Ordinance and that the defendants had infringed the plaintiffs' copyright in these drawings.

2.3 The Parties' Duty to Inform

There is no statutory duty to inform on the part of the employee.

2.4 Employee Remuneration Right

Section 14 of the Copyright Ordinance provides that unless there is any agreement to the contrary, the employer shall pay an award to the employee upon the exploitation of the copyright work *beyond the initial contemplation by both parties*. Other than the foregoing, the Copyright Ordinance does not provide for the remuneration of an employee for their copyright work.

3 EMPLOYEES' SEMICONDUCTOR CHIPS

3.1 The Legal Framework

The relevant legal framework is principally governed by statutory provisions stipulated under the Layout-Design (Topography) of Integrated Circuits Ordinance (Cap. 445).

However, there has been no reported instance of dispute in relation to employees' layout-design (topography) rights in Hong Kong as yet.

3.2 Ownership of Rights

According to Section 2 of the Layout-Design (Topography) of Integrated Circuits Ordinance, the ownership of a layout-design shall be determined as follows subject to any agreement to the contrary:

(i) where the layout-design is not created in pursuance of a commission or in the course of employment, the designer of the layout-design is the owner;
(ii) where the layout-design is created in pursuance of a commission, the person who commissioned the layout-design is the owner; and
(iii) where the layout-design is not created in pursuance of a commission but is created by an employee in the course of their employment, the employer is the owner.

The above is applicable to both the private and public sectors.

3.3 The Parties' Duty to Inform

There is no statutory duty to inform on the part of the employee.

3.4 Employee Remuneration Right

The employee remuneration right is not codified in the relevant statutes.

4 EMPLOYEES' INVENTIONS

4.1 The Legal Framework

The relevant legal framework is principally governed by statutory provisions stipulated under the Patents Ordinance (Cap. 514). In particular, Sections 57-61 of the Patents Ordinance contain provisions concerning employee's inventions under the following section headings:

Section 57 Right to employees' inventions
Section 58 Compensation of employees for certain inventions
Section 59 Amount of compensation
Section 60 Enforceability of contracts relating to employees' inventions
Section 61 Supplementary.

There has been no reported instance for dispute in relation to employees' standard patent rights in Hong Kong as yet. The recent passage of Patents (Amendment) Ordinance 2016 on 10 June 2016 has not included any reform in this regard.

4.2 Ownership of Rights

4.2.1 Ownership of Employee's Inventions

Section 57 of the Patents Ordinance (Cap. 514) provides that the employer will be the owner of an invention made by their employee if:

(i) it was made in the course of the normal duties of the employee or in the course of duties falling outside their normal duties, but specifically assigned to them, and the circumstances in either case were such that an invention might reasonably be expected to result from the carrying out of their duties; or

(ii) the invention was made in the course of the duties of the employee and, at the time of making the invention, because of the nature of their duties and the particular responsibilities arising from the nature of their duties they had a special obligation to further the interests.

Other than the above, any other invention made by an employee shall be regarded as belonging to the employee.

Recently, the Court of First Instance has made a detailed interpretation of Section 57 of the Patents Ordinance. In *Acron International Technology Ltd v Chan Yiu Wai* HCA 1709/2010 *[2016] HKEC 170*, the Plaintiff claimed against his two former employees (the '**Defendants**') for the ownership of an invention developed by the Defendants during the course of their employment with the Plaintiff. The Court, following LIFFE Administration and Management v Pinkava [2007] 4 AER 981, confirmed that there are two primary limbs to be satisfied in section 57 of the Patents Ordinance, namely,

(i) the Invention was made in the course of the Defendants' normal duties; and
(ii) the circumstances were such that an invention might reasonably be expected to result from the carrying out of duties.

In summary, the determination of whether there is a breach under section 57 of the Patents Ordinance depends entirely on the fact, e.g:

(i) whether the invention in disputes is similar to patents owned by the Plaintiff's and derived from them;
(ii) the date of the creation of the invention in disputes;
(iii) what were the Defendants' normal duties;
(iv) whether the invention made in the course of normal duties in the course of employments; and
(v) whether an invention, regardless of whether it is the invention in question or an invention similar to the invention in question, might reasonably be expected from the carrying out of a defendants' duties taking into account the qualities of the defendants.

4.2.2 Two Additional Criteria in Determining Ownership

To determine the ownership for employees' inventions, there are two additional criteria to be taken into consideration:

(i) whether the invention was made before or after the commencement of the Patents Ordinance; and
(ii) the employee's connection to Hong Kong.

According to Section 61 of the Patents Ordinance (Cap. 514), the sections covering employee's rights, i.e. Sections 57-60, will not apply to an invention made by an employee before the commencement of the Patents Ordinance, i.e. 27 June 1997, nor to an invention made by an employee unless at the time they made the invention, one of the following conditions was satisfied:

(i) they were mainly employed in Hong Kong; or
(ii) they were not mainly employed anywhere or their place of employment could not be determined, but their employer had a place of business in Hong Kong to which the employee was attached, whether or not they were also attached elsewhere.

4.2.3 Employee's Rights in Jointly-Made Inventions

Section 61(3) of the Patents Ordinance provides that the employee's rights conferred by Sections 57-60 refer to the invention made by the employee alone or jointly with any other person. However, it will not be regarded as an invention made by the employee where the employee merely contributes advice or other assistance in the making of an invention by another employee.

4.2.4 Employee's Rights Arising from Patents Outside Hong Kong

Section 61(4) of the Patents Ordinance further provides that those patents or patent applications relating to employee's inventions under Sections 57-60 cover patents, patent applications or other forms of protection for an invention under the law of Hong Kong or the law in force in any other country, territory or area or under any treaty or international convention.

4.2.5 Employee's Rights Shall Not Be Diminished

Section 60 of the Patents Ordinance renders any contractual terms or arrangements to be unenforceable if they diminish an employee's rights in inventions of any description made by the employee after the commencement of the Patents Ordinance, i.e. 27 June 1997 and the date of the contract, or in or under patents for those inventions or applications for such patents. For example, the employee's rights may be deemed diminished in case their employment agreement seeks to exclude their rights to be reasonably compensated as conferred by Section 58 of the Patents Ordinance.

The contractual terms or arrangements referred to by Section 60 include:

(i) a term in a contract entered into by an employee with the employer;
(ii) a term in a contract entered into by an employee with some other person at the request of the employer or in pursuance of the employer's contract of employment; and
(iii) arrangement as to performance of duties by a public officer of the government.

The above is applicable to both the private and public sectors.

4.3 The Parties' Duty to Inform

There is no statutory duty to inform on the part of the employee.

4.4 Employee Remuneration Right

Employee remuneration right is conferred on an employee under Section 58 of the Patents Ordinance on conditions that:

(i) an invention made by an employee on or after the commencement of the Patents Ordinance, i.e., 27 June 1997[2]; and
(ii) it is required that at the time they made the invention, the employee was mainly employed in Hong Kong, or they did not have a main place of employment anywhere but their employer had a place of business in Hong Kong to which they were attached, whether or not they were also attached elsewhere[3].

4.4.1 The Invention Has Always Belonged to the Employer[4]

Where the ownership of the invention is vested in the employer by operation of law as set forth in Section 57 (see section 4.2.1 above), the employee may apply to court to seek compensation from the employer if the following conditions are met:

(i) a patent has been granted for the invention;
(ii) the patent is of 'outstanding benefit' to the employer; and
(iii) it is just that the employee should be awarded compensation.

2. Section 61(1) of the Patents Ordinance.
3. Section 61(2) of the Patents Ordinance.
4. Section 58(1), Patents Ordinance (Cap. 514).

4.4.2 *The Invention Originally Belonging to the Employee but Is Subsequently Assigned or Exclusively Licensed to the Employer*[5]

The employee may apply to court to seek compensation from the employer if the following conditions are met:

(i) a patent has been granted for the invention;
(ii) the employee's rights in the invention, or in any patent or application for a patent for the invention, have been assigned or exclusively licensed to the employer since the commencement of Section 58 of the Ordinance, i.e. 27 June 1997;
(iii) having regard to all the circumstances of the case, including the terms of the contract of assignment or grant or any ancillary contract ('the relevant contract'), the benefit derived by the employee from the relevant contract is inadequate in relation to the benefit derived by the employer from the patent; and
(iv) it is 'just' (fair) that the employee should be awarded compensation to be paid by the employer in addition to the benefit derived from the relevant contract.

4.4.3 *Limitation Period for Seeking Compensation by Employee*

The application to court for compensation by the employee under Section 58 has to be made within the prescribed period defined under Section 53 of the Patents (General) Rules (Cap. 514A) as 'that period which begins when the relevant patent is granted and expires one year after the relevant patent has ceased to have effect'.

Where the patent has ceased to have effect by reason of a failure to pay any renewal fee and an application for restoration is made under Section 40 or Section 127 of the Patents Ordinance, then if restoration is ordered, the prescribed period will continue as if the patent had remained continuously in effect. But if restoration is refused, the prescribed period will expire one year after the patent ceased to have effect or six months after the refusal, whichever is the later.

4.4.4 *Amount of Compensation*

The amount of compensation is determined by Section 59 of Patents Ordinance as a 'fair share' of the benefit the employer derives from the patent or patent application for the invention. Accordingly, the benefit, meaning benefit in money or money's worth,[6] covers the following:

(i) both future and present benefit are taken into account;[7]

5. Section 58(2), Patents Ordinance (Cap. 514).
6. Section 61(7), Patents Ordinance (Cap. 514).
7. Section 59(1), Patents Ordinance (Cap. 514).

(ii) benefit derived or expected to be derived by an employer from the assignment or grant of property or right in the invention, the patent application or the patent to a person connected with them[8] even if that person had not been connected with them;[9] and
(iii) benefit derived or expected to be derived from the patent by their personal representatives or by any person in whom the patent was vested by their assent.[10]

When an employee applies to court for compensation under Section 58, the court shall, among other things, take the following matters into account:

(i) where the invention has always belonged to the employer:[11]
 (a) the nature of the employee's duties, their remuneration and the other advantages they derive or have derived from their employment or has derived in relation to the invention under the Ordinance;
 (b) the effort and skill which the employee has devoted to making the invention;
 (c) the effort and skill which any other person has devoted to making the invention jointly with the employee concerned, and the advice and other assistance contributed by any other employee who is not a joint inventor of the invention; and
 (d) the contribution made by the employer to the making, developing and working of the invention by the provision of advice, facilities and other assistance, by the provision of opportunities and by their managerial and commercial skill and activities.
(ii) where the invention originally belonged to the employee but is subsequently assigned or exclusively licensed to the employer:[12]
 (a) any conditions in a licence or licences granted under the Ordinance or otherwise in respect of the invention or the patent;
 (b) the extent to which the invention was made jointly by the employee with any other person; and
 (c) the contribution made by the employer to the making, developing and working of the invention by the provision of advice, facilities and other assistance, by the provision of opportunities and by their managerial and commercial skill and activities.

According to Section 59(5) of the Patents Ordinance, the compensation can be paid in one lump sum or periodically, or both.

8. Section 61(8), Patents Ordinance (Cap. 514) defines the meaning of 'person connected with' in relation to an employer.
9. Section 59(2), Patents Ordinance (Cap. 514).
10. Section 61(5), Patents Ordinance (Cap. 514).
11. Section 59(3), Patents Ordinance (Cap. 514).
12. Section 59(4), Patents Ordinance (Cap. 514).

According to Section 59(6) of the Patents Ordinance, the refusal of the court to make any such order on an application made by an employee under Section 58 will not be a bar to a further application for compensation under Section 58 made by the employee or any successor in title.

According to Section 59(7) of the Patents Ordinance, the court may on the application of either the employer or the employee vary, discharge or suspend any provision of the order.

5 EMPLOYEES' UTILITY MODELS

5.1 The Legal Framework

There is another type of patent available in Hong Kong and it is known as a short-term patent. However, there has been no reported instance of dispute in relation to employees' short-term patent rights in Hong Kong as yet.

Similar to utility models which are available in other jurisdictions, a short-term patent has a shorter patent term, which is eight years, and does not require substantive examination for grant of patent. Short-term patents are also governed by the Patents Ordinance (Cap. 514) and the foregoing discussion in relation to patents is equally applicable to short-term patents.

5.2 Ownership of Rights and Termination

The foregoing discussion in relation to patents is equally applicable to short-term patents.

5.3 The Parties' Duty to Inform

There is no statutory duty to inform on the part of the employees.

5.4 Employee Remuneration Right

The foregoing discussion in relation to patents is equally applicable to short-term patents.

6 EMPLOYEES' DESIGN RIGHTS

6.1 The Legal Framework

The relevant legal framework is principally governed by statutory provisions stipulated under the Registered Designs Ordinance (Cap. 522). Nevertheless, there has been no reported instance of dispute in relation to employees' registered design rights in Hong Kong.

6.2 Ownership of Rights and Termination

In relation to employee works, according to Section 3 of the Registered Designs Ordinance, in general, the designer will be the original owner of the design subject to the following exceptions:

(i) where a design is created in pursuance of a commission for money or money's worth, the person commissioning the design shall, subject to any contrary agreement between the parties, be treated as the original owner of the design;

(ii) where, in a case of design not concerning a commission for money or money's worth, a design is created by an employee in the course of their employment, the ownership of the design belongs to their employer;

(iii) where a design, or the right to apply a design to any article, becomes vested, whether by assignment, transmission or operation of law, in any person other than the original owner, either alone or jointly with the original owner, that other person or, as the case may be, the original owner and that other person, shall be treated for the purposes of this Ordinance as the owner of the design or as the owner of the design in relation to that article;

(iv) in the case of a design generated by computer in circumstances such that there is no human designer, the person by whom the arrangements necessary for the creation of the design are made shall be deemed to be the designer.

Section 2 of Registered Designs Ordinance provides that an 'employee' means:

> a person who works or (where the employment has ceased) worked under a contract of employment (whether with the government or with any other person).

The above is applicable to both the private and public sectors.

6.3 The Parties' Duty to Inform

There is no statutory duty to inform on the part of the employee.

6.4 Employee Remuneration Right

The employee remuneration right is not codified in the relevant statutes.

7 EMPLOYEES' PLANT BREEDER'S RIGHTS

7.1 The Legal Framework

The relevant legal framework is principally governed by statutory provisions stipulated under the Plant Varieties Protection Ordinance (Cap. 490). The registration of Plant

Variety Rights is administered by the Agriculture, Fisheries and Conservation Department of the Hong Kong SAR Government. However, there has been no reported instance of dispute in relation to employees' plant variety rights in Hong Kong as yet.

7.2 Ownership of Rights and Termination

Section 2 of the Plant Varieties Protection Ordinance (Cap. 490) provides that an 'owner of any plant variety' means:

(i) a person who bred or discovered and developed that variety;
(ii) an agent of that person;
(iii) a successor of that person.

Accordingly, for an application for plant varieties protection, the applicant shall be the original breeder of the variety. If the applicant is not the original breeder of the variety, a letter of authorisation or other documentation for proof of ownership must be provided with the application.

The above is applicable to both the private and public sectors.

7.3 The Parties' Duty to Inform

There is no statutory duty to inform on the part of the employee.

7.4 Employee Remuneration Right

The employee remuneration right is not codified in the relevant statutes.

8 INSTANCES FOR DISPUTES

There is no court specialising in the handling of intellectual property (IP) disputes in Hong Kong. In general, the 'court' referred to in most IP-related laws including the Patents Ordinance (Cap. 514), the Plant Varieties Protection Ordinance (Cap. 490), the Registered Designs Ordinance (Cap. 522) and Sections 20-23 of the Layout-Design (Topography) of Integrated Circuits Ordinance (Cap. 445) means 'the Court of First Instance'. Lower courts such as the District Court and the Magistrates' Courts may also have jurisdiction to adjudicate on certain disputes under the Copyright Ordinance (Cap. 528).[13]

13. Jurisdiction of District Court: e.g. Copyright Ordinance (Cap. 528) ss. 134 and 232; the court referred in the Prevention of Copyright Piracy Ordinance (Cap. 544) includes magistrates.

8.1 Copyright Tribunal

The Copyright Ordinance confers power upon the Copyright Tribunal to hear and determine cases[14] including determining the award to employees for using their work outside reasonable contemplation as stipulated under Section 14 of the Copyright Ordinance.

8.2 Labour Tribunal and Minor Employment Claims Adjudication Board

The Labour Tribunal in Hong Kong hears claims arising from non-compliance of the Employment Ordinance (Cap. 57) or the Apprenticeship Ordinance (Cap. 47). The Tribunal also deals with cases involving breaches of a term of a contract of employment, whether for performance in Hong Kong or under a contract to which the Contracts for Employment Outside Hong Kong Ordinance (Cap. 78) applies, and claims arising from breach of a term of a contract of apprenticeship.

For claims lodged by not more than ten claimants for a sum of money not exceeding HKD 8,000 per claimant, they will be heard by the Minor Employment Claims Adjudication Board instead of the Labour Tribunal.

9 CONCLUDING REMARKS

The development of Hong Kong towards a knowledge-based economy has created more and more jobs for local population as knowledge workers. As the number of knowledge workers grows, employees' IP has become increasingly important. Furthermore, since the local administration has been determined to promote the development of Hong Kong as an IP trading hub and to boost the competitiveness of its creative industries, employees' IP will become even more important in terms of attracting world-class talents and companies to Hong Kong and it will be interesting to see if and how employees' IP will evolve in near future to cope with the development of Hong Kong.

BIBLIOGRAPHY

Table of Legislation

Apprenticeship Ordinance (Cap. 47).
Contracts for Employment Outside Hong Kong Ordinance (Cap. 78).
Copyright Ordinance (Cap. 528).
Employment Ordinance (Cap. 57).
Layout-Design (Topography) of Integrated Circuits Ordinance (Cap. 445).

14. S. 173 and s. 233, Copyright Ordinance (Cap. 528).

Patents Ordinance (Cap. 514).
Patent (Amendment) Ordinance (not yet in force)
Patents (General) Rules (Cap. 514C).
Plant Varieties Protection Ordinance (Cap. 490).
Prevention of Copyright Piracy Ordinance (Cap. 544).
Registered Designs Ordinance (Cap. 522).

Table of Cases

Acron International Technology Ltd v Chan Yiu Wai HCA 1709/2010 *[2016] HKEC 170*
Emagist Entertainment Ltd v Nether Games (Hong Kong) Ltd and Others HCA1659/2012 (2013) 1 HKLRD 898.
L & T (Asia Pacific) Limited and Others v Wong Hong Lai Trading As Chong Yip Concrete Mixer Engineering (A Firm) and Others HCA 4204/2002 [2005] HKEC 246
Total Lubricants Hong Kong Ltd and Others v Christophe De La Cropte De Chanterac and Others HCA1694/2008 [2010] HKEC 270.

CHAPTER 26

Japan

Tsugizo Kubo

1 BACKGROUND

There exists no general or fixed principle in Japan whereupon it is the employer who is to be owner from the outset of the intellectual property rights over creations made by its employees in the context of their employment contract and in the performance of their tasks or duties, or that it is the employee who must conserve their rights. The situation differs with regard to the various intellectual property rights.

The definitions of 'intellectual property' and 'intellectual property right' in Japan are stipulated in Article 2(1) and (2) of the Intellectual Property Basic Act (Act No. 122 of 2002) as follows:

(1) The term 'intellectual property' as used in this Act shall mean (i) inventions, devices (utility models), new plant varieties, industrial designs, works and other property that is produced through creative activities by human beings (including discovered or solved laws of nature or natural phenomena that are industrially applicable), (ii) trademarks, trade names and other marks that are used to indicate goods or services in business activities, and (iii) trade secrets and other technical or business information that is useful for business activities.

(2) The term 'intellectual property right' as used in this Act shall mean a patent right, a utility model right, a plant breeder's right, an industrial design right, a copyright, a trademark right, a right that is stipulated by laws and regulations or pertaining to an interest that is protected by laws on other intellectual property.

At present, in Japan, intellectual property laws are composed of the following eight Acts:

(i) Copyright Act (Act No. 48 of 1970) - works
(ii) Act on the Circuit Layout of a Semiconductor Integrated Circuit (Act No. 43 of 1985) - microchip circuit layout-designs
(iii) Patent Act (Act No. 121 of 1959) - inventions
(iv) Utility Model Act (Act No. 123 of 1959) - utility models
(v) Industrial Design Act (Act No. 125 of 1959) - industrial designs
(vi) Plant Variety Protection and Seed Act (Act No. 83 of 1998) - new plant varieties
(vii) Trademark Act (Act No. 127 of 1959) - marks, indications
(viii) Unfair Competition Prevention Act (Act No. 47 of 1993)

With regard to the last two Acts there are no substantial legal disputes of relations between employers and employees regarding ownership of trademark rights or protected IP rights enforced by the Unfair Competition Prevention Act. It is traditionally understood that these rights are all granted to the employers or the legal/business entities from the outset. Both Acts are absent of specific provisions concerning this issue.

With regard to copyright and layout of semiconductor integrated circuits, the author or the creator of a work or a microchip circuit layout-design made by an employee shall be generally the employer or the legal/business entity to which the employee belongs, unless otherwise agreed specifically.

Rights over inventions, utility models, industrial designs and new plant varieties made by an employee in the course of their duties under an employment relationship shall originally belong to the employee, unless an agreement has been made in advance with the employer[1]. If the employer wants to obtain these IP rights entirely or partly from the employee, the employer shall pay 'adequate remuneration' to the employee.

2 EMPLOYEES' COPYRIGHT

Japan, as a Member State of the Berne Convention, adopts the non-formality principle, without requiring any application or registration for the establishment of copyright. Therefore, in contrast with the Patent Act and other Acts relating to IPR registration, the Copyright Act does not provide for a 'right to register copyright'.

The Copyright Act grants the following rights to authors of works:

- moral rights, which include (i) the right of making the work available to the public (publication, Article 18(1)), (ii) the right of determining the indication of the author's name (authorship, Article 19(1)), (iii) the right of preserving the integrity (integrity, Article 20(1)), and (iv) the right of maintaining the honour and/or reputation of the author (reputation, Article 113(6)); and
- economic rights, which include (i) the right of reproduction (Article 21), (ii) the right of presentation and performance (Article 22), (iii) the right of screen

1. This new exception, expressly vesting a right to obtain a patent of an 'employee invention' in the employer, entered into force on 1 April 2016. See further section 4 below.

presentation (Article 22*bis*), (iv) the right of public transmission etc. (Article 23), (v) the right of recitation (Article 24), (vi) the right of exhibition (Article 25), (vii) the right of distribution (Article 26), (viii) the right of transfer of copied works (Article 26*bis*), (ix) the right of rental (Article 26*ter*), (x) the right of translation and adaptation etc. (Article 27), and (xi) the right of the original author in the exploitation of a derivative work (Article 28).

2.1 The Legal Framework

Article 15 of the Copyright Act concerns the authorship of a work created by an employee in the course of their employment relationship. More specifically, the author of a work created by an employee in the course of employment shall be the employer or the legal/business entity to which the employee belongs, unless otherwise agreed specifically. Therefore, all kinds of copyrights including 'moral rights of the author' and 'economic rights' mentioned above shall originally belong to the employer or the legal/business entity who shall be deemed to be the author. In other words, 'employer works' may exist in Japan.

The definition of a work is stipulated in Article 2(1)(i) of the Copyright Act as 'a production in which thoughts or sentiments are expressed in a creative way and which falls within the literary, scientific, artistic or musical domain'.

Article 15 (Authorship of a work created by an employee in the course of employment relationship) of the Copyright Act stipulates as follows:

> (1) The authorship of a work (except a computer program) which, on the initiative of a legal/juridical entity or other employer (hereinafter in this Article referred to as an "'employer"' collectively), is made by its employee in the course of employment relationship in connection with the employer's business operations, and is or will be made public under the name of the employer as the author, shall be attributed to the employer, unless otherwise specifically stipulated by a contract, employment regulation or any other stipulation in force at the time of the making of the work.
> (2) The authorship of a computer program which, on the initiative of the employer is made by its employee in the course of employment relationship in connection with the employer's business operations, shall be attributed to the employer, unless otherwise specifically stipulated by a contract, employment regulation or any other stipulation in force at the time of the making of the work.'

2.2 Ownership of Rights

2.2.1 *Conditions for Enabling the Authorship of Employees' Works to Be Attributed to the Employer*

Conditions for enabling the authorship of employees' works to be attributed to the employer practically mean conditions for enabling employees' works to be regarded as the legal/juridical entity's works. Article 15 of the Copyright Act provides for different requirements for enabling the authorship of works made by employees to be attributed to the employer, between 'works other than computer programs' and 'computer programs'. The definition of a 'computer program' is 'an expression of a combination of instructions to cause a computer to function in order to be able to obtain a certain result' (Article 2(1) (x*bis*)).

2.2.2 *Works other than Computer Programs*

As provided in Article 15(1), a 'work other than a computer program' should satisfy the following requirements in order to be recognised as an 'employer work':

(1) the work has been made on the initiative of the 'employer',
(2) the work has been made by an employee of the 'employer' in the course of an employment relationship (more exactly, in the course of the performance of their duties/tasks under employment relationship),
(3) the work is or will be made public under the name of the 'employer', and
(4) there exists no special provision on authorship in a contract, employment regulation or any other stipulation in force at the time of creation of the work.

2.2.3 *Computer Programs*

In corporate practice, computer program are not usually made public under the name of the employer even if they are made by its employee in the course of employment relationship, because the computer program works are, in general, mainly developed for in-house use only in many cases. Therefore, in order for the employer to be regarded as the author of 'computer programs' created by its employees in the course of the employment relationship, the requirements (1) to (4) in 2.2.2 above shall be satisfied, except (iii).

2.2.4 *Influence of the Status of a Legal Entity's Works*

Where a work made by an employee is recognised as an 'employer work' under the provision of Article 15 of the Copyright Act, the authorship and any right of the work shall belong not to the employee, but to the employer. The employer shall not need the consent of the employee who has created the work to make the work public under the name of the employer or altering the contents of the work and so on, because the employer, as author, has moral rights such as the right of making the work public, the right of determining the author's name, the right of preserving integrity, and the right of maintaining the honour and/or reputation.

On the other hand, where a work made by an employee fails to satisfy the Article 15 requirements of an 'employer work', the authorship and the rights of the work shall exclusively belong to the employee. However, the employer may, by contract or employment regulation, assign or transfer the rights to an 'employee work' to themselves. In such case, as in typical copyright transactions, the employer may only receive the transfer of the economic rights over the work and shall not receive the moral rights because these rights belong exclusively to the employee alone and are deemed non-transferable.

2.3 The Parties' Duty to Inform

With regard to an express positive duty to give notice to an employer of the creation of any intellectual property by the employee (creator), no statute regulating intellectual property rights in Japan expressly contains such a duty. However, an employee is by implication still required to give notice to their employer by virtue of a general duty of loyalty towards the employer. Such duty may be expressly prescribed in the internal employment regulations or similar created by the employer.

2.4 Employee Remuneration Right

Special rules regulating an employee's right to remuneration regarding a work made by the employee do not exist in the Copyright Act or other statutes in Japan. If and to what extent employee-creators are entitled to remuneration depends instead on any agreement that has been voluntarily entered into by the employer and the employee, or an award to be granted by the employer at its discretion.

3 EMPLOYEES' SEMICONDUCTOR CHIPS

3.1 The Legal Framework

The general principle in Article 15 of the Copyright Act is mirrored in the law regulating microchip circuit layout-designs. Article 5 of the Act on the Circuit Layout of a Semiconductor Integrated Circuits (hereinafter the Semiconductor Chip Act) stipulates as follows:

> With respect to a circuit layout-design created by an employee in the course of employment relationship, who performs his or her duties for a legal/juridical entity or other employer (hereinafter in this Article referred to as an "'employer"' collectively), such employer shall be the creator of the circuit layout-design, unless otherwise specifically stipulated by a contract, employment regulation or any other stipulation in force at the time of the creation of the circuit layout-design.

3.2 Ownership of Rights

When an employee creates a microchip circuit layout-design in the course of an employment relationship, the employer shall be automatically a creator of the circuit layout-design, and obtain a 'circuit layout-design right' thereof. The basic condition for the application of Article 5 of the Semiconductor Chip Act is that no other contract has been entered into between the employer and the employee. Where the parties have concluded another contract, it will invalidate the presumption stipulated in Article 5 of the Semiconductor Chip Act, and the attribution of the 'circuit layout-design right' will be determined by the contract instead.

3.3 The Parties' Duty to Inform

Concerning an express positive duty to give notice to an employer of the creation of any intellectual property by the employee (creator), no statute regulating intellectual property rights in Japan expressly contains such a duty. However, an employee is by implication still required to give notice to the employer by virtue of a general duty of loyalty towards the employer. Such duty may be expressly prescribed in the internal employment regulations or similar created by the employer.

3.4 Employee Remuneration Right

Special rules regulating an employee's right to remuneration regarding a microchip circuit layout-design made by the employee do not exist in the Semiconductor Chip Act or any other statute in Japan. Instead, if, and to what extent, the employee-creators are entitled to remuneration depends on any contract that has been entered into by the employer and the employee, or an award to be granted by the employer at its discretion.

4 EMPLOYEES' INVENTIONS

4.1 The Legal Framework

Even though the legal issue of the employee invention system is not only restricted to patent law and is deeply related to other legal areas, such as labour law and the law of contracts, the relevant provisions in Japan are stipulated in the legal framework under 'only one' Article of the Patent Act – Article 35 – which provides a basic legal framework for the employee invention system. The Utility Model Act and the Industrial Design Act apply *mutatis mutandis* this Article, and the Plant Variety Protection and Seed Act essentially stipulates a similar scope.

In 2015 Article 35 of the Patent Act was revised and a special exception was introduced for employee inventions. These amendments are applicable to inventions created on or after 1 April 2016. The following sections will therefore first consider rules applicable to inventions created prior to 1 April 2016 and subsequently discuss the new provisions as applicable to inventions created on or after this date.

4.1.1 *History of the Employee Invention System*

Rules regarding 'employee inventions' were not present in either the Provisional Regulations for Monopoly of 1871 or the Patent Monopoly Act of 1885 in Japan. The first provisions on 'employee inventions' were established in the Patent Act of 1909. This Act adopted the inventor system in principle, but stipulated that unless otherwise provided, rights to 'employee inventions' belong to the employer from the outset.

However, an innovative system that was almost the same as the current one was adopted in the Patent Act of 1921. That is, under that system the employer could obtain a licence free of charge (non-exclusive licence under the current law) for an invention made in the course of the employee's duties ('employee invention' under the current law), and if stipulated in advance, the invention could be assigned to the employer in return for an 'adequate remuneration' (Section 14). Thus, a legal system that is almost the same as the current one was already established in 1921, but the actual practices of the system are not quite clear due to lack of related documents, statistics, and court decisions.

The current law (the Patent Act of 1959) follows mostly the basic structure of the Act of 1921, and the original Article 35(4) of the 1959 Act was revised to form the current Article 35(4) and (5) in 2004. During the last decade, dramatic changes have been observed in Japan regarding the actual situation and people's awareness in dealing with the employee invention system.

4.1.2 *Basic Structure of Article 35 of the Patent Act*

If an employee makes an invention, all of the rights over the invention originally belong to the employee, not to the employer. The invention made by the employee that comes within the scope of their employer's business operations and is made as part of the employee's present or past duties/tasks resulting from the employment relationship is called the 'employee invention', and is subject to some special rules stipulated in Article 35 of the Patent Act.

The employer may justifiably and automatically obtain a non-exclusive licence free of charge for such an 'employee invention'. Moreover, it is possible for the employer to one-sidedly stipulate, in advance, the transfer (acquisition) of the right to obtain a patent, and the patent rights, or the grant of an exclusive licence. Usually, such a stipulation is prepared or arranged in the internal employment regulations or similar documents by the employer.

However, in case such a transfer is made or an exclusive licence is granted, the employee has the legal, and mandatory, right to receive a payment of 'adequate remuneration' thereof from its employer. Court cases relating to legal issues of 'employee invention systems' concern, in a significant majority of cases, the interpretation of 'adequate remuneration.'

On the other hand, the employer does not have a justifiable right to demand a transfer of the right or granting of an exclusive licence relating to the 'employee invention' from its employee after the invention is completed, if no stipulation exists in advance. To put it another way, an 'employee invention' may pass to the employer, in advance, in accordance with the provision(s) of the contract, employment regulation or any other stipulation. In many cases, the employment regulations or the like established or arranged by the employer provides for 'advance succession' of an 'employee invention', so the employer will be the owner of all of rights over an 'employee invention' when the invention is completed.

4.1.3 Provisions of Article 35 of the Patent Act on the Employee Invention System as Applicable to Inventions Created Prior to 1 April 2016

Article 35 (Inventions by Employees) of the Patent Act stipulates as follows:

(1) An employer, a legal/juridical entity, or a national or local government/public-entity (hereinafter referred to as an 'employer' collectively) shall have a non-exclusive licence on a patent right concerned, (i) where an employee, an officer of the legal/juridical entity, or a national or local public official (hereinafter referred to as an 'employee' collectively) has obtained a patent for an invention which falls within the scope of the business operations of his or her employer by the nature of the invention, and an act or acts resulting in the invention were part of the present or past duties/tasks of the employee in the course of employment relationship with the employer (hereinafter referred to as an 'employee invention'), or (ii) where a successor to the right to obtain a patent for the employee invention has obtained a patent therefor.

(2) In the case of an invention made by an employee, unless the invention is an employee invention, any provision in the contract, employment regulation or any other stipulation providing in advance (i) that the right to obtain a patent or the patent right for such an invention made by the employee shall be transferred to his or her employer, or (ii) that a provisional exclusive licence or an exclusive licence on such an invention shall be granted to his or her employer, shall be null and void.

(3) An employee shall have the right to receive the payment of adequate remuneration (i) where, in accordance with the provision(s) of the contract, employment regulation or any other stipulation, he or she transfers the right to obtain a patent or the patent right of the employee invention to his or her employer or grants an exclusive licence therefor to his or her employer, or (ii) where, in the case that he or she has granted a provisional exclusive licence for the employee invention to his or her employer in accordance with the provision(s) of the contract, employment regulation or any other stipulation, an exclusive licence is deemed to have been established under Article 34*bis* (2).

(4) [The newly introduced provision by the 2004 amendment, in force from 'patent applications filed on 1 April 2005']

Where the provision(s) of the contract, employment regulation or any other stipulation provides for the remuneration stipulated in the preceding paragraph, the payment of the remuneration to the employee in accordance with the provision(s) shall not be what is recognised unreasonable in light of circumstances including: (i) whether there were negotiations between the employer and the employee to set standards for the determination of the remuneration; (ii) whether the set standards had been disclosed to the employees; (iii) whether the opinions from the employees and others on the calculation of the amount of the remuneration had been heard; and (iv) any other relevant factor.

(1) [The original provision of the 1959 Act, in force until 'patent applications filed on 31 March 2005']

The amount of such remuneration as prescribed in the preceding paragraph shall be decided by reference to (i) the profit to be received by the employer from the employee invention and (ii) the degree of the contribution made by the employer to the making of the employee invention.

(2) [The revised provision by the 2004 amendment, corresponding to the above original provision (4), in force from 'patent applications on 1 April 2005']

Where there is no provision setting forth the remuneration as stipulated in the preceding paragraph, or where under the preceding paragraph, the payment of the remuneration in accordance with the provision(s) is recognised as unreasonable, the amount of the adequate remuneration referred to in Paragraph (3) shall be determined (by a court) in light of (i) the amount of the profits to be received by the employer from the working of the employee invention, (ii) the employer's burden and contribution to the employee invention, (iii) the benefits or treatment of the employee in relation to the employment invention, and (vi) any other relevant circumstance to be considered.

4.2 Ownership of Rights

4.2.1 Patent Rights Originally Belong to Employees

Under the Japanese Patent Law, rights over an invention made by an employee in the course of the employment relationship who has performed their duties/tasks for the employer shall originally belong to the 'employee'.

When an invention is completed, the original right over the invention is recognised as a right to file a patent application, or a 'right to obtain a patent', which arises from the principal sentence of Article 29(1) of the Patent Act: 'Any person who has made an invention which is industrially applicable may obtain a patent thereof.' Accordingly, 'the person entitled to the right to obtain a patent' shall be the 'inventor'. The term 'person' in this provision is traditionally construed as referring to a natural person. Consequently, the 'right to obtain a patent' with respect to the employee invention shall belong to the employed-inventor who is a natural person, not to the employer. It is not possible to forcibly attribute an employee invention to the employer without any consideration in accordance with the provision(s) of a relevant contract, employment regulation or any other rules. Hence, there is no such case where all of the rights of the employee invention belong to the employer from the outset under the Japanese Patent Law. Thus, under this basic concept, Article 35 of the Patent Act practically plays a very important role for both the employer and the employee.

4.2.2 Conditions for Transferring Employee Inventions to Employers

Following Article 35(1) of the Patent Act, an 'employee invention' is defined as an invention that:

(i) falls within the scope of the business operations of the employer; and
(ii) results from an act or acts that were part of the present or past duties/tasks of the employee in the course of the employment relationship with the employer.

Article 35(2) of the Patent Act stipulates that rights to inventions *other than* 'employee inventions' may not be transferred to the employer under the provision(s) of a contract, employment regulation or any other stipulation in advance. Conversely, therefore, the provision(s) of a contract, employment regulation or any other stipulation may validly provide in advance that an 'employee invention' may be transferred to the employer. This may be one-sidedly established or arranged by the employer without any prior consultation with the employees. In other words, an internal regulation set by the employer in advance indicating the transfer of rights without the conclusion of a contract with the employees is also effective, and it is not permissible for the employee to refuse to cooperate in the procedure for the transfer of the right for the reason that the remuneration is not paid concurrently, or is insufficient/deficient.

However, and perhaps most importantly, Article 35(3) stipulates that the employee shall have the legal right to receive 'adequate remuneration' when the employer has facilitated the transfer of the right to obtain a patent or the patent rights to the 'employee invention', or an exclusive licence thereof for the benefit of the employer, in accordance with the provision(s) of a contract, employment regulation or any other stipulation. Therefore, a payment of 'adequate remuneration' shall be required in order for the employer to obtain the rights to the employee invention from the employee, or an exclusive licence thereof.

As a rule, a voluntary agreement between both parties presumes 'adequate remuneration', unless the remuneration is deemed unreasonable by virtue of the internal regulations. Where the remuneration is not defined in the provision(s) of a contract, employment regulation or any other stipulation, or where it is defined but the remuneration is deemed unreasonable by virtue of the internal regulations, the amount of 'adequate remuneration' shall be determined by the court, if such claim is submitted by the employee, with reference to factors such as the amount of profits to be received by the employer from the employment invention, the burden borne by the employer, the contribution made by the employer, the benefits received by the employee in relation to the employment invention, and any other relevant factors.

The amount of profits that the employer receives from the employee invention is not the amount of profits that is received by merely working or exploiting the employee invention, because the employer automatically enjoys a free, non-exclusive licence on the patent right by virtue of Article 35(1) of the Patent Act. Hence, if the employer adopts a complete open policy relating to employee inventions, the employed inventor may receive only some part of the received royalties paid by the licensees thereof as 'adequate remuneration'. Due to the ambiguity of Article 35(4) and (5) of the Patent Act, Japanese industrial companies (i.e. employers) have their own regulations concerning employees' inventions that are specific to the company, such as the type of business, the scale of operations, and the importance of inventions in each company.

4.3 The Parties' Duty to Inform

Concerning an express positive duty to give notice to an employer of the creation or completion of any intellectual property by the employee (inventor, etc.), no statute regulating intellectual property rights in Japan expressly contains such a duty. However, an employee is by implication still required to give notice to the employer by virtue of a general duty of loyalty towards the employer. Such duty may be expressly prescribed in the internal employment regulations or similar document created by the employer.

4.4 Employee Remuneration Right

In the *Olympus Optical Pick-up Equipment* case, in which an ex-employee inventor sued for the shortfall amount of an 'adequate remuneration', the Supreme Court (Supreme Court at the Third Petty Bench) decision of 22 April 2003 substantially affirmed both lower instance decisions of the Tokyo District Court of 16 April 1999 and the Tokyo High Court of 22 May 2001, and had, for the first time, the opportunity to review the interpretation of Article 35 and 'adequate remuneration'. The Supreme Court clearly addressed that 'even if the amount of the remuneration was to be determined under the defendant's employment regulations on employees' inventions, where the amount of the remuneration actually paid to the employee is less than the amount of the adequate remuneration stipulated under Article 35(4) (the 1959 original provision), the court may calculate an amount of the remuneration independently and order the defendant (employer) to pay the shortfall of the remuneration'. In this case, the employee received the amount of JPY 211,000 (about USD 2,110) during the employment time, and the Supreme Court ordered the Olympus company to pay the additional shortfall amount of JPY 2,289,000 (about USD 22,890).

Thus, this Supreme Court judgment has established a judicial rule that, even where the amount of the remuneration is determined under the provision(s) of the internal employment regulations, or other similar document, concerning employees' inventions, the court may review the regulations and decide an amount of the remuneration independently. This means that Article 35(3) is substantially a mandatory or compulsory provision that cannot be superseded by a separate internal agreement between the employer and the employees.

Following this decision, the number of lawsuits concerning the amount of adequate remuneration has drastically increased. More than one hundred cases have been litigated during the last decade (2004 to 2013). In almost all cases, after retirement, ex-employee inventors demand the shortfall amount of the remuneration which they received from the employer. In reality, it is quite difficult for employees to institute proceedings before a court during their employment.

The most publicised lawsuit over 'adequate remuneration' to date was filed in 2001 by Mr. Shuji Nakamura who is an ex-employee of Nichia Corp., and serves as a professor of the University of California, Santa Barbara in the US, and won the 2014 Nobel Prize for Physics. He instituted proceedings against Nichia Corp. in Japan for his inventions (about 300 inventions including Japan Patent No. 2628404) that paved the way for blue light-emitting diodes to be put to practical use for the first time in the world. Blue LEDs are now a vital part of large-screen TVs, Blu-ray discs, lighting and traffic signals. After resigning from Nichia, he claimed JPY 20 billion (about USD 200 million) in compensation for his invention against the company. The Tokyo District Court ruled in favour of Mr. Nakamura, noting that his contribution is estimated to be worth 50% of the profits the company gained from his invention and ordered Nichia to pay that amount in a decision handed down in January 2004. The Court calculated, at first, the amount of JPY 60 billion (about USD 600 million), but, as Mr. Nakamura was claiming only JPY 20 billion, the Court ordered the payment of JPY 20 billion. However, Nichia subsequently appealed to the Tokyo High Court. Following a decisive court recommendation by the Tokyo High Court in January 2005, the parties agreed to settle the case with Nichia paying JPY 840 million (about USD 8.4 million) to Mr. Nakamura. The Tokyo High Court factored in Nichia's contribution to the completion of his inventions, sales promotion efforts and other related factors.

In the aftermath of this extraordinary Tokyo District Court decision, Article 35 was amended by the National Diet, and a new provision was inserted (Amendment by the Law No. 79 of 2004, enacted on 28 May 2004, in force from 1 April 2005). Henceforth 'adequate remuneration' under the current Article 35(4) and (5) of the Patent Act will be decided as follows:

(1) By virtue of Article 35(4), an 'adequate' remuneration can, as a principal rule, be decided by 'voluntary agreement' between the employer and the employee. Therefore, where the amount of remuneration for transfer of the right relating to an employee invention to the employer is regulated by the provision(s) of a contract, employment regulation or any other stipulation, that remuneration shall be regarded as an 'adequate remuneration', unless the remuneration is deemed unreasonable.

(2) However, if a remuneration relating to an employee invention has not been provided or agreed on in the provision(s) of a contract, employment regulation or any other stipulation, or where the payment of remuneration in accordance with the provision(s) of the contract, employment regulation or any other stipulation is recognised unfairly unreasonable according to Article 35(4), the remuneration evaluated based on the stipulation of Article 35(5) of the Patent Act shall be 'adequate'.

(3) The original wording of Article 35(4) in the 1959 Act refers to 'the degree of the contribution made by the employer to the making of the employee invention', which merely considers the factors before completion of the invention. The (newly inserted) Article 35(5) takes into account, however, contributions after the completion of the invention, such as executing the patent application procedure, developing technologies for exploitation of the invention, conducting sales, advertising and undertaking licence negotiations, as well as contributions to creation, and any other relating factors.

As of December 2014, there has been no newly reported case or court decision concerning the issues of 'adequate remuneration' of the revised Article 35(4) or (5) of the Patent Act. The court settlement (January 2005) of the Blue LED case by the Tokyo High Court mentioned above was proceeding between the enactment (28 May 2004) of the 2004 amendment of Article 35 and its entry into force (1 April 2005).

Even though more than one hundred lawsuits similar to the Blue LED case were filed by frustrated employed-inventors under the 1959 original provisions of Article 35 of the Patent Act, there has been no court decision ordering the payment of such massive remuneration as in the Blue LED case.

The following exemplifies some of the major and massive-amount court cases concerning the interpretation of Article 35 of the Patent Act and determination of 'adequate remuneration' within the context of employee inventions:

Olympus Optical Pick-up Equipment Case (filed in 1995)
The Supreme Court decision of 22 April 2003 affirmed the decisions of the Tokyo High Court on 22 May 2001 and the Tokyo District Court on 16 April 1999. The Court essentially established that if the remuneration under Article 35(3) paid by the employer is not sufficient, the employee who transferred the right to obtain a patent to the employer is entitled to require of the employer the payment of the deficient/shortfall amount. This is the turning point court decision of the employee invention system in Japan as it encouraged frustrated employee-inventors to institute proceedings asking the court to review the amount of remuneration which had been received.

Hitachi Magnetic Optical Disk Case (filed in 1998)
The Supreme Court decision of 17 October 2006 affirmed the decision of the Tokyo High Court of 29 January 2004. The Supreme Court held that when adequate remuneration for the employee invention is being calculated, the rights to obtain a patent in foreign countries as corresponding patents should be included in Article 35 of the Patent Act and ordered the payment of about JPY 162 million (about USD 1.6 million) by Hitachi Ltd. The former employee-inventor (plaintiff) had only received about JPY 2 million (about USD 20,000) before instituting proceedings.

Ajinomoto Aspartame (an artificial sweetener) Case (filed in 2002)
The Tokyo District Court in its decision of 24 February 2004 ordered the payment of about JPY 190 million (about USD 1.9 million) by the Ajinomoto Co., Inc. holding that Article 35 of the Patent Act is to be applied to the right to obtain a patent in a foreign country, so such right should be included in 'a right to obtain a patent' in the provisions of Article 35. The case was appealed to the Tokyo High Court and settled in 2004 resulting in a payment of JPY 150 million (about USD 1.5 million). The former employee-inventor (plaintiff) had only received JPY 10 million (about USD 100,000) before instituting proceedings.

Toshiba Flash Memory Case (filed in 2003)
This case was filed before the Tokyo District Court and settled in 2006 resulting in a payment of JPY 87 million (about USD 0.87 million) by Toshiba Corporation.

Canon Laser Printer Case (filed in 2003)
The Intellectual Property High Court in its decision of 26 February 2009 ordered Canon Inc. the payment of about JPY 70 million (about USD 0.7 million). The Intellectual Property High Court had disagreed with the Tokyo District Court which ordered the payment of about JPY 34 million (about USD 340,000) as remuneration, and this was upheld by the Supreme Court on 19 October 2010. The former employee-inventor (plaintiff) had only received JPY 876,000 (about USD 8,800) before instituting proceedings.

Brother Label Writer Case (filed in 2005)
The Intellectual Property High Court in its decision of 25 June 2009 ordered Brother Industries, Ltd. the payment of about JPY 56 million (about USD 0.56 million). The Intellectual Property High Court disagreed with the Tokyo District Court which ordered the payment of about JPY 37 million (about USD 370,000) as remuneration), and this was upheld by the Supreme Court on 24 October 2011. The two employee-inventors (plaintiffs) had jointly received only JPY 237,700 (about USD 2,400) before instituting proceedings.

Please note that these seven major and massive-amount cases all relate to the original provision of Article 35(4) of the Patent Act, and not the revised provisions of Article 35(4) and (5) in the 2004 amendment which is in force for patent applications filed from 1 April 2005. As of December 2014, there is no reported court decision concerning the issues of 'adequate remuneration' of the 2004 amendment of Article 35(4) and (5).

4.5 The Latest Amendment of Article 35 of the Patent Act in 2015 as Applicable to Inventions Made on or after 1 April 2016

After the long-heated debate in Japan concerning the employee invention system relating to Article 35 of the patent law, the National Diet finally approved and passed the bill partially revising Article 35 of the patent law on 3 July 2015, and the amended law was promulgated on 10 July 2015. The amendment came into effect on **1 April 2016**.

In that debate, industrial circles and business enterprises insisted that patents relating to employee inventions be ascribed to companies without any condition to prevent them from being left at a disadvantage. Also, business enterprises asserted that the process of the determining 'reasonable remuneration' for an employee invention is becoming quite more complicated, and gradually and potentially increasing the employer's risk of being sued by their employee-inventors again. But, on the other hand, mainly labour unions and academics have objected, saying that giving companies carte blanche would erode employee morale and trigger a brain drain from Japan to overseas.

This latest amendment changed the long-established rule since 1921, which is the right to obtain a patent inherently belongs to the employee-inventor even if that invention is an employee invention. And, this amendment newly introduced the unique system that the employer may **optionally** select in advance whether the right to obtain a patent for an employee invention **inherently belongs to the employer** when the employee invention is made or completed. However, even in this new option selected, the employer shall or must pay '**reasonable remuneration or so (reasonable benefits)**' to the employee-inventor, as almost same in the old law.

4.5.1 *Outline of the Amendment of Article 35 (Paragraphs 1 to7)*

(1) The right for employee invention is optionally enabled to inherently belong to the employer

The revised law firmly maintains the old law's basic-principle that a right to obtain a patent for an employee invention is inherently vested in the employee-inventor (**Paragraph 1** of Article 35 is not changed), but prescribes an exception for an 'employee invention' that a right to obtain a patent for an 'employee invention' is inherently vested in the employer only when the employee and employer make, in advance, such an agreement, contract or in-house employment service rules. Unless such an agreement is made, a right to obtain a patent for an 'employee invention' is inherently vested in the employee, as the same as under the old law. The instability in the ownership of a right to obtain a patent will be eliminated under the latest amendment. However, some small, family-controlled, or midsize companies, universities, or research institutions are expected to select and keep the old system.

By way of reference, **Paragraph (1)** of Article 35 of the patent law stipulates as below:

> An employer, a legal entity or a state or local public entity (hereinafter referred to as the 'employee') shall have a non-exclusive license on the patent right concerned, (i) where an employee, an executive officer of a legal entity or a national or local public official (hereinafter referred to as the 'employee') has obtained a patent for an invention which by reason of its nature falls within the scope of the business of the employer and an act or acts resulting in the invention were part of the present or past duties of the employee performed on behalf of the employer (hereinafter referred to as an 'employee's invention'), or (ii) where a successor in title to the right to obtain a patent for an employee's invention has obtained a patent therefor.

The newly added **Paragraph (3)** of Article 35 of the patent law stipulates as follows:

> In the case of an employee's invention made by the employee, where any contractual provision, service regulation or other stipulation provides in advance that the right to obtain a patent for such an invention shall be acquired by the employer, the right to obtain a patent for such an invention shall devolve on the employer, as from the time when it comes into existence.
> *Note: Hereinafter, the italicized parts show the amended portions introduced by the 2015 revision.*

Corresponding to the newly added **Paragraph (3)** above, **Paragraph (2)** of Article 35 was amended as follows:

> In the case of an invention made by an employee which is not subject to an employee's invention, any contractual provision, service regulation or other stipulation providing in advance that the right to obtain a patent such an invention *shall be acquired by the employer*, or that the patent right granted for such an

invention shall pass to the employer, or that the employer shall have a provisional exclusive license or exclusive license on such an invention, shall be null and void.

(2) The right to receive 'reasonable remuneration or other economic profits (reasonable benefits)'

An employee-inventor has a right to receive 'reasonable remuneration or other economical profits (reasonable profits)'. The latest amendment clarified that the reward for an 'employee invention' should not be limited to monetary remuneration, but also may include any economical profits or fringe benefits, such as providing stock options, job promotions with a pay raise, paid leave, or opportunities for studying abroad etc., but excludes honorary titles and appreciation certificates. An employee-inventor enjoys this right evenly not only in the case where the right to obtain a patent for an 'employee invention' is inherently vested in the employer in the new system, but also in the case where the employer succeeds to any employee's right relating to the employee's invention under the old law system.

The revised **Paragraph (4)** of Article 35 (amended from the old Paragraph (3) of Article 35) stipulates as follows:

> The employee shall have the right to receive an adequate *money and other financial benefit (referred to as 'reasonable benefits' in Paragraphs (5) and (7))* (i) where he has, with respect to an employee's invention, *made the employer to acquire* the right to obtain a patent, (ii) where he has made the employer to succeed the patent right, (iii) where he has granted the employer an exclusive license on such invention with respect to an employee's invention in accordance with the contractual provision, service regulation or other stipulation, or (iv) where, in the case where he has granted the employer a provisional exclusive license on such invention with respect to an employee's invention in accordance with the contractual provision, service regulation or other stipulation, it is deemed that an exclusive license is granted under Article 34bis (34-2)(2).

(3) Official guideline for procedures to determine the details of 'reasonable benefits'

In determining whether the 'reasonable benefits' for the employee invention are deemed to be unreasonable, (i) negotiation, (ii) disclosure and (iii) hearing shall be considered (so stipulated **Paragraph (5)** (slightly changed from the corresponding old paragraph (4)). Further, the Minister of Economy, Trade and Industry (METI) shall publish the guideline for (i) the procedures concerning negotiation between the employer and employee, (ii) disclosure of the criteria established, and (iii) hearing from the employees.

This guideline was published on 22 April 2016. If there is no such an official guideline for determining the reasonable profits as a reward for an employee's invention, the reasonable profits determined by employers may largely vary from company to company, and in some cases result in lack of employee-inverters' incentive for making inventions. It is true in the general observation that the employers' superiority over their employers may lead to an imbalance in terms of power between employers and employees.

The slightly amended **Paragraph (5)** of Article 35 stipulates as follows:

> *The grant of the reasonable benefits in the preceding paragraph*, as provided for in the contract, service regulation or other stipulation shall not be considered to be unreasonable, in view of (i) the situation under which a negotiation is carried out between the employer and his employees in the course of establishing the criteria for determining the <u>contents of the reasonable benefits</u>, (ii) the situation

under which the criteria established are disclosed, and (iii) the situation under which the views or opinions of employees are heard for determining the contents of the reasonable benefits, etc.

The newly added **Paragraph (6)** of Article 35 of the patent law stipulates as follows:

The Minister of Economy, Trade and Industry shall, for the purpose of encouraging inventions by promoting their protection and utilization, upon hearing the views of the Industrial Structure Council, provide a guideline with respect to matters that shall be considered as provided for in the preceding paragraph, and publish the same.

Although the guideline mentioned in **Paragraph (6)** is not legally binding, the employer should understand and follow the procedures prescribed in the guideline, so that its policy and formula of 'reasonable benefits' can be strongly defended in court if necessary. The outcomes of disputes with employee-inventors over 'reasonable benefits' will hinge on the employer's execution and documentation of those procedures. That is, the establishment of a clear process for claiming the right to patent employee inventions will lower employers' risk of losing lawsuits brought by employee-inventors due to conflict over 'reasonable benefits' terms and conditions. Anyway, setting up in-house clear procedures will make businesses more likely prevail in court.

(4) What happens when the 'reasonable benefits' to be considered to be unreasonable

If the 'reasonable benefits' granted by the employer is considered to be unreasonable, the employee-inventor may institute a lawsuit before the Tokyo or Osaka District Court based on **Paragraph (7)** (slightly amended from the corresponding old Paragraph (5) based on the revised new **Paragraph (5)**). In this case, the Court will decide the reasonable benefits in accordance with (i) the amount of the profits that the employer gained or will gain from the invention, (ii) the burden, the contribution and the treatment upon the employee assumed and made or to be assumed and made by the employer, and (iii) any other circumstances relating to the invention.

The revised **paragraph (7)** of Article 35 of the patent law stipulates as follows:

Where there is no stipulation with respect to *reasonable benefits* or where *the grant of the reasonable benefits* as provided for in the stipulation shall be considered to be unreasonable according to the provision of ***Paragraph (5)***, *the contents of the reasonable benefits to be received under* ***Paragraph (4)*** shall be determined, taking into consideration (i) the amount of profits that the employer made or will make from the invention, (ii) the burden, the contribution and the treatment upon the employee assumed and made or to be assumed and made by the employer in connection with the invention, and (iii) any other circumstances relating to the invention.

5 EMPLOYEES' UTILITY MODELS

Under the Utility Model Act in Japan, the attribution of rights over employees' utility models (devices) is provided by applying *mutatis mutandis* Article 35 of the Patent Act (Article 11(3) of the Utility Model Act), including the new amendments. Therefore, relating to employees' utility model rights, the situation is analogous. Please see section 4 above.

6 EMPLOYEES' INDUSTRIAL DESIGN RIGHTS

Under the Industrial Design Act in Japan, the attribution of rights over employees' industrial designs is provided by applying *mutatis mutandis* Article 35 of the Patent Act (Article 15(3) of the Industrial Design Act), including the new amendments. Therefore, relating to employees' design rights, the situation is analogous. Please see section 4 above.

7 EMPLOYEES' NEW PLANT VARIETIES

7.1 The Legal Framework

Article 3(1) of the Plant Variety Protection and Seed Act ('PVP Act') provides that any person who has bred or developed a protectable plant variety may obtain a registration of the new plant variety ('plant variety registration'). The term 'person' is limited to mean a 'natural person' as in the case of Article 29(1) of the Patent Act, and where a new plant variety is bred or developed by an employee in the course of their duties/tasks under the employment relationship with the employer, the 'right to obtain a plant variety registration' over the 'employee's plant variety' shall originally belong not to the employer, but to the employee that is a natural person, as in patent law.

7.2 Ownership of Rights

7.2.1 Article 8 of the PVP Act

Article 8 of the PVP Act concerns 'plant varieties bred by employees' and is quite similar to Article 35 of the Patent Act. It stipulates as follows:

(1) 'With respect to a new plant variety bred by an employee, an officer of a legal/juridical entity, or a national or local public official (hereinafter referred to as an 'employee' collectively), any provision of the contract, employment regulation or any other stipulation providing in advance that the employer, the legal/juridical entity, or national or local public entity (hereinafter referred to as an 'employer' collectively) shall file an application for the plant variety registration, or have a title of the applicant transferred to the employer if the employee filed an application for the plant variety registration, or have, if the employee obtained a plant variety registration, the breeder's right thereof transferred or an exclusive licence therefor granted to the employer, shall be null and void. However, this shall not apply where the breeding of the plant variety, by its nature, falls within the scope of the business operations of the employer and, moreover, the acts which have led to the breeding fall

within the duties/tasks of the employee under employment relationship (hereinafter referred to as an 'employee-bred plant variety').[2]

(2) An employee may make a claim on the employer for the payment of remuneration to be determined in consideration of the profits to be received by the employer from the 'employee-bred plant variety' and the extent to which the employer has contributed to the breeding of the 'employee-bred plant variety', where the employer filed an application for the 'employee-bred plant variety' registration in accordance with provision(s) of the contract, employment regulation or any other stipulation, or had the title of the applicant transferred to the employer if the employee filed an application for the 'employee-bred plant variety' registration, or had, after the employee obtained an 'employee-bred plant variety' registration, the breeder's right transferred or an exclusive licence therefor granted to the employer.[3]

(3) Where an employee or his or her successor in title has obtained the 'employee-bred plant variety' registration, the employer or its general-successor shall be granted a non-exclusive licence on the breeder's right of the 'employee-bred plant variety'.'[4]

7.2.2 Breeder's Rights Originally Belong to Employees

The first condition for transferring rights to plant varieties to the employer is that the new plant variety is an 'employee-bred' plant variety. This is defined as:

(i) falling within, by its nature, the scope of the business operations of the employer; and
(ii) results from an act or acts in the course of the employee's duties/tasks under the employment relationship with the employer.

As a second condition, Article 8(2) of the PVP Act also provides that, where a new plant variety is to be bred or developed by an employee, the 'right to obtain a plant variety registration' of the 'employee-bred plant variety' shall be transferred to the employer, prior to the filing of an application for registration thereof in accordance with provision(s) of a contract, employment regulation or any other stipulation. The employer may arrange to set up such a regulation or stipulation one-sidedly.

Thirdly, when an employee transfers the 'right to obtain a plant variety registration' or the breeder's right to the 'employee-bred plant variety' to the employer, the employee may claim the payment of remuneration to be decided by reference to the profits that the employer gains or will gain from the plant variety, as well as the extent of contribution that the employer had made to the breeding of the new plant variety.

2. This paragraph corresponds to Art. 35(2) of the Patent Act.
3. This paragraph corresponds to Art. 35(3) of the Patent Act.
4. This paragraph corresponds to Art. 35(1) of the Patent Act.

Accordingly, the third condition for transferring an employee-bred plant variety to the employer, as in the case of employee inventions, is the payment of remuneration to the employee.

There is no restriction or limitation arising from the nature of moral rights in the case where the employee transfers the right to obtain a plant variety registration relating to the employee-bred plant variety or the breeder's right thereof to the employer.

7.3 The Parties' Duty to Inform

Concerning an express positive duty to give notice to an employer of the creation of any intellectual property by the employee (breeder), no statute regulating intellectual property rights in Japan expressly contain such a duty. However, an employee is by implication still required to give notice to the employer by virtue of a general duty of loyalty towards the employer. Such duty may be expressly prescribed in the internal employment regulations or similar document created by the employer.

7.4 Employee Remuneration Right

As mentioned above, Article 8(2) of the PVP Act provides that the employee may claim the payment of remuneration when they have transferred the right or granted an exclusive licence thereof to the employer with respect to the new plant variety bred in the course of their duties/tasks under the employment relationship with the employer, as in the case of employees inventions.

Up to now, there has been no reported court decision concerning the payment of remuneration under Article 8(2) of the PVP Act.

8 INSTANCES FOR DISPUTES

As the diagram below shows, in Japan, there exists one Supreme Court, eight High Courts, and 50 District Courts.

The Intellectual Property High Court was established in April 2005 as a special branch of the Tokyo High Court and handles cases relating only to intellectual property and has exclusive jurisdiction over appeals filed against judgements rendered by the Tokyo District Court and the Osaka District Court in the first instance, both of which exclusively handle civil cases relating to patent rights, and have territorial jurisdiction over the East and West part of Japan. It means that 48 District Courts other than the Tokyo and Osaka District Courts have no jurisdiction over civil cases relating to patent rights.

The Intellectual Property High Court has also exclusive jurisdiction over actions challenging trial decisions made by the Japan Patent Office ('JPO'). For this reason it seems that the Trial and Appeal Board of the JPO is equivalent to the general District Court in terms of power.

The Family Court has, in the first place, jurisdiction over all disputes concerning family matters. The Summary Court has first instance jurisdiction over civil cases involving claims for an amount not exceeding JPY 1,400,000 (which corresponds to about USD 14,000).

In Japan, there exists no Labour Court or Tax Court at present. Article 76 of the Constitution of Japan states that 'No extraordinary tribunal shall be established, nor shall any organ or agency of the Executive be given final judicial power.'

Figure 26.1

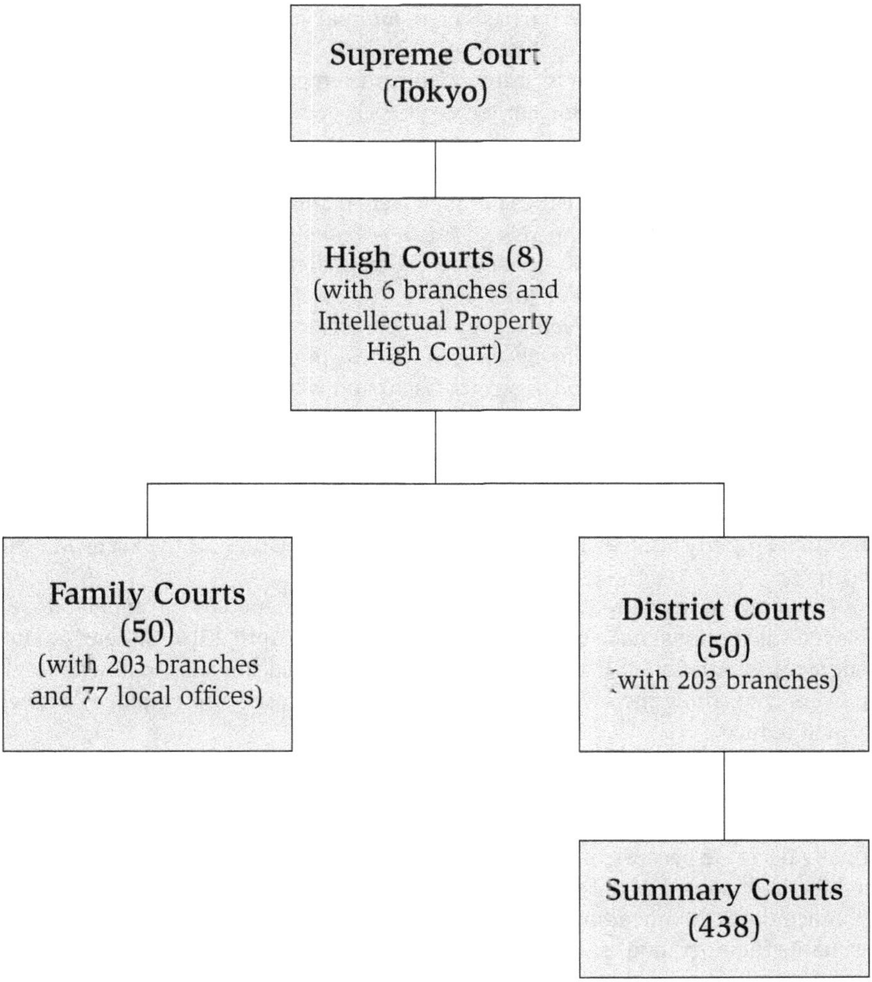

9 CONCLUDING REMARKS

Prior to July 2015, the biggest feature of the 2004 amendment of Article 35 of the Patent Act, the centrepiece of which was the introduction of a reasonable or fair process for deciding the amount of 'adequate remuneration' for employee inventions, is the enhancement of transparency. Although the concept of 'adequate remuneration' will remain, the amount of remuneration will be decided on the basis of the employer's internal regulations and not by the courts, provided that employees play a part in deciding the regulations so that their opinions are sufficiently reflected. This is what is called a 'voluntary agreement' in the 2004 amendment. Furthermore, disclosure of the voluntary agreement with employee inventors is necessary.

The amount calculated on the basis of an internal voluntary agreement was to be deemed 'adequate remuneration'. However, Article 35(5) of the Patent Act maintained that if the voluntary agreement is unreasonable, the employee inventor may, as in the past, file a lawsuit against their employer to seek compensation to reflect that the remuneration be 'adequate'.

While since 2004 there had been no reported court case concerning Article 35(5), some large businesses and industries in Japan were maintaining that it is quite difficult to determine how to lower the risk of such lawsuits, as, after all, the 'adequate remuneration' standard had been subject to litigation, which had created some uncertainty for employers. For this reason, the Keidanren (Japanese Business Federation)/JIPA (Japan Intellectual Property Association), representing major Japanese industries, had been lobbying for a reform of the employee invention system taking the form of a serious and vigorous discussion with respect to the form such a reform could take.

Bowing to these strong pressures from the business/industries lobby, the Government of Japan proposed a policy change relating to the employee invention system in the 'Intellectual Property Strategic Program 2013' released on 25 June 2013 by the Intellectual Property Strategy Headquarters of the Cabinet Office for the Prime Minister of Japan.

The **new paragraph (3)** of Article 35 of the patent law is applied to the employee's inventions made or completed on and after 1 April 2016. Under this latest amendment of Article 35, it is expected that the potential litigation risk against the employers concerning the ownership of the rights with respect to employee's inventions will reduce.

Also, it is expected that the potential litigation risk regarding 'reasonable benefits' will be diminished. However, in this point, there will be a real possibility or misgiving that the value-level of 'financial benefits' granted to the employee-inventors will gradually decrease because of short-sighted profit supremacy policy by the employers generally, in disregard of the spirit of the official guideline mentioned above.

Since the 2004 amendment of Article 35 established a standard for determining 'reasonable remuneration' as a reward for an employee invention, there has been no substantial instance of litigation or lawsuit seeking 'reasonable remuneration' for damages relating to an employee invention neither under the 2004-revised patent law, nor the 2015 latest amendment, as of May 2017.

While Article 35 of the patent law does not explicitly address a duty to notify an employer of an employee invention, whether an employee has such a duty depends on the provisions of the employment regulation or other contract. On a practical level, it is often the case that regulations on employee inventions or so require its employees to notify its employer of any inventions made by the employees.

Following this latest amendment of Article 35 of the Patent Act, both of the Utility Model Act and the Design Act which are also managed and handled by the Japan Patent Office (JPO) were revised simultaneously in the same manner. However, the Plant Variety Protection and Seed Act is not revised which follows the patent-approach concept and scheme. This Act is instead handled by the Ministry of Agriculture, Forestry and Fisheries (MAFF) and Article 8 thereof is the almost same provision as the original Article 35 of the Patent Act of 1960.

BIBLIOGRAPHY

Table of Sources

Nobuhiro Nakayama, *Patent Law* (2nd rev. and enl. ed., 2000), http://www.iip.or.jp/translation/nakayama/index.html (English translation, File 3, Subsection 1, Item 3 Employees' Inventions, 28 pages).

AIPPI-JAPAN, *AIPPI Report Q183, Employers' rights to intellectual property* (2003), https://www.aippi.org/download/commitees/183/GR183japan.pdf (English, 20 pages).

Patent System Subcommittee, Intellectual Property Policy Committee, Industrial Structure Council of the Ministry of Economy, Trade and Industry (METI), JAPAN, *A Final Report on Improvement of Employee-Invention System* (December 2003), http://www.jpo.go.jp/shiryou_e/toushin_e/shingikai_e/pdf/employee-invention.pdf (English translation, 38 pages).

The Japan Patent Office (JPO), *The Case Studies of the Procedures under the New Employee Invention System* (November 2004), https://www.jpo.go.jp/shiryou_e/s_sonota_e/pdf/case_studies.pdf (English translation, 61 pages).

Article by j-cast.com on November 1, 2004, *Japan Revises its Patent System related to Employee Inventions Voluntary Agreements between Employers and Employees to Be Honored from April 2005*, http://en.j-cast.com/2004/11/01000033.html (English translation).

Takeshi Ueno, President, Japan Intellectual Property Association (JIPA), *An Employee Invention System to be Reformed for Promoting Growth-Accelerating Innovation* (26 April 2013). http://www.jipa.or.jp/english/opinion/pdf/employee_invention130610.pdf (English translation, 13 pages).

Keidanren proposal, *Proposal on Revision of the Employee Inventions System – To Strengthen Industrial Competitiveness of Japanese Companies* (13 May 2013), http://www.keidanren.or.jp/en/policy/2013/046.html (English translation).

Asahi Shimbun, *Giving Patent Rights to Companies would Demoralise Creative*

Employees (19 June 2013) http://digital.asahi.com/articles/TKY201306260384.html?ref=comkiji_txt_end (English translation).

Osamu Uchiyama & Norifumi Tanaka, Article of The Asahi Shimbun, *Government Seeks to Review Patent Ownership to Favor Businesses* (7 June 2013) http://ajw.asahi.com/article/business/AJ201306070088 (English translation).

The Intellectual Property Strategy Headquarters of the Cabinet Office for the Prime Minister of Japan, *Intellectual Property Strategic Program 2013* (pages 13 to 14) released on 25 June 2013, http://www.kantei.go.jp/jp/singi/titeki2/ (English translation).

Table of Legislation

Provisional Regulations for Monopoly of 1871.
Patent Monopoly Act of 1885.
Patent Ordinance of 1888.
Patent Act of 1899.
Patent Act of 1909.
Patent Act of 1921 (Act No. 23 of 1921).
Patent Act (Act No. 121 of 1959) Article 35.
Utility Model Act (Act No. 123 of 1959) Article 11.
Industrial Design Act (Act No. 125 of 1959) Article 15.
Copyright Act (Act No. 48 of 1970) Article 15.
Act on the Circuit Layout of a Semiconductor Integrated Circuit (Act No. 43 of 1985) Article 5.
Revision of Article 15 of the Copyright Act (Act No. 62 of 1985).
Plant Variety Protection and Seed Act (Act No. 83 of 1993) Article 8.
Revision of Article 35 of the Patent Act (Act No. 79 of 2004).
Revision of Article 35 of the Patent Act (Act No. 55 of 2015).

Table of Cases

Olympus Optical Pick-up Equipment Case, Supreme Court Decision of 22 April 2003.
Ajinomoto Aspartame Case, Tokyo District Court Decision of 24 February 2004.
Nichia Blue LED Case, Tokyo District Court Decision of 30 January 2004.
Nichia Blue LED Case, Tokyo High Court Recommendation for Settlement of 11 January 2005.
Hitachi Magnetic Optical Disk Case, Supreme Court Decision of 17 October 2006.
Canon Laser Printer Case, Supreme Court Decision of 19 October 2010.
Brother Label Writer Case, Supreme Court Decision of 24 October 2011.

CHAPTER 27

Republic of Korea

Gyooho Lee

1 BACKGROUND

As can be seen from Table 27.1, the total number of applications for patents, utility models, trademarks, and designs was 475,802 in 2015 and increased 7% from the previous year, when the total number of applications for patents, utility models, trademarks, and designs was 444,552.[1]

Table 27.1 The Number of Application of Intellectual Property Rights Before Korean Intellectual Property Office

	Patent	Utility Model	Trademark	Design	Total
2011	178,924 (5.2)	11,854 (–13.2)	134,234 (3.9)	56,524 (–1.2)	381,536 (3.1)
2012	188,305 (5.2)	12,422 (4.8)	142,176 (5.9)	63,135 (11.7)	406,650 (6.6)
2013	204,589 (8.3)	10,968 (–11.7)	159,217 (12.0)	66,940 (6.0)	441,714 (8.6)
2014	210,292 (2.8)	9.184 (–16.3)	160,663 (0.9)	64,413 (–3.8)	444,552 (0.6)
2015	213,694 (1.6)	8,711 (–5.2)	185,443 (15.4)	67,954 (5.5)	475,802 (7.0)

Source: Korean Intellectual Property Office, 2015 White Paper on Intellectual Property, p. 502 (2016).
Note: The brackets show the change in application rate in comparison with the previous year.

The record of intellectual property applications in Korea ranks No. 4 in the world. Most employees' inventions have been made by companies, research institutions, and universities rather than by individuals.[2] For example, between 2005 and 2015, patent

1. Korean Intellectual Property Office, 2015 White Paper on Intellectual Properties, p. 502 (2016).
2. Yong Ho Phi, *A Study on Assessment of Fair Compensation for Employees' Invention in Accordance with Industrial Classification* (Report submitted to Korean Intellectual Property Office), 9 (December 2013); WIPO World IP Indicators (December 2015).

applications made by legal persons held approximately 80% of the annual patent applications as shown in Table 27.2. Also, as far as a copyrighted works for hire are concerned, substantial portions of copyrighted materials have been created by governmental employees, municipalities, or legal persons.[3] Hence, in terms of employees' intellectual creations, to strike a balance between employers' and employees' interests under the Korean legal regime it is crucial to promote employees' intellectual creation to develop a creative economy. In this regard, it should be noted that patent applications made by universities held about 10% of the annual patent applications made by legal persons as shown in Table 27.3 and the number of the former had increased gradually from 2011 to 2015.

Table 27.2 Total Number of Patent Applications Made by Legal Entities from 2005 to 2012

	Patent applications made by individuals (A)	Patent applications made by legal persons (B)	Total (C)	Ratio of B to C
2005	24,368	136,553	160,921	84.9%
2006	27,062	139,127	166,189	83.7%
2007	32,189	140,280	172,469	81.3%
2008	33,215	134,689	167,904	80.2%
2009	35,588	127,935	163,523	78.2%
2010	33,267	136,834	170,101	80.4%
2011	35,424	143,500	178,924	80.2%
2012	36,956	151,349	188,305	80.4%
2013	38,433	166,156	204,589	81.2%
2014	39,200	171,092	210,292	81.4%
2015	42,148	171,546	213,694	80.3%

Source: Korean Intellectual Property Office, 2011 Yearbook of Statistical Data on Intellectual Properties; Korean Intellectual Property Office, 2015 White Paper on Intellectual Properties, p. 504 (2016); Korean Intellectual Property Office, 2014 White Paper on Intellectual Properties, p. 576 (2015).

Table 27.3 Patent Applications Made by Universities from 2011 to 2015

Year	Patent applications made by universities (A)	Patent applications made by legal persons (B)	Ratio of A to B
2011	13,561	143,500	9.45%
2012	14,695	151,349	9.7%
2013	16.039	166,156	9.65%

3. Yong-Sik Song, Sang-Jeong Lee, and Byun-Il Kim, *Intellectual Property Law* 262 (2013).

Year	Patent applications made by universities (A)	Patent applications made by legal persons (B)	Ratio of A to B
2014	18.010	171,092	10.53 %
2015	19,913	171,546	11.61 %

Source: Korean Intellectual Property Office, 2014 White Paper on Intellectual Properties, p. 586 (2015); Korean Intellectual Property Office, 2015 White Paper on Intellectual Properties, p. 504 (2016).

2 EMPLOYERS' COPYRIGHT

2.1 The Legal Framework

Disputes between employers and employees concerning the ownership of copyright are governed by the Korean Copyright Act ('KCA'). Prior to the amendment of the KCA in 2009,[4] the Act prescribed that 'When a work is made public in the name of a legal person, etc. by its employee, its author shall be the relevant legal person, etc., unless otherwise stipulated in the contract or work regulation, etc.' Also, prior to the amendment of the KCA in 2009, the former Computer Program Protection Act,[5] which only dealt with copyrights in computer programs, stipulated that 'When a computer program for hire is created under the scheme of the State, a legal person, or other employer (hereinafter "legal person, etc.") in his/her name, its author shall be the legal person, etc., unless otherwise stipulated in the contract or work regulation, etc.' Essentially, as far as a work for hire is concerned, the former Korean Computer Program Protection Act had a provision identical to the one under the former KCA, except for the scope of their application and applicability to undisclosed works.

A work for hire is primarily governed by the KCA and secondarily by the Korean Civil Code.

2.2 Ownership of Rights

When it comes to the current KCA,[6] which has been revised several times since the revision of 2009, the main sentence of Article 9 of the KCA prescribes that 'When a work is made (or expected to be made) public in the name of a legal person, etc. by its employee, its author shall be the relevant legal person, etc., unless otherwise stipulated in the contract or work regulation, etc.'

4. Act No. 9625 on 22 April 2009, effective since 23 July 2009. The Korean Copyright Act was enacted as Act No. 432 on 28 January 1957 and effective since that date. Since then, it has been revised 26 times.
5. It was enacted on 31 December 1986 by Act No. 3920, effective since 1 July 1987. Then, following several revisions, it was repealed on April 22, 2009 by Act No. 9625 and effective since 23 July 2009 because it was merged into the Korean Copyright Act of 2009
6. Act No. 14634, amended on 21 March 2017 and effective since the same date.

The proviso of Article 9 of the KCA of 2009,[7] which states that 'Provided, that in cases of a computer program work (hereinafter referred to as "program"), its public disclosure is not required,' was added when the provisions of the former Computer Program Protection Act were integrated into the KCA in 2009.

Under Article 9 of the KCA, authorship of a work expected to be made public as well as one made public in the name of a legal person shall be vested in that legal person, unless otherwise stipulated in a contract, work regulation, or the like. Therefore, an employee is considered as the author if the work is published under their name. In addition, the provision eliminates the debate on the authorship and copyright ownership of a work expected to be published in the name of an employer.

The work for hire provision will be applied to undisclosed computer programs which are not made public or is not expected to be made public by virtue of the proviso of Article 9 of the KCA.

In sum, within limits prescribed by the KCA, the ownership of copyrights may be established or modified by contractual agreement between an employer and an employee, employment regulations set by the employer, and the like. The default rule under the KCA is that an employer owns a copyright on an employee's works.

Article 9 of the KCA establishes the rule as to when an employer is deemed the author, even in the absence of an assignment of copyright or work regulations or the like. The copyright will be vested in the employer at the outset if the subject matter of the copyright is created by an employee within the scope of their employment. The default rule under the KCA is that the employer, whether they are an individual, a legal entity, a municipality, or the state, will have economic rights and moral rights in the copyrighted works made by the employee when they are made within the scope of the employment. However, the subject matter will be owned by the employee if it is created in their name even if still within the scope of their employment.

A copyright in a work for hire shall commence from the time of its creation, and shall not require fulfilment of any procedures or formalities.[8] Registration of a work before the Korea Copyright Commission is worthwhile for employers because it shifts the burden of proof to the person who challenges the authorship of the work. The person whose real name is registered as the author is presumed to be the author of the registered work, and the work the date of creation of which or the date on which it has been made public for the first time is registered is presumed to have been created or made public for the first time on the date it was registered. However, where the date, month and year of creation has been registered after one year passed from the time when a work had been created, it shall not be presumed to have been created on the date, month and year registered.[9] The following matters may be registered, and shall not bind third parties without their registration:

7. Act No. 9625, amended on 22 April 2009 and effective since 23 July 2009.
8. Art. 10(2) of the KCA.
9. Art. 53(3) of the KCA.

(i) transfer by assignment of author's economic right (excluding that by inheritance or other successions in general), or restriction on the disposal of author's economic right;
(ii) restriction on establishment, transfer, alteration, extinction or disposal of the right of exclusive publication (Article 57 of the KCA) or the publishing right (Article 63 of the KCA); or
(iii) restriction on establishment, transfer, alteration, extinction or disposal of the right of pledge, the object of which is author's property right, the right of exclusive publication (Article 57 of the KCA) and the publishing right (Article 63 of the KCA).[10]

As far as a licence of a work for hire is concerned, Articles 46, 57-63*bis*, of the KCA will be applied. Article 46(1) of the KCA prescribes that 'The holder of author's economic right may grant another person authorisation to exploit the work.' Article 46(2) of the KCA states that 'The person who obtained such authorisation pursuant to paragraph 1 shall be entitled to exploit the work in such a manner and within the limit of such conditions so authorised.' In addition, Article 46(3) of the KCA stipulates that 'The right of exploitation as authorised under paragraph 1 may not be transferred by assignment to the third party without the consent of the holder of author's economic right.' Based on Article 46 of the KCA, the licensee has a contractual right to exploit the copyrighted work of the licensor.[11] In addition to the said provision, there are special provisions on the right of exclusive publication (Articles 57-62 of the KCA) and publishing right (Articles 63-63*bis* of the KCA). A contract to establish a right of exclusive publication is a special type of exclusive licensing recognised in America. The right of exclusive publication was adopted to implement the Free Trade Agreement between Korea and US ('KORUS FTA'), which was finalised on 2 April 2007 and effectively entered into force on 15 March 2012. Article 18(10)(4) of the KORUS FTA states that 'Each Party shall make available to right holders[27] civil judicial procedures concerning the enforcement of any intellectual property right.' Pursuant to the footnote 27 of Article 18(10)(4) of the KORUS FTA, it makes clear that it includes 'a federation or an association having legal standing and authority to assert such rights, and also includes *a person that exclusively has any one or more of the intellectual property rights* encompassed in a given intellectual property [emphasis added]'. However, the scope within which exclusive publication is allowed in Korea is narrower than that of exclusive licensing in America.

According to Article 57 of the KCA, the owner of an economic right can establish a right of exclusive publication to publish[12] or reproduce and forward[13] their copy-

10. Art. 54 of the KCA.
11. Seung-Jong Oh, *Copyright Law*, 604 (4th ed., Pakyoung Publishing Co. 2016); Hae Wan Lee, *Copyright Law*, 521-522 (3rd ed., Pakyoung Publishing Co. 2015).
12. See Art. 2(24) of the KCA, stating that 'The term "publication' means a reproduction and distribution of the works or music records for the demand of the public.'
13. See Art. 2(10) of the KCA, prescribing that 'The term "interactive transmission" means, among types of public transmission, to make works, etc. available for the public so that the members of the public may have access at the time and place of their own choice, including transmission to be done accordingly.' See cf. Art. 2(7), stating that 'The term "public transmission" means

righted work. In other words, the right of exclusive publication covers only either the right of reproduction and distribution or the right of reproduction and forwarding, so that it is prohibited to encompass the right of public performance, display, or rental or to make derivative work. As far as a publishing right is concerned, a person who holds the right to reproduce and distribute a copyrighted work may establish the right to publish such work for a person who intends to do so in documents or pictures by printing them or by other method similar thereto.[14] The holder of a publishing right may hold the right to publish the original copy of the work that is the object of the publishing right as prescribed by a publishing contract.[15] Since the right of exclusive publication was introduced into the KCA, provisions on the publishing right seem to be redundant. However, taking into account domestic publishers' strong voice to object to the deletion of provisions on the publishing right, the provisions on the publishing right survived. Suppose that a copyright infringer pirates two pages of a 1000-page book. The holder of the publishing right on the 1000-page book will not win the case brought by their against its infringer because the holder of the publishing right does not hold the right to publish some portions of the entire original copy of the work, but has the right to publish the original copy of the work. Nonetheless, the holder of exclusive publication can successfully bring an action against the copyright infringer.

In this regard, it should be discussed whether the general principles of contract law under the Korean Civil Code are applied to a copyright licensing contract. Korean case law tends to rely on the general principle of contract law under the Korean Civil Code to determine whether copyright-related contract is a copyright assignment contract or a copyright licensing contract.[16] According to the Judgment rendered by the Korean Supreme Court on 30 July 1996 (Case No. 95 Da 29130) (*Jeong Sung Shin et al. v Jiku Corp.*): 'The interpretation of a legal act generally makes clear the objective meaning which parties bestow on its representative act. When its objective meaning is not clear by its plain meaning only, the court must interpret it reasonably in accordance with logics, doctrine of experience, common sense of society in general and general perception of its business, and in line with social justice and ideology of equity, by comprehensively taking into account the contractual terms, the motive for and the circumstances of the legal act, the purpose for and genuine intent with which the parties want to accomplish by the legal act, and trade practices, etc.' and thus went on to state that 'When it is unclear whether a copyright-related contract is an assignment contract or a licence contract, the court should presumably interpret it in favour of its author by reserving the copyright to the author if the contractual term does not apparently state whether it is based on assignment or license. If the contractual term is ambiguous, the court must interpret it by comprehensively taking into account trade practices or the knowledge of both parties, parties' behaviours, etc.'

---

transmitting works, stage performances, music records, broadcasting or database (hereinafter referred to as "works, etc.") by means of radio communication or wire communication so that the public may receive them or have access to them.'

14. Art. 63(1) of the KCA.
15. Art. 63(2) of the KCA.
16. Judgment Rendered by the Korean Supreme Court on 30 July 1996, Case No. 95 Da 29130; Seung-Jong Oh, *supra* note 11, at 630.

The employer's economic rights to a work for hire shall continue to exist for a period of 70 years after it has been made public.[17] If it has not been made public within 50 years after its creation, the employer's economic right shall continue to exist for a period of 70 years after its creation.[18]

The employer's moral rights to a work for hire is not transferable to another by assignment or inheritance. The employer's moral rights will be terminated when they die in cases where they are a natural person, or when it is dissolved in cases where it is a legal person.[19]

For copyrights, the original jurisdiction of disputes that arise under Article 9 of the Korean Copyright Act is vested in District Courts in Korea.

2.2.1 Ownership of Rights at Universities or Research Institutions

Unless otherwise agreed between an employer (a university or a research institution) and an employee (professor, researcher, etc.), the general rule is that a copyright in a work for hire, which is, or is expected to be, made public in the name of an employer under scheme of the employer, will be vested in the employer. Work regulations on employees' inventions in universities usually include provisions on ownership of works for hire. Nonetheless, university professors do not author their books or articles under the scheme of universities, so that Article 9 of the KCA may not be applied to their works. In some cases, where universities subsidise their professors to publish their works in the name of the universities, the ownership on the works can be transferred to the universities or legal persons organised by the universities for cooperation between businesses and universities by an assignment contract or the regulations of the university. However, even in these cases, the moral rights in the copyrighted works made by professors are vested in them.

2.2.2 Case Law Related to Employee's Copyright

The provision concerning a work for hire must be construed narrowly because it is an exception to the basic rule that a creator of a work is an author. The proviso of Article 9 of the KCA must be interpreted more stringently than the main sentence of Article 9 of the KCA because the former prescribes that authorship and ownership of a computer program made for hire is vested in an employer even though it was not made public in the name of the employer. So, in interpreting the phrase, 'under the scheme' of an employer, in a case related to a computer program, the Korean Supreme Court held that the employer could give instructions to employees expressly or impliedly to plan the development of a new computer program. However, the Court went on to hold that there must be circumstances from which the employer's implied intent to plan the

17. Main sentence of Art. 41 of the Korean Copyright Act.
18. Proviso of Art. 41 of the Korean Copyright Act.
19. Gyooho Lee, *Copyright Law: Cases and Explanations* 436 (2017).

development of the computer program can be inferred so that it can be deemed to be identical to the employer's express intent to do so.[20]

The Korean Supreme Court held that Article 5 of the former Computer Protection Act[21] would not be applied to a contract for work which is purported to pay the price for the development of a computer program unless an employer exclusively lays a plan to develop a new computer program, invest their capital in its development, the employer commissions its development to a developer of the computer program by using the latter's employees and its developer develops the computer program and supplies the employer with it only for the employer.[22]

The court decisions rendered by the Korean Supreme Court focused on the definition of the term 'under the scheme of an employer', in terms of a work for hire. In addition, as far as copyright on works for hire at universities or research institutions is concerned, the important factor will be the name which appears on front or rear cover sheet of works because Article 9 of the KCA cannot be applied to the cases where an employer's name is not stated on a cover sheet of works for hire.

2.3 The Parties' Duty to Inform

These are no statutory provisions which require employees to inform employers of the creation of works for hire. In some cases, it may be stated in an agreement between employer and employee.

2.4 Employee Remuneration Right

As far as employees' works for hire are concerned, employees usually do not have a remuneration right.

3 EMPLOYERS' SEMICONDUCTOR CHIPS

3.1 The Legal Framework

The Act on the Layout-Designs of Semiconductor Integrated Circuits ('Layout-Designs Act') was enacted on 8 December 1992,[23] which aims to contribute to the sound development of the national economy through the promotion of industries and technologies relating to semiconductors by protecting the rights of persons who create a layout-design for a semiconductor integrated circuit and by encouraging the fair use of such a layout-design.[24] The Layout-Designs Act has been amended several times

20. Judgment rendered by the Korean Supreme Court on 14 January 2010, Case No. 2007 Da 61168.
21. It corresponds to proviso of Art. 9 of the KCA.
22. Judgment rendered by the Korean Supreme Court on 9 May 2013, Case No. 2011 Da 69725.
23. Act No. 4526, effective since 1 September 1993.
24. Art. 1 of Layout-Designs Act.

since its enactment in 1992.[25] However, Article 5 of the Layout-Designs Act, which regulates ownership of layout-designs created in the course of employment, maintains the original version of the 1992 enactment even though some minor changes were made to the wording of Article 5.[26]

For layout-designs of semiconductor chips, the original jurisdiction of disputes that arise in connection with ownership, transfer, infringement of layout-design right is vested in District Courts in Korea. However, there are no Supreme Court decisions related to a layout-design of semiconductor integrated circuits. Also, to date, The Lay-Design Review and Mediation Committee, which was established in 1995 by the Korean Intellectual Property Office pursuant to the Layout-Designs Act, has had no chance to hear any dispute related to layout-design rights.[27]

3.2 Ownership of Rights

The default rule is that an employer is the creator of a layout-design of a semiconductor integrated circuit when it is created by an employee in the course of their employment. Unlike copyrights, the Layout-Designs Act requires registration for layout-design rights to be valid. Article 5 of the Layout-Designs Act of 2014, entitled 'Creator of Layout-Design in Service,' prescribes that 'With respect to a layout-design created in the course of service by a person employed by the State, a corporation, an organisation, or any other employer (hereinafter referred to as "corporation or a similar employer"), the corporation or similar employer shall be deemed the creator of the layout-design, unless otherwise provided for by an agreement, employment regulations, or the like.'

Article 2(4) of the Layout-Designs Act, protects a layout-design,[28] a semiconductor integrated circuit manufactured according to a layout-design, or a product manufactured by applying a semiconductor integrated circuit. 'Layout-design right' means rights created when the creation of a layout-design is registered with the Commissioner

25. It was amended on 5 January 1995 (Act No. 4890, effective since 1 July 1995), 13 December 1997 (Act No. 5453, effective since 1 January 1998 and revised in accordance with the amendment of the Act on Administrative Proceedings), 13 December 1997 (Act No. 5454, effective since 1 January 1999 and revised in accordance with the amendment of the Act on Government's Organisation), 28 December 1998 (Act No. 5599, effective since 1 January 1999), 26 January 2002 (Act No. 6626, effective since 1 July 2002 and revised in accordance with the amendment of Civil Procedure Act), 27 April 2007 (Act No. 8397, effective since 28 October 2007), 29 February 2008 (Act No. 8852, effective since 29 February 2008 and revised in accordance with the amendment of the Act on Government's Organisation), 26 December 2008 (Act No. 9183, effective since 27 March 2009), 23 March 2013 (Act No. 11690, effective since 23 March 2013 and revised in accordance with the amendment of the Act on Government's Organisation), and 21 January 2014 (Act No. 12289, effective since 21 January 2014) and 3 February 2015 (Act No. 13150, effective since 4 August 2015).
26. Act No. 9183, revised on 26 December 2008 and effective since 27 March 2009.
27. Gyooho Lee, Keon-Hyun Ahn and Jacques de Werra, *Euro-Korean Perspectives on the Use of Arbitration and ADR Mechanisms for Solving Intellectual Property Disputes*, 30(1) Arbitration International 100 (2014).
28. According to Art. 2(2) of the Layout-Designs Act, the term 'layout-design' means a design of laying out various circuit elements and wires connecting such elements in two or three dimensions for manufacturing a semiconductor integrated circuit.

of the Korean Intellectual Property Office.[29] Hence, layout-design rights are created when an original layout-design is registered before the Korean Intellectual Property Office.[30] The duration of a layout-design right is ten years from the date of registration of the layout-design which meets the originality requirement.[31] However, the duration of a layout-design right shall not exceed ten years from the date on which the layout-design is initially used for making a profit or 15 years from the date on which it is created.[32]

A person who has registered a layout-design or a person who has succeeded to the rights from such person shall have the exclusive right to use the layout-design for profit. However, if an exclusive licence is granted for the layout-design, the right of the exclusive licensee takes precedence over the right of the licensor to the extent that the exclusive licensee holds the exclusive licence for the layout-design.[33] The effects of a layout-design right shall not extend to any of the following lay-out designs:

(i) a layout-design reproduced directly or vicariously, for education, research, analysis, evaluation, or any other similar activities or for personal use for any non-profit purpose;
(ii) a creative layout-design produced as a result of research, analysis, evaluation, or any other similar activities; or
(iii) a layout-design produced by any person other than the right-holder of another layout-design, which is creative, albeit identical with the latter layout-design.[34]

The layout-design right has no effect on any person who transfers, leases, displays (limited to cases for the purpose of transfer or lease), or imports a layout-design, a semiconductor integrated circuit manufactured according to a layout-design, or a product manufactured by applying a semiconductor integrated circuit, for profit with any semiconductor integrated circuit or similar product legally manufactured and delivered to them.[35] Also, the layout-design right has no effect on any person who has accepted, in good faith and without negligence on their part, a semiconductor integrated circuit or similar product, manufactured by illegally reproducing a semiconductor integrated circuit or similar product registered by any other person and delivered to them, and who transfers, leases, displays (limited to cases for the purpose of transfer or lease), or imports a layout-design, a semiconductor integrated circuit manufactured according to a layout-design, or a product manufactured by applying a semiconductor integrated circuit, for profit with the semiconductor integrated circuit or similar product so delivered.[36]

29. Art. 2(5) of the Layout-Designs Act.
30. Art. 6 of the Layout-Designs Act.
31. Art. 7(1) of the Layout-Designs Act.
32. Art. 7(2) of the Layout-Designs Act.
33. Art. 8 of the Layout-Designs Act.
34. Art. 9(1) of the Layout-Designs Act.
35. Art. 9(2) of the Layout-Designs Act.
36. Art. 9(3) of the Layout-Designs Act.

Layout-design rights are transferable.[37] Rights to a layout-design created jointly by two or more persons shall be jointly owned by the persons who created it, and each joint holder's share is deemed to be equal, unless there is a specific agreement among the joint creators to the contrary.[38] If a layout-design right is owned by joint owners, neither joint owner may transfer their share nor pledge their share, without the other joint holder's consent.[39] If a layout-design right is owned by joint owners, each joint owner has a right to use the layout-design without the other joint holder's consent, unless there is a specific agreement to the contrary.[40] If a layout-design right is owned by joint owners, neither joint owner may grant any person an exclusive licence, or a non-exclusive licence, of the layout-design without the other joint owners' consent.[41] The holder of a layout-design right may grant another person a right to use the layout-design exclusively.[42] A person who has been granted an exclusive licence has an exclusive right to use the layout-design for making a profit within the scope of the licence granted.[43]

An exclusive licensee may transfer their exclusive licence only when any of the following events occurs:

(i) when the exclusive licence is transferred along with the business to which the layout-design has been applied;
(ii) when the exclusive licence is transferred by inheritance or any other ordinary succession; or
(iii) when the holder of the layout-design rights consents to the transfer.[44]

No exclusive licensee may pledge an exclusive licence without the consent of the holder of the layout-design rights.[45] If an exclusive licence is jointly owned, neither joint owner may grant any person a non-exclusive licence without the other joint holder's consent.[46]

The holder of a layout-design right or an exclusive licensee for a layout-design may grant another person a right to use the layout-design non-exclusively.[47] Provided that an exclusive licensee who intends to grant another person a non-exclusive licence shall obtain consent from the holder of the layout-design right. A person who has been granted a non-exclusive licence has a right to use the layout-design for making a profit within the scope of the licence granted.[48] A non-exclusive licensee may transfer their non-exclusive licence only when any of the following events occurs:

37. Art. 10(1) of the Layout-Designs Act.
38. Art. 10(2) of the Layout-Designs Act.
39. Art. 10(3) of the Layout-Designs Act.
40. Art. 10(4) of the Layout-Designs Act.
41. Art. 10(5) of the Layout-Designs Act.
42. Art. 11(1) of the Layout-Designs Act.
43. Art. 11(2) of the Layout-Designs Act.
44. Art. 11(3) of the Layout-Designs Act.
45. Art. 11(4) of the Layout-Designs Act.
46. Art. 11(5) of the Layout-Designs Act.
47. Art. 12(1) of the Layout-Designs Act.
48. Art. 12(2) of the Layout-Designs Act.

(i) when the licence is transferred along with the business to which the layout-design has been applied;
(ii) when the licence is transferred by inheritance or any other ordinary succession; or
(iii) when the holder of the layout-design right, which refers to the holder of the layout-design right and the exclusive licensee, if the licence is a non-exclusive licence granted by the exclusive licensee, consents to the transfer.[49]

No non-exclusive licensee may pledge their non-exclusive licence without the consent of the holder of the layout-design right.[50]

A layout-design right becomes ineffective if:

(i) a corporation or an organisation that has a layout-design right is dissolved, and the layout-design right escheats to the state pursuant to the Civil Act or any other Act; or
(ii) an individual with a layout-design right dies without any inheritor and the layout-design right escheats to the state pursuant to the Civil Act or any other Act.[51]

No holder of a layout-design right may surrender their layout-design right without the consent from exclusive and non-exclusive licensees, and pledgees.[52] No exclusive licensee may surrender their exclusive licence without consent from the persons to whom the exclusive licensee granted a non-exclusive licence, or from pledgees thereof.[53] No non-exclusive licensee may surrender their non-exclusive licence without consent from pledgees thereof.[54] Once a layout-design right or an exclusive or non-exclusive licence is surrendered, the right thereto shall become ineffective immediately upon such surrender.[55]

3.2.1 Ownership of Rights at Universities or Research Institutions

The basic rule is that universities or research institutions own layout-design rights when the layout-design is created by their employees in the course of employment. Universities generally do not have regulations on ownership of layout-design rights in terms of layout-designs created in the course of employment because they can rely on the default rules of Article 5 of the Layout-Designs Act and because they do not have financial resources sufficient for their employees to create layout-designs of semiconductor chips. Also, lack of public awareness on this provision can contribute to absence of work regulations on layout-design rights in universities.

49. Art. 12(3) of the Layout-Designs Act.
50. Art. 12(4) of the Layout-Designs Act.
51. Art. 17 of the Layout-Designs Act.
52. Art. 18(1) of the Layout-Designs Act.
53. Art. 18(2) of the Layout-Designs Act.
54. Art. 18(3) of the Layout-Designs Act.
55. Art. 18(4) of the Layout-Designs Act.

3.3 The Parties' Duty to Inform

There are no statutory provisions which govern an employee's duty to inform an employer of the creation of a layout-design. The parties are autonomous in agreeing on this matter.

3.4 Employee Remuneration Right

There are no statutory provisions which stipulate employees' remuneration rights under the Layout-Designs Act. The parties are autonomous in agreeing on this matter.

4 EMPLOYERS' INVENTIONS

4.1 The Legal Framework

Employees' inventions and utility models are protected under the Korean Invention Promotion Act ('KIPA')[56] if they are created in connection with their duties within the scope of employment. Employees' inventions used to be regulated by the Korean Patent Act ('KPA')[57] until the amendment of the KPA in 2006.[58] Therefore, prior to the amendment of the KPA in 2006, the provisions of Articles 39-40 of the former KPA in 2005,[59] had covered employees' inventions only. The KPA of 2006 deleted the provisions on employees' inventions. These provisions were moved to the KIPA in 2006.[60] The application of the amendment to the KIPA of 2006 was extended to employees' utility models and designs. Also, Article 3 to 14*quater* of the KIPA of 2006 included employees' duty to inform and clarified the criteria and proceedings for fair compensation.

Now, the term 'invention' means an invention, design, or creation eligible for protection under the KPA, the Utility Model Act, or the Design Protection Act in accordance with Article 2(1) of the KIPA. Article 2(2) of the KIPA goes on to state that 'The term "employee invention" means an invention that an employee, executive of a corporation, or public official (hereinafter referred to as "employee, etc.") makes in connection with his/her duties, where it falls within the scope of business of the employer, the corporation, the State, or the competent local government (hereinafter referred to as "employer, etc.") and the activities that have led to the invention fall within the present or past duties of the employee, etc.'

Transfer or termination of, or remedies for, a patent right or a utility model right are governed by the KPA or the Korean Utility Model Act, respectively, unless otherwise prescribed in the KIPA.

56. Act No. 14687, revised on 21 March 2017 and effective since 22 September 2017.
57. Most recent amendment to the KPA was made as Act No. 14035, which was amended on 29 February 2016 and effective since 1 March 2017.
58. Act No. 7871, amended on 3 March 2006 and effective since that date.
59. Act No. 7554, amended on 31 May 2005 and effective since 1 December 2005.
60. Act No. 7869, amended on 3 March 2006 and effective since 4 September 2006.

4.2 Ownership of Rights

Rules governing the ownership of patent rights between employers and employees can be found in Article 10 of the KIPA.

According to Article 10(1) of the KIPA 'If the employee, etc. acquires a patent or completes the registration of a utility model or design (hereinafter referred to as "patent, etc."), or succeeds to the right to acquire a patent, etc. and thus acquires a patent, etc., with regard to an employee invention, the employer shall have a non-exclusive licence for the patent, utility model right, or design right (hereinafter referred to as "patent right, etc.").'

However, in cases where an employer is a company which does not constitute a small and medium-sized enterprise in accordance with Article 2 of the Framework Act on Small and Medium-Sized Enterprises, the employer cannot have a non-exclusive licence for patent rights, etc. unless any of the following contract or work regulations are drawn up in advance following consultation with the employee, etc.:

(i) A contract or work regulations which succeed right to acquire patent, etc. or patent right, etc. to employer, etc. in terms of employees' invention; or
(ii) A contract or work regulations which grant exclusive licence for employer, etc. in terms of employees' invention.[61]

The revised KIPA of 2013[62] reflected economic democratisation in Korea. Hence, businesses can be granted a non-exclusive licence only if they draw up a contract or work regulations, which succeed right to acquire patents, etc. or patent rights, etc. to them or grant exclusive licence to the employees, in advance after consultation with them. The purpose of the revised KIPA of 2013 is to promote employees' bargaining power to negotiate with employers in terms of rights to inventions. However, it is highly debatable in that it can encourage businesses not to innovate. It is silent as to what will happen to businesses if they fail to consult with employees prior to drawing up a contract or work regulations which succeed patent rights or utility model rights to employees.

There are special rules which are applied to the state or municipalities with respect to employees' inventions and utility models. The state or a local government shall succeed to a public official's right to invention and utility model, and thus a public official's patent rights, etc. for an invention and utility model, to which the state or a local government succeeds, shall be vested in the state or municipalities.[63] As far as public schools are concerned, special-purpose organisations established by them succeed the rights to employees' inventions and utility models.[64]

61. Art. 10(1) of the KIPA.
62. Act. No. 11960, amended on 30 July 2013 and effective since 31 January 2014.
63. The main sentence of Art. 10(2) of the KIPA.
64. The proviso of Art. 10(2) of the KIPA prescribes that 'an exclusively responsible organisation under the latter part of Art. 11(1) of the Technology Transfer and Commercialisation Promotion Act (hereinafter referred to as "exclusively responsible organisation") shall succeed to rights of an employee, who works at a national or public school under Art. 3 of the Higher Education Act (hereinafter referred to as "national or public school"), to an employee invention, and thus

A clause of a contract or employment regulations shall be void in cases where the clause requires an employee to transfer the right to acquire a patent, etc., or a patent right, etc., to the employer, or requires an employee to grant an exclusive licence for a patent, etc. to the employer, in advance in connection with any invention made by the employee which does not fall within the definition of an employee invention under the KIPA.[65] The Commissioner of the Korean Intellectual Property Office is in charge of the disposal and management (including waiver of patent rights, etc.) of patent rights etc. that become state property pursuant to the provision on ownership of employees' inventions and utility models.[66]

As mentioned above, within limits prescribed by the KIPA, the ownership of patent rights may be established or modified by contractual agreement between an employer and an employee, employment regulations set by the employer. The default rule under the KIPA is that an employee owns a patentable right or a patent right in their invention or a utility model right or a right to acquire registration of their utility models unless they are not a public servant of the central government or municipalities. An inventor (employee) can enter into an express agreement to assign their rights in an invention to an employer. In the vast majority of situations, the express assignment is contained in an employment contract between the inventor (employee) and the employer or in work regulations, by which the inventor assigns inventions relating to the business of the employer to the employer in return for compensation payments. In the absence of an express agreement, work regulations or similar documents, which would provide for a transfer of patent rights, patentable rights, utility model rights, or rights to acquire the registration of utility models from an employee to an employer so that the employers can have such rights, the ownership of rights on employees' inventions and utility models will be vested in employees. In cases where an employee's invention and utility model has been made jointly by an employee and a third party and an employer succeeds to rights to the invention or utility model in accordance with a contract or employment regulations, the employer shall acquire the employee's share of the rights to the invention.[67]

The duration of a patent right or a utility model right is provided by the KPA or the Korean Utility Model Act respectively. The duration of patent rights shall commence upon registration of the patent and last for 20 years from the filing date of the patent application.[68] The term of a utility model right in connection with employees' utility models shall commence on the registration date of the establishment of the utility model right and shall expire on the tenth anniversary of the filing date of the application for the utility registration.[69]

patent rights, etc. of an employee of a national or public school to an employee invention, to which the exclusively responsible organisation succeeds, shall be owned by the exclusively responsible organisation'.

65. Art. 10(3) of the KIPA.
66. Art. 10(4) of the KIPA.
67. Art. 14 of the KIPA.
68. Art. 88(1) of the KPA.
69. Art. 22(1) of the Korean Utility Model Act.

For patents, jurisdiction of disputes that arise under Article 10 of the KIPA is exclusively vested in the Seoul Central District Court, Busan District Court, Daegu District Court, Gwangju District Court, or Daejeon District Court, where constitutes a district court in the jurisdictional area of a High Court which has jurisdiction over the location of a competent court pursuant to general or special jurisdiction.[70] Nonetheless, a party may bring a lawsuit concerning patents to Seoul Central District Court even though Daejeon District Court has exclusive jurisdiction for a patent dispute because Seoul Central District Court has a special division which deals with intellectual property rights.[71]

4.2.1 *Case Law*

In a case where an employee, who had engaged in handicrafts, mould making, and jig and fixture building as a skilled technician at a musical instrument company, created a device which effectively decompressed bushing in the hole of the flange, a component of the piano, the Korean Supreme Court held that 'the activities that have led to the invention fall within the present or past duties of the employee' and that the requirement under Article 17(1) of the former KPA (currently Article 2(2) of the KIPA) is satisfied when an employee is prearranged to or expected to engage in the creation of an invention in light of their duties and responsibilities.[72] Thus, the Court acknowledged that the employee's utility model constituted an employee invention under the former KPA.

In terms of the date when an employer has a non-exclusive licence for employees' rights to acquire the registration of utility model, the Korean Supreme Court in the *Thermostats for Heating Equipment* case held that the employer obtains a non-exclusive licence for the rights to a utility model on the date of completion of an employee's utility model, rather than on the date of registration of a utility model.[73] The facts of the case were that between 21 July 1993 and September 1994, the defendant, the CEO of company A, produced and sold thermostats on heating equipment, which had been registered by Dong-Won Lee and Kyu-Yong Hwang as a utility model on 6 November 1990. The defendant argued that the utility model was created by Dong-Won Lee and Kyu-Yong Hwang, who were employees of company B, within the scope

70. Article 24 of the Korean Civil Procedure Act prescribes as follows:
    Article 24 (Special Jurisdiction for Intellectual Property Rights, etc.)
    (1) A lawsuit concerning an intellectual property right, except for patent rights, utility model rights, design rights, trademark rights, and protected plant variety rights(hereinafter 'patent rights, etc.), and an international transaction may be brought to a district court in the jurisdictional area of a high court which has jurisdiction over the location of a competent court pursuant to Articles 2 through 23.
    (2) Exclusive jurisdiction for a lawsuit concerning patent rights, etc. is exercised by a district court in the jurisdictional area of a high court which has jurisdiction over the location of a competent court pursuant to Articles 2 through 23
    (3) In spite of Paragraph 2, a party may bring the lawsuit concerning patent rights, etc. to Seoul Central District Court.
71. Article 24 (3) of the Korean Civil Procedure Act.
72. Judgment rendered by the Korean Supreme Court on 27 December 1991, Case No. 91 Hu 1113.
73. Judgment rendered by the Korean Supreme Court on 27 June 1997, Case No. 97 Do 516.

of their employment, and that company B had acquired a non-exclusive licence for the utility model from the two employees and that subsequently the non-exclusive licence was assigned to company A, and that therefore production and distribution of the thermostats for heating equipment made by company A was legitimate based on the non-exclusive licence.

The Seoul District Court established that: (i) company B applied for registration of a utility model right on the thermostat for heating equipment on 13 July 1987; (ii) company B assigned the right to acquire the utility model right to the two employees of company B approximately in 1989 and the two employees modified the application by adding their names, and this modified application was approved and registered by the Korean Intellectual Property Office on 31 March 1989; (iv) company B became subsequently insolvent, and the assets of company B were transferred to its creditors, while the utility model right was excluded from the list of assets of company B; (v) the assets were transferred from the creditors to company A, and entered into a free, non-exclusive licence contract which allowed company A to produce, use, sell, and distribute the thermostats from 20 July 1991 to 20 July 1993; and (vi) company A registered the non-exclusive licence for the utility model right before the Korean Intellectual Property Office on 23 July 1991. The Korean Supreme Court affirmed the lower court's finding, holding that the non-exclusive licence for the utility model right was not transferred to company A even though the utility model was created by employees within the scope of employment.[74]

The Korean Supreme Court has held[75] that where an employee (inventor) assigns their rights to acquire a patent to another (the joint inventor) without notifying the employer of the completion of an employee invention, such assignment will not constitute a disclosure of the employer's trade secret, but an occupational breach of trust with respect to the employer.

Where an invention which an employer exploits is known to the public at the time of the patent application of the employee invention, so that the employer is free to exploit it and their competitor could have been easily aware of such circumstance, the employer is not obliged to pay remuneration for a licence to the employee.[76]

In *LG Life Sciences Co. Ltd.*,[77] the Korean Supreme Court dealt with the criteria to determine whether a person is considered to be a joint inventor, and with the statute of limitation for a remuneration claim to an employee invention and the point in time when the period begins to run. In *LG Life Sciences Co. Ltd.*,[78] the Court opined that:

> In order to become joint inventors, the (potential) joint inventor must have actually been in relationship of mutual cooperation for completion of the invention. Accordingly, the status of a joint inventor is acknowledged when a person has contributed to completion of the invention by more than merely providing basic tasks or ideas for the invention, providing general administration for researchers,

74. Judgment rendered by the Korean Supreme Court on 27 June 1997, Case No. 97 Do 516.
75. Judgment rendered by the Korean Supreme Court on 15 November 2012, No. 2012 Do 6676.
76. Judgment rendered by the Korean Supreme Court on 8 September 2011, Case No. 2009 Da 91507.
77. Judgment rendered by the Korean Supreme Court on 28 July 2011, Case No. 2009 Da 75178.
78. Judgment rendered by the Korean Supreme Court on 28 July 2011, Case No. 2009 Da 75178.

> organising data and performing experiments upon instruction by a researcher, or supporting or commissioning completion of the invention by providing funds or equipment. Instead, a joint inventor must have contributed to the invention by newly presenting or adding or supplementing concrete conceptions with respect to technical problems within the invention or by providing actual advice and/or instructions that make the invention possible. In other words, a person must have contributed creative activity to the technical idea to an actual extent in order to be recognised as a joint inventor. Meanwhile, in the case of chemical inventions, which are made in a field frequently referred to as "the science of experiments," notwithstanding certain disparities that may be caused by the contents and technical level of an invention in this field, there are many cases where the invention substantially lacks predictability or implementability and thus can not be seen as a complete invention without a trial experiment to present experiment data. In such cases, the status of a joint inventor must be determined by taking into account whether the person in question made actual contributions to specifically defining and completing the invention in the process of the experiment.

Also, in *LG Life Sciences Co. Ltd.*,[79] the Court went on to hold that 'A remuneration claim for an employee invention lapses if not exercised for ten years, just as with general claims. Generally speaking, the statute of limitation runs when the employer succeeds from the employee the right to acquire a patent for the invention. However, in cases where a company has internal regulations defining a time period to pay remuneration for employee inventions, the prescribed time of payment becomes the starting point of the statute of limitation.'

4.2.2 *Ownership of Rights at Universities or Research Institutions*

The standard form of university regulations on employees' inventions provided by the Korean Intellectual Property Office, only covers employees' inventions and utility models.[80] Hence, its application is not extended to employees' designs. It prescribes the type of fair compensation by virtue of Article 15(1) of the KIPA. In cases where the Office of Industry-Academy Cooperation succeed rights on employees' inventions and utility models, Article 15(1) of the standard model states fair compensation for application, registration, licensing, disposal, reservation of application. Also, Article 17 of the standard model prescribes that an inventor or a creator of utility model is compensated even if they retire, and that the heir of an inventor or of a creator of a utility model receives remuneration if the inventor or the creator of a utility model dies. In practice, most universities have work regulations on employees' inventions and utility models. If a professor of a national university makes an invention or a utility model, Article 10(2) of the KIPA is applied to them.

79. Judgment rendered by the Korean Supreme Court on 28 July 2011, Case No. 2009 Da 75178.
80. Korean Intellectual Property Office, Standard Model for University Regulations on Employees' Inventions, p. 21 (November 2011) ('KIPO, Standard Model').

4.3 The Parties' Duty to Inform

An employee who creates an employee's invention and a utility model shall notify the employer in writing and without delay. In cases where an employee's invention or utility model is created jointly by two or more employees, such notice shall be given jointly by the employees.[81] An employer (excluding the state or a local government) shall, upon receiving such notice, notify the employee as to whether he or she succeeds to rights to the invention within four months from the date when they receive such notice.[82] However, no employer may assert their succession to rights to an invention or a utility model against an employee, if there is no clause in a contract or employment regulation that requires the employee to transfer the right to acquire a patent, etc., or a patent right, etc., to the employer in advance or requires the employee to grant an exclusive licence to such patent, etc. to the employer.[83] An employer shall be deemed to have succeeded to the rights to the invention or utility model if they give notice of their intent to succeed to the rights to the invention or utility model within four months from the date when they received the notice from the employee.[84] If an employer does not give notice of whether they succeed to the rights to the invention or utility model within four months from the date of receipt of the notice from the employee, the employer shall be deemed to have given up the succession to the rights to the invention. In such cases, the employer shall not be entitled to a non-exclusive licence without the consent of the employee who made the invention.[85]

4.4 Employee Remuneration Right

An employee shall have a right to obtain fair compensation when they agree with their employer under a contract or employment regulations to have the employer succeed to their rights to the patent, etc. for an employee invention or to grant an exclusive licence for an employee invention.[86] An employer must draft work regulations which expressly include the type of compensation, the criteria to determine an amount of compensation, the method of payment, and inform employees of them.[87] An employer must consult with the majority of employees who are, or will be, subject to such work regulations in terms of their drafting or modification.[88] However, if such work regulations are to the detriment of the employees, they require consent of the majority of employees who will be bound by such contract or work regulations.[89] An employer must inform the employees who qualify for the compensation of the details of the

81. Art. 12 of the KIPA.
82. Main sentence of Art. 13 of the KIPA; Art. 7 of the Enforcement Decree of the KIPA, revised on 30 December 2016 in accordance with Enforcement Decree No 27751 and effective since 1 January 2017
83. Proviso of Art. 13(1) of the KIPA.
84. Art. 13(2) of the KIPA.
85. Art. 13(3) of the KIPA.
86. Art. 15(1) of the KIPA.
87. Art. 15(2) of the KIPA.
88. Art. 15(3) of the KIPA; Art. 7*bis*(1)(1) of the Enforcement Decree of the KIPA.
89. Art. 15(3) of the KIPA; Art. 7*bis*(1)(2) of the Enforcement Decree of the KIPA.

amount of the compensation.[90] An employer must inform employees of the type of compensation, the criteria to determine the amount of compensation, and the method of compensation payment not later than 15 days before the date when the newly drafted or modified work regulations are effective.[91]

In cases where a contract or employment regulations provide for compensation with respect to employees' inventions or utility models pursuant to Article 15(2)-(4) of the KIPA, the compensation shall be deemed fair unless it does not take into account the benefits the employer is anticipated to obtain with the invention or the utility model and the degree of contribution of the employer and the employee to the completion of the invention or the utility model.[92]

The state or a local government shall pay fair compensation when it succeeds to rights to an employee invention made by a public servant. In such cases, matters necessary for the payment of compensation shall be prescribed by a Presidential Decree or Municipal Ordinance.[93] The employer shall pay fair compensation pursuant to Article 15, even in cases where the employer has succeeded to rights to an employee invention or utility model, but gives up filing a patent application or application for registration of a utility model or voluntarily revokes a patent application or application for registration of a utility model. In such cases the economic benefits that the employee could have obtained if the invention had been protected as industrial property shall be considered in determining the amount of compensation for the invention.[94]

An employer may establish, and operate, an organisation to deliberate on the following regarding employee inventions:

(i) matters concerning the establishment, amendment, and enforcement of regulations on employee inventions;
(ii) matters concerning the mediation of differences between employees and the employer regarding compensation for employee inventions; and
(iii) other necessary matters concerning employee inventions.[95]

An organisation for deliberation on employee inventions shall comprise the same number of respective representative(s) of an employer and employees. Experts in appropriate fields may be commissioned as advisors of the organisation for deliberation, if necessary.[96]

If a dispute arises in connection with an employee's invention, the employer or employee may file an application for mediation with the committee for mediation of disputes over industrial property rights. Employees shall keep details of an employee's invention or utility model confidential until the employer files a patent application for

90. Art. 15(4) of the KIPA.
91. Art. 7*bis*(1)(2) of the Enforcement Decree of the KIPA.
92. Art. 15(6) of the KIPA.
93. Art. 15(7) of the KIPA.
94. Art. 16 of the KIPA.
95. Art. 17 of the KIPA.
96. Art. 17(2) of the KIPA.

the invention or an application for registration of a utility model unless the employer confirms that they will not succeed to rights to an invention.[97]

4.4.1 *Case Law*

In determining the amount of the proper remuneration which the employee is expected to receive, the Korean Supreme Court held that:

> In determining the amount of the proper remuneration which the employee is expected to receive pursuant to Article 40 (2) of the Patent Act prior to the revision of 3 March 2006 (Act No. 7869) (hereinafter "the former Patent Act") [98] in cases where the employer succeeds the employee's invention from the employee, the court should take into account the benefits the employer, etc. is anticipated to obtain with an employee's invention and the degree of contribution by the employer, etc. and employee, etc. to the completion of the invention. The benefits which the employer is expected to obtain refers to the benefit which he/she can obtain by acquiring the status to exploit the employee's invention exclusively in

97. Art. 19 of the KIPA.
98. The Article 15(6) of the KIPA corresponds to Article 40(2) of the former Patent Act. Article 15 (6) of the KIPA prescribes as follows:
Article 15 (Compensation for Employee's Inventions)
(1) An employee, etc. shall be entitled to fair compensation where the employer, etc. succeeds, under a contract or employment regulations, to the right to acquire patent, etc., or a patent, right, etc., for an employee's invention, or are to be granted an exclusive license in the patent right, etc.
(2) The employer, etc. shall prepare compensation regulations stipulating the criteria for determining the type and amount of compensation referred to in paragraph (1), the method of payment thereof, etc., and inform the employee, etc. thereof in writing. < Amended by Act No. 11960, July 30, 2013 >
(3) The employer, etc. shall consult with the employee, etc. regarding the preparation or modification of the compensation regulations under paragraph (2): Provided, That where modifying the compensation regulations disadvantageously to the employee, etc., the employer, etc. shall obtain consent of a majority of the employees, etc. to whom the relevant contract or compensation regulations apply. < Amended by Act No. 11960, July 30, 2013 >
(4) The employer, etc. shall give written notice to an employee, etc. who will receive compensation under paragraph (1) of the details of compensation, including the amount of compensation determined pursuant to the compensation regulations under paragraph (2). < Newly Inserted by Act No. 11960, July 30, 2013 >
(5) Necessary matters, such as the scope of employees, etc. consultation with whom or consent from whom is required of the employer, etc. pursuant to paragraph (3) and procedures, shall be prescribed by Presidential Decree. < Newly Inserted by Act No. 11960, July 30, 2013 >
(6) Compensation paid by the employer, etc. to the employee, etc. pursuant to paragraphs (2) through (4) shall be deemed fair compensation: Provided, That the foregoing shall not apply to cases where the amount of compensation excludes the benefits the employer, etc. is anticipated to obtain with an employee's invention and the degree of contribution by the employer, etc. and employee, etc. to the completion of the invention. < Newly Inserted by Act No. 11960, July 30, 2013 >
(7) The State or a local government shall pay fair compensation, when it succeeds to rights to an employee's invention of a public official pursuant to Article 10 (2). In such cases, matters necessary for the payment of compensation shall be prescribed by Presidential Decree or Municipal Ordinance. < Amended by Act No. 11960, July 30, 2013 >

that the employer can have non-exclusive license of the patent even though the employer does not succeed the employee' invention according to Article 39(1) of the former Patent Act. [99] [100]

If the invention which the defendant-employer exploited had been known to the public at the time when the patent application for the employee' invention was filed, the defendant is not obliged to pay the remuneration for its exploitation to the plaintiff-employee from whom the defendant succeed the patentable right because anyone could exploit it freely and their competitors could have known the situation easily.[101]

Unless the employer cannot obtain any exclusive benefit due to the patent for the employee invention practically because there is a ground for invalidation of the patent for the employee' invention, such as lack of novelty or non-obviousness, and the competitors can easily notice the situation, the mere fact that there is a reason for invalidation of the patent for the employee' invention does not exclude the employer from paying the remuneration for the employee' invention to the employee. The employer can take into account the reason for invalidation of the patent in determining exclusive benefits attributable to the patent.[102]

In general, the extinctive prescription of a claim for the remuneration for the employee' inventions like a general claim shall become complete if not exercised for a period of ten years from the time when the employer succeeded the patentable right for the employee' invention from the employee. However, its extinctive prescription runs from the time of payment of the remuneration stated under the work regulations of the employer's company if its work regulations set the time of payment of the remuneration.[103]

---

99. Art. 10(1) of the KIPA corresponds to Art. 39 (1) of the former Patent Act.
Article 10(1) of the KIPA stipulates as follows:
(1) If an employee, etc. acquires a patent or completes registration of a utility model or design (hereinafter referred to as "patent, etc."), or succeeds to the right to acquire a patent, etc., for an employee's invention, the employer, etc. shall have a non-exclusive license in the right to a patent, utility model, or design (hereinafter referred to as "patent right, etc."): Provided, That the foregoing shall not apply to an employer, etc. which is an enterprise other than a small and medium business under Article 2 of the Framework Act on Small and Medium Enterprises unless it concludes or prepares, in advance, any of the following contracts or employment regulations subject to consultation with an employee, etc.: < Amended by Act No. 11960, July 30, 2013 >
1. A contract or employment regulations that requires the employer, etc. succeed to the right to acquire a patent, etc. or a patent right, etc. for an employee's invention made by the employee, etc.;
2. A contract or employment regulations that requires the grant of an exclusive license to the employer, etc. for an employee's invention made by the employee, etc.
100. Judgment rendered by the Korean Supreme Court on 25 January 2017, Case No. 2014 Da 220347; Judgment rendered by the Korean Supreme Court on 8 September 2011, Case No. 2009 Da 91507.
101. Judgment rendered by the Korean Supreme Court on 8 September 2011, Case No. 2009 Da 91507.
102. Judgment rendered by the Korean Supreme Court on 25 January 2017, Case No. 2014 Da 220347.
103. Judgment rendered by the Korean Supreme Court on 28 July 2011, Case No. 2009 Da 75178.

5 EMPLOYEES' UTILITY MODELS

The same rules apply for employee utility models as for employee inventions. See section 4 above.

6 EMPLOYERS' DESIGN RIGHTS

6.1 The Legal Framework

The term 'invention' means an invention, design, or creation eligible for protection under the Patent Act, the Utility Model Act, or the Design Protection Act in accordance with Article 2(1) of the KIPA. Article 2(2) of the Invention Promotion Act goes on to state that 'The term "employee invention" means an invention that an employee, executive of a corporation, or public official (hereinafter referred to as "employee, etc.") makes in connection with his/her duties, where it falls within the scope of business of the employer, the corporation, the State, or the competent local government (hereinafter referred to as "employer, etc.") and the activities that have led to the invention fall within the present or past duties of the employee, etc.'

Rules governing the ownership of design rights between employers and employees can be found in Article 10 of the KIPA. Within the limits prescribed by the KIPA, the ownership of design rights may be established or modified by contractual agreement between an employer and an employee, or employment regulations set by the employer.

Transfer, termination of, and remedies for, a design right are regulated by the Korean Design Protection Act unless prescribed otherwise in the KIPA. For designs the jurisdiction of disputes that arise under Article 10 of the KIPA is vested in the District Courts of Korea. However, there have been no reported court decisions in connection with employees' designs so far.

6.2 Ownership of Rights

The ownership of rights to employees' designs will be determined by Article 10 of the KIPA because the application of the KIPA is extended to employees' designs. The duration of a design right shall commence upon registration of the design right and last for 20 years from the filing date of the application of registration of the design right.[104] For design rights, jurisdiction of disputes that arise under Article 10 of the KIPA is exclusively vested in the Seoul Central District Court, Busan District Court, Daegu District Court, Gwangju District Court, or Daejeon District Court, where constitutes a district court in the jurisdictional area of a High Court which has jurisdiction over the location of a competent court pursuant to general or special jurisdiction.[105]

104. Art. 91 of the Korean Design Protection Act (Act No. 14686, revised on 21 March 2017 and effective since 22 September 2017).
105. Article 24(2) of the Korean Civil Procedure Act.

Nonetheless, a party may bring a lawsuit concerning design rights to Seoul Central District Court even though Daejeon District Court has exclusive jurisdiction for a dispute on a design right because Seoul Central District Court has a special division which deals with intellectual property rights.[106]

6.2.1 Ownership of Rights at Universities or Research Institutions

The standard form of university regulations on employees' inventions provided by the Korean Intellectual Property Office, only covers employees' inventions and utility models.[107] Hence, its application is not extended to employees' designs. In practice, most universities have work regulations on employees' inventions and utility models. In addition, they do cover intellectual property other than inventions and utility models. Hence, they include designs, trademarks, copyrights, geographical indications, and semiconductor chips.

6.3 The Parties' Duty to Inform

The parties' duty to inform in terms of employees' designs will be determined by Articles 12 and 13 of the KIPA because the application of the KIPA is extended to employees' designs.

6.4 Employee Remuneration Right

The remuneration right on employees' designs will be determined by Articles 15 and 16 of the KIPA because the application of the KIPA is extended to employees' designs.

7 EMPLOYERS' PLANT BREEDER'S RIGHTS

7.1 The Legal Framework

Rights to new plant variety rights are governed pursuant to the Act on Protection of New Plant Varieties in Korea ('ANPV'). Prior to its amendment in 2012,[108] they used to be prescribed in the Korean Seed Industry Act, which included substantive provisions on breeding, distribution and management of seeds, and procedural provisions on protection of new plant varieties. There have been no court decisions reported in terms of new plant varieties so far.

106. Article 24 (3) of the Korean Civil Procedure Act.
107. KIPO, Standard Model, p. 21.
108. Act No. 11458, comprehensively amended on 1 June 2012 and effective since 2 June 2013.

7.2 Ownership of Rights

A plant breeder or their successor has a right to entertain protection of their plant breeding.[109] The ANPV does not have any provision on employees' new plant varieties in connection with the private sector, but instead only contains provisions regarding employees' new plant varieties bred by public servants-plant breeders.

A public servant's right to a new plant variety shall be succeeded to by the state or a local government in cases where the new plant variety which has been bred by the public servant belongs by reason of its nature to the scope of the business of the State or a local government, and where the breeding of (new) varieties comes within present or past duties of the public servant. However, the public servant's right to a new variety shall be succeeded to by an exclusively responsible organisation established pursuant to Article 11(1) of the Technology Transfer and Commercialisation Promotion Act where the variety falls under the breeding as part of duties of the school personnel of national schools and public schools under the Higher Education Act.[110]

For plant variety rights, jurisdiction of disputes that arise under Article 10 of the KIPA is exclusively vested in the Seoul Central District Court, Busan District Court, Daegu District Court, Gwangju District Court, or Daejeon District Court, where constitutes a district court in the jurisdictional area of a High Court which has jurisdiction over the location of a competent court pursuant to general or special jurisdiction.[111] Nonetheless, a party may bring a lawsuit concerning plant variety rights to Seoul Central District Court even though Daejeon District Court has exclusive jurisdiction for a dispute on a plant variety right because Seoul Central District Court has a special division which deals with intellectual property rights.[112]

7.2.1 Ownership of Rights at Universities or Research Institutions

Many private universities or private research institutions do not have work regulations on ownership of employees' new plant variety rights. This is most likely attributable to the fact that the ANPV only regulates the public servants' new plant variety rights. As far as national universities or public universities are concerned, Article 28(1) of the ANPV will be applied. However, there is no reason for the ANPV to be silent as to employees' new plant varieties in private universities. Hence, Article 28(1) of the ANPV needs to be modified to extend its application to private universities.

A new plant variety right is established when it is registered.[113] The duration of new plant variety rights is 20 years from the date when the right is registered. However,

109. Art. 17(1) of the ANPV, which is revised on 29 February 2016 in accordance with Act No. 14035 and effective since 1 March 2017.
110. Art. 28(1) of the ANPV.
111. Article 24(2) of the Korean Civil Procedure Act.
112. Article 24 (3) of the Korean Civil Procedure Act.
113. Art. 54 of the ANPV.

the protection term of new plant variety rights on fruit trees and mountain trees is extended to 25 years from the date when the right is registered.[114]

7.2.2 *Case Law*

In a declaratory claim to confirm the existence of a non-exclusive licence based on prior use, the Korean Supreme Court held that the plaintiff could not acquire non-exclusive licence based on prior use because he/she came to know the protected plant variety from the defendant who was the breeder of the protected plant variety whose application for its protection was filed, saying as follows:

> Article 64 of the ANPV prescribes that "A person who runs or prepares to run a business of exploiting a protected plant variety in the Republic of Korea as he/she learns of it from another person who breeds or has bred the protected plant variety without the knowledge of the details of the protected plant variety being applied for plant variety protection as at the time an application for the plant variety protection is filed, he/she shall have a non-exclusive license of the plant variety protection of which application is filed, within the scope of the purpose of such business." This provision is intended to strike a balance between a prior user of a plant variety right, who runs or prepares to run a business of exploiting a protected plant variety at the time an application for the plant variety protection is filed, and a holder of the plant variety right under the first-to-file rule.[115]
>
> In light of the gist and the text of the provision, the prior user who can acquire non-exclusive license based on prior use in accordance with the provision means an independent breeder of a new plant variety whose origin is different from the breeder of the protected plant variety whose application for its protection is filed or the person who comes to know it from the independent person who breeds or has bred the protected plant variety. Hence, the plaintiff cannot acquire non-exclusive license based on prior use because he/she came to know the protected plant variety from the defendant who is the breeder of the protected plant variety whose application for its protection is filed.[116]

7.3 The Parties' Duty to Inform

A public servant who breeds a new plant variety within the scope of their duties must immediately report the matter about the new plant variety to the head of the organisation to which they belong.[117] The head of the organisation must subsequently decide whether the new plant variety falls under the category of a new plant variety bred by a public servant plant breeder.[118] The head of the organisation must, in writing,

114. Art. 55 of the ANPV.
115. Article 25(1) of the ANPV prescribes that 'Where at least two applications for plant variety protection for the same plant variety are filed on different days, only the person who first files an application for plant variety protection shall be entitled to plant variety protection.'
116. Judgment rendered by the Korean Supreme Court on 11 June 2015, Case No. 2014 Da 79488.
117. Art. 4 of the Enforcement Decree of the ANPV, which is Presidential Decree No. 26702 revised on 10 December 2015 and effective since the same date.
118. Art. 5(1) of the Enforcement Decree of the ANPV.

notify the public servant plant breeder of their decision.[119] In cases where the public servant plant breeder receives from the head of the organisation a notice that the new plant variety bred by them falls under a new plant variety for hire, they must assign their right to the new plant variety to the head of the organisation immediately.[120]

The parties' duty to inform needs to be prescribed in the ANPV rather than in the Enforcement Decree of the ANPV because it is substantial for the ownership of rights.

7.4 Employee Remuneration Right

In cases where a public servant (a plant breeder) breeds a new plant variety, they will be entitled to an employee's remuneration right which is recognised by Article 29 of the ANPV. The details of such remuneration are prescribed in the Enforcement Decree of the ANPV.

8 INSTANCES FOR DISPUTES

The Intellectual Property Trial and Appeal Board ('IPTAB') was launched concurrently with the Patent Court on 1 March 1998. The IPT is the tribunal of the first instance for industrial property rights-related disputes and is independently operated within the Korean Intellectual Property Office. The IPTAB was established under the jurisdiction of the Commissioner of the Korean Intellectual Property Office to be responsible for trials and retrials regarding parents, utility models, designs and trademarks. It does not hear copyright cases. As far as employees' inventions are concerned, the IPTAB can hear and decide on: (i) a trial against ruling of refusal of registration of a patent or against ruling of refusal to extend duration of protection of a patent;[121] (ii) a trial against decision to reject the amendment when the decision to reject amendment of application for registration of a patent is disputed in a trial against ruling of refusal of its registration;[122] (iii) a trial for invalidation of correction of the specifications or drawings of a patent invention;[123] (iv) a ruling which cancel a patent, followed by an application for cancellation of the patent,[124] (v) an invalidation trial of a patent;[125] (vi) a trial to confirm the scope of a patent right;[126] (vii) invalidation trial of registration for extension of term of a patent right;[127] and (viii) a trial for granting non-exclusive licence.[128]

119. Art. 5(2) of the Enforcement Decree of the ANPV.
120. Art. 6 of the Enforcement Decree of the ANPV.
121. Art. 132 ter of the KPA.
122. Art. 136 of the KPA.
123. Art. 137 of the KPA.
124. Art. 132-13 (1) of the KPA.
125. Art. 133 of the KPA.
126. Art. 135 of the KPA.
127. Art. 134 of the KPA.
128. Art. 138 of the KPA.

Figure 27.1 Original and Appellate Jurisdictions of Disputes Related to Intellectual Property Rights

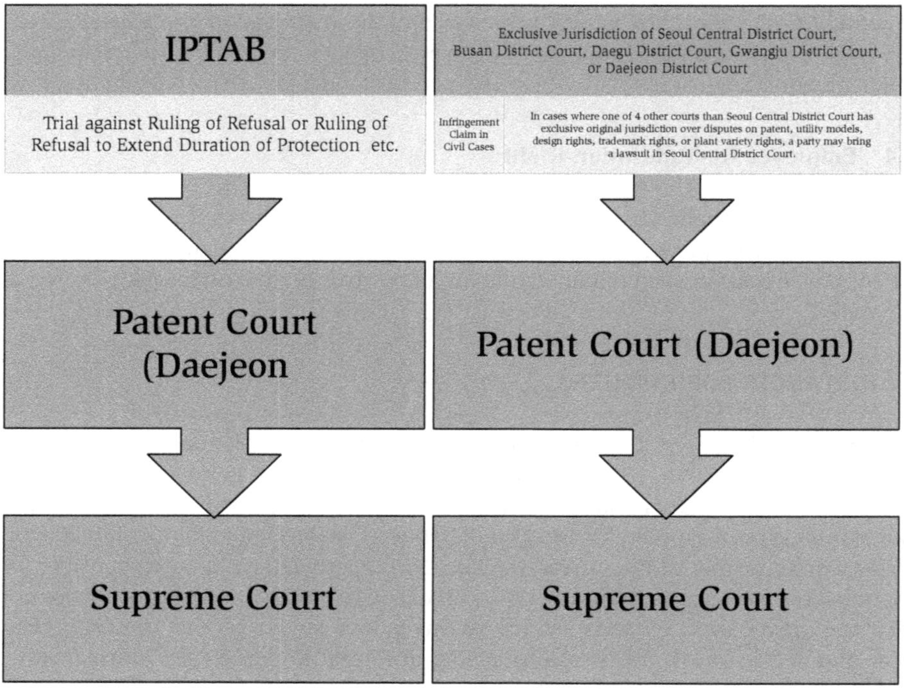

The Patent Court, which is located in Daejeon, Republic of Korea, has exclusive jurisdiction over any action against decisions rendered by the IPTAB or against dismissal of a request for a trial or retrial before the IPTAB. The adjudicative power of the Patent Court is exercised by a collegiate panel of the Patent Court composed of three judges.

As far as infringement actions over patents, utility models, design rights, trademark rights, or plant variety rights are concerned, jurisdiction of disputes that arise under Article 10 of the KIPA is exclusively vested in the Seoul Central District Court, Busan District Court, Daegu District Court, Gwangju District Court, or Daejeon District Court, where constitutes a district court in the jurisdictional area of a High Court which has jurisdiction over the location of a competent court pursuant to general or special jurisdiction.[129] Nonetheless, a party may bring a lawsuit concerning patents to Seoul Central District Court even though Daejeon District Court has exclusive

129. Article 24(2) of the Korean Civil Procedure Act.

jurisdiction for a patent dispute because Seoul Central District Court has a special division which deals with intellectual property rights.[130] In terms of the appellate jurisdiction of infringement claims over patents, utility models, design rights, trademark rights, or plant variety rights, the Korean Patent Court has exclusive appellate jurisdiction over them.[131]

The Supreme Court reviews and judges as the court of the last resort the final appeal cases against a judgment of a High Court, an appellate court, the Patent Court, the re-appeal cases against a ruling or an order of a court of appeal on a ruling, a High Court, an appellate court, or the Patent Court, and the cases falling under the jurisdiction of the Supreme Court pursuant to other Acts. Therefore, any person who has received a judgment from the Patent Court may appeal the Supreme Court. The hierarchy of original and appellate jurisdictions of intellectual property rights-related cases is shown in Figure 27.1.

9 CONCLUDING REMARKS

As mentioned above, big businesses in Korea can be granted a non-exclusive licence only if they draw up a contract or work regulation, which succeed right to acquire a patent, etc. or a patent right, etc. to them or grant exclusive licence for them, in advance after consultation with the employees under Article 10(1) of the revised KIPA of 2013. This provision aims to promote employees' bargaining power to negotiate with employers in terms of employees' invention. The provision, however, is believed to give them a disincentive for innovation. Therefore, I propose that this provision be revised to encourage businesses to take an innovative step.

BIBLIOGRAPHY

Table of Sources

Gyooho Lee, Copyright Law: Cases and Explanations 406 (2017).

Gyooho Lee, Keon-Hyun Ahn and Jacques de Werra, Euro-Korean Perspectives on the Use of Arbitration and ADR Mechanisms for Solving Intellectual Property Disputes, 30(1) Arbitration International 100 (2014).

Hae Wan Lee, Copyright Law, 521-522 (3rd ed., Pakyoung Publishing Co. 2015).

Korean Intellectual Property Office, 2011 Yearbook of Statistical Data on Intellectual Properties.

Korean Intellectual Property Office, 2014 White Paper on Intellectual Properties (2015).

Korean Intellectual Property Office, 2015 White Paper on Intellectual Properties (2016).

130. Article 24 (3) of the Korean Civil Procedure Act.

131. Article 28 quarter, subparagraph 2 of the Korean Court Organisation Act.

Korean Intellectual Property Office, Standard Model for University Regulations on Employees' Inventions, p. 21 (November 2011).
Seung-Jong Oh, Copyright Law, 604 (4th ed., Pakyoung Publishing Co. 2016).
WIPO World IP Indicators (December 2015).
Yong-Sik Song, Sang-Jeong Lee, and Byun-Il Kim, Intellectual Property Law 262 (2013).
Yong Ho Phi, A Study on Assessment of Fair Compensation for Employees' Invention in Accordance with Industrial Classification (Report submitted to Korean Intellectual Property Office), 9 (December 2013).

Table of Legislation

Act No. 14634, amended on 21 March 2017 and effective since the same date.
Act No. 14687, revised on 21 March 2017 and effective since 22 September 2017.
Act No. 14686, revised on 21 March 2017 and effective since 22 September 2017.
Enforcement Decree of the KIPA, revised on 30 December 2016 in accordance with Enforcement Decree No. 27751 and effective since 1 January 2017.
Act No. 14035, amended on 29 February 2016 and effective since 1 March 2017.
Enforcement Decree No. 26702 revised on 10 December 2015 and effective since the same date.
Act No. 13150, revised on 3 February 2015 and effective since 4 August 2015.
Act. No. 11960, amended on 30 July 2013 and effective since 31 January 2014.
Act No. 11458, comprehensively amended on 1 June 2012 and effective since 2 June 2013.
Act No. 9625 on 22 April 2009, effective since 23 July 2009.
Act No. 9183, revised on 26 December 2008 and effective since 27 March 2009.
Act No. 7871, amended on 3 March 2006 and effective since that date.
Act No. 7869, amended on 3 March 2006 and effective since 4 September 2006.
Act No. 7554, amended on 31 May 2005 and effective since 1 December 2005.
Act No. 4526, enacted on 8 December 1992 and effective since 1 September 1993.

Table of Cases

Judgment rendered by the Korean Supreme Court on 25 January 2017, Case No. 2014 Da 220347.
Judgment rendered by the Korean Supreme Court on 11 June 2015, Case No. 2014 Da 79488.
Judgment rendered by the Korean Supreme Court on 9 May 2013, Case No. 2011 Da 69725.
Judgment rendered by the Korean Supreme Court on 15 November 2012, No. 2012 Do 6676.
Judgment rendered by the Korean Supreme Court on 8 September 2011, Case No. 2009 Da 91507.

Judgment rendered by the Korean Supreme Court on 8 September 2011, Case No. 2009 Da 91507.

Judgment rendered by the Korean Supreme Court on 28 July 2011, Case No. 2009 Da 75178.

Judgment rendered by the Korean Supreme Court on 28 July 2011, Case No. 2009 Da 75178.

Judgment rendered by the Korean Supreme Court on 14 January 2010, Case No. 2007 Da 61168.

Judgment rendered by the Korean Supreme Court on 27 June 1997, Case No. 97 Do 516.

Judgment rendered by the Korean Supreme Court on 30 July 1996, Case No. 95 Da 29130.

Judgment rendered by the Korean Supreme Court on 27 December 1991, Case No. 91 Hu 1113.

CHAPTER 28

Singapore

Susanna Leong

1 BACKGROUND

The law in Singapore gives effect to the notion that a person is entitled to the fruits of their labour. Therefore, we see in copyright that the author of a work is, in general save for certain exceptions, regarded by the law as the first owner of any copyright in it. Similarly in patents, only the inventors or joint inventors of an invention are entitled to a patent grant. The position is the same for the other intellectual property rights such as design rights, semiconductor chips and plant varieties.

A notable exception to this general rule is the employee. In the absence of any express agreement to the contrary, if an employee in pursuance of their terms of employment produces an item of intellectual property in the course of their employment, the right to that intellectual property vests in their employer and not the employee. In all other circumstances, the employee who is the author or inventor is entitled to the right of their intellectual creations.

At the heart of the discussion of an employee's entitlement to inventions and works is: (a) the distinction between a contract 'of services' (an employee) and a contract 'for services' (an independent contractor); and (b) the concept of work done 'in the course of employment'. These concepts and distinctions will be discussed in greater detail in the ensuing paragraphs particularly in the context of copyright and patents.

With regard to the employee's obligation to disclose or inform their employer of an invention, the duty of good faith and fidelity requires an employee to impart all inventions to their employer and to hold any resultant patent on trust for their employer. Complicated issues such as whether the employee is the inventor (especially in cases where the employee works in a team), whether the invention was made in the course of the normal duties of the employee or if the duties were specifically assigned to them can arise and will be discussed below. As we will see in certain cases, the terms of the contract of employment may be a relevant factor which the courts will consider in the deliberation of the employee's entitlement to the invention and the duty of good faith owed to their employer.

It should also be noted that the law does not confer upon an employee in Singapore a statutory right to remuneration in respect of inventions created in the course of employment and has instead left the matter to be decided between the parties by private contractual arrangement. Indeed, as we shall see in the discussions below, the employment contract between employee and employer play a significant role in the ownership and remuneration of works and inventions of an employee.

Finally, by way of background, it should be noted that throughout this chapter, reference will be made to relevant UK case law. Singapore's colonial ties with the UK coupled with the fact that certain intellectual property legislation such as the Patents Act is modelled on that in the UK have made references to UK case law useful and persuasive in fleshing out the relevant Singapore provisions, especially in areas where Singapore case law is lacking.

2 EMPLOYEES' COPYRIGHT

2.1 The Legal Framework[1]

Under the Copyright Act,[2] the general rule is that the author of a work (who is the *creator*) is also the *first owner* of any copyright in it, subject to the following exceptions:[3]

(a) works produced in the course of employment; and
(b) works produced in pursuant to contracts varying first ownership.

Under Section 30(6) of the Copyright Act if the work is made by the author in pursuance of the terms of their employment by another person under a contract of service or apprenticeship, copyright does not vest in the author but in their employer. Within this general rule are two further exceptions:

(a) works of an employed journalist;[4] and
(b) commissioned photographs, portraits and engravings.[5]

1. The following section was taken from pages 141-146 of *Authorship and Ownership* in Susanna Leong, *Intellectual Property Law of Singapore* (Academy Publishing, 2013). <http://www.sal.org.sg/Lists/BookTitles/DispForm.aspx?ID = 219>. Reproduced with permission.
2. Cap 63, 2006 Rev Ed.
3. S. 30(4)-(5) Copyright Act.
4. S. 30(4) Copyright Act.
5. S. 30(5) Copyright Act.

2.2 Ownership of Rights[6]

2.2.1 Works Produced in the Course of Employment

In order for Section 30(6) of the Copyright Act to apply, the work must be made by the author in the course of employment pursuant to a *contract of service* as opposed to a *contract for services*. In the latter case, the author would not be regarded as an employee but as an independent contractor and, as such, copyright will vest in the author in respect of the work created. However, it is not always easy to distinguish a contract of service from a contract for services, as explained by Denning LJ, as he then was, in *Stevenson Jordan v McDonnell & Evans*:[7]

> 'It is often easy to recognise a contract of service when you see it, but difficult to say where the difference lies. A ship's master, a chauffeur, and a reporter on the staff of a newspaper are all employed under a contract of service; but a ship's pilot, a taxi-man, and a newspaper contributor are employed under a contract for services. One feature which seems to run through the instances is that, *under a contract of service, a man is employed as part of the business, and his work is done as an integral part of the business*; whereas, *under a contract for services, his work, although done for the business, is not integrated into it but is only accessory to it.'* (Emphasis added)

Frequently, the employment contract is a 'mixed contract'; partly a contract *of service* and partly a contract *for services*. The employment contract in *Stevenson Jordan* was an example of such a 'mixed contract'. An ex-employee was the author of a book based mainly on information acquired while he was an employee of the company. One part of the book comprised lectures delivered publicly by the author whilst in the employment of the company and the other part consisted of materials acquired while the author was on a particular assignment for an individual client of the company. The court held that the author's employment contract with the company had been a mixed contract; partly a contract of service and partly a contract for the provision of services beyond the contract of employment. The material in the book comprising the public lectures was not made under a contract of service whilst the materials acquired while the author was on a special assignment were made under a contract for service. The court held that in such a case the work done by the author was severable and copyright in the work vested accordingly.

Similarly, in *Byrne v Statist* [8] the court held that the copyright in a Portuguese translation of a speech undertaken by an employee of a newspaper vested in the employee as the work had been undertaken by the employee in his own time and independently of his ordinary duties.

6. The following section was taken from pages 141-146 of *Authorship and Ownership* in Susanna Leong, *Intellectual Property Law of Singapore* (Academy Publishing, 2013). <http://www.sal.org.sg/Lists/BookTitles/DispForm.aspx?ID = 219>. Reproduced with permission.
7. [1952] 1 TLR 101 at [111].
8. [1914] 1 KB 622.

Applying the test set out by the English Court of Appeal in *Stevenson Jordan*, the English High Court in *Noah v Shuba* [9] held that the copyright in the work 'A Guide to Hygienic Skin Piercing' written by the plaintiff, who was a consultant epidemiologist at the Communicable Disease Surveillance Centre of the Public Health Laboratory Service ('PHLS'), vested in him and not his employer, PHLS, as it was not written in the course of his employment. The doctor/author in *Noah v Shuba* wrote the guide at home in the evenings and weekends. Although he had had discussions with colleagues about the contents of the guide and had made use of the library facilities at PHLS as well as the secretarial support, this did not render the guide a work done by the doctor/author in the course of his employment with PHLS.

Even as an individual employed under a contract of service, a person may sometimes perform services outside the contract which are useful accessories to their contracted work but not part of their contract of employment. The copyright, if any, in a work which has resulted from any of these outside services would thus vest in the employee.

According to Section 30(3) of the Copyright Act,[10] the position of the employer and the employee *vis-à-vis* the issue of copyright ownership may be excluded or modified by agreement. However, this 'agreement' must satisfy two requirements if it was to obviate the application of the statutory rule that copyright in a work done in the course of employment vests in the employer:

(a) it must be an unequivocal agreement that, notwithstanding the existence of a contract of employment, the title to copyright in works created during the course of the contract shall not vest in the employer; and
(b) the agreement must be a legally effective one.[11]

There was another compelling reason why the court ruled in favour of the doctor/author in *Noah v Shuba*.[12] This was the fact that for a long time it had been the practice at PHLS for employees to retain copyright in work written by them. PHLS had never claimed copyright in these works (such as journal articles) which were written by its employees. Therefore, the doctor/author still retained copyright, even if the work was made in the course of employment. The court held that there was an *implied term* of his contract of service which excluded the operation of the statutory rule. The statutory rule usually vested the copyright in such works in the employer. In this case, copyright vested instead in the doctor/author employee.

2.2.2 Employed Journalists

The situation concerning a journalist employed under a contract of service with a newspaper, magazine or similar periodical is specially provided for under Section 30(4)

9. [1991] FSR 14.
10. (Cap 63, 2006 Rev Ed).
11. See *Noah v Shuba* [1991] FSR 14.
12. [1991] FSR 14.

of the Copyright Act. The copyright in a work created by a journalist for the purpose of publication in a newspaper, magazine or similar periodical is split. Insofar as the copyright relates to the publication of the work in a newspaper, magazine or similar periodical or the reproduction of the work for the purpose of its being so published, the copyright vests in the proprietor of the newspaper, magazine or similar periodical. All other copyrights subsisting in the work of the employed journalist vest in the journalist. Thus, a journalist at a newspaper who has conducted a series of interviews with famous politicians or business magnates and has written interview reports for publication in the newspaper would be able to collate their works and publish them subsequently in a book. Copyright in the book would vest in the journalist and not their employer.

Issues of copyright ownership between the proprietor of a newspaper, magazine and similar periodical and the employed journalist may be modified by agreement as provided for in Section 30(3) of the Copyright Act.

2.2.3 Commissioned Photographs, Portraits and Engravings

In the case of commissioned photographs, portraits and engravings, copyright does not vest in the author of the works (the photographer, painter or engraver) but in the person who has commissioned the making of the works.[13] Two conditions must be satisfied in order for the copyright in the works to be vested in the commissioner:

(a) the existence of an agreement for valuable consideration for the creation of the work; and
(b) the relevant work is made in pursuance of the agreement.

If the work involved was commissioned for a particular purpose and this purpose had been communicated to the creator, then, under copyright, the creator retains the right to prevent the work from being used for purposes other than that for which it was commissioned. For example, if a photographer has been commissioned to take a series of photographs of the new skyline of Singapore in the Marina Bay area for the particular purpose of a tourism advertisement campaign to market Singapore as a choice tourist destination and this purpose had been communicated to the photographer, then applying Section 30(5) Copyright Act the copyright in the commissioned photographs vests in the commissioner and not the photographer/author. However, if the commissioner were to include the photographs in a book entitled 'Scenes of Singapore', under the exception provided in Section 30(5) Copyright Act, the photographer/author would have a right to stop it as it is outside the particular purpose agreed between the parties under the contract of commission.

It should be noted that a commissioner is free to enter into an agreement with the author that excludes or modifies the legal position under Section 30(3) Copyright Act.

13. S. 30(5) Copyright Act.

In 2016, the Singapore Ministry of Law ('MinLaw') and the Intellectual Property Office of Singapore ('IPOS') launched a public consultation on a number of proposed changes to Singapore's copyright regime. One such proposed change in the law involves the situation of commissioned photographs, portraits and engravings. If the proposed change is accepted by the Singapore government, creators of commissioned photographs, portraits and engravings would have first ownership of the copyright over the work he or she has created. This will bring the legal position of commissioned photographs, portraits and engravings in Singapore to align with the general position adopted in other jurisdictions such as the United Kingdom, the United States of America, Japan and Canada. A Copyright Amendment Bill containing the proposed changes is expected to be tabled before Singapore parliament sometime in the third quarter of 2017.

2.2.4 Works Produced in Pursuant to Contracts Varying First Ownership

Apart from photographs, portraits and engravings, the Copyright Act does not contain special provisions for commissioned works. Therefore, first principles apply and the law says that for all other commissioned works (and in the absence of an agreement to the contrary) the *author* is the first owner of copyright subsisting in the work.[14] The mere fact that the author has been commissioned to produce the work is insufficient to entitle the commissioner to the copyright. The commissioner would have to establish the entitlement under some express or implied term of the agreement. Consequently, if the commissioner wishes to own copyright in the work, they must obtain an assignment of it from the author to them.

Where there exists an implied term for the author to assign copyright in the commissioned work to the commissioner pursuant to a contract of services between the author and the commissioner, the courts have held the copyright to belong in equity to the commissioner. The case of *Massine v de Basil* [15] is a case on point. The issue before the court was whether the copyright in the plaintiff's choreography for a ballet was intended to form part of the repertoire of the defendant's ballet company. The English Court of Appeal found that the contract between the plaintiff and the defendant was one of employment, thus copyright in the choreography vested in the defendant employer. More importantly, the court held that given the nature of a ballet, which was necessarily a composite work made up of the various elements such as music, story, choreography, costumes and settings, it must have been intended by the parties that the copyright in the whole ballet and each of its elements should vest in the defendant who commissioned the choreography.

2.3 The Parties' Duty to Inform

There are no provisions governing the parties' duty to inform.

14. *Ray v Classic FM* [1998] FSR 622 at [641], *per* Lightman J.
15. [1936–45] Mcgregor's Copyright Cases 223.

2.4 Employee Remuneration Right

There are no provisions governing the employee's remuneration rights regarding copyrights.

3 EMPLOYEES' SEMICONDUCTOR CHIPS

3.1 The Legal Framework[16]

An integrated circuit is simply an electronic circuit. It consists of a connected arrangement of transistors in a three-dimensional design on a surface of thin substrate[17] of semiconductor material. Integrated circuits incorporated in chips are found in virtually all electronic equipment today such as computers, mobile phones, calculators, digital watches, cars and 'smart cards'. The creation of new computer chips often involves extensive research and development particularly in the complex layout-design of how the transistors are connected and arranged to form the circuit and the integration of these transistors into a small chip. Although integrated circuits are complex and time-consuming to create, they are capable of mass production once created and can be easily copied. Layout-designs of integrated circuits are also referred to as semiconductor topographies and it is on these layout-designs that the law confers protection.

TRIPS[18] obliges WTO[19] Member States to pass legislation to protect integrated circuits *sui generis*. As part of fulfilling its obligations under TRIPS, Singapore passed the Layout-Designs of Integrated Circuits Act[20] (Layout Designs Act) on 15 February 1999.

3.2 Ownership of Rights[21]

Under the Layout Designs Act, a *sui generis* right is created for the protection of layout-designs of integrated circuits. Section 2 of the Layout Designs Act defines 'integrated circuit' to mean:

16. The following section was taken from pages 631-637 of *Integrated Circuits Designs* in Susanna Leong, *Intellectual Property Law of Singapore* (Academy Publishing, 2013). <http://www.sal.org.sg/Lists/BookTitles/DispForm.aspx?ID=219>. Reproduced with permission.
17. In this context, substrate refers to the solid material such as silicon or aluminium oxide that serves as the foundation upon which the integrated circuits are deposited.
18. The TRIPS Agreement is the Agreement on Trade-related Aspects of Intellectual Property Rights in Annex 1 C of the Marrakesh Agreement Establishing the World Trade Organization signed in Marrakesh, Morocco on 15 April 1994. See Art. 35, 36, 37 and 38 TRIPS.
19. World Trade Organization. See n 18 above.
20. (Cap 159A, 2000 Rev Ed).
21. The following section was taken from pages 631-637 of *Integrated Circuits Designs* in Susanna Leong, *Intellectual Property Law of Singapore* (Academy Publishing, 2013). <http://www.sal.org.sg/Lists/BookTitles/DispForm.aspx?ID=219>. Reproduced with permission.

> ... a product, in its final form or an intermediate form, in which the elements, at least one of which is an active element, and some or all of the interconnections are integrally formed in and on, or in or on, a piece of material and which is intended to perform an electronic function.

'Layout-design' is defined to mean:

> ... the 3-dimensional disposition, however expressed, of the elements of an integrated circuit (at least one of which is an active element), and of some or all of the interconnections of an integrated circuit, or such a 3-dimensional disposition prepared for an integrated circuit intended for manufacture.

The basic principle is that the *creator* is the owner of the protected layout-design. This is subject to the following exceptions:

(a) where the layout-design is created in pursuance of a commission: the person who commissioned the layout-design is the owner; and
(b) where the layout-design is created in by an employee in the course of their employment: the employer is the owner.

A layout-design may be owned by more than one person and all of them can be qualified owners. Any requirement of the consent of a qualified owner requires the consent of all the qualified owners.[22]

3.3 The Parties' Duty to Inform

There are no provisions governing the parties' duty to inform.

3.4 Employee Remuneration Right

There are no provisions governing employees' remuneration rights regarding semiconductor chips.

4 EMPLOYEES' INVENTIONS

4.1 The Legal Framework[23]

Section 19(1) of the Patents Act provides that '*any person* may make an application for a patent either alone or jointly with another' (emphasis added). However, according to Section 19(2) of the Patents Act, a patent may be granted *only* to a few groups of individuals and *to no other person*.

22. S. 6(3) Layout-Designs of Integrated Circuits Act.
23. The following section was taken from pages 533-554 of *Entitlement* in Susanna Leong, *Intellectual Property Law of Singapore* (Academy Publishing, 2013). <http://www.sal.org.sg/Lists/BookTitles/DispForm.aspx?ID=219>. Reproduced with permission.

As a general rule, a patent for an invention may be granted to 'the inventor or joint inventors'[24] of the claimed invention. Besides inventors and joint inventors, the categories of person entitled to a patent grant also include persons who by law, treaty or international convention, or prior contractual arrangements with the inventor, are entitled to the whole of the property in the invention in Singapore, and the successor or successors in title of the first two categories.[25]

Examples of persons who by law are entitled to a patent for an invention are the employers of employee-inventors under Section 49 of the Patents Act. By the same token, a person or a company may be able to claim an entitlement to a patent grant on account of prior contractual arrangements with the inventor. For example, a person who is engaged as a paid consultant to a company may enter into an agreement with the company that any inventions which they make during the term of their consultancy will belong to the company. In such a case, the contractual obligation between the consultant-inventor and the company will displace the consultant-inventor's primary claim as regards any inventions which they make within the contractual period. The company which engages them is entitled to claim a grant of the patent by virtue of the contractual obligation (although the consultant would remain the inventor).

It is important to note that a patent for an invention may be revoked if it was granted to a person who was not entitled to be granted that patent.[26] There is a rebuttable presumption[27] that the person or persons making the application for patents are persons so entitled under Section 19(2) of the Patents Act.

In general, the correct approach in any dispute over entitlement is to first ascertain who is the inventor or joint inventors of the claimed invention under Section 19(2)(*a*) of the Patents Act. Only when this issue is settled can one proceed to decide if an individual or individuals are entitled to a patent grant under the other categories provided in Sections 19(2)(*b*) and 19(2)(*c*) of the Patents Act.

4.2 Ownership of Rights[28]

4.2.1 Inventor or Joint Inventors: Who is an 'Inventor'?

Section 2(1) of the Patents Act defines an 'inventor', in relation to an invention, to mean 'the *actual deviser* of the invention and "joint inventor" shall be construed accordingly' (emphasis added). How does one go about determining who is 'the *actual deviser* of the invention' and thus, the inventor of the claimed invention? The identity of the inventor is easy to ascertain if only one individual is responsible for the invention. However, in practice, most inventions are the result of a team of researchers or collaborators working closely with one another. Frequently, more than one

24. S. 19(2)(*a*) Patents Act.
25. S. 19(2)(*b*) and 19(2)(*c*) Patents Act.
26. S. 80(1)(*b*) Patents Act.
27. S. 19(3) Patents Act.
28. The following section was taken from pages 533-554 of *Entitlement* in Susanna Leong, *Intellectual Property Law of Singapore* (Academy Publishing, 2013). <http://www.sal.org.sg/Lists/BookTitles/DispForm.aspx?ID=219>. Reproduced with permission.

researcher may be working on similar ground-breaking works and it is not uncommon for each of them to have claimed to have independently created the said invention. However, in a first-to-file system, it is the inventor who puts in the patent application first that is granted the patent for the invention. The other independent inventors who file their applications later will have their patent applications defeated on the ground of a lack of novelty. Thus, issues of entitlement to grant are often complicated and require careful analysis by the courts in each case.

The phrase 'the *actual deviser* of the invention' has been interpreted to mean 'the natural person who came up with the inventive concept'[29] excluding those who merely '*contributed* to the claim' especially if the contribution is in the form of '*non-patentable integers* derived from prior art'.[30] Only persons who have made contributions to 'the formulation of the inventive concept'[31] may be regarded as 'actual devisers' of the invention.

The starting point is the proper identification of the 'inventive concept' of the said invention, i.e., the 'heart' of the invention.[32] This is achieved by examining the patent specification consisting of the claims, the description and any other drawings. However, courts have cautioned against 'dividing the information in a patent into a myriad of sub-concepts'[33] and taking a 'mechanistic, element by element approach to inventorship',[34] as such an approach would not produce a fair result.

It should always be remembered that there is supposed to be only *one* inventive concept in a patent.[35] It is accepted that the identification of the 'inventive concept' of the claimed invention is not always an easy task and it is not possible to be very specific about how this is done.[36] Indeed, Lord Hoffman in *Rhone-Poulenc Rorer Inc v Yeda Ltd* [37] stated:

> In some cases this may be quite complex because the inventive concept is a relationship of discontinuity between the claimed invention and the prior art. Inventors themselves will often not know exactly where it lies.

However, having identified the 'inventive concept' of the claimed invention, courts proceed to determine who has contributed to the formulation of it. It is

29. *University of Southampton's Applications* [2005] RPC 220 at 234, per Laddie J. See also Dien Ghin Electronic (S) Pte Ltd *v* Khek Tai Ting [2011] 3 SLR 227 at [13].
30. *Rhone-Poulenc Rorer Inc v Yeda Ltd* [2008] RPC 1 at [20], per Lord Hoffmann. See also Henry Bros (Magherafelt) Ltd *v* Ministry of Defence [1997] RPC 693 at 706.
31. *University of Southampton's Applications* [2005] RPC 220 at 234, *per* Laddie J.
32. *Markem Corp v Zipher Ltd* [2005] RPC 31 at 798, *per* Jacob LJ: 'what one is normally looking for is "the heart" of the invention. There may be more than one "heart" but each claim is not to be considered as a separate "heart" on its own'.
33. *IDA v University of Southampton* [2006] RPC 21 at [43], *per* Jacob LJ. In fact, the judge advised that in identifying the 'inventive concept', one should 'proceed more like a hedgehog than a fox'.
34. *Markem Corp v Zipher Ltd* [2005] RPC 31 at 798, *per* Jacob LJ citing with approval the judgment of Christopher Floyd QC in *Stanelco Fibre Optics Ltd's Applications* [2005] RPC 15.
35. See s. 25(5)(*d*) Patents Act and s. 14(5)(d) of the Patents Act 1977 (c 37) (United Kingdom).
36. *Markem Corp v Zipher Ltd* [2005] RPC 31 at 798, *per* Jacob LJ.
37. [2008] RPC 1.

important to remember that the onus is on the person who is asserting entitlement to show that on a balance of probabilities they had made a contribution to the essential elements of the invention.[38]

In general, non-creative or non-technical contributions are not considered meaningful contributions to the formulation of the 'inventive concept' of the claimed invention and are disregarded in the courts' assessments. Non-creative contributions generally refer to those by persons who merely assisted in carrying out the experiments as per the instructions of the principal investigators or researchers; cleaned up the utensils in the laboratories or supplied the necessary starting materials.

Similarly, non-technical contributions are also not taken into consideration and these typically include those by persons who merely provide administrative support such as funding, staffing, and facilities to the actual researchers to carry out their research and experiments.

4.2.2 Employee-Inventors

As discussed above, although any person can apply for a patent either alone or jointly with another person, only very limited categories of persons, for instance, inventors or joint inventors in title, are entitled to be granted a patent.[39] Granting patent rights to the inventors or joint inventors is consistent with the broad objective of the patent system, which is to provide incentives and to reward people who have laboured over their inventions.

Within this basic framework of patent ownership and protection, there is a major exception for inventions produced by employees. In Singapore, as is the case in many other jurisdictions, ownership of inventions made by employees in the course of employment vests with the employers and not the employee-inventors.

4.2.2.1 Who Is an 'Employee'?

Section 2(1) of the Patents Act defines an 'employee' to mean 'a person who works or (where the employment has ceased) worked under a contract of employment or in employment under or for the purpose of a Government department'.

In common law, if an employee makes an invention in the course of their employment, they hold their interest in the invention and in any resulting patent on trust for their employer, unless they can show that they have a beneficial interest which the law recognises.[40]

38. See *Viziball Ltd's Application* [1988] RPC 213.
39. S.19(2) Patents Act. Besides inventors and joint inventors the categories also include persons who by law, treaty or international convention, or prior contractual arrangements with the inventor, were entitled to the whole of the property in the invention in Singapore, and the successor or successors in title of the first two categories. There is also a presumption that the person or persons making the application for patents are persons so entitled under s. 19(2) Patents Act. See s. 19(3) Patents Act.
40. *Sterling Engineering Co Ltd v Patchett* [1955] 1 All ER 369 at 373, *per* Viscount Simonds.

The key statutory reformulation establishing the rules on ownership for employee-inventors' inventions is Section 49 of the Patents Act, which is identical to Section 39 of the Patents Act 1977 (UK). Section 49 of the Patents Act provides:

> (1) Notwithstanding anything in any rule of law, an invention made by an employee shall, as between him and his employer, be taken to belong to his employer for the purposes of this Act and all other purposes if:
> (a) the invention was made in the course of the normal duties of the employee or in the course of duties falling outside his normal duties, but specifically assigned to him, and the circumstances in either case were such that an invention might reasonably be expected to result from the carrying out of his duties; or
> (b) the invention was made in the course of the duties of the employee and, at the time of making the invention, because of the nature of his duties and the particular responsibilities arising from the nature of his duties he had a special obligation to further the interests of the employer's undertaking.
> (2) Any other invention made by the employee shall, as between him and his employer, be taken for those purposes to belong to the employee.
> (3) Where by virtue of this Section an invention belongs, as between him and his employer, to an employee, nothing done:
> (a) by or on behalf of the employee or any person claiming under him for the purposes of pursuing an application for a patent; or
> (b) by any person for the purpose of performing or working the invention, shall be taken to infringe any copyright or design right to which, as between him and his employer, his employer is entitled in any model or document relating to the invention.
> (4) In subsection (3), the reference to application for a patent includes an application for other protection for an invention, and includes an application for a patent or any other protection under the law of a country other than Singapore or under any treaty to which Singapore is a party.

Section 49 of the Patents Act only governs the rights and liabilities between the employer and the employee. Third party rights are not affected.

Under Section 49 of the Patents Act an invention belongs to the employer and not to the employee-inventor in only two situations.

(i) When the invention was made in the course of the normal duties of the employee or if the duties were specifically assigned to them (s. 49(1)(a) Patents Act)

In general, the employee-inventor who falls into this category is someone who has been employed to design or invent and the invention arises in the course of their duties. Specifically, the employer claiming entitlement to the invention under Section 49(1)(*a*) of the Patents Act must prove:

(a) the existence of an employment relationship between the parties;
(b) the invention must have been made during the course of the employee's *normal duties* or their *specifically assigned duties*; and
(c) an invention might reasonably have been expected to result from the carrying out of such duties.

It has been said that there is no single, general test that was determinative of whether a person was an employee of another or otherwise. In the Singapore context, the court takes guidance of the following non-exhaustive factors in its analysis and the outcome of which is dependent on the circumstances of each case:[41]

(a) whether the work of the alleged employee was done as an integral part of the business of the alleged employer;
(b) whether the alleged employee was paid a regular salary or commission, the former is more likely to be treated as an employee than the latter;
(c) whether there were stipulations as to working hours;
(d) whether the alleged employee was entitled to overtime pay;
(e) whether the alleged employer contributed to the Central Provident Fund[42] account of the alleged employee;
(f) whether the alleged employee was entitled to leave and holidays;
(g) whether the alleged employee was entitled to medical leave;
(h) whether the alleged employer had the power to dismiss the alleged employee from his service.

An employee's 'normal duties' are the actual duties which they are employed to do.[43] What these duties may entail may be gleaned from the terms of the employment contract,[44] the job description[45] and the nature of the business involved.[46] Any extra or different duties to be undertaken by the employee beyond the 'normal duties' may be regarded as duties 'specifically assigned' when they were first assigned to the employee but it is also possible that in the course of time these 'specifically assigned' duties may become 'normal duties' of the employee.[47] This point was demonstrated by *LIFFE Administration v Pinkova* [48] in which the English Court of Appeal found that over a period of time, the design of an exchange-tradable credit derivative had become part of

41. *National University Hospital (Singapore) Pte Ltd* v *Cicada Cube Pte Ltd* [2017] SGHC 53, *per* Aedit Abdullah JC at [84].
42. The Central Provident Fund ('CPF') is a mandatory comprehensive savings scheme for working Singaporeans and permanent residents. The CPF was set up in 1955 with the objective of helping Singaporeans and permanent residents to fund their retirement, healthcare, and housing needs. It is an employment based scheme with the employer and the employee each contributing a mandated quantum into the CPF account held by the employee.
43. *Harris' Patent* [1985] RPC 19.
44. *Greater Glasgow Health Board's Application* [1996] RPC 207 at 221, *per* Jacob LJ.
45. *Ibid.*, at 222, *per* Jacob LJ.
46. *Harris' Patent* [1985] RPC 19 at 30, *per* Falconer J.
47. *LIFFE Administration v Pinkova* [2007] EWCA Civ 217.
48. *Ibid.*

the employee's normal duties when this was not what the employee was first hired to do. Therefore, it is important to remember that the duties of an employee are not necessarily 'cast in stone' and may not be restricted to the terms originally found in the written contract of employment. These duties are capable of expansion and contraction in the course of time 'by a continuous process of subtle variation'.

As for the condition that *the circumstances* in either of the two situations (normal duties or specifically assigned duties) are such that an invention might reasonably have been expected to result from the carrying out of such duties, the courts in the UK have held that the test is an objective one. The circumstances referred to in the Section are those surrounding the making of the invention by the employee and not the general circumstances of employment. At the end of the day, the court should be satisfied that considering the 'objective circumstances in which the invention was made',[49] i.e. that the employee is someone who has been hired to invent or innovate, it is reasonable to expect an invention to result from the discharge of these duties. If so, the invention rightfully belongs to the employer.

Even if the employee is someone who has not been employed by the employer as an inventor or researcher, and even if there are no express terms in the employee's letter of appointment or contract of employment that imposed any contractual duty for the employee to research, innovate or invent for the employer, the court may still conclude that the invention was reasonably expected to result from the carrying out of the employee's normal duties in the course of his employment. This was the conclusion which the Singapore High Court came to in the recent decision of *National University Hospital (Singapore) Pte Ltd* v *Cicada Cube Pte Ltd* [2017] SGHC 53. The case centred upon the ownership and recognition as inventors of a patented process for the collection of laboratory specimens in a hospital. The defendant, Cicada Cube Pte Ltd, was a software engineering company which argued that its employees came up with the invention whilst the plaintiff, National University Hospital (Singapore) Pte Ltd claimed that its employees, including one senior doctor, Dr Sethi, had developed the invention after years of testing. One of the issues contended by the defendant was that even if the court found the inventor Dr Sethi to be the plaintiff's employee, he did not conceive of the invention during the course of his duties, and further, the invention could not reasonably be expected to result during the course of Dr Sethi's normal duties which primarily concerned that of treatment of patients and not to develop technology that might assist the treatment of patients. Disagreeing with the defendant, the High Court judge found that the invention was indeed one which reasonably expected to result from Dr Sethi's carrying out of his normal duties as the Chief of the Department of Laboratory Medicine of the hospital. In coming to this conclusion, Judicial Commissioner Aedit Abdullah opined:

> The courts needed to recognize that there was a strong innovative role inherent in many jobs these days, even if it was not explicitly mentioned within the normal or day-to-day job scope. All employees, whether in the public or private sector,

49. *Ibid.*

> whether managerial or clerical, were expected to contribute to the organization by proposing innovations or improvements. What counted as being within the scope of employment had to be seen in that light.

Within the factual matrix of the case, the conclusion of the Singapore High Court must be correct as Dr Sethi was undoubtedly a 'key' employee of the plaintiff and it was his duty to look into ways to improve the hospital's operations and workflow processes for the management of laboratory specimen collection. Nevertheless, the High Court's observation that *all* employees regardless of their job scope are expected to contribute to his or her organisation by way of proposing innovations or making improvements does have the possible effect of significantly enhancing the position of an employer in staking claims over inventions created by his employees. For firms to stay competitive, employees must have an innovative role in their job scope, this is both necessary and important. Whilst this is sound public policy and should be supported, a distinction should still be made between key employees who are in senior managerial positions and junior clerical staff as a failure to do so may, it is submitted, lead to unjust outcomes and may result in the unintended effect of disincentivising disclosures of ground breaking innovations or improvements by employees.

(ii) When the invention was made in the course of the duties of the employee and because of the nature of their duties they had a special obligation to further the interests of the employer's undertaking (s. 49(1)(b) Patents Act)

To come within the situation provided in Section 49(1)(*b*) Patents Act the employer must show:

(a) the existence of a relationship of employment between the parties;
(b) the invention was made during the course of the duties of the employees; and
(c) the nature of the employee's duties and responsibilities were such that the employee had *a special obligation* to further their employer's interests.

The ambit of the UK's equivalent of Singapore's Section 49(1)(*b*) Patents Act – Section 39(1)(b) Patents Act 1977 (UK) – was considered in *Harris' Patent*.[50] Falconer J considered 'that the extent and nature of the "special obligation to further the interests of the employer's undertaking" will depend upon the status of the employee and attendant duties and responsibilities of that status'.[51] Therefore, normally a managing director's obligation to further the interests of their employer's undertaking cannot be the same as the obligation to be expected from a junior sales manager. Further guidance on how courts have applied the legal principles embodied in Section 39(1) (b) of the Patents Act 1977 (UK) can be extracted from older as well as more recent English case law.

50. [1985] RPC 19.
51. *Ibid.*, at [37]–[38].

In *Worthington Pumping Engine Co v Moore*,[52] a case which took place at the turn of the twentieth century, the defendant was a senior Vice-President in the employment of the plaintiff, an American pump corporation. The defendant was charged with the sole control of the plaintiff's business outside the United States and was paid USD 10,000 per annum (a large sum of money at the time). It was evident from the facts that as a result of the defendant's senior status in the company, he was privy to important information regarding new ideas or methods of construct relevant to the plaintiff's business. Using information which was privy to him as a result of his senior status in the company, the defendant developed improvements in the plaintiff's pumps which resulted in the defendant making two patent applications in his own name. The court expected the defendant not to misuse such critical information that was the subject of patents for his own personal benefit at the expense of the plaintiff corporation.

Essentially, this meant that the defendant, by virtue of his status in the plaintiff corporation as well as the nature of his obligations and responsibilities, had a *special obligation to further the interests* of his employer (the plaintiff corporation). Accordingly, the court held that the two patents which the defendant had registered in his own name whilst being in the employment of the plaintiff were to be held by him on trust for the plaintiff.

More recently, in *Staeng Limited's Patents*,[53] a case before the UK Patent Office,[54] a senior executive of a company, who had the right to participate in his company's profit bonus scheme, was held to be in a position in which he had a special obligation to further the interests of his employer's undertaking. The patents in suit were for securing electric cable sheathing to the body portion of a 'connector backshell adaptor'[55] and were registered in the name of S and referred to R as the inventor. N sought an order to name him as the sole inventor and to grant him all rights in the UK in respect of these patents. R was the employee of S and N was the employee of H. H was a much larger company than S and S was subsequently acquired by H. It was not disputed that prior to the acquisition, S and H worked closely on inventions. It was also not disputed that as a result of the close collaboration and frequent visits by R to N's office showing N a backshell adaptor which S has developed, N came up with the inventive concept of a constant spring tension to hold the cable screen to the adaptor without the use of specialised machinery. This was the nub of the inventive concept in the patents in suit.

---

52. (1903) 20 RPC 41.
53. [1996] RPC 183.
54. The Patents Law Amendment Act 1852 (c 83) (UK) brought the Patent Office into operation on 1 October 1852. The Patent Office was initially responsible for patents but has also taken on registered designs and registered trade marks. On 2 April 2007, the Intellectual Property Office became an operating name of The Patent Office.
55. See [1996] RPC 183 at 187 where Dr P Ferdinando, the hearing officer in this case, explained the invention:

    The adaptor is used to form a cable termination for connection to a complementary connector, for example in forming cable harnesses. The sheathing can either be the braid of a cable or a heat-shrinkable or flexible plastics sleeve. The sheathing is placed over the rear end of the backshell adaptor and is rigidly secured there by means of a spiral spring wrapped around the adaptor over the sheathing. As disclosed the spring engages in a recessed portion between tow flanges on the adaptor and prevents the sheathing being pulled off.

Therefore, it was accepted that N was an inventor and the only issue before the UK Patent Office was whether he was the sole inventor. It was accepted by the registrar that N's duties and obligations did not require him to invent and that his responsibilities within H were not concerned with backshell adaptors. Nevertheless, N's job description assigned to him the creative role of searching for new products and to think of novel uses for existing products.

The Patent Office held that as a senior executive in the company with the right of participation in his employer H's profit bonus scheme, N must have recognised that an invention in the field of backshell adaptors would be advantageous to his employer H and as a result of his status, N had a special obligation to further the interests of his employers. Therefore, the patents were held to belong to H, the employers of N.

4.2.2.2 Relevance of the Terms in the Employment Contract

It is common to find express terms in an employment contract the effect of which is to vest ownership of any invention or other intellectual property rights created by the employee in the course of employment in the employer. What is the effect of such contractual terms on the issue of ownership of an invention made by an employee in Singapore? More importantly, can parties contract out of the statutory rules governing ownership of employee-inventors' inventions under Section 49 of the Patents Act? Although the rights to an employee's invention are essentially governed by Section 49 of the Patents Act, parties are free to enter into contracts which may alter the statutory rules on rights to an invention. This is clear on a reading of Section 50(4) of the Patents Act which provides that:

> Nothing in this Part [IX] shall be construed as precluding the operation of an agreement or a contract in relation to the right to an invention.

Therefore, parties in Singapore are able to contract out of the statutory rules governing rights to an invention under Section 49 of the Patents Act provided the relevant contractual terms entered into are enforceable in law and properly cover the issues of ownership.[56]

In this regard, *Electrolux v Hudson*,[57] decided before the enactment of the Patents Act 1977 (UK) is instructive. The first and second defendants were the inventors in this case. They were a married couple. In their spare time, the first defendant and his wife devised an adaptor for use in vacuum cleaners which allowed any open-mouthed bag to fit into a vacuum cleaner for the purpose of collecting the dust sucked up by the

56. By contrast, certain contracts entered into by an employee in contravention of s. 42 of the Patents Act 1977 (UK) are not enforceable. In particular, 'any term in a contract ... which *diminishes the employee's rights in inventions of any description made by him* after the appointed day and the date of the contract ... *shall be unenforceable against him to the extent that it diminishes his rights in an invention of that description so made*, or in or under a patent for such an invention or an application for any such patent' (emphasis added). See s. 42(2) Patents Act 1977 (UK).
57. [1977] FSR 312.

cleaner. The significance of the defendants' invention was that the adaptor was capable of use on any vacuum cleaner which used an integral part, a disposable bag.

The first and second defendants applied for a patent in respect of their invention. At the time of the making of the invention, the first defendant was employed by the plaintiffs, Electrolux Limited, as a senior storekeeper. The plaintiffs claimed that the invention was made by the first defendant in the course of his employment and in accordance with clause 8 of their 'Standard Conditions of Employment for Staff', the plaintiffs were entitled to the benefit of the invention. The defendants argued that the invention was created during their spare time without using any knowledge acquired by the first defendant as a result of his employment and that even if clause 8 of the 'Standard Conditions of Employment for Staff' was to apply, it was too wide to be enforceable.

The outcome thus turned on the proper interpretation of clause 8 a part of which stated:

> If at any time during the period of your employment, you shall invent, discover or ascertain any process, invention or improvement (whether patentable or not) relating to any articles manufactured and/or marketed by the company or its associated companies or undertakings or firms in the United Kingdom or elsewhere, you shall forthwith, at the expense of the company, disclose to the company or any agent appointed by it, a full and complete description of the nature of the said process invention, discovery or improvement and the mode of performing it, and the said process, invention, discovery or improvement shall remain the sole and absolute property of the company; and you shall, if and when required by the company, at the expense of the company, join with the company in applying for letters patent for the said process, invention, discovery or improvement (if patentable) and/or similar protection in any part of the world and shall on the request by, and at the cost of, the company, execute and do all such instruments and things necessary to test the said process, invention, discovery or improvement, and any letters patent and or other similar protection that may be obtained relating thereto, in the company or otherwise as the company may direct.

Whitford J held the above clause was far too wide to be enforceable. He confirmed the appropriateness and legitimacy of employers inserting such express terms governing ownership in employees' inventions in contracts of employment. In fact, if there are no express terms in the contract of employment, courts have been ready to imply appropriate terms and conditions to safeguard the interests of the employer. However, 'what sort of clause may be appropriate must necessarily depend upon the nature of the work done by the employee'.[58]

In this case, the court found that the clause was far too wide, even in the case of research workers, as it purported to cover 'the discovery or ascertainment of any process, invention or improvement relating to any articles manufactured and/or marketed not only by the company, in this case Electrolux Limited, but its associated companies or undertakings or firms in the United Kingdom or elsewhere'.[59] Thus, the

58. *Electrolux v Hudson* [1977] FSR 312 at 322, *per* Whitford J.
59. *Ibid.*, *per* Whitford J.

application of such a clause on all other members of the staff including a storekeeper like the first defendant would undoubtedly give rise to absurd results.

Another area in which contractual terms play a significant role in the ownership of employees' inventions is in academia. As a general rule, it is entirely legitimate for a university to employ a faculty member on condition that the faculty member assigns to the university all inventions and intellectual property rights arising out of the member's employment with the university.[60] This must be the case too for universities in Singapore given the effect of Section 50(4) of the Patents Act. However, it must be appreciated that university academics differ from ordinary employees of private commercial entities in that they are not merely employees of the university but are also members of the university's faculties.

As far as undertaking research is concerned, a university faculty member is not obliged to further the commercial interests of the university and can decide on their own what research to pursue and how to pursue it. Consequently, courts are slow to imply into employment contracts of university faculty members terms and conditions that call for ownership of inventions.

In general, the duty to disclose patentable inventions by members of the faculty in a university and issues of ownership in patents are governed by the university's intellectual property policies which are usually incorporated into the employment contracts.

4.2.2.3 *Putting the Patent Grant in Issue*

Section 19 of the Patents Act governs the right of entitlement to be granted a patent and Section 19(3) of the Patents Act provides a rebuttable presumption that the person who applies for a patent is entitled to a grant. Section 19 of the Patents Act corresponds to Section 7 of the Patents Act 1977 (UK).

Issues on entitlement to a patent grant can arise in the following factual situations:

(a) a patent is granted to 'B' who is an employee of 'A'. The employer 'A' claims entitlement to the patent grant under Section 49 of the Patents Act.
(b) a patent is granted solely to 'B'. 'A' is of the view that he has contributed to the 'inventive concept' of the claimed invention.
(c) a patent is granted to 'B'. 'A' asserts that he has independently created the claimed invention ahead of 'B'.
(d) a patent is granted to 'B'. 'A' contends that 'B' has created the claimed invention using confidential information communicated to 'B' by 'A'.
(e) a patent is granted to 'B', who is an ex-employee of 'A'. The former employer 'A' argues that 'B' has used trade secrets which belonged to 'A' as 'B's' former employer.

60. *University of Western Australia v Gray* [2008] FCA 498; (2008) 246 ALR 603.

The procedures for putting a patent grant in issue are provided in Sections 20, 21 and 47 of the Patents Act. Section 20 of the Patents Act (that corresponds to Section 8 of the Patents Act 1977 (UK)) sets out the procedures for the determination of disputes on issues of entitlement referred to the registrar before the grant of a patent. If a reference on issues of entitlement has been made to the registrar before the grant of a patent under Section 20 of the Patents Act but these are not determined before the time of grant, Section 21 of the Patents Act (that corresponds to Section 9 of the Patents Act 1977 (UK)) operates to allow the patent application to proceed to grant but the outstanding issues on entitlement will be determined in accordance with Section 47 of the Patents Act in the same manner as disputes on issues of entitlement referred to the registrar after grant. It is important to note that there is a two year time limit from the date a patent is granted for any reference to the registrar on issues of entitlement.[61]

What does a person claiming entitlement to a patent registered in another's name have to prove under Section 19 of the Patents Act in order to successfully put the patent grant in issue?

In the past, it was thought that to successfully put a patent grant in issue, the only thing which a challenger needed to show was simply that they are entitled to the patent grant under Section 19(2) of the Patents Act (Section 7 of the Patents Act 1977 (UK)). Nothing further was required. However, in a controversial re-appraisal of the law, the English Court of Appeal in *Markem Corp v Zipher Ltd* [62] held that to successfully put a patent grant in issue in a first-to-file system like the UK, it is not enough for the challenger to merely show that they are the inventor and the registered owner of the patent is not the inventor. The challenger must go one step further: to establish that the registered owner of the patent is *not* entitled to the patent because they had procured the grant by some other breaches of the law, e.g., breach of confidence, contract or statutory duties.

The facts in *Markem Corp v Zipher Ltd* were uncomplicated. Markem and Zipher both made thermal printers. Zipher was formed by former employees of Markem. Shortly afterwards, the ex-employees solved a problem in thermal printing, the goal of which was identified whilst they were still working for Markem. The ex-employees filed for a patent. Markem claimed that it was entitled to the patent because it had helped identify the goal. The Court of Appeal held that Markem could put the patent grant in issue if it could show that:

(a) it was entitled as joint inventor; and
(b) there existed a reason why the ex-employees were *not* entitled to apply for the patent, either collectively or individually.

This essentially meant that Markem needed to show that the ex-employees realised the inventive concept whilst employed by Markem and that the ex-employees acted in breach of confidence or contrary to a reasonable restraint of trade provision in their employment contracts when applying for the patent. Markem failed to do so.

61. S. 47(5)(*b*) Patents Act.
62. [2005] 31 RPC 761.

The Court of Appeal was concerned that if this was not the rule 'there would be great uncertainty and damage to freedom of employment and research'[63] and Jacob LJ, delivering the leading judgment, explained:[64]

> We have in mind particularly the case of ex-employees. Whether whilst still employed or thereafter they cannot use 'chicken 3'[65] material, whether for themselves or for their new employer. That is a rule founded in conscience and enforced by equity. But information which they are free to use is different. If the law imposed any sort of impediment on that the consequences would be serious. So if they use part of their 'permitted' information to make an invention, they (or their new employer) should be entitled to apply for a patent.
>
> ...
>
> If the subsequent patent consists of material which is not the confidential information of the former employer it can hardly be patentable. It will be old or obvious. The former employer's remedy is not to claim entitlement to such a claim, but to have it revoked. The 'unfairness' only arises in relation to such a case. If the employee has taken secret information there will be no unfairness – for the action for breach of confidence as well as an entitlement under s. 8 [cf the Patents At 1977 (UK)] will lie at the suit of the employer.

According to Jacob LJ, the truly complicated cases on entitlement would involve 'an invention [which] is made partly using information in breach of confidence and partly information added to that – where the applicant has added his own ingenuity to robbery'[66] and *Markem Corp v Zipher* was not such a case.

One year later, the same court was again confronted with issues on inventorship and entitlement in the case of *IDA v University of Southampton*.[67] Professor H, a professor with the University of Southampton, patented a device for killing insects by the use of poison talcum powder given an electrostatic charge so that it stuck to the insects' legs. An employee of IDA, having read a report of Professor H's invention in the local newspapers, telephoned Professor H and suggested an improved invention whereby the poison talcum powder was given a magnetic charge instead. The suggestion worked. Consequently, Professor H and the University of Southampton applied for and were granted a patent. IDA applied to have the patent put in its name.

The English Court of Appeal held that the inventive concept in the patent was the idea to use magnetic charge rather than electrostatic charge (all other elements of the claimed invention were not new because of the prior patent held by Professor H) and thus, the IDA employee was the inventor. The employee of IDA had imparted the inventive concept during the telephone conversation with Professor H and that conversation was subject to an implied duty of confidentiality. Applying *Markem Corp*

63. *Ibid.*, at [80], *per* Jacob LJ.
64. *Ibid.*, at [80].
65. In *Faccenda Chicken Ltd v Fowler* [1985] FSR 105, Goulding J described the classes of trade secret which the law would protect by an action for breach of confidence after the employee had left his employment and 'chicken 3' was the third class of trade secret mentioned: 'specific trade secrets so confidential that, even though they may necessarily have been learnt by heart, and even though the servant may have left the service, they cannot lawfully be used for anyone's benefit but the masters'.
66. *Ibid.*, at [86].
67. [2006] RPC 21.

v Zipher, IDA had shown both that it was entitled to be granted the patent under Section 7 of the Patents Act 1977 (UK) and had provided a separate legal reason why Professor H was not entitled, because he was in breach of confidence. Therefore, IDA's challenge to the patent grant was successful and the patent was put in IDA's name.

It is important to note that Jacob LJ's formulation in *Markem Corp v Zipher*, that in a first-to-file system a challenger should not be entitled simply because they independently arrived at the same invention unless they were also able to provide a separate legal reason why the person registered with the patent was *not* entitled to the patent was overruled by the House of Lords in the UK in the more recent case of *Rhone-Poulenc Rorer International Holdings Inc v Yeda Research and Development Co Ltd* [68] ('*Rhone-Poulenc*'). The House of Lords held subsequently that the challenger needed only show that the patent could have been granted to them under Section 7(2) of the Patents Act 1977 (UK) and if the challenger claimed sole entitlement to the patent, they must also prove that the person registered with the patent was *not* entitled to be granted the patent under Section 7(2) of the Patents Act 1977 (UK), i.e. they were not an inventor of the claimed invention. There was no further need to prove the second element in *Markem Corp v Zipher* that by reason of a breach of confidence or contract the person granted was not entitled.

The House of Lords found that the Court of Appeal was wrong to think that there was a problem when the challenger 'A' is an independent creator of the inventive concept in the patent registered in the name of 'B'. In particular, Lord Hoffmann considered that the Court of Appeal's reasoning in *Markem Corp v Zipher* 'involved a confusion of rules which go to the question of who is *entitled* to a patent and rules which go to the *validity* of a patent', the former 'are exhaustively stated in Section 7(2) [of the Patens Act 1977 (UK)]'[69] whilst the latter 'are well known and include the requirement of novelty'.[70]

His Lordship explained why the broad principle laid down by Jacob LJ in *Markem Corp v Zipher* was wrong and that proof of the second element was unnecessary even if the challenger was an independent creator of the claimed invention with the help of the following three scenarios:[71]

(i) If 'B' (the person with the patent grant) arrived independently at the same invention, 'B' is entitled because he filed first. 'A' (also an independent creator of the invention) cannot apply under Section 37 of the Patents Act 1977 (UK) [the equivalent of Section 47 of the Patents Act] to take over a patent in which he had no involvement. Furthermore, 'A' cannot apply for a separate patent, because once 'B' has applied for his patent 'A's' application lacks novelty. 'A' may however rely on the defence of prior user in an infringement proceeding. In this scenario, 'B' is entitled.

68. [2008] RPC 1.
69. *Rhone-Poulenc Rorer International Holdings Inc v Yeda Research and Development Co Ltd* [2008] RPC 1 at [26], per Lord Hoffmann.
70. *Ibid.*
71. *Ibid.*, at [28].

(ii) If 'B' used an inventive concept which 'A' communicated to him without imposing a duty of confidence, the information will be deemed to have been communicated to the public. 'B's' patent is invalid for lacking novelty. If 'A's' patent was filed after the communication to 'B', 'A's' patent may also be invalidated due to lack of novelty. Thus, neither 'B' nor 'A' is entitled to the patent. Both can use the invention freely.

(iii) If 'B' used an inventive concept which 'A' communicated to him in confidence, the patent is valid and 'A' can claim entitlement. The reason 'A' is entitled is because he was the 'actual deviser', i.e. the inventor, and not because 'B' was in breach of confidence. The duty of confidence goes to the question of invalidity and not entitlement. 'A' can claim sole entitlement if he can show that 'B' was a mere pirate and was not a joint inventor. In this scenario, 'A' is entitled as an inventor under Section 7 of the Patents Act 1977 (UK) [the equivalent of Section 19 of the Patents Act].

On the topic of entitlement, it is understood that the position in the UK may be summarised as follows:

(a) Section 7 of the Patents Act 1977 (UK) exhaustively governs the right of a person to be granted a patent. There is no need to refer to other principles of law;
(b) the correct approach in an entitlement dispute is to identify the inventor or inventors; and
(c) it is important to distinguish issues on entitlement from issues on validity. Breach of confidence affects the validity of the patent and it may shed light on whether the party granted the patent was truly the 'actual deviser' of the claimed invention. Mere pirates who misuse others' confidential information to create inventions are entitled to file patents for their creations. However, these individuals may find themselves accountable for their actions under the law of confidence.

These principles are equally applicable in the Singapore context and serve as useful and persuasive guidelines for the courts.

4.3 The Parties' Duty to Inform[72]

4.3.1 Obligations of Employees

In general, employees are expected in law to always act in good faith in furtherance of the interests of their employers. Therefore, an employee must not disclose or use confidential information belonging to the employer for their own personal profit during

72. The following section was taken from pages 533-534 of *Entitlement* in Susanna Leong, *Intellectual Property Law of Singapore* (Academy Publishing, 2013). <http://www.sal.org.sg/Lists/BookTitles/DispForm.aspx?ID = 219>. Reproduced with permission.

the period of their employment and thereafter. However, the law does not extend this protection further than is necessary to protect the employer's proprietary interests. As such, after employment has ceased, an employee is free to make use of general knowledge acquired during their employment.

The law takes a similar attitude towards inventions created by an employee during their employment. All inventions created by an employee during their employment, as well as the patent rights associated with the inventions, belong to their employer. An employee is under a duty to impart these inventions to their employer and to hold any resultant patent on trust for their employer. This is the position even if the employer may not intend to use or exploit the invention until a much later time after the cessation of the employee's contract of service. The employee continues to hold these inventions resulting from any innovative ideas and activities on trust for their employer and can be asked to assign them to their employer. Whilst it is possible that there may be circumstances which amount to a release of their rights by the employer, courts in the UK have held that the mere fact that an employer did not think it necessary or desirable to use the invention created by their employee until some years after the termination of the employment was not in itself adequate evidence of such a release or waiver of the employer's right to the said invention.[73]

4.4 Employee Remuneration Right[74]

4.4.1 Employee-Inventors' Compensation Rights

Although the provision on employees' inventions in Singapore is similar to that in the UK, Singapore did not adopt the employee compensation scheme[75] which gives a statutory right to the employee to 'a special bonus'.

In the UK, this right only arises when:

(a) the patent for the invention is of outstanding benefit to the employer;[76] and
(b) it is just that compensation be given.

It seems that the scheme is an attempt by the UK legislature to reduce the unfairness in the basic legal framework of patent ownership and monopoly. In Singapore, the Patents Act has instead left the matter to be decided between the parties by private contractual arrangement.

As discussed above, under Section 50(4) of the Patents Act parties are free to enter into contractual arrangements which may determine issues of ownership in employee-inventors' inventions. Given the unequal bargaining power of employer and

73. *Triplex v Scorah (*1937) 55 RPC 21.
74. The following section was taken from pages 533-554 of *Entitlement* in Susanna Leong, *Intellectual Property Law of Singapore* (Academy Publishing, 2013). <http://www.sal.org.sg/Lists/BookTitles/DispForm.aspx?ID=219>. Reproduced with permission.
75. Ss. 40-41 Patents Act 1977 (UK).
76. It is important to note that it must be the patent that is of outstanding benefit to the employer and *not* the invention.

employee, governmental intervention may be considered desirable, if not necessary. If the matter is left to private contractual arrangements, the employee, for economic reasons or otherwise, may have no choice but to accept the standard terms of the employer. Unless the employee could prove that the arrangement was entered into under undue influence or duress at common law, there is little that they can do to protect their own interests.

As a matter of fairness and justice, some kind of employees' compensation scheme ought to have been adopted by the Patents Act. Fundamentally, the enactment of a statutory right for the employee to request compensation also effectively strengthens their bargaining position. Consequently, a more equitable arrangement can perhaps be reached between the parties.

On the other hand the employees' compensation scheme is not free of problems. Many of the ambiguities in the wording of in Section 40 of the Patents Act 1977 (UK), e.g., the meaning of 'outstanding benefit', have yet to be resolved.

Disputes of this kind can be protracted and expensive. Since many employees may not have the resources to litigate, such disputes are likely to be settled out of court. Furthermore, the statutory scheme may be successfully avoided by other arrangements and the conduct of the parties.[77]

It is accepted that although the UK statutory employee compensation scheme may appear generous on paper, it is in fact very difficult to engage in practice.[78] Given these competing considerations, Singapore has favoured a simple, albeit a less generous, approach on the issue.

5 EMPLOYEES' UTILITY MODELS

Singapore does not have a regime for registration of utility models.

6 EMPLOYEES' DESIGN RIGHTS

6.1 The Legal Framework[79]

Legal protection for industrial designs in Singapore has a very short history. In fact, until the year 2000, Singapore did not have her own laws on industrial designs nor the

77. See Cornish, Llewelyn and Taplin, *Intellectual Property: Patents, Copyright, Trade Marks and Other Allied Rights* [7-07]–[7-15] (7th ed., Sweet and Maxwell 2010). The compensation scheme may be replaced by collective agreements made by or on behalf of either the employee's or the employer's union with the other, individual agreements between the parties and by the employer in not patenting the invention.
78. See British Steel's Patent [1992] RPC 117; Memco-Med's Patent [1992] RPC 403 and GEC Avionics' Patent [1992] RPC 107.
79. The following section was taken from pages 593–630 *Industrial Designs* in Susanna Leong, *Intellectual Property Law of Singapore* (Academy Publishing, 2013). <http://www.sal.org.sg/Lists/BookTitles/DispForm.aspx?ID=219>. Reproduced with permission.

legal mechanism to grant design protection rights. For more than half a century,[80] the governing Act in Singapore with regard to protection of designs was the United Kingdom Designs (Protection) Act ('UKDPA').[81]

The UKDPA is a short Act comprising five sections under which owners of designs had first to seek registration of their designs in the UK under the RDA 1949 (UK)[82] or any Act amending or substituted for that Act. Once the design was successfully registered in the UK, Section 2 of the UKDPA in Singapore then granted to the registered proprietor of the design the right to 'enjoy in Singapore the like privileges and rights as though the certificate of registration in the United Kingdom had been issued with an extension to Singapore'. Like the old patent system, this system of registered designs in Singapore is often referred to as the 'piggy back system'. However, this has all changed.

On 25 August 2000, the Registered Designs Act 2000[83] ('RDA 2000') was passed by the Singapore Parliament and came into effect on 13 November 2000. Under the RDA 2000, a system for registration of designs was established in Singapore, wherein owners of designs are able to obtain protection for their designs without having to first seek registration in the UK as required under the old law.

6.2 Ownership of Rights[84]

The basic rule of ownership of a registered design is that the *designer* is owner of the design.[85] In turn, a 'designer' in relation to a design means:[86]

> [T]he person who *creates* it or, if there are 2 or more such persons, each of those persons. (Emphasis added)

This means that the designer is someone who is responsible for the creative process which ultimately yields the design. They are the person who conceives of the idea in the design, in particular the features of the design such as the shapes, textures, colours, patterns, lines as well as the juxtaposition or configuration of all these various elements, pulling them all together to form the design in question.

This basic rule is however subject to a number of exceptions:

80. The United Kingdom Designs (Protection) Act (Act 17 of 1938) came into effect on 16 September 1938.
81. United Kingdom Designs (Protection) Act (Cap 339, 1985 Rev Ed).
82. Registered Designs Act 1949 (c 88) (UK).
83. Registered Designs Act (Cap 266, 2005 Rev Ed).
84. The following section was taken from pages 593-630 *Industrial Designs* in Susanna Leong, *Intellectual Property Law of Singapore* (Academy Publishing, 2013). <http://www.sal.org.sg/Lists/BookTitles/DispForm.aspx?ID=219>. Reproduced with permission.
85. S. 4(1) RDA 2000.
86. 2(1) RDA 2000. See also s. 2 of the Registered Designs Act 1949 (UK): the original proprietor of a registered design is the *author* of the design which in turn is defined to mean the person who *creates* the design.

(a) if the design is created in pursuance of a commission for money or money's worth, the ownership of the design resides with the person commissioning the design;[87]
(b) if the design is created by an employee in the course of their employment, the ownership of the design resides with their employer.[88]

The legal positions of the parties in (a) and (b) above may be altered by contractual agreement:[89]

(a) in situations where a design or the right to apply a design to an article becomes vested in persons other than the owner by way of an assignment, transmission or operation of law, then the ownership of the design resides in the persons so vested;[90] and
(b) for computer-generated designs where there is no human designer, the law deems the person by whom the arrangements necessary for the creation of the design are made as the designer of the design.[91]

The issue of ownership or proprietorship of a registered design is an important one as it affects the plaintiff's *locus standi* to sue for infringement.[92] Under the RDA 2000 an infringement of a registered design is actionable by the registered owner. In other words, a plaintiff who has not established their title or ownership of the design does not have the right to sue. In an infringement proceeding, the alleged infringer can plead, *inter alia*, that the proprietorship of the design lies with a party who is not asserting it and, consequently, the plaintiff has no right to sue.

The RDA 2000 has since undergone a comprehensive review in 2016 and following from this review, Singapore's law on designs will be updated to provide a conducive legal and policy framework for businesses to capitalise on the value of their intellectual property through designs in service and product innovation. One such change in the law pertains to the ownership of design rights. When the amendments to the RDA 2000 are in force after parliament passes the Registered Designs (Amendment) Bill 2017,[93] the rights to a design of a commissioned design will vest in the designer automatically (unless assigned to the commissioner by written agreement), a reversal of the current position in which the commissioner of the design owns the rights.[94] The proposed change is to encourage the growth of Singapore's local design industry and to align Singapore's position with international best practice.

87. S. 4(2) RDA 2000.
88. S. 4(3) RDA 2000.
89. S. 4(4) RDA 2000.
90. S. 4(5) RDA 2000.
91. S. 4(6) RDA 2000.
92. *Hunter Manufacturing Pte Ltd v Soundtex Switchgear and Engineering Pte Ltd* [1999] 3 SLR(R) 1108.
93. Bill No. 23/2017. Read the first time and the second time in parliament on 3 and 8 April 2017 respectively.
94. See section 3 of the Registered Designs (Amendment) Bill 2017.

6.3 The Parties' Duty to Inform

There are no provisions governing the parties' duty to inform.

6.4 Employee Remuneration Right

There are no provisions governing the employee's remuneration rights regarding designs.

7 EMPLOYEES' PLANT VARIETIES

7.1 The Legal Framework[95]

As part of Singapore's obligations under the USA-Singapore Free Trade Agreement signed in 2003,[96] Singapore acceded to the UPOV[97] and became a member on 30 July 2004. In June 2004, the Singapore Parliament passed the Plant Varieties Protection Act[98] ('PVPA') which came into force on 1 July 2004.

The primary objective of the PVPA is 'to provide greater incentive to encourage the development of new plant varieties, and to spur innovation amongst plant breeders in Singapore'.[99]

7.2 Ownership of Rights[100]

Any breeder, whether an individual or a company, can apply for plant variety protection under the PVPA.[101]

A breeder in relation to any plant variety is defined as a person who bred, or discovered and developed, the plant variety.[102] However, if the plant variety was bred, or discovered and developed by a person in the course of performing their duties or functions as an employee of another person, that other person is considered the breeder under the PVPA.

95. The following section was taken from pages 579-589 *Protection of Plant Varieties in Singapore* in Susanna Leong, *Intellectual Property Law of Singapore* (Academy Publishing, 2013). <http://www.sal.org.sg/Lists/BookTitles/DispForm.aspx?ID=219>. Reproduced with permission.
96. The USA-Singapore Free Trade Agreement was signed 6 May 2003.
97. The International Union for the Protection of New Varieties of Plants which came into force on 10 August 1968.
98. Plant Varieties Protection Act (Cap 232A, 2006 Rev Ed).
99. Second Reading Speech on Plant Varieties Bill by Senior Minister of State Associate Professor Ho Peng Kee on 15 June 2004.
100. The following section was taken from pages 579-589 *Protection of Plant Varieties in Singapore* in Susanna Leong, *Intellectual Property Law of Singapore* (Academy Publishing, 2013). <http://www.sal.sg/Lists/BookTitles/DispForm.aspx?ID=219>. Reproduced with perm ission.
101. S. 13(1) PVPA.
102. S. 2 PVPA.

7.3 The Parties' Duty to Inform

There are no provisions governing the parties' duty to inform.

7.4 Employee Remuneration Right

There are no provisions governing the employee's remuneration rights regarding plants varieties.

8 INSTANCES FOR DISPUTES

The courts system in Singapore is made up of the State Courts,[103] the High Court[104] and the Court of Appeal.[105] Civil actions for infringement of copyright may be commenced in the State Courts whilst the appropriate forum for commencement of infringement actions under the Trade Marks Act and the Patents Act is the High Court. Appeals may be made from the judgments of the State Courts to the High Court and the Court of Appeal.

Since 2002, a specialised Intellectual Property Court has been established within the High Court and designated High Court judges, with considerable expertise in intellectual property cases, are appointed to hear intellectual property disputes. There is however a dearth of cases involving ownership of intellectual property, whether copyright, patents or other allied rights, in Singapore.[106]

9 CONCLUDING REMARKS

Singaporean law makes a distinction between the concepts of authorship/inventorship and ownership in respect of rules governing entitlement of intellectual creations such as copyrighted works and patentable inventions. The author/inventor is the person who has created the work or invented the invention whereas the owner is the person entitled to exploit the work or invention created.

In general, the author/inventor is also the owner of the intellectual property created. This is the general rule. However, it is clear from the foregoing discussions that

103. Formerly known as the Subordinate Courts. It has only original jurisdiction.
104. The High Court has both original and appellate jurisdiction.
105. The Court of Appeal only has appellate jurisdiction.
106. See *Alteco Chemical Pte Ltd v Chong Yean What t/a Yamayo Stationary Manufacturer* [1999] 2 SLR(R) 915 (High Court decision on the concept of authorship in copyright) and *Asia Pacific Publishing Pte Ltd v Pioneers and Leaders (Publishers) Pte Ltd* [2011] 4 SLR 381 (Court of Appeal decision on the concept of authorship in copyright). See also *Dien Ghin Electronic (S) Pte Ltd v Khek Tai Ting (t/a Soon Heng Digitax)* [2011] 3 SLR 227 (High Court decision on who is the actual devisor of an invention); *Trek Technology (Singapore) Pte Ltd v FE Global Electronics Pte Ltd* [2005] 3 SLR(R) 389 (High Court decision on the exceptions to the general rule that the patent for an invention shall be granted to the inventor); and *Nguyen Michael Anh v Serial Systems Ltd* [2004] SGHC 264 (High Court decision on employer's right to an invention when it was made by the employee while in the service of the employer).

authorship/inventorship and ownership may not always vest in the same person. The employee is such a case in point. Thus, Singapore's approach is to create *an exception* to the general rule. Such an approach may be explained by the economic incentive perspective of intellectual property laws. Since the purpose of intellectual property laws is encourage the creation of works and inventions, accordingly the person entitled to the right of ownership is the person who provided the funding for the creation of the works and inventions. In the context of the employee-author/ inventor, this person is the employer as the employee is but a mere instrument who serves to fulfil the objective of their employer who has invested resources in the creation of the work or invention.

Singapore's approach may be contrasted with that of other countries which adopt a more 'pro-author' approach. In these countries, the author is treated as an indispensable foundation upon which the work or invention is brought to fruition. Therefore, no exception is made to the general rule and the employee-author/inventor is also the owner of the intellectual property created.

The divergence in the way countries approach the issue of employee's intellectual property rights is expected to persist as countries hold differing views on the explanations why intellectual property rights should be upheld. Harmonisation in this area of the law, whilst desirable, would be difficult to achieve.

BIBLIOGRAPHY

Table of Sources

Second Reading Speech on Plant Varieties Bill by Senior Minister of State Associate Professor Ho Peng Kee (15 June 2004).

Cornish, Llewelyn & Taplin, *Intellectual Property: Patents, Copyright, Trade Marks and Other Allied Rights* (7th ed., Sweet & Maxwell 2010).

Table of Legislation

Copyright Act, Cap 63 (2006 Rev Ed).

Layout-Designs of Integrated Circuits Act, Cap 159A (2000 Rev Ed).

Patents Act (1977) and Cap 221 (2005 Rev Ed).

Patents Law Amendment Act 1852, 15 & 16 Vict (1 October 1852).

Plant Varieties Protection Act, Cap 232A (2006 Rev Ed).

Registered Designs Act, Cap 266 (2005 Rev Ed).

The International Union for the Protection of New Varieties of Plants (10 August 1968).

The United Kingdom Designs (Protection) Act, Act 17 of 1938 (16 September 1938) and Cap 339 (1985 Rev Ed).

The USA-Singapore Free Trade Agreement (6 May 2003).

TRIPS Agreement (Marrakesh, Morocco on 15 April 1994).

Table of Cases

Alteco Chemical Pte Ltd v Chong Yean What t/a Yamayo Stationary Manufacturer [1999] 2 SLR(R) 915
Asia Pacific Publishing Pte Ltd v Pioneers & Leaders (Publishers) Pte Ltd [2011] 4 SLR 381
British Steel's Patent, RPC 117 (1992).
Dien Ghin Electronic (S) Pte Ltd v Khek Tai Ting (t/a Soon Heng Digitax) [2011] 3 SLR 227
Electrolux v Hudson, FSR 312 at 322 (1977).
Faccenda Chicken Ltd v Fowler, FSR 105 (1985).
GEC Avionics' Patent, RPC 107 (1992).
Greater Glasgow Health Board's Application, RPC 207 at 221 (1996).
Harris' Patent, RPC 19 at [37]-[38] (1985).
Hunter Manufacturing Pte Ltd v Soundtex Switchgear & Engineering Pte Ltd, 3 SLR(R) 1108 (1999).
IDA v University of Southampton, RPC 21 at [43] (2006).
LIFFE Administration v Pinkova, EWCA Civ 217 (2007).
Markem Corp v Zipher, RPC 31 at [86] (2005).
Memco-Med's Patent, RPC 403 (1992).
Noah v Shuba, FSR 14 (1991).
Nguyen Michael Anh v Serial Systems Ltd [2004] SGHC 264
Ray v Classic FM, FSR 622 at [641] (1998).
Rhone-Poulenc Rorer International Holdings Inc v Yeda Research and Development Co Ltd, RPC 1 at [26] (2008).
Stanelco Fibre Optics Ltd's Applications, RPC 15 (2005).
Sterling Engineering Co Ltd v Patchett, 1 All ER 369 at 373 (1955).
Trek Technology (Singapore) Pte Ltd v FE Global Electronics Pte Ltd [2005] 3 SLR(R) 389
Triplex v Scorah, 55 RPC 21 (1937).
University of Southampton's Applications, RPC 220 at 234 (2005).
University of Western Australia v Gray, FCA 498, 246 ALR 603 (2008).
Viziball Ltd's Application, RPC 213 (1988).

C In Oceania

CHAPTER 29

Australia

Rights in Intellectual Property Created in Employment

Ann Monotti

1 BACKGROUND

In Australia, the rights of employers in intellectual property ('IP')[1] that arise from the work of their employees are governed by a mix of statutory provisions, the express terms in any contract of employment, and any terms that are implied by law or by fact into that contract. The implication of terms in law follows common law principles developed in employment law cases and then applied to IP disputes involving both copyright and patents. Terms may also be implied by fact to give effect to the parties' presumed or hypothetical intention. In some circumstances of employment, fiduciary obligations not to profit from their position at the expense of their employer and to avoid conflicts of interest and duty may be relevant in the context of IP created by the employee.[2] Finally, it is possible for collective agreements which fall within the operation of the Fair Work Act 2009 (Cth) to deal with matters of employee IP. However, these issues are not generally seen as industrial issues.[3]

1. This term refers to the statutory regimes that control copyright, patents, designs, circuit layouts, plant breeder's rights and trade marks. It also includes the doctrines of passing off and breach of confidence. This chapter only addresses copyright, patents, designs, circuit layouts, and plant breeder's rights.
2. E.g., *Victoria University of Technology v Wilson* (2004) 60 IPR 392, [149]. See 4.2.3 below.
3. Marilyn Pittard, 'The innovative worker: genius, accidental inventor or thief?' in Marilyn Pittard, Ann L Monotti and John Duns, *Business Innovation and the Law: Perspectives from Intellectual Property, Labour, Competition and Corporate Law*, 164-167 (Edward Elgar 2013) (Pittard 2013); John Howe and Andrew Newman, 'Collective bargaining and the ownership of employee

A prerequisite for considering an employer's rights in IP is a finding that the originator was an employee when they created the subject matter. The common law of employment developed principles to clarify the position where the nature of the contractual relationship between the parties is contested. These principles distinguish between a *contract of service* (employees) and a *contract for services* (independent contractors). This inquiry may be relevant to ownership of all forms of IP. In practice, however, it has arisen for judicial consideration only in the context of Section 35(6) of the Copyright Act 1968 (Cth) where the common law distinction is expressly maintained.[4]

An inquiry as to whether a person is an employee under a *contract of service* involves the consideration and balancing of a range of factors, none of which is conclusive. Commonly, evidence will point in both directions so it is the 'totality of the relationship between the parties which must be considered'.[5] Relevant factors that support employee status include the right of the employer to exercise control in the manner of doing work,[6] whether the performance of the work is an integral part of the business,[7] and whether the person is presented to the public as an emanation of the principal.[8] Other factors include 'the mode of remuneration, the provision and maintenance of equipment, the obligation to work, the hours of work and provision for holidays, the deduction of income tax, and the delegation of work by the putative employee'.[9]

Factors that point towards the person being an independent contractor include whether the person was in business on their own account, whether they invoiced for each job they performed and charged the principal goods and services tax ('GST') on the invoices, whether they had their own Australian Business Number ('ABN'),[10] concern of the principal that proper workers' compensation and public liability insurance were arranged,[11] and whether they could continue to perform services for other organisations.[12]

All statutory regimes, except the Patents Act 1990 (Cth), make express provision for employers to own the relevant IP rights ('IPRs') in their employees' creations. However, the specific provisions differ in their language and effect. In some cases the

---

creation' 26 AJLL 273 (2013); A. Monotti, *Who Owns my Research and Teaching Materials – My University or Me?* 19 The Syd LR 425, 455 (1997) (Monotti 1997).

4. See 2.2.1 below: 'Employee or independent contractor'.
5. *Stevens v Brodribb Sawmilling Co Pty Ltd* (1986) 160 CLR 16, 29, (Mason J).
6. *Zuijs v Wirth Brothers Pty Ltd* (1955) 93 CLR 561, 571; *Hollis v Vabu Pty Ltd* (2001) 207 CLR 21, 41; *Sweeney v Boylan Nominees Pty Ltd* (2006) 226 CLR 161, [30].
7. *Beloff v Pressdram Ltd* [1973] 1 All ER 241, 253; *Stevenson, Jordan and Harrison Ltd v Macdonald and Evans* (1952) 69 RPC 10, 22 (Denning LJ); *Stevens v Brodribb Sawmilling Co Pty Ltd* (1986) 160 CLR 16, 26–29 (Mason J), 35–37 (Wilson and Dawson JJ); *Redrock Holdings Pty Ltd v Hinkley* (2001) 50 IPR 565, [22].
8. *Hollis v Vabu Pty Ltd* (2001) 207 CLR 21, [50].
9. *Stevens v Brodribb Sawmilling Co Pty Ltd* (1986) 160 CLR 16, 24, (Mason J), 36–37, (Wilson and Dawson JJ); *Sweeney v Boylan Nominees Pty Ltd* (2006) 226 CLR 161, [31].
10. *Centrestage Management Pty Ltd v Riedle* (2008) 77 IPR 550, [14].
11. *Sweeney v Boylan Nominees Pty Ltd* (2006) 226 CLR 161, [31].
12. *Hollis v Vabu Pty Ltd* (2001) 207 CLR 21; *Centrestage Management Pty Ltd v Riedle* (2008) 77 IPR 550, [14].

rights vest initially in the employer when the prescribed conditions are satisfied. In other cases, the rights of the employer are derivative and require the assignment of the creator's rights in law or equity.

No statutory regime for IP makes special provision for rights in IP created by academic employees. However, all Australian universities modify the statutory regimes to some extent by agreement to reflect the different duties and relationships that apply in this workplace environment. The detailed provisions that relate to allocation of rights in IP generally as well as to their management and commercial exploitation[13] are contained in institutional IP statutes, regulations[14] or policies. While there are broadly common approaches, the precise terms differ according to the objectives each university seeks to achieve. The operation of any terms will be subject to any express agreements that govern particular projects such as collaborative research, commissioned teaching or course materials.

In addition to asserting ownership of IP that is created in pursuance of the duties of employment, universities commonly extend the circumstances in which they assert ownership. This might include where the creation of IP arose using assets, special funding, facilities, resources, IP owned by the university or another member of the university, or that was created or developed in collaboration. As these are claims to subject matter that may be outside the scope of the employer's statutory or common law entitlement, their validity is dependent upon an enforceable agreement with the employee to assign any future rights.[15] Although universities expressly claim ownership of most IPRs that are created by their employees they provide a balance of rights in internal legislation and employment contracts that includes mandatory compensation. In contrast, no law provides for employees generally to be paid mandatory special compensation for the IPRs they create for their employer.

2 EMPLOYEES' COPYRIGHT

2.1 The Legal Framework

The legal framework in Australia that governs relations between employers and employees in the field of rights in copyright is contained in the Copyright Act 1968 (Cth), judicial decisions that apply and interpret the relevant provisions and the general principles of contract, equity, and employment law.

The statutory provisions that govern the ownership of copyright between employers and employees differ according to the nature of the subject matter under

13. All university IP management policies provide for some form of profit distribution where inventions and some other forms of IPR are exploited commercially. A. L. Monotti, with S. Ricketson, *Universities and Intellectual Property: Ownership and Exploitation*, Ch. 7 (Oxford University Press, 2003) (Monotti with Ricketson 2003).
14. Statute and regulations apply by their own force: *Ex parte Forster; Re University of Sydney* [1964] NSWR 1000, 1007; Monotti 1997, 429; Monotti with Ricketson 2003.
15. *Copyright Act 1968* (Cth) s. 197(1); Monotti 1997; *University of Western Australia v Gray* [2009] 179 FCR 346.

consideration. The Copyright Act distinguishes between 'original works'[16] which are the creations of human authors and related subject matter (films, sound recordings, television and sound broadcasts, and published editions)[17] where the claimants are the investors or producers who have brought that subject matter into existence. Other rights that are protected under the Act include moral rights[18] and performers' rights.[19]

2.2 Ownership of Rights

Subject to three exceptions, it is the author of a literary, dramatic, musical, or artistic work who is the first owner of any copyright subsisting in the work[20] and who has the power to exercise all the exclusive rights in a copyright work and to authorise others to do so.[21] The Copyright Act does not define 'author', but courts interpret the elements of authorship as they relate to works as requiring a contribution to its protectable elements – the expression as opposed to the ideas that underlie the work.[22] The author is regarded as the person who creates those elements and who is responsible for first reducing the work into a material form.[23] To be considered as an author of a literary work, e.g., the person or persons must have exercised 'independent intellectual effort'[24] and/or 'sufficient effort of a literary nature'.[25]

The three exceptions to first ownership vesting in the author relate to:

- certain literary, dramatic or artistic works created by journalists;[26]
- certain commissioned artistic works;[27] and
- works created by employees when neither of these preceding exceptions apply.[28]

The operation of each of these three statutory exceptions can be modified by express or implied agreement.[29]

16. *Copyright Act 1968* (Cth) Pt III ('Copyright in Original Literary, Dramatic, Musical and Artistic Works').
17. *Ibid.* Pt IV ('Copyright in Subject Matter other than Works').
18. *Ibid.* Pt IX.
19. *Ibid.* Pt XIA.
20. *Copyright Act 1968* (Cth) s. 35(2).
21. *Ibid.* s. 13.
22. For a detailed examination of persons entitled to copyright see: M. J. Davison, A. L. Monotti and L. Wiseman, *Australian Intellectual Property Law* (3rd ed., 2016).
23. *Copyright Act 1968* (Cth) ss. 22(1), 10(1); *Sands and McDougall Pty Ltd v Robinson* (1917) 23 CLR 49, 55.
24. *IceTV Pty Ltd v Nine Network Australia Pty Ltd* (2009) 239 CLR 458, [33] [48] (French CJ, Crennan and Kiefel JJ).
25. *Ibid.* [99] (Gummow, Hayne and Heydon JJ).
26. *Copyright Act 1968* (Cth) s. 35(4).
27. *Ibid.* s. 35(5).
28. *Ibid.* s. 35(6).
29. *Ibid.* s. 35(3).

The main exception to the principle that the author is first owner of copyright applies to works created by employees.[30] Subject to a contrary agreement, and to special provisions for employee journalists that appear below, Section 35(6) provides that an employer owns copyright in literary, dramatic, artistic and musical works made by an author 'in pursuance of the terms of his or her employment ... under a contract of service or apprenticeship.' The resolution of this issue in each case depends upon the answers to the following two questions:

- Was the author an employee at the time the work was made?
- If so, was the work created 'in pursuance of the terms of his or her employment'?

For copyright in a work of joint authorship to vest in an employer it is necessary that all the authors be employees of that employer.[31]

2.2.1 Employee or Independent Contractor

In some cases there may be doubt as to whether the author was an employee at the time the work was created or was instead engaged as an independent contractor under a contract for services.[32] This mixed question of law and fact is determined with reference to the actual terms of the contract along with the circumstances of each case. As mentioned above, these include the right of the employer to exercise control in the manner of doing work and whether the performance of the work is an integral part of the business.

The case of *Redrock Holdings Pty Ltd v Hinkley*[33] provides an example of how this question might be resolved within the context of Section 35(6) of the Copyright Act. The Victorian Supreme Court was asked to decide the nature of the engagement of Mr. Hinkley as a computer programmer for Redrock. Mr. Hinkley was paid a salary from which tax was deducted, and had entitlement to sick leave, annual leave and long service. Redrock provided the necessary equipment for his work and made contributions to superannuation on his behalf. The work he performed was integrated into the business of Redrock. The Court recognised that Mr. Hinkley exercised a high degree of professional skill and expertise in the performance of his duties that necessarily reduced the ability of the employer to exercise control in the manner of doing such work. Although this factor might suggest an engagement under a *contract for services*, it was outweighed by the other circumstances which influenced the Court to conclude

30. *Ibid.* s. 35(6); *Beloff v Pressdram Ltd* [1973] 1 All ER 241; *Stevens v Brodribb Sawmilling Co Pty Ltd* (1986) 160 CLR 16; *Zuijs v Wirth Brothers Pty Ltd* (1955) 93 CLR 561, 572; *Hollis v Vabu Pty Ltd* (2001) 207 CLR 21; *Sweeney v Boylan Nominees Pty Ltd* (2006) 226 CLR 161.
31. *Acohs Pty Ltd v Ucorp Pty Ltd* (2010) 86 IPR 492, [62]. The trial decision was overturned on appeal but not on this point; *Acohs Pty Ltd v Ucorp Pty Ltd* (2012) 201 FCR 173 [41].
32. *Redrock Holdings Pty Ltd v Hinkley* (2001) 50 IPR 565; *EdSonic Pty Ltd v Cassidy* (2010) 189 FCR 271, [42]–[44].
33. (2001) 50 IPR 565, [20]–[24].

that Mr. Hinkley was employed under a *contract of service*. Consequently, the computer programs he wrote during this period vested in Redrock as the employer pursuant to Section 35(6).

2.2.2 In Pursuance of the Terms of His or Her Employment

Assuming that there is a contract of service, and that the relevant work is created during the term of that employment, the issue then arises as to whether the particular work 'is made by the author in pursuance of the terms of his or her employment'. The terms of the contract of employment must be identified with reference to the principles of employment and labour law. They may be written, oral or implied or a combination of both and the terms may have changed over the period of employment.[34]

Two decisions have considered the meaning of the expression 'in pursuance of the terms of his or her employment'. The case of *Edsonic Pty Ltd v Cassidy* [35] involved a dispute as to ownership of copyright in certain vocational and educational materials in respect of which Ms Cassidy received remuneration from Edsonic in the form of royalties. The amount of those royalties bore no relationship to the time she spent creating the works but instead reflected the price paid for the work. Edsonic separately paid Ms Cassidy a salary (from which tax was deducted) to create other training and educational materials for a third party, the Property Council of Australia.

In the absence of direct authority and drawing on notions that apply in the context of patent law,[36] Moore J commented that 'in pursuance of the terms of his or her employment' requires consideration of:

> whether the relevant work is made in furtherance of the contract of employment with the employer. That is, did the employee make the work because the contract of employment expressly or impliedly required or [sic] least authorised the work to be made.[37]

In this case, the entirely different mode of remuneration for the production of the disputed works imbued them with a different character that excluded them from being made in pursuance of the terms of ... her employment.

The second decision is *Insight SRC IP Holdings Pty Ltd v Australian Council for Educational Research Ltd,*[38] an infringement action in which the ownership of the allegedly infringing copyright work was challenged. The approach of Besanko J in the trial was to refer to the Shorter Oxford Dictionary definition of the 'related expression

34. *Insight SRC IP Holdings Pty Ltd v Australian Council for Educational Research Ltd* (2012) 211 FCR 563, [40]; (2013) 101 IPR 484 (FFC). In the context of patents for inventions see: *Victoria University of Technology v Wilson* (2004) 60 IPR 392, [120].
35. *EdSonic Pty Ltd v Cassidy* (2010) 189 FCR 271.
36. See: *University of Western Australia v Gray* [2009] 179 FCR 346.
37. *EdSonic Pty Ltd v Cassidy* (2010) 189 FCR 271, [41].
38. (2012) 96 IPR 495, [40], [98].

"Pursuant to"'. These meanings were 'following upon', 'consequent upon and conformable to' and 'in accordance with'. His conclusion on the facts was 'He was not required to produce it.'[39]

Other circumstances that have influenced a decision in favour of the employee are the absence of an express term of employment to create the contested subject matter, evidence that the work was created for the employee's own purposes, and evidence of an agreement that the employer could use the work in return for providing the employee with departmental data for the purposes of their studies and research.[40]

Journalists and other authors employed by the proprietor of a newspaper, magazine or similar periodical

The general principal that the employer owns copyright in works created by its employees in pursuance of their employment is qualified in the case of authors employed by proprietors of newspapers, magazines or similar periodicals to make literary, dramatic or artistic works for the purpose of inclusion in those publications.[41] Special provision is made under Section 35(4) of the Copyright Act for carving out certain rights for the employed author so that the rights in copyright in those works are distributed between them in ways that divide the copyright ownership. Unless otherwise agreed,[42] the journalist is the owner of copyright for the reproduction of the work for the purpose of including it in a book. They are also the owner of copyright in the work to reproduce it in the form of a hard copy facsimile[43] that is made from a paper edition of the newspaper, magazine or periodical in which it is published.[44] The proprietor is the owner of copyright in the work for all other purposes.[45]

2.2.3 Producers and Broadcasters

Sound recordings, films, broadcasts[46] and published editions are not recognised as works of authorship in Australia. Instead, the rights stem from the person who is the 'maker' (except where the sound recording or film is commissioned),[47] or the 'publisher' in the case of published editions.[48] This may be an individual or a company.[49]

Performers have no copyright in their performances but have limited protection under Part X1A of the Copyright Act in the form of rights against those who make unauthorised use of a performance. These rights are personal to the performer and are

39. The appeal court endorsed this decision without expressly addressing the meaning of the expression 'in pursuance of'. *Insight SRC IP Holdings Pty Ltd v Australian Council for Educational Research Ltd* (2013) 101 IPR 484, [33]-[36].
40. *Insight SRC IP Holdings Pty Ltd v Australian Council for Educational Research Ltd* (2012) 211 FCR 563, [92]-[99].
41. *Copyright Act 1968* (Cth) s. 35(4) – where the employer is the proprietor of a newspaper, magazine or similar periodical.
42. *Ibid.* s. 35(3).
43. *Ibid.* s. 35(7).
44. *Ibid.* s. 35(4)(c).
45. *Ibid.* s. 35(4)(d).
46. *Copyright Act 1968* (Cth) s. 22(5).
47. *Ibid.* ss. 97(3), 98(3).
48. *Ibid.* ss. 97–100.
49. *Ibid.* s. 84.

separate from and additional to copyright in the material that is performed and in any copyright subject matter arising from the performance itself.

Sound recordings

There is no express provision for ownership of copyright in sound recordings created by employees. Such a provision is unnecessary because ownership does not vest in the employee creators but in the maker of the sound recording.[50] In the absence of an agreement to the contrary, an employer will own copyright in a sound recording that is first embodied in discs or tapes[51] owned by the employer.[52] Ownership in such a case is not dependent upon any employment relationship or express duties, but simply rests on the ownership of the physical medium in which the recording is first embodied. Furthermore, a statement on a label on the recording that a specified person is the maker of the recording is prima facie evidence of that fact[53] although it does not imply that this person was the only owner of copyright.[54] Where sound recordings are made for valuable consideration pursuant to a contract of commission, the commissioner is the owner.[55]

Sound recordings of live performances

Special provisions apply to ownership of sound recordings of live performances. The makers of the sound recording are both the person who owned the record on which the recording is made and the performers who appeared in the recorded performance.[56] However, when the performer engages in the performance under the terms of their employment under a contract of service or apprenticeship, the employer of a performer is taken to be a maker of the recording.[57]

Cinematograph films

The maker of a cinematograph film is the person who undertook the arrangements necessary for the production of the first copy of the film.[58] This is generally considered to be the film's producer.[59] Following statutory amendment in 2005,[60] this includes each director of a film that is not commissioned.[61] If the film is made for valuable consideration under agreement, the person who commissions the film owns its copyright.[62]

---

50. *Ibid.* s. 97(2).
51. *Ibid.* s. 10(1): 'record' means a disc, tape, paper, electronic file or other device in which sounds are embodied.
52. *Ibid.* s. 22(3)(b).
53. *Ibid.* s. 130(2).
54. *Ibid.* s. 130(4).
55. *Ibid.* s. 97(3).
56. *Ibid.* ss. 22(3A), 97(2)(A), 100AA–100AH.
57. *Ibid.* s. 100AD(2).
58. *Ibid.* s. 22(4).
59. *Seven Network (Operations) Ltd v TCN Channel Nine Pty Ltd* (2005) 146 FCR 183, [12]; *Wills v Australian Broadcasting Corporation (No 3)* (2010) 89 IPR 252, [29]–[30].
60. *Copyright Amendment (Film Directors' Rights) Act 2005* (Cth).
61. *Copyright Act 1968* (Cth) s. 98(4).
62. *Ibid.* s. 98(3); *Avtex Airservices Pty Ltd v Bartsch* (1992) 107 ALR 539.

As directors of non-commissioned films since 19 December 2005 are co-owners of copyright in their films with the film's producers, an express provision deals with the position of an employed director who directs the film under the terms of their employment under a contract of service or apprenticeship. In this case, unless otherwise agreed, the employer is substituted for the director as a maker of the film.[63]

2.2.4 *Moral Rights*

Part IX of the Copyright Act provides a framework for individuals[64] to have non-economic rights in the form of moral rights in literary, dramatic, musical or artistic works, and cinematograph films and in performances (so far as they consist of sounds) (together defined as 'works').[65] The rights are in addition to any other rights that the individual or anyone else has under the Act in relation to the performances[66] or works.[67]

Authors of works are given a right of attribution of authorship,[68] a right not to have authorship falsely attributed and a right of integrity of authorship.[69] In the case of cinematograph films, the moral rights are given to the maker of the film, namely its principal director, producer and screenwriter.[70] In the case of a joint authorship of a literary, dramatic, musical or artistic work, each author enjoys their own moral rights in the work.[71] Where there is more than one principal director, producer or screen-writer, each person has their own separate and independent moral rights in the film.[72] Performers – each person who contributed to the sounds in the performance as well as the conductor of a live performance of a musical work[73] – are given a right of attribution of performership, a right not to have performership falsely attributed and a right of integrity of performership.

Moral rights are personal to the author or performer and cannot be assigned, given away or otherwise alienated even though another person may own the economic rights in the copyright.[74] This principle operates irrespective of whether the works or performances are made in the course of employment or otherwise.

However, the Copyright Act provides for circumstances in which an author or performer can give written consent to acts or omissions that might otherwise infringe

63. *Copyright Act 1968* (Cth) s. 98(5).
64. *Ibid.* s. 190.
65. Moral rights in performances and works apply in relation to the whole or a substantial part of the performance or work; *Copyright Act 1968* (Cth) ss. 195AZH, 195AZP.
66. *Copyright Act 1968* (Cth) s. 192(2).
67. *Ibid.* ss. 192(1), 189.
68. *Ibid.* s 193; *Corby v Allen & Unwin Pty Ltd* (2013) 297 ALR 761.
69. *Copyright Act 1968* (Cth), Part IX.
70. *Ibid.* ss. 189, 191, 195AZJ, 195AZK, 195 AZL.
71. *Ibid.* s. 195AZI.
72. *Ibid.* ss. 195AZJ, 195AZK, 195 AZL.
73. *Ibid.* ss. 189, 191B.
74. *Ibid.* ss. 195AN (3), 195ANB(3).

moral rights.[75] A consent given for the benefit of the copyright owner is presumed to extend to their licensees and successor in title and to anyone who is authorised to do acts comprised in the copyright.[76] Any consent, whether it be for an employer or someone else, that is obtained through the application of duress or false or misleading statements has no effect.[77]

Employers may seek comprehensive written consent for their benefit from their employees that are not subject to the restrictions that affect an author's or a performer's written consent given to others.[78] In the case of films or literary, dramatic, musical or artistic works as included in a film, an employee can consent to all acts and omissions in relation to all works made or to be made by the employee in the course of his or her employment.[79] A consent can have both retrospective and prospective operation in relation to all works and all acts or omissions.

In the case of literary, dramatic, musical and artistic works (other than as included in a film),[80] the provision relates to a consent that is 'genuinely given'[81] in relation to all or any acts or omissions (whether occurring before or after the consent is given) in relation to all works made or to be made by the employee in the course of his or her employment.

A consent by a performer is similarly comprehensive, but its application to 'all performances in which the employee is to be a performer in the course of his or her employment' assumes that the consent is obtained before any performance has taken place.[82]

This wider protection available for employers from otherwise infringing their employees' moral rights no doubt recognises the impracticality in many industries for adhering to those moral rights. It is common practice for employment contracts to include some form of consent in relation to acts or omissions that may otherwise infringe the employee's moral rights.

2.2.5 University (Academic) Employees' Copyright

Australian universities generally assert ownership of all copyright subject matter that their academic employees create in pursuance of the terms of their employment. An employee who creates copyright works that do not vest initially in the university employer under the Copyright Act can agree to a contractual assignment of those rights to the employer, either in the employment contract or by separate negotiation. It is

75. *Ibid.* ss. 195AW (cinematograph films and literary, dramatic, musical and artistic works in films), 195AWA (literary, dramatic, musical and artistic works not used in films), 195AXJ (performances); Elizabeth Adeney, *The Moral Rights of Authors and Performers: An International and Comparative Analysis* (Oxford University Press, 2006).
76. *Copyright Act 1968* (Cth) ss. 195AW(5), 195AWA(5), 195AXJ(5).
77. *Ibid.* ss. 195AWB; 195AXK.
78. *Ibid.* ss. 195(AW)(3) (films or works in films), 195(AWA)(3) (work that is not a film or included in a film), 195AXJ(3) (performances).
79. *Ibid.* s. 195AW(4).
80. *Ibid.* s. 195AWA(4).
81. *Ibid.* s. 195AWA(2).
82. *Ibid.* s. 195AXJ(4).

common for universities to make such extended claims in the employment contract in circumstances that are specified in the relevant IP regulation or policy.[83] These may include such things as the use of assets, funding, facilities, or university resources, or IP owned by the university or another member of the university.

In the case of teaching and course materials,[84] a term that generally is defined widely to cover all materials produced for use in teaching in any form, some universities will grant the creators a non-exclusive licence. This might be, for instance, to use all university teaching materials to 'fulfil their teaching, research and training responsibilities' at the university[85] or at other educational institutions.[86]

As with all other forms of IP created by academic employees, a university employer may agree to assign its rights of ownership to the creator or creators. Such an agreement may include provisions such as the retention of the right to use the subject matter for education, academic and research purposes, assumption of liability for on-going costs associated with IP protection, appropriate indemnities, for recovery of funds expended by the university and a royalty payable from commercialisation of the IP.[87]

There is a notable exception to university ownership of copyright subject matter. In recognition of the long established principle that academics are entitled to own copyright in their scholarly works, no Australian university retains all rights it may have as employer under Section 35(6) of the Copyright Act. All universities relinquish rights in favour of the author in certain categories of works whose scope varies among institutions. This may be by way of waiver or assignment. In each case, it is common for a university to claim some form of perpetual royalty free non-exclusive and irrevocable licence in those works that is usually expressed to be for the university's teaching, research and educational purposes. In some cases, this licence acknowledges the practical reality that academics often must assign copyright to publishers. The Australian National University, e.g., provides that a licence in respect of educational materials, scholarly works and creative works 'is subject to any overriding contractual obligations the creator owes to third parties (e.g., a publisher of a textbook or a journal article)'.[88]

Although a broadly common approach among universities to own all copyright subject matter except for scholarly works can be identified, the breadth of the scope of exempted subject matter varies. Three examples demonstrate a spectrum of

83. Monotti with Ricketson 2003, Ch. 5. See 4.2.4. below for a more detailed discussion of the university internal legislation and policies that govern IPRs.
84. This phrase is generally defined widely to include not only traditional materials such as lecture notes and material, syllabi, handouts, study guides but also course software and multi-media. See, e.g., University of Queensland, *Intellectual Property for Staff, Students and Visitors – Policy*, cl. 2 (4.10.13a, 2016) (UQ Policy 2016).
85. See, e.g., QUT, D/3.1 *Intellectual Property Policy*, 3.1.6 (30 September 2015) (QUT Policy 2015).
86. See, e.g., UQ Policy 2016, cl. 5.8.
87. See, e.g., Australian National University, *Procedure: IP Protection and Commercialisation*, cls 25 and 26 (ANUP_003605, 1 January 2017) (ANU Procedure 2017); University of Sydney Intellectual Property Policy 2016, cl. 12(8) (Sydney IP Policy 2016).
88. Australian National University, *Intellectual Property Policy* (ANUP_003603, 2017), cl. 11b (ANU Policy 2017).

approaches. First, a recently revised policy at The Australian National University presents a 'generous approach'. The university makes a broad claim to own IP,[89] but does not assert ownership of copyright in 'education materials,[90] and scholarly[91] or creative works[92] ' unless they are specifically commissioned.[93]

Secondly, the University of Melbourne adopts a relatively 'middle of the road' approach that only limits its rights in relation to 'scholarly works'. This term is defined quite narrowly to exclude teaching material 'or any other work required to be disclosed under policy and this statute'.[94] This latter provision ensures that there is common ownership of rights in all IP associated with an invention or other reported subject matter.

A third example at the other end of the spectrum is the Queensland University of Technology. It makes a broad claim to own IP created by staff members in the course of their employment but will assign the right to publish scholarly works to the author, while reserving a non-exclusive licence for its 'educational, research and commercialisation purposes and make it available via its institutional repository'.[95]

Moral rights

The treatment of moral rights in copyright subject matter created by academic employees varies considerably. The most common approach involves recognition of the moral rights of authors and their express exclusion from the operation of the IP policy. No broad assertions are made as to seeking consents from their employees for otherwise infringing acts.[96]

However, perhaps as a reflection of increasing collaborations with sponsors and third parties for the creation and exploitation of copyright subject matter, recently revised IP regulations and policies often include specific provisions. First, the Australian National University provides that it may request a creator to provide a written consent in relation to their moral rights in relation to certain works that might be created during a project between the university and an external sponsor or other third

89. ANU Policy 2017, cl. 2.
90. ANU Policy 2017, Definitions: Briefly, it 'means all materials produced by staff ... in the course of, or for use in, teaching at the University (for example, lecture notes and material, syllabi, handouts, study guides, course software and assessment materials) regardless of format (for example, printed, digital, electronic, multi-media presentations and web content) ...'.
91. ANU Policy 2017, Definitions: 'Scholarly works means copyright works that are intended for academic publication (for example, any article, book, manuscript, or manual) and have been peer reviewed, regardless of format (for example, printed, digital or electronic versions) created by staff, students ... but excludes education Materials.'
92. ANU Policy 2017, Definitions: 'Creative work means artistic, literary, musical, dramatic, film, sound or multimedia work, performance, diagram, map, chart, photograph or other creative work in which copyright subsists (irrespective of the medium in which the work is recorded).'
93. ANU Policy 2017, cls 11 and 12.
94. University of Melbourne Statute (21 July 2016), cl. 5: '"Scholarly works" means any article, book, musical composition, creative writing or like publication or any digital or electronic version of these works that contains material based on the creator's scholarship, learning or research, ...'
95. QUT Policy 2015, cl. 3.1.5.
96. University of New South Wales, *Intellectual Property Policy*, cl. 11 (effective 6 March 2013) (UNSW Policy 2013).

party. That consent would be sought before the project commences and the university would protect the moral rights wherever possible 'particularly where consent to infringement is sought which may affect the academic standing or proper attribution of academic work intended for open publication'.[97]

Secondly, the University of Queensland similarly warns that participation in a project with an external sponsor or other third party might require a prior written consent in relation to the creator's moral rights in relation to certain works created during the project.[98] The purpose of such a clause is to allow the university to ask for moral rights consent if doing so is required by a commercialisation partner. In contrast, the University of Sydney Intellectual Property Policy 2016 deals expressly only with moral rights of students who wish to participate in any third party activity.[99] The policy also provides that the University will take all reasonable steps to ensure that any originator is acknowledged in any documentation relating to university owned IP (other than copyright). Specific provisions relate to the University's obligations in the case of copyright material owned by the University. [100]

A final example is the recently revised Monash University IP regulations. The University recognises moral rights of the creator and acknowledges the obligation to attribute authorship in University owned copyright works. However, there is no limitation on its right to 'request that a creator sign a consent with respect to his or her moral rights subsisting in a copyright work'.[101]

Persons employed by the Commonwealth or a state

The Copyright Act provides specifically for ownership of copyright created in public service employment, where persons are employed under a law of the Commonwealth or of a state. For example, the Public Service Act 1999 (Cth) provides that all persons engaged on behalf of the Commonwealth of Australia as employees to perform the functions in a Department or Executive Agency must be engaged under that Act or under authority of another Act.[102] Section 17 of the Copyright Act treats persons so employed as if they were employed under a contract of service or under a contract of apprenticeship. Consequently, under Section 35(6) the Commonwealth or state as employer will own copyright in any work that is created by that person in pursuance of the terms of their employment.[103]

97. ANU Policy 2017, cl. 20.
98. UQ Policy 2016, cl. 11.
99. Sydney IP Policy 2016, cl. 8.4(c).
100. *Ibid.* cl.15.
101. *Monash University (Vice-Chancellor) Regulations 2016*, Part 5 Intellectual Property, cl. 26(3).
102. *Public Service Act 1999* (Cth) ss. 6(1), 22.
103. Copyright in an original literary, dramatic, musical or artistic work, sound recording or cinematograph film that is otherwise made by, or under the direction or control of the Commonwealth or a State is owned by the Commonwealth or the State as the case may be; *Copyright Act 1968* (Cth), Pt VII, s. 176(2); *Insight SRC IP Holdings Pty Ltd v Australian Council for Educational Research Ltd* (2012) 211 FCR 563, [38]–[39].

2.3 The Parties' Duty to Inform

There is no statutory duty to inform an employer of the creation of a copyright work to which the employer is entitled. The existence of a duty to inform or disclose is one that might arise under express or implied terms in a particular contract of employment or pursuant to the obligation of a fiduciary where that relationship exists.

Australian universities commonly require their academic staff to disclose the creation of specified categories of IP in respect of which the university asserts ownership ('university IP'). The broad purpose of an obligation to inform or disclose is to alert the university to its creation and to give the relevant staff an opportunity to assess the identity of all potential Creators, ownership, commercial potential and protection that might be applicable to the IP Disclosure The report is generally made to officers within an office of commercialisation or technology transfer who will have a certain period of time within which to decide, in consultation with the creator and relevant school, whether or not to proceed with some form of commercialisation. In some cases the reporting obligations are comprehensive across all forms of IP for which the university asserts ownership and thus include copyright subject matter. This obligation may be otherwise unqualified or may be limited to IP that the creator believes may have value and is contemplating that commercialisation[104] or believes may be capable of either or both protection or commercialisation,[105] or merely that objectively has the potential to be commercialised.[106]

2.4 Employee Remuneration Right

The Copyright Act 1968 makes no express provision for an employee to claim remuneration for the creation of a copyright work or other subject matter that is owned by the employer. Any right to compensation becomes a matter for negotiation. In the case of academic employees, all universities provide for sharing the net proceeds of commercial exploitation of university IP in accordance with their regulations and policies in effect at the time.[107] A common approach involves sharing the net proceeds equally between the creator, their department or school, and the university central administration.[108] In some cases, the university's share may be retained by the university commercialisation company when that company manages the commercialisation of the IP.[109]

104. ANU Procedure 2017, cl. 3.
105. Sydney IP Policy 2016, cl. 11(1).
106. *Monash University (Vice-Chancellor) Regulations 2016*, Pt 5 – *Intellectual Property*, cl. 29.
107. See 4 below: 'Employees' inventions'.
108. ANU Procedure 2017, cl. 16.
109. UQ Policy 2016, cl. 9.1.

3 EMPLOYEES' SEMICONDUCTOR CHIPS

3.1 The Legal Framework

The legal framework governing relations between employers and employees in the field of semiconductor chips comprises the Circuit Layouts Act 1989 (Cth) and general principles of contract, equity and employment law. The legislation provides unregistered protection for eligible,[110] original[111] designs of three-dimensional integrated circuits that are recorded in some material form ('eligible layouts').[112] There has been minimal litigation in this field and no judicial consideration of the provisions that relate to rights of employers in eligible layouts created by their employees.

3.2 Ownership of Rights

3.2.1 Employees Generally

The maker of the eligible layout ('EL') is usually its first owner.[113] However, Section 16(2) of the Circuit Layouts Act deems the employer to be the maker and hence the owner of EL rights where the employee makes the EL 'under the terms of his or her employment by another person under a contract of service or apprenticeship'.[114] Therefore, EL rights are vested in the employer in the first instance. The common law principles that distinguish a contract of service from a contract for services and their application under Section 35(6) of the Copyright Act 1968 (Cth),[115] are likely to be applicable to construing this phrase in a dispute as to the employment status of the creator of an EL.

The application of Section 16(2) to rights 'in a particular eligible layout' can be excluded or modified by agreement.[116] If the intention of an employer and employee is that the latter will own all EL rights in eligible layouts that they create in the course of employment, there are means to achieve this aim. The employer can sign an agreement which assigns future EL rights either wholly or partly to the employee or another person. When the rights come into existence they vest in the assignee (employee) by virtue of the Act.[117]

110. *Circuit Layouts Act 1989* (Cth) s. 5: eligibility is determined with reference to the maker's citizenship, residency or place of incorporation or with reference to the place of first commercial exploitation (eligible person).
111. *Ibid.* s. 11.
112. *Ibid.* ss. 5 and 10(b).
113. *Ibid.* s. 16(1); *Avel Pty Ltd v Wells* (1992) 108 ALR 97.
114. *Circuit Layouts Act 1989* (Cth) s. 16(2).
115. *Redrock Holdings Pty Ltd v Hinkley* (2001) 50 IPR 565; *EdSonic Pty Ltd v Cassidy* (2010) 189 FCR 271, [42]-[44] (Moore J).
116. *Circuit Layouts Act 1989* (Cth) s. 16(3).
117. *Ibid.* s. 44(1).

3.2.2 *University Researchers' EL Rights*

Australian universities generally assert ownership of all rights in circuit layouts which vest in them initially under the Circuit Layouts Act.[118] An employee who creates an EL to which the employer is not entitled under the Circuit Layouts Act can agree to a contractual assignment of those rights to the employer, either in the employment contract or by separate negotiation. It is common for universities to make extended claims to rights in ELs that their academic employees create in specified circumstances. These might include such activities as the use of assets, funding, facilities, or resources, and IP owned by the university or another member of the university, or when the EL is commissioned.

A university employer may agree to assign its rights of ownership to the creator or creators. Such an agreement may include provisions such as the retention of the right to use the EL for education, academic and research purposes, assumption of liability for on-going costs associated with IP protection, appropriate indemnities, for recovery of funds expended by the university and a royalty payable from commercialisation of the IP.[119]

3.3 The Parties' Duty to Inform

There is no statutory duty to inform an employer of the creation of an EL to which the employer is entitled. The existence of a duty to inform or disclose is one that might arise under express or implied terms in a particular contract of employment or pursuant to the obligation of a fiduciary where that relationship exists.

Australian universities commonly require their academic staff to disclose the creation of specified categories of university IP, including circuit layouts. The broad purpose of an obligation to inform or disclose is to alert the university to its creation and to give the relevant staff an opportunity to the identity of all potential creators, ownership, commercial potential and protection that might be applicable to the IP Disclosure.[120]

3.4 Employee Remuneration Right

The Circuit Layouts Act vests EL rights in the employer in the first instance. Any right to compensation becomes a matter for negotiation. In the case of academic employees, all universities provide for sharing the net proceeds of commercial exploitation of university IP in accordance with their policies in effect at the time.[121]

118. Monotti with Ricketson 2003, Ch. 5. See 4.2.4. below for more detailed discussion of the university internal legislation and policies that govern IPRs.
119. See 2.2.5 above. e.g., ANU Procedure 2017, cls 25 and 26.
120. See 2.3 above.
121. See 4.4 below.

4 EMPLOYEES' INVENTIONS

4.1 The Legal Framework

The Patents Act 1990 (Cth) provides for registration of two kinds of patents for inventions. Standard patents have a potential term of 20 years, require the presence of an inventive step and require examination prior to registration. Innovation patents, on the other hand, have a maximum eight year term, require a lesser standard of inventiveness - innovative step - and are registered without examination. The Patents Act contains no provision for rights of an employer in employee inventions of either kind. Instead, the legal framework that governs relations between employers and employees is contained in the express or implied terms of any employment contract as construed using the general principles of contract and employment law.[122] In some circumstances of employment, the law relating to fiduciary duties will also be relevant.[123]

4.2 Ownership of Rights

Australian legislation focuses upon a person's entitlement to be granted a patent. At present, a patent grant can be made only to the inventor in the first instance,[124] or to a person who derives title to the invention from the inventor.[125] In some circumstances, the employer may be the inventor, but generally entitlement will be derived from an employee inventor who assigns the right to apply for a patent.[126] The legislation recognises an employer's rights in Section 15(1)(b) and Section 15(1)(c) as someone who is entitled to an assignment of that patent on grant,[127] or as a person who derives title to the invention from the inventor.[128]

In the absence of an express contractual stipulation with the inventor, the existence of an employer's rights is determined according to the common law and the application of the principles governing the employment relationship.[129] As with copyright, an essential threshold inquiry is to establish that the inventor is an employee

122. *University of Western Australia v Gray No. 20* (2008) 246 ALR 603, [144-157]; *Spencer Industries Pty Ltd v Collins* (2002) 54 IPR 434; *Victoria University of Technology v Wilson* (2004) 60 IPR 392, [107]-122]; *Sterling Engineering Co. Ltd v Patchett* [1955] AC 534, 543–544.
123. *Victoria University of Technology v Wilson* (2004) 60 IPR 392, [141]ff.
124. *Patents Act 1990* (Cth) s. 15(1)(a).
125. *Patents Act 1990* (Cth) s. 15(1)(b), (c) and (d); *JMVB Enterprises Pty Ltd v Camoflag Pty Ltd* (2006) 154 FCR 348, [70], [71].
126. *George C. Warner Laboratories Pty Ltd v Chemspray Pty Ltd* (1967) 41 ALJR 75; *Davies Shephard Pty Ltd v Stack* (2001) 108 FCR 422, [29]; *Sigma Pharmaceuticals (Australia) Pty Ltd v Wyeth* [2011] FCAFC 132, [270]-[291].
127. *Patents Act 1990* (Cth) s. 15(1)(b).
128. *Ibid.* s. 15(c).
129. *University of Western Australia v Gray No. 20* (2008) 246 ALR 603, [144-157]; *Spencer Industries Pty Ltd v Collins* (2003) 58 IPR 425; *Victoria University of Technology v Wilson* (2004) 60 IPR 392, [107]-122]; *Sterling Engineering Co Ltd v Patchett* [1955] AC 534, 543–544.

when the invention is created. However, the mere existence of a contract of service does not necessarily result in the entitlement of an employer to the employee's inventions.[130] Whether an invention made by an employee is the property of the employer will depend upon many factors. These include:

- the nature of the invention;
- the duties which the employee is engaged to perform;
- the position that the employee occupies in the employer's operations; and
- the circumstances in which it is made. Relevant circumstances include whether the invention was made during the employer's time, whether there is a relationship of confidence, whether the invention will be useful to the employer's business and whether the employee was responding to the employer's instructions in making a decision on the facts of the particular case at hand.[131]

It is important to note that the fact that the employee made the invention using the employer's materials and time does not necessarily lead to the conclusion that the employer should own the invention.[132] This entitlement will arise under an express or implied term in the employment contract.

4.2.1 Express Provisions in the Employment Contract

In most cases, the employment contract will contain an express term that governs ownership of employee inventions and entitlement will depend upon the interpretation of that term. An express agreement may require the employee to assign to the employer all inventions that are made during employment.[133] Indeed, if a company fails to insert any condition concerning the making of inventions in its employment contracts, courts are willing to imply a condition that is appropriate to safeguard its interests.[134]

The nature of the work done by the employee will suggest the type of clause that is appropriate. While an express or 'pre-assignment' clause is the most sensible approach to take when an employer expects its employees will invent, such clauses will attract the doctrine of restraint of trade and will be construed strictly against the employer.[135] The restrictions imposed must be reasonably necessary for the protection

130. *Fine Industrial Commodities Ltd v Powling* (1954) 71 RPC 253, 257; *Worthington Pumping Engine Co. v Moore* (1903) 20 RPC 41, 48–9; *Triplex Safety Glass Co. Ltd v Scorah* [1938] Ch. 211; *British Reinforced Concrete Engineering Co. Ltd v Lind* (1917) 34 RPC 101.
131. *Fine Industrial Commodities Ltd v Powling* (1954) 71 RPC 253, 257; *Worthington Pumping Engine Co v Moore* (1903) 20 RPC 41, 48–9; *Reiss Engineering Co Ltd v Harris* [1985] 14 IRLR 232.
132. *Mellor v William Beardmore and Co Ltd* (1927) 44 RPC 175, 191; *Kwan v The Queensland Corrective Services Commission* (1994) 31 IPR 25, 33; *In the Matter of Charles Selz Ltd's Application* (1954) 71 RPC 158, 164; *Fine Industrial Commodities Ltd v Powling* (1954) 71 RPC 253, 257; *Victoria University of Technology v Wilson* (2004) 60 IPR 392, [104].
133. *Electrolux Ltd v Hudson* [1977] FSR 312.
134. *Ibid.* 321–322. See below 4.2.2.
135. *Herbert Morris Ltd v Saxelby* [1916] 1 AC 688; *Triplex Safety Glass Co. Ltd v Scorah* [1938] Ch. 211. For the common law doctrine of restraint of trade, see e.g., *Thorsten Nordenfelt v Maxim*

of the legitimate interests of the person imposing the restrictions and reasonable in reference to the interests of the public. They 'must afford no more than adequate protection to the party in whose favour (the restraint) is imposed'.[136]

If the covenant fails to meet the test of reasonableness, by being more than is adequate for the protection of the covenantee, it will not be enforced.[137] There is authority that restrictions will be enforceable where the employee accepts the restriction in return for an adequate level of remuneration during employment.[138] Adequacy of protection for the legitimate interests of the employer is measured at the time of entry into the restriction.[139]

4.2.2 *Implied Duty to Assign Inventions*

A number of obligations of employees to their employers are implied by law into the contract of employment. Those which are relevant in determining ownership of IP are the duty of good faith and fidelity and the duty to obey lawful and reasonable orders.

There is necessarily to be implied into an employee's contract of service a term that the employee 'will serve his employer with good faith and fidelity'.[140] However, as Lord Greene MR stated in *Hivac Limited v Park Royal Scientific Instruments Limited*,[141] 'the practical difficulty in any given case is to find exactly how far that rather vague duty of fidelity extends'. This must be a 'question on the facts of each particular case'.[142] This duty requires the employee 'to carry out faithfully the work the employee is employed to do to the best of his ability'.[143] Even if an express covenant is declared unenforceable as an unreasonable restraint,[144] the employee remains bound by a duty of good faith to their employer.

An important aspect of this duty is reflected in a term implied at common law that the employer owns all inventions created in the performance of employment duties. This implied term is invoked in the absence of an express term or where an express

---

Nordenfelt Guns and Ammunition Co [1894] AC 535; *Amoco Australia Pty Ltd v Rocca Bros Motor Engineering Co Pty Ltd* (1973) 133 CLR 288; *Peters (WA) Ltd v Petersville Ltd* (2001) 205 CLR 126; J. D. Heydon, *The Restraint of Trade Doctrine* (3rd ed, Sydney: LexisNexis Butterworths, 2008) (Heydon 2008).

136. *Herbert Morris Ltd v Saxelby* [1916] 1 AC 688, 707; *Quadramain Pty Ltd v Sevastapol Investments Pty Ltd* (1976) 133 CLR 390; C. Arup, 'What/Whose Knowledge? Restraints of Trade and Concepts of Knowledge' 36 *MULR* 369 (2012).
137. *Maggbury Pty Ltd v Hafele Australia Pty Ltd* (2001) 210 CLR 181 (confidential information); but see *Restraints of Trade Act 1976* (NSW).
138. See Heydon 2008, 174–84.
139. *Lindner v Murdock's Garage* (1950) 83 CLR 628, 653; *Amoco Australia Pty Ltd v Rocca Bros Motor Engineering Co Pty Ltd* (1973) 133 CLR 288, 318.
140. *Electrolux Ltd v Hudson* [1977] FSR 312, 326; *Wessex Dairies Ltd v Smith* [1935] 2 KB 80; *British Syphon Company Ltd v Homewood* [1956] 1 WLR 1190; *Hivac Ltd v Park Royal Scientific Instruments Ltd* [1946] Ch. 169; *Robb v Green* [1895] 2 QB 315, 320 (A. L. Smith LJ); *Lancashire Fires Ltd v S. A. Lyons and Co. Ltd* [1996] FSR 629, 648.
141. [1946] Ch. 169, 174.
142. See also *Christopher Simon French v Paul Julian Mason* [1999] FSR 597.
143. *Harris' Patent* [1985] RPC 19, 29; *Adamson v Kenworthy* (1932) 49 RPC 57.
144. *Electrolux Ltd v Hudson* [1977] FSR 312; *Triplex Safety Glass Co. Ltd v Scorah* [1938] Ch. 211.

term is found to be invalid.[145] This implied term extends to inventions that an employee creates in performance of their employment duties or in following their employer's directions. Hence, it turns critically on what the employee has been engaged to do. The established position at common law[146] is that 'any product of the work which the servant is paid to do belongs to the master'.[147]

The law is well settled upon the position of an officer or employee who makes an invention affecting the business of their employer. It is an implied term of employment that any invention or discovery made in the course of the employment of the employee in doing that which they are engaged and instructed to do during the time of their employment, and during working hours, and using the materials of their employers, is the property of the employer and not of the employee. Having made a discovery or invention in course of such work, the employee becomes a trustee for the employer of that invention or discovery, and they are therefore as a trustee bound to give the benefit of any such discovery or invention to their employer.[148]

This principle applies to the products of all work including patentable inventions.[149] The employee will hold the inventions on trust for the employer[150] and would be expected to assign the title to the inventions to the employer. However, practical difficulties arise in determining what the employee is paid to do. In particular, the ultimate question is

> to ascertain whether, if at all, it was part of an employee's engagement with his or her employer to utilise his or her 'inventive faculty' ... in an agreed way or for an agreed purpose for the benefit of, or to further the purposes of, the employer. ... was the employee 'employed to make or discover inventions at all?', or, as French J put it, did the employee have 'a duty to invent'?[151]

While the general duty of good faith imposes an obligation to carry out faithfully the work they were employed to do to the best of their ability, it 'does not assist in the formulation of the actual duties which the employee is employed to do'.[152] These must be determined as a question of fact.[153] A duty to conduct research will not automatically vest ownership of an employee's inventions in the employer. It all depends upon the nature of the research that the employee is retained to perform; this is taken from the nature of the employer's business and the nature of the research.[154]

145. *Sterling Engineering Co. Ltd v Patchett* [1955] AC 534, 543–544.
146. *University of Western Australia v Gray* [2009] 179 FCR 346, [151].
147. *Sterling Engineering Co. Ltd v Patchett* [1955] AC 534, 547 (Lord Reid); 543-544 (Viscount Simonds).
148. *Victoria University of Technology v Wilson* (2004) 60 IPR 392, [104].
149. *British Reinforced Concrete Engineering Co. Ltd v Lind* (1917) 34 RPC 101; *Triplex Safety Glass Co. Ltd v Scorah* [1938] Ch. 211.
150. *Triplex Safety Glass Co. Ltd v Scorah* [1938] Ch. 211; *Spencer Industries Pty Ltd v Collins* (2002) 54 IPR 434, [74].
151. *University of Western Australia v Gray* (2009) 179 FCR 346, [152].
152. *Harris' Patent* [1985] RPC 19, 29–30; *Adamson v Kenworthy* (1932) 49 RPC 57, 68.
153. See *Greater Glasgow Health Board's Application* [1996] RPC 207; *Vokes Ltd v Heather* (1945) 62 RPC 135, 136.
154. *In the Matter of Charles Selz Ltd's Application* (1954) 71 RPC 158.

Hence, where a person is employed to solve a particular problem and creates an invention to achieve a solution, the employer will own that invention.[155] Similarly, this will occur where an employee is employed to make a technical design or proposal,[156] to make an improvement to the type of goods their employer manufactured[157] or to make or discover inventions.[158] In these circumstances, the invention flows directly from carrying out the duties that the employee is engaged to perform. The employment duties can change over the course of employment and rights in any invention must be assessed with reference to the work performed at the relevant time.[159]

The employee has a duty at common law to obey lawful and reasonable orders of the employer.[160] Therefore, although a person may not be employed to invent from the outset, it is open to an employer to change those duties by 'verbal instruction'.[161] The contractual validity of any directions to perform duties that involve a duty to invent will depend upon whether the order was reasonable and lawful as being within the scope or course of employment.[162]

4.2.3 *Fiduciary Obligations*

It is necessary to distinguish between both the employee's contractual duty of good faith and the implied duty to assign inventions when they are created in pursuance of the employment duties (on the one hand) and such fiduciary duty (if any) as the employee may owe to an employer arising from the position occupied in the organisation.[163] Not all employees will owe fiduciary duties to their employers,[164] but where a duty exists it must accommodate itself to the terms of the contract of employment so that it is consistent with it and conforms to it. The fiduciary relationship cannot be

155. *Adamson v Kenworthy* (1932) 49 RPC 57.
156. *British Reinforced Concrete Engineering Co. Ltd v Lind* (1917) 34 RPC 101; *Barnet Instruments Ltd v Overton* (1949) 66 RPC 315; *Aneeta Window Systems (Vic) Pty Ltd v K. Shugg Industries Pty Ltd* (1996) 34 IPR 95, 106.
157. *Sterling Engineering Co. Ltd v Patchett* [1955] AC 534, 543-4.
158. *Triplex Safety Glass Co. Ltd v Scorah* [1938] Ch. 211.
159. *British Reinforced Concrete Engineering Co. Ltd v Lind* (1917) 34 RPC 10, 109; *University of Western Australia v Gray* (2009) 179 FCR 346, [155]; *The Royal Children's Hospital* (2011) 213 IR 94.
160. *R v Darling Island Stevedoring and Lighterage Co Ltd; Ex parte Halliday and Sullivan* (1938) 60 CLR 601, 621-2 (Dixon J).
161. *Victoria University of Technology v Wilson* (2004) 60 IPR 392, 430.
162. Pittard 2013, 175-176.
163. *University of Nottingham v Fishel* [2000] ICR 1462, 1490; *Hospital Products Ltd v United States Surgical Corporation* (1984) 156 CLR 41, 97; *Victoria University of Technology v Wilson* (2004) 60 IPR 392, 438; *Hivac Ltd v Park Royal Scientific Instruments Ltd* [1946] Ch. 169; as to fiduciary relationships see John Glover, *Equity, Restitution and Fraud* (LexisNexis Butterworths 2004); Patrick Parkinson, *Fiduciary Obligations* in Patrick Parkinson, (ed.), *The Principles of Equity* (2d ed., Law Book Company 2003), Ch. 10.
164. E.g., *Aneeta Window Systems (Vic) Pty Ltd v K. Shugg Industries Pty Ltd* (1996) 34 IPR 95, 106.

superimposed on the contract so that it alters the operation which the contract was intended to have on its true construction.[165]

An employer will be entitled to own inventions when 'the employee's managerial and inventive responsibilities combined are such that he or she, as a fiduciary, is obliged to give the employer the benefit of any invention developed in the course of his or her employment that is germane to the employer's business'.[166]

In the course of its analysis of employee inventions, the Banks Committee in the United Kingdom made the following comments about the extent of this duty:

'It seems to be generally true to say that directors of companies and managers of businesses to whom no specific duties are allocated have a general duty to forward the interests of the company or business and will accordingly hold any inventions made during their employment (at least so far as they relate to the relevant business) in trust for the employer.'[167]

To avoid a breach of duty and liability, a person occupying a fiduciary position who wishes to enter into a transaction that would otherwise amount to a breach of duty must make a full disclosure to the person to whom the duty is owed of all relevant facts known to the fiduciary and that person must consent to the fiduciary's proposal.[168] In providing a remedy for a breach of fiduciary duty, it seems that the court would be prepared to recognise a constructive trust over the inventions up until innocent third-party interests became involved.[169]

Finally, where a term is implied to the effect that an employee holds an invention in trust for their employer, this obligation continues after employment.[170] Being confidential information until a patent application is made publicly available, this obligation is effectively another way of expressing an ex-employee's obligation to respect the confidentiality of their former employer's trade secrets.

4.2.4 University Employee Researchers' Rights

In the absence of an express contractual stipulation with the inventor, the established position at common law[171] is that an implied term in law applies in employment contracts to the effect that 'any product of the work which the servant is paid to do

165. *Victoria University of Technology v Wilson* (2004) 60 IPR 392, 438; *Hospital Products Ltd v United States Surgical Corporation* (1984) 156 CLR 41, 97; *University of Nottingham v Fishel* [2000] ICR 1462, 1491–2.
166. *University of Western Australia v Gray* (2009) 179 FCR 346, [155]; *Fine Industrial Commodities Ltd v Powling* (1954) 71 RPC 253 (managing director); *British Syphon Company Ltd v Homewood* [1956] 1 WLR 1190.
167. Board of Trade, *The British Patent System: Report of the Committee to Examine the Patent System and Patent Law* (Cmnd 4407, 1970) (Banks Committee), [443].
168. *Victoria University of Technology v Wilson* (2004) 60 IPR 392, 437; R. P. Meagher, J. D. Heydon and M. D. Leeming, *Meagher, Gummow and Lehane's Equity: Doctrines and Remedies* (4th ed., Butterworths 2002).
169. *Victoria University of Technology v Wilson* (2004) 60 IPR 392.
170. *Triplex Safety Glass Co. Ltd v Scorah* [1938] Ch. 211.
171. *University of Western Australia v Gray* [2009] 179 FCR 346, [151].

belongs to the master'.[172] However, following the decision of the Full Federal Court in *University of Western Australia v Gray*, [173] it is now established that a university cannot rely upon this implied term in law as the legal basis for a claim to own academic inventions. The court made it clear that academic employment is far removed from the types of employment relationships that gave rise to the common law implied term.[174] The distinctiveness of universities and academic employment within them were central to the court's decision to classify academic employment as a sub-class of employment contracts which do not attract the common law implication.[175]

Hence, in the absence of an enforceable express term in the employment contract or a contract governing a particular research project, an academic employee's duty to undertake unspecified research will not require them to assign to their employer any inventions that result from the performance of that research. In the absence of an implied term in law, universities who require their academic employees to assign inventions must achieve this outcome with express contractual provisions that are both valid and enforceable. If an academic employee is bound to assign inventions that result from the performance of their research, there is some *obiter* to suggest that the nature of the research that an academic is engaged to perform will be judged according to the norm for their department. It is not informed by what happens in other parts of the university. Therefore, there is no duty to invent if the sort of research undertaken within the school in question is directed to the preparation and presentation of peer-reviewed learned papers.[176]

Academic employees occupying senior and specific employment positions at the university have been found in one case to owe fiduciary duties to their university employer.[177] In *Victoria University of Technology v Wilson*, Professor Wilson and Dr. Feaver respectively occupied positions of head of the School of Applied Economics and head of a research centre. An opportunity to design an educational on-line system was presented to them because of their university positions and they commenced this work in these capacities. Subsequently they decided to continue work on the project in their personal capacities with third party involvement without informing the university of their actions and in honest ignorance of the university's interests. In doing so, they excluded the university from any interest in the IPRs in the project outcomes that it

172. *Sterling Engineering Co. Ltd v Patchett* [1955] AC 534, 547 (Lord Reid); 543–4 (Viscount Simonds).
173. *University of Western Australia v Gray* [2009] 179 FCR 346.
174. *Ibid.* [173], [206].
175. *Ibid.* [206]. *University of Western Australia v Gray No. 20* (2008) 246 ALR 603, [158–164]; See Monotti with Ricketson 2003, Ch. 6; W. van Caenegem, 'University of Western Australia *v* Gray' 23 *AJLL* 327 (2010); C. Arup, 'Employee Inventions: Labour Law Meets Intellectual Property' 21 *AJLL* 208 (2008); A. L. Monotti, 'Establishing clear rights in academic employee inventions: lessons learnt from University of Western Australia *v* Gray' in Marilyn Pittard, Ann L Monotti and John Duns, *Business Innovation and the Law: Perspectives from Intellectual Property, Labour, Competition and Corporate Law*, 74 (Edward Elgar 2013); A. L. Monotti, 'Is it time to codify principles for ownership of academic employee inventions? The disconnect between policy and the law' 38 *Monash ULR* 102 (2012).
176. *Victoria University of Technology v Wilson* (2004) 60 IPR 392, 423.
177. *Ibid.* 433.

would have enjoyed if the work had remained a university project. Those project outcomes included an inventions the subject of a provisional patent application.

Nettle J considered that the failure to make a full and true disclosure of their invention to the university and to seek permission to continue the project in their personal capacities amounted to a breach of fiduciary duties. If there had been no third party involvement, Nettle J would have ordered Professor Wilson and Dr. Feaver to hold the invention and hence the provisional patent application upon constructive trust for the university to 'give effect to the rule that a fiduciary must account for benefits which have flowed to him in breach of his duty'.[178]

The legal framework for ownership of rights created by academic employees is established under the terms of the contract of employment in force at the time the invention is created. All Australian universities have internal legislation or policies that govern IPRs that are created within the university. These include matters such as ownership and allocation of rights in IP generally as well as provisions relating to their commercial exploitation. Internal legislation of the university is binding of its own force, provided that its terms are within legislative power of the university. Policies will bind the employee only when these form part of the employment contract and it is common practice for academic employment contracts to incorporate the terms of all internal university policies as amended from time to time.

The common approach in these policies is to claim ownership of all patentable inventions that are created by academic employees, not only in pursuance of their employment duties, but also in additional certain specified circumstances. These include such things as creation using facilities, resources and funding provided by the university, creation under specific direction, and creation using university owned IP. In particular cases an express agreement to which the employee is a party may be supplemented or modified with contracts that govern IP ownership in particular research projects.

As the rights to patentable inventions vest in the inventor in the first instance under the Patents Act, a university that asserts broad claims to ownership that extend beyond their common law rights must have the agreement of the employee. The governing university legislation might grant them statutory powers to deal with their IPRs but does not grant them power to appropriate IPRs that are not otherwise vested in them. Therefore, when the internal legislation and policies purport to deal with IPRs that vest first in the creator, such as patentable inventions, it can do so validly only if the inventor has assigned or agreed to assign those rights to the university.[179]

A university employer may agree to assign its rights of ownership to the creator or creators. Such an agreement may include provisions such as the retention of the right to use the IP for education, academic and research purposes, assumption of liability for on-going costs associated with IP protection, appropriate indemnities, for recovery of funds expended by the university and a royalty payable from commercialisation of the IP.[180]

178. *Ibid.* 448–9.
179. Monotti 1997; *University of Western Australia v Gray* [2009] 179 FCR 346.
180. See, e.g., ANU Procedure 2017, cls 25 and 26.

4.3 The Parties' Duty to Inform

There is no statutory duty to inform an employer of the creation of a patentable invention to which the employer is entitled. The existence of a duty to inform or disclose is one that might arise under express or implied terms in a particular contract of employment or pursuant to the obligation of a fiduciary where that relationship exists.

Australian universities commonly require their academic staff to disclose the creation of specified categories of university IP that include inventions. The broad purpose of an obligation to inform or disclose is to alert the university to its creation and to give the relevant staff an opportunity to the identity of all potential creators, ownership, commercial potential and protection that might be applicable to the IP Disclosure. The report is generally made to officers within an office of commercialisation or technology transfer who will have a certain period of time within which to decide, in consultation with the creator and relevant school, whether or not to proceed with protection. This obligation may be otherwise unqualified or may be limited to IP that the creator believes may have value and is contemplating that commercialisation,[181] or believes may be capable of either or both protection or commercialisation,[182] or merely that objectively has the potential to be commercialised.[183]

4.4 Employee Remuneration Right

The Patents Act makes no express provision for an employee to claim remuneration for the creation of a patentable invention or other subject matter. Any right to compensation becomes a matter for negotiation between the employee and the employer. In the case of academic employees, all universities provide for sharing the net proceeds of commercial exploitation of IP owned by the university in accordance with their policies in effect at the time. A common approach involves sharing the net proceeds equally between the creator, their department or school, and the university central administration.[184] In some cases, the university's share may be retained by the university commercialisation company when that company manages the commercialisation of the IP.[185]

5 EMPLOYEES' UTILITY MODELS

Australian law does not protect utility models under this name but does provide for the grant of an innovation patent under the Patents Act. Innovation patents provide monopoly protection for the owner of the patent that is granted for a maximum of eight years without prior examination and that does not require the presence of an inventive

181. ANU Procedure 2017, cl. 3.
182. Sydney IP Policy 2016, cl. 11(1).
183. *Monash University (Vice-Chancellor) Regulations 2016*, Pt 5 – *Intellectual Property*, cl. 29.
184. ANU Procedure 2017, cl. 16.
185. UQ Policy 2016, cl. 9.1.

step. The level of inventiveness is less than that required for a standard patent and is defined as an 'innovative step'.[186] As the legal framework for governance of employer rights and principles for ownership of rights are the same for both standard patents and innovation patents they are not repeated here.

6 EMPLOYEES' DESIGN RIGHTS

6.1 The Legal Framework

Australia provides registered monopoly protection for the visual appearance of the finished article under the Designs Act 2003 (Cth).[187] This legislation includes a provision that governs relations between employers and employees in this field. As with other forms of IP, the legal framework also includes judicial decisions that apply and interpret the relevant statutory provisions as well as the general principles of contract, equity and employment law.

There is no statutory regime that specifically protects unregistered designs. However, the Copyright Act 1968 (Cth) will continue to govern the relationship between employees and employers in those instances where an artistic work that is applied industrially to products continues to have copyright protection.[188]

6.2 Ownership of Rights

The Designs Act draws a distinction between entitlement to apply for registration of a design and entitlement 'to be entered on the register as the registered owner of a design that has not yet been registered'.[189] While anyone can apply to have a design registered,[190] the registered owner or owners[191] can only be a person who falls within one or more of the listed categories in s. 13(1). These have been construed by Gordon J in *LED Technologies Pty Ltd v Elecspess Pty Ltd* [192] to comprise the following:

- the designer, who is the person who 'created' the relevant design;
- the employer, if the design was made in the course of employment, whether or not there is a specific contract;[193]

186. *Patents Act 1990* (Cth) ss. 7 (4)-(5) (innovative step), 18(1A) (requirements for patentability), 62 (grant) and 68 (term).
187. *Designs Act 2003* (Cth) ss. 5 and 7.
188. *Copyright Act 1968* (Cth) Div 8: limits copyright protection in relation to 'corresponding designs' that are applied industrially.
189. *Designs Act 2003* (Cth) s. 13(1); *Foster's Australia Ltd v Cash's (Australia) Pty Ltd* (2013) 219 FCR 529, 159 [118].
190. *Designs Act 2003* (Cth) s. 21.
191. *Ibid.* s. 14(1): the owner of the registered design is the person (or persons) who is entered as such in the register.
192. (2008) 80 IPR 85, 97 [29]-[30].
193. In a case under the under the *Designs Act 1906* (Cth), the payment of a regular salary for the performance of services from which the designs arose was sufficient to vest ownership in the

- a person who contracted the designer to make the design,[194] whether or not there is an employment relationship; or
- another person who derives title from the designer, employer or commissioner by assignment, will or operation of law.

The designer, namely the creator or author[195] of the design, is the 'person whose mind conceives the relevant shape, configuration, pattern or ornamentation applicable to the article in question and reduces it to visible form'.[196]

Therefore, unless there is a contrary agreement,[197] the employer is the person who is entitled to be entered on the register as the registered owner of a design if the designer created the design 'in the course of employment'.[198] In the absence of precedent, Spender J in *Courier Pete Pty Ltd v Metroll Queensland Pty Ltd* [199] accepted that the correct approach in answering the question whether the design was 'created in the course of employment' is to ask 'what the employee was employed to do?' In adopting this approach, he drew indirectly on the principles articulated by Lord Reid in *Patchett v Stirling Engineering Co Ltd* [200] and followed in Australian patent cases.[201] If this involves the creation of the type of designs that were ultimately produced, then the employer is entitled to the benefit of those designs from the time of their creation. In the absence of any 'duty of creativity', the employer's claim can arise only when it gave a specific direction to the employee to perform the work that resulted in the production of the design.[202]

6.2.1 University Researchers' Rights in Registered Designs

As with other forms of IPRs, a university's entitlement might rest upon either initial vesting or under an assignment of rights. Australian universities generally assert ownership of all rights in registered designs, registrable and unregistered designs in employment contracts.[203] In the case of a design that has not yet been registered and that is created 'in the course of employment', rights will vest initially in the university as owner under the Designs Act. The 'assertion' of ownership merely confirms the

employer as being, although there was no employment or other written contract; *Gram Engineering Pty Ltd v Bluescope Steel Ltd* (2013) 106 IPR 1.

194. *Foster's Australia Ltd v Cash's (Australia) Pty Ltd* (2013) 219 FCR 529.
195. This term was used in *Designs Act 1906* (Cth) s. 19(1) and its judicial interpretation was applied in the context of s. 13(1) of the *Designs Act 2003* (Cth).
196. *Chris Ford Enterprises Pty Ltd v BH and JR Badenhop Pty Ltd* (1985) 7 FCR 75, 80; *LED Technologies Pty Ltd v Elecspess Pty Ltd* (2008) 80 IPR 85, 96 [9], [26]; *Allen Hardware Products v Tclip* [2008] ADO 8, [19].
197. A provision introduced on the recommendation 66 of the ALRC Report 74 *Designs*.
198. *LED Technologies Pty Ltd v Elecspess Pty Ltd* (2008) 80 IPR 85.
199. (2010) 87 IPR 397.
200. (1955) 72 RPC 50.
201. *Spencer Industries Pty Ltd v Collins* (2002) 54 IPR 434.
202. *Courier Pete Pty Ltd v Metroll Queensland Pty Ltd* (2010) 87 IPR 397, 400 [19]; *Gram Engineering Pty Ltd v Bluescope Steel Ltd* (2013) 106 IPR 1, [403] (managing director).
203. Monotti with Ricketson 2003, Ch. 5. See 4.2.4 above for more detailed discussion of the university internal legislation and policies that govern IPRs.

statutory vesting of those rights when the terms (or duties) of employment are construed to encompass the creation of designs. However, the definition of a 'design' is one that is 'in relation to a product'. Hence, there will be a limited class of academic employees whose terms of employment would extend to their creation.

Nevertheless, it is common for universities to make extended claims to rights in all IP, including designs, that their academic employees create in specified circumstances. These might include such activities as the use of assets, funding, facilities, or resources, and IP owned by the university or another member of the university, or when the design is commissioned. The validity of these claims will depend upon the particular circumstances of the case under consideration.[204]

A university employer may agree to assign its rights of ownership to the creator or creators. Such an agreement may include provisions such as the retention of the right to use the IP for education, academic and research purposes, assumption of liability for on-going costs associated with IP protection, appropriate indemnities, for recovery of funds expended by the university and a royalty payable from commercialisation of the IP.[205]

6.3 The Parties' Duty to Inform

There is no statutory duty to inform an employer of the creation of a design to which the employer is entitled. The existence of a duty to inform or disclose is one that might arise under express or implied terms in a particular contract of employment or pursuant to the obligation of a fiduciary where that relationship exists.

Australian universities commonly require their academic staff to disclose the creation of specified categories of university IP including designs. The broad purpose of an obligation to inform or disclose is to alert the university to its creation and to give the relevant staff an opportunity to the identity of all potential creators, ownership, commercial potential and protection that might be applicable to the IP Disclosure. [206]

6.4 Employee Remuneration Right

The Designs Act vests rights to be registered as the owner of a design in the employer in the first instance. Any right to compensation becomes a matter for negotiation between the employee and the employer. In the case of academic employees, all universities provide for sharing the net proceeds of commercial exploitation of university IP in accordance with their policies in effect at the time.[207]

204. *University of Western Australia v Gray* (2009) 179 FCR 346.
205. See, e.g., ANU Procedure 2017, cls 25 and 26.
206. See 4.3 above.
207. See 4.4 above.

7 EMPLOYEES' PLANT VARIETIES

7.1 The Legal Framework

The Plant Breeder's Rights Act 1994 (Cth) ('PBRA') provides an applicant with registered protection for a new plant variety. The grant of a plant breeder's right ('PBR') to the applicant provides him with proprietary rights in relation to propagating material of a protected plant variety.[208] The PBRA includes express provisions that govern the rights of employers when new plant varieties are bred by their employees under the defined circumstances.[209] The general principles of contract, equity and employment law are also relevant to a determination of rights in PBR bred by an employee.

7.2 Ownership of Rights

The right to apply for PBR is given to the breeder of the plant variety.[210] On acceptance of the application, the breeder is granted PBR in that variety. 'Breeder' is defined broadly to mean the person or persons who bred the variety.[211] When a person breeds the new variety 'in the course of performing duties or functions as ... [an] employee ...', the employer is the 'breeder'. As the employer is the person entitled to make the application for and to be granted the PBR,[212] the rights will vest initially in the employer.

7.2.1 University Researchers' Plant Breeder's Rights

As with other forms of IPRs, a university's entitlement might rest upon either initial vesting or under an assignment of rights. Australian universities generally assert ownership of all rights in plant varieties.[213] In the case of a new plant variety that has not yet been registered and that is created 'in the course of performing duties or functions' as an employee of the university, rights to apply for and to be granted ownership of PBR will vest initially in the university under the PBRA. In this case, the 'assertion' of ownership merely confirms the statutory vesting of those rights when the duties and functions of employment are construed to encompass the creation of new plant varieties. However, the specialised nature of this intellectual output would limit the class of academic employees whose terms of employment would extend to their creation.

208. PBRA 1994 (Cth) ss. 11, 24 and 44(1).
209. *Ibid.* ss. 3 and 24.
210. *Ibid.* ss. 24 and 44.
211. *Ibid.* s. 3(1).
212. *Ibid.* s. 3.
213. See generally Monotti with Ricketson 2003, Ch. 5. See 4.2.4 above for more detailed discussion of the university internal legislation and policies that govern IPRs.

Nevertheless, it is common for universities to make extended claims to rights in all IP, including plant varieties, that their academic employees create in specified circumstances. These might include such activities as the use of assets, funding, facilities, or resources, and IP owned by the university or another member of the university or commissioned plant varieties. The validity of these claims will depend upon the particular circumstances of the case under consideration.[214]

A university employer may agree to assign its rights of ownership to the employee who created the new plant variety. Such an agreement may include provisions such as the retention of the right to use the IP for education, academic and research purposes, assumption of liability for on-going costs associated with IP protection, appropriate indemnities, for recovery of funds expended by the university and a royalty payable from commercialisation of the IP.[215]

7.3 The Parties' Duty to Inform

There is no statutory duty imposed on an employee to inform an employer when they breed a new plant variety. The existence of a duty to inform is one that would arise under express or implied terms in the contract of employment. Universities commonly require their academic staff to disclose the creation of any IPRs including new plant varieties. This duty may apply to all IPRs to which the university is entitled or might be limited to rights that have potential commercial value.

7.4 Employee Remuneration Right

The PBRA vests PBR in the employer in the first instance when the new plant variety is bred by a person in the course of performing duties or functions as an employee. Any right to compensation becomes a matter for negotiation between the employee and the employer.

In the case of academic employees, all universities provide for sharing the net proceeds of commercial exploitation of university IP in accordance with their policies in effect at the time. A common approach involves sharing the net proceeds equally between the creator, their department or school, and the university central administration.[216] In some cases, the university's share may be retained by the university commercialisation company when that company manages the commercialisation of the IP.[217]

214. Monotti 1997; *University of Western Australia v Gray* (2009) 179 FCR 346.
215. See, e.g., ANU Procedure 2017, cls 25 and 26.
216. *Ibid.* cl. 16.
217. UQ Policy 2016, cl. 9.1.

8 INSTANCES FOR DISPUTES

8.1 Jurisdiction Conferral

A dispute relating to employees' IPR usually arises in the context of infringement actions although this is not necessarily the case. Infringement actions are generally heard in the Federal Court of Australia ('Federal Court') that has jurisdiction with respect to:

(1) all matters arising under the Patents Act and the Designs Act; [218] and
(2) infringement actions and some other matters arising under the Copyright Act, Circuit Layouts Act and PBRA.[219]

State and Territory Supreme Courts have concurrent jurisdiction over statutory IPR with that of the Federal Court except in the case of PBR.[220] The PBRA confers jurisdiction with respect to infringement actions and some other matters on the Federal Court that is exclusive of the jurisdiction of all other courts.[221]

In cases brought in the Federal Court, its jurisdiction extends to matters that are *associated* with the *core* matters under consideration but are not otherwise within its jurisdiction.[222] Hence, in a matter that concerns ownership of IPRs created by employees, such matters might include common law and equitable rights that are otherwise within the exclusive jurisdiction of the State and Territory Supreme Courts.

The Federal Circuit Court of Australia ('Federal Circuit Court'),[223] established to hear shorter and simpler matters, may also hear disputes concerning ownership of employee copyright and designs as its jurisdiction extends to civil matters under the Copyright Act and all matters arising under the Designs Act.[224] This jurisdiction also extends to associated matters not otherwise within its jurisdiction.[225] The jurisdiction of the Federal Circuit Court of Australia will also extend to include plant breeder's rights matters on 25 August 2015.[226] Both the Federal Circuit Court and the Federal

218. *Patents Act 1990* (Cth) s. 154; *Designs Act 2003* (Cth) s. 83; *Federal Court of Australia Act 1976* (Cth) Part 111.
219. *Copyright Act 1968* (Cth) s. 131C; *Circuit Layouts Act 1989* (Cth) s. 47; *PBRA 1994* (Cth) ss. 56(1)-(2).
220. *PBRA 1994* (Cth) s. 56(1)-(2); *Patents Act 1990* (Cth) s. 155; *Copyright Act 1968* (Cth) ss. 131A, C and D; *Designs Act 2003* (Cth) s. 84; *Circuit Layouts Act 1989* (Cth) ss. 40 and 47; *Federal Court of Australia Act 1976* (Cth) Part III.
221. The jurisdiction of the High Court under s. 75 of the *Constitution* is maintained.
222. *Federal Court of Australia Act 1976* (Cth) s. 32(1).
223. Established by the *Federal Magistrates Act 1999* (Cth) - now known as the *Federal Circuit Court of Australia Act 1999* (Cth) – as the Federal Circuit Magistrates Court of Australia, but renamed on 12 April 2013.
224. *Copyright Act 1968* (Cth) s. 131D; *Designs Act 2003* (Cth) s. 83A.
225. *Federal Circuit Court of Australia Act 1999* (Cth) s. 18.
226. *PBRA 1994* (Cth) s 56A.

Court have discretion to order the transfer of a pending proceeding to the Federal Court or the Federal Circuit Court as the case may be.[227]

Apart from proceedings initiated in a Federal, State or Territory court, the issue of rights in employee inventions, designs and PBR might arise in the context of establishing entitlement to be granted those rights. In the case of a patent or design application, the Commissioner of Patents or Registrar of Designs has authority to make decisions in disputes between applicants that arise under the Patents or Designs Acts. The Registrar of Designs may also revoke a registration on grounds relating to entitlement.[228] The position with PBR differs in that no appeal to a court is possible from decisions of the Secretary. Decisions to grant, or refuse to grant PBR in a plant variety are reviewable by the Administrative Appeals Tribunal ('AAT').[229]

Finally, it is theoretically possible that ownership issues might come before the Fair Work Commission,[230] the body that functions as a tribunal for hearing matters raised under the Fair Work Act 2009 (Cth). In a case involving an unfair dismissal application for wrongly using or exploiting employer IPRs, the dismissed employee might argue that they are the owner of those rights. In addition, the Industrial Relations Commission in the States of New South Wales and Queensland have jurisdiction under their respective Industrial Relations Acts to vary or declare void employment contracts that are unfair.[231] While there may be scope for issues relating to ownership of IPR to arise within this context, this is uncommon.[232]

8.2 Appeal Jurisdiction

Appeals provisions are relatively complex. Appeals from decisions of a single judge in the Federal Court are heard by the Full Federal Court and from decisions of a single judge in a State or Territory Supreme Court are heard in the respective Court of Appeal. A party may also seek special leave of the High Court of Australia to appeal the full court decision.

The Federal Court also has jurisdiction to hear appeals from the Federal Circuit Court,[233] from a judgment of another court exercising jurisdiction under the relevant Act,[234] from decisions of the Commissioner of Patents and Registrar of Designs[235] and from the AAT on questions of law.[236] Where appellate jurisdiction from decisions of the Commissioner of Patents or the Registrar of Designs is exercised, any further appeal

227. *Federal Circuit Court of Australia Act 1999* (Cth) s. 39; *Federal Court of Australia Act 1976* (Cth) s. 32AB.
228. *Patents Act 1990* (Cth) ss. 32 and 59; *Designs Act 2003* (Cth) ss. 29 and 52.
229. *PBRA 1994* (Cth) ss. 44 and 77(1)(b)(xv); *Administrative Appeals Tribunal Act 1975* (Cth).
230. *Fair Work Act 2009* (Cth) Part 5-1; formerly known as Fair Work Australia.
231. *Industrial Relations Act 1996* (NSW) s. 106; *Industrial Relations Act 2016* (Qld) s. 471.
232. *Elkins v Full Bloom Pty Ltd* [2000] NSWIRComm 223.
233. *Federal Court of Australia Act 1976* (Cth) s. 24(1)(d).
234. *Patents Act 1990* (Cth) s. 158.
235. *Designs Act 2003* (Cth) s. 83(2) (Federal Court) and s. 83A(2) (Federal Circuit Court); *Patents Act 1990* (Cth) s. 154(2); *Federal Court of Australia Act 1976* (Cth) s 19(2) (original jurisdiction).
236. *Administrative Appeals Tribunal Act 1975* (Cth) s. 44.

requires leave of the Court.[237] Special provisions may apply in relation to each of the statutory IP disputes. For instance, in the case of the Copyright and Circuit Layouts Acts, an appeal from a decision of a single judge of a State and Territory Supreme Courts lies to the Federal Court or directly to the High Court with special leave. [238]

9 CONCLUDING REMARKS

In Australia, the ownership of IPRs that are created by employees in the course of their employment may be determined with reference to IP legislation, the express and implied terms in the employment contract, perhaps fiduciary obligations, or a combination of all of these. In the same way that employers own the tangible products that their employees create in the course of employment, the underlying policy is that rights in IPRs associated with the products of that work are also to be owned by the employer. In the case of copyright, plant breeder's rights and circuit layouts, rights in the employee's creation vest initially in the employer provided that the specified criteria are satisfied. In contrast, employee designs and inventions vest initially in the designer or inventor. The employer's rights are derivative and rely on an assignment by the employee that is enforceable in law or equity.

The discussion in the sections above considered the rights of employers in each category of IP without consideration of the fact that work undertaken in employment might produce outcomes that are suitable for protection under more than one IP regime. For instance, computer software may have copyright protection but also be patentable, a new plant variety may have protection under both the Patents Act and the PBRA, and a patentable drug may also be the subject of published articles. Furthermore, patentable inventions may be commercially exploited as products that are suitable for registered design protection. The statutes do not exclude ownership of overlapping rights in relation to different aspects of the same subject matter. Instead, they deal with this in the context of infringement.[239] Therefore, it is important for employers and employees to consider in their contractual negotiations whether IPRs in the same product should vest in the same ownership.

As a general proposition, a contract of employment will include an express term to provide that the employer owns all IPR that arise from the work that the employee is paid to perform. An exception applies in the context of academic employment contracts where university employers provide for academic employees to own copyright in 'scholarly works' within the terms of their IP policies. However, express provision is now generally included to deal with the problems that can arise with

237. *Patents Act 1990* (Cth) s. 158(2); *Designs Act 2003* (Cth) s. 87(2); *Federal Court of Australia Act 1976* (Cth) ss. 24 and 25.
238. *Copyright Act 1968* (Cth) s. 131B; *Circuit Layouts Act 1989* (Cth) s. 41(2).
239. E.g., s. 226 of the *Patents Act 1990* (Cth) provides that copyright in any literary or artistic work contained in a published specification made available for public inspection is not infringed by the two dimensional reproduction of the whole or a part of that work; Division 8 of the *Copyright Act* deals with the overlapping rights that arise when an artistic work is industrially applied to a product as a design. *Circuit Layouts Act 1989* (Cth) s. 24 addresses the possible overlap between EL rights and copyright in imported integrated circuits.

divided rights. One such problem is the potential for the owner of copyright in an article to destroy or devalue the rights of the owner of any invention by premature public disclosure of its details.[240]

When the contract is silent as to ownership of IPRs or when a clause is unenforceable, the entitlement is determined with reference to the relevant statutory provisions. These provisions use largely equivalent but not identical language to define circumstances in which employers are entitled to the intellectual products of the work that their employees are paid to perform. However, different drafting means that a common operation and outcomes arising from those provisions can arise only when courts interpret the language to mean the same thing. With the exception of the Patents Act that identifies the rights of an employer with reference to well trodden common law and equitable principles, and without express reference to employer's rights, the Copyright, Circuit Layouts, Designs and Plant Breeder's Rights Acts expressly define entitlement with reference to:

(1) '*in pursuance of the terms* of his or her employment ... under a contract of service or apprenticeship' (copyright);
(2) '*under the terms* of his or her employment by another person under a contract of service' (circuit layouts);
(3) 'if the designer created the design *in the course of employment* ... with another person' (designs); and
(4) 'in the *course of performing duties or functions* as ... [an] employee of a body (whether corporate or unincorporate)' (PBR).

Judicial consideration of these issues appears only in relation to disputes involving copyright, designs and patents. Despite the linguistic differences, a common or closely equivalent approach can be identified in the most recent decisions. In *Courier Pete Pty Ltd v Metroll Queensland Pty Ltd*,[241] Spender J approached the meaning of 'course of employment' in the Designs Act with resort to the common law principles that courts apply to determine entitlement to employee inventions.[242] In the context of the Copyright Act, Moore J in *Edsonic Pty Ltd v Cassidy* [243] adopted an approach that 'comprehends notions similar to those addressed by the Full Court in *Gray* in the context of patent law though plainly I must ultimately focus on and give effect to the words used in the Act itself'.[244]

While the lack of uniform statutory provisions potentially raises the possibility of different outcomes among IPR regimes, the approaches taken in these cases by single judges suggest a willingness to turn to the patent law for a uniform approach in spite

240. E.g., University of New South Wales *Procedure for Disclosing and Exploiting IP* (2013), cl. 3.3; Monotti with Ricketson 2003, [6.180]-[6.189].
241. (2010) 87 IPR 397.
242. (2010) 87 IPR 397, 400-401; *Allen Hardware Products v Tclip* [2008] ADO 8, [19]; *Patchett v Stirling Engineering Co Ltd* (1955) 72 RPC 50, 58 (Lord Reid); *Spencer Industries Pty Ltd v Collins* (2003) 58 IPR 425.
243. *EdSonic Pty Ltd v Cassidy* (2010) 189 FCR 271,. See 2.2.1 above.
244. *Ibid.* [41].

of any linguistic differences in the statutes themselves. Even with uniformity of approach, this apparent simplicity does not overcome the fact that it is often very difficult to apply these principles in different and frequently complex factual circumstances.[245] This lack of clarity is particularly problematic in the context of academic employment where crafting an express contractual regime to meet the many and varied circumstances of employment is challenging.[246]

Another issue that may be problematic in particular cases arises when the statutory provisions or common law principles determine ownership of employee IPRs. Ownership will vest absolutely in either the employer or the employee according to the circumstances of each case as a default position. It is an 'all or nothing' approach that in some cases may be neither fair nor desirable. This is especially true of the university sector. An outcome that recognises some balancing of rights can be achieved only through an agreement that modifies the operation of those legal principles. There are clear limitations in the current paradigm that can arise especially from the unclear scope of duties of employment under the express or implied terms of the contract of employment. Nevertheless, there is little evidence of increasing numbers of disputes that require adjudication in the courts.

BIBLIOGRAPHY

Table of Sources

Elizabeth Adeney, *The Moral Rights of Authors and Performers: An International and Comparative Analysis* (Oxford University Press 2006).

C. Arup, *Employee Inventions: Labour Law Meets Intellectual Property*, 21 *AJLL* 208 (2008).

C. Arup, *What/Whose Knowledge? Restraints of Trade and Concepts of Knowledge?*, 36 MULR 369 (2012).

M.J. Davison, Ann L. Monotti & L. Wiseman, *Australian Intellectual Property Law* (3rd ed., 2016).

Mark Freedland & Jeremias Prassl, *Resolving invention ownership disputes: limitations of the contract of employment* in Marilyn Pittard, Ann L Monotti & John Duns, *Business Innovation and the Law: Perspectives from Intellectual Property, Labour, Competition and Corporate Law* (Edward Elgar 2013).

John Glover, *Equity, Restitution and Fraud* (LexisNexis Butterworths 2004).

J. D. Heydon, *The Restraint of Trade Doctrine* (3rd ed., Sydney: LexisNexis Butterworths 2008).

245. Mark Freedland and Jeremias Prassl, 'Resolving invention ownership disputes: limitations of the contract of employment' in Marilyn Pittard, Ann L Monotti and John Duns, *Business Innovation and the Law: Perspectives from Intellectual Property, Labour, Competition and Corporate Law*, 74 (Edward Elgar 2013), 144.
246. *University of Western Australia v Gray* [2009] 179 FCR 346; Monotti with Ricketson (2003); Monotti (2012) and (2013); Arup (2008); van Caenegem (2010)

John Howe & Andrew Newman, *Collective bargaining and the ownership of employee creation*, 26 AJLL 273 (2013).

R.P. Meagher, J.D. Heydon & M. D. Leeming, *Meagher, Gummow and Lehane's Equity: Doctrines and Remedies* (4th ed., Butterworths 2002).

Ann L. Monotti, *Is it time to codify principles for ownership of academic employee inventions? The disconnect between policy and the law*, 38 Monash ULR 102 (2012).

Ann L. Monotti, with S. Ricketson, *Universities and Intellectual Property: Ownership and Exploitation*, Ch. 7 (Oxford University Press 2003).

Ann L. Monotti, *Who Owns my Research and Teaching Materials – My University or Me?* 19 The Syd LR 425 (1997).

Ann L. Monotti, *Establishing clear rights in academic employee inventions: lessons learnt from University of Western Australia v Gray*, in Marilyn Pittard, Ann L Monotti & John Duns, *Business Innovation and the Law: Perspectives from Intellectual Property, Labour, Competition and Corporate Law* (Edward Elgar 2013).

Patrick Parkinson, *Fiduciary Obligations* in *The Principles of Equity*, ch 10 (ed. Patrick Parkinson, 2d ed., Law Book Company 2003).

Patrick Parkinson, *The Principles of Equity* (ed. Patrick Parkinson, 2d ed., Law Book Company 2003).

Marilyn Pittard, Ann L Monotti & John Duns, *Business Innovation and the Law: Perspectives from Intellectual Property, Labour, Competition and Corporate Law* (Edward Elgar 2013).

Marilyn Pittard, *The innovative worker: genius, accidental inventor or thief?* in Marilyn Pittard, Ann L Monotti & John Duns, *Business Innovation and the Law: Perspectives from Intellectual Property, Labour, Competition and Corporate Law* (Edward Elgar 2013).

W. van Caenegem, 'University of Western Australia *v* Gray' 23 *AJLL* 327 (2010).

Board of Trade, *The British Patent System: Report of the Committee to Examine the Patent System and Patent Law* (Cmnd 4407, 1970) (Banks Committee).

Copyright Agency of Australia, *Moral Rights*: http://www.copyright.com.au/get-information/other-rights/moral-rights (accessed 16 March 2015).

Table of Legislation

Administrative Appeals Tribunal Act 1975 (Cth).
Circuit Layouts Act 1989 (Cth).
Copyright Act 1968 (Cth).
Copyright Amendment (Film Directors' Rights) Act 2005 (Cth).
Designs Act 1906 (Cth).
Designs Act 2003 (Cth).
Fair Work Act 2009 (Cth).
Federal Circuit Court of Australia Act 1999 (Cth).
Federal Court of Australia Act 1976 (Cth).

Industrial Relations Act 1996 (NSW).
Industrial Relations Act 2016 (Qld).
Intellectual Property Laws Amendment Act 2015 (Cth).
Patents Act 1990 (Cth).
Plant Breeder's Rights Act 1994 (Cth).
Public Service Act 1999 (Cth).
Restraints of Trade Act 1976 (NSW).

Table of University Legislation, Policies and Procedures

Australian National University, *Intellectual Property Policy* (ANUP_003603, 2017): https://policies.anu.edu.au/ppl/document/ANUP_003603 (accessed 18 July 2017).
Australian National University, *Procedure: IP Protection and Commercialisation,* (ANUP_003605, 1 January 2017): https://policies.anu.edu.au/ppl/document /ANUP_003605 (accessed 18 July 2017).
Monash University (Vice-Chancellor) Regulations 2016, Pt 5 - Intellectual Property: https://www.monash.edu/legal/legislation/current-statute-regulations-and-related-resolutions (accessed 18 July 2017).
Queensland University of Technology, QUT, D/3.1 *Intellectual Property Pol icy,* (30 September 2015): http://www.mopp.qut.edu.au/D/D_03_01.jsp (accessed 18 July 2017).
University of Melbourne Statute (21 July 2016): http://www.unimelb.edu.au/ governance/statutes (accessed 18 July 2017).
University of New South Wales, *Intellectual Property Policy*, (effective 6 March 2013): https://www.gs.unsw.edu.au/policy/ippolicy.html (accessed 18 July 2017).
University of New South Wales *Procedure for Disclosing and Exploiting IP* (effective 13 February 2013): https://www.gs.unsw.edu.au/policy/ipdisc losingprocedure.html (accessed 18 July 2017).
University of Queensland, *Intellectual Property for Staff, Students and Visitors – Policy,* (4.10.13a, 2016): https://ppl.app.uq.edu.au/content/4.10.13-intellectual-property-staff-students-and-visitors (accessed 18 July 2017).
University of Sydney Intellectual Property Policy 2016: http://sydney.edu.au/ legal/regulations/ip.shtml (accessed 18 July 2017).

Table of Cases

Acohs Pty Ltd v Ucorp Pty Ltd, 86 IPR 492 (2010).
Acohs Pty Ltd v Ucorp Pty Ltd, 201 FCR 173 (2012).
Adamson v Kenworthy, 49 RPC 57 (1932).
Allen Hardware Products v Tclip, ADO 8 (2008).
Amoco Australia Pty Ltd v Rocca Bros Motor Engineering Co Pty Ltd, 133 CLR 288 (1973).

Aneeta Window Systems (Vic) Pty Ltd v K. Shugg Industries Pty Ltd, 34 IPR 95 (1996).
Avel Pty Ltd v Wells, 108 ALR 97 (1992).
Avtex Airservices Pty Ltd v Bartsch, 107 ALR 539 (1992).
Barnet Instruments Ltd v Overton, 66 RPC 315 (1949).
Beloff v Pressdram Ltd, 1 All ER 241 (1973).
British Reinforced Concrete Engineering Co. Ltd v Lind, 34 RPC 101 (1917).
British Syphon Company Ltd v Homewood, 1 WLR 1190 (1956).
Centrestage Management Pty Ltd v Riedle, 77 IPR 550 (2008).
Chris Ford Enterprises Pty Ltd v BH & JR Badenhop Pty Ltd, 7 FCR 75 (1985).
Christopher Simon French v Paul Julian Mason, FSR 597 (1999).
Corby v Allen & Unwin Pty Ltd (2013) 297 ALR 761.
Courier Pete Pty Ltd v Metroll Queensland Pty Ltd, 87 IPR 397 (2010).
Davies Shephard Pty Ltd v Stack, 108 FCR 422 (2001).
EdSonic Pty Ltd v Cassidy, 189 FCR 271 (2010).
Electrolux Ltd v Hudson, FSR 312 (1977).
Ex parte Forster; Re University of Sydney, NSWR 1000 (1964).
Fine Industrial Commodities Ltd v Powling, 71 RPC 253 (1954).
Foster's Australia Ltd v Cash's (Australia) Pty Ltd, 219 FCR 529 (2013).
George C. Warner Laboratories Pty Ltd v Chemspray Pty Ltd, 41 ALJR 75 (1967).
Gram Engineering Pty Ltd v Bluescope Steel Ltd, 106 IPR 1 (2013).
Greater Glasgow Health Board's Application, RPC 207 (1996).
Harris' Patent, RPC 19 (1985).
Herbert Morris Ltd v Saxelby, 1 AC 688 (1916).
Hivac Ltd v Park Royal Scientific Instruments Ltd, Ch 169 (1946).
Hollis v Vabu Pty Ltd, 207 CLR 21 (2001).
Hospital Products Ltd v United States Surgical Corporation, 156 CLR (1984).
IceTV Pty Ltd v Nine Network Australia Pty Ltd, 239 CLR 458 (2009).
In the Matter of Charles Selz Ltd's Application, 71 RPC 158 (1954).
Insight SRC IP Holdings Pty Ltd v Australian Council for Educational Research Ltd, 211 FCR 563 (2012).
JMVB Enterprises Pty Ltd v Camoflag Pty Ltd, 154 FCR 348 (2006).
Kwan v The Queensland Corrective Services Commission, 31 IPR 25 (1994).
Lancashire Fires Ltd v S. A. Lyons & Co. Ltd, FSR 629 (1996).
LED Technologies Pty Ltd v Elecspess Pty Ltd, 80 IPR 85 (2008).
Lindner v Murdock's Garage, 83 CLR 628 (1950).
Maggbury Pty Ltd v Hafele Australia Pty Ltd, 210 CLR 181 (2001).
Mellor v William Beardmore & Co Ltd, 44 RPC 175 (1927).
Peters (WA) Ltd v Petersville Ltd, 205 CLR 126 (2001).
Quadramain Pty Ltd v Sevastapol Investments Pty Ltd, 133 CLR 390 (1976).
R v Darling Island Stevedoring and Lighterage Co Ltd; Ex parte Halliday and Sullivan, 60 CLR 601 (1938).
Redrock Holdings Pty Ltd v Hinkley, 50 IPR 565 (2001).
Reiss Engineering Co Ltd v Harris, 14 IRLR 232 (1985).
Robb v Green, 2 QB 315 (1895).

Sands & McDougall Pty Ltd v Robinson, 23 CLR 49 (1917).
Seven Network (Operations) Ltd v TCN Channel Nine Pty Ltd, 146 FCR 183 (2005).
Sigma Pharmaceuticals (Australia) Pty Ltd v Wyeth, FCAFC 132 (2011).
Spencer Industries Pty Ltd v Collins, 54 IPR 434 (2002).
Sterling Engineering Co. Ltd v Patchett, AC 534 (1955).
Stevens v Brodribb Sawmilling Co Pty Ltd, 160 CLR 16 (1986).
Stevenson, Jordan and Harrison Ltd v Macdonald and Evans, 69 RPC 10 (1952).
Sweeney v Boylan Nominees Pty Ltd, 226 CLR 161 (2006).
The Royal Children's Hospital 213 IR 94 (2011).
Thorsten Nordenfelt v Maxim Nordenfelt Guns and Ammunition Co, AC 535 (1894).
Triplex Safety Glass Co. Ltd v Scorah, Ch 211 (1938).
University of Nottingham v Fishel, ICR 1462 (2000).
University of Western Australia v Gray, 179 FCR 346 (2009).
University of Western Australia v Gray No. 20, 246 ALR 603 (2008).
Wessex Dairies Ltd v Smith, 2 KB 80 (1935).
Victoria University of Technology v Wilson, 60 IPR 392 (2004).
Wills v Australian Broadcasting Corporation (No 3), 89 IPR 252 (2010).
Vokes Ltd v Heather, 62 RPC 135 (1945).
Worthington Pumping Engine Co. v Moore, 20 RPC 41 (1903).
Zuijs v Wirth Brothers Pty Ltd, 93 CLR 561 (1955).

D In North America

CHAPTER 30
Canada

Norman Siebrasse

1 BACKGROUND

Canada is a federal state with a common law legal tradition in all provinces except Quebec, which has a civil law tradition. The law relating to employee rights in intellectual property are generally governed by both the federal intellectual property statutes, which provide basic rules and presumptions regarding initial ownership of a work, and provincial statutory laws relating to the employment relationship, as well as the case law interpreting and supplementing those statutory sources. It is difficult to generalize about whether the rights belong initially to the employer or employee, because different statutes have different provisions; so, under the Copyright Act,[1] an employer is generally (though not always) the first owner of a work, while in patent law, the employee/inventor is generally (though not always) the first owner.[2] Moreover, these presumptive rules may be displaced by employment agreements.

2 EMPLOYEES' COPYRIGHT

2.1 The Legal Framework

Copyright is governed by the federal Copyright Act, which also sets out default rules governing initial ownership of copyright both in general and in the case of a work created by an employee.[3] The default ownership rights set out in the Act may be altered

1. *Copyright Act*, R.S.C. 1985, c. C-42.
2. David Vaver, *Intellectual Property Law*, 125, 368 (2nd ed., Irwin Law 2011).
3. *Copyright Act*, s. 13(1).

by contract, except for moral rights, which may be waived but not assigned.[4] The contractual rights between the employer and employee are governed by provincial contract law.

2.2 Ownership of Rights

2.2.1 Basic Rule

The basic rules relating to ownership of copyright are set out in Section 13 of the Copyright Act:

> '13.
> (1) Subject to this Act, the author of a work shall be the first owner of the copyright therein.
> ...
> (2) Where the author of a work was in the employment of some other person under a contract of service or apprenticeship and the work was made in the course of his employment by that person, the person by whom the author was employed shall, in the absence of any agreement to the contrary, be the first owner of the copyright ...'

The primary rule set out in Subsection 13(1) is that the author of a work is the first owner of the copyright therein. Subsection 13(3) sets out an exception to this rule, with three points of note: the employer will be the first owner of the copyright if: (1) the author was employed 'under a contract of service or apprenticeship' and (2) the work was made 'in the course of' that employment, (3) subject to any 'agreement to the contrary'. The same principle is also generally applicable to officers, directors, and key employees who create a work for the benefit of the corporation.[5] These rules also apply to neighbouring rights.[6] It is also statutorily presumed that the author is the first owner of the copyright,[7] which implies that an employer wishing to establish that it was the first owner bears the burden of proving that the work was created in the course of employment and under a contract of service.[8] The presumption is not a strong one and is determinative only in the absence of direct evidence.[9]

2.2.1.1 Contract 'of Service'

As noted, for the employer to be first owner, the author must be employed in an apprenticeship or a contract 'of service.' The latter is contrasted with a contract 'for

4. *Ibid.* at s. 14.1(2).
5. *Harmony Consulting Ltd. v G.A. Foss Transport Ltd.*, 2011 FC 340 230 aff'd 2012 FCA 226 57.
6. See s. 2 of the *Copyright Act* defining 'copyright' to include neighbouring rights.
7. *Copyright Act*, s. 34.1(b).
8. Planification-Organisation-Publications Système (POPS) Ltée v. 9054-8181 Québec Inc. 2014 FCA 185, [76]; *Dynabec Ltée c. Société d'informatique RDG Inc.*, 6 C.P.R. (3d) 299, 313 (Q.S.C. 1984); *Underwriters' Survey Bureau Ltd. v Massie and Renwick Ltd.*, [1940] S.C.R. 218, 233.
9. *Harmony Consulting Ltd. v G.A. Foss Transport Ltd.*, 2011 FC 340 62.

services,' under which the author will be first owner. In general terms, if a party is hired as an independent contractor for the specific purpose of creating a work, first ownership will rest with the independent contractor, and ownership will have to be expressly assigned to the counter-party, notwithstanding that the purpose of the contract was for the contractor to create the work for the benefit of the counter-party.

There is relatively little Canadian case law directly on the interpretation of a contract 'of service', and the cases which have addressed the issue have done so briefly and by reference to the particular facts of the case, without setting out any tests or considerations which would provide more general guidance.[10] However, essentially the same distinction arises in other areas of law, such as tort law, employment law, taxation law and unemployment insurance law, and the factors identified in those contexts may be considered in copyright cases to determine whether an employment relationship exists.[11] The definition of an apprenticeship does not appear to have attracted any case law.

2.2.1.2 In the Course of Employment

For the employer to be the first owner, it is not enough that the employment is under a contract 'of service.' The work must also be created 'in the course of his employment'.[12] As with the question of whether the relationship is under a contract 'of service,' case law from other areas of law may be considered in determining what is meant by 'in the course of employment.'[13] The main question is whether creating the work was within the duties of the employee.[14] It may be asked whether 'the work [was] created as an integral part of the workings of the enterprise, or did the author create it on his own account?'[15] Similarly, it may be asked whether the employee is 'proceeding generally about his master's affairs'.[16]

Whether the work was created by an employee during their free time is one factor that may be considered,[17] but it is not determinative. If the work is in the course of the employee's duties, copyright will belong to the employer, even if it is created largely

10. *See Planification-Organisation-Publications Système (POPS) Ltée v. 9054-8181 Québec Inc.* 2014 FCA 185, [74], [84]; *Hutton v Canadian Broadcasting Corp.*, 29 C.P.R. (3d) 398, 446-447 (A.B.Q.B. 1989); *Goldner v Canadian Broadcasting Corp.*, 7 C.P.R (2d) 158, 162 (F.C.T.D. 1972); *Wiebe v Saskatchewan Institute of Applied Science and Technology*, 2007 SKQB 60, 38; *Netupsky v Dominion Bridge Co.*, 58 C.P.R. 7, 28-29 (B.C.C.A. 1969); *Dynabec Ltée*, 6 C.P.R. (3d) at 312-313; *Underwriters' Survey Bureau Ltd.*, [1940] S.C.R. at 23?; *Andrews v. McHale* 2016 FC 624, [91]-[92].
11. *Saxon c. Communications Mont-Royal Inc.*, J.E. 2001-91, 79 (C.S.); *Amusements Wiltron Inc. c. Mainville*, (1991) 40 C.P.R. (3d) 521 (Q.S.C.); *Harmony Consulting Ltd. v G.A. Foss Transport Ltd.*, 2012 FCA 226, 56 aff'g 2011 FC 340 229-41.
12. *Copyright Act*, s. 13(3).
13. *Mejia v LaSalle College International Vancouver Inc.*, 2014 BCSC 1559, 205.
14. *Mejia v LaSalle College International Vancouver Inc.*, 2014 BCSC 1559, 207.
15. *Fénollar c. PRB Média*, 2006 QCCS 1956, 30, quoting Normand Tamaro, The 2005 Annotated Copyright Act, Toronto, Carswell, 2005, 352, quoting in turn *Beloff v Pressdram Ltd.* [1973] 1 All E.R. 241, 250 (Ch. D.).
16. *Mejia v LaSalle College International Vancouver Inc.*, 2014 BCSC 1559, 207.
17. *Fénollar c. PRB Média*, 2006 QCCS 1956, 31.

outside of normal working hours.[18] Conversely, copyright may belong to the employee even if the work was done partly during normal working hours, if it was not in the course of the employee's duties.[19] Another relevant factor is whether it would be contrary to the principle of good faith in employment to use the employer's materials and resources to benefit from copyright in the work.[20] The determination does not depend on the employee's labelling the work as a personal project.[21] In *Lachance c. Productions Marie Eykel Inc.* the trial judge found the following factors relevant: presence of a direct superior; membership in a union; whether the work was integrated with the work done at the organization; whether the work was carried out according to a mandate given by a superior.[22] The creativity of the author is not a relevant consideration.[23] Whether the person claiming ownership is the sole author, or one of several authors, matters little.[24]

In determining whether the work was done in the course of employment Professor Vaver suggests two relevant questions:

> (1) Would the worker have broken their contract by not producing the work in the form they did?
> (2) Would the acquisition or retention of copyright by the worker be inconsistent with her duty of good faith and loyalty to their employer? If either question is answered yes, the employer should own the copyright. If both are answered no, the copyright belongs to the worker.[25]

This two-fold test has not been expressly adopted by Canadian courts, but is consistent with the case law and is likely to be influential.

2.2.1.3 Agreement to the Contrary

Even if the work is created by an employee under a contract of service in the course of their employment, the employee will be the first owner of the copyright if there is an agreement to that effect.[26] It is well established that such an agreement need not be express or in writing,[27] and may be implied. A general understanding in the field that copyright will belong to the employee may suffice.[28] An implied agreement may of

18. *Corso v Nebs Business Products Ltd.*, 72 C.C.E.L. (3d) 110, 128 (Ont. S.C. 2009); *Seanix Technology Inc. v Ircha*, 78 C.P.R. (3d) 443, 449 (B.C.S.C. 1998).
19. *Mejia v LaSalle College International Vancouver Inc.*, 2014 BCSC 1559, 206.
20. *Corso v Nebs Business Products Ltd.*, 72 C.C.E.L. (3d) 110, 128 (Ont. S.C. 2009).
21. *Hanis v Teevan*, 81 C.P.R. (3d) 496, 510 (Ont. C.A. 1998).
22. *Lachance c. Productions Marie Eykel Inc.*, 2012 QCCS 1012, 48-50.
23. *Lachance c. Productions Marie Eykel Inc.*, 2014 QCCA 158, 14.
24. *Planification-Organisation-Publications Système (POPS) Ltée v. 9054-8181 Québec Inc.* 2014 FCA 185, [84].
25. Vaver, *supra* at 127.
26. *Saxon c. Communications Mont-Royal Inc.*, J.E. 2001-91 (C.S.)
27. See e.g. *Andrews v. McHale* 2016 FC 624, [92]-[96].
28. *Dolmage v Erskine*, 23 C.P.R. (4th) 495, 517 (Ont. Sm. Cl. 2003).

course also be altered by an express contract. An agreement between employer and employee may provide for joint ownership of the copyright.[29]

2.2.1.4 *Academic Employees*

One example of an implied 'agreement to the contrary' within the meaning of Subsection 13(3) is what is sometimes referred to as the 'academic exception.' It is generally understood in the university community that copyright in academic articles written by a faculty member belongs to the faculty member, notwithstanding that the faculty member is employed under a contract of service and that writing articles is within the normal course of those services.[30] This implied agreement can be modified by collective agreement provisions between the university and the faculty. If copyright ownership in materials authored by faculty members is not mentioned in such an agreement, a university could institute a policy requiring faculty members to assign copyright in materials prepared by them.[31]

This implied agreement is not available to all employees of academic institutions; rather the role of the employee within the university will determine whether or not there is an implied agreement to the contrary. Work written by employees hired by the university was held to be property of the university in *Hanis v Teevan*.[32] A subsequent court, holding that copyright belonged to the employee, distinguished *Hanis* by drawing a distinction between administrative staff and faculty.[33]

2.2.2 *Newspaper Articles*

Subsection 13(3) deals specifically with newspaper articles, providing that:

> (3) ... where the work is an article or other contribution to a newspaper, magazine or similar periodical, there shall, in the absence of any agreement to the contrary, be deemed to be reserved to the author a right to restrain the publication of the work, otherwise than as part of a newspaper, magazine or similar periodical.

This is on the face of it a right of veto, which only allows the employee to restrain publication of the work (other than as part of a newspaper, magazine or similar periodical). It does not make the journalist who has created a work in the course of employment under a contract of service the first owner of the copyright, nor does it give

29. *Kennedy v. Ruminski* 2014 FC 526.
30. *Ibid.* at 517-519. See also Vaver, *supra* at 128-129, criticizing the term 'academic exception,' while indicating that the result in *Dolmage* was correct. His criticism of the term being misleading however is not reflected in the case law, as the court in *Dolmage* was very careful in stating that the academic exception was simply an understood implied agreement to the contrary (517-519).
31. *University of British Columbia (Re)*, [2006] B.C.L.R.B.D. No. 56, 2006 CanLII 6155, 42, 57.
32. *Hanis v Teevan*, [1995] O.J. No. 981, 1995 CarswellOnt 5181, 313-314 (Ont. S.C.), *aff'd on this point in* 81 C.P.R. (3d) 496 (Ont. C.A. 1998). See Vaver, *supra* at 129 criticizing the decision in *Hanis* on the facts.
33. *Dolmage*, 23 C.P.R. (4th) at 517.

the journalist any independent right to publish the work. Importantly, a staff writer will not have a cause of action unless they previously exercised their right to restrain publication.[34]

Freelance authors, like other independent contractors, retain copyright in their work and have the right to reproduce their individual works.[35] The publisher of a newspaper or magazine may reproduce articles as part of its right to reproduce the newspaper or magazine. Thus the right turns on the definition of a 'newspaper.' While the precise definition is unclear, a CD-ROM version of the newspaper falls within the definition of a 'newspaper' and an electronic database comprising newspaper articles does not.[36]

2.2.3 Government Works

Section 12 of the Copyright Act provides that where a work is created 'by or under the direction or control of Her Majesty or any government department', the copyright in the work belongs to Her Majesty, subject to any agreement to the contrary. This provision is evidently broader than that applicable to private parties, in that copyright in a work prepared by an independent contractor under the direction of the government, will belong to the government. There is currently no Canadian case law on what constitutes 'direction or control'.[37]

2.2.4 Reversion

Generally, copyright may be assigned in whole or in part, and for the entirety of the term or only a part.[38] However, where the author of a work is the first owner, the last 25 years of the copyright revert to the author's estate on the author's death, notwithstanding any agreement to the contrary.[39]

2.2.5 Moral Rights

The author of a work, and not the first owner, has certain moral rights in the work.[40] Moral rights may not be assigned, even to an employer, though they may be waived.[41] Waiver of moral rights is common.

34. *Robertson v Thomson*, 2006 SCC 43, 62.
35. *Ibid.* at 6-7.
36. *Ibid.* at 62-63.
37. The issue has been raised in *Keatley Surveying Ltd. v Teranet Inc.*, 2014 ONSC 1677 in which the court certified a class of land surveyors, however there is yet no decision on the merits. See also Vaver, *supra* at 134, discussing case law from other jurisdictions which may provide guidance in the Canadian context.
38. *Copyright Act*, s. 13(4).
39. *Ibid.* at s. 14(1).
40. *Ibid.* at s. 14.1.
41. *Ibid.* at s. 14.1.

2.3 The Parties' Duty to Inform

There is no statutory duty to give an employer notice of the creation of copyrighted works. However, employees owe their employers a general duty of good faith as a matter of employment law.[42] This duty of good faith relates to copyrighted works created by an employee,[43] and may create a duty to give notice in particular circumstances. Disclosure may also be required by contract.

2.4 Employee Remuneration Right

There is no right to remuneration, either at common law or under the Copyright Act. Any right to remuneration in respect of particularly valuable works must stem from the employment contract.

3 EMPLOYEES' SEMICONDUCTOR CHIPS

3.1 The Legal Framework

The federal Integrated Circuit Topography Act ('ICTA'),[44] provides protection to integrated circuit topographies – semiconductor chip designs – on registration under the Act, for a ten-year term. There has been no litigation under the Act, so there is no judicial elaboration on the provisions of the ICTA.

3.2 Ownership of Rights

Section 2(4) of the ICTA provides that:

> Where a topography is created in the course of employment or pursuant to a contract, the employer or party to the contract for whom the topography was created shall be deemed to be the creator of the topography unless the employer and employee or the parties to the contract, as the case may be, otherwise agree.

The exclusive right in a topography is created on registration of the topography,[45] and the creator of the topography, or their successor in title, is the person entitled to apply for registration.[46] The combined effect is that the employer is normally the first owner of the topography right.

42. *57134 Manitoba Ltd. v Palmer*, 26 C.P.R. (3d) 8, 14 (B.C.C.A. 1989).
43. *Corso*, 72 C.C.E.L. (3d) at 128 citing *Lamb v Evans*, [1893] 1 Ch. 218 (Eng. C.A.).
44. *Integrated Circuit Topography Act*, S.C. 1990, c. 37.
45. *Ibid.* at s. 3(1).
46. *Ibid.* at s. 16(1).

3.3 The Parties' Duty to Inform

There is no statutory duty to give notice to an employer of the creation of an integrated circuit topography, though, as with copyright, such a duty may arise under the employee's duty of good faith towards his or her employer which arises under general employment law.

3.4 Employee Remuneration Right

There is no employee right to remuneration. Any entitlement to remuneration for creation of a particular topography will depend on the contractual agreement that has been entered into by the employer and the employee.

4 EMPLOYEES' INVENTIONS

4.1 The Legal Framework

Rights in inventions are purely a creature of the federal Patent Act.[47] There are no specific provisions in the Act governing employee inventions generally, but a number of principles relating to employee inventions have emerged from the case law. These are considered in sections 4.2-4.4, below. Inventions made by public servants are governed in detail by the Public Servants Inventions Act ('PSIA'),[48] which is considered in section 4.2.1.

4.2 Ownership of Rights

The general presumption is that an employee is the owner of any invention made by them.[49] This presumption applies even if the employee has made use of the employer's time and material in developing the invention.[50] There are two exceptions to this presumption. Rights to the invention will belong to the employer if

- there is an agreement to that effect, which may be express or implied;[51]
- the employee was engaged to invent or innovate, and the invention was created in the course of those duties as an innovator.[52]

47. *Patent Act*, R.S.C. 1985, c. P-4.
48. *Public Servants Inventions Act*, R.S.C. 1985, c. P-32.
49. *Comstock Canada v Electec Ltd.*, 38 C.P.R. (3d) 29, 52-53 (F.C.T.D. 1991); *Techform Products Limited v Wolda*, 15 C.P.R. (4th) 44, 60 (Ont. C.A. 2001); *Spiroll v Putti*, 22 C.P.R. (2d) 261, 274 (B.C.S.C. 1975) *aff'd in* 29 C.P.R. (2d) 260 (B.C.C.A. 1976); *Groves v Canasonics Inc*, 2015 ABQB 314, [48] (express agreement).
50. *Comstock*, 38 C.P.R. (3d) at 52-53.
51. *Comstock*, 38 C.P.R.(3d) at 53; *Techform*, 15 C.P.R. (4th) at 60; *Spiroll*, 22 C.P.R. (2d) at 274; *Nature-Control Technologies Inc. v Li*, 2014 BCSC 1868 at 258-63.
52. *Comstock*, 38 C.P.R. (3d) at 53-54; *Spiroll*, 22 C.P.R.(2d) at 274-275; *Devoe-Holbein Inc. v Yam*, 3 C.P.R. (3d) 52, 66 (Q.S.C. 1984).

It is possible that the rights may also belong to the employer if ownership of the invention by the employee would be inconsistent with the employee's general duty of good faith towards the employer,[53] but it is clear that the mere fact that a duty of good faith is owed does not in itself prevent the employee from claiming the invention as their own.[54]

In the absence of an express or implied agreement, factors the court may consider in determining whether the rights to the invention belong to the employer include the following:

(1) whether the employee was hired for the express purpose of inventing;
(2) whether the employee at the time they were hired had previously made inventions;
(3) whether an employer had incentive plans encouraging product development;
(4) whether the conduct of the employee once the invention had been created suggested ownership was held by employer;
(5) whether the invention is the product of the problem the employee was instructed to solve, i.e. whether it was their duty to make inventions;
(6) whether the employee's invention arose following their consultation through normal company channels (i.e. was help sought?).
(7) whether the employee was dealing was highly confidential information or confidential work;
(8) whether it was a term of the servant's employment that they could not use the ideas which they developed to their own advantage.[55]

If the rights to the invention belong to the employer, that entitlement will survive the end of the employment relationship.[56]

4.2.1 *Government Inventions*

Inventions made by public servants are governed by the PSIA, which provides that the following inventions belong to the Crown:

(a) an invention made by a public servant while acting within the scope of their duties or employment, or made by a public servant with facilities, equipment or financial aid provided by or on behalf of Her Majesty; and
(b) an invention made by a public servant that resulted from or is connected with their duties or employment.[57]

53. See Vaver, *supra* at 369, citing Australian authority.
54. *Comstock*, 38 C.P.R. (3d) at 53-54; *Spiroll v Putti*, 29 C.P.R. (2d) 260, 262(B.C.C.A. 1976).
55. *Comstock*, 38 C.P.R. (3d) at 53-54.
56. *Techform*, 15 C.P.R. (4th) at 61.
57. *Public Servants Inventions Act*, s. 3.

This is contrary to the general presumption that the employee owns their invention. A public servant who makes an invention is required to inform the appropriate minister.[58] The PSIA authorizes the appropriate minister to make a payment to the public service inventor in respect of inventions which vest in the Crown, 'in such amount as the appropriate minister and the public servant may agree on or as the appropriate minister determines'.[59] A public servant is defined to include a member of the Canadian Forces.[60] This includes members of the Reserve Force, even if they are not employed and do not receive benefits.[61]

4.3 The Parties' Duty to Inform

There is no general statutory duty to give an employer notice of the creation of an invention. Disclosure may be required by contract.[62] The general duty of good faith may give rise to a duty to inform in particular circumstances, but, as noted, it does not in itself give the employer any entitlement to the invention, and in the absence of any entitlement, there seems no obvious reason why the employer should be informed.

Under PSIA, because the Crown is the owner of any invention made by a public servant (in the circumstances described above), public servants are required to inform the appropriate minister of the invention; they are prohibited from filing an application for a patent outside Canada without permission; and they must disclose that they are a public servant in any application for a patent in Canada.[63] Failure to disclose is an offence with a potential penalty of a fine and imprisonment.[64] The invention may also be vested in the Crown, but this question remains open.[65] Failure to disclose does not, however, result in invalidation of the patent.[66]

4.4 Employee Remuneration Right

There is no employee right to remuneration. Any entitlement to remuneration for creation of a particular invention will depend on the contractual agreement that has been entered into by the employer and the employee.

5 EMPLOYEES' UTILITY MODELS

Canadian law does not protect utility models.

58. *Ibid.* at s. 4(1)(a).
59. *Ibid.* at s. 10.
60. *Public Servants Inventions Act*, s. 2. The definition also includes members of the Royal Canadian Mounted Police.
61. *Brown v. Canada* 2016 FCA 37, [27].
62. See e.g. *Techform*, 15 C.P.R. (4th) at 50-51.
63. *Public Servants Inventions Act*, s. 4(1).
64. *Public Servants Inventions Act*, s. 11.
65. *Brown v. Canada* 2016 FCA 37. [50].
66. *Brown v. Canada* 2016 FCA 37, [50].

6 EMPLOYEES' DESIGN RIGHTS

6.1 The Legal Framework

Canada governs the ownership of design under the federal Industrial Design Act.[67] The term design includes any combination of features of shape, configuration, pattern, or ornament in a finished work that are aesthetic in nature.[68] Rights under the Act are acquired by registration.[69] There has been little litigation under the Act, and there has been little judicial consideration of the rights of employees in designs.

6.2 Ownership of Rights

The Act provides that:

> The author of a design is the first proprietor of the design, unless the author has executed the design for another person for a good and valuable consideration, in which case the other person is the first proprietor.[70]

The salary paid to an employee who creates a design in the course of their employment constitutes good and valuable consideration within the meaning of this provision.[71] This means that the person ordering the work to be done will be the first owner, whether the author of the design is an employee or an independent contractor.[72] However, it does not appear that the employer will be the first owner of every design made by an employee. The design must be executed for consideration, and this requirement and the case law imply that the employer will be the first owner only if the creation of the design is part of the employee's duties, so that it can be said that the salary is consideration for the design.

6.3 The Parties' Duty to Inform

There is no statutory duty on an employee to give notice to an employer of the creation of an industrial design, though, as with copyright, such a duty may arise under the employee's duty of good faith towards their employer which arises under general employment law. If a design is created in the course of the employee's duties, so that the employer is the first owner, the duty of good faith may well be a duty to disclose the creation of the design to the employer.

---

67. *Industrial Design Act*, R.S.C. 1985, c. 1-9.
68. *Ibid.* at s. 2.
69. *Ibid.* at s. 9.
70. *Ibid.* at s. 12(1).
71. *Angelstone Ltd. v Artistic Stone Ltd.*, 33 C.P.R. 155, 160-161 (Ex. Ct. 1960); *Ribbons (Montreal) Ltd. v Belding Corticelli Ltd.*, 36 C.P.R. 65, 71 (Ex. Ct. 1961); *Renewal Manufacturing Co. v Reliable Toy Co.*, 9 C.P.R. 67, 74-75 (Ex. Ct. 1949).
72. *Uniformes Town and Country Inc. v Labrie*, 24 C.P.R. (3d) 162 (F.C.T.D. 1989), *rev'd in* 44 C.P.R. (3d) 514 (F.C.A. 1992) (the appellate court disagreed that the designs were executed for 'good and valuable consideration').

6.4 Employee Remuneration Right

There is no employee right to remuneration. Any entitlement to remuneration for creation of a particular design will depend on the contractual agreement that has been entered into by the employer and the employee.

7 EMPLOYEES' PLANT VARIETIES

7.1 The Legal Framework

New plant varieties are eligible for protection under the federal Plant Breeders' Rights Act ('PBRA').[73] Rights under the PBRA are gained after application and registration of the new variety.[74]

7.2 Ownership of Rights

The breeder of a new plant variety is entitled to apply for the grant of plant breeder's rights respecting that variety,[75] and the PBRA defines a 'breeder' in the following terms:

> 'breeder', in respect of a plant variety, means
>
> (a) any person who originates or who discovers and develops the plant variety, or
>
> (b) any person in respect of whom an officer, servant or employee, while acting within the scope of his or her duties as the officer, servant or employee of that person, originates or discovers and develops the plant variety; ...'

Under paragraph (b) of this definition, it is clear that the employer, not the employee, will be considered the breeder, if the employee originates (or discovers and develops) the plant variety in the course of their employment. By implication, it is clear that the mere fact that the originator of a new variety is an employee does not mean that their employer will be considered the breeder, if the origination of the variety was unrelated to their duties.

It appears, from paragraph (a), that an independent contractor who originates a new variety under a contract 'for services,' as copyright law would put it, the independent contractor will be considered the breeder, with the consequence that an express or implied contractual term will be required to vest rights to the variety in the party who ordered the work to be done.

73. *Plant Breeders' Rights Act*, S.C. 1990, c. 20.
74. *Ibid.* at ss. 7(1), 27(1).
75. *Ibid.* at s. 7(1).

7.3 The Parties' Duty to Inform

There is no statutory duty on an employee to give notice to an employer of the origination of a new variety, though such a duty may arise under the employee's duty of good faith towards their employer which arises under general employment law. As with industrial designs, if the variety is originated in the course of the employee's duties, so that the employer is the first owner, the general duty of good faith may well give rise to a duty to disclose the creation of the design to the employer.

7.4 Employee Remuneration Right

There is no employee right to remuneration. Any entitlement to remuneration for the origination of a particular variety will depend on the contractual agreement that has been entered into by the employer and the employee.

8 INSTANCES FOR DISPUTES

There are no specialized tribunals for dealing with disputes between employers and employees over ownership of intellectual property rights. The dispute will therefore be heard by the provincial superior courts, which are the general courts of inherent jurisdiction, unless the dispute falls within the exclusive or concurrent jurisdiction of the Federal Court. The Federal Court of Canada has exclusive jurisdiction in all cases in which it is sought to impeach or annul a patent or alter any register of federal intellectual property rights, including copyrights, trademarks, industrial designs or integrated circuit topographies, or for the registration of any of those rights.[76] If such a remedy is sought by an employee or employer in a dispute over ownership of rights, the action must be brought in the Federal Court. Conversely, the Federal Court does not have jurisdiction with respect to ownership rights in intellectual property matters that relate only to the interpretation of contracts. If the dispute turns on whether a contract exists, or its meaning, it may not be brought in the Federal Court. If the dispute involves both types of questions, the Federal Court will take jurisdiction in order to address matters within its exclusive jurisdiction, and it will also adjudicate matters of ownership within that context of the matters within its jurisdiction.

9 CONCLUDING REMARKS

The variety of statutory provisions makes it difficult to generalize about Canadian law relating to employee ownership of inventions, but in broad terms, and subject to the many exceptions discussed above, the employer will be the first owner of a work

76. *Federal Courts Act*, R.S.C. 1985, c. F-7, s. 20(1).

created by an employee within the course of their employment. Employees are not entitled to any remuneration in such cases, except to the extent provided by the employment contract.

BIBLIOGRAPHY

Table of Sources

Normand Tamaro, *The 2005 Annotated Copyright Act*, (Toronto, Carswell 2005).
David Vaver, *Intellectual Property Law* (2nd ed., Irwin Law 2011).

Table of Legislation

Copyright Act, R.S.C. 1985, c. C-42.
Federal Courts Act, R.S.C. 1985, c. F-7.
Industrial Design Act, R.S.C. 1985, c. 1-9.
Integrated Circuit Topography Act, S.C. 1990, c. 37.
Patent Act, R.S.C. 1985, c. P-4.
Plant Breeders' Rights Act, S.C. 1990, c. 20.
Public Servants Inventions Act, R.S.C. 1985, c. P-32.Table of Cases
57134 Manitoba Ltd. v Palmer, 26 C.P.R. (3d) 8 (B.C.C.A. 1989).
Amusements Wiltron Inc. c. Mainville, (1991) 40 C.P.R. (3d) 521 (Q.S.C.)
Andrews v. McHale 2016 FC 624
Angelstone Ltd. v Artistic Stone Ltd., 33 C.P.R. 155 (Ex. Ct. 1960).
Beloff v Pressdram Ltd. [1973] 1 All E.R. 241, 250 (Ch. D.).
Brown v. Canada 2016 FCA 37
Comstock Canada v Electec Ltd., 38 C.P.R. (3d) 29 (F.C.T.D. 1991).
Corso v Nebs Business Products Ltd., 72 C.C.E.L. (3d) 110 (Ont. S.C. 2009).
Devoe-Holbein Inc. v Yam, 3 C.P.R. (3d) 52 (Q.S.C. 1984).
Dolmage v Erskine, 23 C.P.R. (4th) 495 (Ont. Sm. Cl. 2003).
Dynabec Ltée c. Société d'informatique RDG Inc., 6 C.P.R. (3d) 299 (Q.S.C. 1984).
Fénollar c. PRB Média, 2006 QCCS 1956.
Goldner v Canadian Broadcasting Corp., 7 C.P.R. (2d) 158 (F.C.T.D. 1972).
Groves v Canasonics Inc, 2015 ABQB 314 *Hanis v Teevan*, [1995] O.J. No. 981, 1995 CarswellOnt 5181 (Ont. S.C.).
Hanis v Teevan, 81 C.P.R. (3d) 496 (Ont. C.A. 1998).
Harmony Consulting Ltd. v G.A. Foss Transport Ltd., 2012 FCA 226 aff'g 2011 FC 340.
Hutton v Canadian Broadcasting Corp., 29 C.P.R. (3d) 398 (A.B.Q.B. 1989)
Keatley Surveying Ltd. v Teranet Inc., 2014 ONSC 1677.
Kennedy v. Ruminski 2014 FC 526
Lachance c. Productions Marie Eykel Inc., 2012 QCCS 1012.
Lachance c. Productions Marie Eykel Inc., 2014 QCCA 158.

Nature-Control Technologies Inc. v Li, 2014 BCSC 1868.
Netupsky v Dominion Bridge Co., 58 C.P.R. 7 (B.C.C.A. 1969).
Planification-Organisation-Publications Système (POPS) Ltée v. 9054-8181 Québec Inc. 2014 FCA 185
Renewal Manufacturing Co. v Reliable Toy Co., 9 C.P.R. 67 (Ex. Ct. 1949).
Ribbons (Montreal) Ltd. v Belding Corticelli Ltd., 36 C.P.R. 65 (Ex. Ct. 1961).
Robertson v Thomson, 2006 SCC 43.
Saxon c. Communications Mont-Royal Inc., J.E. 2001-91 (C.S.)
Seanix Technology Inc. v Ircha, 78 C.P.R. (3d) 443 (B.C.S.C. 1998).
Spiroll v Putti, 22 C.P.R. (2d) 261 (B.C.S.C. 1975).
Spiroll v Putti, 29 C.P.R. (2d) 260 (B.C.C.A. 1976).
Techform Products Limited v Wolda, 15 C.P.R. (4th) 44 (Ont. C.A. 2001).
U.A. v Société Radio-Canada/Canadian Broadcasting Corp., 1 C.L.R.B.R. 129, 1982 CarswellNat 564 (N.S.).
Underwriters' Survey Bureau Ltd. v Massie & Renwick Ltd., [1940] S.C.R. 218.
Uniformes Town & Country Inc. v Labrie, 24 C.P.R. (3d) 162 (F.C.T.D. 1989).
Uniformes Town & Country Inc. v Labrie, 44 C.P.R. (3d) 514 (F.C.A. 1992).
University of British Columbia (Re), [2006] B.C.L.R.B.D. No. 56, 2006 CanLII 6155.
Wiebe v Saskatchewan Institute of Applied Science and Technology, 2007 SKQB 60.

CHAPTER 31

United States of America

Joseph Jeffrey Hawley, Marketa Trimble & Darryl C. Wilson

1 BACKGROUND

Aspects of intellectual property in the United States are governed by either federal or state law. For intellectual property rights that are governed by federal law, federal statutes apply to initial ownership, while subsequent ownership issues, including issues arising under licences and other contracts, including employee contracts, are resolved under state law.

The states gave the federal government the power to establish intellectual property rights. 'The Congress shall have power ... To promote the progress of science and useful arts, by securing for limited times to authors and inventors the exclusive right to their respective writings and discoveries... .' Thus, as Congress used this power, rights flow 'to authors and inventors'. For example, in the Patent Act, 35 U.S.C. § 261 provides that 'patents shall have the attributes of personal property', and § 262 provides for dispositions between joint owners. In 17 U.S.C § 201, the Copyright Act provides: '(a) Initial Ownership – Copyright in a work protected under this title vests initially in the author or authors of the work. The authors of a joint work are co-owners of copyright in the work'. Congress's power to regulate interstate business transactions through the commerce clause provides the basis for federal trademark law. Trademark rights are conferred upon those who utilize distinctive identifications to indicate the source of their goods and services.

The most common legal relationship that might alter this ownership exists where the author or inventor is employed. If the author or inventor is not the subject of an employment legal relationship, parties can always enter into contracts that can alter ownership rights.

Most ownership issues at the state level are resolved through either the state's employment laws or through the state's contract laws. Only a few states have specific statutes that relate to ownership of intellectual property that is otherwise regulated by federal law. Those states are: (1) California, (2) Delaware, (3) Illinois, (4) Kansas, (5) Minnesota, (6) Nevada, (7) North Carolina, (8) Utah, and (9) Washington. Most of these state provisions alter the enforceability of an employment agreement that purports to require assignment of all intellectual property to the employer. In the typical statute, such provisions are not enforceable if they relate to intellectual property that is produced on the employee's own time not using resources of the employer.

Most disputes regarding subsequent ownership issues are resolved in state courts of general jurisdiction. However, where there is sufficient diversity of parties, the dispute might also be resolved in federal courts. Federal courts may also decide about ownership as an incidental question in a case subject to federal question jurisdiction, e.g., to determine a party's standing to sue in a patent infringement case. Federal and state courts use the same procedures as the courts use, respectively, for other types of business disputes, with some courts adopting special court rules for patent cases. Federal courts apply the appropriate state law to the issues governed by state law, including issues of subsequent ownership.

2 EMPLOYEES' COPYRIGHT

2.1 The Legal Framework

Copyright is governed by federal and state law; state law has played a lesser role in copyright since the 1976 Copyright Act, which extended copyright protection to unpublished works. In the Copyright Act of 1976, Congress pre-empted most state copyright laws;[1] however, some state statutes concerning copyright continue to exist. For example, statutes in some states protect pre-1972 sound recordings that are not protected under the federal copyright statute. Some other states are still contesting the issue as reflected by the Florida Supreme Court's recent agreement to hear arguments on that matter in the case of Flo & Eddie, Inc. v Sirius XM Radio Inc.

2.2 Ownership of Rights

Under the Copyright Act, the initial ownership of copyright vests in the author, and where a work is created jointly by two or more authors, the authors become co-owners of the copyright in their jointly created work (17 U.S.C. §201(a)). The author of each individual contribution to a collective work, such as a periodical issue or encyclopaedia, owns the copyright in that contribution, while the compiler owns the copyright in the selection and arrangement of the individual works.[2]

Under the 'work made for hire' doctrine codified in the Copyright Act, initial ownership of copyright vests in the employer who employed the author or the commissioning party who commissioned the work.[3] As for the employment, the US

1. 17 U.S.C. 301.
2. 17 U.S.C. §201(c).
3. 17 U.S.C. §201(b).

Supreme Court has held that for the purposes of the doctrine the term 'employee' should be interpreted according to common law agency principles.[4] Among the factors that a court will consider to determine whether an author created a work as an employee are 'the hiring party's right to control the manner and means by which the product is accomplished', 'the skill required, the source of the instrumentalities and tools, [and] the location of the work'. As for commissioned works, only some works 'specially ordered or commissioned' will be considered 'works made for hire', including a contribution to a collective work, a part of a motion picture or other audiovisual work, and a translation.[5]

Parties can contract around the work made for hire doctrine; an agreement to the contrary must be in a written and signed instrument.[6] There are strong incentives for employers and parties commissioning works to have works created as works made for hire; the fact that a work is a work made for hire has important legal implications. The limited moral rights that are available under the Copyright Act to works of visual art are not available to works made for hire, which are excluded from the definition of a 'work of visual art'.[7] Works made for hire are also excluded from the provisions about termination of transfers and licences.[8]

The scheme with respect to the ownership of rights does not vary as between the private sector, the public sector, and research universities. In each instance, the initial rights are determined by the Copyright Act. Works by the US Government enjoy no copyright protection under the US Copyright Act;[9] state laws vary in their treatment of copyright to works by the respective state governments.

A special provision for ownership of mask works by officers or employees of the Federal government is noted below.

2.3 The Parties' Duty to Inform

As with other aspects, the duty to inform is governed by the contract between the parties.

2.4 Employee Remuneration Right

Employees have no remuneration rights above and beyond their contract.

4. *Community for Creative Non-Violence v Reid*, 490 U.S. 730 (1989).
5. 17 U.S.C. §101.
6. 17 U.S.C. §201(b).
7. 17 U.S.C. §101 and §106A.
8. 17 U.S.C. §203 and §304(c).
9. 17 U.S.C. §105.

3 EMPLOYEES' SEMICONDUCTOR CHIPS

3.1 The Legal Framework

The United States enacted the Semiconductor Chip Protection Act of 1984 ('SCPA') to protect the topography of semiconductor chips. The SCPA is part of the Copyright Act and is found in 17 U.S.C. §§ 901-914.

3.2 Ownership of Rights

Ownership is determined according to the provisions of the Copyright Act, discussed above. The scheme with respect to the ownership of rights does not vary as between the private sector, the public sector or research universities, except as noted just below with respect to ownership by an officer or employee of the United States Government.

The ownership of a 'mask work' is determined by section 201 of the Copyright Act discussed above. Section 903 (d) states: 'Mask works prepared by an officer or employee of the United States Government as part of that person's official duties are not protected under this chapter.'

3.3 The Parties' Duty to Inform

As with other aspects, the duty to inform is governed by the contract between the parties.

3.4 Employee Remuneration Right

Employees have no remuneration rights, unless it is through contract.

4 EMPLOYEES' INVENTIONS

4.1 The Legal Framework

In the case of intellectual property rights in new, non-obvious, and useful processes, machines, manufactures or composition of matter, the federal law governs initial ownership and is embodied in the Patent Act, Title 35 of the United States Code.

4.2 Ownership of Rights

Absent a special statutory provision to the contrary, the inventor (or inventors) is (or are) the initial owner(s) of the rights to the intellectual property that they create. For example, 35 U.S.C. § 261 provides that 'patents shall have the attributes of personal property'; 35 U.S.C. § 101 provides that 'Whoever invents or discovers ...'; 35 U.S.C. § 102 provides that 'A person shall be entitled to a patent unless ...'; 35 U.S.C. § 115 requires that an application shall include 'the name for any inventor for any invention claimed in the application'. The statute contemplates joint owners in 35 U.S.C. § 262. Further regulations require that joint owners must act together as a composite entity when seeking to engage in certain activities involving the Patent Office.

Although initial ownership of an invention vests in the inventor(s), it is common practice for employers to require employee-inventors to sign an invention assignment agreement that assigns inventions made by the employee-inventor to the employer. While the language of these assignment agreements varies, a typical assignment agreement would be limited to inventions made within the scope of the employee's duties. As noted *supra*, assignment agreements that purport to obligate the inventor to assign all inventions, including those made outside the scope of employment and not using the employer's resources, may run afoul of certain state laws. Assignments generally need to comply with §§ 261 and 262 in order to be enforceable in a subsequent ownership dispute.

In the absence of an invention assignment agreement, unless the employee was specifically hired to 'invent,' when an employee makes an invention within the scope of his or her employment, a common law doctrine known as the 'shop rights doctrine' may apply. Under the shop rights doctrine, the inventor owns the invention, but the employer has an equitable right to use the invention without additional compensation to the employee. Essentially, the employer's right is limited to a non-exclusive, royalty-free, and largely non-transferable right to use the invention.

The scheme with respect to the initial ownership of rights does not vary as between the private sector, the public sector, and research universities.

Chapter 18 of 35 U.S.C. concerns the 'Patent Rights in Inventions Made with Federal Assistance.' This chapter deals with a wide variety of subjects including the duty of a contractor to inform the government of inventions made under government contract, 'march-in rights,' etc. However, none of these provisions alter the basic initial ownership of the invention.

4.3 The Parties' Duty to Inform

As with other aspects, the duty to inform is governed by the contract between the parties.

4.4 Employee Remuneration Right

Employees have no remuneration rights, unless it is through contract.

5 EMPLOYEES' UTILITY MODELS

The Congress of the United States has not chosen to provide for utility models.

6 EMPLOYEES' DESIGN RIGHTS

Design patents are included in the federal statute relating to patents in general. 35 U.S.C. § 171 provides that 'Whoever invents any new, original and ornamental design...'; 35 U.S.C. § 171 further provides that 'The provisions of this title [35 U.S.C. §§ 1 et seq.] relating to patents for inventions shall apply to patents for designs, except as otherwise provided.' No alteration of ownership is 'otherwise provided'.

7 EMPLOYEES' PLANT VARIETIES

There are two statutes that relate to the protection of plants. The Plant Variety Protection Act ('PVPA') protects sexually reproduced and tuber propagated plants. The Plant Patent Act of 1930 is limited to asexually reproduced plants, excluding tuber propagated plants and is now embodied in 35 U.S.C. Chapter 15. 35 U.S.C. § 161 provides that 'Whoever invents or discovers and asexually reproduces any distinct and new variety of plant ...'; 35 U.S.C. § 161 further provides that 'The provisions of this title [35 U.S.C. §§ 1 et seq.] relating to patents for inventions shall apply to patents for plants, except as otherwise provided.' No alteration of ownership is 'otherwise provided'.

The Plant Variety Protection Act is embodied in 7 U.S.C. Chapter 57.

7 U.S.C. § 2421 (a) provides 'An application for a certificate of Plant Variety Protection may be filed by the owner of the variety sought to be protected.' While this begs the question as to who is an 'owner' it is clear from the context of the statute that the breeder is the initial owner. For example, 7 U.S.C. § 2423 (a) provides for application by joint breeders ('When two or more persons are the breeders, one person (or the successor of the person) may apply, naming the others'). 7 U.S.C. § 2422 (5) requires 'A statement of the basis of applicant's ownership.'

8 INSTANCES FOR DISPUTES

In the event of disputes, resolution is through a civil lawsuit in state court. This can be altered by the nature of the contract between the parties. For example, the parties can agree to arbitration or mediation. If there is sufficient 'diversity', the action might be brought in federal court applying state law.

BIBLIOGRAPHY

Table of Legislation

California Statute

Cal. Labor Code § 2870 (West 2011) – Employment agreements; assignment of rights:

(a) Any provision in an employment agreement which provides that an employee shall assign, or offer to assign, any of his or her rights in an invention to his or her employer shall not apply to an invention that the employee developed entirely on his or her own time without using the employer's equipment, supplies, facilities, or trade secret information except for those inventions that either:
   (1) Relate at the time of conception or reduction to practice of the invention to the employer's business, or actual or demonstrably anticipated research or development of the employer; or
   (2) Result from any work performed by the employee for the employer.

(b) To the extent a provision in an employment agreement purports to require an employee to assign an invention otherwise excluded from being required to be assigned under subdivision (a), the provision is against the public policy of this state and is unenforceable.

Cal. Labor Code § 2871 (West 2011) – Conditions of employment or continued employment; disclosure of inventions
No employer shall require a provision made void and unenforceable by Section 2870 as a condition of employment or continued employment. Nothing in this article shall be construed to forbid or restrict the right of an employer to provide in contracts of employment for disclosure, provided that any such disclosures be received in confidence, of all of the employee's inventions made solely or jointly with others during the term of his or her employment, a review process by the employer to determine such issues as may arise, and for full title to certain patents and inventions to be in the United States, as required by contracts between the employer and the United States or any of its agencies.
Cal. Labor Code § 2872 (West 2011) – Notice to employee; burden of proof
If an employment agreement entered into after 1 January 1980, contains a provision requiring the employee to assign or offer to assign any of his or her rights in any invention to his or her employer, the employer must also, at the time the agreement is made, provide a written notification to the employee that the agreement does not apply

to an invention which qualifies fully under the provisions of Section 2870. In any suit or action arising thereunder, the burden of proof shall be on the employee claiming the benefits of its provisions.

Delaware Statute

Del. Code Ann. tit. 19 § 805 (West 2011) - Employee's right to certain inventions

Any provision in an employment agreement which provides that the employee shall assign or offer to assign any of the employee's rights in an invention to the employee's employer shall not apply to an invention that the employee developed entirely on the employee's own time without using the employer's equipment, supplies, facility or trade secret information, except for those inventions that:

(i) relate to the employer's business or actual or demonstrably anticipated research or development; or
(ii) result from any work performed by the employee for the employer.

To the extent a provision in an employment agreement purports to apply to the type of invention described, it is against the public policy of this State and is unenforceable. An employer may not require a provision of an employment agreement made unenforceable under this section as a condition of employment or continued employment.

Illinois Statute

765 Comp. Stat. Ann. 1060/2 (West 2011) - Employee rights to inventions—conditions:

(1) A provision in an employment agreement which provides that an employee shall assign or offer to assign any of the employee's rights in an invention to the employer does not apply to an invention for which no equipment, supplies, facilities, or trade secret information of the employer was used and which was developed entirely on the employee's own time, unless (a) the invention relates (i) to the business of the employer, or (ii) to the employer's actual or demonstrably anticipated research or development, or (b) the invention results from any work performed by the employee for the employer. Any provision which purports to apply to such an invention is to that extent against the public policy of this State and is to that extent void and unenforceable. The employee shall bear the burden of proof in establishing that his invention qualifies under this subsection.
(2) An employer shall not require a provision made void and unenforceable by subsection (1) of this Section as a condition of employment or continuing employment. This Act shall not preempt existing common law applicable to any shop rights of employers with respect to employees who have not signed an employment agreement.

(3) If an employment agreement entered into after 1 January 1984, contains a provision requiring the employee to assign any of the employee's rights in any invention to the employer, the employer must also, at the time the agreement is made, provide a written notification to the employee that the agreement does not apply to an invention for which no equipment, supplies, facility, or trade secret information of the employer was used and which was developed entirely on the employee's own time, unless (a) the invention relates (i) to the business of the employer, or (ii) to the employer's actual or demonstrably anticipated research or development, or (b) the invention results from any work performed by the employee for the employer.

Kansas Statute

Kan. Stat. Ann. § 44-130 (West 2011) – Employment agreements assigning employee rights in inventions to employer; restrictions; certain provisions void; notice and disclosure:

(1) Any provision in an employment agreement which provides that an employee shall assign or offer to assign any of the employee's rights in an invention to the employer shall not apply to an invention for which no equipment, supplies, facilities or trade secret information of the employer was used and which was developed entirely on the employee's own time, unless:
  (a) The invention relates to the business of the employer or to the employer's actual or demonstrably anticipated research or development; or
  (b) the invention results from any work performed by the employee for the employer.

(2) Any provision in an employment agreement which purports to apply to an invention which it is prohibited from applying to under subsection (a), is to that extent against the public policy of this state and is to that extent void and unenforceable. No employer shall require a provision made void and unenforceable by this section as a condition of employment or continuing employment.

(3) If an employment agreement contains a provision requiring the employee to assign any of the employee's rights in any invention to the employer, the employer shall provide, at the time the agreement is made, a written notification to the employee that the agreement does not apply to an invention for which no equipment, supplies, facility or trade secret information of the employer was used and which was developed entirely on the employee's own time, unless:
  (a) The invention relates directly to the business of the employer or to the employer's actual or demonstrably anticipated research or development; or

(b) the invention results from any work performed by the employee for the employer.

(4) Even though the employee meets the burden of proving the conditions specified in this section, the employee shall disclose, at the time of employment or thereafter, all inventions being developed by the employee, for the purpose of determining employer and employee rights in an invention.

Minnesota Statute

Minn. Stat. Ann. § 181.78 (West 2011) – Agreements; terms relating to inventions
Subdivision 1. Inventions not related to employment. Any provision in an employment agreement which provides that an employee shall assign or offer to assign any of the employee's rights in an invention to the employer shall not apply to an invention for which no equipment, supplies, facility or trade secret information of the employer was used and which was developed entirely on the employee's own time, and (1) which does not relate (a) directly to the business of the employer, or (b) to the employer's actual or demonstrably anticipated research or development, or (2) which does not result from any work performed by the employee for the employer. Any provision which purports to apply to such an invention is to that extent against the public policy of this state and is to that extent void and unenforceable.
Subd. 2. Effect of subdivision 1. No employer shall require a provision made void and unenforceable by subdivision 1 as a condition of employment or continuing employment.
Subd. 3. Notice to employee. If an employment agreement entered into after 1 August 1977 contains a provision requiring the employee to assign or offer to assign any of the employee's rights in any invention to an employer, the employer must also, at the time the agreement is made, provide a written notification to the employee that the agreement does not apply to an invention for which no equipment, supplies, facility or trade secret information of the employer was used and which was developed entirely on the employee's own time, and (1) which does not relate (a) directly to the business of the employer, or (b) to the employer's actual or demonstrably anticipated research or development, or (2) which does not result from any work performed by the employee for the employer.

Nevada Statute

Nev. Rev. Stat. Ann. § 600.500 (West 2011) – Employer is sole owner of patentable invention or trade secret developed by employee
Except as otherwise provided by express written agreement, an employer is the sole owner of any patentable invention or trade secret developed by his employee during the course and scope of the employment that relates directly to work performed during the course and scope of the employment.

North Carolina Statute

N.C. Gen. Stat. Ann. § 66-57.1 (West 2011) – Employee's right to certain inventions
Any provision in an employment agreement which provides that the employee shall assign or offer to assign any of his rights in an invention to his employer shall not apply to an invention that the employee developed entirely on his own time without using the employer's equipment, supplies, facility or trade secret information except for those inventions that (i) relate to the employer's business or actual or demonstrably anticipated research or development, or (ii) result from any work performed by the employee for the employer. To the extent a provision in an employment agreement purports to apply to the type of invention described, it is against the public policy of this State and is unenforceable. The employee shall bear the burden of proof in establishing that his invention qualifies under this section.
N.C. Gen. Stat. Ann. § 66-57.2 (West Supp. 2016) – Employer's rights

(a) An employer may not require a provision of an employment agreement made unenforceable under G.S. 66-57.1 as a condition of employment or continued employment. An employer, in an employment agreement, may require that the employee report all inventions developed by the employee, solely or jointly, during the term of his employment to the employer, including those asserted by the employee as nonassignable, for the purpose of determining employee or employer rights.
(b) An employer's ownership of an employee's invention, discovery, or development that has or becomes vested in the employer by contract or by operation of law shall not be subject to revocation or rescission in the event of a dispute between the employer and employee concerning payment of compensation or benefits to the employee, subject to any contrary provision in the employee's written employment agreement. The foregoing provision shall not apply where the employee proves that the employer acquired ownership of the employee's invention, discovery, or development fraudulently.
(c) If required by a contract between the employer and the United States or its agencies, the employer may require that full title to certain patents and inventions be in the United States.

Utah Statute

Utah Code Ann. § 34-39-3 (West 2011) – Scope of act—When agreements between an employee and employer are enforceable or unenforceable with respect to employment inventions—Exceptions

(1) An employment agreement between an employee and his employer is not enforceable against the employee to the extent that the agreement requires

the employee to assign or licence, or to offer to assign or licence, to the employer any right or intellectual property in or to an invention that is:
(a) created by the employee entirely on his own time; and
(b) not an employment invention.

(2) An agreement between an employee and his employer may require the employee to assign or licence, or to offer to assign or licence, to his employer any or all of his rights and intellectual property in or to an employment invention.

(3) Subsection (1) does not apply to:
(a) any right, intellectual property or invention that is required by law or by contract between the employer and the United States government or a state or local government to be assigned or licensed to the United States; or
(b) an agreement between an employee and his employer which is not an employment agreement.

(4) Notwithstanding Subsection (1), an agreement is enforceable under Subsection (1) if the employee's employment or continuation of employment is not conditioned on the employee's acceptance of such agreement and the employee receives a consideration under such agreement which is not compensation for employment.

(5) Employment of the employee or the continuation of his employment is sufficient consideration to support the enforceability of an agreement under Subsection (2) whether or not the agreement recites such consideration.

(6) An employer may require his employees to agree to an agreement within the scope of Subsection (2) as a condition of employment or the continuation of employment.

(7) An employer may not require his employees to agree to anything unenforceable under Subsection (1) as a condition of employment or the continuation of employment.

(8) Nothing in this chapter invalidates or renders unenforceable any employment agreement or provisions of an employment agreement unrelated to employment inventions.

Washington Statute

Wash. Rev. Code Ann. § 49.44.140 (West 2011) – Requiring assignment of employee's rights to inventions—Conditions

(1) A provision in an employment agreement which provides that an employee shall assign or offer to assign any of the employee's rights in an invention to the employer does not apply to an invention for which no equipment, supplies, facilities, or trade secret information of the employer was used and which was developed entirely on the employee's own time, unless (a) the

invention relates (i) directly to the business of the employer, or (ii) to the employer's actual or demonstrably anticipated research or development, or (b) the invention results from any work performed by the employee for the employer. Any provision which purports to apply to such an invention is to that extent against the public policy of this state and is to that extent void and unenforceable.

(2) An employer shall not require a provision made void and unenforceable by subsection (1) of this section as a condition of employment or continuing employment.

(3) If an employment agreement entered into after 1 September 1979, contains a provision requiring the employee to assign any of the employee's rights in any invention to the employer, the employer must also, at the time the agreement is made, provide a written notification to the employee that the agreement does not apply to an invention for which no equipment, supplies, facility, or trade secret information of the employer was used and which was developed entirely on the employee's own time, unless (a) the invention relates (i) directly to the business of the employer, or (ii) to the employer's actual or demonstrably anticipated research or development, or (b) the invention results from any work preformed [performed] by the employee for the employer.

Wash. Rev. Code Ann. § 49.44.150 (West 2011) - Requiring assignment of employee's rights to inventions—Disclosure of inventions by employee
Even though the employee meets the burden of proving the conditions specified in RCW 49.44.140, the employee shall, at the time of employment or thereafter, disclose all inventions being developed by the employee, for the purpose of determining employer or employee rights. The employer or the employee may disclose such inventions to the department of employment security, and the department shall maintain a record of such disclosures for a minimum period of five years.

Table of Cases

Community for Creative Non-Violence v Reid 490 U.S. 730 (1989)
JustMed v Byce 600 F. 3d 1118 (9th Cir. 2010)
Stanford University v Roche Molecular Systems 131 S. Ct. 2188 (2011)
Banks v Unisys Corp 228 F.3d 1357 (Fed. Cir. 2000)

E In Latin America

CHAPTER 32
Argentina

Martín Bensadon

1 BACKGROUND

Taking into account that nowadays most development and research work, and therefore, inventive work, is done in labour relationships, one of the most important practical issues in patent law is labour inventions.

In fact, the sole inventor who is not part of an organisation where their intellectual capacity can be facilitated, and from where they could receive material and financial support and the possibility of working with other inventors, is a figure of decreasing importance in the actual technology world.

On the other hand, today's fast-paced and technologically advanced society also poses a challenge to authors whose copyrighted works are very fundamental in terms of business, and plays a significant role in incentivising investment in the development of new content, and ultimately, underpin economic growth and innovation across the world. When such copyrightable works are somewhat connected to a labour relationship, the need to determine who owns the copyright to materials developed is of essence to avoid complex problems.

2 EMPLOYEES' COPYRIGHT

2.1 The Legal Framework

Ownership of a copyrighted work is governed by Section 4 of Law 11,723 which states that the author of a work is the original owner of the copyright to such work.

As authorship of a work implies intellectual creation and this, in turn, requires human activity, it has been stated that only individuals may be the original owners of copyright. There are certain exceptions to this rule that allow corporations to be the original copyright owners, namely anonymous works (that are considered to belong to the institutions, corporations or other legal entities that organised such collective works), and also works resulting from an employment relationship under certain circumstances. Strictly speaking, the original owner of the copyright is still the individual who created the work, but the economic content of such copyright is immediately transferred to the employer.

2.2 Ownership of Rights

There are no specific provisions under Argentine law with regard to the ownership of copyright to works made for hire.

If works have been developed or created in the course of employment, the rules of the Employment Contract Law No. 20,744 apply. Although these rules do not make a specific reference to copyright, they do provide a specific regime for inventions and discoveries obtained in the course of employment which, some authors claim, should be extended to works covered by copyright.

On the basis of these rules, if the employee has been hired to develop or create certain works, the economic rights on such works would belong to the employer, and the employee would retain the moral rights. The employer would also have the right to modify or adapt the works it has thus obtained; if works have been developed on the basis of intellectual property previously belonging to the employer, the employer would also be entitled to copyright on such works, with the limits described in the preceding paragraph. All the other works developed or created by employees would be the property of such employees and would only be transferred to employers through specific transactions accepted by the employees. If works are made for hire in circumstances that do not involve employment relationships, the property rights on such works will depend on the contractual terms agreed by the parties. The original right on these works will still belong to their authors, except in the case of anonymous works.

Moral rights will always belong to the author. Economic rights will also belong to the author, unless express or implied covenants between the parties provide otherwise. Moral rights are considered to be an essential constituent element of copyright under Argentine law. Moral rights to a given work together with the creation of such work are part of the original copyright granted to the author and consist of the following elements:

(a) the right to preserve the text, title and other contents of the work, even if the economic rights on such work have been assigned;
(b) the right to have the author named and identified as such together with the work, regardless of who holds economic rights in connection with such work;
(c) the right to decide when the work will be published (this would only apply for as long as the author has decided to published the work); and
(d) the right to modify the work, even after it has been published.

Argentine authorities have generally been reluctant to infer the transfer of copyright from agreements under which certain works will be supplied by their author with specific purposes. In the absence of express provisions, it will normally be inferred that the acquirer has the right to put the work to their own use, but not to reproduce or otherwise exploit it in connection with third parties. Similarly, if a given work has been developed for a specific project of the hirer, the author's consent is necessary if the work is to be put into a different use by the hirer.

Law 25,036 has amended Section 4 of Law 11,723, which now includes a sub-section d) providing that, in the case of computer programs produced by persons hired with that purpose, in the course of their employment, the employer owns the copyright over such programs, except if the parties have agreed otherwise. The result does not differ from what follows from the general rules included in the Employment Contract Law, described previously in this section.

2.3 The Parties' Duty to Inform

Please see section 2.2.

2.4 Employee Remuneration Right

Please see section 2.2.

3 EMPLOYEES' SEMICONDUCTOR CHIPS

3.1 The Legal Framework

There are no specific regulations regarding semiconductor chips in Argentina. However, Argentina has adhered to the TRIPS Agreement (Law No. 24,425) the provisions of which are directly applicable.

In this regard, Part II Section 6 of the TRIPS Agreement (Articles 35 to 38) regulates the protection of semiconductor chips, incorporating some of the provisions of the Treaty on Intellectual Property in Respect of Integrated Circuits (Washington, 1989), which establishes:

35 Article: Relation to the IPIC Treaty

Members agree to provide protection to the layout-designs (topographies) of integrated circuits (referred to in this Agreement as 'layout-designs') in accordance with Articles 2 through 7 (other than paragraph 3 of Article 6), Article 12 and paragraph 3 of Article 16 of the Treaty on Intellectual Property in Respect of Integrated Circuits and, in addition, to comply with the following provisions.'

36 Article: Scope of the Protection

Subject to the provisions of paragraph 1 of Article 37, Members shall consider unlawful the following acts if performed without the authorisation of the right holder: (9)

importing, selling, or otherwise distributing for commercial purposes a protected layout-design, an integrated circuit in which a protected layout-design is incorporated, or an article incorporating such an integrated circuit only insofar as it continues to contain an unlawfully reproduced layout-design.'

37 Article: Acts Not Requiring the Authorisation of the Right Holder

1. Notwithstanding Article 36, no Member shall consider unlawful the performance of any of the acts referred to in that Article in respect of an integrated circuit incorporating an unlawfully reproduced layout-design or any article incorporating such an integrated circuit where the person performing or ordering such acts did not know and had no reasonable ground to know, when acquiring the integrated circuit or article incorporating such an integrated circuit, that it incorporated an unlawfully reproduced layout-design. Members shall provide that, after the time that such person has received sufficient notice that the layout-design was unlawfully reproduced, that person may perform any of the acts with respect to the stock on hand or ordered before such time, but shall be liable to pay to the right holder a sum equivalent to a reasonable royalty such as would be payable under a freely negotiated licence in respect of such a layout-design.
2. The conditions set out in subparagraphs (a) through (k) of Article 31 shall apply *mutatis mutandis* in the event of any non-voluntary licensing of a layout-design or of its use by or for the government without the authorisation of the right holder.'

38 Article: Term of Protection

1. In Members requiring registration as a condition of protection, the term of protection of layout-designs shall not end before the expiration of a period of ten years counted from the date of filing an application for registration or from the first commercial exploitation wherever in the world it occurs.
2. In Members not requiring registration as a condition for protection, layout-designs shall be protected for a term of no less than ten years from the date of the first commercial exploitation wherever in the world it occurs.
3. Notwithstanding paragraphs 1 and 2, a Member may provide that protection shall lapse 15 years after the creation of the layout-design.'

3.2 Ownership of Rights

Bearing in mind that the above-mentioned specific provisions do not regulate the ownership of semiconductor chips developed during a labour relationship, general provisions regarding patents or copyright should be applied. However, there is no case law in Argentina regarding this issue.

3.3 The Parties' Duty to Inform

For reasons mentioned in section 3.1 and 3.2, please see sections 2 and 4.

3.4 Employee Remuneration Right

For reasons mentioned in section 3.1 and 3.2, please see sections 2 and 4.

4 EMPLOYEES' INVENTIONS

4.1 The Legal Framework

Ownership of inventions developed during a labour relationship is essentially governed by Section 10 of the Argentine Patent and Utility Models Law No. 24,481, as amended ('APL') and Sections 82 and 83 Argentine Labour Law No. 20,744 as amended ('ALL'). Section 10 of the APL provides the following: Inventions developed during a labour relationship:

(a) Inventions made by an employee during the life of their labour or work or service contract with the employer, the total or partial purpose of which is the performance of inventive activities, shall belong to the employer.
(b) The employee, as the author of the invention under the above-mentioned relationship, shall be entitled to a supplementary remuneration for their inventive work, if their personal contribution to the invention and the importance thereof to the enterprise and employer, evidently exceed the explicit or implicit content of their labour contract or relationship. If the conditions set forth in subsection a) above do not exist, when the employee makes an invention related to their professional activity within the enterprise, where the knowledge acquired within the enterprise or the use of the facilities thereof, have predominantly influenced on its obtention, the employer shall be entitled to the ownership of the invention or to reserve for himself the right to use it. The employer shall have to exercise such an option within 90 days after the invention was made.
(c) When the employer assumes the ownership of an invention or reserves for himself the right to use it, the employee shall be entitled to a fair monetary compensation to be established according to the industrial and commercial importance of the invention, taking into account the value of the means or knowledge facilitated by the enterprise, and the contributions by the employee; in the event that the employer grants a licence to third parties, the inventor shall be entitled to claim from the patentee, the payment of up to 50% of the royalties effectively received by the patentee.
(d) An industrial invention shall be considered as developed during the fulfilment of a labour contract or of a contract for services, when the patent

application has been filed within one year after the date the inventor left their job in the field of activity in which they made the invention.

(e) The labour inventions not carried out under the circumstances established by subsections a) and b) above, shall exclusively belong to their author.

(f) Any anticipated waiver by the employee to the rights granted by this section shall not be valid.

Furthermore, Section 10 of National Decree No. 260/96, which regulates the APL, provides the following:

> It will be understood that the right to apply for a patent belongs to the employer, when the performance of inventive activities has been established as a total or partial purpose of the employee's duties.

For the purposes of the second paragraph, subsection b), Section 10 of the Law, it shall only be understood that the development of the invention has been predominantly influenced by the knowledge acquired within the company or by the use of facilities provided by the latter, when the invention concerns to the employer's activities or is related to the specific tasks being or having been performed by the inventor in the service of said employer.

Once an invention is made under the conditions established in the second paragraph, subsection b), Section 10 of the Law, if the employer fails to exercise their right to opt within the term established in the last paragraph of the same subsection, the right of ownership to the patent shall be with the employee-inventor.

Should the invention have been made by an employee under the conditions set forth in the second paragraph, subsection b), Section 10 of the Law and before the patent is granted, the inventor may file a founded written petition under sealed envelope with the National Patent Office or with the provincial delegations authorised by the National Institute of Industrial Property, alleging their ownership rights thereto. In this event, the parties shall be notified by the Commissioner of Patents in order that they file their pertinent allegations in writing within an inextensible term of 15 days as from respective notifications. Within 30 days after the above-mentioned petitions, or the submission of the evidence offered, as the case may be, the National Institute of Industrial Property shall issue a duly founded decision determining who is entitled to apply for the patent. Such decision shall be notified to both parties through reliable means.

In case of disagreement between the employer and their employee on the amount of supplementary remuneration or monetary compensation as provided by the first paragraph of subsection b) and in subsection c) of Section 10 of the Law, respectively, either of them may at any time request the intervention of the National Institute of Industrial Property for the settlement of their dispute, by filing duly founded allegations. Notice of such a petition shall be served upon the other party for ten days from such notification. Within 20 days after the answer to the notification or the submission of the evidence offered, as applicable, the National Institute of Industrial Property shall issue a duly founded resolution establishing the supplementary remuneration or monetary compensation which, at its discretion, is equitable. Such a resolution shall be duly notified to both parties through reliable means.

The decisions of the National Institute of Industrial Property as referred to in the two preceding paragraphs may be appealed to a Federal Court of Justice in Civil and Commercial Matters with territorial jurisdiction over the domicile of the work place, within 20 business days as from such notice. The appeal shall have no staying effect.

In addition, Section 82 of ALL contains similar provisions to Section 10 of the APL.

> 82 Section. – Employee's inventions.
> The employee's personal inventions or discoveries are of his/her property, even if the employee used instruments which he/she did not own
> The inventions or discoveries that may derive from industrial procedures, methods or installations of the facility or from try outs, investigations, improvements or development of the procedures already followed, shall be the employer's property.
> The employer shall also be the owner of the inventions or discoveries, formulas, designs, materials and combinations which may be obtained by hiring the employee for such purpose.'
> 83 Section. – Employer's priority – Prohibition – Secret.
> The employer will have priority among third parties provided that they were under equal conditions, should the employee decided to assign the rights on his/ her invention or discovery, under the assumption stated in the first paragraph of section 82 of this law.
> The parties are obliged to keep secret the inventions or discoveries achieved under any of the referred methods.'

4.2 Ownership of Rights

According to the APL and the ALL, an employee's invention may be owned either by the employer or the employee, and may also be assigned by the employer or employee to a third party.

The employer would be entitled to the invention if the invention is made by an employee during the time of their labour or work or service contract with the employer, and the total or partial purpose of such contract is the performance of inventive activities ('ordinary inventions'). Furthermore, the invention will be owned by the employer, but the employer will be entitled to extra compensation, when the invention is made in the same situation as the above-described, but where the employee adds a personal contribution to the invention and the importance thereof to the employer exceeds the explicit or implicit content of the labour relation or contract ('extraordinary inventions').

The employer shall be entitled to the ownership of the invention or to reserve for himself the right to use it, and shall have to exercise such option within 90 days after the invention was made, when the total or partial purpose of the labour contract is other than performing inventive activities, and the employee performs an invention related to their professional activity within the company, where the knowledge acquired within the enterprise or the use of the facilities thereof, have predominantly influenced on its obtention ('non-related inventions').

The employee is entitled to the ownership of 'non-related inventions', provided that the employer refuses to claim the employee's invention, or of inventions not related at all to the employee's work (i.e. the invention was not made among any of the conditions explained in point a. (i), (ii) or (iii)), ('personal inventions').

According to Section 10 d) of the APL, the invention shall be considered as developed during the fulfilment of a labour contract or of a contract for services, when the patent application has been filed within one year after the date the inventor left their job in the field of activity where they made the invention.

According to Section 83 of the ALL, the employer will have priority among third parties provided that they are under equal conditions, should the employee decide to assign the rights on their personal inventions. Further, according to Section 83 of the ALL the parties are obliged to keep secrecy about the inventions or discoveries achieved under these circumstances.

4.3 The Parties' Duty to Inform

In the case of non-related inventions, the employer shall exercise the option between being entitled to the ownership of the invention or reserve for themselves the right to use the invention within 90 days after the invention was made. This presumes the employee has fulfilled their duty to notify the employer about the existence of the invention; otherwise the deadline cannot begin.

4.4 Employee Remuneration Right

In the case of ordinary inventions, the employee is not entitled to any remuneration. On the other hand, in the case of extraordinary inventions, the employee shall be entitled to a 'supplementary remuneration' if their personal contribution to the invention and the importance thereof to the enterprise and employer evidently exceed the explicit or implicit content of their labour relationship.

The APL does not define 'supplementary remuneration'. However, according to Section 10 of National Decree No. 260/96, in case of disagreement between the employer and their employee on the amount of supplementary remuneration either of them may at any time request the intervention of the National Institute of Industrial Property ('INPI') for the settlement of their dispute, by filing duly founded allegations. Notice of such a petition shall be served upon the other party for ten days of such notification. Within 20 days after the answer to the notification or the submission of the evidence offered, as applicable, the INPI shall issue a duly founded decision establishing the supplementary remuneration which, at its discretion, is equitable. Such a resolution shall be duly notified to both parties through reliable means.

The decision of the INPI may be appealed to a Federal Court of Justice in Civil and Commercial Matters with territorial jurisdiction over the domicile of the work place, within 20 business days from such notice. The appeal shall have no staying effect.

Finally, in case of 'non-related inventions', the employer has to pay a fair compensation ('monetary compensation') to the employee. The compensation shall be established according to the industrial and commercial importance of the invention, taking into account the value of the means and knowledge facilitated by the enterprise, and the contributions of the employee. In the event the employer grants a licence to third parties, the inventor shall be entitled to claim from the patentee the payment of up to 50% of the royalties effectively received by the patentee.

In case of disagreement between the employer and their employee on the amount of the monetary compensation, Section 10 of National Decree No. 260/96 establishes the same administrative proceeding as for the case of disagreement over supplementary remuneration.

5 EMPLOYEES' UTILITY MODELS

The legal framework described in section 4 for employees' inventions is applicable for employees' utility models.

6 EMPLOYEES' DESIGN RIGHTS

6.1 The Legal Framework

Ownership of industrial models and designs developed during a labour relationship is essentially governed by Section 1 of Industrial Models and Designs Decree-Law No. 6673 ('IML'). In this regard, Section 1(2) of the IML establishes:

> Industrial designs created by persons employed by other persons shall belong to their authors, who shall have the exclusive right of exploitation, except where the author was specially contracted to create them or if the author was merely carrying out instructions received from the persons for whom he or she works. If the design was a joint work of the employer and the employee, it shall belong to both, unless otherwise agreed.

6.2 Ownership of Rights

The general principle is that all industrial models or designs – or the right to obtain their registration – shall belong to their authors, unless the total or partial purpose of the service agreement with the employer is the creation of industrial models or designs. This case may be configured when the author has executed a work or service lease agreement or a labour agreement, so the industrial model or design belongs to the principal, lessor or employer, even when they are not its author or the employee is the mere performer of instructions given by the employer. In this case the employee is not a real author but a performer of a third person's instructions, and the author would be the person who gives the instructions.

If the model or design was a joint work of the employer and the employee, it shall belong to both, unless otherwise agreed.

6.3 The Parties' Duty to Inform

There is no regulation regarding the parties' duty to inform in the IML.

6.4 Employee Remuneration Right

There is no regulation regarding the employees' remuneration right in the IML.

7 EMPLOYEES' PLANT VARIETIES

7.1 The Legal Framework

Section 40 of Regulatory Decree of the PVP Law No. 2183/91 establishes:

> 'If the new variety has been obtained by two or more people, provisions of the Civil Code regarding co-ownership shall be applied in order to determinate its ownership. If the person that obtained the new variety is an employee, Section 82 of the Labour Law No. 20, 744 as amended shall be applied.'

7.2 Ownership of Rights

If the breeder is an employee, the employer itself is entitled to file and own a PVP application/right. Section 40 of the regulatory decree of the PVP Law sets forth that the ALL applies if a person obtained the new variety as an employee. As explained above, according to the ALL when the invention is made by an employee who has been employed for the performance of inventive activities, such invention belongs to the employer.

7.3 The Parties' Duty to Inform

There is no regulation of this issue.

7.4 Employee Remuneration Right

There is no regulation of this issue.

8 INSTANCES FOR DISPUTES

8.1 Competent Court

Section 10 of the APL says nothing with respect to the competent court to hear disputes regarding ownership of patents rights over employee's inventions.

However, the last paragraph of Section 10 of the National Decree No. 260/96 states that the decisions of the INPI regarding supplementary remuneration and compensatory indemnification may be appealed at the Federal Court on Civil and Commercial Matters with territorial jurisdiction over the domicile of the work place. This means that the jurisdiction over the Federal Courts on Civil and Commercial Matters is determined by a decree and not a law. Therefore, the provision of the National Decree No. 260/96 is an extension of competence in favour of the Federal Justice on Civil and Commercial Matters.

Notwithstanding the above, the extension of competence set by the National Decree No. 260/96 is prohibited countrywide and in the City of Buenos Aires according to Section 19 of Law of Organisation and Procedure of the National Labour Courts No. 18,345 ('OL'), which provides: 'The jurisdiction of the National Labour Courts, even territorial, cannot be extended.'

On the other hand, Section 20 of the OL sets forth the competence of the National Labour Courts as follows:

> 'The following will fall under national labour courts' competence: in general, contentious cases on individual conflicts of law, whoever the parties were, including the nation, its decentralised departments, the local government of the City of Buenos Aires and any public entity; for claims or counterclaims based on employment agreements, collective labour agreements, awards effectively in collective agreements, or labour laws or regulations; and the claims between employees and employers resulting from employment agreements, even when such claims were grounded on common law provisions.

Furthermore, Section 21 of the OL states:

> The following will in particular be of national labour courts' competence: a) cases in which the determination of issues directly related to individual and collective aspects of the labour law has a decisive influence … .

In the case of ownership of patents rights over employees' inventions, although employees' inventions are regulated by the APL, the action will be grounded in the labour law, since in order to determine various issues such as the existence of the employment relationship itself the labour law and not the industrial property law should be applied.

Therefore, it can be concluded that Section 10 of the Decree No. 260/96 would probably be considered unconstitutional since its provisions are contrary to the provisions of Section 19, 20 and 21 of the OL.

8.2 Procedure

The lawsuit may follow the rules of labour proceedings, which are governed mainly by the OL. Furthermore, Section 155 of the OL establishes that the Code of Civil Procedure is additionally applicable as far as its rules are compatible with the OL.

Litigation in Argentina is mostly written. The case is mainly decided on the basis of written submissions by the parties, although during the proceeding there are opportunities to meet the judge and explain the case orally. The main procedural stages in a lawsuit are: (a) bill of complaint and answer thereto; (b) evidence period; (c) final arguments; and (d) judgment. It is possible to appeal subsequently to the National Labour Court of Appeals, and in exceptional cases there may be an extraordinary appeal before the Supreme Court.

A pre-trial conciliation proceeding is mandatory before being entitled to start labour legal actions in which National Labour Courts are competent.

If the conciliation proceeding is concluded without agreement, the next step is filing a bill of complaint. The plaintiff must serve notice of the bill of complaint upon the defendant, who has ten working days to file an answer. In the same term, the defendant can also file a counterclaim.

Parties must produce all relevant documentary evidence upon submitting their claim or their answer and they must also indicate all other means of evidence they intend to rely upon. If the matter in dispute involves only questions of law (i.e., there are no contested facts), the proceedings will skip the evidence period and will proceed directly to the judgment phase.

Once the production of the evidence has been completed, the parties can file final arguments. Once the final arguments have been filed, or if the term to do so has expired, the judicial docket remains with the Court, which will study it and render its decision.

9 CONCLUDING REMARKS

Labour inventions have not become a relevant issue in Argentina. In any event, it should be acknowledged that the Argentine Patent Law provides a detailed regulation for labour inventions with a proper balance between the rights of the employees and the rights of the employers.

BIBLIOGRAPHY

Table of Sources

Iván Alfredo Poli, *El Modelo Industrial* (Abeledo Perrot, Buenos Aires 2008).

Iván Alfredo Poli, *Comentario al proyecto de ley de patentes de invención* (Talleres Gráficos Dorrego, Buenos Aires 1992).

Martín Bensadon, *Ley de Patentes Comentada y Concordada con el ADPIC y el*

Convenio de Paris (Lexis Nexis, Buenos Aires 2007).
Martín Bensadon, *Derecho de Patentes* (Abeledo Perrot, Buenos Aires 2012).

Table of Legislation

Patent and Utility Models Law No. 24,481 and its Regulatory Decree No. 260/96.
Labour Law No. 20,744.
Industrial Models and Designs Decree-Law No. 6673.
Law of Organisation and Procedure of the National Labour Justice No. 18,345.
Intellectual Property Law No. 11,723.
TRIPS Agreement (approved by Law No. 24,425).
PVP Law No. 20,247 and its Regulatory Decree No. 2183/91.

Table of Cases

Mittelman, Susana Raquel v Laboratorios Bacón S.A., Court of Appeals on Labour Matters, Division I (26 December 2007).
Fernandez, Eduardo Carlos v Massalin Particulares S.A., Court of Appeals on Labour Matters, Division IV (31 August 2011).

CHAPTER 33

Brazil

*Elisabeth Kasznar Fekete**

1 BACKGROUND

Litigating employees' or service providers' intellectual property rights cases in Brazil can take a long time, starting with procedural incidents challenging court competence aspects, going through the evidence phase at the heart of the classification analysis stage, which often includes hearing witnesses, and finally, discussing at length the assessment of accounting criteria and award quantification, that can affect 50% of the company's net profits.

The relationship between an employee's invention and the employer's field of activities is not foreseen as a classification criterion in statutory law, based predominantly on a two-factor system: existence of the employee's inventive or creative duties in the contract and their use of the company's resources.

The definition of who is considered an 'employee' or an 'intern' and the extent of their duties are regulated by the Brazilian Labour Laws Consolidation ('CLT', of 1943), while the intellectual property rights of the parties have their framework in different acts, depending on the contemplated technological, plant variety, software, literary, artistic or scientific creation.

Key statutory provisions present some ambiguities and have long not been updated. In an era of open innovation, fairly assigning intellectual property rights to the parties directly involved in the creative process presents constant challenges in Brazil. Involving a complex set of rules, three main fields of law require consideration: labour,

* With special gratitude to Nathalia de Oliveira Maranhão and Juliana Dias Avezum for their efficient cooperation and support with researches for the first edition of this chapter.

intellectual property and contract, not to mention the changes that the country is experiencing in its economic, cultural and educational objectives.[1]

Public policies in Brazil take account of the needs of its growing market, e-commerce, developing local R&D, and innovation funding by the government. At the same time, a model attracting private investment in new technologies must be continuously encouraged.

Although all employees are potential inventors, regardless of their contractual function, the subject matter of this chapter finds no guidelines in international treaties and it was introduced in the Brazilian legislation in 1943, by the country's above-mentioned CLT, that contained a general rule regarding employment contracts in its Article 454, establishing that in the course of an employment contract, inventions developed by an employee using the employer's equipment belonged to both parties, except if the employee was hired for research. In this case, the invention title was given to the employer, who had one year to start using it, as from the date of the granting of the patent. Otherwise, the employee would gain the exclusive rights to the invention.[2]

The first Brazilian Industrial Property Law ('BIPL') was entered into force in 1945. As from this Act, employees'/employers' rights were transferred from the Labour Act and received closer attention. In time, with the country's development, new acts on industrial property, copyright, plant variety, software and semiconductors were enacted, all containing provisions on this matter.[3]

The resulting scenario is a set of rules based on labour law principles, spread out in different acts, most of which have similar provisions.

As a general provision, The Brazilian Federal Constitution of 1988, currently in force, provides a frame for the employees' social rights, among which their participation in the profits or results of the companies, independently of their remuneration, is foreseen in Article 7, XI as an open possibility, not related to intellectual property creations or innovations. The support to the companies that invest in research, training and remuneration systems that grant to the employees participation in the economic gains resulting from their productivity is also encompassed by the Federal Constitution, in paragraph 4 of its Article 218.

Inventions, utility models and designs have had the greater persistence, detail and organisation with regard to employees' rights, addressed in a general rule set forth by Article 5, XXIX of the Federal Constitution, which ensures that the creator of the invention is entitled to obtain the patent, considered by scholars as the most important

1. Professor Jeremy Philips highlighted employee inventions as a 'truly multidisciplinary problem' and outlined the non-legal aspects involved in his *Employees' Inventions: an Analysis of the Nature of the Subject*, in *Employees' Inventions – A Comparative Study* 7-40 (Fernsway Publications).
2. Decree-Law No. 5,452 of 1 May 1943. This Labour Act is still in force, but its Art. 454 was revoked by the subsequent Industrial Property Law.
3. In the Industrial Property area: Ordinance No. 7,903 of 27 August 1945 followed by: Ordinance No. 253 of 28 February 1967 (as from which the subject deserved a broader regulation), Ordinance No. 1,005 of 21 October 1969, Law No. 5,772 of 21 December 1971 and Industrial Property Law No. 9,279 of 14 May 1996, currently in force.

faculty within the material rights to the invention.[4] The Brazilian system adopts the 'inventor principle' (*Erfinderprinzip*), which establishes that the privilege to protect an invention belongs to its creator (natural person). In practice, however, they often turns out to be a different person, not the inventor, who is allowed to apply for the patent either by a legal or a contractual provision.[5] The first legal standard regarding ownership is therefore the rule that the inventor's heirs or successors, legal entities and assignees,[6] provided that they prove such, may acquire the patent for protection of the invention. Furthermore, whenever an invention or utility model is created jointly by several persons, the patent may be claimed by all or any of the co-inventors, being granted to all in co-ownership, by naming and qualifying the others to guarantee their respective rights.[7]

Chapter XIV of Title I of BIPL on employees' inventions applies also to patents resulting from any other contractual relationship, such as R&D contracts between natural or legal persons, and also to public servants. It classifies inventions into three categories, which are further described below.

While the inventor's right to have their name mentioned on the letters patent[8] is a moral right and as such cannot be renounced, independently of the invention's category, material rights can be waived to different degrees, according to the corresponding category.

In view of the existing laws applicable to the different intellectual property assets that can be generated, it is highly recommended, in Brazil, in order to avoid disputes and to secure the parties' rights, to draw up a comprehensive and clear contract when hiring employees, interns and outsourced service providers in general, and not only those who deal with R&D work, considering that jurisprudence on the interpretation of the legal classification is under constant construction when the contracts lack definition. However, the parties' will is subject to limited flexibility, since many labour law rules are mandatory (of public order) and labour principles are, as mentioned above, at the root of the Brazilian system. Moreover, without a written contract dealing with employees' intellectual property rights, the ownership and remuneration solutions imposed by law will automatically apply.

4. João da Gama Cerqueira, *Tratado da Propriedade Industrial (Industrial Property Treaty)*, 244 (Rio de Janeiro, Forense, updated 2010. v.2. 6.2.). See also: Denis Borges Barbosa, *Inventor Empregado ou Prestador de Serviços*, in *Tratado da Propriedade Intelectual*, Tomo II, 1308 et seq. (Rio de Janeiro, Lumen Júris 2010).
5. Art. 6 of the Industrial Property Law, complementing the constitutional provision, establishes the general rules regarding inventorship and ownership of industrial creations.
6. BIPL – regulates the rights and obligations relating to industrial property (Art. 6, § 2 (14 May 1996), Industrial Property Law).
7. BIPL.
8. On the employee's naming right, see Carlos Eduardo Frank Fischer, *Critérios de Nomeação de Inventores e Autores segundo a Prática Internacional e a Legislação Brasileira*, 95 ABPI 47-50 (July/August 2008).

2 EMPLOYEES' COPYRIGHT

2.1 The Legal Framework

The current Brazilian Copyright Act (Law No. 9,610/1998) reduced the provisions regarding employees' and employers' rights over a creation, when compared to the previous Act (Law nº 5.988/1973), which Article 36 established the co-ownership between the employer and the employed author on the latter's works, and was not replaced by any equivalent rule concerning software, such provisions can be found in Article 4 of the Brazilian Software Act (Act No. 9,609/1998).[9]

2.2 Ownership of Rights

2.2.1 Copyright

Before addressing ownership, a few words are needed with respect to the authorship of copyright works. According to Article 11 of the Brazilian Copyright Act, the author is *the individual who creates a literary, artistic or scientific work*. In other words, the author of a work is considered to be the person who, by any of the modes of identification provided by the law, has, according to the normal usage, indicated or announced their authorship.[10] Copyright arises through the creation of the work and is independent of registration.

The person who adapts, translates, arranges or orchestrates a work that has fallen into the public domain owns the copyright of the new version, but cannot object to another adaptation, translation, arrangement, or orchestration, unless it is a copy of such version.[11]

Co-authorship occurs when two or more people work together in the creation of the same work. The person who simply assists the author in producing the literary, artistic or scientific work by revising, updating, monitoring or managing their publication or presentation by any means is not considered a co-author. The same applies to the other types of intellectual property rights in Brazil.

Bearing in mind that copyrighted works are indivisible, the co-authorship regime implies in a certain restriction of the rights of each individually considered author[12]

---

9. In accordance with Art. 10 of the Agreement on Trade Related Aspects of Intellectual Property Rights ('TRIPS'), computer programs are protected as literary works under the Berne Convention.
10. Brazilian Copyright Law, Art. 13 (19 Feb. 1998).
11. Brazilian Copyright Law, Art. 14 (19 Feb. 1998).
12. Eliane Y. Abrão, *Direitos de Autor e Direitos Conexos (Copyrigths and Neighbouring Rights)* (São Paulo, Migalhas, 2014), 27.

However, concerning audio-visual works, a co-author is anyone who develops the literary, musical or musical-literary plot or script, as well as the director. Finally, it is clear that the co-author of an animated cartoon is anyone who creates the cartoon drawings used in the audio-visual work.[13]

Ownership is addressed in the Brazilian Copyright Act only in one paragraph, that deals with 'collective works': Article 17(2) sets forth that the property rights on the entire collective work belong to its organiser. This rule is often applied to copyrights generated in corporations in general and to companies working with publications, editors in particular, depending, of course, on the circumstances of each case.

Article 13 of Law n° 6.533/1978, that foresees the rule according to which the copyrights regarding professional services of artists and technicians on entertainment spectacles cannot be assigned caused national repercussion, mainly with the Supreme Federal Court's understanding that such article Is constitutional, rendered in 1980 on Anti-constitutionality claim n° 1.031/DF[14].

Analysing some of the relevant judgments of Brazilian Labour Courts on employed authors' works, scholars point out that in view of the labour relationship between the parties, Intellectual Property Acts are not always sufficiently clear and whenever a labour principle is violated, courts apply the interpretation that is more favourable to the employee[15].

Unless the 'collective' work rule can be applied, Courts often understand as it happened in the *Forum TV* Case, that under the current Brazilian model, copyrights belong exclusively to the employee, independently of his or her labour relationship, and that even if the work was made in the context of his or her professional duties, the employer must, whenever it wishes to exploit it, obtain his or her express authorisation, in view of the Copyright Act´s Articles 11, 28 and 29.[16]

Following the same reasoning, in the *Caldas Junior* Case, the court stated that the employer newspaper cannot assign the employed photographer journalist's photographs to other communication vehicles in a gratuitous way and therefore granted damages to the employee. It also ruled that the violation of copyrights causes damages *in re ipsa*, because authorship is a personality right and the unfair profit obtained on the intellectual work constitutes an invasion.[17]

13. The provisions on co-authorship are established in the Brazilian Copyright Law, Art. 15 and 16 (19 February 1998).
14. Maurício Kioshi Kanashiro, *A Proteção do Autor Empregado sob a Perspectiva da Função Social do Direito Autoral (The Protection of the Employed Author Under the Perspective of the Copyright´s Social Function)*, ABPI Magazine n° 131, 3-26 (July/August 2014), 25.
15. Suzete da Silva Reis, *Os Direitos Trabalhistas do Autor Empregado: Uma Análise do Entendimento dos Tribunais do Trabalho (The Employed Author´s Labour Rights: an Analysis on the Understanding of the Labour Courts)*, PIDCC, Aracajú, Brazil, Year III, Edition n° 06/2014, 23-39 (June 2014), www.pidcc.com.br, 38.
16. *João Paulo R. N. da Gama v. Forum TV Mais Ltda.*, Superior Court of Justice´s decision on Special Appeal n° 1.322. 325-DF (2011/0026518-1), of 18 February 2014. In this particular case, the work was anonymous.
17. *Roberto Vinicius da Silva* v. *Empresa Jornalística Caldas Júnior Ltda.*, the 4th Region´s Labour Court decision N° 0000 428-13.2010.5.04.0028-00, on the company´s appeal, of 1 August. 2013.

2.2.2 *Computer Programs*

Article 4, head paragraph, of the Brazilian Software Act states that the employer, service contracting party, or public body has the total and exclusive rights over a computer program developed and produced during the validity of an employment contract or statutory relationship, expressly intended for research and development, or when the activity of the employee, service provider or public servant is foreseen in the contract or if it was created as a result of the employee's activities during the relationship.

According to the second paragraph of the same Article, the employee, service provider or public servant exclusively owns the rights to the software program developed with no connection to the employment contract, statutory relationship, or by-law obligation, and without the use of resources, technological information, trade secrets, materials, facilities or equipment of the employer, company or entity to which the employee is bound by a service agreement or similar contract.

Since the Software Act is silent when it comes to computer programs created by the employee who was not hired to develop them, but who used the employer's resources, the ownership is that envisaged in the contract; otherwise, if nothing has been agreed, it will be decided by the court.

2.2.3 *Jurisprudence Concerning Computer Programs*

Daniel José Elias v Borgwarner Brasil Ltda

According to the elements described by the Honourable Judge Ana Cláudia Torres Vianna in the decision, 'the known Software Act deals with only two situations regarding the invention's ownership. That is, it either belongs to the employer, in the cases shown in the *caput* of Article 4, or it belongs to the employee, in the situations shown in the second paragraph. It does not deal with the invention created jointly, a hypothesis known in the doctrine as casual invention, in which the rights to exploit belong exclusively to the employer and a fair remuneration as a share in the distribution of the gains of the invention is assured to the employee.' Majority decision accompanied by The Honourable Judge Eurico Cruz Neto (author's translation).

The legislator did not conceive of extending the treatment set by Act No. 9,279/1996 to computer inventions because such inventions, due to increasingly rapid and frequent innovation, have turned into work tools, used to enhance and speed up the productive systems in all areas of expertise, having no reason to exist outside of the workplace to which they are related. This way, according to Honorable Judge Ana Cláudia Torres Vianna, 'the intellectual property will only belong to the employee when he developed a project that is not related to his employment contract, using his own resources. It is obvious that the knowledge acquired in the workplace will influence the invention because the human being has only one brain that cannot be compartmentalised. However, the Act expressly highlights that if technological factors, the employer's trade secrets or business contribute to the invention, then the invention's ownership will belong to the employer or to the contracting public body, and not to the employee.' Majority decision accompanied by the Honorable Judge Eurico Cruz Neto (author's translation).[18]

2.3 The Parties' Duty to Inform

Since it is not expressly mentioned in the law, it is recommended to put in the employment contract a clause whereupon the employee has the duty to inform the employer of any invention or creation, collaborating as necessary to enable the protection and enforcement of the rights on said creation.

Although it is not a statutory provision, the employee should inform the employer of any invention or creative work connected to the activities developed by the latter. Eliane Ribeiro do Prado points out that granting the employer the right of preference to a licence is expected in view of the employee's loyalty duty, avoiding a potential conflict of interests if the invention is negotiated with a competitor of the employer.[19]

2.4 Employee Remuneration Right

The Copyright Act is silent on a remuneration right, but it is stated in the Software Act that remuneration should be agreed between the parties when the employee is hired for a creative duty.

3 EMPLOYEES' SEMICONDUCTOR CHIPS

3.1 The Legal Framework

Articles 27 and 28 of Brazilian Act No. 11,484/2007 on Digital Television, Semiconductors and Intellectual Property Protection to Integrated Circuits Topographies establish the employee's and employer's rights over their work specifically related to integrated circuit topography. Therefore, these provisions are analogously interpreted and applied to semiconductor components.

3.2 Ownership of Rights

According to Article 27 of the above-mentioned Act, the applicant will be considered the creator of the semiconductor component or the integrated circuit topography, unless agreed otherwise. When two or more people create a semiconductor component or the integrated circuit topography, any or all of them may apply for the registration.

18. *Daniel José Elias v Borgwarner Brasil Ltda.*, Judicial chapter of the Daily Gazette, vol. 8, 17 (TRT. 2005).
19. *Gestão e Justiça no Trabalho Inovador: o Direito do Trabalho na Propriedade Intelectual (Management and Justice in Innovative Work: Labour Law in Intellectual Property)* 46 (Lumen Juris 2011).

If the registration is applied for by only one person, the applicant shall name the other creators and their respective rights. The Act also provides that the inventor's successors, the assignee or whom the law or the employment contract or statutory relationship determines to be the owner, may file the application.

Unless otherwise stipulated by the parties, the rights over the semiconductor component or the integrated circuit topography developed during the employment contract or statutory relationship term, in which the creative activity results from the nature of the services rendered during such relation or when the person uses the resources, technological information, trade secrets, materials, facilities or equipment of the employer, service contracting party or entity, will belong exclusively to the employing party.

On the other hand, the rights will belong to the employee when the semiconductor component or the integrated circuit topography were developed during their spare time and with no relation to their employment contract or with the use of their employer's resources, technology information, trade secrets, materials, facilities or equipment.

The Act provides no provisions regarding the creation of the semiconductor component or the integrated circuit topography by an employee who was not contracted to develop such topography, but made it with the use of the employer's resources. In this case, the dispute will be decided by the court.[20]

3.3 The Parties' Duty to Inform

The same applies as described in section 2.3 above.

3.4 Employee Remuneration Right

The employee only has the right to remuneration agreed in the contract when the rights belong exclusively to the employer, unless otherwise has been agreed, by virtue of Article 28(1) of Act No. 11,484/2007.

4 EMPLOYEES' INVENTIONS

4.1 The Legal Framework

In a specific chapter on the matter, Articles 88 through 93 of the Brazilian Industrial Property Law (Act No. 9,279/1996, 'BIPL') regulate employers' and employees' rights on inventions, utility models and designs.[21]

20. *Ibid.*, p. 62.
21. According to Art. 121 of BIPL, the rights of the employee or supplier of services with regard to designs are governed by the same provisions than those applicable to inventions and utility models.

4.2 Ownership of Rights

The general principles on ownership of patents, utility models and designs are ruled in Articles 6 and 7 of BIPL, the essential elements of which are integrated with the specific chapter on employees' rights.

Under Article 6, unless otherwise agreed, the applicant is presumed to have the right to obtain the patent. This right does not always belong to the creator of the invention. It is possible that the application is filed by the inventor's heirs or successors, as well as by an assignee or by whoever the law or a labour or service contract determines to be the owner. When the invention is created by more than one person, the patent may be applied for by any or all of the co-inventors. In such circumstances, the patent will be granted to all of them and in case one of the co-inventors is willing to transfer their rights, the other co-inventors equally have the preference to the patent.

If two or more inventors have independently devised the same invention, utility model or design, the right to obtain the corresponding patent or registration will go to whoever proves the earliest filing, independently of the dates of the creation of the invention, as addressed by Article 7.

Relying upon Article 92, the provisions regarding employees' and employers' rights are also applied to interns, as well as to the relationship between two legal entities;[22] this is the case, for instance, for technology research contracts between two companies.

In order to determine to which party the ownership rights on an innovation resulting from a contract to be executed in Brazil are granted, one must analyse the employment contract and the conditions that contributed to the invention, utility model or design, to determine to which of the three categories envisaged in the law the invention belongs: service inventions, free inventions, 'composite' inventions. The names of the categories are not found in the statute, but they are mentioned as such by most scholars.

4.2.1 Service Inventions

If the invention, utility model or design is developed by an employee who was specifically hired to perform inventive activities, it belongs to the employer. Also in accordance with Article 88 of BIPL, whenever the employee applies for a patent within one year of the termination of their employment contract, it will be presumed that the invention was developed while their employment contract was still in force. This legal provision aims to prevent the employee not disclosing an invention to the employer, achieving this purpose by means of an inversion of the burden to proof: if, within one

22. With regard to inventions created by directors or shareholders of corporations, with no Labour Law relationship, see the author's Ch. *As invenções nas sociedades anônimas: Questões Societárias e Concorrenciais e Ações Reivindicatórias de Patente (Inventions in Companies: Competition Issues and Actions to Claim the Property of a Patent)*, in *Reivindicando a Criação Usurpada (Claiming the Usurped Creation)* 23-43 (Lumen Juris, Rio de Janeiro 2010).

year of the termination of the contract, the employee creates a new invention, utility model or design that is in no way connected to their former employer and employment contract, they will have to prove this fact; while after the lapse of this term, the employer will have the burden of proving that the invention is in some way connected to the former employment contract.

4.2.2 Free Inventions

In case the invention, utility model or design is not connected to the employment contract and the employee did not use any resources, means, data, materials, premises, or equipment belonging to the employer to develop it, the innovation belongs exclusively to the employee, in view of BIPL's Article 90.

4.2.3 Composite Inventions

Article 91 sheds light on inventions created by an employee who was not hired to invent or to perform R&D functions but used the resources, means, data, materials, premises, or equipment belonging to the employer. For this category, also known as 'dependant inventions' and often referred to by international literature as 'occasional inventions', Brazilian law applies the co-ownership solution:[23] the invention will belong to both parties, in equal parts, but without prejudice to (different) contractual provisions. Thus, in my view, it is possible for the parties to agree differently on ownership. Although the possibility of co-ownership in differentiated proportions has been rarely applied in practice, it is not disputed that the employee must always receive fair remuneration for their work, as explicitly stated in paragraph 2 of the same Article.

The key factor is therefore to know what is deemed to be the 'essence' of the object of a contract, 'which is research or the exercise of inventive activity, or when this results from the nature of the services for which the employee was contracted' (referred to at the end of the heading of BIPL's Article 88), since this factor can determine the classification of the invention as a service or composite invention, which in turn highly impacts on ownership and remuneration aspects.

Although the solution can generally be assumed from the description of duties in the contract, often employers encounter difficulties in drafting such description, since according to Article 456 of the Labour Code, when it is not expressly mentioned in the contract, it shall be understood that the employee has agreed to perform any and all duties compatible with their personal condition.[24] When it comes to industrial property, this provision cannot always be applied: although the duty to invent may be compatible with the employee's conditions, it should be at least implicitly mentioned

23. Regarding co-ownership on invention or utility model patents, see Frank Fischer, *O Regime de Co-Propriedade em Patentes*, 76 ABPI Magazine 3-15 (May/June 2005).
24. Ordinance No. 5,452, Art. 468 (1 May 1943): In individual employment contracts, it is only allowed to change the conditions by mutual consent, and yet, only if they do not harm directly or indirectly the employee, under penalty of the nullity of the infringing provision of this warranty.

in the employment contract, or result from the contracted activity, for an invention to be classified as a service invention. In order to avoid any kind of confusion regarding the existence, or not, of a duty to invent in a labour relationship, it is recommended for the employer to provide adequate mention in the contract of the innovative activities that shall be performed by the employee.

In the *Automotiva Usiminas* case, the Third Region's Labour Court granted to the employee remuneration based on Article 91 of BIPL, not based on the 50% criterion. The invention was not patented and the employee had obtained judicial fees exemption for poverty and had recognised to be one of the co-authors of the invention within the company.[25]

The reasoning that the employed inventor has the right to the remuneration established in Article 91 of BIPL because 'the intellectual activity performed by him surpassed the activities foreseen in the labour contract' was followed by the Second Region´s Labour Court in the *Paranepanema* case, in which the company argued that the employee, in his capacity as a sales engineer, had cooperated only with operationalising the project, not with its creation.[26]

4.2.4 *Jurisprudence*

Among the many judgments rendered by Brazilian Courts on employees' inventions cases, the following examples can be added to those mentioned in this chapter, quoting some of their core sentences:

> Court of Justice of the State of Rio de Janeiro:
> '[Patent concerning] improvements introduced in machines that produce bottles. State court's competence to decide upon the use of the patent. Invention created by former employee during employment contract. Application of Articles 40, §1 and 23, sole paragraph, of the Industrial Property Code. Granting of monthly compensation from the date of the patent application until the date of sale of the machinery, corresponding to the salary that the author received when he resigned his job, duly corrected and with interests. Appeal partially upheld. (Civil Action No. 2.868/87, Reporting Judge Pedro Américo Rios Gonçalves, First Chamber, decided on 15 December 1987)'. (Author's translation).[27]
>
> Court of Justice of the State of São Paulo:
> 'Plaintiff at the time of the facts – at the end of the 1960s – was chief of art of the Veja magazine and had as essential duties of his employment contract, for which he was remunerated, to follow and develop the execution and visual presentation of the magazine, including any graphical patterns. It is, therefore, clear that if the initial issues of the magazine had any type of layout or visual presentation problems, the chief of art would be called to remove such obstacles. It results from this that the development of a new layout process, including the columns

25. *Cremansul Fernandes* v. *Automotiva Usiminas S.A.*, Fisrt Chamber of the 3rd Region´s Labour Court's (Minas Gerais State) decision on Ordinary Appeal TRF-001 39-2012-129-03-00-9-RO, of 23 April 2014.
26. *Amilcar Barros Peres v. Paranapanema S/A*, Third Labour Court of Santo Andre, São Paulo Regional Court, The Hon. Judge Pedro Rogério dos Santos, decision, court action n° 1001057-94.2015.5.02.0433, 22 August 2016.
27. Law Gazette of the State of Rio de Janeiro, vol. 5, 195 (TJRJ, 1988).

measurements, the spaces and new format constitute the core of the plaintiff labour activity. It is clear that the creation of a new format for the magazine, independent of how creative it is, even it being given the plaintiff's name, does not grant him the patent of invention, once it was developed as a request and in benefit of the editor, and in exchange for wage compensation. (Decision on Appeal 9941.01.061498-2/50001, Reporting Judge Teixeira Leite, Fourth Private Law Chamber, decided on 8 April 2010).' (Author's translation).[28]

Superior Labour Court:

'Employee's invention in the course of the employment contract. It is the worker's right to receive from the employer participation in the income resulting from the use of the invention by the company independently of the granting of the patent by the correspondent administrative organ. (Appeal to the Higher Courts No. 1513/ 74, Reporting Minister Orlando Coutinho, Second Chamber, Official Gazette of September 17, 1974).' (Author's translation)[29]

4.3 The Parties' Duty to Inform

The same applies as described in section 2.3 above.

4.4 Employee Remuneration Right

In cases where the invention, utility model or design belongs to the employer (service inventions), the employer is guaranteed the right to an exclusive licence for exploitation and the employee will only receive the agreed salary. It is possible to agree in the employment contract a bonus wage in such cases, as a way of stimulating the employees to innovate. Article 93 of BIPL states that the employees' and employers' rights are applied to both public and private entities. However, the bonus wage is mandatory for public workers, as a method of motivating the researchers in public universities and research centres.

Free inventions belong to the employee, for which reason no special remuneration right applies.

With regard to composite inventions, Brazilian Law grants a 'fair remuneration' to the employee, and a right to exploit composite inventions to the employer (Article 91(2) of BIPL).[30]

As per Article 91(1) of BIPL, when an invention or utility model is created by more than one employee jointly, each of them will have an equal share, except when agreed otherwise. Thus, their remuneration can be agreed, for instance, as proportional to their contribution.

28. *George Benigno Jatahy Duque Estrada v Editora Abril SA*, Judicial Volume of the Daily Gazette Year IV Edition 869, pg. 836 (TJSP, 2011).
29. Official Gazette (TST, 1974).
30. Júlio Emílio Abranches Mansur highlights the technical inadequacy of the term 'remuneration' under the law, in view of the civil right, without any salary nature, arising from co-ownership on a patent in: *A Retribuição Econômica devida ao Empregado pela Exploração de Invenção Mista (The Economic Retribution payable to the Employee for the Exploitation of Joint Invention)*, 82 ABPI Magazine 12 (May/June 2006).

5 EMPLOYEES' UTILITY MODELS

As regards employees' utility models, see section 4 above.

6 EMPLOYEES' DESIGN RIGHTS

As regards employees' design rights see section 4 above.

7 EMPLOYEES' PLANT VARIETIES

7.1 The Legal Framework

The rules on employees' rights to plant varieties are regulated by Articles 38 and 39 of the Brazilian Plant Variety Protection Act (Act No. 9,456/1997).

7.2 Ownership of Rights

According to Article 38 of the Brazilian Plant Variety Protection Act, ownership belongs to the person (individual) or legal entity that obtains the new plant variety.

Additionally, two specific situations are regulated in the Act.

If a new plant variety or an essentially derived plant variety is developed or obtained by an employee who was hired to perform such duty ('service plant variety'), the work will belong exclusively to the employer. The name of the plant breeder shall be included in the application and in the certificate.

On the other hand, if the innovative variety was not foreseen in the employee's contract nor were they hired with an R&D purpose, and it resulted from their personal contribution and from the use of resources, data, means, materials, premises, or equipment of the employer ('composite plant variety'), it belongs to both parties.

A third category ('free plant varieties') is not mentioned in the law; plant varieties obtained by an employee not hired for R&D work and who did not use their employer's intellectual or material resources belong exclusively to the employee.

If the employee applies for the protection certificate within 36 months of the termination of their employment contract, it will be presumed that the new plant variety or essentially derived plant variety was developed while their employment contract was still in force.

These rules do not apply to legal entities that provide services, to interns or to grant holders.

Article 11 states that the protection will be granted for a 15-year period to new plant varieties and essentially derived plant varieties. Vines, fruit trees, forest trees and ornamental trees, including their rootstock, will be protected for a period of 18 years.

The application may be filed by the plant breeder, but also by their successors, as well as by an assignee. If two or more people jointly conducted the process, any or all of them may apply for the protection,

7.3 The Parties' Duty to Inform

The same applies as described in section 2.3 above.

7.4 Employee Remuneration Right

With regard to a 'service plant variety', payment to the employee shall be limited to the salary or the value agreed by the parties, unless otherwise stated in the contract.

The employer has the right to exclusively exploit the new plant variety or essentially derived plant variety, the remuneration agreed between both parties being assured to the employee, notwithstanding the payment of salary or the remuneration agreed.

8 INSTANCES FOR DISPUTES

8.1 Labour Relationships

For decades, the issue of the competent court to solve conflicts involving employees' intellectual property rights has been discussed in several instances, because of the hybrid nature of this matter.

The competence of Labour Courts to decide disputes related to employment contracts is granted by the Brazilian Federal Constitution[31], but it is not always clear to the parties to what extent disputes over intellectual property assets could be treated as having a different nature, and it is not rare to have one of them pleading that a specific approach be dictated.

Two conciliation stages are available at Labour Courts: before the defendant's reply and before the final briefs; neither are specialised in intellectual property disputes.

The Brazilian Patent and Trademark Office ('BPTO') is competent to decide any matter concerning the issuance of patents, such as their granting, cancellation or maintenance. However, it is the Federal Justice's competence to assess the validity of such administrative acts. When it comes to issues such as the classification of the invention into one of the three employees' inventions categories, allocating the ownership and compensations, BPTO has repeatedly declined to analyse such matters, claiming that these do not belong to its area of expertise and that it has neither structure to analyse evidence nor to conduct hearings.

31. See The Summary of seven decisions rendered by Brazilian Labour Courts concerning employees´ creations protected as inventions, copyrights or software, in Suzete da Silva Reis, *Os Direitos Trabalhistas do Autor Empregado: Uma Análise do Entendimento dos Tribunais do Trabalho (The Employed Author´s Labour Rights: an Analysis on the Understanding of the Labour Courts)*, PIDCC, Aracajú, Brazil, Year III, Edition n° 06/2014, 23-39 (June 2014), www.pidcc.com.br.

As recognised in the last years by Labour Courts, the parties may choose to resolve such conflicts through mediation or arbitration process, Brazil does not have a specified governmental alternative dispute resolution body for these matters.

8.2 Relationships Other Than Labour

Moreover, the nature of the issues arising from contracts between two legal entities or between a legal entity and an independent contractor is civil and they are decided therefore by Civil Courts. Consistently with the nature of lawsuits in which the BPTO is a party or an assistant participant, Federal Courts are competent to rule on patent ownership claim actions.

Notably, since there is no harmonised paradigm with regard to instances for the issues at stake, AIPPI's Resolution on Q183 recommends that the same court competent to decide intellectual property issues should also be competent to judge all disputes between employers and employees on such rights.[32]

9 CONCLUDING REMARKS

Seeking to stimulate innovation, the Brazilian legal regime acknowledges, by means of a certain flexibility, the material and moral rights that derive from the protection granted by patents or registrations of industrial or intellectual creations. Such flexibility has become increasingly necessary since in the work environment it is especially hard to balance these rights and even to identify the exact participation of each author.

Employees' and employers' intellectual property rights are alike, but not identical, in the different Acts mentioned in this chapter. Labour law aspects should always be considered by employers when drafting employment agreements and internal regulations, since freedom of contract in this area is limited by labour principles and specific rules found in each Act referred to herein.

Some aspects were modernised: for instance, the obligation that the application on an employee's invention should be filed first in Brazil, existing in the previous statute,[33] was revoked by the current one. The penalty of nullity had never been applied.

However, other critical issues remain unresolved, as for instance, the statute of limitations for payment of the employee's compensation for a composite invention, which has no labour law aspect. According to BPIL's Article 225, the limitation for actions for repairing damages caused to industrial property rights is five years to apply therefor, instead of the two-year statute of limitations of Labour Law.

32. See AIPPI – International Association for the Protection of Industrial Property – recommendation in item 2 of the Resolution adopted at Geneva on 23 June 2004, on the Q183 question, concerning 'Employer's Rights to Intellectual Property.' The same item of the Resolution also recommends that in such disputes, although a conciliatory phase may be desirable in certain circumstances, it should not be compulsory in all cases, http://www.aippi.org.
33. Law No. 5,772/71, revoked in 1996 by current Law No. 9,279/96.

As a general tendency, if the company has a Code of Conduct/Ethics, also addressing the company's rights on intellectual property assets and the employee's non-compete duties, courts are inclined to analyse it together with the employment contract.

The protection of employees' rights by Brazilian Labour Courts has always tended to be strong, in assistance to the weaker party, developing case law with a trend to, depending on the circumstances, classify inventions as composite and allocate 50% of the net profits to the employee-inventor, both in the private and public sectors, resulting in challenges for commercial exploitation and tax issues with economic consequences for the country that have not been specifically quantified.

As far as inventions, utility models and designs are concerned, the same BIPL rules apply to the private and public sectors, however, the latter has many complementary rules, set forth in acts that were implemented within the scope of Brazilian public policies to increase innovation and creativity.[34] In particular, for public university students, researchers and professors, inventions must be analysed together with the Brazilian Federal Innovation Act (Law No. 10,973, of 2 December 2004, in particular its Article 13 and the State Innovation Acts[35]. This is essential in particular in the context of collaboration between industries and universities, licensing, innovation funding, secrecy in R&D and tax incentives.

---

34. Regarding innovation in the public sector, see Claudia Inês Chamas, Marilyn Nogueira, and Simone Henriqueta Cossetin Scholz (coord.), *Scientia 2000: Propriedade Intelectual para. a Academia (Intellectual Property to the Academy)*, (Rio de Janeiro: Oswaldo Cruz Foundation, MCT, Konrad Adenauer Foundation, 2003); Gert Egon Dannemann, *Direitos de Propriedade Industrial. Pesquisa e Desenvonvimento (Industrial Property Rights. Research and Development)*, 56 ABPI Magazine 3-5 (January-February 2002); the lectures of Oswaldo Massambani and Denis Borges Barbosa on *Lei da Inovação: Entrosamento (ou Falta de) entre Universidade e Empresa (Innovation Law: rapport (or lack of) between University and Enterprise)*, in *Anais do XXVIII Seminário Brasileiro da Propriedade Intelectual da ABPI (Anais of the XXVIII Brazilian Seminar on Intellectual Property of ABPI)*, 159 et seq. (2008); Roberto de Alencar Lotufo, *A Experiência da Agência de Inovação* da UNICAMP (*The Experience of Unicamp Innovation Agency*), in *Anais do XXV Seminário Nacional de Propriedade Intelectual da ABPI (Anais of the XXV National Seminar on Intellectual Property of ABPI)* 63 et seq. (2005); the series *Encontros de Propriedade Intelectual e Comercialização de Tecnologia* (*Meetings of Intellectual Property and Technology Commercialization*) of REPICT - Technology Network of Rio de Janeiro, http://www.redetec.org.br; Elisabeth Kasznar Fekete, *Considerações sobre o Projeto de Lei da Inovação à Luz do Direito da Propriedade Intelectual (Considerations on the Innovation Law Project based on the Intellectual Property Law)*, in *Anais do XXIV Seminário Nacional da Propriedade Intelectual da ABPI (Anais of the XXIV Brazilian Seminar on Intellectual Property of ABPI)* 57-63 (Brasilia, August 2004); and the same author, *A Lei da Inovação Tecnológica (Lei No. 10.973/04) e os respectivos Incentivos Fiscais (Lei No. 11.196/05) (The Technological Innovation Law (Law No. 10,973/04) and the related Tax Incentives (Law No. 11,196/05)*, in Eliane Y. Abrão (org.), *Propriedade Material – Direitos Autorais, Propriedade Industrial e Bens de Personalidade (Intangible Property – Copyright, Industrial Property and Personality Property)*, 71-89 (Ed Senac, São Paulo 2006).

35. See in Carla Eugenia Caldas Barros and Murilo Soares Tavares, *A Propriedade Intelectual Derivada da Criação e do Trabalho Intelectual (The Intellectual Property Derived from the Creation and Intellectual Work)*, PIDCC, Aracajú, Brazil, Year III, edition N° 05/2014, 256-322 (February 2014), www.pidcc.com.br, the list of 13 State Innovation Acts currently in force, a comparative analysis with federal and municipal legislation and a broad study of labour relationships and social rights in the context of innovation.

Besides such Acts, public universities and research centres must have their own internal rules on ownership and remuneration of their researches and professors, under coordination of an Intellectual Property Department of their own or shared with other public R&D centres.

BIBLIOGRAPHY

Table of Sources

AIPPI's - International Association for the Protection of Industrial Property - recommendation in item 2 of the Resolution adopted at Geneva on 23 June 2004 on the Q183 question, concerning 'Employer's rights to intellectual property'. The same item of the Resolution also recommends that in such disputes, although a conciliatory phase may be desirable in certain circumstances, it should not be compulsory in all cases, http://www.aippi.org.

Carla Eugênia Caldas Barros and Murilo Soares Tavares, *A Propriedade Intelectual Derivada da Criação e do Trabalho Intelectual (The Intellectual Property Derived from the Creation and Intellectual Work)*, PIDCC, Aracajú, Brazil, Year III, edition N° 05/2014, 256-322 (February 2014), www.pidcc.com.br.

Claudia Inês Chamas, Marilyn Nogueira & Simone Henriqueta Cossetin Scholz (coord.), *Scientia 2000: Propriedade Intelectual para a Academia (Scientia 2000: Intellectual Property for the Academy)* (Fundação Oswaldo Cruz, Rio de Janeiro, MCT, Fundação Konrad Adenauer 2003).

Denis Borges Barbosa, *Tratado da Propriedade Intelectual: Inventor Empregado ou Prestador de Serviços (Intellectual Property Treaty: Employee Inventor or Third Party)*, vol. II, 1308 et seq. (Lumen Júris, Rio de Janeiro 2010).

Denis Borges Barbosa, *Uma Introdução à Propriedade Intelectual (An Introduction to Intellectual Property)* (2nd ed., Lumen Júris, Rio de Janeiro, 2003).

Eliane Ribeiro do Prado, *Gestão e Justiça no Trabalho Inovador: O Direito do Trabalho na Propriedade Intelectual (Management and Justice in the Innovative Work: the Labour Law in the Intellectual Property)* (Lumen Júris, Rio de Janeiro 2011).

Eliane Y. Abrão, *Direitos de Autor e Direitos Conexos (Copyrigths and Neighbouring Rights)* (São Paulo, Migalhas, 2014).Elisabeth Kasznar Fekete, *Anotações sobre a Legislação e a Infraestrutura de Patentes e o Futuro da Inovação Tecnológica no Brasil (Notes on the Legislation and Infrastructure of Patents and the Future of Technology Innovation in Brazil)*, in *A Revisão da Lei de Patentes: Inovação em Prol da Competitividade Nacional (The Law of Patents Review: Innovation to promote National Competitiveness)*, Estudos Estratégicos 1, 165-178 (Brasília: Câmara dos Deputados, Edições Câmara 2013). Speaker articles in CD-ROM.

Elisabeth Kasznar Fekete, *Arbeitnehmererfindungen: Recht und Wirklichkeit in Brasilien*, 726-734 (12th ed., GRUR International 1984).

Elisabeth Kasznar Fekete, *As Invenções nas Sociedades Anônimas: Questões Societárias e Concorrenciais e Ações Reivindicatórias de Patente*, 23-43 *(Inventions in Compa-*

nies: Corporate and Competition Issues and Actions to Claim the Property of a Patent), in: *Reivindicando a Criação Usurpada (Claiming the Usurped Creation)*, 23-43, organised by Denis Borges Barbosa (Lumen Júris, Rio de Janeiro 2010).

Elisabeth Kasznar Fekete, *Considerações sobre o Projeto de Lei da Inovação à luz do Direito da Propriedade Intelectual (Considerations on the Innovation Law Project based on the Intellectual Property Law)*, in *Anais do XXIV Seminário da Propriedade Intelectual da ABPI (Anais of the XXIV National Seminar on Intellectual Property of ABPI)*, 57-63 (16-18 August 2004).

Elisabeth Kasznar Fekete, *Direitos, Obrigações e Perspectivas dos Inventores-Empregados no Brasil (Rights, Obligations and Perspectives of the Employee-Inventors in Brazil)*, in *Anais do II Seminário sobre Propriedade Industrial da ABPI (Anais of the II Seminar on Industrial Property of ABPI)*; and *Transferência de Tecnologia do INPE – Instituto de Pesquisas Espaciais (Technology Transfer of INPE – Institute of Space Researches)*, São José dos Campos, Brazil, 31-41, (24-25 October 1988).

Elisabeth Kasznar Fekete, *Notas sobre a Cessão de Invenções Futuras e seus Efeitos sobre o Direito de Obtenção da Patente (Notes on the Cease of Future Inventions and its Effects on the Patent Obtaining Right)*, http://www.esaoabsp.edu.br/revista/edicao 09/index.swf (accessed 19 May 2014).

Elisabeth Kasznar Fekete, *O Regime Jurídico das Criações Empresariais e Terceirizadas (The Legal Regime of the Corporate Creations and Outsourcing)*, in: Anais do XX Seminário da Propriedade Intelectual da ABPI (Anais of the XX Intellectual Property Seminar of ABPI), 35-45 (21-22 August 2000).

Elisabeth Kasznar Fekete, *O Regime Jurídico do Segredo de Indústria e Comércio no Direito Brasileiro (The legal regime of Trade Secret in the Brazilian Law)*, 489 (Rio de Janeiro, Forense 2003).

Elisabeth Kasznar Fekete, *Quem é o Titular do Resultado da Pesquisa Financiada com Recursos Públicos? (Who is the Owner of the Search Result financed with Public Funds?)*, in *Anais do IV Encontro de Propriedade Intelectual e Comercialização de Tecnologia (Anais of the IV Intellectual Property and Technology Commercialization Meeting)*; 45-65, Debates, 71-78 - REPICT - Rede de Tecnologia do Rio de Janeiro (July 2001).

Elisabeth Kasznar Fekete, *Relação entre Empregado e Empregador à Luz da Nova Lei de Propriedade Industrial (Relationship between Employee and Employer based on the new Law of Industrial Property)*, in *Anais do I Seminário Internacional de Inovação: da Ideia ao Mercado (Anais of the I International Innovation Seminar: from the Idea to the Market)*, Porto Alegre, FIERGS/MICT (Federação das Indústrias do Estado do Rio Grande do Sul e Ministério da Indústria, do Comércio e do Turismo (Industrial Federation of Rio Grande do Sul State, and the Ministry of Industry, Commerce and Tourism), 129-139 (1998).

Frank Fischer, Carlos Eduardo Eliziario de Lima and Paula Santos Silva, *Critérios de Nomeação de Inventores e Autores segundo a Prática Internacional e a Legislação Brasileira (Appointment Criteria of Inventors and Authors according to the International Practice and to the Brazilian Legislation)*, ABPI Magazine No. 95, 47-50 (July/August 2008).

Frank Fischer, Carlos Eduardo Eliziario de Lima, and Paula Santos Silva, *O Regime de Co-Propriedade em Patentes (The Scheme of Co-Ownership in Patents)*, 76 ABPI Magazine 3-15 (May/June 2005).

Jeremy Philips, *Employees' Inventions: an analysis of the nature of the subject*, in: *Employees' Inventions – A Comparative Study* 7-40 (Fernsway Publication).

João da Gama Cerqueira, *Tratado da Propriedade Industrial (Industrial Property Treaty)* (Lumen Júris, Rio de Janeiro, updated 2010).

Júlio Emílio Abranches Mansur, *A Retribuição Econômica devida ao Empregado pela Exploração de Invenção Mista (The Economic Retribution payable to the Employee by the Exploitation of Joint Invention)*, 82 ABPI Magazine 12 (May/June 2006).

Maurício Kioshi Kanashiro, *A Proteção do Autor Empregado sob a Perspectiva da Função Social do Direito Autoral (The Protection of the Employed Author Under the Perspective of the Copyright's Social Function)*, ABPI Magazine n° 131, 3-26 (July/August 2014).

Series: *Encontros de Propriedade Intelectual e Comercialização de Tecnologia (Encounters of Intellectual Property and Technology Commercialization)*, REPICT – Rede de Tecnologia do Rio de Janeiro, http://www.redetec.org.br.

Suzete da Silva Reis, *Os Direitos Trabalhistas do Autor Empregado: Uma Análise do Entendimento dos Tribunais do Trabalho (The Employed Author's Labour Rights: an Analysis on the Understanding of the Labour Courts)*, PIDCC, Aracajú, Brazil, Year III, Edition n° 06/2014, 23-39 (June 2014), www.pidcc.com.br

Table of Legislation

Brazilian Copyright Law: Law No. 9,610 (19 February 1998).

Brazilian Federal Constitution (5 October 1988).

Brazilian Industrial Property Law: Law No. 9,279 (14 May 1996).

Brazilian Plant Variety Protection Law: Law No. 9,456 (25 April 1997).

Brazilian Semiconductor Components and Topographies of Integrated Circuits: Law No. 11,484 (30 May 2007).Brazilian Software Law: Law No. 9,609 (19 February 1998).

Labour Act: Decree-Law No. 5,452 (1 May 1943).Labour Act: Decree-Law No. 5,452 (1 May 1943).

CHAPTER 34

Chile

Cristóbal Porzio

1 BACKGROUND

The ownership of creations and inventions resulting from or developed under the existence of an employment relationship needs to be examined under the prism of intellectual and industrial property law and also under the scope of labour law.

Although both the Intellectual Property Law and the Industrial Property Law contain express provisions related to the attribution of the rights of property that apply to inventions and creations obtained under the frame of a labour relationship, in most cases these provisions will need to be interpreted in harmony with the regulations and the principles that govern labour law in Chile.

In Chile, the expression 'employee' is directly associated with the concepts of:

(i) individual (the worker or employee);
(ii) provision of personal services;
(iii) relationship of subordination and dependence with the employer;
(iv) determined labour conditions; and
(v) payment of a remuneration or salary for the services provided.[1]

It is relevant to consider the definition contained in Article 3 of the Chilean Labour Code, which provides that the 'employer' is 'the natural or legal person who uses the intellectual or material services of one or more persons under a contract of employment'. In the same manner the same Article 3 of the Labour Code states that the 'worker' (or employee)[2] is 'any natural person or individual who provides personal,

1. These rules are found in Art. 7 of the Chilean Labour Code, which defines the individual employment contract.
2. Chilean legislation recognises the concept of 'worker' and not the concept of 'employee' to refer to a natural person or individual who provides personal services at the risk and on account of

intellectual or material services'. Likewise, Chilean Law defines 'company', for labour purposes, as 'any organisation of personal, tangible and intangible means, sorted under a direction, for the achievement of economic, social, cultural or charitable purposes, provided with a certain legal individuality'.[3] Thus, it is only possible to speak about 'employees' in those cases where there is a labour relationship linking the parties, regardless of whether or not there is a written labour contract. We need to stress the expression 'written' since even in those cases in which a labour contract has not been signed by the parties, if the elements numbered (i) to (v) mentioned above exist, the relationship will be considered as a labour contract. In addition, in this last case, consisting in the lack of a written agreement, the declaration made by the worker/employee as to their relationship, will be considered as a valid labour contract, containing the clauses declared by the employee.[4]

On the other hand, local practice indicates that in Chile most of the inventions and intellectual creations arise in the context of a labour-dependent relationship or at universities or research and development centres. Depending on the case, the state may assist with different funds that can be claimed by inventors and/or by applicants. This case is very common for universities.

It is not common in Chile that investments in research and inventions be conducted by private individuals, since the financing of materials and human resources required by research and development activities are rarely available for individuals. These means are mainly concentrated in companies, research centres and universities, and these entities are the ones that make the investments in resources for the development of new inventions and creations. As a result, and to allow these entities to obtain the benefits from their investments, in many cases the law permits them to exploit the property titles. For completeness, it should be noted that in the last ten years individuals and private companies have created funds for capital risk. Some of these funds will assist individuals, small companies and universities in research projects and in projects consisting of the 'transformation of an idea' into a viable invention, and thereafter in the protection of such invention, and in some cases also in the finding of the best manner to put such an invention on the market.

Notwithstanding the above, the Law provides for some specific provisions to protect individuals that may be in a weak position in relation to their employers.

Though the regulation of employees' rights exists in Chile as a legal framework, it has to be said that some particular cases may not be specifically regulated. As an example, there is an absence of rules governing the information duties of employees, or the non-existence of parameters to determine the retribution that the employee is entitled to receive for their work. Depending on the focus given to the question, these

someone else. The concept of 'employee' is commonly associated with government employees and as individuals who provide personal services to state institutions. Notwithstanding this, in this presentation we will use both concepts indistinctly.

3. In our opinion this concept should be extended to the provisions of the Industrial and Intellectual Property Laws that regulate the attribution of property rights in works, inventions and creations that are obtained under the frame of a labour relationship, since these are rules that regulate matters that affect these kind of relationships.
4. Art. 9 of the Chilean Labour Code.

omissions in the Chilean legislation can be considered as an attribution that the law gives to the parties in order to regulate certain aspects of the relationship between them, or as deficiencies in the law, which does not provide an appropriate solution or elements to implement the existing rules.

As we will see in the present chapter, the fact that the employee provides their creative and inventive services within a labour relationship, does not affect their constitutional right consisting in 'the right to create freely'.

According to the above, the employer's faculty to indicate how the provision of services should be established shall be exercised only in general instructions, leaving the employee free to develop their activity. If this is not the case, and if the employer indicates to the employee all the aspects pursuant to which they must provide their services, leaving no room for their liberty to create, the employee will not be considered an inventor or author of the inventive or creative result. In such case the employee is a mere executor of the employer's instructions.[5]

2 EMPLOYEES' COPYRIGHT

2.1 The Legal Framework

In Chile the regulation of employees' creations is to be found in the Intellectual Property Law No. 17,336. Some additional provisions can also be found in the Chilean Labour Code and in the Civil Code.

The provisions related to employees' copyrights shall apply in cases where the creations are generated as part of an employment contract. This contract must have as its object the provision of creative services as a means obligation, and not a result obligation. On the other hand, in cases where the contract's object is the execution of one or more specific works, without a working relationship between the parties, the rules applicable in conjunction with the provisions of the Copyright Law will not be those that are found in the Labour Code, but the Chilean Civil Code will apply instead. In this respect, it is to be noted that the ninth paragraph, Title XXVI, Book IV, which rules on 'the rental of intangible services' will apply.

Chilean law provides for special provisions related to intellectual property rights of employees in cases where there is an employment contract in which the provision of services contracted by the employer, as the contract's object, is the creation of intellectual works. It is not relevant to analyse those cases in which the provision of services by the employee is not intended to be the creation of intellectual works. In such cases there are no doubts that the rights are vested in the employee, regardless of place or occasion in which the creation is done. So, the general provisions contained in the Intellectual Property Law regarding the ownership and enjoyment of copyright would be applicable.

5. It is to be noted that since the first edition of the present chapter, the Chilean Labour Law has gone through many important changes. However, it is to be said that the provisions commented in the present chapter, have not been the subject of modifications.

In addition, it is relevant to consider Article 89 of the Law No. 17,336 which establishes that the rights recognised by the Intellectual Property Law for the benefit of the copyright holder do not affect the protection that the Industrial Property Law (No. 19,039) or other legal rules may recognise.

2.2 Ownership of Rights

2.2.1 Preliminary Considerations

Under Chilean law the author is always an individual because only individuals have the intellectual capacity to create. This situation is recognised by Article 7 of the Intellectual Property Law, which states that, 'the original owner of copyright is the author of the work'. In consequence, only the employee can be considered the author of the work.

It should be noted that the existence of an employment contract in which the object is creative services does not limit the constitutional protection of freedom to create of the employee, but only submits the employee to the direction and instructions of the employer, limiting the scope and the extent of their activity but not their creative capacity and expression. This situation is expressly recognised in the final section of Article 145-A of the Chilean Labour Code, included in the rules that govern the employment contract of arts and entertainment workers. This Article points out that, 'In the case of the creation of a work, the employment contract, in any case, can affect the freedom to create of the artist under contract, notwithstanding its obligation to comply with the services according to the terms specified in the contract.' Although this is a rule that refers to a special type of employment contract, we can extend the sphere of its application to all employment contracts that have as their object the provision of creative services by the employee. This is because this rule is uniform and complies with the constitutional protection of freedom to create established in Article 19 No. 25 of the Chilean Constitution, which provides that, 'The Constitution guarantees to all persons: No. 25 the freedom to create and disseminate the arts, as well as their copyright in intellectual and artistic creations of any kind ... .' As a result, the authorship of the employee and their freedom in the provision of its creative services are recognised at a legal and constitutional level. However, these acknowledgments have a limitation given by the employers' property rights regarding the result of the creative services, considering that it is they who, under the employment contract, provides to the employee the conditions, materials, media, and all that is necessary for the creation.

Chilean legislation does not ignore the rights of the employer. Even though it does not recognise employers' authorship, according to certain legal provisions it is possible to attribute to a legal entity – the employer – the ownership of copyright. This is possible according to Article 7(2) which recognises as a secondary holder of the copyright to, '... [the one] who acquired it from the author by any title'. Thus, employers acquire from their employees the copyright, as a secondary owner.

On the other hand, if the employer is also an individual, authorship should be allocated alike to the employee, because Article 7 clearly distinguishes between the original and the secondary ownership of copyright. The secondary ownership will correspond to the employer, since they will acquire such right directly from the employee.

The recognised ownership right of the employer is not absolute, because it does not cover all the rights covered by the copyright according to Chilean legislation. Thus, according to the Intellectual Property Law, copyright is composed of two separate rights, namely the moral rights and the economic right. In this respect, we can state that the Chilean system 'follows the French School or Theory'. The employer could be the owner of the economic rights only because the moral rights are non-transferable and are attached to the creator/author/employee.

2.2.2 Moral Rights of the Employee-Author

Moral rights correspond to one type of rights recognised by copyright. They maintain a link between the creator and the created work. Moral rights are regulated in Chapter IV of the Intellectual Property Law.

According to the Chilean legislation moral rights belong only and exclusively to the original author and are inalienable from the work and non-transferable to third parties. They may, however, be transmitted to the heirs of the author. The author preserves the moral rights even if they have assigned the economic rights to another person.

Moral rights are listed in Article 14 of Law No. 17,336, which provides that the author, as exclusive holder of the moral right, during their lifetime has the following powers/rights:

(i) to claim the paternity of the work, associating to the same their name or known pseudonym;
(ii) to oppose or fight against any deformation, mutilation or other modification done without their express and previous written assent – the acts of conservation, reconstitution or restoration of the works that may have suffered damage that alter or reduce its artistic value, will not be considered as alterations of the piece of work;
(iii) to keep the work unpublished;
(iv) to authorise to third parties to finish the incomplete work, with a prior consent of the editor or of the assignee, if any of them were to exist; and
(v) to claim respect of their will to keep the work anonymous or pseudonymous while it does not belong to the common cultural common heritage.

The moral rights are inviolable rights that can only belong to the author, for this reason when the author is an employee, these rights will stay with them. Needless to say, it is not possible for the employee to transfer the moral rights to the employer.

Considering this situation in light of Article 7 of the Law, the only rights that can be transferred by the author to another are the economic rights, so the employer can be the secondary holder of the economic rights.

Notwithstanding the above, there are certain cases expressly regulated in the Law, in which the employer is recognised as the original owner of the economic right. Considering the special situation of an author who is the subject to a labour contract, it is necessary to bear in mind certain elements in relation to the author's powers and rights as recognised by the Law in relation to the moral rights. The author-worker maintains their right to claim paternity of the work. Such an author-worker is not allowed nor entitled to transfer nor to assign nor to attribute the above-mentioned paternity to the employer. In addition, the employer will have the obligation to recognise the paternity of the work of the employee. With regard to the right to object any distortion, mutilation or other modification made to the work without the prior express consent of the author, it is necessary to ascertain who may make these alterations to the work. If the alteration is made by a person who is a stranger to the labour relationship, the faculty of the author-worker will remain intact and they will be entitled to oppose to any alteration. In the case the alteration of the work is performed by the employer or a third party authorised by the employer, it will be necessary to check the terms on which the worker has agreed with the employer the provision of their creative service. Additionally, it is necessary to consider whether the employee authorised in the contract with the employer to make changes in the work for the purposes of adequacy or updating, or if according to the nature of the work it is essential and necessary to make such changes. For example, if the work is one of those that requires constant updates or was created from creative elements that complement each other and have been contributed by a number of people, as it could be the case of a collective work.

In relation to the right of maintaining the work unpublished, it is necessary to consider such a provision in the contract. Thereafter, it will be possible to determine or conclude if the employee has authorised the employer to publish the work, or if they have limited such faculty. In general, it would be possible to conclude that this authorisation is given to the employer when the employee accepts the provision of creative services, since it implies that the worker places the work available to the employer in order to be published by them, without prejudice to the worker's authorship. Another alternative would consist of having the worker agreeing in the contract to give this authorisation when the employer intends to require same.

The right to authorise third parties to finish an incomplete work must be understood as granted by the worker to the employer at the time the hiring took place. In fact, in case that the employee is unable to finish the work, the employer must be able to continue the development of the creative work by means of a different worker from the one originally hired to do it.

Finally, the right of the author to demand the respect of their decision to keep the work anonymous or pseudonymous, if such a work does not belong to the cultural common heritage, shall be exercised by the employee in accordance with the provisions of the employment contract, respecting the worker's will to keep the work anonymous or pseudonymous as long as this is possible according to the contract or when the completion of it does not imply any harm to the employee.

2.2.3 Economics Rights of the Employee-Author

It was mentioned above that copyright in Chile is composed of two rights. We have already developed some ideas regarding the 'moral rights'. We therefore need to discuss what the Chilean law states regarding 'economic rights'. The economic rights allow the author to obtain economic remuneration for the use made by third parties of their work and it also allows them to prevent other persons from using their work without their permission.

Economic rights are regulated in Chapter V of the Intellectual Property Law, which regulates both their exercise and their limitations. Article 17 of the Law describes the 'economic rights' generally by reference to the 'copyright holder'. Such a holder could be the same person as the owner of the moral rights or a third party. This is in line with Article 7 of the Law, which distinguishes between the original owner and the secondary owner of the copyright. Consequently, only the original owner of the copyright, the author or their heirs in their capacity as representatives of the deceased author, will hold the moral rights, as indicated above.

On the other hand, both the original and the secondary owners can be holders of the economic right, since the provisions that regulate this right are not limited only to the author. Moreover, secondary ownership is possible because the original owner is entitled to transfer or assign to third parties the economic dimension of the copyright, thus the secondary holder acquires ownership directly from the author, who transfers in a total or partial manner their economic rights on the work to another person. In order for a third party to become a secondary holder of the economic right, an act through which the original owner assigns or transfers their right to a third party will be necessary.

Article 73 of the Intellectual Property Law states that:

> the total or partial transfer of copyright or connected rights, in any title, must be recorded with the Registry, within a term of 60 days starting from the date of execution of the respective act or contract. The transfer must be made by a public or notarised private document.

From the aforementioned Article we can raise the following questions: is the employment contract sufficient in order to transfer the economic right to the employer? Must the contract comply with the above-mentioned formalities? In my opinion the labour contract is not a sufficient convention to transfer the economic right to the employer, since it only defines the relationship between the parties. So, it is not necessary to comply with the formalities required by Article 73. If we demand these formalities as necessary in all labour contracts in which service provision is the development of a creative activity, it would imply the imposition of additional requirements that are not contemplated expressly in the Labour Code for this type of contracts. Consequently, in order to assign the economic rights, it would be necessary to have a specific Act. The same should comply with the formalities of Article 73, according to which the employee states their intention to yield the economic rights, specifying the works that are subject of the assignment and the conditions under which such assignment is made.[6] This solution is suitable for the situation of employment contracts in which the object of the services provision by the employee is a creative service and not the execution of a certain work.

Thus, the employee under a labour contract is required to: (i) comply with the provision of the hired creative service; and (ii) in the event that the result of their creative service is the conclusion of a work, transfer to the employer the economic rights of the work, fulfilling the formalities established under Article 73. In consequence, this second obligation contracted by the worker will depend on the existence of a result and will allow omission of the formalities established in Article 73, at the time of the execution of the employment contract in which the object is the provision of creative services.

It should be noted that as a general rule, Article 8 of the Intellectual Property Law establishes as a simple legal presumption, absent proof to the contrary, that the author of a work is the person who, according to the respective registration, owns the registered copy of the work, or who appears as such on the work, by indication of their name, pseudonym, signature or usual identification mark.

The Intellectual Property Law contains special rules relative to different types of creations, in which the Law establishes simple legal presumptions (that admit proof to the contrary) in relation to rights transfers that the employee would make to the employer. Likewise, there are certain cases in which the Law grants the original ownership directly to the employer. These are further discussed below.

2.2.4 Computer Programs/Software

Article 3(16) of the Intellectual Property Law includes, within the types of creations specially protected by law, computer programs. These are defined as 'any mode or form of expression, such as source or object program, even preparatory documentation, its technical description and manuals'. In relation to the copyright ownership of these kind of works, Article 8(2) states that in the case of computer programs, 'unless otherwise stipulated in writing, the natural or legal persons whose dependents, in the performance of their labour functions, have produced it, will be the holders of the respective copyright'.

As a consequence, in the case of computer programs the Law presumes that the ownership of the copyright belongs to the employer, but admits that this presumption may be altered by means of a written stipulation contained in the labour contract or on a separate document, giving the ownership of the economic rights to the employee.

Furthermore, in relation to the duration of the protection granted to the employer, it is necessary to consider the provisions of Article 10 of the Law, which distinguishes

6. In this sense Schuster Santiago, *Derechos de Autor en las Relaciones Laborales y su Vínculo con el Tratado de Libre Comercio entre Chile y Estados Unidos*, in *Estudios de Derecho y Propiedad Intelectual: Homenaje a Arturo Alessandri Besa* 374 (Jurídica de Chile ed., 1st ed., Santiago of Chile).

the case when the employer is an individual or legal entity. Thus, if the employer is an individual, the protection that the Law provides is extended throughout the entire life of the author and extends to 70 years after the date of death of the author. If the employer is a legal entity, the protection will last for 70 years, counted from the date of the first publication.

2.2.5 *Works Created by Employee-Authors of Media Companies, News and Information Agencies or Television or Broadcaster Stations*

The rules applicable to these kinds of employees are contemplated in Article 24(c), (d) and (e) of the Intellectual Property Law.

Article 24(c) rules the situation of works created by authors who are employees of newspapers, magazines and other media companies, under an employment contract that binds the employee with the employer. The first sentence of Article 24(c) states that:

> the media company acquires the right to publish in the newspaper, magazine or journal which the author provides his services including the articles, drawings, photographs and other productions made by personnel subject to a labour contract, keeping the authors all other rights that the law protects.

Thus, the rule only entitles the employer to publish the employee's creation in a media publication of the employer company under the terms of the Law No. 17,336, which defines publication in its Article 5(o) as 'public offer of the work ... with the consent of the holder of the right, as long as tangible copies are offered to the public in reasonable quantity'.

I believe that the consent of the right owner required by Article 5 in order to publish the work is given by the employee to the employer at the time of the conclusion of the labour contract, even if an express clause in this sense is not included in the contract, since by virtue of the agreement on the provision of creative services by the worker, they assume the legal consequences set out in Article 24(c)(1) that the benefit may have, if there is a concrete result of their services (as a creative work). In relation to the extension of this authorisation granted by the employee, I believe that it is limited, and only applicable, to works produced by the author within the framework of the provision of services defined in the employment contract, not being extendable to works of the employee produced outside of the labour relationship. Moreover, this rule is restrictive and indicates what aspects of the copyright the employee grants to the employer, while maintaining with the employee the remaining rights or the rights reserved to the author.

Notwithstanding the above, the Law does not prevent the employer from publishing these creations in other newspapers, magazines or journals for the same company, other than that or those in which the employee provides their services. However, these publications entitle the authors to receive an additional payment of fees, from the employer, in accordance with the tariff of the Chilean Journalists Association.

It should be mentioned that at present the tariff of the Chilean Journalists Association is for reference purposes to the free exercise of the profession of a journalist. I consider that this reference to the tariff of the Journalist Association is a minimum established by Law for the author, and nothing prevents the parties freely agreeing to a greater amount or alternative benefits. On the other hand, if the publication is made by a different media company not related to the employer, this third company that publishes the creations must pay the author the fees fixed by the Journalist Association.

The right to receive an additional compensation must be fulfilled within one year from the publication of the creative production, but is suspended in favour of the authors in respect of the employer's media company while the employment contract is in force. This suspension is only applicable in the first case of publishing in a media company of the employer, because only in this case shall the payment of additional compensation be made by the employer.

Under Article 24(d) and (e), the rules that apply to authors who provide services to newspapers, magazines and other media companies, also apply to those who work in news and information agencies regarding articles, drawings, photographs and other productions protected by the Law, as well as those employees of broadcaster or television stations.

2.2.6 Databases and Other Compilations

Article 3(1) of Law No. 17,336 specifically protects 'books, pamphlets, article and writings, whatever their form and nature, including encyclopaedias, handbooks, dictionaries, anthologies, and compilations of all kinds'. In turn, Article 3(17) extends the protection of the Law to:

> ... data compilations or other readable materials in machine or other form, which by selection or arrangement reason of their contents, constitute intellectual creations. This protection does not cover the material data itself and is without prejudice to any subsisting copyright in respect to the data or materials contained in the compilation.

These provisions attribute copyright to the maker of the compilations of data or other material, which is referred in the Law as 'organiser'. Said person will own the copyright to the compilation, without prejudice to the copyright of the particular authors of the works that are part of the collection.

In my opinion, the particular nature of this type of collection work, in which several individuals take part in the final creation as participants, being directed by a person, whether natural or legal, makes it possible to consider this work as a collective one, and which would therefore fall under Article 5(c) of the Law which defines a collective work as 'the work that is produced by a group of authors, by the initiative and under the guidance of a natural or legal person who coordinates, reports and publishes the final work under his name'.

These rules of Articles 3 and 5(c), are linked and consistent with the provisions of Article 24(a) and (b), which regulate two cases:

(a) *Anthologies, chrestomathies and similar compilations*: in respect to these compilations the owner of the copyright is the organiser. However, they will be forced to obtain the free or paid authorisation of the author of the each particular work included in the compilation. If such authorisation is granted for free, it is necessary to expressly recognise that.
(b) *Encyclopaedias, dictionaries and similar compilations*: in respect of these works, made on the order of the organiser, the copyright regarding the compilation and the individual contributions belong to the organiser.

Consequently, it is possible to conclude that in those cases in which the organiser of the work is the employer, we are witnessing an exception to the general rule in the Intellectual Property Law (which states that the employee is the original owner of the copyright and the employer the secondary one), since according to the rules concerning databases and compilations, the Law attributes to the employer, as the organiser, the original ownership of the copyright, irrespective of whether the employer is a natural or legal person.

Finally, it is important to remember that although the Law defines collective works in Article 5(c), it does not establish rules relating to the ownership of the copyright of these works, except for the particular case of works mentioned in Article 24(a) and (b), which expressly grants the original ownership to the organiser (natural or legal person).

2.2.7 Cinematographic Works

With regard to cinematographic works, Article 25 of the Law states that copyright belongs to its producer. In turn, Article 26 defines a cinematographic producer as 'a person, individual or company, who takes the initiative and responsibility to produce the cinematographic work'. However, Article 27 of the Law recognises as authors of a cinematographic work the individuals who perform the intellectual creation of the cinematographic work. Consequently, workers who are hired to provide their creative services for the realisation of a film are recognised by the law as co-authors.

Article 28 states that those authors who abandon their participation in the realisation of a cinematographic work, do not lose the rights that correspond to them for their participation in the film. Hence, they will be able to freely exploit and use their personal contribution. However, they cannot oppose the use of their contribution in the conclusion of the work.

In relation to the rights that authors involved (as workers) in the creation of a cinematographic work may have, Article 29 of the Law expressly states that:

> ... the contract between the authors of the cinematographic work and the producer include the assignment to him of all the work rights, and entitles the producer to show the work in public, present it on television, reproduce it in copies, transfer and lease it, without prejudice of the rights that the Law grants to authors of the used works and other collaborators.

That notwithstanding, Article 31 states that:

> ... the authors of the plot, of the music, of the lyrics, dubbing and of the work that eventually may have been the object of adaptation, will retain the right to use, separately, their respective contributions, as long as they have not agreed to use it exclusively for the production of the film.

Finally, the Law recognises a right of authors involved in the creation of a cinematographic work to be named in the cinematographic work. This particularly includes the name of the director, the authors of the staging of the original work, of the script adaptation, of the music and lyrics, and the main interpreters and performers.

2.2.8 Works Produced by Public Employees in the Performance of Their Duties

Article 88 recognises the ownership of the copyright to works produced by public employees in the performance of their duties, to the state, municipalities, government corporations, semi-public or autonomous institutions and other legal state persons, in which such public employees may provide their services. Additionally, Article 88 empowers the state entity, owner of the copyright, to release, by means of a resolution, any of these works, with the purpose that such works may become part of the common cultural heritage, so that they can be exploited freely.

However, this provision is not applicable to 'works developed in the context of the owned activity of public companies or in which the state participates, when the work has a strategic direction for their purposes or when it is expressly stated by the Law that established and regulates the entity'.

2.3 The Parties' Duty to Inform

In Chilean intellectual property legislation there is no express positive obligation to inform an employer of the creation of a work protected by copyright. However, such potential obligation may be governed by the employment contract and by the will of the parties.

2.4 Employee Remuneration Right

Regarding the employee's right to receive remuneration for the creations made in the context of an employment contract, the only express provision is the one that we have reviewed in the case of works by employee-authors of media companies, news agencies and news or radio stations or television stations, when the creations are used in publications of media companies related to the employer, but distinct from the latter, or by a third company.

In general, this is an issue that is not expressly regulated in Chilean legislation. Hence, the regulation of the employee remuneration will be found the agreement between the employer and the employee, considering that the Chilean Constitution guarantees the employees the right to receive a fair remuneration for their work.

3 EMPLOYEES' SEMICONDUCTOR CHIPS

Protection of semiconductors (integrated circuits) is to be found in Chile in the Industrial Property Law (No. 19,039), under the general provisions of Title VI, 'Inventions in Service', Articles 68 to 72.

4 EMPLOYEES' INVENTIONS

4.1 The Legal Framework

As noted throughout this presentation, the majority of inventions and investigations that occur in different fields in Chile are developed by private and public companies, or universities, by persons specifically hired for this purpose.

Industrial Property Law No. 19,039 was the first Chilean law to regulate this matter, and is currently governed by Title VI 'Inventions in Service', by Articles 68 to 72. These rules govern both the case of dependent workers, who maintain a relationship of subordination and dependence with the employer, and those who are responsible for the development of an invention through an intangible lease of services. This second case is not the object of this chapter and will therefore be omitted.

Given that Law No. 19,039 contains a complete section devoted to this subject, which is not framed within a specific type of industrial property protected by the Law, it is possible to conclude that these rules apply to all types of IP rights contained or provided for in the mentioned Law. Thus, these rules apply to all IPRs indicated in Article 1 of Law No. 19,039, i.e., trademarks, (invention) patents, utility models, industrial drawings and designs and layout-designs or topographies of integrated circuits.

For other types of industrial property rights that are not regulated in Law No. 19,039, such as plant breeders' rights, it would be possible to apply the provisions of Title VI, as long as the particular law that rules these property titles makes express reference to the provisions of Title VI.

4.2 Ownership of Rights

The general rule regarding the ownership of intellectual property rights is contained in Article 3(1) of the Industrial Property Law No. 19,039 Regulations,[7] which states that 'the power to obtain an industrial property right belongs to its true creator or inventor, to his heirs or assignees...'. In order to determine the ownership of the industrial property rights in relation to inventions developed under an employment contract, called *inventions in service*, the Chilean legislation provides for four different situations:

7. Decree No. 236 of the Ministry of Economy, Development and Reconstruction; Secretary of Economics, Building and Reconstruction, of 25 August 2005.

(a) inventions resulting from the inventive or creative activity of an employee, where the provision of services is specifically contracted by the employer.
(b) own inventions of the employee or *free inventions*.
(c) inventions not related to the provision of services contracted by the employer, but obtained with occasion of the provision of such services or by means of taking advantage of the means that the employer makes available to the employee.
(d) inventions of employees of universities or research institutions.

The Industrial Property Law does not distinguish, in contrast to the Copyright Law, between ownership of moral and the economic rights to the invention. Nevertheless, Article 11 of the Regulation of Law No. 19,039 requires to include 'the full name of the inventor', among the mandatory information to be included in all patent or utility model applications. In addition, if the applicant is not the inventor, it will be necessary to file the necessary document (assignment document) to support such a situation. On the other hand, Article 71 of the Law states that the rights established in favour of the employee in the articles of Title VI are waived before the grant of the patent, utility model, layout-design or integrated circuit topography, as appropriate, if there is no written contrary clause.

The following considers the specific types of inventions regulated by Title VI.

4.2.1 Inventions Resulting from the Inventive or Creative Activity of an Employee, Where the Provision of Services is Contracted by the Employer

The situation of these inventions is regulated in Article 68 of the Industrial Property Law, which states that:

> ... in employment and provision services contracts, the nature of which is the development of an inventive or creative activity, the right to apply for registration as well as any industrial property rights, shall belong exclusively to the employer or the person that has commissioned the service, unless there is a specific stipulation to the contrary.

The rule states that in employment contracts in which the object of the contract, i.e., the provision of services contracted by the employer, is actually the realisation of an inventive or creative activity, the ownership of the industrial property rights arising from the invention resulting from that activity and the faculty to request the protection of the invention is with the employer. Notwithstanding the foregoing, an exception is set in the case where the parties expressly agree that all or part of these rights correspond to the employee inventor.

The solution provided by the Law is consistent with the elements of alienation and dependence, characteristic of the employment contract. The employee actually provides their services on account and under the risk of the employer who provides their employees with the economic and technical means to perform their activities. Although the Law provides the ownership of the industrial property rights to the employer by the mere fact of agreeing to the provision of inventive or creative services, without further acts or provisions under which the employee assigns the ownership to the employer, Article 11 of the Law Regulations requires that the patent or utility model application should indicate the inventor's full name and the documents that confirm the transfer or assignment of rights, in those cases when the applicant is a person different than the inventor.

Finally, it has to be noted that Article 10(3) of the Chilean Labour Code requires among the employment contract, the details of the personal services to be provided by the employee. To fulfil this requirement it is necessary to define with certain detail the types of services to be provided by the employee, rather than establishing it generally and in broad terms. Thus, if the service is described as a creative or inventive activity, the agreement should indicate the specific kind of creations which will be object of the employee's activity.

Consequently, the ownership of the industrial property right by the employer exists only in respect of the employee's inventions resulting from the creative services defined in the contract and cannot be extended to other inventions made by the employee. This approach allows the existence of the other categories of inventions, as they are qualified by employee's scope of degree of inventive freedom and autonomy.

4.2.2 *Inventions Owned by the Employee or Free Inventions*

These inventions are regulated in Article 69(1) of the Industrial Property Law, which states that:

> ... a worker that, according to his employment contract, is not obliged to make an inventive or creative function, will be entitled to apply for the registration/ protection of his invention as well as any industrial property rights arising from his inventions, which belong exclusively to him.

In this case the inventions are made by an employee, who is not required to perform an inventive or creative function, and that takes place outside the field of business of the company and is unrelated to the employer. For this reasons, the Law recognises the employee as the exclusive owner of any industrial property right. This kind of invention is known in doctrine as a *free invention.*[8] Finally, it needs to be noted that to consider an invention of an employee as a free invention, the employee must not have used or benefited from the means and expertise available to them in the company. Additionally, the inventor must create their invention out of their normal scope of activity and for which they were hired, without any relation with the employer or the provision of services.

8. Alfredo Sierra Herrero, *Trabajadores Inventores: Titularidad de sus Creaciones y Régimen Retributivo*, 18(2) *Revista de Derecho Universidad Católica del Norte* 162, 171 (2011).

4.2.3 *Inventions Not Related to the Provision of Services Contracted by the Employer, but Produced in the Framework of it or Taking Advantage of the Media that the Employer Has Made Available*

Article 69(2) of the Industrial Property Law regulates this kind of invention. The Article states that an invention produced in a service of this kind requires the employee 'to carry out the invention has obviously benefited of the knowledge acquired in the company and has used the means provided by it'. Such inventions are known in doctrine as *mixed* or *exploitation inventions*[9] and are characterised by the fact that, although the employee has not been hired to the provision of inventive services, the employee has used for their invention the knowledge and means provided by the employer to the fulfilment of their ordinary duties. Therefore, for an invention to fall into this category, which turn out to be an intermediate category between the other two types, it is necessary that:

(1) the employee has not been hired to provide creative or inventive services;
(2) the employee has obviously benefited from the knowledge acquired in the company and has used the means that the employer has provided them with in order to fulfil their ordinary duties.

This last requirement implies that the employee must simultaneously be equipped with the knowledge and the means of the company, because if they only use one of these elements, the invention will belong entirely to them as a free invention.

With regard to knowledge, it must be based on new records, originally or unknown by the employee, acquired by them on occasion and during their provision of services in the company, and that are relevant to the obtaining of the invention as is required by Law.

In relation to the means, it is understood that it is the effective use and exploitation of instruments, tools, materials and other items that the employer offers to the employee in connection with the execution of their duties, and that obviously benefits them in the inventive process or as part of the final invention.

Regarding the ownership of intellectual property rights in these inventions, the rule provides that the faculties and rights belong to the employer as the original holder, in which case the employer must give the employee an additional remuneration as agreed by the parties.

Another relevant point in relation to mixed inventions is that - unlike the other two types of inventions - to determine than an invention falls into this category, it is necessary to assess the degree of utilisation and eventual relevance of the knowledge and resources of the employer for obtaining the invention. This appreciation may lead to conflicts between the employee and the employer.

9. Sierra Herrero, *supra* n. 8, at 163.

Finally, although the Law considers the employer as the original holder of the industrial property rights, it also establishes a benefit for the employee, who will be entitled to receive an additional remuneration from the employer, which is further discussed below.

The final paragraph of Article 69 extends this right of the employee to workers who have obtained an invention that exceeds the framework of the creative services hired by the employer.

4.2.4 Inventions of Employees of Universities or Investigation Institutions

This matter is regulated by Article 70 of the Industrial Property Law, which provides that:

> ... the right to request the relevant register as well as any industrial property rights resulting from the inventive and creative activity of employees hired in a dependent or independent relation by universities or research institutions included in Law Decree No. 1,263,[10] of 1975, belong to these entities, or who they may determine.

This notwithstanding, it is normal that the by-laws of these entities will regulate the ways in which the inventor or creator can benefit from the profits received from the exploitation of their work.

This Article grants universities and research institutions included in the Decree-Law No. 1,263 the right to apply for the title to inventions made by their employees. This situation is justified by the fact that universities and research institutions are the entities that mainly conduct research or create innovations in Chile, which become the object of protection by industrial property rights. In return, and in order to establish an incentive for the development of creative functions by employees of such institutions, the Article states that these entities must establish the manners in which the inventor or creator should obtain a share in the benefits obtained from the work. It has to be said that the main universities of Chile have internal rules that regulate these situations.

4.3 The Parties' Duty to Inform

Chilean law does not regulate an obligation to inform the employer about the creation of an invention. Notwithstanding this, the parties may agree to include such an obligation in the labour contract.

10. Organic Law Decree No. 1,263 of Financial Management of the State, this law lists a number of services and institutions that make up the financial management system of the State, within which there are: Geological Research Institute, Natural Resources Institute, Technology Research Institute, National Commission for Scientific and Technological Research, Agricultural Research Institute, among others.

4.4 Employee Remuneration Right

To analyse the employee's right to be paid for the result of their creative services it is necessary to distinguish the type of service invention in question. It should be noted that free inventions are excluded from this analysis, since the ownership of the invention belongs exclusively to the employee.

4.4.1 Employee's Remuneration for Inventions Resulting from the Inventive or Creative Activity of an Employee, Where Provision of Services Is Contracted by the Employer

The provisions governing such inventions do not refer to a payment for the employee's rights. In principle, the employee hired to provide inventive services, which have as their object the creation of an invention, will have the right to receive only the original remuneration agreed with the employer in the employment contract for the provision of the services, regardless of the result of this activity.

Notwithstanding the above, the parties may agree in the individual or collective employment contract that if the employee obtains the expected inventive result, the employee will be entitled to receive an additional remuneration. On the other hand, if the employee hired to develop an inventive activity obtains a result that exceeds the framework of their original activity, they will be entitled to receive an additional retribution, as the last section of Article 69 provides. The employee will have the right to receive this additional remuneration only if the invention resulting from their work exceeds the result expected by the employer or if they perform an inventive activity that exceeds the agreed services and which was not required under the employment contract, irrespective of whether this additional result benefits the employer.

The determination of the amount of this additional compensation is not regulated by the Law, so it has to be agreed by the parties. In my opinion, a possible criterion to consider in order to determine the amount of the compensation is the benefit and the importance that this 'unexpected' result has for the employer.

Finally, in relation to the nature of this additional compensation it is necessary to consider that it falls within the definition of remuneration given by the Chilean Labour Code in Article 41: as 'money compensations and additional compensations in species quantifiable in money that the employee must receive from the employer based on the employment contract'. The remuneration for this kind of invention is precisely a monetary compensation to be paid by the employer to the employee based on the employment contract, particularly for the result of the provision of the agreed services.

4.4.2 *Employee's Remuneration for Inventions Not Related to the Provision of Services Contracted by the Employer, but Produced in the Framework of It or Taking Advantage of the Media That the Employer Makes Available to the Employee*

In respect to this type of invention the employee is entitled to receive additional compensation, which has to be determined by the parties. The Law does not provide specific criteria to estimate the amount of the additional compensation.

5 EMPLOYEES' UTILITY MODELS

Employees' utility models are regulated under the general provisions of Title VI of the Industrial Property Act, Articles 68 to 72.

6 EMPLOYEES' DESIGN RIGHTS

Employees' design rights are regulated under the general provisions of Title VI of the Industrial Property Act, Articles 68 to 72.

7 EMPLOYEES' PLANT BREEDERS' RIGHTS

In Chile there are no express rules governing intellectual property rights of employees in relation to plant varieties obtained in the performance of their duties. As we saw earlier, it is not possible to apply the general provisions of the Industrial Property Law to this case. New plant varieties are regulated in the Law No. 19,342. Thus, the rights an employee may eventually have on new plant varieties must be agreed by the parties.

8 INSTANCES FOR DISPUTES

In order to analyse the instances for dispute it is necessary to consider that even though Chilean legislation contemplates rules in order to regulate this matter, it is not common that legal disputes arise in relation to the attribution of ownership of industrial or intellectual property rights between employees and employers. Consequently, there are few cases on this subject that have come to the attention of the Courts of Justice. However, it is necessary to distinguish conflicts that are generated by the attribution of ownership of intellectual property rights (i.e. copyright) from those generated by the attribution of industrial property rights (i.e. patents, trademarks).

8.1 Resolution of Conflicts Related to the Attribution of Intellectual Property Rights

There is not a specific procedure established in Chilean legislation, nor is the jurisdiction of a specific court imposed by the law for these matters. However, considering that these conflicts will be generated in the context of the existence of an employment relationship, which in Chile is guided by principles of special guardianship and protection of employees' rights, I believe that this matter should be resolved by the Labour Court.

8.2 Resolution of Conflicts Related to the Attribution of Industrial Property Rights

According to Article 72 of the Industrial Property Law, the competent court to hear this kind of dispute is the Industrial Property Court. However, the Law does not appear to determine the relevant type of procedure.

In order to remedy the omission of the Law, the Industrial Property Court issued an Agreed Order dated 26 October 2012, which provides that the prosecution of these issues must be subject to the rules governing the procedure of oppositions to a trademark registration, regulated under Part I, Paragraph 2 of Law No. 19,039. Additionally, it establishes that the Industrial Property Court will hear these matters in one instance, i.e. special complaint being only possible in limited circumstances and procedure before the Supreme Court.

Subsequently, on 10 June 2013, the Industrial Property Court revoked the Agreed Order on prosecution, and provided a new interpretation of Article 72 of Law No. 19,039. Under the new interpretation, the Court held that disputes regarding service inventions should be resolved in accordance with the general procedure of opposition, under which the competent Court of First Instance should be the Industrial Property National Institute ('INAPI'), and the Court of Second Instance should be the Court of Industrial Property. This new interpretation makes it possible to appeal a decision of first and second instance to the Supreme Court by filing an annulment, which seeks to correct errors of the law or interpretation of legal rules.

Without prejudice to the competent body established in Law to resolve conflicts in such cases, it is possible that the employment relationship may establish elements involving invention services with other labour issues, with the result that the Labour Courts will be deemed more appropriate. This can happen, e.g., if issues related to service inventions are linked to contractual breaches of the obligation of confidentiality or exclusivity established in the labour contract or other issues related to it.

9 CONCLUDING REMARKS

As we have seen, the situation of employees' creations and inventions is regulated in Chilean legislation. The law provides specific solutions in order to determine the ownership of industrial and intellectual property rights in different cases in the context of the employment relationship of the employer and the employee. The solution provided for by the law will differ depending on the specific characteristics of every case, and also depending on the existence or non-existence of a labour contract. Cases that are not expressly regulated by the law, will be governed by the principle of the autonomy of the will of the parties. On this last point, practice of the last years indicate that in the case of employment contracts concerning employees of the R&D groups of companies, and also in the case of employees of laboratories of universities, the labour contracts are every day more sophisticated. In addition, several institutions of this nature have drafted internal codes of conduct/rulings for IP, that establish clearly rights and obligations of the parties involved. In the case of Universities, some are deciding to apply these internal rulings for IP to the students who may have participation in new developments. Sometimes, this kind of documents simply 'repeat' the text of the Law, while in other cases, these texts are stricter than the provision of the Law.

BIBLIOGRAPHY

Table of Sources

Alfredo Sierra Herrero, *Trabajadores Inventores: Titularidad de sus Creaciones y Régimen Retributivo (Inventors Worker: Ownership of their Creations and Remuneration System), Revista de Derecho Universidad del Norte, Sección: Estudios* 151, 188 (2011).

Biblioteca del Congreso Nacional, Historia de la Ley 19,039.

Biblioteca del Congreso Nacional, Historia de la Ley 17,336.

Santiago Schuster, *Derechos de Autor en las Relaciones Laborales y su Vínculo con el Tratado de Libre Comercio entre Chile y Estados Unidos (Copyright in Labor Relationships and its Link with the Free Trade Agreement Between Chile and the United States)*, in *Homenaje a Arturo Alessandri Besa: Estudios de Derecho y Propiedad Intelectual (Tribute to Arturo Alessandri Besa: Law and Intellectual Property Studies)* 369, 384.

Instituto Nacional de Propiedad Industrial, *Derecho de Autor: consideraciones generales (Copyright: general considerations)*, at: http://www.inapi.cl/portal/institucional/600/w3-article-842.html (accessed 27 July 2017).

Instituto Nacional de Propiedad Industrial, *Derecho Moral (Moral Right)*, at: http://www.inapi.cl/portal/institucional/600/w3-article-844.html (accessed 27 July 2017).

Instituto Nacional de Propiedad Industrial, *Derecho Patrimonial (Economic Right)*, at: http://www.inapi.cl/portal/institucional/600/w3-article-843.html (accessed 27 July 2017).

Instituto Nacional de Propiedad Industrial, *Invenciones y creaciones surgidas en el seno de la relación laboral y durante la prestación de servicios (Inventions and creations arising within employment relationship and during the service provision)*, at: http://www.inapi.cl/portal/institucional/600/w3-article-837.html (27 July 2017).

Table of Legislation

Chilean Civil Code
Chilean Labour Code
Decree No. 236 of the Ministry of Economy, Development and Reconstruction; Secretary of Economics, Building and Reconstruction, of 25 August 2005.
Intellectual Property Law No. 17,336.
Intellectual Property Law No. 17,336 Regulation.
Industrial Property Law No. 19,039.
Industrial Property Law No. 19,039 Regulation.
Industrial Property Court Agreed Order dated 26 October 2012.
Industrial Property Court Agreed Order dated 10 June 2013.
Organic Law Decree No. 1,263 of Financial Management of the State.

CHAPTER 35
Colombia

José Roberto Herrera Díaz

1 BACKGROUND

Before entering into analysis of intellectual property aspects applicable to employees, it is necessary to understand the concept of 'employee' in Colombia. Colombian law does not provide a definition of 'employee'. This term has a very close relationship with the legal definition of a labour law contract. In this regard, the International Labour Organization ('ILO') has identified three common elements in labour law contracts around the world. These concepts are not always defined by the law or case law of a country.[1] Not only the ILO, but also the labour law doctrine in the majority of countries has established a consensus on the existence of three elements. The first element, called the *personal element*, means that only a natural person can be considered an employee.[2] Second is the *subordination element* by which the employer is entitled to give orders to the employees; this element involves a *hierarchical aspect*.[3] The third and final element is the *salary* that employers pay their employees as a remuneration for their work. Colombia is one of the countries in which said elements can be clearly found.[4] Article 22 of the Colombian Labour Law Code defines a labour law contract as: 'one in which a natural person promises to give his or her services to another legal or

1. OIT. Contratos de Trabajo, at: http://www.ilo.org/public/spanish/dialogue/themes/ce.htm#ci.
2. Efren Cordova, *Naturaleza y Elementos del Contrato de Trabajo México*. UNAM, Série B, EstudiosDoctrinales No. 188, ch. 16 at 302 (1997).
3. Likewise, the employee has to be free in their decision to work; there cannot be any form of forced labour. On this point see *id.*, at 306.
4. Cordova, *supra* note 2 at 306.

natural person, under continuous subordination [i.e., receiving orders from the employer], with remuneration in exchange for the services [translated by author]'.[5]

When these three elements are present, there is a presumption of the existence of a labour law contract, despite what the parties may have agreed in a different contract.[6] In Colombia the main distinction of a labour law contract and other contracts is the *subordination* element. In the course of services provided under civil law contracts, there are orders given to the person providing the services, but the autonomy that the provider of services has is broader than that in an employment contract. The term *employment* is not only linked to works created under a labour law contract.[7] In most jurisdictions the term employment is a generic term by which not only labour contracts are included, but also contracts by which a person agrees to provide services to another are included. Although the focus of this study is the employee's intellectual property rights in the narrow sense, meaning labour law contracts, some considerations will be made in relation to other kinds of agreements that are included within the generic concept of employment.[8] For instance, works created under contracts to provide services in Colombia are clear examples that show the broad understanding of the term employment. These other kinds of relationships can sometimes provide answers in ownership issues in the employees' intellectual property rights scenario.

Colombian law belongs to the continental law system by which the rules established in statutes are considered primary sources of law. Colombia belongs to the Andean Community and many of the provisions relating to intellectual property rights are established in the Andean Community Decisions No. 486/2000 and 351/1993 that are currently applicable for the member countries (Bolivia, Colombia, Ecuador and Peru). Many provisions are harmonised for the member countries, but the Community law leaves to the national law the regulation on the ownership of employees' intellectual property rights.

The main provisions concerning employees' intellectual property rights are found in intellectual property laws, rather than labour laws. Similar to other jurisdictions, determining the ownership of the intellectual property created by employees has been an issue open to different interpretations by the courts and legal doctrine. In relation to the ownership of works created by employees or independent contractors, Article 20 of the Author's Right Act of 1982 is the norm that establishes the rules on this matter. Regarding the industrial property rights created by employees, the Commercial Code was the applicable law in relation to employees' invention patents and utility models,

5. Original text of Art. 22 reads: 'Art. 22. – *Definición. 1. Contrato de trabajo es aquél por el cual una persona natural se obliga a prestar un servicio personal a otra persona natural o jurídica, bajo la continuada dependencia o subordinación de la segunda y mediante remuneración (Código sustantivo del trabajo)*'.
6. The theory by which the presence of the three said elements form a labour law contract was born in Germany in the first quarter of the previous century, soon after the first labour laws were enacted. For further information See El derecho del Trabajo en España. *Ideologías Jurídicas y Contrato de Trabajo* 127, 128, 129 and 130 (Instituto de Estudios Sociales, Madrid 1981).
7. For further information see *supra* note 2, at 309.
8. José Roberto Herrera Díaz, *Ownership of Copyright in Works Created in Employment Relationships: Comparative Study of the Laws of Colombia, Germany and the United States of America*, at: http://papers.ssrn.com/sol3/papers.cfm?abstract_id=1705516.

however there was uncertainty in relation to the ownership of other industrial property rights created by employees, and in 2011 the topic of industrial property rights created by employees or independent contractors was harmonised and is regulated in Colombia with Law No. 1450.

2 EMPLOYEES' COPYRIGHT

2.1 The Legal Framework

The theoretical foundation of the applicable system in Colombia, in relation to the protection of works, is based on the author rights system of the civil law tradition, by which legal persons are not considered authors and only flesh and blood authors may have this attribute. Regardless of that, Colombia pertains to the dualistic author rights system by which economic and moral rights are treated differently, whereupon the former may be assigned or licensed from the author to third parties and the latter cannot be transferred or licensed and are considered inalienable to the author. The economic rights to a work created by an employed author may be transferred to an employer by an explicit agreement or by a legal presumption in case of absence of a written agreement. This general legal framework is established in the Andean Community Decision No. 351 of 1993 and the Colombian Author's Rights' Law No. 23 of 1982.

2.2 Ownership of Rights

The general rule concerning the ownership of the moral and economic rights is that the flesh and blood author is the only one who may have this quality. Legal persons can only own authors economic rights, if these rights have been transferred to them. Therefore, the general rule is that the employer and employee may agree on the ownership of the economic rights exclusively, but not on the moral rights as they always pertain to the employed author.

Regarding collective works, Article 92 of the Author's Right Law No. 23 from 1982, established that the ownership of copyright in collective works created under an employment contract or a contract for services in which it is impossible to distinguish the individual contribution of each of the natural persons, shall belong to the publisher or the legal entity or the natural person for whose account and at whose risk the contributions have been made. Concerning works created by employees of the state, Article 91 of the Author's Rights Law establishes that the copyright in works created by public employees or officials in the exercise of the constitutional and legal obligations incumbent on them, shall be the property of the public body concerned. This provision shall not apply to lectures or talks given by professors. Moral rights shall be exercised by authors as long as such exercise is not incompatible with the rights and obligations of the public bodies concerned. There has been no major issue concerning the understanding of the ownership of the copyright in works made by public servants or collective works created in the course of employment. Disputes in Colombia regarding the ownership of the economic rights concern events where a private sector employer and employee did not agree on the ownership of said rights. Before June 2011 there were three different interpretations concerning the application of the previous Article 20 of the Author's Rights law that stated:

> When one or more authors during a contract to provide services make a work under the directions of a natural or legal person, the author or authors will only receive the fees that they agreed on by contract. Just by this fact, the author or authors transfer their rights on the work, without prejudice to Article 30 letters a) and b). (Author's translation)[9]

The following three interpretations were made of the previous Article 20 of the Author's Rights Law:

(1) Article 20 of the Author's Rights Law No. 23, 1982, is only addressed to independent contractors that were commissioned to create works. Therefore, employed authors still maintained their economic rights and a separate written agreement different from the labour law contract was required to assign those rights.[10]

(2) Article 20 addressed the hypothesis of the ownership of works created by employees and independent contractors. Thus, the economic rights are transferred to the employer (or commissioning party) in case of silence when there is a written agreement and the work is created in execution of employees or independent contractors duties.[11] This position was also supported with the existence of Article 10 of the Andean Community Decision that established that unless there is evidence to the contrary, and in accordance with national law, legal persons may exercise the economic rights of works created under employment or civil services agreements.

(3) Article 20 does not establish any legal presumption of economic rights transferral, and therefore, independent contractors and employees must sign a separate contract certified by a public notary in order to transfer these rights. This position was pronounced due to the existence of Article 183 of the Author's Rights Law that established that economic rights sale agreements must be recognised before a public notary.[12]

With the enactment of Law No. 1450 of 2011, which modified Articles 20 and 183 of the Author's Right Law, the mentioned controversy on the application of the legal presumption of the economic rights transferral was solved in the following way:

---

9. It is important to note that Art. 20 starts with the word 'authors'. This is without doubt an example of an author's right provision where there is no need to distinguish the transfer of the moral rights as it is implied that the moral rights are inalienable and the author or authors will retain them.
10. Opinion No. 1-2004-21716 of 31 December 2004 by the Colombian Copyright Office.
11. Opinion No. 1,538 of 23 October 2003, by the Colombian State Advisor Court (*Consejo deEstado*)
12. Opinion of Fernando Zapata Lopez, *El derecho de autor y la marca*, 2 *Revista la PropiedadInmaterial del Centro de Estudios de Propiedad Intelectual de la Universidad Externado deColombia* 19, 20 and 21.

Economic rights will be assigned to the employer if there is a written labour law agreement and if the work created by the employee is related to the usual activities that the employer was engaged in when the work was created. Despite the effort of the lawmaker to regulate this issue, the current provision is unfortunate for the economic interests of employed authors because it does not take into account if the work was created in execution of the employees' duties, as was stated in the previous Article 20. Instead, there is a broader requirement based on the usual activities of the employer, which in most cases include various activities and can be enough to transfer the economic rights in case of silence. One interpretation that may serve the interests of the employed author is related to the existence of a principle in the 'author's rights' theory by which the author may only transfer the economic rights that are agreed in a contract. Consequently, the legal presumption of Article 20 may be understood as a limited transferral of the economic rights that the employer exclusively needs for the execution of their usual activities and therefore not necessarily of all the economic rights.

The most recent revision of Article 20 does not refer to the exercise of the paternity and integrity moral rights of the employed author against the employer. On the contrary, in the new Article 20 this part is deleted and it does not establish any rule on the matter. The silence on the limitation of the moral rights may be understood as an implicit waiver of the moral rights and a new possibility for the employer to exercise these rights without the authorisation of the employee. The general principle concerning moral rights regulated in Article 30 of the Author's Right Law may prevent this situation as the Article establishes that moral rights cannot be waived. Perhaps the lawmaker deleted this part of Article 20 as it was enough that Article 30 established that moral rights cannot be waived. This inalienable characteristic of moral rights may even limit the exercise of economic rights that have been transferred to an employer. For instance, if the employee has the moral right of divulgation, they can prevent the employer from publishing the work with a probable indemnification. In certain types of works that are constantly modified (e.g. software), employees may prevent modifications by the employer. Due to these practical problems, it would be recommended to allow, in Colombian legal framework, implied limitations on the exercise of moral rights by employees against employers in certain types of works, such as software or media journals. Otherwise, the application of economic rights that are transferred to the employer would be impractical for certain types of works. The reality shows that employers sometimes exercise the moral rights of the employee as they publish or alter their works without the employee authorisation. It is important to note that after transferring the economic rights, the employee is not interested in claiming their moral rights or at least suing the employer for a moral right infringement. Therefore and in spite of the existence of the inalienable character of the moral rights, the lawmaker should authorise waivers for certain types of works that must be frequently updated or altered.

If we compare the regulation of works created by public servants and employees from the private sector, the first provision has an implicit waiver on the moral rights as Article 91 states: 'Moral rights shall be exercised by authors insofar as such exercise is not incompatible with the rights and obligations of the public bodies concerned'. This would be an additional reason for allowing the waiver of the moral rights in this scenario, even with the existence of Article 30 of the author rights law that clearly establishes that moral rights are inalienable.

Finally, regarding employees' duties to inform and remuneration rights, Colombian law and Andean Community Decisions do not establish any provisions. Practice shows that employees do not have remuneration different from their salary when they create copyrighted-protected works and the employer exercises the associated economic rights.

2.3 The Parties' Duty to Inform

There are no provisions governing the parties' duty to inform.

2.4 Employee Remuneration Right

There are no provisions governing the employee's remuneration rights regarding copyrights.

3 EMPLOYEES' SEMICONDUCTOR CHIPS

3.1 The Legal Framework

Article 87 of Andean Community Decision No. 486 of 2000 recognises the protection of the layout-designs of integrated circuits. The requirement for protection is that the layout-design must be original and in order to have this attribute, the layout-design must be:

> the result of intellectual effort on the part of its creator and is not commonplace in the integrated circuit industry. Where a layout-design consists of one or more elements that are commonplace in the integrated circuit industry, it shall be considered original if the combination of those elements, as a whole, meets that condition.

3.2 Ownership of Rights

Article 88 of the Andean Community Decision establishes that in case the designer has not agreed with the employer on the ownership of the layout-design of the integrated circuit, the respective right will belong to the employer:

> Where the layout-design has been created under a commission contract or service contract entered into for that purpose, or in the course of employment relations where the designer has it among his duties, the right to protection shall belong to the person who contracted the work or service, or to the employer, unless otherwise provided by contract. (Translation provided by WIPO)[13]

13. http://www.wipo.int/edocs/lexdocs/laws/en/can/can012en.pdf.

In June 2011 a new law (No. 1450/2011) was enacted in Colombia that established rules regarding the ownership of intellectual property rights. The changes regarding copyright have been described in section 2 above. Concerning the changes regarding industrial property rights, it is important to mention that Article 29 of the new Law establishes that unless otherwise agreed, the ownership of the industrial property rights created in the course of employment belongs to the employer, with the only condition that the labour law agreement is in written and not verbal form. This rule is compatible with the wording of Article 88 of the Andean Decision, with the only additional condition that the labour law agreement must be in written form. Law No. 1450 of 2011 harmonised all the previous provisions of the Andean Community Decision and the Commercial Code, that established separate rules on the ownership of industrial property rights created under the course of employment. The new Law created one exclusive rule of automatic transfer of the industrial property rights to the employer when there is a written labour law agreement. The Andean Community Decision 486 of 2000 establishes the rights that are considered industrial property and therefore all these types of rights should fall within the new legal presumption of Article 29 of Law No. 1450 of 2011. The provision has received some criticism among industrial property practitioners who believe that the fact that all industrial property rights must be owned by the employer is too broad. Some consideration in section 4 on patents will be made on this point as there should be special rules for this scenario.

There is no employee's remuneration right or employee's duty to inform regarding layout-designs of integrated circuits.

3.3 The Parties' Duty to Inform

There are no provisions governing the parties' duty to inform.

3.4 Employee Remuneration Right

There are no provisions governing the employees' remuneration rights regarding Semiconductor Chips.

4 EMPLOYEES' INVENTIONS

4.1 The Legal Framework

The following three provisions contain rules regarding employees' inventions:

> Article 23 of the Andean Community Decision No. 486 of 2000:
> Without prejudice to the provisions of the national legislation of each member country, in the case of inventions occurring in the course of employment relations, the employer, whatever its form and nature, may transfer part of the economic benefits deriving from the inventions to the employee inventors with the purpose of promoting research activity.

It is important to note that this provision essentially leaves it to the Member States to regulate the topic, and that is also applicable to utility models by virtue of Article 85 of the Andean Decision.

Article 539 of the Commercial Code and Article 29 of Law No. 1450 of 2011:

> Employee's inventions and utility models were regulated by Article 539 of the Commercial Code of 1971. Said norm established that unless otherwise agreed, inventions created by employees belong to the employer. This provision was in accordance with the later Article 23 of the Andean Decision which stated that unless otherwise agreed, the invention belongs to the employer. The condition that the Commercial Code established for considering the property of the employer was that the employee must be hired for investigation or the invention must be related to the function for which the employee was hired. This Article created the possibility of remunerating the employee based on the following three aspects: employee's salary, importance of the invention in the market, and economic benefits reported to the employer after exploiting the invention in the market. Finally, the norm determined that in case the employer and the employee did not agree on the amount of the remuneration, said quantity must be fixed by a judge.

There is an issue regarding the scope of the mentioned provision, as Article 29 of Law No. 1450 of 2011 has established that unless otherwise agreed, the industrial property created by an employee belongs to the employer if there is a written and non-verbal labour law agreement. Invention patents are a type of industrial property right and one interpretation may be that the new Article 29 replaced the old provision of Article 539 of the Commercial Code, despite there not being an express provision by which the new Article derogated the previous one.

Another possible interpretation that would be more accurate may be that Article 29 only replaced the old Article 539 in the aspects that were regulated by the new provision. In other words, the new provision abolishes the requirement that the employee must be hired for investigation or the invention must be related to the function by which the employee was hired. With the current provision only a labour law agreement and silence of the parties in relation to the ownership of the invention will be needed in order to declare the employer as the patent owner. An aspect that was established by Article 539 in relation to the possibility to remunerate the employee would be maintained, as the new Law remains silent with regard to the employee's remuneration right and did not expressly derogate Article 539.

4.2 Ownership of Rights

Given the introduction on applicable law, it may be stated that the current situation in Colombia is that the ownership of invention patents and utility models belong to the employer if:

- the employee and employer have not agreed on the contrary;
- there is a written labour law agreement.

The previous conditions established by the Commercial Code are only applicable in relation to the possibility to remunerate the employee for creating the invention patent or the utility model.

4.3 The Parties' Duty to Inform

There are no current provisions or bills related to the parties' duty to inform.

4.4 Employee Remuneration Right

Article 539 of the Commercial Code establishes a remuneration right in which the following aspects should be studied in order to calculate the amount of remuneration:

- the employee's salary;
- importance of the invention on the market;
- economic benefits reported to the employer after exploiting the invention on the market.

The regular practice in Colombia is that the employees who create invention patents or utility models do not receive payments additional to their salaries. There is no significant case law by which employees can claim the payment of the remuneration right from the employer. A reason why the remuneration right is not currently in existence in the legal framework may be that there are no legal sanctions against the employer (patent owner) when the remuneration right has not been recognised to the employee. There are no special actions to protect the right, which is due to the fact that the norm that established this provision was the Commercial Code and not an industrial property right statute, such as Andean Decision No. 486 of 2000. A new bill under an industrial property law should be created to establish clear rules on the remuneration rights and more importantly, legal rights to protect it against the employer.

5 EMPLOYEES' UTILITY MODELS

See section 4 above.

6 EMPLOYEES' DESIGN MODELS

6.1 The Legal Framework

Andean Community Decision No. 486 establishes in Title V the provisions regarding industrial designs. Article 114 provides that the rights on the industrial design belong to the designer and that legal and natural persons may own the rights on the design.

There are no provisions related to the designs created by employed authors in this Decision. Article 29 of Law No. 1450 of 2011 establishes a general the rule on this matter.

6.2 Ownership of Rights

Article 29 of Law No. 1450 of 2011 establishes a general rule by which all industrial property rights created by employees belong to the employer if there is a written labour law contract. Because industrial designs fall within industrial property rights and therefore it is enough in Colombia that there is a written labour law contract despite the (potential) absence of an agreement on the ownership of the design between the employer and the employee, in order to declare the employer as the owner of the design.

6.3 The Parties' Duty to Inform

There is no party duty to inform.

6.4 Employee Remuneration Right

No remuneration right has been established.

7 EMPLOYEES' PLANT VARIETIES

7.1 The Legal Framework

Plant varieties are recognised by Andean Decision No. 345 of 1993 to protect the rights of breeders of new plant varieties by the grant of breeders' certificates. There are various national provisions that provide various rules on plant varieties. The main provisions are Decree No. 533 of 1994, Resolution No. 1893 of 1995 and the UPOV Convention from 1978.

Article 14 of the mentioned Andean Decision establishes that a natural person or legal entity can be the owner of a plant variety. However, there are no provisions in the Decision specifically oriented on plant varieties created under a labour law agreement. The only provision that may be found in the Decision concerning this is Article 15, which establishes the possibility for the state employer to transfer the benefits of the registered plant variety to its breeder employees in order to stimulate research activity. However, this is only an option that the state employer has, but it does not bring a clear answer regarding the ownership of a variety created under a labour law agreement.

Article 1 of Resolution No. 1893 of 1995 establishes certain definitions in line with the UPOV Convention of 1978, in which the employer may be considered the owner of a plant variety created by an employee. This is only a possibility and there is not a specific rule establishing the ownership for the employee or employer. For certain commentators the employer is always the owner of a plant variety created by an employee,[14] both in the public and private sector; however I do not share this opinion as the law is not clear on this point.

7.2 Ownership of Rights

There are no clear provisions that establish the ownership of rights of plant varieties created by employees. One possible interpretation may be that Article 29 of Law 1450 of 2011, which refers to a transfer of industrial property rights, may be applicable to plant varieties. This would not be the most adequate interpretation as plant varieties are not considered as a type of industrial property right or a patent in Colombia, but as a *sui generis* right. There are no clear rules on the ownership issue on plant varieties created by employees in Colombia.

7.3 The Parties' Duty to Inform

There is no party duty to inform.

7.4 Employee Remuneration Right

There is no mandatory remuneration right. The only possibility that the law establishes refers to plant varieties created by public servants. In this case the state may give some economic benefits to the public servant as Article 15 of Andean Decision No. 345 provides.

8 INSTANCES FOR DISPUTES

Intellectual property disputes are decided in Colombia by civil, criminal or administrative courts. Labour law matters are decided by labour law courts. In 2012, new judicial attributions were given to the Copyright Office and the Industrial Property Office in Colombia. The parties that have civil litigation concerning intellectual property can choose between special judges from the national intellectual property law offices, or civil court judges (which are, however, not always knowledgeable on intellectual property matters as they handle all type of civil law disputes). Another instance for dispute resolution concerning intellectual property matters in Colombia is the Arbitration Panel for Intellectual Property Matters from the Chamber of Commerce of Bogota. In this panel there is a list of arbitrators exclusively dealing with intellectual property matters. Conflicts concerning employees' intellectual property rights do not have relevant case law because civil and criminal IP litigation in Colombia has mostly been related to counterfeiting. There are no special panels for employee invention

14. See Pablo Felipe Robledo Del Castillo, *Protección de los derechos de obtentor de variedades vegetales en Colombia*, 3 *Revista la propiedad Inmaterial* 8 (2001).

cases like it exists in other jurisdictions. The new attributions given to the intellectual property offices in Colombia may help in the future to increase the IP litigation and probably to solve conflicts regarding employees' intellectual property rights. The only case and recommendation that have been decided in Colombia, which are related to the topic under study, are from the Constitutional Court and State Adviser Court, were certain topics considering the legitimacy of laws on employees copyrights were decided[15] and recommended[16] respectively.

Finally, it is important to mention that Colombian law does not provide a specific provision that indicates whether a labour or civil law court will be the competent institution to decide a dispute related to employees' intellectual property rights. Despite these two branches of law being much focused on legal theory, the intellectual property provisions are considered more special and technical. This is why the civil law courts or intellectual property offices acting in a judicial capacity will be the competent institutions to decide a conflict of this nature.

9 CONCLUDING REMARKS

As regards employees' copyright, as in many other different jurisdictions of the common law or civil law system, the topic of employees' intellectual property rights has not been conciliatory in the legal doctrine. Different rules and different interpretations have been created around this topic without clarity on the current state of issues. There have been some advances concerning employees' copyright rules in Colombia and recent provisions from 2011 have diminished the rights of employed authors. One of these provisions relates to the case in which the employer and employee remain silent in a written labour law agreement on the ownership of the copyright created by an employee. In this event it is enough that the work relates to the usual activities at the time of creation of the work in order to transfer the author economic rights to the employer. This is an unfavourable provision to the employed author as it does not take into account if the work relates to the functions for which they were hired, but instead to a very broad concept – the usual activities of the employer. The silence of the new Law No. 1450 of 2011 regarding the moral rights that the employee may exercise against the employer that owns the economic rights is a way in which the employee may stop the employer from exploiting their work without authorisation.

The moral rights in Colombia are unalienable and there is no possibility to waive them, but the law maker should allow the moral right waivers at least in certain type of events, such as with employees that produce computer programs or media journals for which it is impractical to allow the exercise of the moral right of divulgation by the employed author. The lawmaker should allow waiver of the moral rights for certain

15. Constitutional Court Judgment No. C-276/1996.
16. Opinion No. 1,538 of 23 October 2003, by the Colombian State Advisor Court (*Consejo de Estado*).

kinds of works in which the employed authors are not interested. Once the economic rights are assigned to the employer, the author loses the moral interest to their work in some type of works.

The 'author's rights' legal tradition establishes that moral rights cannot be waived, but the lawmaker should update the legal system to make it applicable in practice. For instance, countries like Japan have even accepted the transfer of the moral rights to the employer in certain type of works such as software. In theory this type of provision would be completely against the author's rights, but it is highly recommended to create initiatives that at least allow the waivers of some moral rights for certain kind of works. Creative commons licences are becoming popular in civil law countries, based on the inflexible character of the classical author's rights theory. However, with some amendments to the civil law system concerning moral rights, certain situations may be solved without the need to bring creative common rules. This proposal does not mean the elimination of the moral rights that are completely necessary in much type of works, but establishing new rules on a case-by-case basis (e.g. allowing moral rights waivers in computer programs created by employees) may help solve many issues concerning moral rights.

In Colombia there are no provisions in relation to employees' remuneration rights in copyrighted works. There should be rules allowing a remuneration right at least for certain type of works in which the benefits obtained by the employer are clearly significant in comparison to the salary that is paid to the employed author.

Regarding employees' industrial property rights the absence of case law in this topic is a consequence of the fact that all the industrial property rights created by employees belong to the employers. The general rule in case of silence of the employee and employer on the ownership of the industrial property rights is that these rights belong to the employers. There are very few events in which employees agree for the employer to be the owners of the industrial property rights. This is due in part to the classic labour law principle by which all creations of the employee belong to the employer and that the salary is a fair remuneration for the services rendered. This is the general trend, but Colombian law should create at least certain types of incentives to the employed inventors that have been only remunerated by their salary. There are valuable inventions that report amazing economic benefits to employers and employed inventors are only remunerated in many cases with a minimum wage.

Article 539 of Colombia's Commercial Code acknowledges a remuneration right that has not been applicable in practice due to the fact that it does not have any sanction or legal action in case the employer does not recognise it. There should be a new bill by which a remuneration right for creativity is awarded to the employed inventor, at least for invention patents. This right should provide legal action by which the employee may exercise the right against the employer; otherwise it would be difficult to make it a clear incentive. There are different countries that have a specific law on this point such as Germany and Sweden. These laws, which have produced an incentive to innovation, could be a guide to a future bill in Colombia.

It is a big mistake in Colombia that all the industrial property rights created by employees are ruled by the single provision of the 2011 framework (Law No. 1450, Article 29). According to this, all rights are transferred to the employer in case of silence. Invention patents cannot be ruled by this provision as there are many hypotheses that need to be detailed such as the parties' duty to inform that is non-existent in Colombia, or the relation between the invention and the tasks for which the employed inventor was hired. With the current law, a written labour law agreement and that the employee and employer did not agree on the ownership of the industrial property rights, is enough to transfer the rights to the employer. These broad provisions clearly diminish the interest of the employed inventors that in many events do not create inventions related to the functions for which they were hired.

There is no clear answer that determines the ownership of plant varieties created by employees. Part of the doctrine establishes that it belongs to the employer. The law only establishes the possibility that the employer owns this right without a clear rule. A new bill is also suggested to provide a clear framework on this topic.

BIBLIOGRAPHY

Table of Sources

OIT. Contratos de Trabajo, at http://www.ilo.org/public/spanish/dialogue/themes/ce.htm#ci.

Efren Cordova, *Naturaleza y Elementos del Contrato de Trabajo*. México, UNAM, Série B, EstudiosDoctrinales No. 188 (1997).

El derecho del Trabajo en España, *Ideologías Jurídicas y Contrato de Trabajo* (Instituto de Estudios Sociales, Madrid 1981).

José Roberto Herrera Díaz, *Ownership of Copyright in Works Created in Employment Relationships: Comparative Study of the Laws of Colombia, Germany and the United States of America*, at: http://papers.ssrn.com/sol3/papers.cfm?abstract_id=1705516

Opinion No. 1-2004-21716 of 31 December 2004 by the Colombian Copyright Office.

Fernando Zapata Lopez, *El derecho de autor y la marca*, 2 *Revista la Propiedad Inmaterial del Centro de Estudios de Propiedad Intelectual de la Universidad Externado de Colombia*.

Pablo Felipe Robledo Del Castillo, *Protección de los derechos de obtentor de variedades vegetales en Colombia*, 3 *Revista la propiedad Inmaterial* (2001)

Table of Legislation

Commercial Code, Article 539.

Labour Law Code, Article 22.

Andean Community Decision No. 486 of 200, Articles 23, 88, 265.

Andean Community Decision No. 351 of 1993, Article 10.

Author's Right Law No. 23 of 1982, Articles 20, 91, 92, 183.

Law 1450 of 2011, Articles 28, 29 and 30.

Table of Cases

Constitutional Court Judgment No. C-276/1996.
Consejo de Estado, Opinion No. 1,538 of 23 October 2003, by the Colombian State Advisor Court.

CHAPTER 36

Ecuador

*Maria Rosa Fabara Vera, Pablo Fabara Vera & Gerardo Naranjo Ormaza**

1 BACKGROUND

Historically, intellectual property law has been divided into two categories: (i) copyright and related rights, and (ii) industrial property rights. The latter includes, *inter alia*, inventions and distinctive signs. Ecuador adopted this categorisation in its law, adding plant varieties breeding rights as an additional intellectual property branch.

Intellectual property protection in Ecuador includes patents, utility models, industrial designs, trademarks, trade names, commercial slogans, geographical indications, copyright and related rights, integrated circuit designs, plant varieties breeding rights; additionally, the protection of indigenous communities has been included as part of the recently adopted legislation.

In Ecuador intellectual property is regulated both by its domestic legislation and by the Andean Community legislation at a supranational level; it must also comply with a number of international treaties and agreements under the WIPO framework (among others, the Berne Convention for the Protection of Literary and Artistic Works, the Paris Convention for the Protection of Industrial Property, Patent Cooperation Treaty (PCT), etc.), the WTO (Agreement on Aspects of Intellectual Property Rights Aspects regarding Trade - Trade-Related Intellectual Property Agreement (TRIPS Agreement), or Annex 1C of the 1994 Agreement establishing the WTO), as well as within the framework of free trade agreements with certain countries and / or regional blocks (Ecuador - US FTA).

Moreover, it is worth mentioning that Ecuador, by virtue of being an Andean Community member (CAN), shares legislation with other member countries

* This chapter is based upon work prepared by Luis Marín Tobar.

(Colombia, Peru, and Bolivia) in certain matters for which these countries have yielded their legislative competence to the Andean Community Commission; among them, intellectual property.

Thus, in intellectual property matters, essentially there is: Decision 345 of the Cartagena Agreement Commission, to regulate the Common Regime to Protect the Rights of Plant Varieties Breeders, Decision 351 of the Cartagena Agreement Commission regulating The Common Regime on Copyright and Related Rights, and Decision 486 of the Andean Community Commission, which regulates the Industrial Property Common System.

Domestically, Ecuador published in the Supplement of Official Register No. 899 of 9 December 2016, new legislation on intellectual property, known as, the 'Código Orgánico de la Economía Social del Conocimiento' (The Social Economy of Knowledge Code) (hereinafter 'COESC'). This Code repealed the previous Intellectual Property Law (Supplementary to Official Register No. 426 of 28 December 2006) in aspects related to intellectual property[1], and it established a National Competent Authority[2] for matters concerning the rights. As a result, the Ecuadorian Institute of Intellectual Property will be replaced, however it will remain operative until the new Competent National Authority is appointed through an Executive Decree signed by the President. Currently, through Executive Decree 1435 (Organic Code of Social Economy of Knowledge Regulation), the name and the general organisational framework of the new Competent Authority, the National Service of Intellectual Rights (SENADI), was established. A detailed regulation of its tasks and duties, or of its organisation, is not available as of July 2017, nor has it started to operate as such.

This Code anticipates a radical change in the protection of intellectual property. The acquisition and exercise of intellectual property rights will be balanced against the effective enjoyment of rights to health, nutrition, education, information, access to culture, participation in scientific development, economic activities development, freedom to work, access to goods and quality services and the right to other forms of ownership. Intellectual property rights related to goods that guarantee fundamental rights will be of public interest and their protection must maintain societal needs, uses may be authorised without the holder's consent.

Intellectual property rights related to strategic sectors, and its protection will be limited and will allow the Government access to such rights on grounds of public, social or national interests. It also contemplates Government participation in compensation for rights involving biodiversity. It also envisages communities' participation regarding rights that involve traditional knowledge. This Code establishes regulation for all intellectual property matters with various modifications to the regime established in the Intellectual Property Law.

It should however be noted that provisions established in the COESC will be applied in a manner compatible with CAN Decisions, in agreement with the Second General Provision of this regulatory body. This notwithstanding, the domestic legal system must in any case be compliant with Andean Community regulation, since the

1. The previous framework does regulate other aspects.
2. Within the National Secretariat for Higher Education, Science and Technology (SENESCYTP).

Andean Community legal order has certain characteristics imposing: (i) supranationality - body of international law to which sovereign powers are assigned, whose acts have binding effects; (ii) intangibility - Governments cannot ignore CAN competence acts; (iii) binding nature - pursuant Governments decision to surrender part of their sovereignty; (iv) supremacy, prevalence or pre-eminence - community law prevails over domestic legislation; (v) direct application - it does not require incorporation into domestic law for its enforceability; (vi) immediate effect - option to demand compliance with community regulation; and, (vii) autonomy - Andean Community bodies have independence from countries authorities.

With this in mind, it is worth mentioning that Andean Decisions do not as such regulate the acquisition and exercise of intellectual property rights for which reason the provisions of COESC, which regulate the treatment of intellectual property rights in the context of employee-employer relationships, apply in their entirety.

2 EMPLOYEES' COPYRIGHT

2.1 The Legal Framework

By virtue of Article 30 of Decision 351 of the Cartagena Agreement Commission, (the applicable supranational law), provisions related to the assignment or concession of property rights and user licences for protected works shall be governed by domestic legislation of the Member Countries. Consequently, rights to subject matter created in employment contexts are governed by such legislation. In relation to copyright-protected subject matter, these rules are found in Section III Chapter III of Title II COESC.

2.2 Ownership of Rights

Under Article 115 COESC, unless otherwise agreed upon or if there is a special provision, the ownership of works created in employment relationships or upon request, will belong to its author. If the author transfers their rights, they shall retain the possibility to exploit the works in a different way than that which was initially contemplated under the contract, if they do so in good faith and do not unreasonably harm the normal exploitation undertaken by the employer or principal.

2.3 The Parties' Duty to Inform

Under the current legislation, there are no provisions specifically governing the parties' duty to inform; however, in practice, to allow the distribution of economic profits from the exploitation of works, the Authority may, upon a party request and with due justification, through an Administrative Tutelage, request such information by virtue of Article 567 COESC.

2.4 Employee Remuneration Right

The author is entitled to the inalienable right of fair payment for the exploitation of their work. While this right is not applicable to computer programs, it is applicable even in cases of transfer of rights or of ownership of a work created in the context of an employment relationship and by commission.

3 EMPLOYEES' SEMICONDUCTOR CHIPS

3.1 The Legal Framework

By virtue of Article 276 of the Andean Community Commission Decision 486 (the applicable supranational legislation), matters pertaining to Intellectual Property (not included under the previously mentioned Decision) shall be regulated by the domestic legislation of the Member Countries.

Hence the rights to semiconductor chips are governed by Section II Chapter IV of Title III COESC concerning Industrial Property.

3.2 Ownership of Rights

According to Article 329 COESC, industrial property rights to semiconductor chips made by an employee within an employment contract are regulated in the same way as rights to inventions (patents), discussed in section 4 below. Thus, the power to acquire the right to an integrated circuit layout developed in fulfilment of a contract belongs to the principal or the employer, unless otherwise agreed, but the inventors shall be entitled to not less than 25% of the ownership value.

3.3 The Parties' Duty to Inform

Under the current legislation, there are no provisions governing the parties' duty to inform; however, in practice, to distribute the economic benefits of the exploitation of industrial property rights, the Authority may, upon a party request and with due justification, through an Administrative Tutelage, request such information by virtue of Article 567 COESC.

3.4 Employee Remuneration Right

Since inventors cannot receive a percentage lower than 25% of the value over ownership rights, they are entitled to receive royalties for the exploitation of the invention corresponding to that percentege. The same will apply to semiconductor chips.

4 EMPLOYEES' INVENTIONS

4.1 The Legal Framework

By virtue of Article 276 of Decision 486 of the Andean Community Commission, (the relevant supranational legislation), matters regarding Intellectual Property not encompassed in the Decision shall be regulated by the domestic legislation of the Member Countries.

Hence the rights to inventions created in employment contexts are governed by Section II of Chapter II Title III COESC.

4.2 Ownership of Rights

Article 277 COESC establishes that the power to acquire the right to a (patented) invention created under an employment contract belongs to the principal or employer, unless otherwise stipulated. The rights may be transferred or licensed by prior agreement.

The provision further stipulates that when an employment contract does not demand from an employee to undertake their own inventive activity, but instead to make an invention based on the employers' resources or confidential information, the ownership shall belong to the employee. Yet the employer will be entitled to an exclusive, non-transferable licence and free of charge.

Conversely, if an employment contract does not require of the employee to undertake their own inventive activity, and the invention is made without using the employers' resources or confidential information, the ownership will belong to the employee.

Finally, when an employee or commissioner does not directly participate in an inventive process, the ownership will belong only to those who were involved in such process.

4.3 The Parties' Duty to Inform

Under the current law, there are no provisions regulating the parties' duty to inform; however, in practice, with the aim to distribute economic benefits over patents exploitation, the Authority may, upon a party request and with due justification, through an Administrative Tutelage, request such information by virtue of Article 567 COESC.

4.4 Employee Remuneration Right

Inventors are entitled to a percentage over the patent of not less than 25% of the ownership value. Hence inventors are entitled to royalties arising from the exploitation of the rights, which correspond to that percentage.

5 EMPLOYEES' UTILITY MODELS

5.1 The Legal Framework

By virtue of Article 276 of Decision 486 of the Andean Community Commission, (the relevant supranational legislation), matters regarding Intellectual Property not encompassed in the Decision shall be regulated by the domestic legislation of the Member Countries.

Hence the rights to utility models created in employment contexts are governed by Chapter III of Title III COESC.

5.2 Ownership of Rights

By virtue of Article 324 COESC, utility models are subject to the same provisions regulating patents, hence ownership over utility models made by an employee in the context of an employment contract will be regulated in the same way as invention patents. In this regard, the rights to utility models developed pursuant to an employment contract belong to the principal or the employer, unless otherwise agreed.

5.3 The Parties' Duty to Inform

Under the current law there are no provisions regulating the parties' duty to inform; however, in practice, with the aim to distribute economic benefits over the work´s exploitation, the Authority may, upon a party request and with due justification require information through an Administrative Tutelage by virtue of Article 567 COESC.

5.4 Employee Remuneration Right

Since inventors cannot receive a percentage lower than 25% of the value of an invention; inventors are entitled receive the royalties arising from their exploitation in this percentage. The same will apply to utility models.

6 EMPLOYEES' DESIGN RIGHTS

6.1 The Legal Framework

By virtue of Article 276 of Decision 486 of the Andean Community Commission, (the relevant supranational legislation), matters regarding Intellectual Property not encompassed in the Decision shall be regulated by the domestic legislation of the Member Countries.

Hence the rights to industrial designs created in employment contexts are governed Section II Chapter V of Title III COESC.

6.2 Ownership of Rights

As per Article 350 COESC, industrial design rights made by an employee under an employment contract are regulated in the same way as invention patents are. Hence rights to industrial designs created under a contract belong to the principal or employer; unless agreed otherwise.

6.3 The Parties' Duty to Inform

Under the current legislation, there are no provisions governing the parties' duty to inform; however, in practice, to allow the distribution of economic profits from work exploitation, the Authority, with due justification, could, through an Administrative Tutelage, request such information, in agreement with COESC Article 567

6.4 Employee Remuneration Right

Since inventors are entitled to a percentage not lower than 25% of the value of an invention, inventors are entitled receive royalties arising from its exploitation corresponding to that percentege. The same will apply to industrial designs.

7 EMPLOYEES' PLANT VARIETIES

7.1 The Legal Framework

By virtue of the second transitional provision of Decision 345 of the Cartagena Agreement Commission (the applicable supranational legislation), breeders' rights shall be regulated by the domestic legislation of the Member Countries.

Hence, breeding rights to plant varieties created in employment contexts are governed by Section II of Title IV COESC.

7.2 Ownership of Rights

By virtue of Article 482 COESC, breeding rights to plan varieties created by an employee under an employment contract are regulated in the same way as invention patents are. Therefore, the rights to plant varieties created under an employment contract will belong to the principal or employer, unless otherwise agreed.

7.3 The Parties' Duty to Inform

Under the current law there are no provisions regulating the parties' duty to inform; however, in practice, for purposes of distributing economic profits from works exploitation, the Authority may, at a party´s request and with due justification, through and Administrative Tutelage request such information in agreement with Article 567 COESC.

7.4 Employee Remuneration Right

Similarly to inventors, breeders are entitled to a percentage of not less than 25% of the ownership value, and are entitled to receive royalties arising from its exploitation corresponding to that percentege.

8 INSTANCES FOR DISPUTES

Claim procedures or actions, or enforcement of employees' intellectual property rights are the same as that of any rights holder in Ecuador. Among the most common are: (i) initiating an administrative process before IEPI´s National Directorate; i.e. before the National Directorate of Industrial Property, the National Directorate of Copyright, or the National Directorate of Plant Varieties Breeders; (ii) filing a demand for preventive and precautionary measures, and, (iii) filing a cost and damages claim and initiation of civil proceedings.

An employee seeking compensation or damages resulting from infringement of their rights can seek redress of damages either directly through civil court, or through a preceding administrative action whereby the authority recognises the infringement and provides the legal basis for a direct damages claim in civil court. The civil, ordinary damages action is brought under the Ecuadorean Civil Code and the Code of Civil Procedure. A labour judge would not be empowered to settle a labour dispute where it must first settle an IP infringement or ownership of IP rights dispute. A decision from a civil court could be appealed before the Provincial Court of Justice, and finally through a cassation recourse.

It is relevant to note that there is a possibility to use alternative conflict resolution methods, such as: extrajudicial agreements, mediation or arbitration. These alternatives are noticeably faster, however the parties ought to meet a number of requirements; for instance the parties can submit the dispute to alternative resolution either through a pre-conflict arbitral clause or by reaching an agreement to that effect once the conflict has ensued.

9 CONCLUDING REMARKS

With the exception of computer programs, for all types of intellectual property the legal framework provides for remuneration for the exploitation of the relevant subject matter.

However, in copyright the author is deemed to be the owner unless otherwise has been agreed and if there is an agreement to the contrary, authors have the inalienable right to obtain an equitable remuneration for the exploitation of the work … . For the remaining intellectual property, the employer is deemed as the owner, while the inventors and breeders receive a 25% ownership right, and are entitled to royalties corresponding to such a percentage.

As a final remark, it should be mentioned that the Labour Code includes provisions concerning the non-disclosure of technical secrets held by the employer to which the employee has access.

CHAPTER 37

Mexico

Jorge Gomez, Liliana Arellano & Jonathan Rangel

1 BACKGROUND

Employees' work is a significant asset for every company. Therefore, adopting clear and protective measures in connection with the IP created by employees becomes one of the first concerns of every employer, mainly for a company in which innovation is the main business.

In such regard, the ownership of inventions and copyrights created under the scope of an employment relationship involves the convergence of three different bodies of laws: the Mexican Copyright Law, the Mexican Industrial Property Law and the Federal Labour Law.

Moreover, such convergence between the laws, plus the fact that different IP rights are subject to different protection as well as the grey areas of the relevant laws, notably add complexity to these sorts of cases. Thus, it is important to know the difference between IP created by employees, and that resulting from a 'work for hire' relationship; a clarification that may lead to an excellent employment relationship or long and costly litigation.

2 EMPLOYEES' COPYRIGHT

2.1 The Legal Framework

In Mexico employee's copyright is governed by the Federal Law on Copyright (FLC), which provides in Article 84 the rules applicable to works created under an employment relationship,[1] while Article 103 concerns the ownership of rights of computer programs created by employees.[2]

In addition, Article 46 of the Regulations of the Federal Law on Copyright (Regulations) provides the rules to be applied to works created in the official service of the Federation.

2.2 Ownership of Rights

Article 84 of the FLC provides that in case of a work made in the course of an employment relationship (under a written individual employment contract), the rules that have to be applied concerning the ownership of rights are the following:

If there is an intellectual property clause, it will apply. Some individual employment contracts include an intellectual property clause, in which the employer states that the copyright of works created by employees will belong to the employer.

If there is no clause or agreement about the ownership of the relevant copyright in the contract, the FLC provides that the rights will be divided on an equal basis. The employer and the employee will be entitled to own 50% respectively of the relevant copyright. Furthermore, the employer is entitled to disclose the work without the employee's authorisation, but the employee is not authorised to do the same. They will always be required to obtain the employer's prior authorisation. This means that the right of disclosure will belong completely to the employer.

If there is no written individual employment contract, the ownership of rights will belong to the employee.

2.2.1 Computer Programs/Software

Concerning computer programs/software, Article 103 of the FLC provides that, unless otherwise agreed, the economic rights to computer programs created by employees in the course of their duties shall belong to the employer. Therefore, the employer will own the copyright (economic rights) to the computer programs created by its employees only in the following cases:

(a) when the computer program is created by an employee in the course of their duties; or

1. See Art. 84 of FLC at: http://www.indautor.gob.mx/ingles/documentos_normas/ley_f_derecho_autor_ingles.pdf.
2. See Art. 103 of FLC at: http://www.indautor.gob.mx/ingles/documentos_normas/ley_f_derecho_autor_ingles.pdf.

(b) when the computer program is created by the employee as a result of instructions given by the employer.

For cases not falling within the ambit of either, it seems that the copyright to a relevant work will belong to the employee exclusively. In such case, the employer will have to execute an assignment agreement with its employee in order to obtain the copyright of the relevant work. Such assignment agreement would have to meet with the following requirements:

(1) *The contract must be in writing*: the agreement shall be in writing; otherwise it would be null.
(2) *The contract must be temporary*: according to Article 33 of the FLC, the transfer of economic rights can be for a maximum of 15 years, except when the nature of the work or the scale of the relevant investment requires additional time. The legal framework provides cases in which the assignment can be for more than 15 years:
   (a) works which are so lengthy that their publication involves a much higher level of investment than that usually required for other works in the same class;
   (b) musical works that require a longer broadcasting period;
   (c) incidental contributions to a more extensive work, such as prologues, presentations, introductions, prefaces and other similar contributions;
   (d) literary or artistic works, including musical works, that are incorporated in electronic media programs referred to in Article 111 of the Law; and
   (e) other works for which the investment cannot be recouped within 15 years, owing to their nature, the scale of investment involved, the number of copies or the number of performers participating in the performance.
(3) It is important to stress that in case of absence of any express provision, the assignment of economic rights shall be deemed to be for a term of five years. Once the period of assignment ends, the ownership of rights will return to the employee/author.
(4) *The contract must be onerous*: The employee has the right to receive remuneration for the assignment of their work, this remuneration can be:
   (a) a proportional share in the proceeds from the exploitation concerned; or
   (b) an exact amount of remuneration, in which case it is possible to have remuneration in money or payment in kind.

2.2.2 Government Employees

Regarding government employees (federal/state/municipalities), Article 46 FLC Regulations provides that their works shall be considered produced as a hired work unless expressly agreed otherwise in each case.[3] The provision means that the ownership rights of works created in the official service shall belong to the entity, ministry or governmental office which the employee works for. The only right that the governmental employee will have is to be mentioned as author (moral right).

2.3 The Parties' Duty to Inform

Article 84 of the LFC provides the obligation on the employee to get the employer authorisation in order to disclose the work. This obligation is not imposed on the employer.

No other similar provision is present in the FLC or its Regulations.

2.4 Employee Remuneration Right

There are no special rules regulating employees' right to remuneration in the FLC or its Regulations. Nevertheless, from their provisions it is possible to confirm the following:

If the creation of the work is part of the duties of the employee or the creation is by instruction of the employer, the remuneration for that creation may be included in the employee's salary.

In governmental employee's case, given that their works are considered as works created by hire, there has to be remuneration to the employee.

Finally, it is worth mentioning that the Mexican Constitution[4] provides that nobody can be deprived of the fruits of their work, unless there is a court decision which states otherwise. Also, the Mexican Constitution provides that no one can be forced to work without a fair remuneration.

3 EMPLOYEES' SEMICONDUCTOR CHIPS

3.1 The Legal Framework

The general principles concerning layout-designs of integrated circuits are contained in Title V*bis* of the Mexican Industrial Property Law, which was enacted on 26 December 1997 and entered into full force and effect on 1 January 1998.

The incorporation of this new title provides the protection and enforcement of layout-designs of integrated circuits in compliance with the provisions agreed in the TRIPS and NAFTA agreements.

3. See Art. 46 of FLC Regulations at: http://www.indautor.gob.mx/ingles/documentos_normas/reglamento_ing.pdf.
4. See Art. 5 of Mexican Constitution at: http://www.diputados.gob.mx/LeyesBiblio/htm/1.htm.

3.2 Ownership of Rights

No special provisions concerning the ownership of rights within an employment relationship are stated in our legal system.

The ownership of a given layout-design of integrated circuit is granted to the applicant by virtue of *Article 178bis(2) of the Mexican Industrial Property Law.*[5]

3.3 The Parties' Duty to Inform

The Mexican Industrial Property Law as well as any other applicable laws do not expressly provide such a duty.

3.4 Employee Remuneration Right

No special provisions concerning the employee remuneration right in connection with a *layout-design of integrated circuit registration* is provided in our legal system. However, if remuneration applies, it would be determined in the agreement reached by the parties (employer and employee).

4 EMPLOYEES' INVENTIONS

4.1 The Legal Framework

Inventions are covered by Articles 9 to 26[6] of the Mexican Industrial Property Law. The provisions do not differentiate between inventions made by inventors in the private and public sector, but contain a detailed list of rights and obligations for the patent holder and its recorded licensees.

4.2 Ownership of Rights

According to Article 14[7] of the Mexican Industrial Property Law, the provisions of the Mexican Labour Law (mainly Article 163), shall be applied when an invention has been made by a person subject to an employment relationship.

Therefore, under the scope of Article 163[8] of the Mexican Labour Law, the ownership of a given invention is subject to the following rules:

5. See Art. 178*bis*(2) of the Mexican Industrial Property Law at: http://www.wipo.int/wipolex/en/text.jsp?file_id=265590.
6. See Arts. 9 to 26 of the Mexican Industrial Property Law at: http://www.wipo.int/wipolex/en/text.jsp?file_id=265590.
7. See Art. 14 of the Mexican Industrial Property Law at: http://www.wipo.int/wipolex/en/text.jsp?file_id=265590.
8. See Art. 163 of the Mexican Labour Law at: http://www.diputados.gob.mx/LeyesBiblio/pdf/125.pdf.

(i) in any case, the inventor shall be identified as the author of the relevant invention;

(ii) if the services provided by the employee are related to research activities, the invention's ownership as well as the exploitation right shall be held by the employer; and

(iii) if nothing is agreed between the parties, the ownership of the invention would be held by the inventor, but the employer shall have a preferential right - in equal circumstances - to obtain the ownership as well as the exclusive right to use the relevant invention.

4.3 The Parties' Duty to Inform

The Mexican Industrial Property Law as well as any other applicable laws do not expressly provide such a duty.

4.4 Employee Remuneration Right

Following the provisions contained in Article 163 of the Mexican Labour Law, the inventor, *in addition to their salary*, is entitled to receive remuneration, which may be determined by them and the employer. Otherwise, the Mexican Labour Court shall determine such additional remuneration by taking into consideration the importance of the invention and the benefits that the employer may obtain from it.

5 EMPLOYEES' UTILITY MODELS

5.1 The Legal Framework

According to the Mexican Industrial Property Law, general provisions concerning 'inventions' (such as ownership, general concepts, time of protection, etc.) also apply to utility models. In contrast to patents, a utility model is can be registered when it meets the following requirements: (i) is new; and (ii) is industrially applicable. Similarly to patents, Article 163 of the Mexican Labour Law applies when the relevant utility model was created as result of an employment relationship.

5.2 Ownership of Rights

According to Article 14 of the Mexican Industrial Property Law, the provisions of the Mexican Labour Law (mainly Article 163), shall be applied when a utility model was made by a person subject to an employment relationship.

Therefore, under the scope of Article 163 of the Mexican Labour Law, the ownership of a given utility model is subject to the following rules:

(i) in any case, the inventor shall be identified as the author of the relevant invention;
(ii) if the services provided by the employee are related to research activities, the ownership of the utility model as well as the exploitation right shall be held by the employer; and
(iii) if nothing is agreed between the parties, the ownership of the utility model will be held by the creator, but the employer shall have a preferential right – in equal circumstances – to obtain the ownership as well as the exclusive right to use the relevant utility model.

5.3 The Parties' Duty to Inform

The Mexican Industrial Property Law as well as any other applicable laws do not expressly provide such a duty.

5.4 Employee Remuneration Right

Following the provisions contained in Article 163 of the Mexican Labour Law, the creator, *in addition to their salary*, is entitled to receive remuneration, which may be determined by them and the employer. Otherwise, the Mexican Labour Court shall determine such additional remuneration by taking into consideration the importance of the utility model and the benefits that the employer may obtain from such utility model.

6 EMPLOYEES' DESIGN RIGHTS

6.1 The Legal Framework

According to the Mexican Industrial Property Law, general provisions concerning 'inventions' (such as ownership, general concepts, time of protection, etc.) also apply to designs.

In contrast to patents, a design can be registered when it is new and industrially applicable. Similarly to patents, Article 163 of the Mexican Labour Law applies when the relevant design has been created as a result of an employment relationship.

6.2 Ownership of Rights

According to Article 14 of the Mexican Industrial Property Law, the provisions of the Mexican Labour Law (mainly Article 163), shall be applied when a design has been made by a person subject to an employment relationship.

Therefore, under the scope of Article 163 of the Mexican Labour Law, the ownership of a given design is subject to the following rules:

(i) in any case, the designer shall be identified as the author of the relevant design;
(ii) if the services provided by the employee are related to research activities, the ownership of the design as well as the exploitation right shall be held by the employer; and
(iii) if nothing is agreed between the parties, the ownership of the design will be held by the designer, but the employer shall have a preferential right - in equal circumstances - to obtain the ownership as well as the exclusive right to use the relevant design.

6.3 The Parties' Duty to Inform

The Mexican Industrial Property Law as well as any other applicable laws do not expressly provide such a duty.

6.4 Employee Remuneration Right

Following the provisions contained in Article 163 of the Mexican Labour Law, the designer, *in addition to their salary*, is entitled to receive remuneration, which may be determined by them and the employer; otherwise, the Mexican Labour Court shall determine such additional remuneration by taking into consideration the importance of the design and the benefits that the employer may obtain from such design.

7 EMPLOYEES' PLANT VARIETIES

7.1 The Legal Framework

The Mexican Plant Varieties Law[9] provides the general principles concerning the protection and enforcement of plant varieties. Given that Mexico is a party to the UPOV Treaty, plant breeders' rights are basically the same as those granted in other countries.

Unfortunately, no provision concerning the rights that an employee, *within an employment relationship,* may be entitled to exist.

7.2 Ownership of Rights

No provision concerning the ownership of rights derived from an employment relationship concerning plant varieties is provided.

9. See the Mexican Plant Varieties Law at: http://www.upov.int/export/sites/upov/members/en/npvlaws/mexico/mexico_law.pdf.

7.3 The Parties' Duty to Inform

No Mexican provision contains such a duty.

7.4 Employee Remuneration Right

No Mexican provision contains any remuneration right for plant varieties, but it may be determined by the parties.

8 INSTANCES FOR DISPUTES

Conflicts concerning the ownership of intellectual property rights derived from an employment relationship can be heard by the Federal Labour Courts (for employees in the public sector) and the Local Labour Courts (for employees in the private sector).

The ruling issued in such first instance may be challenged by means of an *Amparo appeal*, which is filed and decided by a Federal Circuit Court, specialising in labour matters. The ruling issued by these courts becomes *res judicata*. No further appeals are available.

9 CONCLUDING REMARKS

Intellectual property developed by employees under an employment relationship has always been a controversial topic in our legal system. The limited (and sometimes absent) regulation involves serious complications for the parties and the relevant courts as well, mainly when trying to understand the meaning and scope of the applicable legislation.

Mexican courts are unfamiliar with such kinds of disputes. The mere fact of determining which law to be applied is a subject that is not defined so far. The principles and rules to be observed in every dispute of this nature are still unknown to our Labour Courts. Said circumstances plus the fact that the employees most often decide to give up their job, rather than initiate a costly and lengthy dispute, implies a serious problem that directly affects/violates the employees' IP rights.

Therefore, the need to create accurate legislation that establishes the basic rules to be observed in connection with the IP created within an employment relationship should be one of the main concerns for our Congress. Providing the applicable principles, that now are unknown, may represent a huge step towards a real enforcement of the intellectual property rights derived from an employment relationship in Mexico.

BIBLIOGRAPHY

Table of Legislation

Federal Law on Copyright.
Regulations of the Federal Law on Copyright.
Mexican Industrial Property Law.
Regulations of the Mexican Industrial Property Law.
Mexican Labour Law.
Mexican Plant Varieties Law.

CHAPTER 38
Uruguay

Gustavo Fischer

1 BACKGROUND

The legal system in place in Uruguay is a civil law regime. The legal framework comprises the Constitution and general principles of law, statutes and codes. Case law does not constitute a binding source of law, and the common law doctrines of precedent and *stare decisis* are not applicable in Uruguay. Uruguay is a member of the main international treaties and conventions concerning intellectual property.

Provisions concerning intellectual property rights ('IPR') are not contained in a code or single piece of legislation. There are separate laws for patents, copyrights and plant variety rights, which show no uniformity in treatment of employees' IPR.

As discussed below, there are no general provisions in Uruguayan law governing ownership of IPR over subject matter or works created under an employment relationship. As to existing rules, there are still several areas regarding employees IPR that have not yet been regulated. Even regarding statutes that effectively deal with employees' IPR, there is a degree of legal uncertainty that needs to be addressed.

In the following sections, we discuss the legal framework and main issues in Uruguay concerning employees' rights to copyright, technology innovations (inventions, utility models and industrial designs), and plant variety rights.

2 EMPLOYEES' COPYRIGHT

2.1 The Legal Framework

Applicable legislation in Uruguay is Copyright Law No. 9,739 of 17 December 1937, as amended and updated by Law No. 17,616 passed on 10 January 2003, Law No. 17,805 of 26 August 2004, and Law No. 19,149 of 24 October 2013 ('Copyright Law').

The Copyright Law does not contain general rules in regard to employees' copyright. Nevertheless, the Law has several provisions which deal with employees' copyright for certain types of works. This includes the provisions regarding ownership of computer programs (Article 29 of the Copyright Law), as well as ownership of works in journalism (Article 22).

It is understood that these provisions only apply for the specific types of works to which they refer, and that ownership of protected works other than those mentioned above is governed by the general principles regarding copyright ownership in Uruguay.

Uruguay follows a copyright system which is based on the *droit d'auteur* system, and ownership of a created work pertains in principle to the author, who is always a natural person. Legal entities may hold copyrights as a derivative right obtained from the author.

2.2 Ownership of Rights

There are no general provisions in the Uruguayan Copyright Law governing ownership of rights over works created under an employment relationship. The Copyright Law uses the word 'employer' only once, and that is in relation to the specific case of computer programs.

As has been already mentioned, under the Uruguay legal system, an author is always a living person, and the right to a work is automatically acquired with the creation thereof. It is also worth mentioning that the protection afforded to authors is not subject to compliance with any formality. This had already been recognised for foreign authors as from approval by Uruguay of the Berne Convention back in 1979, and was expressly incorporated into the Copyright Law by the legislative amendment passed in 2003 through Law No. 17,616.

As to the regime for transfer of rights over copyrights, Article 8 of the Copyright Law must be noted:

> The economic rights of the author are transmitted by all means foreseen by the law. For the agreement to be valid, it must be necessarily executed in writing, but may only be enforced against third parties after its recordal in the Register. When the agreement is executed abroad, recordal thereof may be performed before the country's diplomatic or consular authorities.

According to the above, the principle is that transfer of rights must be performed by means of a written document. Otherwise, the agreement may be held to be invalid. However, there has recently been some debate as to whether the requirement for the agreement to be executed in writing applies to all cases of transfer of copyrights or only encompasses the case where the entire ownership over the work is assigned or transferred, but not to the case of a licence or authorisation to use the work (as opposed to absolute transfer of ownership).

In addition, doctrine has also discussed whether it would be possible in Uruguay to assign or transfer in an absolute manner the economic rights over a protected work, or whether this is not allowed (as happens in many other jurisdictions where an absolute transfer is null and void, and only limited licences or authorisations may be granted). However, the language of the Copyright Law seems to authorise such an absolute transfer of copyright ownership from the author to a third party (either an individual or a legal entity).

Articles 2 and 12 use the word 'transfer' (*enajenar, enajenación, enajenada*) as opposed to 'assignment' (*cesión*), thus providing grounds for the position that the law would allow the author to dispose or sell their copyright definitively to a third party.

Moral rights are duly recognised in the Copyright Law and the statute clearly states that, irrespective of the terms and conditions of a transfer or assignment agreement, the author will always retain their moral rights.

2.2.1 The General Rule

Accordingly, for the transfer of copyright ownership of a work created by an employee, the principle is that such transfer must be foreseen in a written document. This may be the written employment agreement entered into between the employer and the employee, or a specific copyright transfer agreement entered into by both parties. In the absence of a contract in writing, Article 8 of the Copyright Law (which for validity requires the transfer to be executed in writing) will apply.

It is important to note that as per Article 33 of the Copyright Law, even in case of an absolute transfer of ownership of copyright over a work, the right to economically exploit the work by the purchaser will be limited to a term of 15 years as from the author's death. Thereafter, ownership will pass on to the author's heirs, which will enjoy ownership for the remainder of the duration of the term of protection (which is up to 50 years from author's death).

Case law addressing the issue of works created under an employment is almost non-existent. However, an isolated 2016 judgement has dealt with this topic. A copyright infringement claim was brought before a First Instance Civil Court by a company alleging that a major retailer had copied their original fashion designs. The First Instance Court ruled for the plaintiff and found copyright infringement. In their appeal, the retailer argued *inter alia* that the ruling should be reversed, since the plaintiff should have evidenced that the fashion designs, created by designers employed by the plaintiff, had been assigned to the plaintiff by their employees in writing (Article 8 of the Copyright Law). The Court of Appeals in Civil Matters confirmed the first instance ruling and dismissed the appeal. Without elaborating, the Court of Appeals expressed that since the fashion designs had been created 'within the framework' of an employment relationship, then copyright ownership should be 'presumed as assigned'. The Court of Appeals ruling erroneously cites as support for this assertion Article 22 of the Copyright Law. This provision only and specifically deals with employees' copyright in journalism, and by no means may be construed as a general rule or extended to apply to situations outside the scope of the statute. There are no other published decisions referring to employees' copyright in general, and we deem this decision as not adequately reflecting how the provisions of the Copyright Law should be construed. As already mentioned, Article 8 of the Copyright Law applies as a general rule, with the exceptions that are expressly foreseen in the law and discussed herein below. Further cases and decisions will undoubtedly help in clarifying and shedding light on the general principle applying to employees' copyrights in general.

2.2.2 Computer Programs

One of the exceptions to the general rule on ownership of created works is the case of computer programs, as provided for in Article 29 of the Copyright Law. In 2003, the Copyright Law was amended to expressly include language encompassing the case of creation, ownership and transfer of computer programs and databases. This was performed taking into consideration the special features of the software industry, the difficulties posed by the chain of ownership for this type of creation, and the need to provide legal certainty with regard to use and exploitation of this kind of works.

Article 29 provides for a presumption that the authors of computer programs and databases have assigned to the 'producer' the economic rights thereon, in an unlimited manner and on an exclusive basis. This includes the authorisation to decide on their disclosure and to exercise the moral rights over the same.

Furthermore, the authors – unless otherwise provided by way of an agreement – may not oppose or prevent the producer from performing or authorising modifications or new versions of the computer programs or databases (e.g., and *inter alia*, 'upgrades').

With specific regard to the employers, the provision in Article 29 envisages that if the computer program or database has been created 'within the framework of an employment relationship, be it public or private, the purpose of which has totally or partially a nature similar to such creations', then it is presumed that the author has 'authorised' the employer the economic rights as well as the exercise of the moral rights, in an unlimited and exclusive manner, unless agreed otherwise.

It is relevant to note that the statute uses a different wording when regulating the rights vested in the producer, and those vested in the employer, pursuant to the presumptions contained in Article 29.

As to the producer, the law presumes that the authors have 'assigned' to the producer in an unlimited and exclusive manner their economic rights over the computer programs. However, as to employers, the provision establishes that it is presumed that the author-employee has 'authorised' the employer the economic rights as well as the exercise of the moral rights, in an unlimited and exclusive manner, unless agreed otherwise.

The law here lacks precision. It is unclear what the scope is of such 'authorisation' of 'economic rights', if use of this expression was unintentional, or if the lawmaker wished to establish any difference between the 'assignment' of rights to software producers, and the 'authorisation' vested in the employer. There is no reference in the records of parliamentary discussion of the 2003 amendment to the Copyright Law.

However, the spirit of the law was always to enable circulation and exploitation of computer programs with a clear title and chain of ownership, and there is no reason to believe that the employer was intended to enjoy lesser rights than those afforded to the software producer.

As to ownership of employee-created computer software, the relevant issue is to establish whether the computer programs have been created *within the framework of the employment relationship*, and also whether such relationship *has totally or partially a nature similar to such creations*.

First, this constitutes an exception to the general rule that transfer of ownership over a copyright must be performed in writing (Article 8 of the Copyright Law). However, establishing whether a given software falls within the scope of an 'employment relationship' is ultimately a matter of evidence, and in case of conflict, it will be up to the Judge to decide upon this issue. It should also be noted that, in addition to the existence of an employment relationship, the statute requires that such relationship has a specific purpose, i.e. *a nature similar to such creations*, which in other words means that the employee must have been hired to create or develop computer programs or databases.

The doctrine in our jurisdiction has emphasised that this presumption not only includes the authorisation for economic rights but also to exercise the moral rights, and that this constitutes a novelty, considering the classic position that moral rights are always vested in the author (an individual) and are not subject to transfer or assignment (Article 31 of the Copyright Law).

2.2.3 Employees' Copyright in Journalism

The Copyright Law was amended in 2004 by Law No. 17,805, to include provisions on copyright in journalism. This amendment contemplates in one of its provisions (Article 22 of the Copyright Law as amended), the rights of employees of journals, magazines or communication companies. A presumption is established that the rights of the authors that have been hired under an employment relationship, have been assigned to be used 'only by the company or means of communication for which the work is being performed'. Use of a journalistic article in a different situation, or with a purpose different to the use for which the author was hired, must have the author's express authorisation. In all cases, the author (either an employee, or a freelance journalist), will keep the publishing rights (Article 23 of the Copyright Law).

2.3 The Parties' Duty to Inform

No express provision regarding employees' duty to inform is envisaged by the Copyright Law.

2.4 Employee Remuneration Right

No express provision regarding employees' remuneration rights is envisaged by the Copyright Law.

3 EMPLOYEES' SEMICONDUCTOR CHIPS

3.1 The Legal Framework

Uruguay has not passed any special legislation addressing protection of semiconductor chips. Neither did the country approve the IPIC.[1]

3.2 Ownership of Rights

The lack of a specific law implementing protection of integrated circuits poses several issues:

> Are the TRIPS provisions that provide protection for layout-designs of integrated circuits, and those incorporated by reference from the IPIC, self-executing?
> What would be the type of protection to be afforded by Uruguay to semiconductor chips? That of copyright, or patents?

These issues remain largely unresolved, and – besides the already noted lack of specific legislation – there is no case law either that could provide guidance as to the nature and scope of protection effectively available.

3.3 The Parties' Duty to Inform

The only express provision establishing a specific duty to inform the employer in regard to development of intellectual property rights is contained in Article 18 of the Patent Law and is described in detail in section 4 below regarding employee inventions.

3.4 Employee Remuneration Right

There is no specific provision regarding employee remuneration rights other than established by Article 17(2) of the Patent Law, as discussed in section 4.4 below.

1. Treaty on Intellectual Property in Respect of Integrated Circuits, done at Washington, D.C., on May 26, 1989.

4 EMPLOYEES' INVENTIONS

4.1 The Legal Framework

Applicable legislation in Uruguay is the Law Regulating Rights and Obligations Relating to Patents, Utility Models and Industrial Designs No. 17.164 of 2 September 1999 ('Patent Law'). The Patent Law, in its Chapter II, Section II, entitled 'Inventions made during an employment relationship', contains a detailed description of the parties' rights and obligations. Indeed, the incorporation provisions regulating the employees' inventions in our patent legislation is considered a great innovation and advancement.

The reasons underlying the inclusion of provisions contemplating employees' inventions are various and comprise the fostering and promotion of research and development, the need to clarify the resulting ownership of patentable subject matter and the ensuing inventions, and to protect the interests of those involved in the innovation process. As in the case of copyrights, there is an intersection between labour law and intellectual property law.

The relevant provisions of the Patent Law are contained in Chapter II, Section II, and provide the following:

> Article 17
> Where an invention has been made in fulfilment of a contract of employment, or a contract for a work made for hire or a service agreement, whose total or partial object is research activity, the right to the resulting patent derived therefrom shall belong to the employer, unless agreed to the contrary.
> In case where the employee's personal contribution to the invention and importance thereof to the company evidently exceed the explicit or implicit terms of his contract or work relationship, the employee shall have the right to a supplementary remuneration.
> Article 18
> Where an employee makes an invention related to his professional activity in the company, without having an obligation to carry out research activity, and if in its obtention the knowledge gained within the company or the use of means provided by the company had a decisive influence, he shall communicate this in writing to his employer.
> If the employer notifies his interest in the invention in writing within 90 days, the right to the patent shall be jointly owned.
> An invention for which a patent application has been filed within one year following cessation of employment shall be deemed to have been developed during the employment relationship.
> Article 19
> Inventions made in the course of an employment relationship not covered by the preceding articles shall exclusively pertain to their author.
> Article 20
> Any contractual provision less favourable to the inventor than those provided for in this Section shall be null and void.

4.2 Ownership of Rights

4.2.1 General Remarks

Article 16 of Patent Law sets forth the general principle in regard to creation of inventions, the so-called 'principle of ownership', i.e., that the patent pertains to its inventor:[2]

> The right to a patent shall belong to the inventor or to his successors in title and it shall be transferable by an act inter vivos or by reason of death.

Contrary to what occurs in regard to other types of intellectual property, the Patent Law contains a set of general provisions governing ownership of rights over inventions created under an employment relationship.

In order for the rules regarding employees' inventions to apply, there are some basic elements that must be present: (i) the existence of an 'invention', as objectively defined by the Patent Law,[3] (ii) that such invention is not excluded from patentability, and that it may effectively give rise to a patent right, and (iii) the existence of an employment relationship, as per the definition thereof provided by labour law doctrine, rules and regulations.

The concept of 'invention' is provided for in Article 8 of the Patent Law, which defines it as:

> Inventions of new products or processes which involve an inventive activity and are susceptible of industrial application shall be patentable.

Articles 13 and 14 of the Patent Law define what is not considered an invention, and the exclusions to patentability. Article 15 refers to second use.

> Article 13
> The following shall not be considered inventions for the purposes of this law:
> A) Discoveries, scientific theories and mathematical methods.
> B) Plants and animals, with the exception of microorganisms and essentially biological processes for the production of plants or animals, except for non-biological or microbiological processes.
> C) Schemes, plans, rules for playing games, business, accounting, financial, educational, publicity, draws or supervision principles or methods.
> D) Literary or artistic works, or any other aesthetic creation, as well as scientific works.
> E) Computer programs considered in an isolated manner.
> F) The various methods of reproducing information.
> G) Biological or genetic material as found in nature.
> Article 14
> The following shall not be patentable:
> A) Diagnostic, therapeutic or surgical methods for the treatment of persons or animals.

2. Beatriz Bugallo, *Propiedad Intelectual*, 397 (1st ed., FCU – Fundación de Cultura Universitaria, 2006).
3. *Ibid.* 397.

> B) Inventions contrary to public order, morality, public health, the population's nutrition, safety or the environment.
>
> Article 15
>
> Pursuant to the provisions of this law, products or processes already patented and included in the state of the art may not be the subject of a new patent simply because the purpose for which they are to be used differs from that comprised in the original patent.

With regard to the existence of an employment relationship, as expressed above, the concept, scope and formalities thereof will be those provided by labour law doctrine, applicable rules and regulations, as well as case law. The employment agreement may contain provisions which are different to those provided for in the Patent Law, with the caveat that the Patent Law establishes a framework of minimum provisions designed to guarantee the employees' rights to their inventions. Clauses contained in an employment contracts that are less favourable for the employee-inventor than those envisaged in the Patent Law, shall be considered null and void (not the entire agreement, but only those provisions which are in breach of Articles 17 to 20 of the Patent Law) by virtue of Article 20:

> Any contractual provision less favourable to the inventor than those provided for in this Section shall be null and void.

4.2.2 *Employee Inventions Situations*

Three main situations concerning employee-developed inventions may be identified:[4]

(a) the *employee-researcher case*, where the invention has been made in fulfilment and within the scope of an employment agreement encompassing a research obligation;
(b) the *company-provided knowledge or means case*, where the invention has been made by the employee in the course of an employment relationship – without having a research obligation – but where use of knowledge acquired in the company or use of means provided by the company have had a decisive influence in the development of the invention;
(c) a *residual criterion*, whereby all other inventions developed or made by an employee fall under the general principle of ownership, and will pertain exclusively to the inventor (irrespective of whether they are an employee or not), unless expressly agreed otherwise.

4. Cf. Bugallo, *supra* n.2, at 397.

4.2.2.1 *The Employee-Researcher Case*

As envisaged by Article 17 of the Patent Law, the title to the patent resulting from a research activity performed because of a labour or service contract pertains to the employer:

> Where an invention has been made in fulfilment of a contract of employment, or a contract for a work made for hire or a service agreement, whose total or partial object is research activity, the right to the resulting patent derived therefrom shall belong to the employer, unless agreed to the contrary.

Where the employee's personal contribution to the invention and importance thereof to the company evidently exceed the explicit or implicit terms of their contract or work relationship, the employee shall have the right to a supplementary remuneration.

In this situation, the employment contract must clearly define the research scope and obligations of the employee, and their remuneration for the work to be performed. The employer is assuming the risk that no patentable subject matter is developed as an outcome of the employment relationship. However, if an invention is made, the right to the invention, and to the resulting patent (and this to exploitation and ownership thereof), will belong exclusively to the employer. In principle, the employee-researcher will only have the compensation that has been agreed upon by virtue of the employment relationship.[5]

The inventor will always keep their moral rights to be recognised as the author of the invention, as set forth in Article 2 of the Patent Law:[6]

> The moral rights of inventors and designers recognised as authors of their inventions or creations shall be inalienable and not subject to limitation, and is transmitted to their heirs.
>
> The economic rights derived from inventions, utility models and industrial designs shall be protected by patents, which shall be guaranteed by the corresponding titles.
>
> The inventor shall be mentioned as such on the patent granted and in publications and official documents relating thereto, unless this right is specifically renounced in writing.
>
> Any agreement by which the inventor renounces their right to be mentioned prior to filing of the application for a patent shall be null and void.

Article 17(2) is further discussed below in section 4.4 with regard to the employee remuneration right.

4.2.2.2 *The Company-Provided Knowledge or Means Case*

Here, the invention has been made by the employee in the course of an employment relationship, which does not include a research obligation, where use of knowledge

5. Bugallo, *supra* n.2, at 398.
6. *Ibid.* 398.

acquired in the company or use of means provided by the company have had a decisive influence in the development of the invention.

The Patent Law provides in Article 18 the case where an employee produces an invention that bears connection with the type of activity they perform within the company or in the course of their employment relationship, but where the employee has not been hired to perform a research and development activity:

Where an employee makes an invention related to their professional activity in the company, without having an obligation to carry out research activity, and if in its obtention the knowledge gained within the company or the use of means provided by the company had a decisive influence, they shall communicate this in writing to their employer.

If the employer notifies their interest in the invention in writing within 90 days, the right to the patent shall be jointly owned.

An invention for which a patent application has been filed within one year following cessation of employment shall be deemed to have been developed during the employment relationship.

In order for this situation to apply, the following elements must be present cumulatively:

(i) that the invention is 'related' to the employee's professional activity in the company;
(ii) that, for development of the invention, the employee-inventor used knowledge that they acquired from the company, or means provided by the company; and
(iii) that the employee-inventor is under no obligation to perform research activities.[7]

It is worth noting that Article 18 does not require that the invention has been made or developed within the company's premises, or within the employee's work timetable. Accordingly, if, e.g., the employee used the company's research facilities or equipment outside normal working hours to achieve the invention, such circumstances would fall within the scope of the situation envisaged by the legal framework. Again, the employee's innovation must fall within the definition of 'invention', and must not be excluded from patentability.

In the situation foreseen by Article 18 of the Patent Law, the employee must compulsorily communicate to the employer, in writing, that they developed an invention. The employer has a 90-day term in order to notify the employee of its interest in the invention. If affirmative, the employee and the employer will jointly own the patent.[8]

7. *Ibid.* 399.
8. *Ibid.* 400.

In the event that the employer does not notify the employee of its interest in the invention (or if it does so beyond the 90-day time limit), it will be considered as uninterested, and the employee will entirely own the invention and the resulting patent(s).[9]

The last paragraph of Article 18 provides that an invention for which a patent application has been filed within one year following cessation of employment shall be deemed to have been developed 'during' the employment relationship.[10] This presumption only bears relevance to the first of the cumulative requirements for this situation, i.e., that the invention is 'related' to the employee's professional activity in the company, since it was performed 'during' the employment relationship. It does not mean that for the development of the invention, the employee-inventor effectively used knowledge that they acquired from the company, or means provided by the company, nor is a proof as to whether the employee-inventor was under an obligation to perform research activities or not. It does not mean either that the employer will automatically become the owner of the invention. These two latter elements (the knowledge and means, and the existence or not of a research obligation) must still be proven by the employer in order for the provisions of Article 18 to apply.

4.2.2.3 The Residual Criterion

All other inventions developed or made by an employee fall under the general principle of ownership, and will pertain exclusively to the inventor (irrespective of whether they are an employee or not), unless expressly agreed otherwise. Article 19 of the Patent Law provides that:

> Inventions made during an employment relationship not covered by the preceding articles shall exclusively pertain to their author.

4.3 The Parties' Duty to Inform

The only express provision establishing a specific duty to inform which appears in the Patent Law is the one contained in Article 18 which provides that:

> Where an employee makes an invention related to his professional activity in the company, without having an obligation to carry out research activity, and if, in its obtention, the knowledge gained within the company or the use of means provided by the company had a decisive influence, he shall communicate this in writing to his employer.

9. *Ibid.* 400.
10. *Ibid.* 400.

It is worth noting that Article 18 does not contain: (i) any specific term for the employee to notify the employer of the development of an invention, nor (ii) any provision outlining the consequences in case of failure by the employee of their duty to inform the employer.

In the absence of an express regulation, I suggest that the general principles of labour law should apply, including the duties of loyalty and fidelity by the employee to the employer, and the obligation to act in good faith. I also suggest that notice to the employer should be given within a reasonable period (the concept of 'reasonableness' to be ultimately judged by a court of law in case of discrepancy, all circumstances considered).

Case law regarding the duty to inform is almost non-existent. A dispute between an employee and an employer concerning the scope of the duty to inform has been decided by the Labour Court and the Civil Court. The employee had developed an invention within the framework of Article 18 of the Patent Law. Furthermore, the employee drafted, prepared and filed with the Uruguayan Patent Office a patent application for his invention, informing the employer neither of the development of the invention nor of the filing of the patent application. Subsequently the employee informed his employer's patent agents of said circumstances and requested them to notify the employer under the terms of Article 18 – i.e. in order to obtain confirmation by the employer of his interest or not in the invention within a 90-day term after the notification. The employer believed that the employee had acted in bad faith and terminated the employment relationship. The employee brought two separate legal actions: one before the Labour Court for compensation derived from termination of the work relationship, including a request for special compensation for abusive termination of employment; the second claim was filed before the Civil Court for damages and prejudices allegedly caused by the information provided by the employer's attorneys to their client, which the employee claimed constituted a breach of professional secrecy (since, in performing the notification to the employer under Article 18, the employer's attorneys not only reported to their employer that the employee had developed an invention, but also had provided their employer with information on the invention itself).

The findings and decisions of the venues were contradictory. The Civil Courts (both the First Instance Court and the Court of Appeals) rejected the employee's claim for damages and prejudice. Analysing the scope of the duty to inform, the Judges stated that the employee is under the obligation not only to report on the existence of the invention, but also to provide to the employer a description thereof; and that therefore the employer's attorneys did not incur any liability and had acted pursuant to Article 18(1) of the Patent Law.

However, the Labour Courts decided otherwise. Both the First Instance Labour Court and the Court of Appeals in Labour Matters understood that the employer had abusively terminated the employment relationship with the employee. The Judges recognised that the employee had failed to provide notice to the employer of the development of the invention and of the filing of a patent application, and that he did only do so with a delay and through the employer's attorneys. However, the Courts ruled that that did not constitute enough cause for termination of the employment relationship and it was not related to the employee's performance in the company. The employer was therefore ordered to pay a severance payment as well as punitive compensation.[11]

4.4 Employee Remuneration Right

Article 17(2) of the Patent Law addresses the case where the result of the professional activity of the employee-researcher exceeds the expectations:

> Where the employee's personal contribution to the invention and importance thereof to the company evidently exceed the explicit or implicit terms of his contract or work relationship, the employee shall have the right to a supplementary remuneration.

This constitutes an exception to the normal circumstances under which the employee-researcher will receive the compensation provided for in the employment contract for their research and development work, irrespective of whether, as an outcome of his professional activity, patentable subject matter is identified and patent rights are ultimately secured by the employer.

Under this particular case, the following must be present in order for the employee to have the right to a supplementary remuneration:

(i) an invention which is of special importance to the company (e.g. if it brings substantial revenue to the company, or for strategic market reasons, or any other reason which sufficiently justifies the importance of the invention);
(ii) a personal contribution by the employee-researcher to the invention; and
(iii) neither the importance of the invention nor of the personal contribution by the employee-researcher to the invention were envisaged and provided for in the employment contract and at the time the employment relationship began (and thus the contribution to the invention and the importance thereof *evidently* exceeded *the explicit or implicit terms of his contract or work relationship*).

On the other hand, if from the outset of the research and development process for which the employee-researcher was hired, it was clear that the potential patentable subject matter and the resulting invention and patent rights would be crucial for the company and of the utmost importance, then the provision of Article 17(2) would not apply.

The solution is the same where the hiring of the employee-researcher was performed taking into account the special skills or abilities of the employee, and in consideration of the researcher's known high-standard of professional work, and expected work product and milestones implicitly or explicitly agreed upon in the employment contract. In such case, the provision of Article 17 for supplementary

11. See Court of Appeals in Labour Matters of the 2nd Circuit, Ruling No. 159/2008, of July 16, 2008, and Court of Appeals in Civil Matters of the 1st Circuit, Ruling No. 1/2012, dated February 1, 2012.

remuneration would apply only if the personal contribution of the employee-researcher *evidently exceeds* such high standards and expectations.

The rationale underlying supplementary remuneration is to provide a means to compensate the employee-researcher in case of an exceptional enrichment of the employer.[12] Surprisingly, the Patent Law does not provide any definition or basis as to what should be understood to constitute an adequate 'supplementary remuneration'.[13]

It is clear that in this scenario, the employee-researcher does not become a joint-owner of the invention and does not have a right on the revenues obtained by the employer through the economic exploitation of the patent. It would appear that the 'supplementary remuneration' should be calculated taking into account the salary payment that the employee-researcher would have received should the employer have known of the importance of the invention and the relevance of the researcher's personal contribution at the time when the employment agreement was entered into.[14]

The 'supplementary remuneration' is an indeterminate legal concept, the content and definition of which will – in case of conflict – ultimately lie with the Judge.

5 EMPLOYEES' UTILITY MODELS

The Patent Law also regulates utility models, in Title III, entitled 'Utility Model Patents'. Article 85 thereof establishes that all the general rules set forth for patents of invention, also apply to utility models:

> Without prejudice to the provisions of this Title, where applicable, the provisions on patents shall apply to utility models.

Accordingly, all provisions regarding employees' inventions discussed above apply equally here.

6 EMPLOYEES' DESIGN RIGHTS

The Patent Law regulates industrial design rights as well, in Title IV, entitled 'Industrial Design Patents'.

Similarly to what is provided for in the case of utility models, Article 98 of the Patent Law establishes that all provisions regarding employees' inventions apply to industrial design rights.

12. Bugallo, *supra* n.2 at 399.
13. Mario Daniel Lamas, *Los derechos de propiedad intelectual en las relaciones laborales. El caso uruguayo*, 179 (Estudios de Derecho y Propiedad Intelectual, Homenaje a Arturo Alessandri Besa, 2009).
14. Lamas, *supra* n.13 at 179.

7 EMPLOYEES' PLANT VARIETIES

7.1 The Legal Framework

The protection of plant varieties is contained in Seed Law No. 16,811 of 21 February 1997, as amended by Law No. 18,467 of 27 February 2009 ('PVR Law'). Implementing regulations are Decree No. 438/004 of 16 December 2004 as amended by Decree No. 219/010 of 14 July 2010. Other relevant provisions are Decree No. 385/009 of 18 August 2009 and modification of Decree No. 453/009 of 28 September 2009 (Decrees defining Small Farmer).

The INASE (*Instituto Nacional de Semillas*) is the PVR Office of Uruguay, responsible for maintaining the registry of ownership of new plant varieties, and of granting registration of new plant varieties in accordance with the above-mentioned laws.

7.2 Ownership of Rights

There are no special rules in the Uruguayan PVR legal framework providing for ownership of rights over employees' plant varieties as that which may be found in the Patent Law.

The PVR Law is structured under the assumption that the ownership of rights over a new plant variety are originally vested in the breeder thereof, as evidenced – among others – by Article 67 of the statute which provides that:

> [T]he National Seed Institute shall keep the Register of Cultivar Ownership, the purpose of which shall be to recognise and guarantee rights to the breeder of a new plant variety by means of the grant and registration of a title of ownership in accordance with the International Convention for the Protection of New Varieties of Plants signed in Paris on 2 December 1961, and amended by additional Acts signed in Geneva on 10 November 1972, and 23 October 1978, as approved by Law No. 16,580 of 21 September 1994.

Absent any express or specific rules regarding ownership of plant varieties bred under an employment relationship, this should be interpreted and decided upon taking recourse to the general principles of labour law concerning employment, and work done for hire.

It should be pointed out, however, that in practice, when applications for registration of new plant varieties in the Register of Cultivar Ownership are filed, the National Seed Institute requires the applicant (if different from the breeder) to expressly provide the identification details of the assignee of the rights. In addition, and most importantly, the assignee must file the documents that prove the transfer of ownership. This may be performed through an assignment of rights document, or otherwise by means of the employment agreement when the new plant variety has been bred by the employee under an employment relationship, the purpose of which is the breeding of cultivars.

There is no set of minimum public order rules providing for employees' joint ownership of rights over new plant varieties, or a supplementary remuneration, as may be found in the Patent Law.

Under normal circumstances, ownership of new plant varieties bred in the course of an employment relationship will be regulated or flow from the nature and scope of the employment agreement. In addition, considering the special features of the plant breeding activity, where substantial time, effort and investment is required until a plant variety qualifies for protection under the PVR system, rarely will an employee undertake the task of developing a plant variety outside of the scope of their employment agreement.

7.3 The Parties' Duty to Inform

The only express provision establishing a specific duty to inform the employer with regard to the development of intellectual property is the one found in Article 18 of the Patent Law and which has been discussed above.

No express duty to inform is contained in the PVR Law.

In the absence of an express regulation, the general principles of labour law should apply, including the duties of loyalty and fidelity by the employee to the employer, and the obligation to act in good faith. I also suggest that notice to the employer should be given within a reasonable period (the concept of 'reasonableness' to be ultimately judged by a court of law in case of discrepancy, all circumstances considered).

7.4 Employee Remuneration Right

The PVR law does not contemplate any remuneration right to the employee, besides the salary or compensation that may have been agreed upon between the employer and the employee in the employment contract.

8 INSTANCES FOR DISPUTES

Depending on the nature of the dispute, Civil Courts or Labour Courts may have jurisdiction over the case. Labour Courts have jurisdiction to decide disputes originating from individual labour conflicts between an employee and the employer, and this may therefore include the dispute concerning the claims arising from termination of the employment relationship related to an invention developed by an employee (this may include deciding whether the termination was justified or not, whether the employee or the employer breached any obligation under the employment relationship, and if compensation is due by the employer to the employee). Decisions of the First Instance Labour Courts may be appealed to the Court of Appeals on Labour Matters. Depending on the case and the contents of the First Instance and Second Instance Rulings, an appeal may be filed before the Supreme Court of Justice.

As to disputes concerning ownership of the employee's intellectual property rights, Civil Courts have jurisdiction thereon. Decisions of the First Instance Civil Courts may be appealed to the Court of Appeals on Civil Matters. Depending on the case and the contents of the First Instance and Second Instance Rulings, an appeal may be filed before the Supreme Court of Justice.

9 CONCLUDING REMARKS

As outlined in the introductory remarks, the legal system of Uruguay is based on civil law. The legal framework comprises the Constitution and general principles of law, statutes and codes. Case law does not constitute a binding source of law, and the common law doctrines of precedent and *stare decisis* are not applicable in Uruguay. Uruguay is a member of the main international treaties and conventions in intellectual property.

Provisions concerning intellectual property rights are not contained in a code or single piece of legislation. There are separate laws for trademarks, patents, copyrights and plant variety rights, which show no uniformity in treatment of employees' IPR. Uruguayan case law on employees' IPR is scarce, and this prevents clarification of legal issues that are obscure in the statutes.

Jurisdiction on conflicts arising from these various laws is similarly not uniform. The administrative authority in charge of granting trademarks and patents in Uruguay is the Uruguayan Patent and Trademark Office (*Dirección Nacional de la Propiedad Industrial*, DNPI), a subordinate body of the Ministry of Industry and Energy (which is part of the Executive branch). As to copyrights, it is the Uruguayan Copyright Office (*Oficina de Derechos de Autor*), of the National Library, within the framework of the Ministry of Education and Culture (which is also a part of the Executive branch). It is discussed whether the Copyright Office is an autonomous entity or is subordinate to the Ministry of Education and Culture. With regard to Plant Variety Rights, the same are administered by the National Seed Institute ('INASE'), which is structured as a public entity not pertaining to the state, having an autonomous nature, but not pertaining to the Executive.

Decisions arising from the DNPI, and the Copyright Office, are reviewable by the Contentious Administrative Court (*Tribunal de lo Contencioso Administrativo*, TCA), which is the highest body in the administrative justice system in Uruguay, and enjoys the same institutional rank of the Supreme Court of Justice (the highest body in the ordinary Judiciary). The Contentious Administrative Court is composed of five Judges, and the rulings of the Courts are not subject to review by any other body or Court.

However, *inter-partes* conflicts concerning the enforcement of IPRs, as well as of the employees' rights afforded by legislation, are brought before and resolved by the Judiciary (typically, Civil First Instance Courts, Court of Appeals, and the Supreme Court of Justice in the case of the cassation appeal, if applicable). There are no special IP Courts.

BIBLIOGRAPHY

Table of Sources

Beatriz Bugallo, *Propiedad Intelectual,* (1st ed., FCU - Fundación de Cultura Universitaria, 2006).

Mario Daniel Lamas, *Los derechos de propiedad intelectual en las relaciones laborales. El caso uruguayo,* (Estudios de Derecho y Propiedad Intelectual, Homenaje a Arturo Alessandri Besa, 2009).

Table of Legislation

Treaty on Intellectual Property in Respect of Integrated Circuits (Washington, D.C., 26 May 1989).

Table of Cases

Court of Appeals in Labour Matters of the 2nd Circuit, Ruling No. 159/2008 (16 July 2008).

Court of Appeals in Civil Matters of the 1st Circuit, Ruling No. 1/2012 (1 February 2012).

Court of Appeals in Civil Matters of the 4th Circuit, Ruling No. 176/2016 (22 November 2016).

F In Africa

CHAPTER 39

Uganda

Ronald Kakungulu-Mayambala

1 BACKGROUND

This section of the chapter sets the background to employees' intellectual property rights (IPRs) in Uganda. The chapter examines employees' intellectual property rights in the five key categories of copyright, semi-conductor chips, inventions, utility models, design rights and plant varieties in Uganda. The study also deals with instances of dispute resolution and offers a general conclusion at the end of the chapter. For purposes of the study, it is imperative to define an employee, a contract of service, a contract of apprenticeship and an employer.

An 'employee' is defined in Section 2 of the Employment Act[1] to mean:

> Any person who has entered into a contract of service or an apprenticeship contract, including, without limitation, any person who is employed by or for the Government of Uganda, including the Uganda Public Service, a local authority or a parastatal organization but excludes a member of the Uganda Peoples' Defence Forces.

From the above definition, it is clear that an employee either be on a contract of apprenticeship, or a contract of service but for purposes of employees' intellectual property rights will not include a 'casual employee' who is 'a person who works on a daily or hourly basis where payment of wages is due at the completion of each day's work.'[2] Since the casual employee is paid on a daily or hourly rate, the general presumption is that they are not in full employment of his or her employer and will ordinarily not create any intellectual property rights during the time of employment, because most of the casual employees are not offering very skilled labour or even if

1. Act No. 6 of 2006.
2. Section 2 of the Employment Act.

they did, their labour will have (already) been paid for on a daily or hourly rate and will be treated as 'work done in the course of employment' or as 'commissioned works' all of which are discussed in detail in this paper.

A key determinant in the employee/employer relationship in Uganda is the contract of apprenticeship. A 'contract of apprenticeship' means a contract of service—

(a) where there is an obligation on the employer to take all reasonable steps to ensure that the employee is taught and acquires, the knowledge and skills of that industry by means of practical training received in the course of the employee's employment; and
(b) where there is a provision for formal recognition of the fact that the employee has acquired the knowledge and skills, intended to be acquired when the employee has done so.

According to Rideout, 'technically a contract of apprenticeship does not give rise to the relationship of employer and employee.'[3] The general view was that the apprentice was after all undergoing training, acquiring skills for a better future in which they will earn a good salary or pay.[4]

However, Deakin and Morris[5], argue for a position similar to that given in Uganda's Employment Act on a contract of apprenticeship; whereby 'the essence of an apprenticeship is that the employer undertakes formally to train and instruct the apprentice in a particular trade or skill; the apprentice, in turn, contracts to serve the employer and to learn from those from whom they receive instruction'.[6] Deakin and Morris[7] seal the above argument with the case of *Dunk v. George Waller & Son Ltd*[8], in which Widgery LJ Stated:

> A contract of apprenticeship secures three things for the apprentice; it secures him, first, a money payment during the period of apprenticeship; secondly, that he shall be instructed and trained and thereby acquire skills which will be of value to him for the rest of his life; and thirdly it gives him status, because ... once a young man, as here, completes his apprenticeship and can show by certificate that he has completed his time with a well-known employer, this gets him off to a good start in the labour market and gives him a status the loss of which may be of considerable damage to him.

It is therefore clear that any forms of intellectual property generated under the contract of apprenticeship will automatically belong to the employer unless and until is evidence adduced to the contrary.

3. Roger W. Rideout, *Principles of Labour Law* (London: Sweet & Maxwell, 1972), at 16.
4. *Id.*
5. Simon Deakin & Gillian S. Morris, Labour Law, 6th Ed (Portland: Hart Publishing, 2012), 172.
6. *Id.*
7. *Id.*
8. *Id.*, [1970] 2 QB 163, 169.

Equally, a contract of service means 'any contract, whether oral or in writing, whether express or implied, where a person agrees in return for remuneration, to work for an employer and includes a contract of apprenticeship'[9], already discussed above. A contract of service needs to be distinguished from a contract for services (also otherwise known as independent contract). Quoting Denning LJ in *Stevenson, Jordan and Harrison Ltd. v. MacDonald and Evans*, Rideout[10] states:

> One feature which seems to run through the instances (of a contract of service and a contract for service) is that, under a contract of service, a man is employed as part of the business, and his work is done as an integral part of the business; whereas, under a contract for services, his work, although done for the business, is not integrated into it but is only necessary to it.[11]

It is therefore imperative that in a contract of service, the fruits of labour, including those of intellectual property, should belong to the employer whereas in a contract for employment, the fruits of labour should belong to the employee as an independent contractor.

Lastly, it is very important the Employment Act states:

> 'Employer' means any person or group of persons, including a company or corporation, a public, regional or local authority, a governing body or an unincorporated association, a partnership, parastatal organization or other institution or organization whatsoever, for whom an employee works or has worked, or normally worked or sought to work, under a contract of service, and includes the heirs, successors, assignees and, transferors of any person or group of persons for whom an employee works, has worked, or normally works.[12]

Thus, in Uganda's legal framework, an employer may be a natural or juristic person or a group of such persons. What matters most is that such a person or group of persons have a valid legal existence. Indeed, it is this existence that is crucial in determining employees' IPRs.

The common law holds sway as a source law in Uganda.[13] The common law was introduced in Uganda on 11 August 1902 using the 1902 Order-In-Council.[14] In the Ugandan context, the common law means 'common law of England.'[15]

9. Section 2 of the Employment Act.
10. Rideout, *supra* n. 3, at 7.
11. *Id.*, [1952] 1 T.L.R. 101.
12. Section 2 of the Employment Act.
13. Section 14(2)(b)(1) of the Judicature Act, Cap. 13, Laws of Uganda.
14. H.F. Morris & James S. Read, Uganda: The Development of its Law and Constitution (London: Stevens & Sons, 1966), 239.
15. Section 2(n) of the Interpretation Act, Cap. 3, Laws of Uganda.

2 EMPLOYEES' COPYRIGHT

2.1 The Legal Framework

The legal framework in Uganda provides for the protection of employees' copyright. An employee is a person who is in the formal employment of another person, both natural and juristic. The Constitution of the Republic of Uganda provides for the protection of copyright and for the right to property.

Article 26 of the Constitution states:

> (1) Every person has a right to own property either individually or in association with others.
> (2) No person shall be compulsorily deprived of property or any interest in or right over property of any description except where the following are satisfied:
> (a) the taking of possession or acquisition is necessary for public use or in the interest of defence, public safety, public order, public morality or public health; and
> (b) the compulsory taking of possession or acquisition of property is made under a law which makes provision for:
> (i) prompt payment of fair and adequate compensation, prior to the taking of possession or acquisition of the property; and
> (ii) a right of access to a court of law by any person who has an interest or right over the property.

It is clear from the above constitutional provision that Uganda's constitution seeks to protect both the tangible (physical) and intangible (non-physical) property including copyright and intellectual property in general. Indeed, this line of thought is fortified by Article 189(1), which states that 'subject to the provisions of this Constitution, the functions and services specified in the Sixth Schedule to this Constitution shall be the responsibility of the Government'. Clause 6 of the Sixth Schedule lists 'copyrights, patents and trademarks and all forms of intellectual property...' as being among the functions and services for which the (central) Government of Uganda is responsible.

In terms of thematic approach, Uganda has the Copyright and Neighbouring Rights Act[16] (CNRA), which repealed the colonial Copyright Act.[17] For the smooth running and operationalisation of the CNRA, the government passed the Copyright and Neighbouring Rights Regulations in 2010. The regulations are detailed on the procedure for registering a copyright and the protection of neighbouring rights in Uganda.

Thus, Uganda has a comprehensive legal framework for the protection of copyright and neighbouring rights. This legal framework comprises of the Constitution, the CNRA and the regulations made thereunder. It can be said that this legal framework is adequate and largely conforms with Uganda's obligations under the World Trade Organization (WTO), the World Intellectual Property Organization (WIPO) and its

16. Act No. 19 of 2006.
17. Cap. 215 of 1964.

Agreement on Trade-Related Aspects of Intellectual Property (TRIPS), the African Regional Intellectual Property Organization (ARIPO) and the East African Community (EAC).

2.2 Ownership of Rights

The CNRA and the regulations made thereunder provide for the ownership of literary, scientific and artistic intellectual works and their neighbouring rights and provide for other related matters and procedures and fees payable. In terms of who owners the copyright in a work in Uganda, the CNRA provides for three major categories viz: the creator of the worker, employees, employers, government or international bodies and any other person, both natural and legal/juristic who commissions such a work.

Ordinarily, 'copyright belongs to the person who created the work, i.e., the author.'[18] Thus, it is the author of the work, who is taken to be the owner of the work. However, where the work was done or created during the course of employment, then shall a work vests in the hands of the employer unless the contrary has been proved. Commissioned works also belong to the entity or person who commissions such a work.

However, computer programs may also be protected under 'section 5(1)(e) of the CNRA which gives works eligible for copyright protection to include "computer programmes and electronic data banks and other accompanying materials"'.[19]

Computer programs are protected under Ugandan law.

2.2.1 Employee Works

Employee works are traditionally the works of the employer.[20] This is more so, especially if the work was created during the course of employment.[21] It is one of the cardinal principles of labour that the employee should work only for the master in the master's time.[22] As Deakin and Morris note:

> Despite the volume of legislation now governing the area of labour law, the common law remains at the core of the system at both the individual and collective levels. At the individual level, with the exception of some public service workers, the employment relationship is constituted by a contract, which is governed, essentially, by principles emanating from the common law. Moreover the existence, or potential existence, of a contract is the basis of access by workers to the vast majority of statutory employment protection rights.[23]

18. Adams & Adams, Practical Guide to Intellectual Property in Africa (Pretoria: Pretoria University Law Press, 2012), 635.
19. David J. Bakibinga & Ronald Kakungulu M., *Intellectual Property Law in East Africa* (Nairobi: lawAfrica, 2016), at 126.
20. Barnett Instruments v. Overton [1994] 69 RPC 315.
21. Patchett v. Steering [1955] RPC.
22. Rideout, *supra* n.3, at 81.
23. Deakin & Morris, *supra* n. 5, at 59.

The common law applies to a wide range of subjects in Uganda.[24] Labour law is one such area to which the common law has made major inroads. According to Rideout, 'so far as the common law is concerned, the employer purchases the services of the employee for so much of that employees' time'.[25] Even 'as far as ownership of copyright is concerned'[26] it belongs to the employer 'as long as it is work produced in the course of the author's employment'.[27] Uganda's statutory law is explicit on ownership of employees' works. Section 8 of the Copyright and Neighbouring Rights Act (CNRA) states that:

> (1) Where a person creates a work:
> (a) in the course of employment by another person;
> (b) on commission by another person or body;
> then in the absence of a contract to the contrary, the copyright in respect of that work shall vest in the employee or the person or body that commissioned the work.
> (2) Where a person creates work under the direction of control of the Government or a prescribed international body, unless agreed otherwise, the copyright in respect of that work shall vest in the Government or international body.
> (3) Vesting of copyright referred to in (1) and (2) above shall apply only to work created within the stipulated schedule of work of an employee.
> (4) The moral right in a work made under this section shall always remain with the actual author of the work.

It is therefore clear that Ugandan legislation follows the common law position that all works created by the employee in the course of her employment belong to her employer.

In *Sikuku Agaitano v. Uganda Baati*[28], the High Court held, that 'the persons who created and did the video shooting or who employed the person who carried out the work of shooting the photos and video is/are the authors or author of the work'.[29] This Ugandan case is key in resolving the dispute between the employee and employer ownership of copyright.

2.2.2 Commissioned Works

One of the most contested areas in respect of employees IPRs is that of commissioned works. According to the Black's Law Dictionary, commission means 'the authority under which a person transacts business for another'.[30] Commissioned works generally refer to works of which a particular person (who can also happen to be an employee) has been specifically asked, usually, at a professional fee to undertake a

24. William Burnett Harvey, An Introduction to the Legal System in East Africa (Nairobi: East African Literature Burea, 1975), 539.
25. Rideout, *supra* n.3, at 89.
26. *Id.*, at 90.
27. *Id.*, at 91.
28. HCCS No. 298 of 2012.
29. *Id.*
30. Bryan A. Garner, Black's Law Dictionary, 8th Ed. (Texas: Thomson West, 1999), 286.

particular work. The work would ordinarily not have been done and if at all by that particular person if it were not for the commission. Section 4 of the CNRA states:

> (1) The author of any work specified in Section 5 (on work eligible for copyright) shall have a right to protection of the work, where the work is original and is reduced to material form in whatever method irrespective of quality of the work or the purpose for which it is created.

In *Sikuku Agaitano v. Uganda Baati*[31], the court emphasised the fact that where a person creates a work on commission by another person or body, then in the absence of a contract to the contrary, in respect of that work, the work shall vest in the employer or the person or body that commissioned the work. Justice Chris Madrama held thus 'the persons who created and did the video shooting or who employed the person who carried out the work of shooting the photos and video is/are the authors or author of the works'.[32] This case underscores the essence of Section 8(1) of the CNRA on commissioned works. The right of enforcement over copyright is thus determined by the entity that is in charge of the expression, which may include the employee or the employer.

Thus, ordinarily, copyright belongs to the author of the work. The author or copyright owner is established through identifying such entity as such. However, under Section 8(1)(b), 'Where a person creates a work on commission by another person or body; then in the absence of a contract to the contrary, the copyright in respect of that work shall vest in the employer or the person or body that commissioned the work.'

Indeed it is only logical and fair that such a commissioned work vests in the person who commissioned the work.[33] It is that person, who is the 'brain' behind the creation of such a work. The commissioned work is the brainchild of the person who commissioned the work. However, 'where person creates work under the direction or control of the Government or a prescribed international body, unless agreed otherwise, the copyright in respect of that work shall vest in the Government or international body'.[34] Again, it is only logical and fair that the Government or international body under whose direction of control the person did the work owns such a work, unless and until the contrary be shown or proved. It is because, if it were not for the direction or control. In a bid to safeguard against abuse and exploitation, the CNRA requires that 'vesting of copyright referred to in (1) and (2) above shall apply only to work created within the stipulated schedule of work of the employee'.[35] For example, whereas it is expected that University Professors should teach, publish and do community service or outreach, it is generally agreed the world over that scholarly works published by the professors remains their (professors') intellectual work or copyright. At the common law, the duty of the servant or employee includes 'to offer personal service and to work

31. HCCS No. 298 of 2012.
32. *Id.*
33. Bakibinga & Kakungulu, *supra* n.19 at 24.
34. Section 8(2) of the CNRA.
35. Section 8(3) of the CNRA.

only for the master or employer in the master's or employer's time'.[36] The duty of the servant or employee to offer personal service and to work only for the master or employer in the master's or employer's time hold sway in Uganda, as a common law country.

2.3 The Parties' Duty to Inform

The employer is under a duty to inform the employee that all copyright emanating from the employee's sweat or work shall belong to the employer. The employer will usually do this through the contract of employment or employment manual. However, even where the employer does not inform the employee, the CNRA stipulates that 'where a person creates a work in the course of employment by another person, then in the absence of any evidence to the contrary, the copyright in respect of that work shall vest in the employer or the person or body that commissioned the work'.[37]

It is very clear that for the case of Uganda, although in practice the employer informs the employee that all copyright created in the course of employment belongs to the employer, all this is mere courtesy or comity since the CNRA has already decreed that such work automatically belongs to the author unless and until evidence to the contrary is given or shown.

2.4 Employee Remuneration Right

The employer is under a duty to remunerate the employee for the latter's time, labour and loyalty. Indeed 'where an employer is unable, or refuses, to pay wages, a labour officer, on application of any employee of that employer, shall declare the contract of service terminated'.[38] It is the duty of the employer to provide work to the employee during the substance of the contract of service.[39] The employee is entitled to wages as long as she remains an employee of her employer.[40] The employee is under a duty to act diligently in the course of her employment or while serving her contract of service, so as not to bring loss or liability to her employer. Two copyright cases in Uganda illustrate these phenomena. In *Al Hajji Nasser Ntege Sebagala v. MTN Uganda Ltd & SMS Media Ltd*[41], 'the plaintiff filed this action for a declaration that the Defendant's use and/or sale of the plaintiff's speeches/addresses as ring tones/caller tunes … constitute an infringement of his copyright'. The key question was 'who was the copyright owner of the caller/ring tunes? 'I am the Minister without fort portal', 'I will put my behind …'? The High Court relying on the United States case of *Falwell v. Penthouse International Ltd*, 215 U.S.P.Q 975 (Vir. Ct. 1981) held that 'copyright in a sound recording vest in the producer … the plaintiff enjoys no copyright in the

36. Rideout, *supra* n. 3, at 81.
37. Section 8(1)(a) of the CNRA.
38. Section 31(1) of the Employment Act.
39. Section 40 of the Employment Act.
40. Section 41 of the Employment Act.
41. HC (Com. Div.) CS No.283 of 2012 at 1.

circumstances, the action for copyright infringement has to fail and is hereby dismissed'.[42] Justice Christopher Madrama Izama further noted that 'it is the third-party who is the author of the recording of the Plaintiff's voice … it carried out the fixation of the plaintiff's voice in a material form and was able to reproduce it and sell it to the Defendant'.[43] Fixation of the copyright is a key determinant in ownership of the copyright between the employer and the employee since owner automatically belongs to the person who puts the work in a material form.

In the case of *Katatumba v. Anti Corruption Coalition Uganda*[44], the plaintiff, an artist, and the copyright holder of the musical production/recording entitled 'Let's Go Green' sued the defendant, a non-governmental organisation, for general and aggravated/exemplary/punitive damages for copyright infringement, when the defendant without prior authorisation from the plaintiff, used portions of the plaintiff's song (above) in the advertisement of her (defendant's) save Namanve Forest campaign. The High Court held that 'the actions of the defendant's officials for which the defendant is vicariously liable was done in blatant disregard of the plaintiff's copyright and the use did not fall within any of the permitted uses under section 15(1) of the CNRA 2006'.[45] Basing on the long-standing common law principle of vicarious liability, the employees of the defendant NGO created liability for their employer.

3 EMPLOYEES' SEMI-CONDUCTOR CHIPS

Semi-conductor chips are not directly protected under Ugandan law.

4. EMPLOYEES' INVENTIONS

4.1 The Legal Framework

Uganda's legal framework is adequate for the protection of inventions including those made in the course of employment as long as the inventions meet the minimum requirements of protection. Indeed, the preamble to the Industrial Property Act[46] (herein after IPA) states:

> An Act to provide for the promotion of inventive and innovative activities, to facilitate the acquisition of technology through the grant and regulation of patents, utility models, industrial designs and technovations and to provide for the designation of a registrar, to provide for the functions of the registrar, and the establishment of a register of industrial property rights and for related matters.

42. *Id.*, at 29-30.
43. *Id.*
44. HC (Com. Div.) CS No.307 of 2011 at 1-3.
45. *Id.*, at 38.
46. Act No. 3 of 2014.

The Industrial Property Act constitutes an unprecedented avenue for standardisation of the legislative framework for the effective regulation and protection of patents in Uganda as well as complying with Uganda's international obligations.

Uganda is a member of the Paris Convention, ARIPO (Harare Protocol), the Patents Corporation Treaty and the WTO/TRIPS Agreement.[47] The IPA defines both an invention and inventor as here below:

> Invention means a new and useful art whether producing a physical effect or not, process, machine, manufacture or composition of matter which is not obvious, or a new and useful improvement of it which is not obvious, capable of being used or applied in trade or industry; and includes an alleged invention.[48]

Thus, an invention must be new, useful, and capable of being used or applied in trade or industry (or industrial application). This is in line with Article 27(1) of the TRIPS Agreement which states that 'patents shall be available for any inventions, whether products or processes, in all fields of technology, provided that they are new, involve an inventive step and are capable of industrial application'.

The inventor is the brain behind the invention and has been defined in the IPA as 'the person who actually devises the invention as defined in section 8; and includes the legal representative of the inventor'. In the context of Uganda, the inventor can be the employee which leads us to the question of ownership of rights.

4.1.1 Protection of Computer Programs

According to Bakibinga and Kakungulu, 'in the Ugandan context, computer programmes may be said to be proscribed from patenting under the Uganda Industrial Property Act section 8(3)(a) and (e) relative to mathematical theories and the presentation of information'.[49] Computer programs that meet the requirements of patentability may be protected as inventions.

4.2 Ownership of Rights

Section 38 of the IPA vests the ownership of rights in an invention or patent in the hands of the owner of the invention, as long as, the inventor has fulfilled the requirements for patentability under the Act. The other rights bestowed upon the inventor and the owner of the invention, which in the context of employment may not be one and the same person or entity include: the right to 'preclude any person from exploiting the patented invention', the right to 'assign or transfer by succession the application for patent or the patent' and the right to 'conclude licences contracts' in line with the invention.[50]

47. Adams & Adams, *supra* n. 18, at 628.
48. Section 2 of the IPA.
49. Bakibinga & Kakungulu, *supra* n. 19, at 126.
50. Section 38 (2)(3) & (4) of the IPA.

The employee who makes computer programs, technical protection measures, databases and circuit drawings, broadcasting rights, designs domain names, in the course of employment, such semi-conductor chips would automatically belong to the employer. The reverse is also true, where the semi-conductor chips are made by an employee but outside the course of employment.

4.2.1 Employees' Technovations

The Industrial Property Act provides for technovations which are defined as follows:

> A solution to a specific problem in the field of technology, proposed by an employee of an enterprise in Uganda for use by that enterprise, and which relates to the activities of the enterprise but which, on the date of the proposal, has not been used or actively considered for use by that enterprise.[51]

A technovator means 'an employee to whom the enterprise has issued a technovation certificate'.[52] Technovations form part of the wider industrial property rights.[53]

4.2.1.1 Infringement of Inventions and Remedies

In case of infringement of the invention, the owner of the invention, may petition the High Court for a host of remedies ranging from an injunction to injuct any further infringing acts, claim damages from the (alleged) infringer, and/or claim for compensation commensurate to the damage caused by the infringement.[54]

In the event that the invention was made in the course of employment, the owner of that invention shall be the employer unless evidence to the contrary is shown. And also, in the event, of an invention made outside the course of employment, the employee shall be the owner of that invention unless evidence is shown to the contrary.

4.2.1.2 Limitations on the Employer's Rights in Respect of the Invention

Several limitations pertain to the employer's rights as the owner of an invention made by an employee in the course of employment. The scope of the protection of a patent shall be limited to only what is stated in the application or claim for the patent.[55] Other limitations of rights includes 'acts which are done for industrial or commercial purposes but do not extend to acts which are done for scientific research', 'articles which have been put on the market in Uganda or in any other country or imported into Uganda by the owner of the patent or with his or her consent', 'the use of articles on

51. Section 81 of the IPA.
52. *Id.*
53. Section 2 of the IPA.
54. Section 40 of the IPA.
55. Section 42 of the IPA.

aircraft, land vehicles or vessels of other countries, which temporarily or accidentally enter the airspace, territory or waters of Uganda'.[56]

4.3 The Parties' Duty to Inform

Either party, be it the employer or the employee who takes or makes or seeks to make any acts which are adverse to the rights of the owner of the patent has a duty to inform the affected party. This is the requirement with compulsory acquisition or licensing[57] and experimental, teaching and medical grounds.[58] There is a right to reasonable remuneration in case of a compulsory licensing by Government.

In the case where employees request for a technovation certificate for the same technovation, the employee who makes the first request will be entitled to the certificate.

Once the technovation certificate has been issued, the technovator shall not communicate his or her technovation to any person other than the enterprise and shall not use it except in collaboration with the enterprise.

An employee who develops a computer program or software is also under a duty to inform the employer.

4.4 Employee Remuneration Right

The employee has a right to be remunerated for his or her labour. However, in the case of inventions, this remuneration may be from either the employer or the invention itself. Thus, it is very important to ascertain whether the invention belongs to the employee or the employer. The IPA is silent on this matter, but recourse may be had to Section 39(1) of the Patent Act of the United Kingdom for guidance. The position in the UK which is of persuasive application to Uganda as a common law country is that 'an employee invention belongs to the employer if made in the course of employment or where the employee had special obligation to further the interests of the employer's undertaking'.[59] Two cases below best illustrate the two positions above.

> In *Electrolux Ltd v. Hudson* [1977] FSR 312, an employee of a company making electrical appliances invented with his wife, at home, in the evening, an adaptor for a vacuum cleaner which allowed the use of any type of disposable bag in any vacuum cleaner taking disposable bags. At the time, the defendant was not employed as a storekeeper. It was held that since the employee was not employed to invent but as a storekeeper, the company was not entitled to the patent. The relevant term in the contract was too wide even for a person employed in a research capacity.[60]

56. Section 43 of the IPA.
57. Compulsory licensing is to be done by only the Government of Uganda.
58. Sections 44, 58, 59 & 66 of the IPA.
59. Bakibinga & Kakungulu, *Supra* n. 19, at 142.
60. *Id.*, at 143.

From the above case, it exemplifies the fact that in order for the invention to belong to the employer, 'it must have been made in the course of duties specifically assigned to the employee as such'.[61] Thus, inventions made when the employee is off duty and not during the employer's time may not belong to the employer.

Similarly, inventions made by the employee but which do not 'pertain' or relate to the employee's job description may also not be considered as the employer's:

> In *Greater Glasgow Health Board's Application* [1996] R.P.C. 207, the issue was whether an employed hospital doctor was entitled to be a proprietor of a patient for the invention of a spacing device for opthalmoscope (eye equipment). The doctor's job description was wide and referred to the use of research facilities, which the doctor was encouraged to avail himself for basic clinical research. However, the employment contract did not mention this as a duty. It was held that the doctor's duty to treat patients did not extend to devising new ways of diagnosis and treatment of patients.[62]

In situations where the employee makes an invention, which is *outstanding benefit* to the employer, the employee may be compensated for the extra work.[63] However, where there is 'pressure or undue influence on the employee by the employer to "surrender" the invention to the latter, then the equitable doctrine of undue influence might apply'.[64]

The employee is entitled to the usual remuneration as an employee and where this is not adequate; they may be given a special compensation by the employer for the invention if the latter deems the invention to be of outstanding benefit.

5 EMPLOYEES' UTILITY MODELS

5.1 The Legal Framework

Part XII of the IPA deals with utility models. A utility model also referred to as a 'Petty Patent' or 'Small Patent' in some jurisdictions is used to protect inventions which would otherwise not meet the stringent requirements of a patent. Thus, the requirements for obtaining a utility model are less stringent as compared to those of patentability. Usually the requirements of newness or novelty, or inventive step (non-obviousness)[65], and industrial application[66], may not be met or are much lower with a utility model. Thus, in practice, protection for utility models is often sought for innovation of a rather incremental character which may not meet the patentability criteria but are nonetheless useful or can be used to solve a specified problem in a given field of technology.[67] It is for this reason that utility models are also designated as innovation patents. According to Section 2 of the IPA, an 'innovation' means 'utility

61. *Id.*, at 142.
62. *Id.*, at 143-144.
63. *Id.*, at 146.
64. *Id.*
65. Section 11 of the IPA.
66. Section 12 of the IPA.
67. Section 8 of the IPA.

models, technovation, models, and industrial designs and any other non-patentable creations or improvements that may be taken as deserving specified intellectual property rights.' Indeed, it follows without mention that utility models can only be obtained for certain fields of technology and only for products but not for processes as opposed to 'inventions' or patents which are generally available across most fields of technology for both products and processes as long as these meet the requirements of patentability.

The IPA provides for the legal framework dealing with specific aspects of the utility model protection, ranging from inventionness, usefulness, subject matter of protection, remedies for infringement, rights of the owner and duration of protection.

5.2 Ownership of Rights

Heralded as key especially for developing countries with low levels of technological development, the ownership of rights in respect of utility models rests with the innovator unless made during the course of employment, in which case the ownership of such a utility model may then vest in the employer. The owner of a utility model in Uganda: be it the employee or the employer, has a right to the exclusive exploitation of that utility model for a limited period of ten (10) years, non-renewable.[68] The owner also has a right to assign or transfer by succession or to conclude licence contracts in respect of the utility model.[69]

5.2.1 Infringement of a Utility Model

Like most intellectual property rights, the proprietor of a registered utility model has a right to sue infringers of his or her product with a view to stopping the infringing activities, destruction of infringing goods, obtaining information as to the source or origin and distribution channels or outlets of the infringing products.

5.2.2 Remedies for Infringement of a Utility Model

The proprietor of a utility model can sue to recover the usual remedies that accrue to a holder of an intellectual property right. These remedies are mainly, listed in Section 40 of the IPA and include: to stop the infringing activities, claim damages, cause the destruction of infringing goods, and obtain information on the origin and distribution channels of the infringing products to the utility model. However, exceptions do pertain to these limitations in respect of, e.g., on private or experimental use.

68. Section 38(1) & (2) and Section 69(3) of the IPA.
69. Section 38(3) & (4) of the IPA.

5.3 The Parties' Duty to Inform

During the subsistence of the utility model, which is the tenth year after the date of the grant of a utility model and is not renewable,[70] there's a duty to inform. Indeed, this duty to inform may even exist prior to the disclosure of the utility model in the case of where the employer/employee relationship exists. Even at registration of the utility model, the duty by the employee to inform the employer exists. Indeed, under Section 68(3) of the IPA, 'a certificate of utility model is prima facie evidence that the holder is the registered owner of a utility model with the registrar'. And 'where the right of a patent conflicts with the right of a utility model certificate in the case referred to in section 17(3), that provision shall apply as if the word "patent" were replaced by the word "utility model certificate"'.[71] This is very important in relation to the applicability of provisions relating to patents with necessary modifications.

There is generally a duty to keep the invention confidential before filing a utility mode application usually within 12 months as the grace period. Otherwise, for an application for a utility model will be rejected or invalidated where it can be proved that there was prior disclosure of the invention before the filing date.

5.4 Employee Remuneration Right

The employee who makes a utility model during the course of employment has remuneration right as per his or her contract of service. This remuneration is usually only limited to the terms and conditions within the employee's contract of employment or engagement. Where the employee makes the utility model in the course of employment, ordinarily, no extra remuneration would be necessary unless the utility model be of an extra ordinary benefit to the employer.

6 EMPLOYEES' DESIGN RIGHTS

6.1 The Legal Framework

The IPA provides for the protection of industrial designs in Part XII. An 'industrial design' is defined as:

> That aspect of a useful article which is ornamental or aesthetic that may consist of three dimensional features like a shape or surface or an article or a three dimensional feature such as patterns, line or colours.[72]

The IPA provides a set of provisions dealing with specific aspects of design rights protection namely; right to industrial design; naming of creator[73], registration of

70. Section 69(3) of the IPA.
71. Section 68(2) of the IPA.
72. Section 71(1) of the IPA.
73. Section 72 of the IPA.

industrial design[74], application and examination[75], duration and renewal of registration of an industrial design[76], restoration of registration and publication of reference to registration[77], graphic representations, specimen and related matters[78], publication of reference to registration[79], rights conferred by registration of industrial designs and related matters[80], and transfer and assignment of industrial designs, licences[81]. Thus, Uganda has an adequate legal framework for the protection of employee's industrial design rights. The only challenge is that the IPA is of 2014 and is yet to be tested.

6.2 Ownership of Rights

The ownership of design rights in Uganda, can vest in the creator, employer, or person commissioning the work, the only requirement being that the actual creator of the work must be named as such.[82] The situation is best captured by Bakibinga who states that:

> The creator of an industrial design (including his/her successor in title) is entitled to the exclusive right to sell, or cause to be sold for commercial or industrial purposes the goods in which the design is incorporated. Similar to the patent regime, an industrial design belongs to the inventor, employer of the inventor or the person who has commissioned the work for an industrial design. Nevertheless the actual inventor has a right to be named as such.[83]

Ownership of employee's design rights in Uganda can thus vest in the employee, as the creator (inventor) of the industrial design, the employer if the industrial design was made during the course of employment, or the person who commissioned the work for an industrial design. However, when the ownership of the industrial design belongs to the employer in the circumstances explained above, the moral rights in the industrial design remain with the creator and the creator has a right to be named as such.

6.2.1 Infringement of Industrial Design

The owner of an industrial design is entitled to sue and recover damages and other remedies if there is unauthorised use, dealing, transfer, assignment, sale or licensing of the design without his or her permission. The duration of the design is 'five (5) years

74. Section 73 of the IPA.
75. Section 74 of the IPA.
76. Section 75 of the IPA.
77. Section 76 of the IPA.
78. Section 77 of the IPA.
79. Section 78 of the IPA.
80. Section 79 of the IPA.
81. Section 80 of the IPA.
82. Section 72 of the IPA.
83. DJ Bakibinga, *Recent Developments in Intellectual Property Law in Uganda*, A paper delivered at the 4th Global Congress on Intellectual Property and Public Interest, New Delhi, India, December 2015.

following the date of registration, renewable for only two further consecutive periods of five years upon payment of the prescribed renewal fee'.[84]

However, as a defence to infringement of a design, Adams and Adams state:

> The registered proprietor of a design shall not be entitled to recover damages in respect of any infringement of design from any defendant who proves that, at the date of the infringement, he was not aware and had no reasonable means of making himself aware of the existence of the registration of the design.[85]

The owner of employee design rights enjoys the economic rights or benefits that accrue to the design whereas the employee retains the moral rights in the design including the right of paternity (to be named as the creator of the design).

6.2.2 Remedies for Infringement of Employees' Design Rights

Similar to the patent regime, infringement of an employee's industrial design gives rise to a number of remedies to the owner to compensate for the injury or loss suffered as a result of the infringement. Such remedies include, injunction, damages, delivery and destruction of infringing copies, inspection, etc.

6.3 The Parties' Duty to Inform

The employee is under a duty to inform the employer of any industrial designs that they make especially during the course of employment and to a full disclosure of the same to his or her master. Similarly, the employer is under a duty to compensate the employee for the time spent in coming up with the design.

6.4 Employee Remuneration Right

The employee has a right to remuneration. An employee who comes up with a design during the course of her employment is entitled to no more remuneration, since she is paid to do what she will have done. However, employees who come up with a design outside the course of employment or one who comes up with a design which is very beneficial to the employer need be given extra remuneration beyond their salary or work related allowed. Commissioned works which result in a design need also be remunerated. Thus, in *Sylvia Katende v. Bank of Uganda*[86], the plaintiff participated in a contest organised by Kampala City Council (KCC), won and was duly paid. She sued the defendant over eight (8) years later claiming the sculpture was made without her knowledge and consent. The High Court found that the plaintiff was commissioned by KCC to do a design for a sculpture, meaning that the model design was made to the order of KCC. Therefore, the work the plaintiff produced was a drawing of the sculpture

84. Section 75(1)&(2) of the IPA.
85. Adams & Adams, *supra* n. 18, at 634.
86. HCCS No. 443 of 2010.

and not the sculpture. The Court further observed that the design and the sculpture are two different works of art, albeit one is based on the other. A drawing is one form of artistic expression just as a sculpture is another form of artistic expression. The sculpture can be described as a derivative work of the art, derived from the drawing.

Those employees who do commissioned designs are remunerated by way of a commission or token payment for the work, and the design will belong to the commissioning entity.

7 EMPLOYEES' PLANT VARIETIES

7.1 The Legal Framework

The Plant Variety Protection Act, 2014 provides for the legal framework on the protection of employees' plant varieties in Uganda. The preamble to the Act states:

> An Act to provide for the promotion of development of new plant varieties and their protection as a means of enhancing breeders innovations and rewards through granting of plant breeders rights and for other related matters.

The Act applies to 'plant varieties under both *in situ* and *ex situ* conditions; derivatives of the plant varieties; plant breeders; and export pf seeds of plant varieties'.[87] The Act is in line with Uganda's obligations under the TRIPS Agreement which obliges its member States to provide for the protection of plant varieties either by patents or by an effective *sui generis* system or by any combination therefore.[88] In Uganda, employees' plant varieties are protected not by patents but by a *sui generis* system as seen in the Plant Variety Protection Act.

7.2 Ownership of Rights

The ownership of the plant variety normally vests in the breeder. A breeder has been defined in Section 3 of the Act as:

> (a) a person who bred or discovered in relation to the use of a plant variety and developed the plant variety;
> (b) the employer of the person to whom paragraph (a) applies or any person who as commissioned the person's work; or
> (c) the successor in title of the first person mentioned in paragraph (a) and (b), as the case may be.

It is therefore clear that the breeder can be the employee, as the person who bred or discovered in relation to the use of a plant variety and developed the plant variety. Similarly, an employee who breeds or discovers a plant variety or another person

87. The Plant Vartiety Protection Act, 2014, Section 1.
88. Article 27(3)(b) of the TRIPS Agreement.

commissioned to develop a plant variety may be called breeders. The only issue will be that at the common law, generally, such a plant variety discovered by an employee during the course of employment will automatically belong to the employer unless evidence to the contrary can be shown. Equally, a plant variety, which is as a result of a commissioned work, shall vest in the commissioning entity.

7.2.1 Rights of Plant Breeders

Section 13 of the Act, gives plant breeders rights, in respect of a new variety as:

(a) the exclusive right to sell, including the right to licence other persons to sell and export plant varieties and reproductive material of plants of that variety; and
(b) the exclusive right to produce, including the right to licence other persons to produce, reproductive material of plants of that variety for sale.

Thus, the main rights of the plant breeder are the exclusive right to sell and to produce the plant variety, and to licence or to export the plant variety.

7.2.2 Infringement of Plant Breeders Rights and Remedies

According to the Act, the plant breeders' rights in respect of a plant variety shall subsist: (a) a period of twenty (20) years in the case of annual crops; or (b) for a period of twenty (25) years for trees, vines and other perennials commencing on the day of filing of the application.[89] Whoever sells or produces a protected plant variety without permission from the breeder is liable to be sued for infringement of the breeders' rights.

Section 20 of the Act provides for the remedies for infringement of rights as follows:

(1) Infringement of plant breeders' rights, shall be actionable at the instance of the holder of the rights.
(2) In any proceedings for the infringement of plant breeders' rights, relief shall be available by way of damages, injunction or any other relief that court may grant.

However, a defendant faced with a suit for infringement of plant breeders' rights may:

(1) A defendant in proceedings for the infringement of plant breeders' rights in respect of a variety shall apply by way of counter-claim for the revocation of plant breeders' rights on the ground that:
    (a) the variety was not new; or

89. Section 19 of the Planty Variety Protection Act.

(b) facts exist which, if they had been known by the office before the grant of those rights, would have resulted in the refusal of the grant.

7.3 The Parties' Duty to Inform

The parties' duty to inform exists on both sides. Where the person who bred or discovered in relation to the use of a plant variety and developed the plant variety, on their own, there's no duty to inform. However, where one breeds or discovers in relation to the use of plant variety and developed the plant variety, as an employee, the employee has a duty to inform the employer of such a remarkable development. And where, the employer seeks to claim the employee's breeder's rights, based on the contract of service, they have a duty to inform the employee.

7.4 Employee Remuneration Right

The breeder has a right to adequate remuneration. Thus, an employee who breeds or discovers in relation to the use of a plant variety and develops the plant variety is entitled to remuneration as an employee if developing new plant varieties is part and parcel of his or her employment. If the development of new plant varieties is merely incidental to their employment, such an employment will have to be compensated.

8 INSTANCES FOR DISPUTES

The TRIPS Agreement provides for dispute prevention and settlement in Part V of the Agreement. In line with this, Uganda as a Member State has transparency, dispute prevention and settlement. All the laws dealing with the different clusters of employees' intellectual property rights in Uganda have dispute prevention and settlement provisions as much as the labour and employment laws do, including a fully fledged and operational division of the High Court dealing with commercial related disputes known as the Commercial Division which handles all intellectual property related disputes in Uganda. Similarly, under the Labour Disputes (Arbitration and Settlement) Act[90], labour and industrial disputes, may be referred to the Industrial Court.

The Employment Act[91] empowers the Labour Officer to settle labour disputes in Uganda.[92] The Labour Officer also has power to investigate and dispose of complaints[93], of any nature including those of employees' intellectual property rights. Similarly, the Labour Officer has power to 'institute criminal or civil proceedings before the Industrial Court in respect of a contravention or alleged contravention of this Act,

90. Cap. 224, Laws of Uganda.
91. Act No. 6 of 2006.
92. *Id.*, Section 12.
93. *Id.*, Section 13.

and may prosecute and appear in his or her own name in respect of the proceedings'.[94] On the question of which court between the High Court and the Industrial Court takes primacy in respect jurisdiction over labour related disputes in Uganda.

In the case of *201 Former Employees of G4S Security Services Uganda Limited v. G4S Security Services Uganda Limited*[95], the Supreme Court of Uganda in dealing with the jurisdiction of the Labour Officer and the Industrial Court noted:

> Clearly, the above provisions intended to oust the jurisdiction of the ordinary civil courts in Uganda by ensuring that employment matters are only handled by Labour Officers and the Industrial Court. It is also evident that these sections conflict with article 139(1) of the Constitution in so far as they limit the unlimited original jurisdiction of the High Court to hear employment matters as a court of first instance. Article 139(1) of the Constitution of Uganda (1995) confers on the High Court unlimited original jurisdiction and appellate jurisdiction ... .[96]

According to Bakayana, 'the *G4S*' case, upheld article 139 of the Constitution (due to the supremacy of the Constitution), where Acts of Parliament, conferred explicit jurisdiction onto both the Labour Officer and the Industrial Court.'[97] However, as of 10 July, 2017, primacy to determine employment disputes in Uganda, will have to revert to the Labour Officers and the Industrial Court, based on the Supreme Court decision in *Rabbo's* Case in which the Court (holding on a similar principle of primacy of jurisdiction but on a taxation dispute) held thus 'the proper procedure therefore is that all tax disputes must be lodged with (the) Tax Appeals Tribunal (TAT) (established under the Tax Appeals Tribunal Act, Cap. 345), and only taken before the High Court on appeal'.[98]

Section 14(1) of the TAT states that 'Any person who is aggrieved by a decision made under a taxing Act by the Uganda Revenue Authority (URA) may apply to the tribunal for a review of the decision.' In Section 14(2), the 'Tribunal has power to review any taxation decision in respect of which an application is properly made.'[99] As of now, and until the decision in *Rabbo's* case is overturned by the Supreme Court under Article 132(4), it remains trite law, that primacy in respect of jurisdiction in employment disputes lies with the Labour Officers and the Industrial Court as opposed to the High Court of Uganda.

This comes on the backdrop of the need for an effective harmonisation and integration of the intellectual property system forms the foundation for the TRIPS Agreement. State parties thereto are mandated to modify their respective domestic legislations in a manner intended to harmonise the international patents systems and the entire intellectual property regime.

94. *Id.*, Section 14.
95. Supreme Court Civil Appeal (SCCA) No. 18 of 2010.
96. *Id.*, cited in Isaac Bakayana, *Thoughts on the Supreme Court decision of Uganda Revenue Authority v. Rabbo Enterprise (U) Limited, Mt. Elgon Hardwares Ltd* (Civil Appeal No. 12 of 2004), July 2017. (Unpublished paper on file with the author).
97. *Id.*, at 3.
98. *Id.*, at 1.
99. *Id.*, at 2.

9 CONCLUDING REMARKS

Uganda's intellectual property laws offer a strong legal framework for the protection of employees' intellectual property rights in the fields of employees' copyright, inventions, utility models, design rights, plant varieties and also provides for avenues of disputes prevention and settlement. However, semi-conductor chips are not directly protected under Ugandan law.

BIBLIOGRAPHY

Adams & Adams, Practical Guide to Intellectual Property in Africa (Pretoria: Pretoria University Law Press, 2012).

Bryan A. Garner, Black's Law Dictionary, 8th Ed. (Texas: Thomson West, 1999).

David J. Bakibinga & Ronald Kakungulu M., *Intellectual Property Law in East Africa* (Nairobi: Law Africa, 2016).

David J. Bakibinga, *Recent Developments in Intellectual Property Law in Uganda*, A paper delivered at the 4th Global Congress on Intellectual Property and Public Interest, New Delhi, India, December 2015.

Glanville Williams, Learning the Law (London: Stevens and Sons, 1982).

H.F. Morris & James S. Read, Uganda: The Development of its Law and Constitution (London: Stevens & Sons, 1966).

Isaac Bakayana, Thoughts on the Supreme Court decision of Uganda Revenue Authority v. Rabbo Enterprise (U) Limited, Mt. Elgon Hardwares Ltd (Civil Appeal No. 12 of 2004), July 2017. (Unpublished paper on file with the author).

Roger W. Rideout, Principles of Labour Law (London: Sweet & Maxwell, 1972).

Simon Deakin & Gillian S. Morris, Labour Law, 6th Ed (Portland: Hart Publishing, 2012).

William Burnett Harvey, An Introduction to the Legal System in East Africa (Nairobi: East African Literature Bureau, 1975).

Legislation

Copyright Act, Cap. 215 of 1964 (repealed by the 2006 Act)

Copyright and Neighbouring Rights Act, Act No. 19 of 2006

Copyrights and Neighbouring Right Regulations, S.1 No. 1 of 2010

Employment Act, Act No. 6 of 2006

Industrial Property Act, 2014

Judicature Act, Cap. 13, Laws of Uganda

Patent Regulations, 1993

Patents (Amendment) Act of 2002

Patents Act, Cap 216 of 1993

Plant Variety Protection Act, 2014

Trade Marks Act, No. 17 of 2010

Trade Marks Act, Cap 217 of 1953 (repealed by the 2010 Act)

Trade Marks Rules (issued under the repealed Trade Marks Act, Cap 217 of 1953 and expressly retained in force)
Trade Secrets Protection Act, 2009
United Kingdom Designs (Protection) Act, Cap 218 of 1937

International/regional conventions

Banjul Protocol (of ARIPO)
Harare Protocol (of ARIPO)
Lusaka Agreement (ARIPO)
Paris Convention
Patent Cooperation Treaty
WIPO Convention
WTO/TRIPS

Cases

Al Hajji Nasser Ntege Sebagala v. MTN Uganda Ltd & SMS Media Ltd, HC (Com. Div.) CS No. 283 of 2012
Barnett Instruments v. Overton [1994] 69 RPC 315
Dunk v. George Waller & Son Ltd [1970] 2 QB 163, 169
Jordan and Harrison Ltd. v. MacDonald and Evans [1952] 1 T.L.R. 101
Katatumba v. Anti Corruption Coalition Uganda, HC (Com. Div.) CS No.307 of 2011
Patchett v. Steering [1955] RPC
Sikuku Agaitano v. Uganda Baati, HCCS No. 298 of 2012
Sylvia Katende v. Bank of Uganda, HCCS No. 443 of 2010
Uganda Revenue Authority v. Rabbo Enterprise (U) Limited, Mt. Elgon Hardwares Ltd (Civil Appeal No. 12 of 2004)
201 Former Employees of G4S Security Services Uganda Limited v. G4S Security Services Uganda Limited, Supreme Court Civil Appeal (SCCA) No. 18 of 2010

Employees' Intellectual Property Rights: An Epilogue

Kacper Szkalej & Sanna Wolk

1 OVERVIEW

Ownership and control of intellectual property rights is mainly regulated by national legislation, and no major international harmonisation efforts are in progress regarding employees' intellectual property rights. However, national laws still vary to a material extent from jurisdiction to jurisdiction, and this even includes the full array of various intellectual property – some jurisdictions do not protect, e.g., utility models. Nevertheless, similarities also exist between national legislation concerning ownership of intellectual property rights created by employees in the course of their employment. For example, in many jurisdictions rules concerning the ownership of rights to inventions apply *mutatis mutandis* to other industrial property such as designs, utility models and topographies of semiconductor chips (where such property is protected). In the European Union, where rights to computer programs and (Community) designs are expressly regulated (Directive 2009/24/ EC on the legal protection of computer programs and Regulation (EC) No. 6/2002 on Community design, respectively) in this context, ownership of rights is initially attributed to the employer. However, whereas clarity in the form of express provisions is generally welcomed, such *lex specialis* results in unnecessary separation of various works protected by copyright, and therefore of default rules, especially where a creation is (or can be) protected by two different IPRs – be it a design protectable by design law and copyright, or a computer program protectable by copyright and, despite Article 52(2)(c) European Patent Convention, patent law (e.g. *Vicom* EPO T 0208/84). Nonetheless, many jurisdictions, while permitting the derivative assignment of the economic rights to the employer, provide that moral rights reside with the employee, such rights being non-transferable and attributable to the physical person who actually created the work.

One particular difference that is worth highlighting is that countries either regulate rights relating to inventions that subsequently become (or have become) the object of a patent application and a patent, and, on the other hand, inventions which do not necessarily become the object of patent applications, i.e. *patentable* inventions. Such differences can create imbalance, especially in cross-border situations, for example. where in one jurisdiction the employee has a right to additional remuneration for a *patentable* invention, but does not have such a right in a jurisdiction where only *patented* inventions are subject to such rules (or where no additional remuneration is payable). The fact that an invention does not become the subject of a patent application does not necessarily mean that it is less valuable, or that it would have not satisfied the requirements of novelty or extended beyond the state of the art, had it become the object of a patent application. In fact, where the employer immediately proceeds to the commercial exploitation of the invention, it might very well evidence that the employer appreciates the value of the invention; or where every attempt is made to maintain an invention as a trade secret, that the invention as such might be valued at more than a 20-year monopoly right is worth.

One common thread for all intellectual property areas is that a properly drafted agreement can help ensure that the party seeking ownership of the relevant intellectual property rights will get what it bargained for and secure the value of its intellectual property assets. Freedom of contract appears to be respected generally in all jurisdictions in relation to IP ownership in employment contexts. However, on the other hand, it is worth noting that by virtue of being an employment relationship, the employer can be accepted to be in a stronger bargaining position, for which reason some jurisdictions do employ various measures for the benefit of the employee, such as additional remuneration based on the profit that is made once the intellectual creation (often an invention) is exploited commercially.

2 ARE UNIFORM RULES DESIRABLE?

The legal situation concerning employee intellectual property rights has been described in Part II. The final issue to be addressed is whether uniform legislation in the field of employee intellectual property rights is necessary, and if so, how it should be formed. This is indeed a challenging question today, for twenty-first century economies.

The existence of differences between national laws on employee intellectual property rights causes complications and problems for cross-national development and research, both within multinational enterprises and for co-operation between companies. Divergent rules concerning employee and employer rights create uncertainty, and the position can be further exacerbated by default, or indeed mandatory, rules prescribed by employment law which relate to the language and applicable law governing an employment contract (as a means to protect the employee). Furthermore, intellectual property rights have shown an increasing tendency to overlap, and a specific product of intellectual creativity may be covered by several and perhaps, regarding ownership, conflicting rights, which in turn can affect the scope of the exclusive rights that are granted. Such complicated uncertainty can arise where, e.g., a computer program is protected in one area by patent law, and by copyright in another – not only does this affect the type and scope of exclusive rights, duration of such rights and their transferability (especially moral rights), but also the initial attribution of ownership by the legal framework. Here it should be stressed that whereas copyright protects the expression of ideas, patents can be said to protect the exploitation of ideas. Consequently, it is desirable to create uniform rules in this area, at least in terms of the initial attribution of ownership, with the caveat that the parties shall be free to agree differently.

There are a number of questions that have to be considered before it is possible to create uniform international rules regarding rights to employees' intellectual creations. Nevertheless, it is important to find a proper balance between the interests of the employee and the employer. On the one hand, a system giving employers control over employee intellectual property rights would be desirable; on the other, it is important to recognise, facilitate and reward creativity. It is suggested that employers should be able to acquire the entire right to employee creations, and employee-creators should be entitled to reasonable remuneration in compensation for the rights transferred to the employer, especially when some employed creators, depending on what intellectual property rights are created, are economically compensated through mandatory national provisions for the rights that are transferred to the employer. As noted above, several countries have already introduced mandatory provisions regarding an employed inventors' right to reasonable remuneration for the rights in the inventions that are transferred to the employer. Statutory employee-inventor compensation schemes have been established in many countries. For other intellectual creations, national solutions vary as regards the employed creator's right to economic compensation and there is no international homogeneity. Nevertheless, it is imperative to have balanced and coherent principles applicable to all intellectual property rights.

AIPPI LAW SERIES

1. Sanna Wolk & Kacper Szkalej (eds), *Employees' Intellectual Property Rights*, Second Edition, 2018 (ISBN 978-90-411-9230-1).
2. Jochen Bühling (ed.), *Patent Protection for Second Medical Uses*, 2016 (ISBN 978-90-411-8253-1).
3. Christopher V. Carani (ed.), *Design Rights: Functionality and Scope of Protection*, 2017 (ISBN 978-90-411-8864-9).